A HISTORY OF MAGIC AND EXPERIMENTAL SCIENCE

VOLUME I

A HISTORY OF MAGIC AND EXPERIMENTAL SCIENCE

DURING THE FIRST THIRTEEN CENTURIES OF OUR ERA

BY LYNN THORNDIKE

VOLUME I

COLUMBIA UNIVERSITY PRESS

NEW YORK

Columbia University Press
Publishers Since 1893
New York Chichester, West Sussex

First published by The Macmillan Company, 1923

Library of Congress Cataloging-in-Publication Data

A complete CIP record is available from the Library of Congress.

ISBN: 978-0-231-08794-0 (cloth)

Columbia University Press books are printed on
permanent and durable acid-free paper.

Printed in the United States of America

CONTENTS

BOOK I. THE ROMAN EMPIRE

BOOK IV. THE TWELFTH CENTURY

BOOK V. THE THIRTEENTH CENTURY

PREFACE

This work has been long in preparation—ever since in 1902-1903 Professor James Harvey Robinson, when my mind was still in the making, suggested the study of magic in medieval universities as the subject of my thesis for the master's degree at Columbia University—and has been foreshadowed by other publications, some of which are listed under my name in the preliminary bibliography. Since this was set up in type there have also appeared: "Galen: the Man and His Times," in *The Scientific Monthly,* January, 1922; "Early Christianity and Natural Science," in *The Biblical Review,* July, 1922; "The Latin Pseudo-Aristotle and Medieval Occult Science," in *The Journal of English and Germanic Philology,* April, 1922; and notes on Daniel of Morley and Gundissalinus in *The English Historical Review.* For permission to make use of these previous publications in the present work I am indebted to the editors of the periodicals just mentioned, and also to the editors of *The Columbia University Studies in History, Economics, and Public Law, The American Historical Review, Classical Philology, The Monist, Nature, The Philosophical Review,* and *Science.* The form, however, of these previous publications has often been altered in embodying them in this book, and, taken together, they constitute but a fraction of it. Book I greatly amplifies the account of magic in the Roman Empire contained in my doctoral dissertation. Over ten years ago I prepared an account of magic and science in the twelfth and thirteenth centuries based on material available in print in libraries of this country and arranged topically, but I did not publish it, as it seemed advisable to supplement it by study abroad and of the manuscript material, and to adopt an arrangement by authors. The result is Books IV and V of the present work.

My examination of manuscripts has been done especially at the British Museum, whose rich collections, perhaps because somewhat inaccessibly catalogued, have been less used by students of medieval learning than such libraries as the

Bodleian and Bibliothèque Nationale. I have worked also, however, at both Oxford and Paris, at Munich, Florence, Bologna, and elsewhere; but it has of course been impossible to examine all the thousands of manuscripts bearing upon the subject, and the war prevented me from visiting some libraries, such as the important medieval collection of Amplonius at Erfurt. However, a fairly wide survey of the catalogues of collections of manuscripts has convinced me that I have read a representative selection. Such classified lists of medieval manuscripts as Mrs. Dorothea Singer has undertaken for the British Isles should greatly facilitate the future labors of investigators in this field.

Although working in a rather new field, I have been aided by editions of medieval writers produced by modern scholarship, and by various series, books, and articles tending, at least, in the same direction as mine. Some such publications have appeared or come to my notice too late for use or even for mention in the text: for instance, another edition of the *De medicamentis* of Marcellus Empiricus by M. Niedermann; the printing of the *Twelve Experiments with Snakeskin* of John Paulinus by J. W. S. Johnsson in *Bull. d. l. société franç. d'hist. d. l. méd.*, XII, 257-67; the detailed studies of Sante Ferrari on Peter of Abano; and A. Franz, *Die kirchlichen Benediktionen im Mittelalter*, 1909, 2 vols. The breeding place of the eel (to which I allude at I, 491) is now, as a result of recent investigation by Dr. J. Schmidt, placed "about 2500 miles from the mouth of the English Channel and 500 miles north-east of the Leeward Islands" (*Discovery*, Oct., 1922, p. 256) instead of in the Mediterranean.

A man who once wrote in Dublin [1] complained of the difficulty of composing a learned work so far from the Bodleian and British Museum, and I have often felt the same way. When able to visit foreign collections or the largest libraries in this country, or when books have been sent for my use for a limited period, I have spent all the available time in the collection of material, which has been written up later as opportunity offered. Naturally one then finds many small and some important points which require verification or further investigation, but which must be postponed until one's next vacation or trip abroad, by which time some of the smaller points are apt to be forgotten.

[1] H. Cotton, *Five Books of Maccabees*, 1832, pp. ix-x.

Of such loose threads I fear that more remain than could be desired. And I have so often caught myself in the act of misinterpretation, misplaced emphasis, and other mistakes, that I have no doubt there are other errors as well as omissions which other scholars will be able to point out and which I trust they will. Despite this prospect, I have been bold in affirming my independent opinion on any point where I have one, even if it conflicts with that of specialists or puts me in the position of criticizing my betters. Constant questioning, criticism, new points of view, and conflict of opinion are essential in the pursuit of truth.

After some hesitation I decided, because of the expense, the length of the work, and the increasing unfamiliarity of readers with Greek and Latin, as a rule not to give in the footnotes the original language of passages used in the text. I have, however, usually supplied the Latin or Greek when I have made a free translation or one with which I felt that others might not agree. But in such cases I advise critics not to reject my rendering utterly without some further examination of the context and line of thought of the author or treatise in question, since the wording of particular passages in texts and manuscripts is liable to be corrupt, and since my purpose in quoting particular passages is to illustrate the general attitude of the author or treatise. In describing manuscripts I have employed quotation marks when I knew from personal examination or otherwise that the Latin was that of the manuscript itself, and have omitted quotation marks where the Latin seemed rather to be that of the description in the catalogue. Usually I have let the faulty spelling and syntax of medieval copyists stand without comment. But as I am not an expert in palaeography and have examined a large number of manuscripts primarily for their substance, the reader should not regard my Latin quotations from them as exact transliterations or carefully considered texts. He should also remember that there is little uniformity in the manuscripts themselves. I have tried to reduce the bulk of the footnotes by the briefest forms of reference consistent with clearness—consult lists of abbreviations and of works frequently cited by author and date of publication—and by use of appendices at the close of certain chapters.

Within the limits of a preface I may not enumerate all the libraries where I have been permitted to work or which have generously sent books—sometimes rare volumes—to Cleveland for my use, or all the librarians who have personally assisted my researches or courteously and carefully answered my written inquiries, or the other scholars who have aided or encouraged the preparation of this work, but I hope they may feel that their kindness has not been in vain. In library matters I have perhaps most frequently imposed upon the good nature of Mr. Frederic C. Erb of the Columbia University Library, Mr. Gordon W. Thayer, in charge of the John G. White collection in the Cleveland Public Library, and Mr. George F. Strong, librarian of Adelbert College, Western Reserve University; and I cannot forbear to mention the interest shown in my work by Dr. R. L. Poole at the Bodleian. For letters facilitating my studies abroad before the war or application for a passport immediately after the war I am indebted to the Hon. Philander C. Knox, then Secretary of State, to Frederick P. Keppel, then Assistant Secretary of War, to Drs. J. Franklin Jameson and Charles F. Thwing, and to Professors Henry E. Bourne and Henry Crew. Professors C. H. Haskins,[1] L. C. Karpinski, W. G. Leutner, W. A. Locy, D. B. Macdonald, L. J. Paetow, S. B. Platner, E. C. Richardson, James Harvey Robinson, David Eugene Smith, D'Arcy W. Thompson, A. H. Thorndike, E. L. Thorndike, T. Wingate Todd, and Hutton Webster, and Drs. Charles Singer and Se Boyar have kindly read various chapters in manuscript or proof and offered helpful suggestions. The burden of proof-reading has been generously shared with me by Professors B. P. Bourland, C. D. Lamberton, and Walter Libby, and especially by Professor Harold North Fowler who has corrected proof for practically the entire work. After receiving such expert aid and sound counsel I must assume all the deeper guilt for such faults and indiscretions as the book may display.

[1] But Professor Haskins' recent article in *Isis* on "Michael Scot and Frederick II" and my chapter on Michael Scot were written quite independently.

ABBREVIATIONS

Abhandl. Abhandlungen zur Geschichte der Mathematischen Wissenschaften, begründet von M. Cantor, Teubner, Leipzig.

Addit. Additional Manuscripts in the British Museum.

Amplon. Manuscript collection of Amplonius Ratinck at Erfurt.

AN Ante-Nicene Fathers, American Reprint of the Edinburgh edition, in 9 vols., 1913.

AS Acta sanctorum.

Beiträge Beiträge zur Geschichte der Philosophie des Mittelalters, ed. by C. Baeumker, G. v. Hertling, M. Baumgartner, et al., Münster, 1891-.

BL Bodleian Library, Oxford.

BM British Museum, London.

BN Bibliothèque Nationale, Paris.

Borgnet Augustus Borgnet, ed. B. Alberti Magni Opera omnia, Paris, 1890-1899, in 38 vols.

Brewer Fr. Rogeri Bacon Opera quaedam hactenus inedita, ed. J. S. Brewer, London, 1859, in RS, XV.

Bridges The Opus Maius of Roger Bacon, ed. J. H. Bridges, I-II, Oxford, 1897; III, 1900.

CCAG Catalogus codicum astrologorum Graecorum, ed. F. Cumont, W. Kroll, F. Boll, et al., 1898.

CE Catholic Encyclopedia.

CFCB Census of Fifteenth Century Books Owned in America, compiled by a committee of the Bibliographical Society of America, New York, 1919.

CLM Codex Latinus Monacensis (Latin MS at Munich).

CSEL	Corpus scriptorum ecclesiasticorum latinorum, Vienna, 1866-.
CU	Cambridge University (used to distinguish MSS in colleges having the same names as those at Oxford).
CUL	Cambridge University Library.
DNB	Dictionary of National Biography.
EB	Encyclopedia Britannica, 11th edition.
EETS	Early English Text Society Publications.
EHR	English Historical Review.
ERE	Encyclopedia of Religion and Ethics, ed. J. Hastings et al., 1908-.
HL	Histoire Littéraire de la France.
HZ	Historische Zeitschrift, Munich, 1859-.
Kühn	Medici Graeci, ed. C. J. Kühn, Leipzig, 1829, containing the works of Galen, Dioscorides, etc.
MG	Monumenta Germaniae.
MS	Manuscript.
MSS	Manuscripts.
Muratori	Rerum Italicarum scriptores ab anno aerae christianae 500 ad 1500, ed. L. A. Muratori, 1723-1751.
NH	C. Plinii Secundi Naturalis Historia (Pliny's Natural History).
PG	Migne, Patrologiae cursus completus, series graeca.
PL	Migne, Patrologiae cursus completus, series latina.
PN	The Nicene and Post-Nicene Fathers, Second Series, ed. Wace and Schaff, 1890-1900, 14 vols.
PW	Pauly and Wissowa, Realencyclopädie der classischen Altertumswissenschaft.
RS	"Rolls Series," or Rerum Britannicarum medii aevi scriptores, 99 works in 244 vols., London, 1858-1896.

TU Texte und Untersuchungen zur Geschichte der altchristlichen Literatur, ed. Gebhardt und Harnack.

DESIGNATION OF MANUSCRIPTS

Individual manuscripts are usually briefly designated in the ensuing notes and appendices by a single word indicating the place or collection where the MS is found and the number or shelf-mark of the individual MS. So many of the catalogues of MSS collections which I consulted were undated and without name of author that I have decided to attempt no catalogue of them. The brief designations that I give will be sufficient for anyone who is interested in MSS. In giving Latin titles, *Incipits,* and the like of MSS I employ quotation marks when I know from personal examination or otherwise that the wording is that of the MS itself, and omit the marks where the Latin seems rather to be that of the description in the manuscript catalogue or other source of information. In the following *List of Works Frequently Cited* are included a few MSS catalogues whose authors I shall have occasion to refer to by name.

LIST OF WORKS FREQUENTLY CITED BY AUTHOR AND DATE OF PUBLICATION OR BRIEF TITLE

For more detailed bibliography on specific topics and for editions or manuscripts of the texts used see the bibliographies, references, and appendices to individual chapters. I also include here some works of general interest or of rather cursory character which I have not had occasion to mention elsewhere; and I usually add, for purposes of differentiation, other works in our field by an author than those works by him which are frequently cited. Of the many histories of the sciences, medicine, and magic that have appeared since the invention of printing I have included but a small selection. Almost without exception they have to be used with the greatest caution.

Abano, Peter of, Conciliator differentiarum philosophorum et praecipue medicorum, 1472, 1476, 1521, 1526, etc.

De venenis, 1472, 1476, 1484, 1490, 1515, 1521, etc.

Abel, ed. Orphica, 1885.

Abelard, Peter. Opera hactenus seorsim edita, ed. V. Cousin, Paris, 1849-1859, 2 vols.

Ouvrages inédits, ed. V. Cousin, 1835.

Abt, Die Apologie des Apuleius von Madaura und die antike Zauberei, Giessen, 1908.

Achmetis Oneirocriticon, ed. Rigaltius, Paris, 1603.

Adelard of Bath, Quaestiones naturales, 1480, 1485, etc.

De eodem et diverso, ed. H. Willner, Münster, 1903.

Ahrens, K. Das Buch der Naturgegenstände, 1892.

Zur Geschichte des sogenannten Physiologus, 1885.

Ailly, Pierre d', Tractatus de ymagine mundi (and other works), 1480 (?).

Albertus Magnus, Opera omnia, ed. A. Borgnet, Paris, 1890-1899, 38 vols.

Allbutt, Sir T. Clifford. The Historical Relations of Medicine and Surgery to the End of the Sixteenth Century, London, 1905, 122 pp.; an address delivered at the St. Louis Congress in 1904.

The Rise of the Experimental Method in Oxford, London, 1902, 53 pp., from Journal of the Oxford University Junior Scientific Club, May, 1902, being the ninth Robert Boyle Lecture.

Science and Medieval Thought, London, 1901, 116 brief pages. The Harveian Oration delivered before the Royal College of Physicians.

Allendy, R. F. L'Alchimie et la Médecine; Étude sur les théories hermétiques dans l'histoire de la médecine, Paris, 1912, 155 pp.

Anz, W. Zur Frage nach dem Ursprung des Gnostizismus, Leipzig, 1897.

Aquinas, Thomas. Opera omnia, ed. E. Fretté et P. Maré, Paris, 1871-1880, 34 vols.

Aristotle, De animalibus historia, ed. Dittmeyer, 1907; English translations by R. Creswell, 1848, and D'Arcy W. Thompson, Oxford, 1910.

Pseudo-Aristotle. Lapidarius, Merszborg, 1473.

Secretum secretorum, Latin translation from the Arabic by Philip of Tripoli in many editions; and see Gaster.

Arnald of Villanova, Opera, Lyons, 1532.

Artemidori Daldiani et Achmetis Sereimi F. Oneirocritica; Astrampsychi et Nicephori versus etiam Oneirocritici; Nicolai Rigaltii ad Artemidorum Notae, Paris, 1603.

Ashmole, Elias, Theatrum chemicum Britannicum, 1652.

Astruc, Jean. Mémoires pour servir à l'histoire de la Faculté de Médecine de Montpellier, Paris, 1767.

Auriferae artis quam chemiam vocant antiquissimi auctores, Basel, 1572.

Barach et Wrobel, Bibliotheca Philosophorum Mediae Aetatis, 1876-1878, 2 vols.

Bartholomew of England, De proprietatibus rerum Lingelbach, Heidelberg, 1488, and other editions.

Bauhin, De plantis a divis sanctisve nomen habentibus, Basel, 1591.

Baur, Ludwig, ed. Gundissalinus De divisione philosophiae, Münster, 1903.

Die Philosophischen Werke des Robert Grosseteste, Münster, 1912.

Beazley, C. R. The Dawn of Modern Geography, London, 1897-1906, 3 vols.

Bernard, E. Catalogi librorum manuscriptorum Angliae et Hiberniae in unum collecti (The old catalogue of the Bodleian MSS), Tom. I, Pars 1, Oxford, 1697.

Berthelot, P. E. M. Archéologie et histoire des sciences avec publication nouvelle du papyrus grec chimique de Leyde et impression originale du Liber de septuaginta de Geber, Paris, 1906.

Collection des anciens alchimistes grecs, 1887-1888, 3 vols.

Introduction à l'étude de la chimie des anciens et du moyen âge, 1889.

La chimie au moyen âge, 1893, 3 vols.

Les origines de l'alchimie, 1885.

Sur les voyages de Galien et de Zosime dans l'Archipel et en Asie, et sur la matière médicale dans l'antiquité, in Journal des Savants, 1895, pp. 382-7.

Bezold, F. von, Astrologische Geschichtsconstruction im Mittelalter, in Deutsche Zeitschrift für Geschichtswissenschaft, VIII (1892) 29ff.

Bibliotheca Chemica. See Borel and Manget.

Björnbo, A. A. und Vogl, S. Alkindi, Tideus, und Pseudo-Euklid; drei optische Werke, Leipzig, 1911.

Black, W. H. Catalogue of the Ashmolean Manuscripts, Oxford, 1845.

Boffito, P. G. Il Commento di Cecco d'Ascoli all' Alcabizzo, Florence, 1905.

Il De principiis astrologiae di Cecco d'Ascoli, in Giornale Storico della Letteratura Italiana, Suppl. 6, Turin, 1903.

Perchè fu condannato al fuoco l'astrologo Cecco d'Ascoli, in Studi e Documenti di Storia e Diritto, Publicazione periodica dell' accademia de conferenza Storico-Giuridiche, Rome, XX (1899).

Boll, Franz. Die Erforschung der antiken Astrologie, in Neue Jahrb. f. d. klass. Altert., XI (1908) 103-26.

Eine arabisch-byzantische Quelle des Dialogs Hermippus, in Sitzb. Heidelberg Akad., Philos. Hist. Classe (1912) No. 18, 28 pp.

Sphaera, Leipzig, 1903.

Studien über Claudius Ptolemaeus, in Jahrb. f. klass. Philol., Suppl. Bd. XXI.

Zur Ueberlieferungsgeschichte d. griech. Astrologie u. Astronomie, in Münch. Akad. Sitzb., 1899.

Boll und Bezold, Sternglauben, Leipzig, 1918; I have not seen.

Bonatti, Guido. Liber astronomicus, Ratdolt, Augsburg, 1491.

Boncompagni, B. Della vita e delle Opere di Gherardo Cremonese traduttore del secolo duodecimo e di Gherardo da Sabbionetta astronomo del secolo decimoterzo, Rome, 1851.

Della vita e delle opere di Guido Bonatti astrologo ed astronomo del secolo decimoterzo, Rome, 1851. Estratte dal Giornale Arcadico, Tomo CXXIII-CXXIV. Della vita e delle opere di Leonardo Pisano, Rome, 1852.

Intorno ad alcune opere di Leonardo Pisano, Rome, 1854.

Borel, P. Bibliotheca Chimica seu catalogus librorum philosophicorum hermeticorum usque ad annum 1653, Paris, 1654.

Bostock, J. and Riley, H. T. The Natural History of Pliny, translated with copious notes, London, 1855; reprinted 1887.

Bouché-Leclercq, A. L'astrologie dans le monde romain, in Revue Historique, vol. 65 (1897) 241-99.

L'astrologie grecque, Paris, 1899, 658 pp.
Histoire de la divination dans l'antiquité, 1879-1882, 4 vols.

Breasted, J. H. Development of Religion and Thought in Ancient Egypt, New York, 1912.
A History of Egypt, 1905; second ed., 1909.

Brehaut, E. An Encyclopedist of the Dark Ages; Isidore of Seville, in Columbia University Studies in History, etc., vol. 48 (1912) 1-274.

Brewer, J. S. Monumenta Franciscana (RS IV, 1), London, 1858.

Brown, J. Wood. An inquiry into the life and legend of Michael Scot, Edinburgh, 1897.

Browne, Edward G. Arabian Medicine (the Fitzpatrick Lectures of 1919 and 1920), Cambridge University Press, 1921.

Browne, Sir Thomas. Pseudodoxia Epidemica, 1650.

Bubnov, N. ed. Gerberti opera mathematica, Berlin, 1899.

Budge, E. A. W. Egyptian Magic, London, 1899.
Ethiopic Histories of Alexander by the Pseudo-Callisthenes and other writers, Cambridge University Press, 1896.
Syriac Version of Pseudo-Callisthenes, Cambridge, 1889.
Syrian Anatomy, Pathology, and Therapeutics, London, 1913, 2 vols.

Bunbury, E. H. A History of Ancient Geography, London, 1879, 2 vols.

Cahier et Martin, Mélanges d'archéologie, d'histoire et de littérature, Paris, 1847-1856, 4 folio vols.

Cajori, F. History of Mathematics; second edition, revised and enlarged, 1919.

Cantor, M. Vorlesungen über Geschichte der Mathematik, 3rd edition, Leipzig, 1899-1908, 4 vols. Reprint of vol. II in 1913.

Carini, S. I. Sulle Scienze Occulte nel Medio Evo, Palermo, 1872; I have not seen.

Cauzons, Th. de. La magie et la sorcellerie en France, 1910, 4 vols.; largely compiled from secondary sources.

Charles, E. Roger Bacon: sa vie, ses ouvrages, ses doctrines, Bordeaux, 1861.

Charles, R. H. The Apocrypha and Pseudepigrapha of the Old Testament, English translation with introductions and critical and explanatory notes in conjunction with many scholars, Oxford, 1913, 2 large vols.

Ascension of Isaiah, 1900, and reprinted in 1917.

The Book of Enoch, Oxford, 1893; translated anew, 1912.

Charles, R. H. and Morfill, W. R. The Book of the Secrets of Enoch, Oxford, 1896.

Charterius, Renatus ed. Galeni opera, Paris, 1679, 13 vols.

Chartularium Universitatis Parisiensis, see Denifle et Chatelain.

Chassang, A. Le merveilleux dans l'antiquité, 1882; I have not seen.

Choulant, Ludwig. Albertus Magnus in seiner Bedeutung für die Naturwissenschaften historisch und bibliographisch dargestellt, in Janus, I (1846) 152ff.

Die Anfänge wissenschaftlicher Naturgeschichte und naturhistorischer Abbildung, Dresden, 1856.

Handbuch der Bücherkunde für die ältere Medicin, 2nd edition, Leipzig, 1841; like the foregoing, slighter than the title leads one to hope.

ed. Macer Floridus de viribus herbarum una cum Walafridi Strabonis, Othonis Cremonensis et Ioannis Folcz carminibus similis argumenti, 1832.

Christ, W. Geschichte der Griechischen Litteratur; see W. Schmid.

Chwolson, D. Die Ssabier und der Ssabismus, Petrograd, 1856, 2 vols.

Clément-Mullet, J. J. Essai sur la minéralogie arabe, Paris, 1868, in Journal asiatique, Tome XI, Sèrie VI.

Traité des poisons de Maimonide, 1865.

Clerval, Hermann le Dalmate, Paris, 1891, eleven pp.
Les écoles de Chartres au moyen âge, Chartres, 1895.

Cockayne, O. Leechdoms, Wortcunning, and Starcraft of Early England, in RS XXXV, London, 1864-1866, 3 vols.
Narratiunculae anglice conscriptae, 1861.

Congrès Périodique International des Sciences Médicales, 17th Session, London, Section XXIII, History of Medicine, 1913.

Cousin, V. See Abelard.

Coxe, H. O. Catalogi Codicum Manuscriptorum Bibliothecae Bodleianae Pars Secunda Codices Latinos et Miscellaneos Laudianos complectens, Oxford, 1858-1885.
Catalogi Codicum Manuscriptorum Bibliothecae Bodleianae Pars Tertia Codices Graecos et Latinos Canonicianos complectens, Oxford, 1854.
Catalogus Codicum Manuscriptorum qui in collegiis aulisque Oxoniensibus hodie adservantur, 1852, 2 vols.

Cumont, F. Astrology and Religion among the Greeks and Romans, 1912, 2 vols. And see CCAG under Abbreviations.

Daremberg, Ch. V. Exposition des connaissances de Galien sur l'anatomie, la physiologie, et la pathologie du système nerveux, Paris, 1841.
Histoire des sciences médicales, Paris, 1870, 2 vols.
La médecine; histoire et doctrines, Paris, 1865.
Notices et extraits des manuscrits médicaux, 1853.

Delambre, J. B. J. Histoire de l'astronomie du moyen âge, Paris, 1819.

Delisle, L. Inventaire des manuscrits latins conservés à la bibliothèque nationale sous les numéros 8823-18613 et faisant suite à la série dont la catalogue a été publié en 1744, Paris, 1863-1871.

Denifle, H. Quellen zur Gelehrtengeschichte des Predigerordens im 13 und 14 Jahrhundert, in Archiv f. Lit. u. Kirchengesch. d. Mittelalters, Berlin, II (1886) 165-248.

Denifle et Chatelain, Chartularium Universitatis Parisiensis, Paris, 1889-1891, 2 vols.

Denis, F. Le monde enchanté, cosmographie et histoire naturelles fantastiques du moyen âge, Paris, 1843. A curious little volume with a bibliography of works now forgotten.

Doutté, E. Magie et religion dans l'Afrique du Nord, Alger, 1909.

Duhem, Pierre. Le Système du Monde: Histoire des Doctrines Cosmologiques de Platon à Copernic, 5 vols., Paris, 1913-1917.

Du Prel, C. Die Magie als Naturwissenschaft, 1899, 2 vols. Occult speculation, not historical treatment; the author seems to have no direct acquaintance with sources earlier than Agrippa in the sixteenth century.

Easter, D. B. A Study of the Magic Elements in the romans d'aventure and the romans bretons, Johns Hopkins, 1906.

Ennemoser, J. History of Magic, London, 1854.

Enoch, Book of. See Charles.

Epiphanius. Opera ed. G. Dindorf, Leipzig, 1859-1862, 5 vols.

Evans, H. R. The Old and New Magic, Chicago, 1906.

Fabricius, Bibliotheca Graeca, 1711.

Bibliotheca Latina Mediae et Infimae Aetatis, 1734-1746, 6 vols.

Codex Pseudepigraphus Veteris Testamenti, 1713-1733.

Farnell, L. R. Greece and Babylon; a comparative sketch of Mesopotamian, Anatolian, and Hellenic Religions, Edinburgh, 1911.

The Higher Aspects of Greek Religion, New York, 1912.

Ferckel, C. Die Gynäkologie des Thomas von Brabants, ausgewählte Kapitel aus Buch I de naturis rerum beendet um 1240, Munich, 1912, in G. Klein, Alte Meister d. Medizin u. Naturkunde.

Ferguson, John. Bibliotheca Chemica, a catalogue of alchemical, chemical and pharmaceutical books in the collection of the late James Young, Glasgow, 1906.

Fort, G. F. Medical Economy; a contribution to the history of European morals from the Roman Empire to 1400, New York, 1883.

Fossi, F. Catalogus codicum saeculo XV impressorum qui in publica Bibliotheca Magliabechiana Florentiae adservantur, 1793-1795.

Frazer, Sir J. G. Folk-Lore in the Old Testament, 3 vols., 1918.

Golden Bough, edition of 1894, 2 vols.

Magic Art and the Evolution of Kings, 2 vols., 1911.

Some Popular Superstitions of the Ancients, in Folk-Lore, 1890.

Spirits of the Corn and of the Wild, 2 vols., 1912.

Garinet. Histoire de la Magie en France.

Garrison, F. H. An Introduction to the History of Medicine, 2nd edition, Philadelphia, 1917.

Gaster, M. A Hebrew Version of the Secretum secretorum, published for the first time, in Journal of the Royal Asiatic Society, London, 1907, pp. 879-913; 1908, pp. 111-62, 1065-84.

Gerland, E. Geschichte der Physik von den ältesten Zeiten bis zum Ausgange des achtzehnten Jahrhunderts, in Königl. Akad. d. Wiss., XXIV (1913) Munich and Berlin.

Gerland und Traumüller, Geschichte der Physikalischen Experimentierkunst, Leipzig, 1899.

Giacosa, P. Magistri Salernitani nondum editi, Turin, 1901.

Gilbert of England, Compendium medicinae, Lyons, 1510.

Gloria, Andrea. Monumenti della Università di Padova, 1222-1318, in Memorie del Reale Istituto Veneto di Scienze, Lettere ed Arti, XXII (1884).

Monumenti della Università di Padova, 1318-1405, 1888.

Gordon, Bernard. Lilium medicinae, Venice, 1496, etc.

Practica (and other treatises), 1521.

Grabmann, Martin. Forschungen über die lateinischen Aristoteles-Uebersetzungen des XIII Jahrhunderts, Münster, 1916.

Die Geschichte der Scholastischen Methode, Freiburg, 1909-1911, 2 vols.

Graesse, J. G. T. Bibliotheca magica, 1843; of little service to me.

Grenfell, B. P. The Present Position of Papyrology, in Bulletin of John Rylands Library, Manchester, VI (1921) 142-62.

Haeser, H. Lehrbuch der Geschichte der Medicin und der Volkskrankheiten, Dritte Bearbeitung, 1875-1882.

Halle, J. Zur Geschichte der Medizin von Hippokrates bis zum XVIII Jahrhundert, Munich, 1909, 199 pp.; too brief, but suggests interesting topics.

Halliwell, J. O. Rara Mathematica, 1839.

Hammer-Jensen. Das sogennannte IV Buch der Meteorologie des Aristoteles, in Hermes, L (1915) 113-36.

Ptolemaios und Heron, Ibid., XLVIII (1913), 224ff.

Hansen, J. Zauberwahn, Inquisition, und Hexenprozess im Mittelalter, Munich and Leipzig, 1900.

Haskins, C. H. Adelard of Bath, in EHR XXVI (1911) 491-8; XXVIII (1913), 515-6.

Leo Tuscus, in EHR XXXIII (1918), 492-6.

The "De Arte Venandi cum Avibus" of the Emperor Frederick II, EHR XXXVI (1921) 334-55.

The Reception of Arabic Science in England, EHR XXX (1915), 56-69.

The Greek Element in the Renaissance of the Twelfth Century, in American Historical Review, XXV (1920) 603-15.

The Translations of Hugo Sanctelliensis, in Romanic Review, II (1911) 1-15.

Nimrod the Astronomer, Ibid., V (1914) 203-12.

A List of Text-books from the Close of the Twelfth Century, in Harvard Studies in Classical Philology, XX (1909) 75-94.

Haskins and Lockwood. The Sicilian Translators of the Twelfth Century and the First Latin Versions of Ptolemy's Almagest, Ibid., XXI (1910), 75-102.

Hauréau, B. Bernard Délicieux et l'inquisition albigeoise, Paris, 1887.

Histoire de la philosophie scolastique, 1872-1880.

Le Mathematicus de Bernard Silvestris, Paris, 1895.

Les œuvres de Hugues de Saint Victor, essai critique, nouvelle édition, Paris, 1886.

Mélanges poétiques d'Hildebert de Lavardin.

Notices et extraits de quelques mss latins de la bibliothèque nationale, 1890-1893, 6 vols.

Singularités historiques et littéraires, Paris, 1861.

Hearnshaw, F. J. C. Medieval Contributions to Modern Civilization, 1921.

Heilbronner, J. C. Historia Matheseos universae praecipuorum mathematicorum vitas dogmata scripta et manuscripta complexa, Leipzig, 1742.

Heim, R. De rebus magicis Marcelli medici, in Schedae philol. Hermanno Usener oblatae, 1891, pp. 119-37.

Incantamenta magica graeca latina, in Jahrb. f. cl. Philol., 19 suppl. bd., Leipzig, 1893, pp. 463-576.

Heller, A. Geschichte der Physik von Aristoteles bis auf die neueste Zeit, Stuttgart, 1882-1884, 2 vols.

Hendrie, R. Theophili Libri III de diversis artibus, translated by, London, 1847.

Hengstenberg, E. W. Die Geschichte Bileams und seine Weissagungen, Berlin, 1842.

Henry, V. La magie dans l'Inde antique, 1904.

Henslow, G. Medical Works of the Fourteenth Century, London, 1899.

Hercher, ed. Aeliani opera, 1864.

ed. Artemidori Oneirocritica, Leipzig, 1864.

ed. Astrampsychi oculorum decades, Berlin, 1863.

Hertling, G. von, Albertus Magnus; Beiträge zu seiner Würdigung, revised edition with help of Baeumker and Endres, Münster, 1914.

Hubert, H. Magia, in Daremberg-Saglio.

Hubert et Mauss, Esquisse d'une Théorie Générale de la Magie, in Année Sociologique, 1902-1903, pp. 1-146.

Husik, I. A History of Medieval Jewish Philosophy, 1916.

Ishak ibn Sulaiman, Opera, 1515.

James, M. R. A Descriptive Catalogue of the McClean Collection of MSS in the Fitzwilliam Museum, 1912.

A Descriptive Catalogue of the MSS in the Fitzwilliam Museum, 1895.

A Descriptive Catalogue of the MSS in the Library of Corpus Christi College, Cambridge, 1912, 2 vols.

A Descriptive Catalogue of the MSS in the Library of Gonville and Caius College, 1907-1908, 2 vols.

A Descriptive Catalogue of the MSS in the Library of Pembroke College, 1905.

A Descriptive Catalogue of the MSS in the Library of Peterhouse, 1899.

A Descriptive Catalogue of the MSS in the Library of St. John's College, Cambridge, 1913.

A Descriptive Catalogue of the MSS in the Library of Sidney Sussex College, Cambridge, 1895.

The Ancient Libraries of Canterbury and Dover, 1903.

The Western MSS in the Library of Emmanuel College, 1904.

The Western MSS in the Library of Trinity College, Cambridge, 1900-1904, 4 vols.

Janus, Zeitschrift für Geschichte und Literatur der Medizin, 1846-.

Jenaer medizin-historische Beiträge, herausg. von T. M. Steineg, 1912-.

Joël, D. Der Aberglaube und die Stellung des Judenthums zu demselben, 1881.

John of Salisbury, Metalogicus, in Migne PL vol. 199.

Polycraticus sive de nugis curialium et vestigiis philosophorum, Ibid. and also ed. C. C. I. Webb, Oxford, 1909.

Joret, Les plantes dans l'antiquité et au moyen âge, 2 vols., Paris, 1897 and 1904.

Jourdain, A. Recherches critiques sur l'âge et l'origine des traductions latines d'Aristote, Paris, 1819; 2nd edition, 1843.

Jourdain, C. Dissertation sur l'état de la philosophie naturelle en occident et principalement en France pendant la première moitié du XIIe siècle, Paris, 1838.

Excursions historiques et philosophiques à travers le moyen âge, Paris, 1888.

Karpinski, L. C. Hindu Science, in American Mathematical Monthly, XXVI (1919) pp. 298-300.

Robert of Chester's Latin translation of the Algebra of al-Khowarizmi, with introduction, critical notes, and an English version, New York, 1915.

The "Quadripartitum numerorum" of John of Meurs, in Bibliotheca Mathematica, III Folge, XIII Bd. (1913) 99-114.

Kaufmann, A. Thomas von Chantimpré, Cologne, 1899.

King, C. W. The Gnostics and their Remains, ancient and medieval, London, 1887.

The Natural History, ancient and modern, of Precious Stones and Gems, London, 1855.

Kopp, H. Beiträge zur Geschichte der Chemie, Brunswick, 1869-1875.

Ueber den Zustand der Naturwissenschaften im Mittelalter, 1869.

Kretschmer, C. Die physische Erdkunde im christlichen Mittelalter, 1889.

Krumbacher, K. Geschichte der byzantinischen Literatur, 527-1453 A. D., 2nd edition, Munich, 1897.

Kunz, G. F. The Curious Lore of Precious Stones, Philadelphia, 1913.

Magic of Jewels and Charms, Philadelphia, 1915.

Langlois, Ch. V. La connaissance de la nature et du monde au moyen âge d'après quelques écrits français à l'usage des laïcs, Paris, 1911.

Maître Bernard, in Bibl. de l'École des Chartes, LIV (1893) 225-50, 795.

Lauchert, F. Geschichte des Physiologus, Strassburg, 1889.

Lea, H. C. A History of the Inquisition of the Middle Ages, New York, 1883, 3 vols.

Le Brun. Histoire critique des pratiques superstitieuses, Amsterdam, 1733.

Lecky, W. E. H. History of European Morals from Augustus to Charlemagne, 1870, 2 vols.

History of the Rise and Influence of the Spirit of Rationalism in Europe, revised edition, London, 1870.

Lehmann, A. Aberglaube und Zauberei von den ältesten Zeiten an bis in die Gegenwart; deutsche autorisierte Uebersetzung von I. Petersen, Stuttgart, 1908. The historical treatment is scanty.

Leminne, J. Les quatre éléments, in Mémoires couronnés par l'Académie Royale de Belgique, vol. 65, Brussels, 1903.

Lévy, L. G. Maimonide, 1911.

Liechty, R. de. Albert le Grand et saint Thomas d'Aquin, ou la science au moyen âge, Paris, 1880.

Lippmann, E. O. von. Entstehung und Ausbreitung der Alchemie, 1919.

Little, A. G. Initia operum Latinorum quae saeculis XIII, XIV, XV, attribuuntur, Manchester, 1904.

ed. Roger Bacon Essays, contributed by various writers on the occasion of the commemoration of the seventh centenary of his birth, Oxford, 1914.

ed. Part of the Opus Tertium of Roger Bacon, including a Fragment now printed for the first time, Aberdeen, 1912, in British Society of Franciscan Studies, IV.

Loisy. Magie, science et religion, in À propos d'histoire des religions, 1911, p.166ff.

Macdonald, D. B. The Religious Attitude and Life in Islam, Chicago, 1909.

Macray, Catalogus codicum MSS Bibliothecae Bodleianae,

V, Codices Rawlinsonianae, 1862-1900, 5 fascs.; IX, Codices Digbeianae, 1883.

Mai, A. Classici Auctores, 1835.

Mâle, E. Religious Art in France in the Thirteenth Century, translated from the third edition by Dora Nussey, 1913.

Mandonnet, P. Des écrits authentiques de S. Thomas d'Aquin, Fribourg, 1910.

Roger Bacon et la composition des trois Opus, in Revue Néo-Scolastique, Louvain, 1913, pp. 52-68, 164-80.

Roger Bacon et la Speculum astronomiae, Ibid., XVII (1910) 313-35.

Siger de Brabant et l'averroïsme latin au XIIIme siècle, Fribourg, 1899; 2nd edition, Louvain, 1908-1910, 2 vols.

Manget, J. J. Bibliotheca Chemica Curiosa, Geneva, 1702, 2 vols.

Manitius, Max. Geschichte der lateinischen Literatur des Mittelalters, Erster Teil, Von Justinian bis zur Mitte des zehnten Jahrhunderts, Munich, 1911, in Müller's Handbuch d. kl. Alt. Wiss. IX, 2, i.

Mann, M. F. Der Bestiaire Divin des Guillaume le Clerc, 1888.

Der Physiologus des Philipp von Thaon und seine Quellen, 1884.

Mappae clavicula, ed. M. A. Way in Archaeologia, London, XXXII (1847) 183-244.

Maury, Alfred. La magie et l'astrologie dans l'antiquité et au moyen âge, 1877. Brief as it is, perhaps the best general history of magic.

Mead, G. R. S. Apollonius of Tyana; a critical study of the only existing record of his life, 1901.

Echoes from the Gnosis, 1906, eleven vols.

Fragments of a Faith Forgotten, 1900.

Pistis-Sophia, now for the first time Englished, 1896.

Plotinus, Select Works of, with preface and bibliography, 1909.

Simon Magus, 1892.

Thrice Great Hermes, London, 1906, 3 vols.

Medicae artis principes post Hippocratem et Galenum Graeci Latinitate donati, ed. Stephanus, 1567.

Medici antiqui omnes qui latinis litteris . . . Aldus, Venice, 1547.

Mély, F. de et Ruelle, C. E. Les lapidaires de l'antiquité et du moyen âge, Paris, 1896. Mély has published many other works on gems and lapidaries of the past.

Merrifield, Mrs. M. P. Ancient Practice of Painting, or Original Treatises dating from the XIIth to XVIIIth centuries on the arts of painting, London, 1849.

Meyer, E. Albertus Magnus, ein Beitrag zur Geschichte der Botanik im XIII Jahrhundert, in Linnaea, X (1836) 641-741, XI (1837) 545.

Meyer, Karl. Der Aberglaube des Mittelalters und der nächstfolgenden Jahrhunderte, Basel, 1856.

Migne, Dictionnaire des Apocryphes, Paris, 1856.

See also under Abbreviations.

Millot-Carpentier, La Médecine au XIIIe siècle, in Annales Internationales d'Histoire, Congrès de Paris, 1900, 5e Section, Histoire des Sciences, pp. 171-96; a chapter from a history of medicine which the author's death unfortunately kept him from completing.

Milward, E. A Letter to the Honourable Sir Hans Sloane, Bart., in vindication of the character of those Greek writers in physick that flourished after Galen. . . particularly that of Alexander Trallian, 1733; reprinted as Trallianus Reviviscens, 1734.

Mommsen, Th. ed. C. Iulii Solini Collectanea rerum memorabilium, 1895.

Moore, Sir Norman, History of the Study of Medicine in the British Isles, 1908.

The History of St. Bartholomew's Hospital, London, 1918, 2 vols.

The Physician in English History, 1913. A popular lecture.

Muratori, L. A. Antiquitates Italicae medii aevi, Milan,

1738-1742, 6 vols. Edition of 1778 in more vols. Index, Turin, 1885.

See also under Abbreviations.

Naudé, Gabriel. Apologie pour tous les grands personnages qui ont esté faussement soupçonnez de Magie, Paris, 1625.

Neckam, Alexander. De naturis rerum, ed. T. Wright, in RS vol. 34, 1863.

Omont, H. Nouvelles acquisitions du départment des manuscrits pendant les années 1891-1910, Bibliothèque Nationale, Paris.

Orr, M. A. (Mrs. John Evershed) Dante and the Early Astronomers, London, 1913.

Paetow, L. J. Guide to the Study of Medieval History, University of California Press, 1917.

Pagel, J. L. Die Concordanciae des Joannes de Sancto Amando, 1894.

Geschichte der Medizin im Mittelalter, in Puschmann's Handbuch der Geschichte der Medizin, ed. Neuburger u. Pagel, I (1902) 622-752.

Neue litterarische Beiträge zur mittelalterlichen Medicin, Berlin, 1896.

Pangerl, A. Studien über Albert den Grossen, in Zeitschrift für katholische Theologie, XXII (1912) 304-46, 512-49, 784-800.

Pannier, L. Les lapidaires français du moyen âge, Paris, 1882.

Payne, J. F. English Medicine in Anglo-Saxon Times, 1904.

The Relation of Harvey to his Predecessors and especially to Galen: Harveian oration of 1896, in The Lancet, Oct. 24, 1896, 1136ff.

Perna. Artis quam chemiam vocant antiquissimi auctores, Basel, 1572.

Perrier, T. La médecine astrologique, Lyons, 1905, 88 pp. Slight.

Petrus de Prussia. Vita B. Alberti Magni, 1621.

Petrus Hispanus. Summa experimentorum sive thesaurus pauperum, Antwerp, 1497.
Philips, H. Medicine and Astrology, 1867.
Picavet, F. Esquisse d'une histoire comparée des philosophies médiévales, 2nd edition, Paris, 1907.
Pico della Mirandola. Opera omnia, 1519.
Pistis-Sophia, ed. Schwartze und Petermann, Coptic and Latin, 1851. Now for the first time Englished, by G. R. S. Mead, 1896.
Pitra, J. B. Analecta novissima, 1885-1888.
Analecta sacra, 1876-1882.
Spicilegium solesmense, 1852-1858.
Poisson, Théories et symboles des Alchimistes, Paris, 1891.
Poole, R. L. Illustrations of the History of Medieval Thought in the Departments of Theology and Ecclesiastical Politics, 1884; revised edition, 1920.
The Masters of the Schools at Paris and Chartres in John of Salisbury's Time, in EHR XXXV (1920) 321-42.
Pouchet, F. A. Histoire des sciences naturelles au moyen âge, ou Albert le Grand et son époque considéré comme point de départ de l'école expérimentale, Paris, 1853.
Ptolemy. Quadripartitum, 1484, and other editions.
Optica, ed. G. Govi, Turin, 1885.
Puccinotti, F. Storia della Medicina, 1850-1870, 3 vols.
Puschmann, Th. Alexander von Tralles, Originaltext und Uebersetzung nebst einer einleitenden Abhandlung, Vienna, 1878-1879.
Handbuch der Geschichte der Medizin, Jena, 1902-1905, 3 vols. Really a cooperative work under the editorship of Max Neuburger and Julius Pagel after Puschmann's death.
A History of Medical Education from the most remote to the most recent times, London, 1891, English translation.
Quetif, J. et Echard J. Scriptores Ordinis Praedicatorum, Paris, 1719.

Rambosson, A. Histoire et légendes des plantes, Paris, 1887.

Rashdall, H. ed. Fratris Rogeri Bacon Compendium Studii Theologiae, 1911.

The Universities of Europe in the Middle Ages, Oxford, 1895, 3 vols. in 2.

Rasis (Muhammad ibn Zakariya) Opera, Milan, 1481, and Bergamo, 1497.

Regnault, J. La sorcellerie: ses rapports avec les sciences biologiques, 1897, 345 pp.

Reitzenstein, R. Poimandres, Leipzig, 1904.

Renzi, S. de. Collectio Salernitana, 1852-1859, 5 vols.

Rose, Valentin. Anecdota graeca et graeco-latina, Berlin, 1864.

Aristoteles De lapidibus und Arnoldus Saxo, in Zeitschrift für deutsches Alterthum, XVIII (1875) 321-447.

Ptolemaeus und die Schule von Toledo, in Hermes, VIII (1874) 327-49.

ed. Plinii Secundi Iunioris de medicina libri tres, Leipzig, 1875.

Ueber die Medicina Plinii, in Hermes, VIII (1874) 19-66.

Verzeichnis der lateinischen Handschriften der K. Bibliothek zu Berlin, Band XII (1893), XIII (1902-1903-1905).

Ruska, J. Das Steinbuch des Aristoteles . . . nach der arabischen Handschrift, Heidelberg, 1912.

Der diamant in der Medizin, in Deutsche Gesell. f. Gesch. d. Mediz. u. d. Naturwiss., Zwanzig Abhandl. z. Gesch. d. Mediz., 1908.

Zur älteren arabischen Algebra und Rechenkunst, Heidelberg, 1917.

Rydberg, V. The Magic of the Middle Ages, 1879, translated from the Swedish. Popular.

Salverte, E. Des sciences occultes, ou essai sur la magie, Paris, 1843.

Sánchez Pérez, J. A. Biografías de Matemáticos Árabes que florecieron en España, Madrid, 1921.

Schanz, M. Geschichte der Römischen Litteratur, Dritter Teil, Munich, 1905; Vierter Teil, Erste Hälfte, Munich, 1914, in Müller's Handbuch d. klass. Alt. Wiss., VIII, 3.

Schepss, G. ed. Priscilliani quae supersunt, 1889.

Schindler. Der Aberglaube des Mittelalters, Breslau, 1858.

Schmid, W. Die Nachklassiche Periode der Griechischen Litteratur, 1913, in Müller's Handb. d. kl. Alt. Wiss., VII, ii, 2.

Schum, W. Beschriebendes Verzeichnis der Amplonianischen Handschriften-Sammlung zu Erfurt, Berlin, 1887.

Sighart, J. Albertus Magnus: sein Leben und seine Wissenschaft, Ratisbon, 1857; French translation, Paris, 1862; partial English translation by T. A. Dixon, London, 1876.

Singer, Charles. Early English Magic and Medicine, 1920, 34 pp.

"Science," pp. 106-48 in "Medieval Contributions to Modern Civilization," ed. F. J. C. Hearnshaw, 1921.

Studies in the History and Method of Science, Oxford, 1917; a second volume appeared in May, 1921.

Stapper, Richard. Papst Johannes XXI, Münster, 1898, in Kirchengesch. Studien herausg. v. Dr. Knöpfler, IV, 4.

Steele, R. Opera hactenus inedita Rogeri Baconi, 1905-1920.

Steinschneider, Moritz. Abraham ibn Ezra, in Abhandl., (1880) 57-128.

Apollonius von Thyana (oder Balinas) bei den Arabern, in Zeitschrift d. deutschen morgenländischen Gesellschaft, XLV (1891) 439-46.

Arabische Lapidarien, Ibid., XLIX (1895).

Constantinus Africanus und seine arabischen Quellen, in Virchow's Archiv für pathologische Anatomie, etc., Berlin, XXXVII (1866) 351-410.

Der Aberglaube, Hamburg, 1900, 34 pp.

Die europäischen Uebersetzungen aus dem Arabischen bis Mitte des 17 Jahrhunderts, in Sitzungsberichte d. kaiserl. Akad. d. Wiss., Philos. Hist. Klasse, Vienna, CXLIX, 4 (1905); CLI, 1 (1906).

Lapidarien, ein culturgeschichtlicher Versuch, in Semitic Studies in memory of Rev. Dr. Alexander Kohut, Berlin, 1897, pp. 42-72.

Maschallah, in Zeitsch. d. deut. morgenl. Gesell., LIII (1899), 434-40.

Zum Speculum astronomicum des Albertus Magnus über die darin angeführten Schriftsteller und Schriften, in Zeitschrift für Mathematik und Physik, Leipzig, XVI (1871) 357-96.

Zur alchimistischen Literatur der Araber, in Zeitsch, d. deut. morgenl. Gesell., LVIII (1904) 299-315.

Zur pseudepigraphischen Literatur insbesondere der geheimen Wissenschaften des Mittelalters; aus hebräischen und arabischen Quellen, Berlin, 1862.

Stephanus, H. Medicae artis principes post Hippocratem et Galenum Graeci Latinitate donati, et Latini, 1567.

Strunz, Franz. Geschichte der Naturwissenschaften im Mittelalter, Stuttgart, 1910, 120 pp. Without index or references.

Studien zur Geschichte der Medizin herausgegeben von der Puschmann-Stiftung an der Universität Leipzig, 1907-.

Sudhoff, Karl. His various articles in the foregoing publication and other periodicals of which he is an editor lie in large measure just outside our period and field, but some will be noted later in particular chapters.

Suter, H. Die Mathematiker und Astronomen der Araber, in Abhandl., X (1900) 1-277; XIV (1902) 257-85.

Die astronomischen Tafeln des Muhammed ibn Musa-al-Khwarizmi, Copenhagen, 1914.

Tanner, T. Bibliotheca Britannico-Hibernica, London, 1748. Still much cited but largely antiquated and unreliable.

Tavenner, E. Studies in Magic from Latin Literature, New York, 1916.

Taylor, H. O. The Classical Heritage, 1901.

The Medieval Mind, 2nd edition, 1914, 2 vols; 3rd edition, 1919.

Theatrum chemicum. See Zetzner.

Theatrum chemicum Britannicum. See Ashmole.

Theophilus Presbyter, Schedula diversarum artium, ed. A. Ilg, Vienna, 1874; English translation by R. Hendrie, London, 1847.

Thomas of Cantimpré, Bonum universale de apibus, 1516.

Thompson, D'Arcy W. Aristotle as a Biologist, 1913.

Glossary of Greek Birds, Oxford, 1895.

Historia animalium, Oxford, 1910; vol. IV in the English translation of The Works of Aristotle edited by J. A. Smith and W. D. Ross.

Thorndike, Lynn. Adelard of Bath and the Continuity of Universal Nature, in Nature, XCIV (1915) 616-7.

A Roman Astrologer as a Historical Source: Julius Firmicus Maternus, in Classical Philology, VIII (1913) 415-35.

Natural Science in the Middle Ages, in Popular Science Monthly (now The Scientific Monthly), LXXXVII (1915) 271-91.

Roger Bacon and Gunpowder, in Science, XLII (1915), 799-800.

Roger Bacon and Experimental Method in the Middle Ages, in The Philosophical Review, XXIII (1914), 271-98.

Some Medieval Conceptions of Magic, in The Monist, XXV (1915), 107-39.

The Attitude of Origen and Augustine toward Magic, in The Monist, XIX (1908), 46-66.

The Place of Magic in the Intellectual History of Europe, Columbia University Press, 1905.

The True Roger Bacon, in American Historical Review, XXI (1916), 237-57, 468-80.

Tiraboschi. Storia della Letteratura Italiana, Modena, 1772-1795.

Tischendorf, C. Acta Apostolorum Apocrypha, Leipzig, 1851.

Evangelia Apocrypha, Leipzig, 1876.

Töply, R. von. Studien zur Geschichte der Anatomie im Mittelalter, 1898.

Unger, F. Die Pflanze als Zaubermittel, Vienna, 1859.

Vacant, A. et Mangenot, E. Dictionnaire de théologie catholique, Paris, 1909-.

Valentinelli, J. Bibliotheca manuscripta ad S. Marci Venetiarum, Venice, 1868-1876, 6 vols.

Valois, Noël. Guillaume d'Auvergne, évêque de Paris, 1228-1249. Sa vie et ses ouvrages, Paris, 1880.

Vincent of Beauvais. Speculum doctrinale, 1472 (?).

Speculum historiale, 1473.

Speculum naturale, Anth. Koburger, Nürnberg, 1485.

Vossius, G. J. De Universae Matheseos natura et constitutione liber, Amsterdam, 1650.

Walsh, J. J. Medieval Medicine, 1920, 221 pp.

Old Time Makers of Medicine; the story of the students and teachers of the sciences related to medicine during the middle ages, New York, 1911. Popular.

The Popes and Science, 1908.

Webb, C. C. I. See John of Salisbury.

Webster, Hutton. Rest Days, 1916.

Wedel, T. C. The Medieval Attitude toward Astrology particularly in England, Yale University Press, 1920.

Wellmann, Max. ed. Dioscorides de materia medica, 1907, 1906.

Die Schrift des Dioskurides Περὶ ἁπλῶν φαρμακῶν, 1914.

White, A. D. A History of the Warfare of Science with Theology in Christendom, New York, 1896, 2 vols.

Wickersheimer, Ernest. Figures médico-astrologiques des neuvième, dixième et onzième siècles, in Transactions of the Seventeenth International Congress of Medicine,

Section XXIII, History of Medicine, London, 1913, p. 313 ff.

William of Auvergne. Opera omnia, Venice, 1591.

Withington, E. T. Medical History from the Earliest Times, London, 1894.

Wright, Thomas. Popular Treatises on Science written during the middle ages in Anglo-Saxon, Anglo-Norman, and English, London, 1841.

ed. Alexander Neckam De naturis rerum, in RS vol. 34, 1863.

Wulf, M. de. History of Medieval Philosophy, 1909, English translation.

Wüstenfeld, F. Geschichte der Arabischen Aerzte und Naturforscher, Göttingen, 1840.

Yule, Sir Henry, The Book of Ser Marco Polo, third edition revised by Henri Cordier, 2 vols., London, 1903.

Zarncke, F. Der Priester Johannes, in Abhandl. d. philol.-hist. Classe, Kgl. Sächs. Gesell. d. Wiss., VII (1879), 627-1030; VIII (1883), 1-186.

Zetzner, L. Theatrum chemicum, 1613-1622, 6 vols.

A

HISTORY OF MAGIC AND EXPERIMENTAL SCIENCE

VOLUME I

A HISTORY OF MAGIC AND EXPERIMENTAL SCIENCE AND THEIR RELATION TO CHRISTIAN THOUGHT DURING THE FIRST THIRTEEN CENTURIES OF OUR ERA

CHAPTER I

INTRODUCTION

Aim of this book—Period covered—How to study the history of thought—Definition of magic—Magic of primitive man; does civilization originate in magic?—Divination in early China—Magic in ancient Egypt—Magic and Egyptian religion—Mortuary magic—Magic in daily life—Power of words, images, amulets—Magic in Egyptian medicine—Demons and disease—Magic and science—Magic and industry—Alchemy—Divination and astrology—The sources for Assyrian and Babylonian magic—Was astrology Sumerian or Chaldean?—The number seven in early Babylonia—Incantation texts older than astrological—Other divination than astrology—Incantations against sorcery and demons—A specimen incantation—Materials and devices of magic—Greek culture not free from magic—Magic in myth, literature, and history—Simultaneous increase of learning and occult science—Magic origin urged for Greek religion and drama—Magic in Greek philosophy—Plato's attitude toward magic and astrology—Aristotle on stars and spirits—Folk-lore in the *History of Animals*—Differing modes of transmission of ancient oriental and Greek literature—More magical character of directly transmitted Greek remains—Progress of science among the Greeks—Archimedes and Aristotle—Exaggerated view of the scientific achievement of the Hellenistic age—Appendix I. Some works on Magic, Religion, and Astronomy in Babylonia and Assyria.

"Magic has existed among all peoples and at every period."—Hegel.[1]

Aim of this book.

THIS book aims to treat the history of magic and experimental science and their relations to Christian thought during the first thirteen centuries of our era, with especial emphasis upon the twelfth and thirteenth centuries. No

[1] Lectures on the Philosophy of Religion; quoted by Sir James Frazer, *The Magic Art* (1911), I, 426.

adequate survey of the history of either magic or experimental science exists for this period, and considerable use of manuscript material has been necessary for the medieval period. Magic is here understood in the broadest sense of the word, as including all occult arts and sciences, superstitions, and folk-lore. I shall endeavor to justify this use of the word from the sources as I proceed. My idea is that magic and experimental science have been connected in their development; that magicians were perhaps the first to experiment; and that the history of both magic and experimental science can be better understood by studying them together. I also desire to make clearer than it has been to most scholars the Latin learning of the medieval period, whose leading personalities even are generally inaccurately known, and on perhaps no one point is illumination more needed than on that covered by our investigation. The subject of laws against magic, popular practice of magic, the witchcraft delusion and persecution lie outside of the scope of this book.[1]

Period covered.

At first my plan was to limit this investigation to the twelfth and thirteenth centuries, the time of greatest medieval productivity, but I became convinced that this period could be best understood by viewing it in the setting of the Greek, Latin, and early Christian writers to whom it owed so much. If the student of the Byzantine Empire needs to know old Rome, the student of the medieval church to comprehend early Christianity, the student of Romance languages to understand Latin, still more must the reader of Constantinus Africanus, Vincent of Beauvais, Guido Bonatti, and Thomas Aquinas be familiar with the Pliny, Galen, and Ptolemy, the Origen and Augustine, the Alkindi and Albumasar from whom they drew. It would indeed be difficult to draw a line anywhere between them. The ancient

[1] That field has already been treated by Joseph Hansen, *Zauberwahn, Inquisition und Hexenprozess im Mittelalter,* 1900, and will be further illuminated by *A History of Witchcraft in Europe,* soon to be edited by Professor George L. Burr from H. C. Lea's materials. See also a work just published by Miss M. A. Murray, *The Witch-Cult in Western Europe,* Oxford, 1921.

authors are generally extant only in their medieval form; in some cases there is reason to suspect that they have undergone alteration or addition; sometimes new works were fathered upon them. In any case they have been preserved to us because the middle ages studied and cherished them, and to a great extent made them their own. I begin with the first century of our era, because Christian thought begins then, and then appeared Pliny's *Natural History* which seems to me the best starting point of a survey of ancient science and magic.[1] I close with the thirteenth century, or, more strictly speaking, in the course of the fourteenth, because by then the medieval revival of learning had spent its force. Attention is centred on magic and experimental science in western Latin literature and learning, Greek and Arabic works being considered as they contributed thereto, and vernacular literature being omitted as either derived from Latin works or unlearned and unscientific.

How to study the history of thought.

Very probably I have tried to cover too much ground and have made serious omissions. It is probably true that for the history of thought as for the history of art the evidence and source material is more abundant than for political or economic history. But fortunately it is more reliable, since the pursuit of truth or beauty does not encourage deception and prejudice as does the pursuit of wealth or power. Also the history of thought is more unified and consistent, steadier and more regular, than the fluctuations and diversities of political history; and for this reason its general outlines can be discerned with reasonable sureness by the examination of even a limited number of examples, provided they are properly selected from a period of sufficient duration. Moreover, it seems to me that in the present stage of research into and knowledge of our subject

[1] Some of my scientific friends have urged me to begin with Aristotle, as being a much abler scientist than Pliny, but this would take us rather too far back in time and I have not felt equal to a treatment of the science of the genuine Aristotle *per se,* although in the course of this book I shall say something of his medieval influence and more especially of the Pseudo-Aristotle.

sounder conclusions and even more novel ones can be drawn by a wide comparative survey than by a minutely intensive and exhaustive study of one man or of a few years. The danger is of writing from too narrow a view-point, magnifying unduly the importance of some one man or theory, and failing to evaluate the facts in their full historical setting. No medieval writer whether on science or magic can be understood by himself, but must be measured in respect to his surroundings and antecedents.

Definition of magic.

Some may think it strange that I associate magic so closely with the history of thought, but the word comes from the *Magi* or wise men of Persia or Babylon, to whose lore and practices the name was applied by the Greeks and Romans, or possibly we may trace its etymology a little farther back to the Sumerian or Turanian word *imga* or *unga,* meaning deep or profound. The exact meaning of the word, "magic," was a matter of much uncertainty even in classical and medieval times, as we shall see. There can be no doubt, however, that it was then applied not merely to an operative art, but also to a mass of ideas or doctrine, and that it represented a way of looking at the world. This side of magic has sometimes been lost sight of in hasty or assumed modern definitions which seem to regard magic as merely a collection of rites and feats. In the case of primitive men and savages it is possible that little thought accompanies their actions. But until these acts are based upon or related to some imaginative, purposive, and rational thinking, the doings of early man cannot be distinguished as either religious or scientific or magical. Beavers build dams, birds build nests, ants excavate, but they have no magic just as they have no science or religion. Magic implies a mental state and so may be viewed from the standpoint of the history of thought. In process of time, as the learned and educated lost faith in magic, it was degraded to the low practices and beliefs of the ignorant and vulgar. It was this use of the term that was taken up by anthropologists and by them applied to analogous doings and notions of primitive men and savages. But we may go too far in regarding magic as a purely social product of tribal society: magicians may be, in Sir James Frazer's words,[1] "the only professional class" among the lowest savages, but note that they rank as a learned profession from the start. It will be chiefly through the writings of learned men that something of their later history and of the growth of interest in experimental science will be traced in this work. Let me add that in this investigation all arts of divination, including astrology, will be reckoned as magic; I have been quite unable to separate the two either in fact or logic, as I shall illustrate repeatedly by particular cases.[2]

Magic of primitive man: does civilization originate in magic?

Magic is very old, and it will perhaps be well in this introductory chapter to present it to the reader, if not in its infancy—for its origins are much disputed and perhaps antecede all record and escape all observation—at least some centuries before its Roman and medieval days. Sir J. G. Frazer, in a passage of *The Golden Bough* to which we have already referred, remarks that "sorcerers are found in every savage tribe known to us; and among the lowest savages . . . they are the only professional class that exists." [3] Lenormant affirmed in his *Chaldean Magic and Sorcery* [4] that "all magic rests upon a system of religious belief," but recent sociologists and anthropologists have

[1] Frazer has, of course, repeatedly made the point that modern science is an outgrowth from primitive magic. Carveth Read, *The Origin of Man,* 1920, in his chapter on "Magic and Science" contends that "in no case . . . is Science derived from Magic" (p. 337), but this is mainly a logical and ideal distinction, since he admits that "for ages" science "is in the hands of wizards."

[2] I am glad to see that other writers on magic are taking this view; for instance, E. Doutté, *Magie et religion dans l'Afrique du Nord,* Alger, 1909, p. 351.

[3] *Golden Bough,* 1894, I, 420. W. I. Thomas, "The Relation of the Medicine-Man to the Origin of the Professional Occupations" (reprinted in his *Source Book for Social Origins,* 4th edition, pp. 281-303), in which he disputes Herbert Spencer's "thesis that the medicine-man is the source and origin of the learned and artistic occupations," does not really conflict with Frazer's statement, since for Thomas the medicine-man is a priest rather than a magician. Thomas remarks later in the same book (p. 437), "Furthermore, the whole attempt of the savage to control the outside world, so far as it contained a theory or a doctrine, was based on magic."

[4] *Chaldean Magic and Sorcery.* 1878, p. 70.

inclined to regard magic as older than a belief in gods. At any rate some of the most primitive features of historical religions seem to have originated from magic. Moreover, religious cults, rites, and priesthoods are not the only things that have been declared inferior in antiquity to magic and largely indebted to it for their origins. Combarieu in his *Music and Magic* [1] asserts that the incantation is universally employed in all the circumstances of primitive life and that from it, by the medium it is true of religious poetry, all modern music has developed. The magic incantation is, in short, "the oldest fact in the history of civilization." Although the magician chants without thought of æsthetic form or an artistically appreciative audience, yet his spell contains in embryo all that later constitutes the art of music.[2] M. Paul Huvelin, after asserting with similar confidence that poetry,[3] the plastic arts,[4] medicine, mathematics, astronomy, and chemistry "have easily discernable magic sources," states that he will demonstrate that the same is true of law.[5] Very recently, however, there has been something of a reaction against this tendency to regard the life of primitive man as made up entirely of magic and to trace back every phase of civilization to a magical origin. But R. R. Marett still sees a higher standard of value in primitive man's magic than in his warfare and brutal exploitation of his fellows and believes that the "higher plane of experience for which *mana* stands is one in which spiritual enlargement is appreciated for its own sake." [6]

Divination in early China.

Of the five classics included in the Confucian Canon, *The Book of Changes* (*I Ching* or *Yi-King*), regarded by

[1] Jules Combarieu, *La musique et la magie*, Paris, 1909, p. v.

[2] *Ibid.*, pp. 13-14.

[3] Among the early Arabs "poetry is magical utterance" (Macdonald (1909) p. 16), and the poet "a wizard in league with spirits" (Nicholson, *A Literary History of the Arabs*, 1914, p. 72).

[4] See S. Reinach, "L'Art et la Magie," in *L'Anthropologie*, XIV (1903), and Y. Hirn, *Origins of Art*, London, 1900, Chapter xx, "Art and Magic." J. Capart, *Primitive Art in Egypt*.

[5] P. Huvelin, *Magie et droit individuel*, Paris, 1907, in *Année Sociologique*, X, 1-471; see too his *Les tablettes magiques et le droit romain*, Mâcon, 1901.

[6] R. R. Marett, *Psychology and Folk-Lore*, 1920, Chapter iii on "Primitive Values."

some as the oldest work in Chinese literature and dated back as early as 3000 B.C., in its rudimentary form appears to have been a method of divination by means of eight possible combinations in triplets of a line and a broken line. Thus, if *a* be a line and *b* a broken line, we may have *aaa, bbb, aab, bba, abb, baa, aba,* and *bab.* Possibly there is a connection with the use of knotted cords which, Chinese writers state, preceded written characters, like the method used in ancient Peru. More certain would seem the resemblance to the medieval method of divination known as geomancy, which we shall encounter later in our Latin authors. Magic and astrology might, of course, be traced all through Chinese history and literature. But, contenting ourselves with this single example of the antiquity of such arts in the civilization of the far east, let us turn to other ancient cultures which had a closer and more unmistakable influence upon the western world.

Magic in ancient Egypt.

Of the ancient Egyptians Budge writes, "The belief in magic influenced their minds . . . from the earliest to the latest period of their history . . . in a manner which, at this stage in the history of the world, is very difficult to understand."[1] To the ordinary historical student the evidence for this assertion does not seem quite so overwhelming as the Egyptologists would have us think. It looks thinner when we begin to spread it out over a stretch of four

[1] E. A. Wallis Budge, *Egyptian Magic,* 1899, p. vii. Some other works on magic in Egypt are: Groff, *Études sur la sorcellerie, mémoires présentés à l'institut égyptien,* Cairo, 1897; G. Busson, *Extrait d'un mémoire sur l'origine égyptienne de la Kabbale,* in *Compte Rendu du Congrès Scientifique International des Catholiques, Sciences Religieuses,* Paris, 1891, pp. 29-51. Adolf Erman, *Life in Ancient Egypt,* English translation, 1894, "describes vividly the magical conceptions and practices." F. L. Griffith, *Stories of the High Priests of Memphis,* Oxford, 1900, contains some amusing demotic tales of magicians. Erman, *Zaubersprüche für Mutter und Kind,* 1901. F. L. Griffith and H. Thompson, *The Demotic Magical Papyrus of London and Leiden,* 1904. See also J. H. Breasted, *Development of Religion and Thought in Ancient Egypt,* New York, 1912.

The following later but briefer treatments add little to Budge: Alfred Wiedemann, *Magie und Zauberei im Alten Ægypten,* Leipzig, 1905, and *Die Amulette der alten Ægypter,* Leipzig, 1910, both in *Der Alte Orient;* Alexandre Moret, *La magie dans l'Egypte ancienne,* Paris, 1906, in *Musée Guimet, Annales, Bibliothèque de vulgarisation,* XX, 241-81.

thousand years, and it scarcely seems scientific to adduce details from medieval Arabic tales or from the late Greek fiction of the Pseudo-Callisthenes or from papyri of the Christian era concerning the magic of early Egypt. And it may be questioned whether two stories preserved in the Westcar papyrus, written many centuries afterwards, are alone "sufficient to prove that already in the Fourth Dynasty the working of magic was a recognized art among the Egyptians." [1]

Magic and Egyptian religion.

At any rate we are told that the belief in magic not only was predynastic and prehistoric, but was "older in Egypt than the belief in God." [2] In the later religion of the Egyptians, along with more lofty and intellectual conceptions, magic was still a principal ingredient.[3] Their mythology was affected by it [4] and they not only combated demons with magical formulae but believed that they could terrify and coerce the very gods by the same method, compelling them to appear, to violate the course of nature by miracles, or to admit the human soul to an equality with themselves.[5]

Mortuary magic.

Magic was as essential in the future life as here on earth among the living. Many, if not most, of the observances and objects connected with embalming and burial had a magic purpose or mode of operation; for instance, the "magic eyes placed over the opening in the side of the body through which the embalmer removed the intestines," [6] or the mannikins and models of houses buried with the dead. In the process of embalming the wrapping of each bandage was accompanied by the utterance of magic words.[7] In "the oldest chapter of human thought extant"—the Pyramid

[1] Budge (1899), p. 19. At pp. 7-10 Budge dates the Westcar Papyrus about 1550 B. C. and Cheops, of whom the tale is told, in 3800 B. C. It is now customary to date the Fourth Dynasty, to which Cheops belonged, about 2900-2750 B. C. Breasted, *History of Egypt*, pp. 122-3, speaks of a folk tale preserved in the Papyrus Westcar some nine (?) centuries after the fall of the Fourth Dynasty.

[2] Budge, p. ix.

[3] Budge, pp. xiii-xiv.

[4] For magical myths see E. Naville, *The Old Egyptian Faith*, English translation by C. Campbell, 1909, p. 233 *et seq.*

[5] Budge, pp. 3-4; Lenormant, *Chaldean Magic*, p. 100; Wiedemann (1905), pp. 12, 14, 31.

[6] So labelled in the Egyptian Museum at Cairo.

[7] Budge, p. 185.

Texts written in hieroglyphic at the tombs at Sakkara of Pharaohs of the fifth and sixth dynasties (c. 2625-2475 B.C.), magic is so manifest that some have averred "that the whole body of Pyramid Texts is simply a collection of magical charms." [1] The scenes and objects painted on the walls of the tombs, such as those of nobles in the fifth and sixth dynasties, were employed with magic intent and were meant to be realized in the future life; and with the twelfth dynasty the Egyptians began to paint on the insides of the coffins the objects that were formerly actually placed within.[2] Under the Empire the famous *Book of the Dead* is a collection of magic pictures, charms, and incantations for the use of the deceased in the hereafter,[3] and while it is not of the early period, we hear that "a book with words of magic power" was buried with a pharaoh of the Old Kingdom. Budge has "no doubt that the object of every religious text ever written on tomb, stele, amulet, coffin, papyrus, etc., was to bring the gods under the power of the deceased, so that he might be able to compel them to do his will." [4] Breasted, on the other hand, thinks that the amount and complexity of this mortuary magic increased greatly in the later period under popular and priestly influence.[5]

Magic in daily life.

Breasted nevertheless believes that magic had played a great part in daily life throughout the whole course of Egyptian history. He writes, "It is difficult for the modern mind to understand how completely the belief in magic penetrated the whole substance of life, dominating popular custom and constantly appearing in the simplest acts of the daily household routine, as much a matter of course as

[1] Breasted (1912), pp. 84-5, 93-5. "Systematic study" of the Pyramid Texts has been possible "only since the appearance of Sethe's great edition,"—*Die Altægyptischen Pyramidentexte*, Leipzig, 1908-1910, 2 vols.

[2] Budge, pp. 104-7.

[3] Many of them are to enable the dead man to leave his tomb at will; hence the Egyptian title, "The Chapters of Going Forth by Day," Breasted, *History of Egypt*, p. 175.

[4] Budge, p. 28.

[5] *History of Egypt*, p. 175; pp. 249-50 for the further increase in mortuary magic after the Middle Kingdom, and pp. 369-70, 390, etc., for Ikhnaton's vain effort to suppress this mortuary magic. See also Breasted (1912), pp. 95-6, 281, 292-6, etc.

sleep or the preparation of food. It constituted the very atmosphere in which the men of the early oriental world lived. Without the saving and salutary influence of such magical agencies constantly invoked, the life of an ancient household in the East was unthinkable." [1]

Power of words, images, amulets.

Most of the main features and varieties of magic known to us at other times and places appear somewhere in the course of Egypt's long history. For one thing we find the ascription of magic power to words and names. The power of words, says Budge, was thought to be practically unlimited, and "the Egyptians invoked their aid in the smallest as well as in the greatest events of their life." [2] Words might be spoken, in which case they "must be uttered in a proper tone of voice by a duly qualified man," or they might be written, in which case the material upon which they were written might be of importance.[3] In speaking of mortuary magic we have already noted the employment of pictures, models, mannikins, and other images, figures, and objects. Wax figures were also used in sorcery,[4] and amulets are found from the first, although their particular forms seem to have altered with different periods.[5] Scarabs are of course the most familiar example.

Magic in Egyptian medicine.

Egyptian medicine was full of magic and ritual and its therapeusis consisted mainly of "collections of incantations and weird random mixtures of roots and refuse." [6] Already we find the recipe and the occult virtue conceptions, the elaborate polypharmacy and the accompanying hocus-pocus which we shall meet in Pliny and the middle ages. The Egyptian doctors used herbs from other countries and preferred compound medicines containing a dozen ingredients to simple medicines.[7] Already we find such magic

[1] Breasted (1912), pp. 290-1.

[2] Budge, pp. xi, 170-1.

[3] Budge, p. 4.

[4] Budge, pp. 67-70, 73, 77.

[5] Budge, pp. 27-28, 41, 60.

[6] From the abstract of a paper on *The History of Egyptian Medicine,* read by T. Wingate Todd at the annual meeting of the American Historical Association, 1919. See also B. Holmes and P. G. Kitterman, *Medicine in Ancient Egypt; the Hieratic Material,* Cincinnati, 1914, 34 pp., reprinted from *The Lancet-Clinic.*

[7] See H. L. Lüring, *Die über die medicinischen Kenntnisse der alten Ægypter berichtenden Papyri*

logic as that the hair of a black calf will keep one from growing gray.[1] Already the parts of animals are a favorite ingredient in medical compounds, especially those connected with the organs of generation, on which account they were presumably looked upon as life-giving, or those which were recommended mainly by their nastiness and were probably thought to expel the demons of disease by their disagreeable properties.

Demons and disease.

In ancient Egypt, however, disease seems not to have been identified with possession by demons to the extent that it was in ancient Assyria and Babylonia. While Breasted asserts that "disease was due to hostile spirits and against these only magic could avail," [2] Budge contents himself with the more cautious statement that there is "good reason for thinking that some diseases were attributed to . . . evil spirits . . . entering . . . human bodies . . . but the texts do not afford much information" [3] on this point. Certainly the beliefs in evil spirits and in magic do not always have to go together, and magic might be employed against disease whether or not it was ascribed to a demon.

Magic and science.

In the case of medicine as in that of religion Breasted takes the view that the amount of magic became greater in the Middle and New Kingdoms than in the Old Kingdom. This is true so far as the amount of space occupied by it in extant records is concerned. But it would be rash to assume that this marks a decline from a more rational and scientific attitude in the Old Kingdom. Yet Breasted rather gives this impression when he writes concerning the Old Kingdom that many of its recipes were useful and rational, that "medicine was already in the possession of much empirical wisdom, displaying close and accurate observation," and that what "precluded any progress toward real science was the belief in magic, which later began to dominate all the

verglichen mit den medic. Schriften griech. u. römischer Autoren, Leipzig, 1888. Also Joret, I (1897) 310-11, and the article there cited by G. Ebers, *Ein Kyphirecept aus dem Papyrus Ebers*, in *Zeitschrift f. ægypt. Sprache*, XII (1874), p. 106. M. A. Ruffer, *Palaeopathology of Egypt*, 1921.

[1] *History of Egypt*, p. 101.

[2] *Ibid*, p. 102.

[3] Budge, p. 206.

practice of the physician." [1] Berthelot probably places the emphasis more correctly when he states that the later medical papyri "include traditional recipes, founded on an empiricism which is not always correct, mystic remedies, based upon the most bizarre analogies, and magic practices that date back to the remotest antiquity." [2] The recent efforts of Sethe and Wilcken, of Elliot Smith, Müller, and Hooten to show that the ancient Egyptians possessed a considerable amount of medical knowledge and of surgical and dental skill, have been held by Todd to rest on slight and dubious evidence. Indeed, some of this evidence seems rather to suggest the ritualistic practices still employed by uncivilized African tribes. Certainly the evidence for any real scientific development in ancient Egypt has been very meager compared with the abundant indications of the prevalence of magic.[3]

Magic and industry.

Early Egypt was the home of many arts and industries, but not in so advanced a stage as has sometimes been suggested. Blown glass, for example, was unknown until late Greek and Roman times, and the supposed glass-blowers depicted on the early monuments are really smiths engaged in stirring their fires by blowing through reeds tipped with clay.[4] On the other hand, Professor Breasted informs me that there is no basis for Berthelot's statement that "every sort of chemical process as well as medical treatment was executed with an accompaniment of religious formulae, of prayers and incantations, regarded as essential to the success of operations as well as the cure of maladies." [5]

Alchemy.

Alchemy perhaps originated on the one hand from the practices of Egyptian goldsmiths and workers in metals, who experimented with alloys,[6] and on the other hand from

[1] *History of Egypt*, p. 101.

[2] *Archéologie et Histoire des Sciences*, Paris, 1906, pp. 232-3.

[3] Professor Breasted, however, feels that the contents of the new Edwin Smith Papyrus will raise our estimate of the worth of Egyptian medicine and surgery: letter to me of Jan. 20, 1922.

[4] Petrie, "Egypt," in EB, p. 73.

[5] Berthelot (1885), p. 235. See E. B. Havell, *A Handbook of Indian Art*, 1920, p. 11, for a combination of "exact science," ritual, and "magic power" in the work of the ancient Aryan craftsmen.

[6] Berthelot (1889), pp. vi-vii.

the theories of the Greek philosophers concerning world-grounds, first matter, and the elements.[1] The words, alchemy and chemistry, are derived ultimately from the name of Egypt itself, Kamt or Qemt, meaning literally black, and applied to the Nile mud. The word was also applied to the black powder produced by quicksilver in Egyptian metallurgical processes. This powder, Budge says, was supposed to be the ground of all metals and to possess marvelous virtue, "and was mystically identified with the body which Osiris possessed in the underworld, and both were thought to be sources of life and power." [2] The analogy to the sacrament of the mass and the marvelous powers ascribed to the host by medieval preachers like Stephen of Bourbon scarcely needs remark. The later writers on alchemy in Greek appear to have borrowed signs and phraseology from the Egyptian priests, and are fond of speaking of their art as the monopoly of Egyptian kings and priests who carved its secrets on ancient steles and obelisks. In a treatise dating from the twelfth dynasty a scribe recommends to his son a work entitled *Chemi,* but there is no proof that it was concerned with chemistry or alchemy.[3] The papyri containing treatises of alchemy are of the third century of the Christian era.

Divination and astrology.

Evidences of divination in general and of astrology in particular do not appear as early in Egyptian records as examples of other varieties of magic. Yet the early date at which Egypt had a calendar suggests astronomical interest, and even those who deny that seven planets were distinguished in the Tigris-Euphrates Valley until the last millennium before Christ, admit that they were known in Egypt as far back as the Old Kingdom, although they deny the existence of a science of astronomy or an art of astrology then.[4] A dream of Thotmes IV is preserved from 1450 B.C. or thereabouts, and the incantations employed by magicians

[1] Berthelot (1885), pp. 247-78; E. O. v. Lippmann (1919), pp. 118-43.
[2] Budge, pp. 19-20.
[3] Berthelot (1885), p. 10.
[4] Lippmann (1919), pp. 181-2, and the authorities there cited.

in order to procure divining dreams for their customers attest the close connection of divination and magic.[1] Belief in lucky and unlucky days is shown in a papyrus calendar of about 1300 B.C.,[2] and we shall see later that "Egyptian Days" continued to be a favorite superstition of the middle ages. Tables of the risings of stars which may have an astrological significance have been found in graves, and there were gods for every month, every day of the month, and every hour of the day.[3] Such numbers as seven and twelve are frequently emphasized in the tombs and elsewhere, and if the vaulted ceiling in the tenth chamber of the tomb of Sethos is really of his time, we seem to find the signs of the zodiac under the nineteenth dynasty. If Boll is correct in suggesting that the zodiac originated in the transfer of animal gods to the sky,[4] no fitter place than Egypt could be found for the transfer. But there have not yet been discovered in Egypt lists of omens and appearances of constellations on days of disaster such as are found in the literature of the Tigris-Euphrates valley and in the Roman historians. Budge speaks of the seven Hathor goddesses who predict the death that the infant must some time die, and affirms that "the Egyptians believed that a man's fate . . . was decided before he was born, and that he had no power to alter it." [5] But I cannot agree that "we have good reason for assigning the birthplace of the horoscope to Egypt," [6] since the evidence seems to be limited to the almost medieval Pseudo-Callisthenes and a Greek horoscope in the British Museum to which is attached the letter of an astrologer urging his pupil to study the ancient Egyptians carefully. The later Greek and Latin tradition that astrology was the invention of the divine men of Egypt and Babylon probably has a basis of fact, but more contemporary evidence is needed if Egypt is to contest the claim of Babylon to precedence in that art.

[1] Budge, pp. 214-5.

[2] Budge, pp. 225-8; Wiedemann (1905), p. 9.

[3] Wiedemann (1905), pp. 7, 8, 11. See also G. Daressy, *Une ancienne liste des décans égyptiens*, in *Annales du service des antiquités de l'Egypte*, I (1900), 79-90.

[4] F. Boll in *Neue Jahrb.* (1908), p. 108.

[5] Budge, pp. 222-3.

[6] Budge, p. 229.

The sources for Assyrian and Babylonian magic.

In the written remains of Babylonian and Assyrian civilization [1] the magic cuneiform tablets play a large part and give us the impression that fear of demons was a leading feature of Assyrian and Babylonian religion and that daily thought and life were constantly affected by magic. The bulk of the religious and magical texts are preserved in the library of Assurbanipal, king of Assyria from 668 to 626 B.C. But he collected his library from the ancient temple cities, the scribes tell us that they are copying very ancient texts, and the Sumerian language is still largely employed.[2] Eridu, one of the main centers of early Sumerian culture, "was an immemorial home of ancient wisdom, that is to say, magic." [3] It is, however, difficult in the library of Assurbanipal to distinguish what is Babylonian from what is Assyrian or what is Sumerian from what is Semitic. Thus we are told that "with the exception of some very ancient texts, the Sumerian literature, consisting largely of religious material such as hymns and incantations, shows a number of Semitic loanwords and grammatical Semitisms, and in many cases, although not always, is quite patently a translation of Semitic ideas by Semitic priests into the formal religious Sumerian language." [4]

Was astrology Sumerian or Chaldean?

The chief point in dispute, over which great controversy has taken place recently among German scholars, is as to the antiquity of both astronomical knowledge and astrological doctrine, including astral theology, among the dwellers in the Tigris-Euphrates region. Briefly, such writers as Winckler, Stücken, and Jeremias held that the religion of the early Babylonians was largely based on astrology and that all their thought was permeated by it, and that they had probably by an early date made astronomical observations and acquired astronomical knowledge which was lost

[1] Some works on the subject of magic and religion, astronomy and astrology in Babylonia and Assyria will be found in Appendix I at the close of this chapter.

[2] Thompson, *Semitic Magic*, pp. xxxvi-xxxvii; Fossey, pp. 17-20.

[3] Farnell, *Greece and Babylon*, p. 102.

[4] Prince, "Sumer and Sumerians," in EB.

in the decline of their culture. Opposing this view, such scholars as Kugler, Bezold, Boll, and Schiaparelli have shown the lack of certain evidence for either any considerable astronomical knowledge or astrological theory in the Tigris-Euphrates Valley until the late appearance of the Chaldeans. It is even denied that the seven planets were distinguished in the early period, much less the signs of the zodiac or the planetary week,[1] which last, together with any real advance in astronomy, is reserved for the Hellenistic period.

The number seven in early Babylonia.

Yet the prominence of the number seven in myth, religion, and magic is indisputable in the third millennium before our era. For instance, in the old Babylonian epic of creation there are seven winds, seven spirits of storms, seven evil diseases, seven divisions of the underworld closed by seven doors, seven zones of the upper world and sky, and so on. We are told, however, that the staged towers of Babylonia, which are said to have symbolized for millenniums the sacred Hebdomad, did not always have seven stages.[2] But the number seven was undoubtedly of frequent occurrence, of a sacred and mystic character, and virtue and perfection were ascribed to it. And no one has succeeded in giving any satisfactory explanation for this other than the rule of the seven planets over our world. This also applies to the sanctity of the number seven in the Old Testament[3] and the emphasis upon it in Hesiod, the Odyssey, and other early Greek sources.[4]

[1] Webster, *Rest Days*, pp. 215-22, with further bibliography. See Orr (1913), 28-38, for an interesting discussion in English of the problem of the origin of solar and lunar zodiac.

[2] Lippmann (1919), pp. 168-9.

[3] Although Schiaparelli, *Astronomy in the Old Testament*, 1905, pp. v, 5, 49-51, 135, denies that "the frequent use of the number seven in the Old Testament is in any way connected with the planets." I have not seen F. von Andrian, *Die Siebenzahl im Geistesleben der Völker*, in *Mitteil. d. anthrop. Gesellsch. in Wien*, XXI (1901), 225-74; see also Hehn, *Siebenzahl und Sabbat bei den Babyloniern und im alten Testament*, 1907. J. G. Frazer (1918), I, 140, has an interesting passage on the prominence of the number seven "alike in the Jehovistic and in the Babylonian narrative" of the flood.

[4] Webster, *Rest Days*, pp. 211-2. Professor Webster, who kindly read this chapter in manuscript, stated in a letter to me of 2 July 1921 that he remained convinced that "the mystic properties as-

Incantation texts older than the astrological.

However that may be, the tendency prevailing at present is to regard astrology as a relatively late development introduced by the Semitic Chaldeans. Lenormant held that writing and magic were a Turanian or Sumerian (Accadian) contribution to Babylonian civilization, but that astronomy and astrology were Semitic innovations. Jastrow thinks that there was slight difference between the religion of Assyria and that of Babylonia, and that astral theology played a great part in both; but he grants that the older incantation texts are less influenced by this astral theology. L. W. King says, "Magic and divination bulk largely in the texts recovered, and in their case there is nothing to suggest an underlying astrological element." [1]

Other divination than astrology.

Whatever its date and origin, the magic literature may be classified in three main groups. There are the astrological texts in which the stars are looked upon as gods and predictions are made especially for the king.[2] Then there are the tablets connected with other methods of foretelling the future, especially liver divination, although interpretation of dreams, augury, and divination by mixing oil and water were also practiced.[3] Fossey has further noted the close connection of operative magic with divination among the Assyrians, and calls divination "the indispensable auxiliary of magic." Many feats of magic imply a precedent knowledge of the future or begin by consultation of a diviner, or a favorable day and hour should be chosen for the magic rite.[4]

Third, there are the collections of incantations, not however those employed by the sorcerers, which were pre-

cribed to the number seven" can only in part be accounted for by the seven planets; "Our American Indians, for example, hold seven in great respect, yet have no knowledge of seven planets." But it may be noted that the poet-philosophers of ancient Peru composed verses on the subject of astrology, according to Garcilasso (cited by W. I. Thomas, *Source Book for Social Origins*, 1909, p. 293).

[1] L. W. King, *History of Babylon*, 1915, p. 299.

[2] Fossey (1902), pp. 2-3.

[3] Farnell, *Greece and Babylon*, pp. 301-2. On liver divination see Frothingham, "Ancient Orientalism Unveiled," *American Journal of Archaeology*, XXI (1917) 55, 187, 313, 420.

[4] Fossey, p. 66.

Incantations against sorcery and demons.

sumably illicit and hence not publicly preserved—in an incantation which we shall soon quote sorcery is called evil and is said to employ "impure things"—but rather defensive measures against them and exorcisms of evil demons.[1] But doubtless this counter magic reflects the original procedure to a great extent. Inasmuch as diseases generally were regarded as due to demons, who had to be exorcized by incantations, medicine was simply a branch of magic. Evil spirits were also held responsible for disturbances in nature, and frequent incantations were thought necessary to keep them from upsetting the natural order entirely.[2] The various incantations are arranged in series of tablets: the *Maklu* or burning, *Ti'i* or headaches, *Asakki marsûti* or fever, *Labartu* or hag-demon, and *Nis kati* or raising of the hand. Besides these tablets there are numerous ceremonial and medical texts which contain magical practice.[3] Also hymns of praise and religious epics which at first sight one would not classify as incantations seem to have had their magical uses, and Farnell suggests that "a magic origin for the practice of theological exegesis may be obscurely traced."[4] Good spirits are represented as employing magic and exorcisms against the demons.[5] As a last resort when good spirits as well as human magic had failed to check the demons, the aid might be requisitioned of the god Ea, regarded as the repository of all science and who "alone was possessed of the magic secrets by means of which they could be conquered and repulsed."[6]

A specimen incantation.

The incantations themselves show that other factors than the power of words entered into the magic, as may be illustrated by quoting one of them.

"Arise ye great gods, hear my complaint,
Grant me justice, take cognizance of my condition.
I have made an image of my sorcerer and sorceress;

[1] Fossey, p. 16.
[2] Lenormant, pp. 35, 147, 158.
[3] Thompson, *Semitic Magic*, pp. xxxviii-xxxix.
[4] *Greece and Babylon*, p. 296.
[5] Lenormant, pp. 146-7.
[6] *Ibid*, p. 158.

I have humbled myself before you and bring to you my cause,
Because of the evil they have done,
Of the impure things which they have handled.
May she die! Let me live!
May her charm, her witchcraft, her sorcery be broken.
May the plucked sprig of the *binu* tree purify me;
May it release me; may the evil odor of my mouth be scattered to the winds.
May the *mashtakal* herb which fills the earth cleanse me.
Before you let me shine like the *kankal* herb,
Let me be brilliant and pure as the *lardu* herb.
The charm of the sorceress is evil;
May her words return to her mouth, her tongue be cut off.
Because of her witchcraft may the gods of night smite her,
The three watches of the night break her evil charm.
May her mouth be wax; her tongue, honey.
May the word causing my misfortune that she has spoken dissolve like wax.
May the charm she had wound up melt like honey,
So that her magic knot be cut in twain, her work destroyed." [1]

Materials and devices employed in the magic.

It is evident from this incantation that use was made of magic images and knots, and of the properties of trees and herbs. Magic images were made of clay, wax, tallow, and other substances and were employed in various ways. Thus directions are given for making a tallow image of an enemy of the king and binding its face with a cord in order to deprive the person whom it represents of speech and will-power.[2] Images were also constructed in order that disease demons might be magically transferred into them,[3] and sometimes the images are slain and buried.[4] In the above incantation the magic knot was employed only by the sorceress, but Fossey states that knots were also used as

[1] Jastrow, *Religion of Babylon and Assyria*, pp. 283-4.
[2] Zimmern, *Beiträge*, p. 173.
[3] *Ibid.*, p. 161.
[4] Fossey, p. 399.

counter-charms against the demons.[1] In the above incantation the names of herbs were left untranslated and it is not possible to say much concerning the pharmacy of the Assyrians and Babylonians because of our lack of a lexicon for their botanical and mineralogical terminology.[2] However, from what scholars have been able to translate it appears that common rather than rare and outlandish substances were the ones most employed. Wine and oil, salt and dates, and onions and saliva are the sort of things used. There is also evidence of the employment of a magic wand.[3] Gems and animal substances were used as well as herbs; all sorts of philters were concocted; and varied rites and ceremonies were employed such as ablutions and fumigations. In the account of the ark of the Babylonian Noah we are told of the magic significance of its various parts; thus the mast and cabin ceiling were made of cedar, a wood that counteracts sorceries.[4]

Greek culture not free from magic.

One remarkable corollary of the so-called Italian Renaissance or Humanistic movement at the close of the middle ages with its too exclusive glorification of ancient Greece and Rome has been the strange notion that the ancient Hellenes were unusually free from magic compared with other periods and peoples. It would have been too much to claim any such immunity for the primitive Romans, whose entire religion was originally little else than magic and whose daily life, public and private, was hedged in by superstitious observances and fears. But they, too, were supposed to have risen later under the influence of Hellenic culture to a more enlightened stage,[5] only to relapse again into magic in the declining empire and middle ages under oriental influence. Incidentally let me add that this notion that in *the past* orientals were more superstitious and fond of

[1] Fossey, p. 83.

[2] *Ibid.*, pp. 89-91. F. Küchler, *Beiträge zur Kenntnis der Assyr.-Babyl. Medizin; Texte mit Umschrift, Uebersetzung und Kommentar*, Leipzig, 1904, treats of twenty facsimile pages of cuneiform.

[3] Lenormant, p. 190.

[4] *Ibid.*, p. 159.

[5] So enlightened in fact that they spoke with some scorn of the "levity" and "lies" of the Greeks.

marvels than westerners in the same stage of civilization and that the orient must needs be the source of every superstitious cult and romantic tale is a glib assumption which I do not intend to make and which our subsequent investigation will scarcely substantiate. But to return to the supposed immunity of the Hellenes from magic; so far has this hypothesis been carried that textual critics have repeatedly rejected passages as later interpolations or even called entire treatises spurious for no other reason than that they seemed to them too superstitious for a reputable classical author. Even so specialized and recent a student of ancient astrology, superstition, and religion as Cumont still clings to this dubious generalization and affirms that "the limpid Hellenic genius always turned away from the misty speculations of magic."[1] But, as I suggested some sixteen years since, "the fantasticalness of medieval science was due to 'the clear light of Hellas' as well as to the gloom of the 'dark ages.' "[2]

Magic in myth, literature, and history.

It is not difficult to call to mind evidence of the presence of magic in Hellenic religion, literature, and history. One has only to think of the many marvelous metamorphoses in Greek mythology and of its countless other absurdities; of the witches, Circe and Medea, and the necromancy of Odysseus; or the priest-magician of Apollo in the *Iliad* who could stop the plague, if he wished; of the lucky and unlucky days and other agricultural magic in Hesiod.[3] Then there were the Spartans, whose so-called constitution and method of education, much admired by the Greek philosophers, were largely a retention of the life of the primitive tribe with its ritual and taboos. Or we remember Herodotus and his childish delight in ambiguous oracles or his tale of seceders from Gela brought back by Telines single-handed because he "was possessed of certain mysterious visible symbols of the powers beneath the earth which were deemed to be of

[1] *Oriental Religions in Roman Paganism,* Chicago, 1911, p. 189.

[2] Thorndike (1905), p. 63.

[3] E. E. Sikes, *Folk-lore in the Works and Days of Hesiod,* in *The Classical Review.* VII (1893), 390.

wonder-working power."[1] We recall Xenophon's punctilious records of sacrifices, divinations, sneezes, and dreams; Nicias, as afraid of eclipses as if he had been a Spartan; and the matter-of-fact mentions of charms, philters, and incantations in even such enlightened writers as Euripides and Plato. Among the titles of ancient Greek comedies magic is represented by the *Goetes* of Aristophanes, the *Mandragorizomene* of Alexis, the *Pharmacomantis* of Anaxandrides, the *Circe* of Anaxilas, and the *Thettale* of Menander.[2] When we candidly estimate the significance of such evidence as this, we realize that the Hellenes were not much less inclined to magic than other peoples and periods, and that we need not wait for Theocritus and the Greek romances or for the magical papyri for proof of the existence of magic in ancient Greek civilization.[3]

Simultaneous increase of learning and occult science.

If astrology and some other occult sciences do not appear in a developed form until the Hellenistic period, it is not because the earlier period was more enlightened, but because it was less learned. And the magic which Osthanes is said to have introduced to the Greek world about the time of the Persian wars was not so much an innovation as an improvement upon their coarse and ancient rites of *Goetia.*[4]

Magic origin urged for Greek religion and drama.

This magic element which existed from the start in Greek culture is now being traced out by students of anthropology and early religion as well as of the classics. Miss Jane E. Harrison, in *Themis, a study of the social origins of Greek religion,* suggests a magical explanation for many a myth and festival, and even for the Olympic games and Greek drama.[5] The last point has been developed in more

[1] Freeman, *History of Sicily,* I, 101-3, citing Herodotus VII, 153.

[2] Butler and Owen, *Apulei Apologia,* note on 30, 30.

[3] For details concerning operative or vulgar magic among the ancient Greeks see Hubert, *Magia,* in Daremberg-Saglio; Abt, *Die Apologie des Apuleius von Madaura und die antike Zauberei,* Giessen, 1908; and F. B. Jevons, "Græco-Italian Magic," p. 93-, in *Anthropology and the Classics,* ed. R. Marett; and the article "Magic" in ERE.

[4] I think that this sentence is an approximate quotation from some ancient author, possibly Diogenes Laertius, but I have not been able to find it.

[5] J. E. Harrison, *Themis,* Cambridge, 1912. The chapter head-

detail by F. M. Cornford's *Origin of Attic Comedy,* where much magic is detected masquerading in the comedies of Aristophanes.[1] And Mr. A. B. Cook sees the magician in Zeus, who transforms himself to pursue his amours, and contends that "the real prototype of the heavenly weather-king was the earthly" magician or rain-maker, that the pre-Homeric "fixed epithets" of Zeus retained in the Homeric poems "are simply redolent of the magician," and that the cult of Zeus Lykaios was connected with the belief in werwolves.[2] In still more recent publications Dr. Rendel Harris [3] has connected Greek gods in their origins with the woodpecker and mistletoe, associated the cult of Apollo with the medicinal virtues of mice and snakes, and in other ways emphasized the importance in early Greek religion and culture of the magic properties of animals and herbs.

These writers have probably pressed their point too far, but at least their work serves as a reaction against the old attitude of intellectual idolatry of the classics. Their views may be offset by those of Mr. Farnell, who states that "while the knowledge of early Babylonian magic is beginning to be considerable, we cannot say that we know anything definite concerning the practices in this department of the Hellenic and adjacent peoples in the early period with which we are dealing." And again, "But while Babylonian magic proclaims itself loudly in the great religious literature and highest temple ritual, Greek magic is barely mentioned in the older literature of Greece, plays no part at all in the hymns, and can only with difficulty be discovered as latent in the higher ritual. Again, Babylonian

ings briefly suggest the argument: "1. Hymn of the Kouretes; 2. Dithyramb, Δρώμενον, and Drama; 3. Kouretes, Thunder-Rites and Mana; 4. a. Magic and Tabu, b. Medicine-bird and Medicine-king; 5. Totemism, Sacrament, and Sacrifice; 6. Dithyramb, Spring Festival, and Hagia Triada Sarcophagus; 7. Origin of the Olympic Games (about a year-daimon); 8. Daimon and Hero, with Excursus on Ritual Forms preserved in Greek tragedy; 9. Daimon to Olympian; 10. The Olympians; 11. Themis."

[1] F. M. Cornford, *Origin of Attic Comedy,* 1914, see especially pp. 10, 13, 55, 157, 202, 233.

[2] A. B. Cook, *Zeus,* Cambridge, 1914, pp. 134-5, 12-14, 66-76.

[3] Rendel Harris, *Picus who is also Zeus,* 1916; *The Ascent of Olympus,* 1917.

magic is essentially demoniac; but we have no evidence that the pre-Homeric Greek was demon-ridden, or that demonology and exorcism were leading factors in his consciousness and practice." Even Mr. Farnell admits, however, that "the earliest Hellene, as the later, was fully sensitive to the magico-divine efficacy of names." [1] Now to believe in the power of names before one believes in the existence of demons is the best possible evidence of the antiquity of magic in a society, since it indicates that the speaker has confidence in the operative power of his own words without any spiritual or divine assistance.

Magic in Greek philosophy.

Moreover, in one sense the advocates of Greek magic have not gone far enough. They hold that magic lies back of the comedies of Aristophanes; what they might contend is that it was also contemporary with them.[2] They hold that classical Greek religion had its origins in magic; what they might argue is that Greek philosophy never freed itself from magic. "That Empedocles believed himself capable of magical powers is," says Zeller, "proved by his own writings." He himself "declares that he possesses the power to heal old age and sickness, to raise and calm the winds, to summon rain and drought, and to recall the dead to life." [3] If the pre-Homeric fixed epithets of Zeus are redolent of magic, Plato's *Timaeus* is equally redolent of occult science and astrology; and if we see the weather-making magician in the Olympian Zeus of Phidias, we cannot explain away the vagaries of the *Timaeus* as flights of poetic imagination or try to make out Aristotle a modern scientist by mutilating the text of the *History of Animals*.

[1] Farnell, *Greece and Babylon*, pp. 292, 178-9.

[2] See Ernest Riess, *Superstitions and Popular Beliefs in Greek Tragedy*, in *Transactions of the American Philological Association*, vol. 27 (1896), pp. 5-34; and *On Ancient superstition*, *ibid.* 26 (1895), 40-55. Also J. G. Frazer, *Some Popular Superstitions of the Ancients*, in *Folk-lore*, 1890, and E. H. Klatsche, *The Supernatural in the Tragedies of Euripides*, in *University of Nebraska Studies*, 1919.

[3] See Zeller, *Pre-Socratic Philosophy*, II (1881), 119-20, for further boasts by Empedocles himself and other marvels attributed to him by later authors.

Plato's attitude toward magic and astrology.

Toward magic so-called Plato's attitude in his *Laws* is cautious. He maintains that medical men and prophets and diviners can alone understand the nature of poisons (or spells) which work naturally, and of such things as incantations, magic knots, and wax images; and that since other men have no certain knowledge of such matters, they ought not to fear but to despise them. He admits nevertheless that there is no use in trying to convince most men of this and that it is necessary to legislate against sorcery.[1] Yet his own view of nature seems impregnated, if not actually with doctrines borrowed from the *Magi* of the east, at least with notions cognate to those of magic rather than of modern science and with doctrines favorable to astrology. He humanized material objects and confused material and spiritual characteristics. He also, like authors of whom we shall treat later, attempted to give a natural or rational explanation for magic, accounting, for example, for liver divination on the ground that the liver was a sort of mirror on which the thoughts of the mind fell and in which the images of the soul were reflected; but that they ceased after death.[2] He spoke of harmonious love between the elements as the source of health and plenty for vegetation, beasts, and men, and their "wanton love" as the cause of pestilence and disease. To understand both varieties of love "in relation to the revolutions of the heavenly bodies and the seasons of the year is termed astronomy," [3] or, as we should say, astrology, whose fundamental law is the control of inferior creation by the motion of the stars. Plato spoke of the stars as "divine and eternal animals, ever abiding," [4] an expression which we shall hear reiterated in the middle ages. "The lower gods," whom he largely identified with the heavenly bodies, form men, who, if they live good lives, return after death each to a happy existence in his proper star.[5] Such a doctrine is not identical with that of nativities

[1] *Laws,* XI, 933 (Steph.).

[2] *Timaeus,* p. 71 (Steph.).

[3] *Symposium,* p. 188 (Steph.); in Jowett's translation, I, 558.

[4] *Timaeus,* p. 40 (Steph.); Jowett, III, 459.

[5] *Ibid.,* pp. 41-42 (Steph.).

and the horoscope, but like it exalts the importance of the stars and suggests their control of human life. And when at the close of his *Republic* Plato speaks of the harmony or music of the spheres of the seven planets and the eighth sphere of the fixed stars, and of "the spindle of Necessity on which all the revolutions turn," he suggests that when once the human soul has entered upon this life, its destiny is henceforth subject to the courses of the stars. When in the *Timaeus* he says, "There is no difficulty in seeing that the perfect number of time fulfills the perfect year when all the eight revolutions . . . are accomplished together and attain their completion at the same time," [1] he seems to suggest the astrological doctrine of the *magnus annus,* that history begins to repeat itself in every detail when the heavenly bodies have all regained their original positions.

Aristotle on stars and spirits.

For Aristotle, too, the stars were "beings of superhuman intelligence, incorporate deities. They appeared to him as the purer forms, those more like the deity, and from them a purposive rational influence upon the lower life of the earth seemed to proceed,—a thought which became the root of medieval astrology." [2] Moreover, "his theory of the subordinate gods of the spheres of the planets . . . provided for a later demonology." [3]

Folk-lore in the *History of Animals.*

Aside from bits of physiognomy and of Pythagorean superstition, or mysticism, Aristotle's *History of Animals* contains much on the influence of the stars on animal life, the medicines employed by animals, and their friendships and enmities, and other folklore and pseudo-science.[4] But

[1] *Timaeus,* p. 39 (Steph.); Jowett, III, 458.

[2] W. Windelband, *History of Philosophy,* English translation by J. H. Tufts, 1898, p. 147.

[3] Windelband, *History of Ancient Philosophy,* English translation by H. E. Cushman, 1899.

[4] For a number of examples, which might be considerably multiplied if books VII-X are not rejected as spurious, see Thorndike (1905), pp. 62-3. T. E. Lones, *Aristotle's Researches in Natural Science,* London, 1912, 274 pp., discusses "Aristotle's method of investigating the natural sciences," and a large number of Aristotle's specific statements showing whether they were correct or incorrect. The best translation of the *History of Animals* is by D'Arcy W. Thompson, Oxford 1910, with valuable notes.

the oldest extant manuscript of that work dates only from the twelfth or thirteenth century and lacks the tenth book. Editors of the text have also rejected books seven and nine, the latter part of book eight, and have questioned various other passages. However, these expurgations save the face of Aristotle rather than of Hellenic science or philosophy generally, as the spurious seventh book is held to be drawn largely from Hippocratic writings and the ninth from Theophrastus.[1]

Differing modes of transmission of ancient oriental and Greek literature.

There is another point to be kept in mind in any comparison of Egypt and Babylon or Assyria with Greece in the matter of magic. Our evidence proving the great part played by magic in the ancient oriental civilizations comes directly from them to us without intervening tampering or alteration except in the case of the early periods. But classical literature and philosophy come to us as edited by Alexandrian librarians [2] and philologers, as censored and selected by Christian and Byzantine readers, as copied or translated by medieval monks and Italian humanists. And the question is not merely, what have they added? but also, what have they altered? what have they rejected? Instead of questioning superstitious passages in extant works on the ground that they are later interpolations, it would very likely be more to the point to insert a goodly number on the ground that they have been omitted as pagan or idolatrous superstitions.

More magical character of directly transmitted Greek remains.

Suppose we turn to those writings which have been unearthed just as they were in ancient Greek; to the papyri, the lead tablets, the so-called Gnostic gems. How does the proportion of magic in these compare with that in the indirectly transmitted literary remains? If it is objected that the magic papyri [3] are mainly of late date and that

[1] See the edition of the *History of Animals* by Dittmeyer (1907), p. vii, where various monographs will be found mentioned.

[2] Perhaps pure literature was over-emphasized in the Museum at Alexandria, and magic texts in the library of Assurbanipal.

[3] A list of magic papyri and of publications up to about 1900 dealing with the same is given in Hubert's article on *Magia* in Daremberg-Saglio, pp. 1503-4. See also Sir Herbert Thompson and

they are found in Egypt, it may be replied that they are as old as or older than any other manuscripts we have of classical literature and that its chief store-house, too, was in Egypt at Alexandria. As for the magical curses written on lead tablets,[1] they date from the fourth century before our era to the sixth after, and fourteen come from Athens and sixteen from Cnidus as against one from Alexandria and eleven from Carthage. And although some display extreme illiteracy, others are written by persons of rank and education. And what a wealth of astrological manuscripts in the Greek language has been unearthed in European libraries by the editors of the *Catalogus Codicum Graecorum Astrologorum!*[2] And occasionally archaeologists report the discovery of magical apparatus[3] or of representations of magic in works of art.

Progress of science among the Greeks.

In thus contending that Hellenic culture was not free from magic and that even the philosophy and science of the ancient Greeks show traces of superstition, I would not, however, obscure the fact that of extant literary remains the Greek are the first to present us with any very considerable body either of systematic rational speculation or of classified collection of observed facts concerning nature. Despite the rapid progress in recent years in knowledge of prehistoric man and Egyptian and Babylonian civilization, the Hellenic

F. L. Griffith, *The Magical Demotic Papyrus of London and Leiden*, 3 vols., 1909-1921; *Catalogue of Demotic Papyri in the John Rylands Library, Manchester, with facsimiles and complete translations*, 1909, 3 vols. Grenfell (1921), p. 159, says, "A corpus of the magical papyri was projected in Germany by K. Preisendanz before the war, and a Czech scholar, Dr. Hopfner, is engaged upon the difficult task of elucidating them."

[1] W. C. Battle, *Magical Curses Written on Lead Tablets*, in *Transactions of the American Philological Association*, XXVI (1895), pp. liv-lviii, a synopsis of a Harvard dissertation. Audollent, *Defixionum tabulae*, etc., Paris, 1904, 568 pp. R. Wünsch, *Defixionum Tabellae Atticae*, 1897, and *Sethianische Verfluchungstafeln aus Rom* (390-420 A.D.), Leipzig, 1898.

[2] Since 1898 various volumes and parts have appeared under the editorship of Cumont, Kroll, Boll, Olivieri, Bassi, and others. Much of the material noted is of course post-classical and Byzantine, and of Christian authorship or Arabic origin.

[3] For example, see R. Wünsch, *Antikes Zaubergerät aus Pergamon*, in *Jahrb. d. kaiserl. deutsch. archæol. Instit., suppl.* VI (1905), p. 19.

title to the primacy in philosophy and science has hardly been called in question, and no earlier works have been discovered that can compare in medicine with those ascribed to Hippocrates, in biology with those of Aristotle and Theophrastus, or in mathematics and physics with those of Euclid and Archimedes. Undoubtedly such men and writings had their predecessors, probably they owed something to ancient oriental civilization, but, taking them as we have them, they seem to be marked by great original power. Whatever may lie concealed beneath the surface of the past, or whatever signs or hints of scientific investigation and knowledge we may think we can detect and read between the lines, as it were, in other phases of older civilizations, in these works solid beginnings of experimental and mathematical science stand unmistakably forth.

Archimedes and Aristotle.

"An extraordinarily large proportion of the subject matter of the writings of Archimedes," says Heath, "represents entirely new discoveries of his own. Though his range of subjects was almost encyclopædic, embracing geometry (plane and solid), arithmetic, mechanics, hydrostatics and astronomy, he was no compiler, no writer of text-books. . . . His objective is always some new thing, some definite addition to the sum of knowledge, and his complete originality cannot fail to strike anyone who reads his works intelligently, without any corroborative evidence such as is found in the introductory letters prefixed to most of them. . . . In some of his subjects Archimedes had no forerunners, *e. g.*, in hydrostatics, where he invented the whole science, and (so far as mathematical demonstration was concerned) in his mechanical investigations." [1] Aristotle's *History of Animals* is still highly esteemed by historians of biology [2] and often evidences "a large amount of personal

[1] T. L. Heath, *The Works of Archimedes,* Cambridge, 1897, pp. xxxix-xl.

[2] On "Aristotle as a Biologist" see the Herbert Spencer lecture by D'Arcy W. Thompson, Oxford, 1913, 31 pp. Also T. E. Lones, *Aristotle's Researches in Natural Science,* London, 1912. Professor W. A. Locy, author of *Biology and Its Makers,* writes me (May 9, 1921) that in his opinion G. H. Lewes, *Aristotle; a Chapter from the History of Science,* London,

observations," [1] "great accuracy," and "minute inquiry," as in his account of the vascular system [2] or observations on the embryology of the chick.[3] "Most wonderful of all, perhaps, are those portions of his book in which he speaks of fishes, their diversities, their structure, their wanderings, and their food. Here we may read of fishes that have only recently been rediscovered, of structures only lately reinvestigated, of habits only of late made known." [4] But of the achievements of Hellenic philosophy and Hellenistic science the reader may be safely assumed already to have some notion.

Exaggerated view of the scientific achievement of the Hellenistic age.

But in closing this brief preliminary sketch of the period before our investigation proper begins, I would take exception to the tendency, prevalent especially among German scholars, to center in and confine to Aristotle and the Hellenistic age almost all progress in natural science made before modern times. The contributions of the Egyptians and Babylonians are reduced to a minimum on the one hand, while on the other the scientific writings of the Roman

1864, "dwells too much on Aristotle's errors and imperfections, and in several instances omits the quotation of important positive observations, occurring in the chapters from which he makes his quotations of errors." Professor Locy also disagrees with Lewes' estimate of *De generatione* as Aristotle's masterpiece and thinks that "naturalists will get more satisfaction out of reading the *Historia animalium*" than either the *De generatione* or *De partibus*. Thompson (1913), p. 14, calls Aristotle "a very great naturalist."

[1] This quotation is from Professor Locy's letter of May 9, 1921.

[2] The quotations are from a note by Professor D'Arcy W. Thompson on his translation of the *Historia animalium*, III, 3. The note gives so good a glimpse of both the merits and defects of the Aristotelian text as it has reached us that I will quote it here more fully:

"The Aristotelian account of the vascular system is remarkable for its wealth of details, for its great accuracy in many particulars, and for its extreme obscurity in others. It is so far true to nature that it is clear evidence of minute inquiry, but here and there so remote from fact as to suggest that things once seen have been half forgotten, or that superstition was in conflict with the result of observation. The account of the vessels connecting the left arm with the liver and the right with the spleen . . . is a surviving example of mystical or superstitious belief. It is possible that the ascription of three chambers to the heart was also influenced by tradition or mysticism, much in the same way as Plato's notion of the three corporeal faculties."

[3] Professor Locy called my attention to it in a letter of May 17, 1921. See also Thompson (1913), p. 14.

[4] Thompson (1913), p. 19.

Empire, which are extant in far greater abundance than those of the Hellenistic period, are regarded as inferior imitations of great authors whose works are not extant; Posidonius, for example, to whom it has been the fashion of the writers of German dissertations to attribute this, that, and every theory in later writers. But it is contrary to the law of gradual and painful acquisition of scientific knowledge and improvement of scientific method that one period of a few centuries should thus have discovered everything. We have disputed the similar notion of a golden age of early Egyptian science from which the Middle and New Kingdoms declined, and have not held that either the Egyptians or Babylonians had made great advances in science before the Greeks. But that is not saying that they had not made some advance. As Professor Karpinski has recently written:

"To deny to Babylon, to Egypt, and to India, their part in the development of science and scientific thinking is to defy the testimony of the ancients, supported by the discoveries of the modern authorities. The efforts which have been made to ascribe to Greek influence the science of Egypt, of later Babylon, of India, and that of the Arabs do not add to the glory that was Greece. How could the Babylonians of the golden age of Greece or the Hindus, a little later, have taken over the developments of Greek astronomy? This would only have been possible if they had arrived at a state of development in astronomy which would have enabled them properly to estimate and appreciate the work which was to be absorbed. . . . The admission that the Greek astronomy immediately affected the astronomical theories of India carries with it the implication that this science had attained somewhat the same level in India as in Greece. Without serious questioning we may assume that a fundamental part of the science of Babylon and Egypt and India, developed during the times which we think of as Greek, was indigenous science."[1]

[1] L. C. Karpinski, "Hindu Science," in *The American Mathematical Monthly,* XXVI (1919), 298-300.

Nor am I ready to admit that the great scientists of the early Roman Empire merely copied from, or were distinctly inferior to, their Hellenistic predecessors. Aristarchus may have held the heliocentric theory[1] but Ptolemy must have been an abler scientist and have supported his incorrect hypothesis with more accurate measurements and calculations or the ancients would have adopted the sounder view. And if Herophilus had really demonstrated the circulation of the blood, so keen an intelligence as Galen's would not have cast his discovery aside. And if Ptolemy copied Hipparchus, are we to imagine that Hipparchus copied from no one? But of the incessant tradition from authority to authority and yet of the gradual accumulation of new matter from personal observation and experience our ensuing survey of thirteen centuries of thought and writing will afford more detailed illustration.

[1] Sir Thomas Heath, *Aristarchus of Samos, the Ancient Copernicus: a history of Greek astronomy to Aristarchus together with Aristarchus's treatise, "On the Sizes and Distances of the Sun and Moon," a new Greek text with translation and notes,* Oxford, 1913, admits that "our treatise does not contain any suggestion of any but the geocentric view of the universe, whereas Archimedes tells us that Aristarchus wrote a book of hypotheses, one of which was that the sun and the fixed stars remain unmoved and that the earth revolves round the sun in the circumference of a circle." Such evidence seems scarcely to warrant applying the title of "The Ancient Copernicus" to Aristarchus. And Heath thinks that Schiaparelli (*I precursori di Copernico nell' antichità,* and other papers) went too far in ascribing the Copernican hypothesis to Heraclides of Pontus. On Aristotle's answer to Pythagoreans who denied the geocentric theory see Orr (1913), pp. 100-2.

APPENDIX I

SOME WORKS ON MAGIC, RELIGION, AND ASTRONOMY IN BABYLONIA AND ASSYRIA

The following books deal expressly with the magic of Assyria and Babylonia:

Fossey, C. La magie assyrienne; étude suivie de textes magiques, Paris, 1902.

King, L. W. Babylonian Magic and Sorcery, being "The Prayers of the Lifting of the Hand," London, 1896.

Laurent, A. La magie et la divination chez les Chaldéo-Assyriens, Paris, 1894.

Lenormant, F. Chaldean Magic and Sorcery, English translation, London, 1878.

Schwab, M., in Proc. Bibl. Archæology (1890), pp. 292-342, on magic bowls from Assyria and Babylonia.

Tallquist, K. L. Die Assyrische Beschwörungsserie Maqlû, Leipzig, 1895.

Thompson, R. C. The Reports of the Magicians and Astrologers of Nineveh and Babylon in the British Museum, London, 1900. Texts and translations—all but three are astrological.

The Devils and Evil Spirits of Babylonia, London, 1904.

Semitic Magic, London, 1908.

Weber, O. Dämonenbeschwörung bei den Babyloniern und Assyrern, 1906. Eine Skizze (37 pp.), in Der Alte Orient.

Zimmern. Die Beschwörungstafeln Surpu.

Much concerning magic will also be found in works on Babylonian and Assyrian religion.

Craig, J. A. Assyrian and Babylonian Religious Texts, Leipzig, 1895-7.

Curtiss, S. I. Primitive Semitic Religion Today, 1902.

Dhorme, P. Choix des textes religieux Assyriens Babyloniens, 1907.

La religion Assyro-Babylonienne, Paris, 1910.

Gray, C. D. The Samas Religious Texts.

Jastrow, Morris. The Religion of Babylonia and Assyria, Boston, 1898. Revised and enlarged as Religion Babyloniens und Assyriens, Giessen, 1904.

Jeremias. Babylon. Assyr. Vorstellungen von dem Leben nach Tode, Leipzig, 1887.

Hölle und Paradies, and other works.

Knudtzon, J. A. Assyrische Gebete an den Sonnengott, Leipzig, 1893.

Lagrange, M. J. Études sur les religions sémitiques, Paris, 1905.

Langdon, S. Sumerian and Babylonian Psalms, Paris, 1909.

Reisner, G. A. Sumerisch-Babylonische Hymnen, Berlin, 1896.

Robertson Smith, W. Lectures on the Religion of the Semites, London, 1907.

Roscher, Lexicon, for various articles.

Zimmern. Babylonische Hymnen und Gebete in Auswahl, 32 pp., 1905 (Der Alte Orient).

Beiträge zur Kenntniss der Babyl. Religion, Leipzig, 1901.

On the astronomy and astrology of the Babylonians one may consult:

Bezold, C. Astronomie, Himmelschau und Astrallehre bei den Babyloniern. (Sitzb. Akad. Heidelberg, 1911, Abh. 2).

Boissier. A. Documents assyriens relatifs aux présages, Paris, 1894-1897.

Choix de textes relatifs à la divination assyro-babylonienne, Geneva, 1905-1906.

Craig, J. A. Astrological-Astronomical Texts, Leipzig, 1892.

Cumont, F. Babylon und die griechische Astrologie. (Neue Jahrb. für das klass. Altertum, XXVII, 1911).

Epping, J., and Strassmeier, J. N. Astronomisches aus Babylon, 1889.

Ginzel, F. K. Die astronomischen Kentnisse der Babylonier, 1901.

Hehn, J. Siebenzahl und Sabbat bei den Babyloniern und im Alten Testament, 1907.

Jensen, P. Kosmologie der Babylonier, 1890.

Jeremias. Das Alter der babylonischen Astronomie, 1908. Handbuch der altorientalischen Geisteskultur, 1913.

Kugler, F. X. Die Babylonische Mondrechnung, 1900.

Sternkunde und Sterndienst in Babel, Freiburg, 1907-1913. To be completed in four vols.

Im Bannkreis Babels, 1910.

Oppert, J. Die astronomischen Angaben der assyrischen Keilinschriften, in Sitzb. d. Wien. Akad. Math.-Nat. Classe, 1885, pp. 894-906.

Un texte Babylonien astronomique et sa traduction grecque par Cl. Ptolémeé, in Zeitsch. f. Assyriol. VI (1891), pp. 103-23.

Sayce, A. H. The astronomy and astrology of the Babylonians, with translations of the tablets relating to the subject, in Transactions of the Society of Biblical Archaeology, III (1874), 145-339; the first and until recently the best guide to the subject.

Schiaparelli, G. V. I Primordi ed i Progressi dell' Astronomia presso i Babilonesi, Bologna, 1908.

Astronomy in the Old Testament, 1905.

Stücken, Astralmythen, 1896-1907.

Virolleaud, Ch. L'Astrologie chaldéenne, Paris, 1905-; to be completed in eight parts, texts and translations.

Winckler, Himmels- und Weltenbild der Babylonier als Grundlage der Weltanschauung und Mythologie aller Völker, in Der alte Orient, III, 2-3.

BOOK I. THE ROMAN EMPIRE

BOOK I. THE ROMAN EMPIRE

FOREWORD

A TRIO of great names, Pliny, Galen, and Ptolemy, stand out above all others in the history of science under the Roman Empire. In the use or criticism which they make of earlier writers and investigators they are also our chief sources for the science of the preceding Hellenistic period. By their voluminousness, their generous scope in ground covered, and their broad, liberal, personal outlooks, they have painted, in colors for the most part imperishable, extensive canvasses of the scientific spirit and acquisitions of their own time. Pliny pursued politics and literature as well as natural science; Ptolemy was at once mathematician, astronomer, physicist, and geographer; Galen knew philosophy as well as medicine. The two latter men, moreover, made original contributions of their own of the very first order to scientific knowledge and method. It is characteristic of the homogeneous and widespread culture of the Roman Empire that these three representatives of different, although overlapping, fields of science were natives of the three continents that enclose the Mediterranean Sea. Pliny was born at Como where Italy verges on transalpine lands; Ptolemy, born somewhere in Egypt, did his work at Alexandria; Galen came from Pergamum in Asia Minor. Finally, these men were, after Aristotle, the three ancient scientists who directly or indirectly most powerfully influenced the middle ages. Thus they illuminate past, present, and future. A trio of great names.

We shall therefore open the present section of our investigation by considering in turn chronologically, Pliny, Ptolemy, and Galen, coupling, however, with our consideration of Ptolemy the work of Seneca on *Natural Questions* Plan of this section.

which shows the same combination of natural science and natural divination. Next we shall consider some representatives of ancient applied science and its relations to magic, and the more miscellaneous writings of Plutarch, Apuleius, and Philostratus's *Life of Apollonius of Tyana*. From the hospitable attitude toward magic and occult science displayed by these last writers we shall then turn back again to consider some examples of literary and philosophical attacks upon superstition, before proceeding lastly to spurious mystic writings of the Roman Empire, Neo-Platonism and its relations to astrology and theurgy, and the works of Aelian, Solinus, and Horapollo.

CHAPTER II

PLINY'S NATURAL HISTORY

I. *Its Place in the History of Science*

Its importance in our investigation—As a collection of miscellaneous information—As a repository of ancient natural science—As a source for magic—Pliny's career—His writings—His own description of the *Natural History*—His devotion to science—Conflict of science and religion—Pliny not a trained naturalist—His use of authorities—His lack of arrangement and classification—His scepticism and credulity—A guide to ancient science—His medieval influence—Early printed editions.

II. *Its Experimental Tendency*

Importance of observation and experience—Use of the word ***experimentum***—Experiments due to scientific curiosity—Medical experimentation—Chance experience and divine revelation—Marvels proved by experience.

III. *Pliny's Account of Magic*

Oriental origin of magic—Its spread to the Greeks—Its spread outside the Graeco-Roman world—Failure to understand its true origin—Magic and divination—Magic and religion—Magic and medicine—Magic and philosophy—Falseness of magic—Crimes of magic—Pliny's censure of magic is mainly intellectual—Vagueness of Pliny's scepticism—Magic and science indistinguishable.

IV. *The Science of the Magi*

Magicians as investigators of nature—The *Magi* on herbs—Marvelous virtues of herbs—Animals and parts of animals—Further instances—Magic rites with animals and parts of animals—Marvels wrought with parts of animals—The *Magi* on stones—Other magical recipes—Summary of the statements of the *Magi*.

V. *Pliny's Magical Science*

From the *Magi* to Pliny's magic—Habits of animals—Remedies discovered by animals—Jealousy of animals—Occult virtues of animals—The virtues of herbs—Plucking herbs—Agricultural magic—Virtue of stones—Other minerals and metals—Virtues of human parts—Virtues of human saliva—The human operator—Absence of medical compounds—Sympathetic magic—Antipathies between animals—Love and hatred between inanimate objects—Sympathy between animate and inanimate objects—Like cures like—The principle of association—Magic transfer of disease—Amulets—Position or direction—The time element—Observance of number—Relation between operator and patient—Incantations—Attitude towards love-charms and birth control—Pliny and astrology—Celestial portents—The stars and the world of nature—Astrological medicine—Conclusion: magic unity of Pliny's superstitions.

"Salve, parens rerum omnium Natura, teque nobis Quiritium solis celebratam esse numeris omnibus tuis fave!"
—Closing words of the Natural History.[1]

I. *Its Place in the History of Science*

Important in our investigation.

WE should have to search long before finding a better starting-point for the consideration of the union of magic with the science of the Roman Empire, and of the way in which that union influenced the middle ages, than Pliny's *Natural History.*[2] The foregoing sentence, with which years ago I opened a chapter on the *Natural History* of Pliny the Elder in my briefer preliminary study of magic in the intellectual history of the Roman Empire, seems as true as ever; and although I there considered his confusion of magic and science at some length, I do not see how I can make the present work well-rounded and complete without including in it a yet more detailed analysis of the contents of Pliny's book.

As a collection of miscellaneous information.

Pliny's *Natural History,* which appeared about 77 A. D. and is dedicated to the Emperor Titus, is perhaps the most

[1] "Farewell, Nature, parent of all things, and in thy manifold multiplicity bless me who, alone of the Romans, has sung thy praise."

[2] For the Latin text of the *Naturalis Historia* I have used the editions of D. Detlefsen, Berlin, 1866-1882, and L. Janus, Leipzig, 1870, 6 vols. in 3; 5 vols. in 3. There is, however, a good English translation of the *Natural History,* with an introductory essay, by J. Bostock and H. T. Riley, London, 1855, 6 vols. (Bohn Library), which is superior to both the German editions in its explanatory notes and subject index, and which also apparently antedates them in some readings suggested for doubtful passages in the text. Three modes of dividing the *Natural History* into chapters are indicated in the editions of Janus and Detlefsen. I shall employ that found in the earlier editions of Hardouin, Valpy, Lemaire, and Ajasson, and preferred in the English translation of Bostock and Riley.

important single source extant for the history of ancient civilization. Its thirty-seven books, written in a very compact style, constitute a vast collection of the most miscellaneous information. Whether one is investigating ancient painting, sculpture, and other fine arts; or the geography of the Roman Empire; or Roman triumphs, gladiatorial contests, and theatrical exhibitions; or the industrial processes of antiquity; or Mediterranean trade; or Italian agriculture; or mining in ancient Spain; or the history of Roman coinage; or the fluctuation of prices in antiquity; or the Roman attitude towards usury; or the pagan attitude towards immortality; or the nature of ancient beverages; or the religious usages of the ancient Romans; or any of a number of other topics; one will find something concerning all of them in Pliny. He is apt both to depict such conditions in his own time and to trace them back to their origins. Furthermore he repeats many detailed incidents of interest to the political or narrative historian of Rome as well as to the student of the economic, social, artistic, and religious life of antiquity. Probably there is no place where an isolated point is more likely to be run down by the investigator, and it is regrettable that exhaustive analytical indices of the work are not available. We may add that, although the work is supposedly a collection of facts, Pliny contrives to introduce many moral reflections and sharp comments on the luxury, vice, and unintellectual character of his times, suggesting Juvenal's picture of degenerate Roman society and his own lofty moral standards.

As a repository of ancient natural science.

Indeed, Pliny's title, *Naturalis Historia,* or at least the common English translation of it, "Natural History," has been criticized as too limited in scope, and the work has been described as "rather a vast encyclopedia of ancient knowledge and belief upon almost every known subject." [1] Pliny himself mentions in his preface the Greek word "encyclopedia" as indicative of his scope. Nevertheless, his work is primarily an account of nature rather than of civili-

[1] Bostock and Riley (1855), I, xvi.

zation, and much of its information concerning such matters as the arts and business is incidental. Most of its books bear such titles as Aquatic Animals, Exotic Trees, Medicines from Forest Trees, The Natures of Metals. After an introductory book containing the preface and a table of contents and lists of authorities for each of the subsequent books, the second book treats of the universe, heavenly bodies, meteorology, and the chief changes, such as earthquakes and tides, in the land and water forming the earth's surface. After four books devoted to geography, the seventh deals with man and human inventions. Four more follow on terrestrial and aquatic animals, birds, and insects. Sixteen more are concerned with plants, trees, vines, and other vegetation, and the medicinal simples derived from them. Five books discuss the medicinal simples derived from animals, including the human body; and the last five books treat of metals and minerals and the arts in which they are employed. It is thus evident that in the main Pliny is concerned with natural science, and that, if his work is a mine of miscellaneous historical information, it should even more prove a rich treasure-house—*"quoniam, ut ait Domitius Piso, thesauros oportet esse non libros"* [1]—for an investigation concerned as intimately as is ours with the history of science.

As a source for magic.

The *Natural History* is a great storehouse of misinformation as well as of information, for Pliny's credulity and lack of discrimination harvested the tares of legend and magic along with the wheat of historical fact and ancient science in his voluminous granary. This may put other historical investigators upon their guard in accepting its statements, but only increases its value for our purpose. Perhaps it is even more valuable as a collection of ancient errors than it is as a repository of ancient science. It touches upon many of the varieties, and illustrates most of the characteristics, of magic. Moreover, Pliny often mentions the *Magi* or magicians and discusses "magic" expressly at some

[1] NH, Preface.

length in the opening chapters of his thirtieth book—one of the most important passages on the theme in any ancient writer.

Pliny's career.

Pliny the Elder, as we learn from his own statements in the *Natural History* and from one or two letters concerning him written by his nephew, Pliny the Younger, whom he adopted, went through the usual military, forensic, and official career of the Roman of good family, and spent his life largely in the service of the emperors. He visited various Mediterranean lands, such as Spain, Africa, Greece, and Egypt, and fought in Germany. He was in charge of the Roman fleet on the west coast of Italy when he met his death at the age of fifty-six by suffocation as he was trying to rescue others from the fumes and vapors from the eruption of Mount Vesuvius.

His writings.

Of Pliny's writings the *Natural History* is alone extant, but other titles have been preserved which serve to show his great literary industry and the extent of his interests. He wrote on the use of the javelin by cavalry, a life of his friend Pomponius, an account in twenty books of all the wars waged by the Romans in Germany, a rather long work on oratory called *The Student,* a grammatical or philological work in eight books entitled *De dubio sermone,* and a continuation of the *History* of Aufidius Bassus in thirty-one books. Yet in the dedication of the *Natural History* to the emperor Titus he states that his days were taken up with official business and only his nights were free for literary labor. This statement is supported by a letter of his nephew telling how he used to study by candle-light both late at night and before daybreak. Pliny the Younger narrates several incidents to illustrate how jealous and economical of every spare moment his uncle was. He would dictate or have books read to him while lying down or in the bath, and on journeys a secretary was always by his side with books and tablets. If the weather was very cold, the amanuensis wore gloves so that his hands might not become too numb to write. Pliny always took notes on what he read, and at his death left his nephew one hundred and sixty notebooks written in a small hand on both sides.

His own description of the *Natural History.*

Such were the conditions under which, and the methods by which, Pliny compiled his encyclopedia on nature. No single writer either Greek or Latin, he tells us, had ever before attempted so extensive a task. He adds that he treats of some twenty thousand topics gleaned from the perusal of about two thousand volumes by one hundred authors.[1] Judging from his bibliographies and citations, however, he would seem to have utilized more than one hundred authors. But possibly he had not read all the writers mentioned in his bibliographies. He affirms that previous students have had access to but few of the volumes which he has used, and that he adds many things unknown to his ancient authorities and recently discovered. Occasionally he shows an acquaintance with beliefs and practices of the Gauls and Druids. Thus his work assumes to be something more than a compilation from other books. He says, however, that no doubt he has omitted much, since he is only human and has had many other demands upon his time. He admits that his subject is dry *(sterilis materia)* and does not lend itself to literary exhibitions, nor include matters stimulating to write about and pleasant to read about, like speeches and marvelous occurrences and varied incidents. Nor does it permit purity and elegance of diction, since one must at times employ the terminology of rustics, foreigners, and even barbarians. Furthermore, "it is an arduous task to give novelty to what is ancient, authority to what is new, interest to what is obsolete, light to what is obscure, charm to what is loathsome"—as many of his medicinal simples undoubtedly are—"credit to what is dubious."

His devotion to science.

It is a great comfort to Pliny, however, in his immense task, when many laugh at him as wasting his time over worthless trifles, to reflect that he is being spurned along with Nature.[2] In another passage [3] he contrasts the blood

[1] NH, Preface.
[2] NH, XXII, 7.
[3] NH, II, 6.

and slaughter of military history with the benefits bestowed upon mankind by astronomers. In a third passage [1] he looks back regretfully at the widespread interest in science among the Greeks, although those were times of political disunion and strife and although communication between different lands was interrupted by piracy as well as war, whereas now, with the whole empire at peace, not only is no new scientific inquiry undertaken, but men do not even thoroughly study the works of the ancients, and are intent on the acquisition of lucre rather than learning. These and other passages which might be cited attest Pliny's devotion to science.

Conflict of science and religion.

In Pliny we also detect signs of the conflict between science and religion. In a single chapter on God he says pretty much all that the church fathers later repeated at much greater length against paganism and polytheism. But his discussion would hardly satisfy a Christian. He asserts that "it is God for man to aid his fellow man,[2] and this is the path to eternal glory," but he turns this noble sentiment to justify deification of the emperors who have done so much for mankind. He questions whether God is concerned with human affairs; slyly suggests that if so, God must be too busy to punish all crimes promptly; and points out that there are some things which God cannot do. He cannot commit suicide as men can, nor alter past events, nor make twice ten anything else than twenty. Pliny then concludes: "By which is revealed in no uncertain wise the power of Nature, and that is what we call God." In many other passages he exclaims at Nature's benignity or providence. He believed that the soul had no separate existence from the body, [3] and that after death there was no more sense left in body or soul than was there before birth. The hope of personal immortality he scorned as "puerile ravings" produced by the fear of death, and he believed still less in the possibility of any resurrection of the body. In short, natural law, me-

[1] NH, II, 46.

[2] NH, II, 5. "Deus est mortali iuvare mortalem. . . ."

[3] NH, VII, 56.

chanical force, and facts capable of scientific investigation would seem to be all that he will admit and to suffice to satisfy his strong intellect. Yet we shall later find him having the greatest difficulty in distinguishing between science and magic, and giving credence to many details in science which seem to us quite as superstitious as the pagan beliefs concerning the gods which he rejected. But if any reader is inclined to belittle Pliny for this, let him first stop and think how Pliny would ridicule some modern scientists for their religious beliefs, or for their spiritualism or psychic research.

Pliny not a trained naturalist.

It is desirable, however, to form some estimate of Pliny's fitness for his task in order to judge how accurate a picture of ancient science his work is. He does not seem to have had much detailed training or experience in the natural sciences himself. He writes not as a naturalist who has observed widely and profoundly the phenomena and operations of nature, but as an omnivorous reader and voluminous note-taker who owes his knowledge largely to books or hearsay, although occasionally he says "I know" instead of "they say," or gives the results of his own observation and experience. In the main he is not a scientist himself but only a historian of science or nature; after all, his title, *Natural History,* is a very fitting one. The question, of course, arises whether he has sufficient scientific training to evaluate properly the work of the past. Has he read the best authors, has he noted their best passages, has he understood their meaning? Does he repeat inferior theories and omit the correcter views of certain Alexandrian scientists? These questions are hard to answer. On his behalf it may be said that he deals little with abstruse scientific theory and mainly with simple substances and geographical places, matters in which it seems difficult for him to go far astray. Scientific specialists were not numerous in those days, anyway, and science had not yet so far advanced and ramified that one man might not hope to cover the entire field and do it substantial justice. Pliny the Younger was perhaps a partial judge, but he described the *Natural History* as "a work remarkable for its comprehensiveness and erudition, and not less varied than Nature herself." [1]

His use of authorities.

One thing in Pliny's favor as a compiler, besides his personal industry, unflagging interest, and apparently abundant supply of clerical assistance, is his full and honest statement of his authorities, although he adds that he has caught many authors transcribing others verbatim without acknowledgment. He has, however, great admiration for many of his authorities, exclaiming more than once at the care and diligence of the men of the past who have left nothing untried or unexperienced, from trackless mountain tops to the roots of herbs.[2] Sometimes, nevertheless, he disputes their assertions. For instance, Hippocrates said that the appearance of jaundice on the seventh day in fever is a fatal sign, "but we know some who have lived even after this." [3] Pliny also scolds Sophocles for his falsehoods concerning amber.[4] It may seem surprising that he should expect strict scientific truth from a dramatic poet, but Pliny, like many medieval writers, seems to regard poets as good scientific authorities. In another passage he accepts Sophocles' statement that a certain plant is poisonous, rather than the contrary view of other writers, saying "the authority of so prominent a man moves me against their opinions." [5] He also cites Menander concerning fish and, like almost all the ancients, regards Homer as an authority on all matters.[6] Pliny sometimes cites the works of King Juba of Numidia, than whom there hardly seems to have been a greater liar in antiquity.[7] He stated among other things in a work which he wrote for Gaius Caesar, the son of Augustus, that a whale six hundred feet long and three hundred and sixty feet broad had

[1] Letter to Macer, Ep. III, 5, ed. Keil. Leipzig, 1896.

[2] NH, VII, 1; XXIII, 60; XXV, 1; XXVII, 1.

[3] XXVI, 76.

[4] XXXVII, 11.

[5] XXI, 88.

[6] XXXII, 24.

[7] Yet C. W. King, *Natural History of Precious Stones*, p. 2, deplores the loss of Juba's treatise, which he says, "considering his position and opportunities for exact information, is perhaps the greatest we have to deplore in this sad catalogue of *desiderata*."

entered a river in Arabia.[1] But where should Pliny turn for sober truth? The Stoic Chrysippus prated of amulets;[2] treatises ascribed to the great philosophers Democritus and Pythagoras[3] were full of magic; and in the works of Cicero he read of a man who could see for a distance of one hundred and thirty-five miles, and in Varro that this man, standing on a Sicilian promontory, could count the number of ships sailing out of the harbor of Carthage.[4]

His lack of arrangement and classification.

The *Natural History* has been criticized as poorly arranged and lacking in scientific classification, but this is a criticism which can be made of many works of the classical period. Their presentation is apt to be rambling and discursive rather than logical and systematic. Even Aristotle's *History of Animals* is described by Lewes[5] as unclassified in its arrangement and careless in its selection of material. I have often thought that the scholastic centuries did mankind at least one service, that of teaching lecturers and writers how to arrange their material. Pliny seems rather in advance of his times in supplying full tables of contents for the busy emperor's convenience. Valerius Soranus seems to have been the only previous Roman writer to do this. One indication of haste in composition and failure to sift and compare his material is the fact that Pliny sometimes makes or includes contradictory statements, probably taken from different authorities. On the other hand, he not infrequently alludes to previous passages in his own work, thus showing that he has his material fairly well in hand.

His scepticism and credulity.

Pliny once said that there was no book so bad but what some good might be got from it,[6] and to the modern reader he seems almost incredibly credulous and indiscriminate in

[1] NH, XXXII, 4.

[2] XXX, 30.

[3] Bouché-Leclercq (1899), p. 519, notes, however, that Aulus Gellius (X, 12) protested against Pliny's credulity in accepting such works as genuine and that "Columelle (VII, 5) cite un certain Bolus de Mendes comme l'auteur des ὑπομνήματα attribués à Démocrite." Bouché-Leclercq adds, however, "Rien n'y fit: Démocrite devint le grand docteur de la magie."

[4] NH, VII, 21.

[5] G. H. Lewes, *Aristotle; a Chapter from the History of Science*, London, 1864.

[6] *Letters of Pliny the Younger*, III, 5, ed. Keil, Leipzig, 1896.

his selection of material, and to lack any standard of judgment between the true and the false. Yet he often assumes an air of scepticism and censures others sharply for their credulity or exaggeration. "'Tis strange," he remarks *à propos* of some tales of men transformed into wolves for nine or ten years, "how far Greek credulity has gone. No lie is so impudent that it lacks a voucher."[1] Once he expresses his determination to include only those points on which his authorities are in agreement.[2]

A guide to ancient science.

On the whole, while to us to-day the *Natural History* seems a disorderly and indiscriminate conglomeration of fact and fiction, its defects are probably to a great extent those of its age and of the writers from whom it has borrowed. If it does not reflect the highest achievements and clearest thinking of the best scientists of antiquity—and be it said that there are a number of the Hellenistic age of whom we should know less than we do but for Pliny—it probably is a fairly faithful epitome of science and error concerning nature in his own time and the centuries preceding. At any rate it is the best portrayal that has reached us. From it we can get our background of the confusion of magic and science in the Hellenistic age, and then reveal against this setting the development of them both in the course of the Roman Empire and middle ages. Pliny gives so many items upon each point, and is so much fuller than the average ancient or medieval book of science, that he serves as a reference book, being the likeliest place to look to find duplicated some statement concerning nature by a later writer. This of course shows that such a statement did not originate with the later writer, but is not a sure sign that he copied from Pliny; they may both have used the same authorities, as seems the case with Greek authors later in the empire who probably did not know of Pliny's work.

His medieval influence.

In the middle ages, however, Pliny had an undoubted direct influence.[3] Manuscripts of the *Natural History* are

[1] NH, VIII, 34.
[2] XXVIII, 1.
[3] Rück, *Die Naturalis Historia des Plinius im Mittelalter*, in *Sitzb. Bayer. Akad. Philos-Philol. Classe* (1908) pp. 203-318. For

numerous, although in a scarcely legible condition owing to corrections and emendations which enhance the obscurity of the text and perhaps do Pliny grave injustice in other respects.[1] Also many manuscripts contain only a few books or fragments of the text, so that it is possible that many medieval scholars knew their Pliny only in part.[2] This, however, can scarcely be argued from their failure to include more from him in their own works; for that might be due to their knowing the *Natural History* so well that they took its contents for granted and tried to include other material in their own works. In a later chapter we shall treat of *The Medicine of Pliny,* a treatise derived from the *Natural History.* Pliny's phrase *rerum natura* figures as the title of several medieval encyclopedias of somewhat similar scope. And his own name was too well known in the middle ages to escape having a work on the philosopher's stone ascribed to him.[3]

citations of Pliny by writers of the late Roman empire and early middle ages, see Panckoucke, *Bibliothèque Latine-Française,* vol. CVI.

[1] Concerning the MSS see Detlefsen's prefaces in each of his first five volumes and his fuller dissertations in Jahn's *Neue Jahrb.,* 77, 653ff, *Rhein. Mus.,* XV, 265ff; XVIII, 227ff, 327.

Detlefsen seems to have made no use of English MSS, but a folio of the close of the 12th century at New College, Oxford, contains the first nineteen books of the *Natural History* and is described by Coxe as "very well written and preserved."

Nor does Detlefsen mention Le Mans 263, 12th century, containing all 37 books except that the last book is incomplete, and with a full page miniature (fol. 10v) showing Pliny in the act of presenting his work to Vespasian. Escorial Q-I-4 and R-I-5 are two other practically complete texts of the fourteenth century which Detlefsen failed to use.

[2] See M. R. James, Eton Manuscripts, p. 63, MS 134, Bl. 4. 7., Roberti Crikeladensis Prioris Oxoniensis excerpta ex Plinii Historia Naturali, 12-13th century, in a large English hand, giving extracts extending from Book II to Book IX.

Of Balliol 124, fols. 1-138, *Cosmographia mundi,* by John Free, born at Bristol or London, fellow at Balliol College, Oxford, later professor of medicine at Padua and a doctor at Rome, also well instructed in civil law and Greek, Coxe writes, "This work is nothing but a series of excerpts from Pliny's *Natural History,* beginning with the second and leaving off with the twentieth." I wonder if John Free may not have used the very MS of the first nineteen books mentioned in the foregoing note, since the second book of the *Natural History* is often reckoned as the first.

In Balliol 146A, 15th century, fol. 3-, the *Natural History* appears in epitome, with a prologue opening, "I, Reginald (*Retinaldus*), servant of Christ, perusing the books of Pliny . . ."

[3] Bologna, 952, 15th century, fols. 157-60, "Tractatus optimus in

That the *Natural History* was well known as a whole at least by the close of the middle ages is shown by the numerous editions, some of them magnificently printed, which were turned off from the Italian presses immediately after the invention of printing. In the Magliabechian Library of Florence alone are editions printed at Venice in 1469 and 1472, at Rome in 1473 and Parma in 1481, again at Venice in 1487, 1491, and 1499, not to mention Italian translations which appeared at Venice in 1476 and 1489.[1] These editions were accompanied by some published criticism of Pliny's statements, since in 1492 appeared at Ferrara a treatise *On the Errors of Pliny and Others in Medicine* by Nicholas Leonicenus of Vicenza with a dedication to Politian.[2] But two years later Pliny found a defender in Pandulph Collenucius.[3]

Early printed editions.

But Pliny's future influence will come out repeatedly in later chapters. We shall now inquire, first, what signs of experimental science he shows, either derived from the past or added by himself. Second, what he defines as magic and what he has to say about it. Third, how much of what he supposes to be natural science must we regard as essentially magic?

II. *Its Experimental Tendency*

It is probably only a coincidence that two medieval manuscripts close the *Natural History* in the midst of the seventy-sixth chapter of the last book with the words, "*Experimenta pluribus modis constant . . . Primum pondere.*"[4] But although from the very nature of his work Pliny makes extensive use of authorities, he not infrequently manifests a realization, as one dealing with the facts of nature should, of the importance of observation and experience as means of

Importance of observation and experience.

quo exposuit et aperte declaravit plinius philosophus quid sit lapis philosophicus et ex qua materia debet fieri et quomodo."

[1] Fossi, *Catalogus codicum saeculo XV impressorum qui in publica Bibliotheca Magliabechiana Florentiae adservantur*, 1793-1795, II, 374-81.

[2] *De erroribus Plinii et aliorum in medicina*, Ferrara, 1492.

[3] *Pliniana defensio*, 1494.

[4] Escorial Q-I-4, and R-I-5, both of the 14th century.

reaching the truth. The claims of many Romans of high rank to have carried their arms as far as Mount Atlas, which Pliny declares has been repeatedly shown by experience to be most fallacious, leads him to the further reflection that nowhere is a lapse of one's credulity easier than where a dignified author supports a false statement.[1] In other passages he calls experience the best teacher in all things,[2] and contrasts unfavorably garrulity of words and sitting in schools with going to solitudes and seeking herbs at their appropriate seasons. That upon our globe the land is entirely surrounded by water does not require, he says, investigation by arguments, but is now known by experience.[3] And if the salamander really extinguished fire, it would have been tried at Rome long ago.[4] On the other hand, we find some assertions in the *Natural History* which Pliny might easily have tested himself and found false, such as his statement that an egg-shell cannot be broken by force or any weight unless it is tipped a little to one side.[5] Sometimes he gives his personal experience,[6] but also mentions experience in many other connections.

Use of the word experimentum.

The word employed most of the time by Pliny to denote experience is *experimentum*.[7] In many passages the word does not indicate anything like a purposive, prearranged, scientific experiment in our sense of that word, but simply the ordinary experience of daily life.[8] We are also told what *experti*,[9] or men of experience, advise. In a number of passages, however, *experimentum* is used in a sense some-

[1] NH, V, 1, 12.

[2] XXVI, 6, "usu efficacissimo rerum omnium magistro"; XVII, 2, 12, "quare experimentis optime creditur."

[3] II, 66.

[4] XXIX, 23.

[5] XXIX, 11.

[6] XXV, 54, "coramque nobis"; XXV, 106, "nos eam Romanis experimentis per usus digeremus."

[7] Sometimes another term, as *usus* in note 2 above, is employed.

[8] See II, 41, 1-2; II, 108; VII, 41; VII, 56; VIII, 7; XIV, 8; XVI, 1; XVI, 64; XVII, 2; XVII, 35; XXII, 1; XXII, 43; XXII, 49; XXII, 51; XXV, 7; XXXIV, 39 and 51. Experience is also the idea in the two following passages, although the word *experimentum* could not smoothly be rendered as "experience" in a literal translation: VII, 50, "Accedunt experimenta et exempla recentissimi census . . ."; XXVIII, 45, "Nec uros aut bisontes habuerunt Graeci in experimentis."

[9] XVI, 24; XXII, 57; XXVI, 60.

what more closely approaching our "experiment." These are cases where something is being tested. For instance, a method of determining whether an egg is fresh or rotten by putting it in water and watching if it floats or sinks is called an *experimentum.*[1] That horses would whinny at no other painting of a horse than that by Apelles is spoken of as *illius experimentum artis,* a test of, or testimony to, his art.[2] The expression *religionis experimento* is applied to a religious test or ordeal by which the virginity of Claudia was vindicated.[3] The word is also used of ways of telling if unguents are good[4] and if wine is beginning to turn;[5] and of various tests of the genuineness of drugs, gems, earths, and metals.[6] It is also twice used of letting down a lighted lamp into a huge wine cask or into wells to discover if there is danger at the bottom from noxious vapors.[7] If the lamp was extinguished, it was a sign of peril to human life. Pliny further suggests purposive experimentation in speaking of *experimenta* to discover water under ground[8] and in grafting trees.[9]

Experiments due to scientific curiosity.

Most of the tests and experiences thus far mentioned have been practical operations connected with husbandry and industry. But Pliny recounts one or two others which seem to have been dictated solely by scientific curiosity. He classifies the following as *experimenta:*[10] the sinking of a well to prove by its complete illumination that the sun casts no shadow at noon of the summer solstice; the marking of a dolphin's tail in order to throw some light upon its length of life, should it ever be captured again, as it was three hundred years later—perhaps the experiment of longest duration on record;[11] and the casting of a man into a pit of

[1] X, 75.
[2] XXXV, 30.
[3] VII, 35.
[4] XIII, 3.
[5] XIV, 25.
[6] XVII, 4; XX, 3 and 76; XXII, 23; XXIX, 12; XXXIII, 19 and 43 and 44 and 57; XXXIV, 26 and 48; XXXVI, 38 and 55; XXXVII, 22 and 76; such phrases as *sinceri experimentum* and *veri experimentum* are used for "test of genuineness."
[7] XXIII, 31; XXXI, 28.
[8] XXXI, 27.
[9] XVII, 26.
[10] II, 75.
[11] IX, 7.

serpents at Rome to determine if he was really immune from their stings.[1]

Medical experimentation.

Experimentum is employed by Pliny in a medical sense which becomes very common in the middle ages. He calls some remedies for toothache and inflamed eyes *certa experimenta*—sure experiences.[2] Later *experimentum* came to be applied to almost any recipe or remedy. Pliny, indeed, speaks of the doctors as learning at our risk and getting experience through our deaths.[3] In another passage he states more favorably that "there is no end to experimenting with everything so that even poisons are forced to cure us." [4] He also briefly mentions the medical sect of Empirics, of whom we shall hear more from Galen. He says that they so name themselves from experiences [5] and originated at Agrigentum in Sicily under Acron and Empedocles.

Chance experience and divine revelation.

Pliny is puzzled how some things which he finds stated in "authors famous for wisdom" were ever learned by experience, for example, that the star-fish has such fiery fervor that it burns everything in the sea which it touches, and digests its food instantly.[6] That adamant can be broken only by goat's blood he thinks must have been divinely revealed, for it would hardly have been discovered by chance, and he cannot imagine that anyone would ever have thought of testing a substance of immense value in a fluid of one of the foulest of animals.[7] In several other passages he suggests chance, accident, dreams,[8] or divine revelation as the ways in which the medicinal virtues of certain simples were discovered. Recently, for example, it was discovered that the root of the wild rose is a remedy for hydrophobia by the mother of a soldier in the praetorian guard, who was warned

[1] XXVIII, 6.

[2] XXVIII, 14.

[3] XXIX, 8. "Discunt periculis nostris et experimenta per mortes agunt." Bostock and Riley translate the last clause, "And they experimentalize by putting us to death." Another possible translation is, "And their experiments cost lives."

[4] XXV, 17. ". . . adeo nullo omnia experiendi fine ut cogerentur etiam venena prodesse."

[5] XXIX, 4 ". . . ab experimentis se cognominans empiricen."

[6] IX, 86.

[7] XXXVII, 15.

[8] According to Galen, as we shall hear later, the Empirics relied a good deal upon chance experience and dreams.

in a dream to send her son this root, which cured him and many others who have tried it since.[1] And a soldier in Pompey's time accidentally discovered a cure for elephantiasis when he hid his face for shame in some wild mint leaves.[2] Another herb was accidentally found to be a cure for disorders of the spleen when the entrails of a sacrificial victim happened to be thrown on it and it entirely consumed the milt.[3] The healing properties of vinegar for the sting of the asp were discovered by chance in this wise. A man who was stung by an asp while carrying a leather bottle of vinegar noticed that he felt the sting only when he set the bottle down.[4] He therefore decided to try the effects of a drink of the liquid and was thereby fully cured.[5] Other remedies are learned through the experience of rustics and illiterate persons, and yet others may be discovered by observing animals who cure their ills by them.[6] Pliny's opinion is that the animals have hit upon them by chance.

Marvels proved by experience.

Pliny represents a number of marvelous and to us incredible things as proved by experience. Divination from thunder, for instance, is supported by innumerable experiences, public and private. In two passages out of the three mentioning *experti* which I cited above, those experienced persons recommended a decidedly magical sort of procedure.[7] In another passage "the experience of many" supports "a strange observance" in plucking a bud.[8] A fourth bit of magical procedure is called "marvelous but easily tested."[9] Thus the transition is an easy one from signs of experimental science in the *Natural History* to our next topic, Pliny's account of magic.

[1] XXV, 6.
[2] XX, 52.
[3] XXV, 20.
[4] XXIII, 27.
[5] Among other virtues of vinegar, besides its supposed property of breaking rocks, Pliny mentions that if one holds some in the mouth, it will prevent one from feeling the heat in the baths.
[6] XXV, 6 and 21 and 50; XXVII, 2.
[7] XVI, 24; XXVI, 60.
[8] XXIII, 59.
[9] XXVIII, 7.

III. *Pliny's Account of Magic.*

Oriental origin of magic.

Pliny supplies some account of the origin and spread of magic [1] but a rather confused and possibly unreliable one, as he mentions two Zoroasters separated by an interval of five or six thousand years, and two Osthaneses, one of whom accompanied Xerxes, and the other Alexander, in their respective expeditions. He says, indeed, that it is not clear whether one or two Zoroasters existed. In any case magic has flourished greatly the world over for many centuries, and was founded in Persia by Zoroaster. Some other magicians of Media, Babylonia, and Assyria are mere names to Pliny; later he mentions others like Apollobeches and Dardanus. Although he thus derives magic from the orient, he appears to make no distinction, as we shall find other writers doing, between the *Magi* of Persia and ordinary magicians, nor does he employ the word magic in two senses. He makes it evident, however, that there have been other men who have regarded magic more favorably than he does.

Its spread to the Greeks.

Pliny next traces the spread of magic among the Greeks. He marvels at the lack of it in the Iliad and the abundance of it in the Odyssey. He is uncertain whether to class Orpheus as a magician, and mentions Thessaly as famous for its witches at least as early as the time of Menander who named one of his comedies after them. But he regards the Osthanes who accompanied Xerxes as the prime introducer of magic to the Greek-speaking world, which straightway went mad over it. In order to learn more of it, the philosophers Pythagoras, Empedocles, Democritus, and Plato went into distant exile and on their return disseminated their lore. Pliny regards the works of Democritus as the greatest single factor in that dissemination of the doctrines of magic which occurred at about the same time that medicine was being developed by the works of Hippocrates. Some

[1] In the opening chapters of Book XXX, unless otherwise indicated by specific citation.

regarded the books on magic ascribed to Democritus as spurious, but Pliny insists that they are genuine.[1]

Its spread outside the Graeco-Roman world.

Outside of the Greek-speaking world, whence of course magic spread to Rome, Pliny mentions Jewish magic, represented by such names as Moses, Jannes, and Lotapes. But he holds that magic did not originate among the Hebrews until long after Zoroaster. He also speaks of the magic of Cyprus; of the Druids, who were the magicians, diviners, and medicine men of Gaul until the emperor Tiberius suppressed them; and of distant Britain.[2] Thus discordant nations and even those ignorant of one another's existence agree the world over in their devotion to magic. From what Pliny tells us elsewhere of the Scythians we can see that the nomads of the Russian steppes and Turkestan were devoted to magic too.

Failure to understand its true origin.

It has been shown that Pliny regarded magic as a mass of doctrines formulated by a single founder and not as a gradual social evolution, just as the Greeks and Romans ascribed their laws and customs to some single legislator. He admits in a way, however, the great antiquity claimed by magic for itself, although he questions how the bulky dicta of Zoroaster and Dardanus could have been handed down by memory during so long a period. This remark again shows how little he thinks of magic as a set of social customs and attitudes perpetuated through constant and universal practice from generation to generation. Yet what he says of its widespread prevalence among unconnected peoples goes to prove this.

Magic and divination.

Pliny has a clearer comprehension of the extensive scope of magic and of its essential characteristics, at least as it was in his day. "No one should wonder," he says, "that its authority has been very great, since alone of the arts it has

[1] Aulus Gellius, X, 12, and Columella, VII, 5, dispute this (Bouché-Leclercq, *L'Astrologie grecque*, p. 519). Berthelot (*Origines de l'alchimie*, p. 145) believes in a Democritan school at the beginning of the Christian era which wrote the works of alchemy attributed to Democritus as well as the books of medical and magical recipes which are quoted in the *Geoponica* and the *Natural History*.

[2] XVI, 95.

embraced and united with itself the three other subjects which make the greatest appeal to the human mind," namely, medicine, religion, and the arts of divination, especially astrology. That his phrase *artes mathematicas* has reference to astrology is shown by his immediately continuing, "since there is no one who is not eager to learn the future about himself and who does not think that this is most truly revealed by the sky." But magic further "promises to reveal the future by water and spheres and air and stars and lamps and basins and the blades of axes and by many other methods, besides conferences with shades from the infernal regions." There can therefore be no doubt that Pliny regards the various arts of divination as parts of magic.

Magic and religion.

While we have heard Pliny assert in general the close connection between magic and religion, the character of the *Natural History,* which deals with natural rather than religious matters, does not lead him to enter into much further detail upon this point. His occasional mention of religious usages in his own day, however, supports our information from other sources that the original Roman religion was very largely composed of magic forces, rules, and ceremonial.

Magic and medicine.

Nearly half the books of the *Natural History* deal in whole or in part with remedies for diseases, and it is therefore of the relations between magic and natural science, and more particularly between magic and medicine, that Pliny gives us the most detailed information. Indeed, he asserts that "no one doubts" that magic "originally sprang from medicine and crept in under the show of promoting health as a loftier and more sacred medicine." Magic and medicine have developed together, and the latter is now in imminent danger of being overwhelmed by the follies of magic, which have made men doubt whether plants possess any medicinal properties.

Magic and philosophy.

In the opinion of many, however, magic is sound and beneficial learning. In antiquity, and for that matter at almost all times, the height of literary fame and glory has been sought from that science.[1] Eudoxus would have it the most noted and useful of all schools of philosophy. Empedocles and Plato studied it; Pythagoras and Democritus perpetuated it in their writings.

Falseness of magic.

But Pliny himself feels that the assertions of the books of magic are fantastic, exaggerated, and untrue. He repeatedly brands the *magi* or magicians as fools or impostors, and their statements as absurd and impudent tissues of lies.[2] *Vanitas,* or "nonsense," is his stock-word for their beliefs.[3] Some of their writings must, in his opinion, have been dictated by a feeling of contempt and derision for humanity.[4] Nero proved the falseness of the art, for although he studied magic eagerly and with his unlimited wealth and power had every opportunity to become a skilful practitioner, he was unable to work any marvels and abandoned the attempt.[5] Pliny therefore comes to the conclusion that magic is "invalid and empty, yet has some shadows of truth, which however are due more to poisons than to magic." [6]

Crimes of magic.

The last remark brings us to charges of evil practices made against the magicians. Besides poisons, they specialize in love-potions and drugs to produce abortions; [7] and some of their operations are inhuman or obscene and abominable. They attempt baleful sorcery or the transfer of disease from one person to another.[8] Osthanes and even Democritus propound such remedies as drinking human blood or utilizing in magic compounds and ceremonies parts of the corpses of men who have been violently slain.[9] Pliny thinks that humanity owes a great debt to the Roman government

[1] XXX, 2. ". . . quamquam animadverto summam litterarum claritatem gloriamque ex ea scientia antiquitus et paene semper petitam."

[2] Examples are: XXV, 59, "Sed magi utique circa hanc insaniunt"; XXIX, 20, "magorum mendacia"; XXXVII, 60, "magorum inpudentiae vel manifestissimum . . . exemplum"; XXXVII, 73, "dira mendacia magorum."

[3] See XXII, 9; XXVI, 9; XXVII, 65; XXVIII, 23 and 27; XXIX, 26; XXX, 7; XXXVII, 14.

[4] XXXVII, 40.

[5] XXX, 5-6.

[6] XXX, 6. "Proinde ita persuasum sit, intestabilem, inritam, inanem esse, habentem tamen quasdam veritatis umbras, sed in his veneficas artis pollere, non magicas."

[7] XXV, 7.

[8] XXVIII, 23.

[9] XXVIII, 2.

for abolishing those monstrous rites of human sacrifice, "in which to slay a man was thought most pious; nay more, to eat men was thought most wholesome." [1]

Pliny's censure of magic is mainly intellectual.

Pliny nevertheless lays less stress upon the moral argument against magic as criminal or indecent than he does upon the intellectual objection to it as untrue and unscientific. Indeed, so far as decency is concerned, his own medicine will be seen to be far from prudish, while he elsewhere gives instances of magicians guarding against defilement.[2] Moreover, among the methods employed and the results sought by magic which he frequently mentions there are comparatively few that are morally objectionable, although they seem without exception false. But many of their recipes aim at the cure of disease and other worthy, or at least admissible, objects. Possibly Pliny has somewhat censored their lore and tried to exclude all criminal secrets, but his censure seems more intellectual than moral. For instance, he fills a long chapter with extracts from a treatise on the virtues of the chameleon and its parts by Democritus, whom he regards as a leading purveyor of magic lore.[3] In opening the chapter Pliny hails "with great pleasure" the opportunity to expose "the lies of Greek vanity," but at its close he expresses a wish that Democritus himself had been touched with the branch of a palm which he said prevents immoderate loquacity. Pliny then adds more charitably, "It is evident that this man, who in other respects was a wise and most useful member of society, has erred from too great zeal in serving humanity."

Vagueness of Pliny's scepticism.

Pliny himself fails to maintain a consistently sceptical attitude towards magic. His exact attitude is often hard to determine. Often it is difficult to say whether he is speaking in sober earnest or in a tone of light and easy pleasantry and sarcasm, as in the passage just cited concerning Democritus. Another puzzling point is his frequent excuse that he will list certain assertions of the magicians in order to

[1] XXX, 4. [2] XXVIII, 19; XXX, 6. [3] XXVIII, 29.

expose or confute them. But really he usually simply sets them forth, apparently expecting that their inherent and patent absurdity will prove a sufficient refutation of them. On the rare occasions when he undertakes to indicate in what the absurdity consists his reasoning is scarcely scientific or convincing. Thus he affirms that "it is a peculiar proof of the vanity of the magicians that of all animals they most admire moles who are condemned by nature in so many ways, to perpetual blindness and to dig in the darkness as if they were buried." [1] And he assails the belief of the *magi* [2] that an owl's egg is good for diseases of the scalp by asking, "Who, I beg, could ever have seen an owl's egg, since it is a prodigy to see the bird itself?" Moreover, he sometimes cites assertions of the magicians without any censure, apology, or expression of disbelief; and there are many other passages where it is practically impossible to tell whether he is citing the magicians or not. Sometimes he will apparently continue to refer to them by a pronoun in chapters where they have not been mentioned by name at all.[3] In other places he will imperceptibly cease to quote the *magi* and after an interval perhaps as imperceptibly resume citation of their doctrines.[4] It is also difficult to determine just when writers like Democritus and Pythagoras are to be regarded as representatives of magic and when their statements are accepted by Pliny as those of sound philosophers.

Magic and science indistinguishable.

Perhaps, despite Pliny's occasional brave efforts to withstand and even ridicule the assertions of the magicians, he could not free himself from a secret liking for them and more than half believed them. At any rate he believed very similar things. Even more likely is it that previous works on nature were so full of such material and the readers of his own day so interested in it, that he could not but include

[1] XXX, 7.

[2] XXIX, 26.

[3] For instance, XXX, 27, he mentions the magi, but not in XXX, 28. Nor are they mentioned in XXX, 29, but in XXX, 30 "plura eorum remedia ponemus" seems to refer to them, although we must look back three chapters for the antecedent of *eorum*.

[4] XXXVII, 14, he says that he is going to confute "the unspeakable nonsense of the magicians" concerning gems, but makes no specific citation from them until the thirty-seventh chapter on jasper.

much of it. Once he explains [1] that certain statements are scarcely to be taken seriously, yet should not be omitted, because they have been transmitted from the past. Again he begs the reader's indulgence for similar "vanities of the Greeks," "because this too has its value that we should know whatever marvels they have transmitted." [2] The truth of the matter probably is that Pliny rejected some assertions of the magicians but found others acceptable; that he gets his occasional attitude of scepticism and ridicule of their doctrines from one set of authorities, and his moments of unquestioning acceptance of their statements from other authors on whom he relies. Very likely in the books which he used it often was no clearer than it is in the *Natural History* whether a statement was to be ascribed to the *magi* or not. Very possibly Pliny was as confused in his own mind concerning the entire business as he seems to be to us. He could no more keep magic out of his *Natural History* than poor Mr. Dick could keep Charles the First's head out of his book. One fact at any rate stands out clearly, the prominence of magic in his encyclopedia and in the learning of his age.

IV. *The Science of the Magi*

Magicians as investigators of nature.

Let us now further examine Pliny's picture of magic, not as he expressly defines or censures it, but as he reflects its own assertions and purposes in his fairly numerous citations from its literature and perhaps its practice. Here I shall rather strictly limit my survey to those statements which Pliny definitely ascribes by name to the *magi* or magic art. The most striking fact is that the magicians are cited again and again concerning the supposed properties, virtues, and effects of things in nature—herbs, animals, and stones. These virtues are, it is true, often employed in an effort to produce wonderful results, and often too they are combined with some fantastic rite or superstitious ceremonial performed by a human agent. But in many cases either no

[1] XXX, 47. [2] XXXVII, 11.

rite at all is suggested or merely some simple medicinal application; and in a few cases there is no mention of any particular operation or result, the magicians are cited simply as authorities concerning the great but unspecified virtues of natural objects. Indeed, they stand out in Pliny's pages not as mere sorcerers or enchanters or wonder-workers, but as those who have gone the farthest and in most detail—too far and too curiously in Pliny's opinion—into the study of medicine and of nature. Sometimes their statements, cited without censure, supplement others concerning the species under discussion;[1] sometimes they are his sole source of information on the subject in hand.[2]

The *magi* on herbs.

Pliny connects the origin of botany rather closely with magic, mentioning Medea and Circe as early investigators of plants and Orpheus among the first writers on the subject.[3] Moreover, Pythagoras and Democritus borrowed from the *magi* of the orient in their works on the properties of plants.[4] There would be little profit in repeating the names of the herbs concerning which Pliny gives opinions of the magicians, inasmuch as few of them can be associated with any plants known to-day.[5] Suffice it to say that Pliny makes no objection to the herbs which they employed. Nor does he criticize their methods of employing them, although some seem superstitious enough to the modern reader. A chaplet is worn of one herb,[6] others are plucked with the left hand and with a statement of what they are to be used for, and in one case without looking backward.[7] The anemone is to be plucked when it first appears that year with a statement of its intended use, and then is to be wrapped in a red cloth and kept in the shade, and, whenever anyone falls sick of tertian or quartan fever, is to be bound on the patient's body.[8] The heliotrope is not to be plucked at all but

[1] XX, 30; XXI, 38, 94, 104; XXII, 24, 29.

[2] XXI, 36; XXIV, 99.

[3] XXV, 5.

[4] XXIV, 99-102.

[5] See XX, 30; XXI, 36, 38, 94, 104; XXII, 9, 24, 29; XXIV, 99, 102; XXV, 59, 65, 80-81; XXVI, 9.

[6] XXI, 38.

[7] XXI, 104; XXII, 24.

[8] XXI, 94.

tied in three or four knots with a prayer that the patient may recover to untie the knots.[1]

Marvelous virtues of herbs.

Pliny does not even object to the marvelous results which the *magi* think can be gained by use of herbs until towards the close of his twenty-fourth book, although already in his twentieth and twenty-first books such powers have been claimed for herbs as to make one well-favored and enable one to attain one's desires,[2] or to give one grace and glory.[3] At the end of his twenty-fourth book [4] he states that Pythagoras and Democritus, following the *magi,* ascribe to herbs unusually marvelous virtues such as to freeze water, invoke spirits, force the guilty to confess by frightening them with apparitions, and impart the gift of divination. Early in his twenty-fifth book [5] Pliny suggests that some incredible effects have been attributed to herbs by the *magi* and their disciples, and in a later chapter [6] he describes the *magi* as so mad about vervain that they think that if they are anointed with it, they can gain their wishes, drive away fevers and other diseases, and make friendships. The herb should be plucked about the rising of the dog-star when there is neither sun nor moon. Honey and honeycomb should be offered to appease the earth; then the plant should be dug around with iron with the left hand and raised aloft. By the time he reaches his twenty-sixth book Pliny's courage has risen, so to speak, enough to cause him at last to enter upon quite a tirade against "magical vanities which have been carried so far that they might destroy faith in herbs entirely." [7] As examples he mentions herbs supposed to dry up rivers and swamps, open barred doors at their touch, turn hostile armies to flight, and supply all the needs of the ambassadors of the Persian kings. He wonders why such herbs have never been employed in Roman warfare or Italian drainage. Pliny's only objection to magic herbs therefore seems to be the excessive powers which are claimed for some of them.

[1] XXII, 29.
[2] XX, 30.
[3] XXI, 38.
[4] XXIV, 99 and 102.
[5] XXV, 5.
[6] XXV, 59.
[7] XXVI, 9.

He adds that it would be strange that the credulity which arose from such wholesome beginnings had reached such a pitch, if human ingenuity observed moderation in anything and if the much more recent system of medicine which Asclepiades founded could not be shown to have been carried even beyond the magicians. Here again we see Pliny failing to recognize magic as a primitive social product and regarding it as a degeneration from ancient science rather than science as a comparatively modern development from it. But he may well be right in thinking that many particular far-fetched recipes and rites were the late, artificial product of over-scholarly magicians. Thus he brands as false and magical the assertion of a recent grammarian, Apion, that the herb cynocephalia is divine and a safeguard against poison, but kills the man who uproots it entirely.[1]

Animals and parts of animals.

In a few cases Pliny objects to the animals or parts of animals employed by the *magi*, as in the passage already cited where he complains that they admire moles more than any other animals.[2] But his assertion is inconsistent, since he has already affirmed that they hold the hyena in most admiration of all animals on the ground that it works magic upon men.[3] Their promise of readier favor with peoples and kings to those who anoint themselves with lion's fat, especially that between the eyebrows, he criticizes by declaring that no fat can be found there.[4] He also twits the *magi* for magnifying the importance of so nasty a creature as the tick.[5] They are attracted to it by the fact that it has no outlet to its body and can live only seven days even if it fasts. Whether there is any astrological significance in the number seven here Pliny does not say. He does inform us, however, that the cricket is employed in magic because it moves backward.[6] A very bizarre object employed by the Druids and other magicians is a sort of egg produced by the hissing or foam of snakes.[7] The blood of the basilisk may also be

[1] XXX, 6.
[2] XXX, 7.
[3] XXVIII, 27.
[4] XXVIII, 25.
[5] XXX, 24.
[6] XXIX, 39.
[7] XXIX, 12.

classed as a rarity. Apparently animals in some way unusual are preferred in magic, like a black sheep,[1] but the logic in the reasons given by Pliny for their selection is not clear in every instance. In some other cases not criticized by Pliny [2] we have plainly enough sympathetic magic or the principle of like cures like, as when the milt of a calf or sheep is used to cure diseases of the human spleen.

Further instances.

The magicians, however, do not scorn to use familiar and easily obtainable animals like the goat and dog and cat. The liver and dung of a cat, a puppy's brains, the blood and genitals of a dog, and the gall of a black male dog are among the animal substances employed.[3] Such substances as those just named are equally in demand from other animals.[4] Minute parts of animals are frequently employed by the magicians, such as the toe of an owl, the liver of a mouse given in a fig, the tooth of a live mole, the stones from young swallows' gizzards, the eyes of river crabs.[5] Sometimes the part employed is reduced to ashes, perhaps a relic of sacrificial custom. Thus for toothache the *magi* inject into the ear nearer the tooth the ashes of the head of a mad dog and oil of Cyprus, while they prescribe for affections of the sinews the ashes of an owl's head in honied wine with lily root.[6] Other living creatures which Pliny mentions as used by the *magi* are the salamander, earthworm, bat, scarab with reflex horns, lizard, tortoise, bed-bug, frog, and sea-urchin.[7] The dragon's tail wrapped in a gazelle's skin and bound on with deer-sinews cures epilepsy,[8] and a mixture of the dragon's tongue, eyes, gall, and intestines, boiled in oil, cooled in the night air, and rubbed on morning and evening, frees one from nocturnal apparitions.[9]

Sometimes the parts of animals are bound on outside the patient's body, sometimes the injured portion of his body

[1] XXX, 6.

[2] XXVIII, 57; XXX, 17.

[3] Use of goat, XXVIII, 56, 63, 78-79; cat, XXVIII, 66; puppy, XXIX, 38; dog, XXX, 24.

[4] XXVIII, 60, 66, 77; XXIX, 26.

[5] XXVIII, 66; XXIX, 15; XXX, 7; XXX, 27; XXXII, 38.

[6] XXX, 8 and 36; see also XXVIII, 60; XXXII, 19 and 24.

[7] XXIX, 23; XXX, 18, 20, 30, 49; XXXII, 14, 18, 24.

[8] XXX, 27.

[9] XXX, 24.

is merely touched with them. Once the whole house is to be fumigated with the substance in question;[1] once the walls are to be sprinkled with it; once it is to be buried under the threshold. Some instances follow of more elaborate magic ritual connected with the use of animals or parts of animals. The hyena is more easily captured by a hunter who ties seven knots in his girdle and horsewhip, and it should be captured when the moon is in the sign of Gemini and without the loss of a single hair.[2] Another bit of astrology dispensed by the *magi* is that the cat, whose salted liver is taken with wine for quartan fever, should have been killed under a waning moon.[3] To cure incontinence of urine one not only drinks ashes of a boar's genitals in sweet wine, but afterwards urinates in a dog kennel and repeats the formula, "That I may not urinate like a dog in its kennel."[4] The magicians insist that the sex of the patient be observed in administering burnt cow-dung or bull-dung in honied wine for cases of dropsy.[5] For infantile ailments the brains of a she-goat should be passed through a gold ring and dropped in the baby's mouth before it is given its milk.[6] After the fresh milt of a sheep has been applied to the patient with the words, "This I do for the cure of the spleen," it should be plastered into the bedroom wall and sealed with a ring, while the charm should be repeated twenty-seven times.[7] In treating sciatica[8] an earthworm should be placed in a broken wooden dish mended with an iron band, the dish should be filled with water, the worm should be buried again where it was dug up, and the water should be drunk by the patient. The eyes of river crabs are to be attached to the patient's person before sunrise and the blinded crabs put back into the water.[9] After it has been carried around the house thrice a bat may be nailed head down outside a window as an amulet.[10] For epilepsy goat's flesh should be

Magic rites with animals and parts of animals.

[1] XXX, 24.
[2] XXVIII, 27.
[3] XXVIII, 66; and see XXIX, 12.
[4] XXVIII, 60.
[5] XXVIII, 68.
[6] XXVIII, 78.
[7] XXX, 17.
[8] XXX, 18.
[9] XXXII, 38.
[10] XXIX, 26.

given which has been roasted on a funeral pyre, and the animal's gall should not be allowed to touch the ground.[1]

Marvels wrought with parts of animals.

Pliny occasionally speaks in a vague general way of his citations from the *magi* concerning the virtues of parts of animals as lies or nonsense or "portentous," but he does not specifically criticize their procedure any more than he did their methods of employing herbs, and he does not criticize their promised results as much as he did before. Indeed, as we have already indicated, the object in a majority of cases is purely medicinal. The purpose of others is pastoral or agricultural, such as preventing goats from straying or causing swine to follow you.[2] The blood of the basilisk, however, is said to procure answers to petitions made to the powerful and prayers addressed to the gods, and to act as a safeguard against poison or sorcery (*veneficiorum amuleta*).[3] Invincibility is promised the wearer of the head and tail of a dragon, hairs from a lion's forehead, a lion's marrow, the foam of a winning horse, a dog's claw bound in deerskin, and the muscles alternately of a deer and a gazelle.[4] A woman will tell secrets in her sleep if the heart of an owl is applied to her right breast, and power of divination is gained by eating the still palpitating heart of a mole.[5]

The *magi* on stones.

In the case of stones the names are again, as in the case of herbs, of little significance for us.[6] The accompanying ritual is slight. There are one or two suspensions from the neck or elsewhere by such means as a lion's mane—the hair of the hyena will not do at all—nor the hair of the cynocephalus and swallows' feathers.[7] There is some use of incantations with the stones, a setting of iron for one stone, burial of another beneath a tree that it may not dull the axe, and placing another on the tongue after rinsing the mouth with honey at certain days and hours of the moon in order to acquire the gift of divination.[8] Indeed, the results promised

[1] XXVIII, 63.
[2] XXVIII, 56; XXIX, 15.
[3] XXIX, 19.
[4] XXIX, 20.
[5] XXIX, 26; XXX, 7.
[6] Pliny ascribes statements concerning stones to the *magi* in the following chapters: XXXVI, 34; XXXVII, 37, 40, 49, 51, 54, 56, 60, 70, 73.
[7] XXXVII, 54 and 40.
[8] XXXVII, 40, 60, 56, 73.

are all marvelous. The stones benefit public speakers, admit to the presence of royalty, counteract fascination and sorcery, avert hail, thunderbolts, storms, locusts, and scorpions; chill boiling water, produce family discord, render athletes invincible, quench anger and violence, make one invisible, evoke images of the gods and shades from the infernal regions.

Other magical recipes.

We have yet to mention a group of magical recipes and remedies which Pliny for some reason collects in one chapter [1] but which hardly fall under any one head. A whetstone on which iron tools are sharpened, if placed without his knowledge under the pillow of a man who has been poisoned, will cause him to reveal all the circumstances of the crime. If you turn a man who has been struck by lightning over on his injured side, he will speak at once. To cure tumors in the groin, tie seven or nine knots in the remnant of a weaver's web, naming some widow as each knot is tied. The pain is assuaged by binding to the body the nail that has been trod on. To get rid of warts, on the twentieth day of the moon lie flat in a path gazing at the moon, stretch the hands above the head and rub the warts with anything that comes to hand. A corn may be extracted successfully at the moment a star shoots. Headache may be relieved by a liniment made by pouring vinegar on door hinges or by binding a hangman's noose about the patient's temples. To dislodge a fish-bone stuck in the throat, plunge the feet into cold water; to dislodge some other sort of bone, place bones on the head; to dislodge a morsel of bread, stuff bits of bread into both ears. We may add from a neighboring chapter a very magical remedy for fevers, although Pliny calls it "the most modest of their promises." [2] Toe and finger nail parings mixed with wax are to be attached ere sunrise to another person's door in order to transfer the disease from the patient to him. Or they may be placed near an ant-hill, in which case the first ant who tries to drag one in-

[1] XXVIII, 12, "Magorum haec commenta sunt. . . ."

[2] XXVIII, 23.

side the hill should be captured and suspended from the patient's neck.

Summary of the statements of the *magi.*

Such is the picture we derive from numerous passages in the *Natural History* of the magic art, its materials and rites, the effects it seeks to produce, and its general attitude towards nature. Besides the natural materials employed and the marvelous results sought, we have noted the frequent use of ligatures, suspensions, and amulets, the observance of astrological conditions, of certain times and numbers, rules for plucking herbs and tying knots, stress on the use of the right or left hand—in other words, on position or direction, some employment of incantations, some sacrifice and fumigation, some specimens of sympathetic magic, of the theory that "like cures like," and of other types of magic logic.

V. *Pliny's Magical Science*

From the *magi* to Pliny's magic

We may now turn to the still more numerous passages of the *Natural History* where the *magi* are not cited and compare the virtues there ascribed to the things of nature and the methods employed in medicine and agriculture with those of the magicians. We shall find many striking resemblances and shall soon come to a realization that there is more magic in the *Natural History* which is not attributed to the *magi* than there is that is. Pliny did not need to warn us that medicine had been corrupted by magic; his own medicine proves it. It is this fact, that virtually his entire work is crammed with marvelous properties and fantastic ceremonial, which makes it so difficult in some places to tell when he begins to draw material from the *magi* and when he leaves off. By a detailed analysis of this remaining material we shall now attempt to classify the substances of which Pliny makes use and the virtues which he ascribes to them, the rites and methods of procedure by which they are employed, and certain superstitious doctrines and notions which are involved. We shall thus find that almost precisely the same factors are present in his science as in the lore of the magicians.

Habits of animals.

Of substances we may begin with animals,[1] and, before we note the human use of their virtues with its strong suggestion of magic, may remark another unscientific and superstitious feature which was very common both in ancient and medieval times. This is the tendency to humanize animals, ascribing to them conscious motives, habits, and ruses, or even moral standards and religious veneration. We shall have occasion to note the same thing in other authors and so will give but a few specimens from the many in the *Natural History*. Such qualities are attributed by Pliny especially to elephants, whom he ranks next to man in intelligence, and whom he represents as worshiping the stars, learning difficult tricks, and as having a sense of justice, feel-

[1] Some works upon animals in antiquity and Greece are:

Aubert und Wimmer, *Aristoteles Thierkunde,* 2 vols., Leipzig, 1868.

Baethgen, *De vi et significatione galli in religione et artibus Graecorum et Romanorum,* Diss. Inaug., Göttingen, 1887.

Bernays, *Theophrasts Schrift über Frömmigkeit.*

Bikélas, O., *La nomenclature de la Faune grecque,* Paris, 1879.

Billerbeck, *De locis nonnullis Arist. Hist. Animal. difficilioribus,* Hildesheim, 1806.

Dryoff, A., *Die Tierpsychologie des Plutarchs,* Progr. Würzburg, 1897. *Über die stoische Tierpsychologie,* in *Bl. f. bayr. Gymn.,* 33 (1897) 399ff.; 34 (1898) 416.

Erhard, *Fauna der Cykladen,* Leipzig, 1858.

Fowler, W. W., *A Year with the Birds,* 1895.

Hopf, L., *Thierorakel und Orakelthiere in alter und neuer Zeit,* Stuttgart, 1888.

Hopfner, T., *Der Tierkult der alten Ægypter nach den griechisch-römischen Berichten und den wichtigen Denkmälern,* in *Denkschr. d. Akad. Wien,* 1913, ii Abh.

Imhoof-Blumer, F., und Keller, O., *Tier- und Pflanzenbilder auf Münzen und Gemmen des klassischen Altertums.* illustrated, 1889.

Keller, O., *Thiere des class. Altertums.*

Krüper, *Zeiten des Gehens und Kommens und des Brütens der Vögel in Griechenland und Ionien,* in Mommsen's *Griech. Jahreszeiten,* 1875.

Küster, E., *Die Schlange in der griechischen Kunst und Religion,* Giessen, 1913.

Lebour, *Zoologist,* 1866.

Lewysohn, *Zoologie des Talmuds.*

Lindermayer, A., *Die Vögel Griechenlands,* Passau, 1860.

Locard, *Histoire des mollusques dans l'antiquité,* Lyon, 1884.

Lorenz, *Die Taube im Alterthume,* 1886.

Marx, A., *Griech. Märchen von dankbaren Tieren,* Stuttgart, 1889.

Mühle, H. v. d., *Beiträge zur Ornithologie Griechenlands,* Leipzig, 1844.

Sundevall, *Thierarten des Aristoteles,* Stockholm, 1863.

Thompson, D'Arcy W., *A Glossary of Greek Birds,* 1895. *Aristotle as a Biologist,* 1913. Also the notes to his translation of the *Historia animalium.*

Westermarck, E., *The Origin and Development of Moral Ideas,* I (1906) 251-60, gives further bibliography on the subjects of animals as witnesses and the punishment of animal culprits.

ing of mercy, and so on.[1] Similarly the lion has noble courage and a sense of gratitude, while the lioness is wily in the devices by which she conceals her amours with the pard.[2] A number of the devices of fishes to escape hooks and nets are repeated by Pliny from Ovid's *Halieuticon,* extant only in fragments.[3] The crocodile opens its jaws to have its teeth picked by a friendly bird; but sometimes while this operation is being performed the ichneumon "darts down its throat like a javelin and eats away its intestines."[4] Pliny also marvels at the cleverness displayed by the dragon and the elephant in their combats with one another,[5] which, however, almost invariably terminate fatally to both combatants, the elephant falling exhausted in the dragon's coils and crushing the serpent by its weight. Others say that in the hot summer the dragons thirst for the blood of the elephant which is very cold; in their combat the elephant falls drained of its blood and crushes the dragon who is intoxicated by the same.

Remedies discovered by animals.

The dragon's apparent knowledge that the elephant is cold-blooded leads us to a kindred topic, the remedies used by animals and often discovered by men only by seeing animals use them. This notion continued in the middle ages, as we shall see, and of course it did not originate with Pliny. As he says himself, "The ancients have recorded the remedies of wild beasts and shown how they are healed even when poisoned." [6] Against aconite the scorpion eats white hellebore as an antidote, while the panther employs human excrement.[7] Animals prepare themselves for combats with poisonous snakes by eating certain herbs; the weasel eats rue, the tortoise and deer use two other plants, while field mice who have been stung by snakes eat *condrion.*[8] The hawk tears open the hawkweed and sprinkles its eyes with the juice.[9] The serpent tastes fennel when it sheds its old

[1] VIII, 1-12.
[2] VIII, 17-21.
[3] XXXII, 5.
[4] VIII, 37.
[5] VIII, 11-12.
[6] XXVII, 2; XVIII, 1.
[7] XXVII, 2; VIII, 41.
[8] XX, 51 and 61; XXII, 37 and 45.
[9] XX, 26.

skin.[1] Sick bears cure themselves by a diet of ants.[2] Swallows restore the sight of their young with chelidonia or swallow-wort,[3] and the historian Xanthus says that the dragon restores its dead offspring to life with an herb called *balis*.[4] The hippopotamus was the original discoverer of bleeding,[5] opening a vein in his leg by wounding himself on sharp reeds along the shore, and afterwards checking the flow of blood by plastering the place with mud.[6] Pliny, however, states in one passage that animals hit upon all these remedies by chance and even have to rediscover them by accident in each new case, "since," he continues in conformity with recent animal psychologists, "reason and practice cannot be transmitted between wild beasts." [7]

Jealousy of animals.

Yet in another passage Pliny deplores the spitefulness of the dog which, while men are looking, will not pluck the herb by which it cures itself of snakebite.[8] Probably Pliny is using different authorities in the two passages. Theophrastus, the pupil of Aristotle, had written a work on *Jealous Animals*. More excusable than the spitefulness of the dog is the attitude of the dragon, from whose brain the gem *draconitis* must be taken while the dragon is alive and preferably asleep. For if the dragon feels that it is mortally wounded, it takes revenge by spoiling the gem.[9] Elephants know that men hunt them only for their tusks, and so bury these when they fall off.[10]

Occult virtues of animals.

Animals have marvelous virtues of their own other than the medicinal uses to which men have put them. For instance, the mere glance of the basilisk is fatal, and its breath burns up vegetation and breaks rocks.[11] But the medicinal effects which Pliny ascribes to animals and parts of animals

[1] VIII, 41; XX, 95.
[2] XXIX, 39.
[3] XXV, 50.
[4] XXV, 5.
[5] VIII, 40; XXVIII, 31.
[6] For further remedies used by animals see VIII, 41; XXIX, 14, 38; XXV, 52-53; XXVIII, 81.
[7] XXVII, 2. ". . . quod certe casu repertum quis dubitet et quotiens fiat etiam nunc ut novom nasci quoniam feris ratio et usus inter se tradi non possit?" Perhaps Pliny would have denied the inheritance of acquired characteristics.
[8] XXV, 51.
[9] XXXVII, 57.
[10] VIII, 4.
[11] VIII, 33.

are well nigh infinite. Many animal substances will have to be introduced in other connections so that we need mention now but a very few: the heads and blood of flies, honey in which bees have died, *cinere genitalis asini,* chicks in the egg, and thrice seven centipedes diluted with Attic honey,[1]—this last a prescription for asthma and to be taken through a reed because it blackens every dish by its contact. Another passage advises eating a rat or shrew-mouse in order to bear a baby with black eyes.[2] These items are enough to convince us that the animals and parts of animals employed by the magicians were not one whit more bizarre and nauseating than the others found in the *Natural History,* nor were the cures which they were expected to work any more improbable. In order to illustrate, however, the delicate distinctions which were imagined to exist not only between the virtues of different parts of the same animal, but also between slightly varied uses of the same part, we may note that scales scraped from the topmost part of a tortoise's shell and administered in drink check sexual desire, considering which, it is, as Pliny remarks, the more marvelous that a powder made of the entire shell is reported to arouse lust.[3] But love turns readily to hatred in magic as well as in romance, and it is nothing very unusual, as we shall find in other authors, for the same thing on slight provocation to work in exactly opposite ways.

The virtues of herbs.

Pig grease, Pliny somewhere informs us, possesses especially strong virtue, "because that animal feeds on the roots of herbs." [4] From the virtues of animals, therefore, let us turn to those of herbs.[5] Pliny met on every hand assertion of their wonderful powers. The empire-builders of Rome employed the sacred herbs *sagmina* and *verbenae* in their embassies and legations. The Gauls, too, use the verbena in

[1] XXIX, 34; XXX, 10, 19; XXVIII, 46; XXIX, 11; XXX, 16.

[2] XXX, 46.

[3] XXXII, 14.

[4] XXVIII, 37.

[5] A recent work on the general theme is Joret, *Les plantes dans l'antiquité,* Paris, 1904; see also F. Mentz, *De plantis quas ad rem magicam facere crediderunt veteres,* Leipzig, 1705, 28 pp.; F. Unger, *Die Pflanze als Zaubermittel,* Vienna, 1859.

lot-casting and prophetic responses.[1] Pliny also states more sceptically that there is another root which diviners take in drink in order to feign inspiration.[2] The Scythians know of a plant which prevents hunger and thirst if held in the mouth, and of another which has the same effect upon their horses, so that they can go for twelve days without meat or drink,[3] —an exaggerated estimate of the hardihood of the mounted Asiatic nomads and their steeds. Musaeus and Hesiod say that one anointed with *polion* will attain fame and dignities.[4]

Pliny perhaps did not intend to subscribe fully to such statements, although he cannot be said to call many of them into question. He did complain that some writers had asserted incredible powers of herbs, such as to restore dragons or men to life or withdraw wedges from trees,[5] yet he seems on the whole in sympathy with the opinion of the majority that there is practically nothing which the force of herbs cannot accomplish. Herophilus, illustrious in medicine, had said that certain herbs were beneficial if merely trod upon, and Pliny himself says the same of more than one plant. He tells us further that binding the wild fig tree about their necks makes the fiercest bulls stand immobile;[6] that another plant subjects fractious beasts of burden to the yoke;[7] while cows who eat *buprestis* burst asunder.[8] Another herb *contacto genitali* kills any female animal.[9] Betony is considered an amulet for houses,[10] and fishermen in Pliny's neighborhood mix a plant with chalk and scatter it on the waves.[11] "The fish dart towards it with marvelous desire and straightway float lifeless on the surface." Dogs will not bark at persons carrying *peristereos*.[12] The "impious plant" prevents any human being who tastes it from having quinsy, while swine are sure to have that disease if they do not eat it.

[1] XXII, 3; XXV, 59; XXVII, 28.

[2] XXI, 105. "Halicacabi radicem bibunt qui vaticinari gallantesque vere ad confirmandas superstitiones aspici se volunt."

[3] XXV, 43-44.

[4] XXI, 21, 84.

[5] XXV, 5.

[6] XXIII, 64.

[7] XXV, 35.

[8] XXII, 36.

[9] XXIV, 94.

[10] XXV, 46.

[11] XXV, 54.

[12] XXV, 78.

Some place it in birds' nests to prevent the voracious nestlings from strangling. Bitter almonds provide another amusing combination of effects. Eating five of them permits one to drink without experiencing intoxication, but if foxes eat them they will die unless they find water near by to drink.[1] There are some herbs which have a medicinal effect, if one merely looks at them.[2] In two cases the masculine or feminine variety of a herb is used to secure the birth of a child of the desired sex.[3]

Plucking herbs.

That the plucking of herbs and digging up of roots was a process very apt to be attended by magical procedure we find abundant evidence in the *Natural History*. Often plants should be plucked before sunrise.[4] Twice Pliny tells us that the peony should be uprooted by night lest the woodpecker of Mars try to pick the digger's eyes out.[5] The state of the moon is another point to be observed,[6] and once an herb is to be gathered before thunder is heard.[7] A common instruction is to pick the plant with the left hand,[8] and once with the thumb and fourth finger of the left hand.[9] Once the right hand should be stretched covertly after the fashion of a pickpocket through the left sleeve in order to pluck the plant.[10] Sometimes one faces east in plucking herbs; sometimes, west; again one is careful not to face the wind.[11] Sometimes the gatherer must not glance behind him. Sometimes he must fast before he takes the plant from the ground;[12] again he must observe a state of chastity.[13] Sometimes he should be barefoot and clothed in white; again he should remove every stitch of clothing and even his rings.[14] Sometimes the use of iron implements is forbidden; again gold or some other material is prescribed;[15] once the herb is to be dug with a nail.[16] Sometimes circles are traced

[1] XXIII, 75.
[2] XXIV, 56-57.
[3] XXV, 18; XXVII, 100.
[4] XX, 14; XXIV, 82; XXV, 92.
[5] XXV, 10; XXVII, 60.
[6] XXIV, 6, 93.
[7] XXV, 6.
[8] XX, 49; XXI, 83; XXIII, 54; XXIV, 63; XXV, 59; XXVI, 12.
[9] XXIII, 59.
[10] XXIV, 62.
[11] XXV, 21, 94.
[12] XXIV, 63 and 118.
[13] XXI, 19.
[14] XXIV, 62; XXIII, 59.
[15] XXIII, 81; XXIV, 6, 62, 116.
[16] XXVI, 12.

about the plant with the point of a sword.[1] Often the plant must not touch the ground again after it is picked,[2] presumably from a fear that its virtue would run off like an electric current. Pliny alludes at least three times [3] to the practice of herbalists of retaining portions of the herbs they sell, and then, if they are not paid in full, replanting the herb in the same spot with the idea that thereby the disease will return to plague the delinquent patient. Frequently one is directed to state why one plucks the herb or for whom it is intended.[4] In one case the digger says, "This is the herb Argemon which Minerva discovered was a remedy for swine who taste it." [5] In another case one should salute the plant and extract its juice before saying a word; thus its virtue will be much greater.[6] In other cases, as an offering to appease the earth, the soil about the plant is soaked with hydromel three months before plucking it, or the hole left by pulling it up is filled with different kinds of grain.[7] Sometimes one sacrifices beforehand with bread and wine or prays to the gods for permission to gather the herb.[8] The customs of the Druids in gathering herbs are mentioned more than once.[9] In gathering the sacred mistletoe on the sixth day of the moon they hold sacrifices and a banquet beneath the tree.[10] Two white bulls are the victims; a priest clad in white cuts the mistletoe with a golden sickle and receives it in a white cloak.[11]

Agricultural magic.

To Pliny's discussion of herbs we may append some specimens of the employment of magic procedure in agriculture and of the superstitions of the peasantry in which his pages abound. To guard against diseases of grain the seeds before planting should be steeped in wine, the juice of a certain herb, the gall of a cow, or human urine, or

[1] XXI, 19; XXV, 21, 94.
[2] XXIII, 71, 81; XXIV, 6; XXVII, 62.
[3] XXI, 83; XXV, 109; XXVI, 12.
[4] XXII, 16; XXIII, 54; XXIV, 82; XXVII, 113.
[5] XXIV, 116.
[6] XXV, 92.
[7] XXI, 19; XXV, 11.
[8] XXIV, 62; XXV, 21.
[9] XXIV, 62-63.
[10] XVI, 95.
[11] See XXIV, 6, for other methods of plucking the mistletoe.

should be touched with the shoulders of a mole [1]—the animal whose use by the *magi* we heard Pliny ridicule. One should sow at the moon's conjunction. Before the field is hoed, a frog should be carried around it and then buried in the center in an earthen vessel. But it should be disinterred before harvest lest the millet be bitter. Birds may be kept away from the grain by planting in the four corners of the field an herb whose name is unfortunately unknown to Pliny.[2] Mice are kept out by the ashes of a weasel, mildew by laurel branches, caterpillars by placing the skull of a female beast of burden upon a stick in the garden.[3] To ward off fogs and storms from orchards and vineyards a frog may be buried as directed above, or live crabs may be burnt in the trees, or a painted grape may be consecrated.[4] Suspending a frog in the granary preserves the corn stored there.[5] To keep wolves away catch one, break its legs, attach it to the ploughshare, and thus scatter its blood about the boundaries of the field; then bury the carcass at the starting-point.[6] Or consecrate at the altar of the Lar the ploughshare with which the first furrow was traced. Foxes will not touch poultry who have eaten the dried liver of a fox or who wear a bit of its skin about their necks. Fern will not spring up again if it is mowed with the edge of a reed or uprooted by a ploughshare upon which a reed has been placed.[7] Of the use of incantations in agriculture we shall treat later.

Virtues of stones.

Pliny appears to have much less faith in the possession of marvelous virtues by gems than by herbs and parts of animals. He not only characterizes the powers attributed to gems by the *magi* and Democritus and Pythagoras as "terrible lies" and "unspeakable nonsense"; [8] but refrains from mentioning many such himself or inserts a cautious "if we believe it" or "if they tell the truth." [9] Of the gem

[1] XVIII, 45.
[2] See also XXV, 6.
[3] XIX, 58.
[4] XVIII, 70.
[5] XVIII, 73.
[6] XXVIII, 81.
[7] XVIII, 8.
[8] XXXVII, 14, 73.
[9] XXXVII, 55-56.

supposed to be produced from the urine of the lynx he says, "I think that this is quite false and no gem of that name has been seen in our time. What is stated concerning its medicinal virtue is also false." [1] To other stones, however, he ascribes various medicinal virtues, either when taken pulverized in drink or when worn as amulets.[2] A few other occult properties are stated without reservation, as that *amiantus* resists all sorceries,[3] that adamant expels idle fears from the mind, that *sideritis* produces discord and litigation, and that *eumeces,* placed beneath one's pillow at night, causes oracular visions.[4] Magnets are said to differ in sex, and the belief of Theophrastus and Mucianus is repeated that certain stones bear offspring.[5]

Other minerals and metals.

Of the metals iron sometimes figures in Pliny's magical procedure, as when he either prescribes or taboos the use of it in cutting herbs or killing animals. In Arcadia the yew-tree is a fatal poison to persons sleeping beneath it, but driving a copper nail into the tree makes it harmless.[6] Pliny says that gold is medicinal in many ways and in particular is applied to wounded persons and to infants as a safeguard against witchcraft.[7] Earth itself is often used to work marvels, but usually some particular portion, such as that between cart ruts or that thrown up by ants, beetles, and moles, or in the right footprint where one first heard a cuckoo sing.[8] However, the rule that an object should not touch the ground is enforced in many other connections [9] than the plucking of herbs, and Pliny twice states that the earth will not permit a serpent who has stung a human being to re-enter its hole.[10] In his discussion of metals Pliny does not allude to transmutation or alchemy, unless it be in his accounts of various fraudulent practices of workers in metal and how Caligula extracted gold from orpiment. But the following directions for preparing antimony show how

[1] XXXVII, 13.

[2] For instance, XXXVII, 12 amber, 37 jasper, 39 aetites, 55 "baroptenus."

[3] XXXVI, 31.

[4] XXXVII, 15, 58, 67.

[5] XXXVI, 25, 39.

[6] XVI, 20.

[7] XXXIII, 25.

[8] XXX, 12, 25.

[9] XX, 3; XXVIII, 6, 9; etc.

[10] II, 63; XXIX, 23.

closely akin to magic the procedure in ancient metallurgy might be. The antimony should be coated with cow-flap and burnt in furnaces, then quenched in woman's milk and pounded in mortars with an admixture of rain-water.[1]

Virtues of human parts.

Various parts and products of the human body are credited with remarkable virtues as the mention just made of woman's milk suggests. Other passages recommend more especially the milk of a woman just delivered of a male child, but most of all that of the mother of twins.[2] *Sed nihil facile reperiatur mulierum profluvio magis monstrificum,* as Pliny proceeds to illustrate by numerous examples.[3] Great virtues are also attributed to the urine, particularly of a chaste boy.[4] A few other instances of remedies drawn from the human body are ear-wax or a powdered tooth against stings of scorpions and bites of snakes,[5] a man's hair for the bite of a dog, the first hairs from a boy's head for gout.[6] Diseases of women are prevented by wearing constantly in a bracelet the first tooth a boy loses, provided it has not touched the ground. Simply tying two fingers or toes together is recommended for tumors in the groin, catarrh, and sore eyes.[7] Or the eyes may be touched thrice with water in which the feet have been washed. Scrofula and throat diseases may be cured by the touch of the hand of one who has died an early death, although some authorities do not insist upon the circumstance of early death but direct that the corpse be of the same sex as the patient and that the diseased spot be touched with the back of the left dead hand.

Virtues of human saliva.

Of all fluids and excretions of the human body the saliva is perhaps used most often in ancient and medieval medicine, as the custom of spitting once or thrice in administering other remedies or performing ceremonies goes to prove. The spittle of a fasting person is the more efficacious. In a chapter devoted particularly to the properties of human

[1] XXXIII, 34.
[2] XX, 51; XXVIII, 21.
[3] VII, 13; XXVIII, 23.
[4] XX, 33; XXII, 30; XXVIII, 18-19.
[5] XXVIII, 8.
[6] XXVIII, 9.
[7] XXVIII, 9-11.

saliva Pliny lists many diseases and woes which it alleviates.[1] In this connection he makes the following absurd assertion which he nevertheless declares is easily tested by experiment. "If a person repents of a blow given from a distance or hand-to-hand, let him spit into the palm of the hand with which he struck, and the person who has been struck will feel no resentment. This is often proved by beasts of burden who are induced to mend their pace by this method after the use of the whip has failed." Pliny adds, however, that some persons try to increase the force of their blows by thus spitting on the hands beforehand. He also mentions as counter-charms against sorcery the practices of spitting into one's urine or right shoe, or when crossing a dangerous spot.

The human operator.

The importance of the human operator as a factor in the performance of marvels, be they medical or magical, is attested by the frequent injunctions of chastity, virginity, nudity, or a state of fasting upon persons concerned in Pliny's procedure. Sometimes they are not to glance behind them, sometimes they are to speak to no one during the operation. Pliny also mentions men who have a special capacity for wonder-working, such as Pyrrhus, the touch of whose toe had healing power,[2] those whose eyes exert strong fascination, whole tribes of serpent-charmers and venom-curers, and others whose mere presence addles the eggs beneath a setting hen.[3] The power of words spoken by men will be considered separately under the head of incantations.

Absence of medical compounds.

While Pliny attributes the most extreme medicinal virtues to simples, he excludes from his *Natural History* the strange and elaborate compounds which were nevertheless so popular in the pharmacy of his age. Of one simple, *laser,* he says that it would be an immense task to attempt to list all the uses that it is supposed to have in compounds.[4] His position is that the simple remedies alone are the direct work of nature, while the mixtures, tablets, pills, plasters,

[1] XXVIII, 7.
[2] VII, 2.
[3] XXVIII, 6.
[4] XXII, 49.

washes are artificial inventions of the apothecaries. Once when he describes a compound called "Hermesias" which aids in the generation of good and beautiful children, it seems to be borrowed by Democritus from the *magi*.[1] Furthermore, Pliny thinks that health can be sufficiently preserved or restored by nature's simple remedies. Compounds are the invention of human conjecture, avarice, and impudence. Such conjecture is often false, not sufficiently taking into account the natural sympathies and antipathies of the numerous ingredients. Often compounds are inexplicable. Pliny also deplores resort to imported drugs from India, Arabia, and the Red Sea, when there are homely remedies at hand for the poorest man.[2]

Sympathetic magic.

We have just heard Pliny refer to the sympathies and antipathies of natural simples, and he often explains the marvelous effects of natural objects upon one another by this relation of love and hatred, friendship or repugnance, discord or concord which exists between them, which the Greeks call sympathy or antipathy, and which Heracleitus was perhaps the first philosopher to insist upon.[3] Some modern students of magic have tried to account for all magic on this theory, and Pliny states that medicine and medicines originated from it.[4]

Antipathies between animals.

This relationship exists between animals,—deer and snakes, for example. So great a force is it that stags track snakes to their holes and extract them thence despite all resistance by the power of their breath. This antipathy continues after death, for the sovereign remedy for snakebite is the rennet of a fawn killed in its mother's womb, while serpents flee from a man who wears the tooth of a deer. But antipathy may change to sympathy, for Pliny adds that in some cases certain parts of deer treated in certain ways attract serpents.[5] This force of antipathy is in-

[1] XXIV, 102.

[2] In this paragraph I have combined views expressed by Pliny in three different passages: XXII, 49 and 56; XXIV, 1.

[3] IX, 88; XXIV, 1; XXVIII, 23; XXXII, 12; XXXVII, 15; etc.

[4] XXIV, 1; XXIX, 17.

[5] VIII, 50; XXVIII, 42.

deed capable of taking the strangest turn. Bed-bugs, foul and disgusting as they are, heal the bite of snakes, especially asps, and sows can eat the poisonous salamander.[1] The antipathy between goats and snakes would seem almost as potent as that between deer and snakes,[2] since we are told that snake-bitten persons recover more quickly, if they frequent the stalls where goats are kept or wear as an amulet the paunch of a she-goat.

Love and hatred between inanimate objects.

There is also "the hatred and friendship of deaf and insensible things." [3] Instances are the magnet's attraction for iron and the fact that adamant can be broken only by the blood of a he-goat, two stock examples of occult influence and natural marvels which continued classic in the medieval period.[4] Pliny indeed regards this last as the clearest illustration possible of the potency of sympathy and antipathy, since a substance which defies iron and fire, nature's two most violent agents, yields to the blood of a foul animal.[5]

Sympathy between animate and inanimate objects.

There is furthermore sympathy and antipathy between animate and inanimate objects. So marvelous is the antipathy of the tamarisk tree for the spleen alone of internal organs, that pigs who drink from troughs of this wood are found when slaughtered to be without spleen, and hence splenetic patients are fed from vessels of tamarisk.[6] The spleenless pig, it may be interpolated, is another commonplace of ancient and medieval science. Smearing the hives with cow dung kills other insects but stimulates the bees who have an affinity for it (*cognatum hoc iis*),[7] probably, although Pliny does not say so, on the theory that they are

[1] XXIX, 17 and 23.

[2] XXVIII, 43.

[3] XX, 1. "Odia amicitiaque rerum surdarum ac sensu carentium . . . quod Graeci sympathiam appellavere." XXIV, 1. "Surdis etiam rerum sua cuique sunt venena ac minimis quoque . . . Concordia valent."

[4] XXVIII, 41; XXXVII, 15. Yet a note in Bostock and Riley's translation, IV, 207, asserts, "Pliny is the only author who makes mention of this singularly absurd notion."

[5] "Nunc quod totis voluminibus his docere conati summus de discordia rerum concordiaque quam antipathiam Graeci vocavere ac sympathiam non aliter clarius intelligi potest."

[6] XXIV, 41.

[7] XXI, 47.

spontaneously generated from it. That the wild cabbage is hostile to dogs is evidenced by the statement of Epicharmus that it cures the bite of a mad dog but kills a dog if he eats it when given to him with meat.[1] Snakes hate the ash-tree so, that if they are hemmed in by its foliage on one side and fire on the other, they flee by preference into the flames.[2] Betony, too, is so antipathetic to snakes that they lash themselves to death when a circle of it is drawn about them.[3] Scorpions cannot survive in the air of Sicily.[4] Perhaps antipathy is also the explanation of Pliny's absurd statement that loads of apples and pears, even if there are only a few of them, are very heavy for beasts of burden.[5] Here, however, the condition may be remedied and perhaps a relationship of sympathy established by showing the beasts how few fruit there really are or by giving them some to eat. That sympathy may even attach to places or religious circumstances Pliny infers from the belief that the priestess of the earth at Aegira, when about to descend into the cave and predict, drinks without injury bull's blood which is supposed to be a fatal poison.[6]

Like cures like.

That like cures like, or more precisely and paradoxically that the cause of the disease will cure its own result, is another notion which Pliny's medicine shares with magic. This is seen in the use of parts of the mad dog to cure its bite,[7] or in rubbing thighs chafed by horse-back riding with the foam from a horse's mouth.[8] The bite of the shrew-mouse, too, is best healed by imposition of the very animal which bit you, but another shrew-mouse will do and they are kept ready in oil and mud for this purpose.[9] The sting of the *phalangium* may be cured by merely looking at another insect of that species, whether it be dead or alive.

From cases in which the cure for the disease is identical with its cause it is but a short step to remedies similar to

[1] XX, 36.
[2] XVI, 24.
[3] XXV, 55.
[4] XXXVII, 54.
[5] XXIII, 62; XXIV, 1.
[6] XXVIII, 41.
[7] XXIX, 32.
[8] XXVIII, 61.
[9] XXIX, 27.

or in some way associated with the ailment. It seems obvious to Pliny that stone in the bladder can be broken by the herb on which grow what look exactly like pearls. "In the case of no other herb is it so evident for what medicine it is intended; its species is such that it can be recognized at once by sight without book knowledge."[1] Similarly *ophites,* a marble with serpentine streaks, is used as an amulet against snake-bite.[2] Mithridates discovered that the blood of Pontic ducks should be mixed in antidotes because they live on poison.[3] Heliotrope seed looks like a scorpion's tail; if scorpions are touched with a sprig of heliotrope they die, and they will not enter ground which has been circumscribed by it.[4] To accelerate a woman's delivery her lover should take off his belt and gird her with it, then untie it, saying that he has bound her and will unloose her, and then he should go away.[5] An epileptic may be cured by driving an iron nail into the spot where his head rested when he fell in the fit.[6]

The principle of association.

Other instances of association are when the remedy employed is some part of an animal who is free from the disease in question or marked by an opposite state of health. Goats and gazelles never have ophthalmia, hence various portions of their bodies are prescribed for eye diseases.[7] Eagles can gaze at the sun, therefore their gall is efficacious in eye-salves.[8] The bird called ossifrage has a single intestine which digests anything; the end of this intestine serves as an amulet against colic, and indigestion may be cured by merely holding the crop of the bird in one hand.[9] But do not hold it too long or your flesh will waste away. The virus of mares is an ingredient in a candle which makes heads of horses seem to appear when it burns;[10] while ink of the *sepia* is used in a candle which causes Ethiopians to be seen when it is lighted.[11] These magic candles are borrowed

[1] XXVII, 74.
[2] XXXVI, 11.
[3] XXV, 3.
[4] XXII, 29.
[5] XXVIII, 9.
[6] XXVIII, 17.
[7] XXVIII, 47.
[8] XXIX, 38.
[9] XXX, 20.
[10] XXVIII, 49.
[11] XXXII, 52.

by Pliny from the works of Anaxilaus, and we shall find them a feature of medieval collections of experiments. Earth from a cart-wheel rut is thought a remedy against the bite of the shrew-mouse because that creature is too torpid to cross such a rut;[1] and Pliny believes that none of the virtues attributed to moles by the magicians is more probable than that they are an antidote to the bite of the shrew-mouse, which shuns even ruts, whereas moles burrow freely through the soil.[2] Pliny finds incredible the assertion made by some that a ship will move more slowly if it has the right foot of a tortoise aboard,[3] but the logic of the magic seems evident enough.

Magic transfer of disease.

In Pliny's medicine there are a number of examples of what may be called magic transfer, in which the aim of the procedure is not to cure the disease outright but to rid the patient of it by transferring it from him to some other animal or object. Intestinal disease may be transferred to puppies who have not yet opened their eyes by pressing them to the body and giving them milk from the patient's mouth. They will die of the disease, when its cause and exact nature may be determined by dissecting them. But finally they must be buried.[4] Griping pains in the bowels will also pass to a duck that is held against the abdomen. One may be rid of a cough by spitting in a frog's mouth or cure catarrh by kissing a mule,[5] although in these cases we are left uninformed whether the disease passes to the animal. But if a person who has been stung by a scorpion whispers the news in the ear of an ass, the ill will be transferred to the ass.[6] A boil may be removed by rubbing nine grains of barley around it, each grain thrice with the left hand, and then throwing them all into the fire.[7] Warts are banished by touching each with a grain of the chickpea and then tying the grains up in a linen cloth and throwing them behind one.[8] If a root of asphodel is applied to sores and then hung

[1] XXIX, 27.
[2] XXX, 7.
[3] XXXII, 14.
[4] XXX, 20 and 14.
[5] XXXII, 29; XXX, 11.
[6] XXVIII, 42.
[7] XXII, 65.
[8] XXII, 72.

up in smoke, the sores will dry up along with the root.[1] To cure scrofulous sores some bind on as many earthworms as there are sores and let them dry up together.[2] A tooth will cease aching if the herb *erigeron* is dug up with iron and the patient thrice alternately touches the tooth with the root and spits, and if he then replaces the herb in the same spot and it lives.[3] If this last is a case of magic transfer, perhaps we may trace the same notion in some of the numerous instances in which Pliny directs that an animal shall be released alive after some part of it has been removed or some other medicinal use made of it.

Amulets.

A common characteristic of magic force and occult virtue is that it will often act at a distance or without any physical contact or direct application. This is manifested in the practice of carrying or wearing amulets, or, what is the same thing, of ligatures and suspensions, in which objects are hung from the neck or bound to some part of the body in order to ward off danger from without or cure internal disease. Instances of such practices in the *Natural History* are well nigh innumerable. Roots are suspended from the neck by a thread;[4] the tongue of a fox is worn in a bracelet;[5] for quinsy the throat is wound thrice with a thong of dog-skin and catarrh is relieved by winding the same about the fingers.[6] A tooth stops aching when worms are taken from a certain prickly plant, put with some bread in a pill-box, and bound to the arm on the same side of the body as the aching tooth.[7] Two bed-bugs bound to the left arm in wool stolen from shepherds are a charm against nocturnal fevers; against diurnal fevers, if wrapped in russet cloth instead.[8] The heart of a vulture is an amulet against snakes, wild beasts, robbers, and royal wrath.[9] The traveler who carries the herb *artemisia* feels no fatigue.[10] Injurious drugs cannot cross one's threshold and do injury in

[1] XXII, 32.
[2] XXX, 12.
[3] XXV, 106.
[4] XX, 81.
[5] XXVIII, 47.
[6] XXX, 12, 15.
[7] XXVII, 62.
[8] XXIX, 17.
[9] XXIX, 24.
[10] XXVI, 89.

one's household, if a sea-star is smeared with the blood of a fox and attached to the lintel or door-post with a copper nail.[1] Not only is a wreath of herbs worn for headache,[2] but a sprig of poplar held in the hand prevents chafing between the thighs.[3] Often objects are placed under one's pillow, especially for insomnia,[4] but any psychological effect is precluded in the case where this is to be done without the patient's knowledge.[5] All sorts of specifications are given as to the color and kind of string, cloth, skin, box, nail, ring, bracelet, and the like in which should be placed, or with which should be bound on, the various gems, herbs, and parts of animals which serve as amulets. But when we are told that a remedy for headache which always helps many consists of a little bone from a snail found between two cart ruts, passed through gold, silver, and ivory, and attached to the body with dog-skin; or that one may bind on the head with a linen cloth the head of a snail decapitated with a reed when feeding in the morning especially at full moon;[6] we feel that we have passed beyond mere amulets, ligatures, and suspensions to more elaborate minutiae of magic procedure.

Position or direction.

Position or direction is often an important matter in Pliny's, as in magic, ceremonial. It perhaps comes out most frequently in his specification of right or left. An aching tooth should be scarified with the left eye-tooth of a dog; a spider which is placed with oil in the ear should be caught with the left hand;[7] epilepsy may be cured if a virgin touches the sufferer with her right thumb;[8] for ophthalmia of the right eye suspend the right eye of a frog from the patient's neck, and the left eye for the left eye;[9] for lumbago tear off an eagle's feet away from the joint, and use the right foot for the right side and the left for pain in the left side.[10] But we have met other examples already, and

[1] XXXII, 16; also XX, 39.
[2] XXII, 30.
[3] XXIV, 32, 38.
[4] XX, 72, 82.
[5] XXVI, 69.
[6] XXIX, 36.
[7] XXX, 8.
[8] XXVIII, 10.
[9] XXXII, 24.
[10] XXX, 18.

also cases of the use of the upper or lower part of this or that according to the corresponding location of an aching tooth in the upper or lower jaw.[1] Tracing circles with and about objects, facing towards this or that point of the compass, the prohibition against glancing behind one, and the stress laid upon finding things or killing animals between the ruts of cart wheels, are other examples of taking into consideration position and direction which we have already met with incidentally to the treatment of other topics. The prescription of a plant which has grown on the head of a statue and of another which has taken root in a sieve thrown into a hedge [2] also seem to take mere position largely into account, more so than the accompanying recommendation of an herb growing on the banks of a stream and of another growing upon a dunghill.[3]

The time element.

The element of time is also important. Operations should be performed before sunrise, early in the morning, at night, and so on. The moon is especially regarded in such directions.[4] When we are informed that sufferers from quartan fever should be rubbed all over with the fat of a tortoise, we are also told that the tortoise will be fattest on the fifteenth day of the moon and that the patient should be anointed on the sixteenth.[5] But this waxing and waning of the tortoise with the moon is primarily a matter of astrology and planetary influence, under which heading we shall also later speak of Pliny's observance of the rising of the dog-star.

Observance of number.

Observance of number is another feature in Pliny's ceremonial, of which we have already met instances. He also alludes to the writings of Pythagoras on the subject and ascribes to Democritus a work on the number four. Pliny's recipes frequently recommend that the operation be thrice repeated. In the case of curing scrofula by the ashes of vipers he prescribes three fingers thereof taken in drink for

[1] See also XXX, 8.
[2] XXIV, 106 and 109.
[3] XXIV, 107 and 110.
[4] Some examples are: XVIII, 75, 79; XXII, 72; XXIII, 71; XXVIII, 47; XXIX, 36; XXXII, 14, 25, 38, 46.
[5] XXXII, 14.

thrice seven days.[1] In another application of a Gallic herb with old axle-grease which has not touched iron, not only must the patient spit thrice to the right, but the remedy is more efficacious if three men representing three different nations anoint the right side with it.[2] The virtue of the number one is not, however, entirely slighted. Importance is attached to the death of a stag from a single wound.[3] Sometimes three and one are joined in the same operation, as when child-birth is aided by hurling through the house a stone or weapon by which three animals, a man, a boar, and a bear, have been killed with single blows. One of the discoveries of Pythagoras which seldom fails is that an odd number of vowels in a child's given name portends lameness, blindness, and like incapacitation on the right side of its body, and an even number, injuries on the left side.[4] In a crown of smilax for headache there should be an odd number of leaves,[5] and in a diet of snails prescribed for stomach trouble an odd number are to be eaten.[6] For a head-wash ten green lizards are boiled in ten *sextarii* of oil,[7] and for an application to prevent eyelashes from growing again when they have been pulled out fifteen frogs are impaled on fifteen bulrushes.[8] The person who has tied on a certain amulet is thereafter excluded from the patient's sight for five days.[9] And so on.

Relation between operator and patient.

This last item suggests a further intangible factor in Pliny's procedure, the doing of things to or for the patient without his knowledge. But this and any other incorporeal relationships existing between operator and patient should perhaps be classed under the head of sympathy and antipathy.

Incantations.

Closely akin to the power of numbers is that of words. Pliny once says of an incantation employed to avert hailstorms that he would not dare in seriousness to insert its

[1] XXX, 12.
[2] XXIV, 112.
[3] VIII, 50.
[4] XXVIII, 6.
[5] XXIV, 17.
[6] XXX, 15.
[7] XXIX, 34.
[8] XXXII, 24.
[9] XXXII, 38.

words, although Cato in his work on agriculture prescribed a similar formula of meaningless words for the cure of fractured limbs.[1] But Pliny does not object to the repetition of incantations or prayers if the words spoken have some meaning. He informs us that *ocimum* is sown with curses and maledictions and that when cummin seed is rammed down into the soil, the sowers pray it not to come up.[2] In another case the sower is to be naked and to pray for himself and his neighbors.[3] In a third case in which a poultice is to be applied to an inflammatory tumor, Pliny says that persons of experience regard it as very important that the poultice be put on by a naked virgin and that both she and the patient be fasting. Touching the sufferer with the back of her hand she is to say, "Apollo forbids a disease to increase which a naked virgin restrains." Then, withdrawing her hand, she is to repeat the same words thrice and to join with the patient in spitting on the ground each time.[4] Indeed, in another passage Pliny states that it is the universal custom in medicine to spit three times with incantations.[5] Perhaps the power of the words is thought to be increased or renewed by clearing the throat. Words were also occasionally spoken in plucking herbs. Ring-worm or tetter is treated by spitting upon and rubbing together two stones covered with a dry white moss, and by repeating a Greek incantation which may be translated, "Flee, Cantharides, a wild wolf seeks your blood."[6] Abscesses and inflammations are treated with the herb *reseda* and a Latin translation which seems irrelevant, if not quite senseless, and which may be translated, "Reseda, make disease recede. Don't you know, don't you know what chick has dug up these roots? May they have neither head nor feet."[7] In the book following this passage Pliny raises the general question of the power of words to heal diseases.[8] He gives many instances of belief in incantations from contemporary popu-

[1] XVII, 47.
[2] XIX, 36.
[3] XVIII, 35.
[4] XXVI, 60.
[5] XXVIII, 7.
[6] XXVII, 75.
[7] XXVII, 106.
[8] XXVIII, 3-4.

lar superstition, from Roman religion, and from the annals of history. He does not doubt that Romans in the past have believed in the power of words, and thinks that if we accept set forms of prayer and religious formulae, we must also admit the force of incantations. But he adds that the wisest individuals believe in neither.

Attitude to love-charms and birth-control.

Pliny's recipes and operations are mainly connected with either medicine or agriculture, but he also introduces as we have seen magical procedure employed in child-birth, safeguards against poisons and reptiles, and counter-charms against sorcery. He more than once avers that love-charms (*amatoria*) lie outside his province,[1] in one passage alleging as a reason that the illustrious general Lucullus was killed by one,[2] but he includes a great many of them nevertheless.[3] Some herbs are so employed because of a resemblance in shape to the sexual organs,[4] another instance of association by similarity. Pliny declared against abortive drugs as well as love-charms,[5] but cited from the *Commentaries* of Caecilius one recipe for birth-control for the benefit of over-fecund women, consisting of a ligature of two little worms found in the body of a certain species of spider and bound on in deer-skin before sunrise. After a year the virtue of this charm expires.[6]

Pliny and astrology.

Pliny devotes but a small fraction of his work to the stars and heavens as against terrestrial phenomena, and therefore has less occasion to speak of astrology than of magic. However, had he been a great believer in astrology he doubtless would have devoted more space to the stars and their influence on terrestrial phenomena. He recognizes none the less, as we have seen, that magic and astrology are in-

[1] XXVII, 35. "Catanancen Thessalam herbam qualis sit describi a nobis supervacuum est, cum sit usus eius ad amatoria tantum." XXVII, 99. "Phyteuma quale sit describere supervacuum habeo cum sit usus eius tantum ad amatoria."

[2] XXV, 7. "Ego nec abortiva dico ac ne amatoria quidem, memor Lucullum imperatorem clarissimum amatorio perisse . . ."

[3] A few examples are: XX, 15, 84, 92; XXIV, 11, 42; XXVI, 64; XXVII, 42, 99; XXVIII, 77, 80; XXX, 49; XXXII, 50.

[4] XXII, 9.

[5] XXV, 7.

[6] XXIX, 27.

timately related and that "there is no one who is not eager to learn his own future and who does not think that this is shown most truly by the heavens."[1] Parenthetically it may be remarked that the general literature of the time only confirms this assertion of the widespread prevalence of astrology; allusions of poets imply a technical knowledge of the art on their readers' part; the very emperors who occasionally banished astrologers from Rome themselves consulted other adepts. In another passage Pliny speaks of men who "assign events each to its star according to the rules of nativities and believe that God decreed the future once for all and has never interfered with the course of events since.[2] This way of thinking has caught learned and vulgar alike in its current and has led to such further methods of divination as those by lightning, oracles, haruspices, and even such petty auguries as from sneezes and shifting of the feet. Furthermore in Pliny's list of men prominent in the various arts and sciences we find Berosus of whom a statue was erected by the Athenians in honor of his skill in astrological prognostication.[3] In another place where he speaks for a moment of "the science of the stars" Pliny disputes the theories of Berosus, Nechepso, and Petosiris that length of human life is ordered by the stars, and also makes the trite objection to the doctrine of nativities that masters and slaves, kings and beggars are born at the same moment.[4] He also is rather inclined to ridicule the enormous figures of 720,000 or 490,000 years set by Epigenes and Berosus and Critodemus for the duration of astronomical observations recorded by the Babylonians.[5] From such passages we get the impression that astrology is widely accepted as a science but that the art of nativities at least is not regarded by Pliny

[1] XXX, 1. On the general attitude to astrology of the preceding Augustan Age and its poets see H. W. Garrod, *Manili Astronomicon Liber II,* Oxford, 1911, pp. lxv-lxxiii, but I think he overestimates the probable effect of the edict of 16 A.D. upon the poem of Manilius.

[2] II, 5. "Astroque suo eventus adsignat nascendi legibus semelque in omnes futuros umquam deo decretum in reliquom vero otium datur."

[3] VII, 37.

[4] VII, 50.

[5] VII, 57.

with favor. But it would not be safe to say that he denies the control of the stars over human destiny. Indeed, in one chapter he declares that the astronomer Hipparchus can never be praised enough because more than any other man he proved the relationship of man with the stars and that our souls are part of the sky.[1] When Pliny disputes the vulgar notion that each man has a star varying in brightness according to his fortune, rising when he is born, and fading or falling when he dies, he is not attacking even the doctrine of nativities; he is denying that the stars are controlled by man's fate rather than that man's life is ordered by the stars.[2]

Celestial portents.

If Pliny thus leaves us uncertain as to the relation of man to the stars, we also receive conflicting impressions from his discussion of various celestial phenomena regarded as portentous. In one passage he speaks of the debt of gratitude owed by mankind to those great astronomical geniuses who have freed men from their former superstitious fear of eclipses.[3] But he explains thunderbolts as celestial fire vomited forth from the planet Venus and "bearing omens of the future." [4] He also gives instances from Roman history of comets which signaled disaster, and he expounds the theory of their signifying the future.[5] What they portend may be determined from the direction in which they move and the heavenly body whose power they receive, and more particularly from the shapes they assume and their position in relation to the signs of the zodiac. Indeed, Pliny even gives examples of ominous eclipses of the sun, although it is true that they were also of unusual length.[6] He also tells us that many of the common people still believed that women could produce eclipses "by sorceries and herbs.[7]

[1] II, 24.
[2] II, 6, "Non tanta caelo societas nobiscum est ut nostro fato mortalis sit ibi quoque siderum fulgor."
[3] II, 9.
[4] II, 18.
[5] II, 23.
[6] II, 30.
[7] XXV, 5.

The stars and the world of nature.

Aside from the question of the control of human destiny by the constellations at birth, Pliny's general theories of the universe and of the influence of the stars upon terrestrial nature are roughly similar to those of astrology. For him the universe itself is God, "holy, eternal, vast, all in all, nay, in truth itself all;" [1] and the sun is the mind and soul of the whole world and the chief governor of nature.[2] The planets affect one another. A cold star renders another approaching it pale; a hot star causes its neighbor to redden; a windy planet gives those near it a lowering appearance.[3] At certain points in their orbits the planets are deflected from their regular course by the rays of the sun,—an unwitting concession to heliocentric theory.[4] Pliny ascribes the usual astrological qualities to the planets.[5] Saturn is cold and rigid; Mars, a flaming fire; Jupiter, located between them, is temperate and salubrious. Besides their effects upon one another, the planets especially influence the earth.[6] Venus, for instance, rules the process of generation in all terrestrial beings.[7] Following the *Georgics* of Vergil somewhat, Pliny asserts that the stars give indubitable signs of the weather and expounds the utility of the constellations to farmers.[8] He tells how Democritus by his knowledge of astronomy was able to corner the olive crop and put to shame business men who had been decrying philosophy; [9] and how on another occasion he gave his brother timely warning of an impending storm.[10] But Pliny does not accept all the theories of the astrologers as to control of the stars over terrestrial nature. He repeats, but without definitely accepting it, the ascription by the Babylonians of earthquakes to three of the planets in particular,[11] and the notion that the gem *sandastros* or *garamantica*, em-

[1] II, 1.
[2] II, 4.
[3] II, 16.
[4] II, 13.
[5] II, 6; and see II, 39.
[6] II, 6. "Potentia autem ad terram magnopere eorum pertinens."
[7] II, 6.
[8] XVIII, 5, 57, 69.
[9] XVIII, 68. Other authorities tell the story of Thales; see Cicero, *De divinatione*, II, 201; Aristotle, *Polit.* I, 7; and Diogenes Laertius.
[10] XVIII, 78.
[11] II, 81.

ployed by Chaldeans in their ceremonies, is intimately connected with the stars.[1] He is openly incredulous about the gem *glossopetra,* shaped like a human tongue and supposed to fall from the sky during an eclipse of the moon and to be invaluable in selenomancy.[2]

Astrological medicine.

Pliny tells how the physician Crinas of Marseilles made a fortune by regulating diet and observing hours according to the motion of the stars.[3] But he does not show much faith in astrological medicine himself, rejecting entirely the elaborate classification of diseases and remedies which the astrologers had by his time already worked out for the revolutions of the sun and moon in the twelve signs of the zodiac.[4] In his own recipes, however, astrological considerations are sometimes observed, as we have already seen, especially the rising of the dog-star and the phases of the moon. Pliny, indeed, states that the dog-star exerts an extensive influence upon the earth.[5] As for the moon, the blood in the human body augments and decreases with its waxing and waning as shell-fish and other things in nature do.[6] Indeed, painstaking men of research had discovered that even the entrails of the field-mouse corresponded in number to the days of the moon, that the ant stopped working during the interlunar days, and that diseases of the eyes of certain beasts of burden also increased and decreased with the moon.[7] But on the whole Pliny's medicine and science do not seem nearly so immersed in and saturated with astrology as with other forms of magic. This gap was for the middle ages amply filled by the authority of Ptolemy, of whose belief in astrology we shall treat in the next chapter.

Conclusion: magic unity of Pliny's superstitions.

We have tried to analyze the contents of the *Natural History,* bringing out certain main divisions and underlying principles of magic in Pliny's agriculture, medicine, and natural science. This is, however, an artificial and difficult

[1] XXXVII, 28.
[2] XXXVII, 59.
[3] XXIX, 5.
[4] XXX, 29.
[5] II, 40.
[6] II, 102.
[7] II, 41.

task, since it is not easy to sever materials from ceremonial or the virtues of objects from the relations of sympathy or antipathy between them. Often the same passage might serve to illustrate several points. Take for example the following sentence: "Thrasyllus is authority that nothing is so hostile to serpents as crabs; swine who are stung cure themselves by this food, and when the sun is in Cancer, serpents are in pain."[1] Here we have at once antipathy, the remedies used by animals, the reasoning, characteristic of magic, from association and similarity, and the belief in astrology. And this confusion, to illustrate which a hundred other examples might be collected from the *Natural History,* demonstrates how indissolubly interwoven are all the varied threads that we have been tracing. They all go naturally together, they belong to the same long period of thought, they represent the same stage in mental development, they all are parts of magic.

[1] XXXII, 19.

CHAPTER III

SENECA AND PTOLEMY: NATURAL DIVINATION AND ASTROLOGY

Seneca's *Natural Questions*—Nature study as an ethical substitute for existing religion—Limited field of Seneca's work—Marvels accepted, questioned, or denied—Belief in natural divination and astrology—Divination from thunder—Ptolemy—His two chief works—His mathematical method—Attitude towards authority and observation—The *Optics*—Medieval translations of *Almagest*—*Tetrabiblos* or *Quadripartitum*—A genuine reflection of Ptolemy's approval of astrology—Validity of Astrology—Influence of the stars not inevitable—Astrology as natural science—Properties of the planets—Remaining contents of Book One—Book Two: regions—Nativities—Future influence of the *Tetrabiblos.*

"When the stars twinkle through the loops of time."
—Byron.

Seneca's *Natural Questions.*

In this chapter we shall preface the main theme of Ptolemy and his sanction of astrology by a consideration of another and earlier ancient writer on natural science who was very favorable to divination of the future, namely, the famous philosopher, statesman, man of letters, and tutor of Nero, Lucius Annaeus Seneca. In point of time his *Natural Questions,* or *Problems of Nature,* is a work slightly antedating even the *Natural History* of Pliny, but it is hardly of such importance in the history of science as the more voluminous works of the three great representatives of ancient science, Pliny, Galen, and Ptolemy. Nevertheless Seneca was well known and much cited in the middle ages as an ethical or moral philosopher, and the title, *Natural Questions,* was to be employed by one of the first medieval pioneers of natural science, Adelard of Bath. Seneca in any case is a name of which ancient science need not be ashamed. He tells us that in his youth he had already written a treatise on earthquakes;[1] and in the present treatise his aim is to inquire into the natural causes of phenomena; he wants to know why things are so. He is aware that his own age has only entered the vestibule of the knowledge of natural phenomena and forces, that it has but just begun to know five of the many stars, that "there will come a time when our descendants will wonder that we were ignorant of matters so evident."[2]

Study of nature as an ethical substitute for existing religion.

In one passage Seneca perhaps expresses his consciousness of the very imperfect scientific knowledge of his own age a little too mystically. "There are sacred things which are not revealed all at once. Eleusis reserves sights for those who revisit her. Nature does not disclose her mysteries in a moment. We think ourselves initiated; we stand but at her portal. Those secrets open not promiscuously nor to every comer. They are remote of access, enshrined in the inner sanctuary."[3] Indeed, he shows a tendency to regard scientific research as a sort of religious exercise or perhaps as a substitute for existing religion and a basis for moral philosophy. He relates physics to ethics. His enthusiasm in the study of natural forces appears largely due to the fact that he believes them to be of a sublime and divine character and above the petty affairs of men. He also as constantly and more fulsomely than Pliny inveighs against the luxury, vice, and immorality of his own day, and moralizes as to the beneficent influence which natural law and phenomena should exert upon human conduct. It is interesting to note that this habit of drawing moral lessons from the facts of nature was not peculiar to medieval or Christian writers.

With such subjects as zoology, botany, and mineralogy Seneca's work has little to do; it does not, like Pliny's

[1] *L. Annaei Senecae Naturalium Quaestionum Libri Septem,* VI, 4, "Aliquando de motu terrarum volumen iuvenis ediderim." The edition by G. D. Koeler, Göttingen, 1819, devotes several hundred pages to a *Disquisitio* and *Animadversiones* upon Seneca's work. I have also used the more recent Teubner edition, ed. Haase, 1881, and the English translation in Clark and Geikie, *Physical Science in the Time of Nero,* 1910. In Panckoucke's *Library,* vol. 147, a French translation accompanies the text.

[2] VII, 25.

[3] VII, 31.

Limited field of Seneca's work.

Natural History, include medicine and the industrial arts; neither does he, like Pliny, cite the lore of the *magi*. The phenomena of which he treats are mainly meteorological manifestations, such as winds, rain, hail, snow, comets, rainbows, and what he regards as allied subjects, earthquakes, springs, and rivers. Perhaps he would not have regarded the study of vegetables, animals, and minerals as so lofty and sublime a pursuit. At any rate, in consequence of the restricted field which Seneca covers we find very little of the marvelous medicinal and magical properties of plants, animals, and other objects, or the superstitious procedure which fill the pages of Pliny.

Marvels accepted, questioned, or denied.

Seneca nevertheless has occasion to repeat some tall stories, such as that the river Alpheus of Greece reappears as the Arethusa in Sicily and there every four years casts up filth from its depths on the very days when victims are slaughtered at the Olympic games.[1] He also affirms that living beings are generated in fire; he believes in such effects of lightning as removing the venom from snakes which it strikes; and he recounts the old stories of floating islands and of waters with the virtue of turning white sheep black.[2] On the other hand, he qualifies by the phrases, "it is believed" and "they say," the assertions that certain waters produce foul skin-diseases and that dew in particular, if collected in any quantity, has this evil property; and he doubts whether bathing in the Nile would enable a woman to bear more children.[3] He ridicules the custom of the city which had public watchmen appointed to warn the inhabitants of the approach of hail-storms, so that they might avert the danger by timely sacrifice or simply by pricking their own fingers so that they bled a trifle. He adds that some suggest that blood may possess some occult property of repelling storm-clouds, but he does not see how there can be such force in a drop or two and thinks it simpler to

[1] III, 26.

[2] V, 6, for animals generated in flames; II, 31, for snakes struck by lightning; III, *passim* for marvelous fountains.

[3] III, 25.

regard the whole thing as false. In the same chapter he states that uncivilized antiquity used to believe that rain could be brought on or driven off by incantations, but that now-a-days no one needs a philosopher to teach him that this is impossible.[1]

Belief in natural divination and astrology.

But while he thus rejects incantations and is practically silent on the subject of natural magic, Seneca accepts natural divination in well-nigh all its branches: sacrificial, augury, astrology, and divination from thunder. He believes that whatever is caused is a sign of some future event.[2] Only Seneca holds that every flight of a bird is not caused by a direct act of God, nor the vitals of the victim altered under the axe by divine interference, but that all has been prearranged in a fatal and causal series.[3] He believes that all unusual celestial phenomena are to be looked upon as prodigies and portents. A meteor "as big as the moon appeared when Paulus was engaged in the war against Perseus"; similar portents marked the death of Augustus and execution of Sejanus, and gave warning of the death of Germanicus.[4] But no less truly do the planets in their unvarying courses signify the future. The stars are of divine nature, and we ought to approach the discussion of them with as reverent an air as when with lowered countenance we enter the temples for worship.[5] Not only do the stars influence the upper atmosphere as earth's exhalations affect the lower, but they announce what is to occur.[6] Seneca employs the statement of Aristotle that comets signify the coming of storms and winds and foul weather to prove that they are stars; and declares that a comet is a portent of bad weather during the ensuing year in the same way that the Chaldeans or astrologers say that a man's natal star determines the whole course of his life.[7] In fact, Seneca's chief, if not sole, objection to the Chaldeans or astrologers would seem to be that in their predictions they take only five

[1] IV, 7.
[2] II, 32.
[3] II, 46.
[4] I, 1.
[5] VII, 30.
[6] II, 10.
[7] VII, 28.

stars [1] into account. "What? Think you so many thousand stars shine on in vain? What else, indeed, is it which causes those skilled in nativities to err than that they assign us to a few stars, although all those that are above us have a share in the control of our fate? Perhaps those which are nearer direct their influence upon us more closely; perhaps those of more rapid motion look down on us and other animals from more varied aspects. But even those stars that are motionless, or because of their speed keep equal pace with the rest of the universe and seem not to move, are not without rule and dominion over us." [2] Seneca accepts the theory of Berosus that whenever all the stars are in conjunction in the sign of Cancer there will be a universal conflagration, and a second deluge when they all unite in Capricorn.[3]

Divination from thunder

It is on thunderbolts as portents of the future that Seneca dwells longest, however.[4] "They give," he declares, "not signs of this or that event merely, but often announce a whole series of events destined to occur, and that by manifest decrees and ones far clearer than if they were set down in writing." [5] He will not accept, however, the theory that lightning has such great power that its intervention nullifies any previous and contradictory portents. He insists that divination by other methods is of equal truth, though possibly of minor importance and significance. Next he attempts to explain how the dangers of which we are warned by divination may be averted by prayer, expiation, or sacrifice, and yet the chain of events wrought by destiny not be broken. He maintains that just as we employ the services of doctors to preserve our health, despite any belief we may have in fate, so it is useful to consult a *haruspex*. Then he goes on to speak of various classifications of thunderbolts according to the nature of the warnings or encouragements which they bring.

Ptolemy.

We pass on from Seneca to a later and greater exponent of natural science and divination, Ptolemy, in the follow-

[1] That is to say, five in addition to the sun and the moon.

[2] II, 32.

[3] III, 29.

[4] II, 31-50.

[5] II, 32.

ing century. He was perhaps born at Ptolemaïs in Egypt but lived at Alexandria. The exact years of his birth and death are unknown, and very little is recorded of his life or personality. The time when he flourished is sufficiently indicated, however, by the fact that his first recorded astronomical observation was in 127 and his last in 151 A. D. Thus most of his work was probably done during the reigns of Hadrian and Antoninus Pius, but he appears to have lived on into the reign of Marcus Aurelius. His strictly scientific style scorns rhetorical devices and literary felicities, and while it is clear and correct, is dry and impersonal.[1]

His two chief works.

Ptolemy's two chief works, the *Geography* in eight books, and ἡ μαθηματικὴ σύνταξις, or *Almagest* (al-μεγίστη) as the Arabs called it, in thirteen books, have been so often described in histories of mathematics, astronomy, geography, and discovery that such outline of their contents need not be repeated here. The erroneous Ptolemaic theories of a geocentric universe and of an earth's surface on which dry land preponderated are equally well known. What is more to the point at present is to note that one of these theories was so well fitted to actual scientific observations and the other was thought to be so similarly based, that they stood the test of theory, criticism, and practice for over a thousand years.[2] It should, however, be said that the *Geography* does not seem to have been translated into Latin until the

[1] A complete edition of Ptolemy's works has been in process of publication since 1898 in the Teubner library by J. L. Heiberg and Franz Boll. They are also the authors of the most important recent researches concerning Ptolemy. See Heiberg's discussion of the MSS in the volumes of the above edition which have thus far appeared; his articles on the Latin translations of Ptolemy in *Hermes* XLV (1910) 57ff, and XLVI (1911) 206ff; but especially Boll, *Studien über Claudius Ptolemäus. Ein Beitrag zur Geschichte der griechischen Philosophie und Astrologie*, 1894, in *Jahrb. f. Philol. u. Pädagogik*, Neue Folge, Suppl. Bd. 21. A recent summary of investigation and bibliography concerning Ptolemy is W. Schmid, *Die Nachklassische Periode der Griechischen Litteratur*, 1913, pp. 717-24, in the fifth edition of Christ, *Gesch. d. Griech. Litt.*

[2] Some strictures upon Ptolemy as a geographer are made by Sir W. M. Ramsay, *The Historical Geography of Asia Minor*, 1890, pp. 69-73.

opening of the fifteenth century,[1] when Jacobus Angelus made a translation for Pope Alexander V, (1409-1410), which is extant in many manuscripts [2] as well as in print.[3] It therefore did not have the influence and fame in the Latin middle ages that the *Almagest* did or the briefer astrological writings, genuine and spurious, current under Ptolemy's name.

His mathematical method.

We may briefly state one or two of Ptolemy's greatest contributions to mathematical and natural science and his probable position in the history of experimental method. Perhaps of greater consequence in the history of science than any one specific thing he did was his continual reliance

[1] Schmid would appear to be mistaken in saying that the *Geography* was already known in Latin and Arabic translation in the time of Frederick II (p. 718, "Seine in erster Linie die Astronomie, dann auch die Geographie und Harmonik betreffenden Schriften haben sich nicht bloss im Originaltext erhalten; sie wurden auch frühzeitig von den Arabern übersetzt und sind dann, ähnlich wie die Werke des Aristoteles, schon zur Zeit des Kaisers Friedrich II, noch ehe man sie im Urtext kennen lernte, durch lateinische, nach dem Arabischen gemachte Übersetzungen ins Abendland gelangt"), for in his own bibliography (p. 723) we read, "*Geographie* . . . Frühste latein. Übersetzung des Jacobus Angelus gedruckt Bologna, 1462." Apparently Schmid did not know the date of Angelus' translation.

However, Duhem, III (1915) 417, also speaks as if the *Geography* were known in the thirteenth century: "les considérations empruntées à la Géographie de Ptolémée fournissent à Robert de Lincoln une objection contre le mouvement de précession des équinoxes tel qu'il est définé dans l'Almageste." See also C. A. Nallino, *Al-Huwarizmi e il suo rifacimento della geografia di Tolomeo,* 1894, cited by Suter (1914) viii-ix, for a geography in Arabic preserved at Strasburg which is based on Ptolemy's *Geography.*

[2] In this Latin translation it is often entitled *Cosmographia.* Some MSS are: CLM 14583, 15th century, fols. 81-215, Cosmographia Ptolomei a Jacobo Angelo translata. Also BN 4801, 4802, 4803, 4804, 4838. Arsenal 981, in an Italian hand, is presumably incorrectly dated as of the 14th century.

This Jacobus Angelus was chancellor of the faculty of Montpellier in 1433 and is censured by Gerson in a letter for his superstitious observance of days.

[3] The several editions printed before 1500 seem to have consisted simply of this Latin translation, such as that of Bologna, 1462, and Vincentiae, 1475, and the Greek text to have been first published in 1507. See Justin Winsor, *A Bibliography of Ptolemy's Geography,* 1884, in *Library of Harvard University, Bibliographical Contributions,* No. 18:—a bibliography which deals only with printed editions and not with the MSS. According to Schmid, however, the *editio princeps* of the Greek text was that of Basel, 1533. C. Müller's modern edition (Didot, 1883 and 1901) gives an unsatisfactory bare list of 38 MSS. See also G. M. Raidel, *Commentatio critico-literaria de Claudii Ptolemaei Geographia eiusque codicibus,* 1737.

upon mathematical method both in his astronomy and his geography. In particular may be noted his important contribution to trigonometry in his table of chords, which modern scholars have found correct to five decimal places, and his contribution to the science of cartography by his successful projection of spherical surfaces upon flat maps.

Attitude towards authority and observation.

Ptolemy based his two great works partly upon the results already attained by earlier scientists, following Hipparchus especially in astronomy and Marinus in geography. He duly acknowledged his debts to these and other writers; praised Hipparchus and recounted his discoveries; and where he corrected Marinus, did so with reason. But while Ptolemy used previous authorities, he was far from relying upon them solely. In the *Geography* he adds a good deal concerning the orient and northern lands from the reports of Roman merchants and soldiers. His intention was to repeat briefly what the ancients had already made clear, and to devote his works chiefly to points which had remained obscure. His ideal was to rest his conclusions upon the surest possible observation; and where such materials were meager, as in the case of the *Geography*, he says so at the start. He also recognized that delicate observations should be repeated at long intervals in order to minimize the possibility of error. He devised and described some scientific instruments and conducted a long series of astronomical observations. He anteceded Comte in holding that one should adopt the simplest possible hypothesis consistent with the facts to be explained.

The *Optics*.

Besides some minor astronomical works and a treatise on music which seems to be largely a compilation an important work on optics is ascribed to Ptolemy.[1] It is the most experimental in method of his writings, although Alexander von Humboldt's characterization of it as the only work in ancient literature which reveals an investigator of nature

[1] *L'ottica di Claudio Tolomeo da Eugenio ammiraglio di Sicilia ridotta in latino*, ed. Gilberto Govi, Turin, 1885.

in the act of physical experimentation[1] must be regarded as an exaggeration in view of our knowledge of the writings of other Alexandrines such as Hero and Ctesibius. As in the case of some of Ptolemy's other minor works, the Greek original is lost and also the Arabic text from which was presumably made the medieval Latin version which alone has come down to us. Yet there are at least sixteen manuscripts of this Latin version still in existence.[2] The translation was made in the twelfth century by Eugene of Palermo, admiral of Sicily, whose name is attached to other translations and who was also the author of a number of Greek poems.[3] Heller states that the *Optics* was lost at the beginning of the seventeenth century but that manuscripts of it were rediscovered by Laplace and Delambre.[4] At any rate the first of the five books is no longer extant, although Bridges thinks that Roger Bacon was acquainted with it in the thirteenth century.[5] It dealt with the relations between the eye and light. In the second book conditions of visibility are discussed and the dependence of the apparent size of bodies upon the angle of vision. The third and fourth books deal with different kinds of mirrors, plane, convex, concave, conical, and pyramidical. Most important of all is the fifth and last book, in which dioptrics and refraction are discussed for the first and only time in any extant work of antiquity,[6] provided the *Optics* has really come down in its present form from the time of Ptolemy. His authorship has been questioned because the subject of refraction is not mentioned in the *Almagest,* although even astronomical refraction is discussed in the *Optics.*[7] De Morgan also

[1] Schmid (1913) still cites it without qualification. Hammer-Jensen has an article, *Ptolemaios und Heron,* in *Hermes,* XLVIII (1913) 224, *et seq.*

[2] Haskins and Lockwood, *The Sicilian Translators of the Twelfth Century,* in *Harvard Studies in Classical Philology,* XXI (1910), 89.

[3] *Ibid.*, 89-94.

[4] A. Heller, *Geschichte der Physik von Aristoteles bis auf die neueste Zeit,* 2 vols., Stuttgart, 1882-1884. The statement sounds a trifle improbable in view of the number of MSS still in existence.

[5] *Opus Maius,* II, 7.

[6] The *Dioptra* of Hero is really geodetical.

[7] Govi (1885), p. 151.

objects that the author of the *Optics* is inferior to Ptolemy in knowledge of geometry.[1] Possibly a work by Ptolemy has received medieval additions, either Arabic or Latin, in the version now extant; maybe the entire fifth book is such a supplement. That works which were not Ptolemy's might be attributed to him in the middle ages is seen from the case of Hero's *Catoptrica,* the Latin translation of which from the Greek is entitled in the manuscripts *Ptolemaei de speculis.*[2]

Medieval translations of *Almagest.*

If there is, as in other parallel cases, the possibility that the medieval period passed off recent discoveries of its own under the authoritative name of Ptolemy, there also is the certainty that it made Ptolemy's genuine works very much its own. This may be illustrated by the case of the *Almagest.* On the verge of the medieval period the work was commented upon by Pappus and Theon at Alexandria in the fourth, and by Proclus in the fifth century. The Latin translation by Boethius is not extant, but the book was in great repute among the Arabs, was translated at Bagdad early in the ninth century and revised later in the same century by Tabit ben Corra. During the twelfth century it was translated into Latin both from the Greek and the Arabic. The translation most familiar in the middle ages was that completed at Toledo in 1175 by the famous translator, Gerard of Cremona. There has recently been discovered, however, by Professors Haskins and Lockwood [3] a Sicilian translation made direct from the Greek text some ten or twelve years before Gerard's translation. There are

[1] *Ptolemy* in Smith's *Dictionary of Greek and Roman Biography.*

[2] It was also so printed in *Sphera cum commentis,* 1518: "Explicit secundus et ultimus liber Ptolomei de Speculis. Completa fuit eius translatio ultimo Decembris anno Christi 1269."

[3] C. H. Haskins and D. P. Lockwood, *The Sicilian Translators of the Twelfth Century and the First Latin Version of Ptolemy's Almagest,* in *Harvard Studies in Classical Philology,* XXI (1910) 75-102.

C. H. Haskins, *Further Notes on Sicilian Translations of the Twelfth Century, Ibid.,* XXIII, 155-66.

J. L. Heiberg, *Eine mittelalterliche Uebersetzung der Syntaxis des Ptolemaios,* in *Hermes* XLV (1910) 57-66; and *Noch einmal die mittelalterliche Ptolemaios-Uebersetzung, Ibid.,* XLVI, 207-16.

two manuscripts of this Sicilian translation in the Vatican and one at Florence, showing that it had at least some Italian currency. Gerard's reputation and his many other astronomical and astrological translations probably account for the greater prevalence of his version, or possibly the theological opposition to natural science of which the anonymous Sicilian translator speaks in his preface had some effect in preventing the spread of his version.

The *Tetrabiblos* or *Quadripartitum.*

Of Ptolemy's genuine works the most germane to and significant for our investigation is his *Tetrabiblos, Quadripartitum,* or four books on the control of human life by the stars. It seems to have been translated into Latin by Plato of Tivoli in the first half of the twelfth century[1] before *Almagest* or *Geography* appeared in Latin. In the middle of the thirteenth century Egidius de Tebaldis, a Lombard of the city of Parma, further translated the commentary of Haly Heben Rodan upon the *Quadripartitum.*[2] In the early Latin editions[3] the text is that of the medieval translation; in the few editions giving a Greek text there is a different Latin version translated directly from this Greek text.[4]

A genuine reflection of Ptolemy's approval of astrology.

In the *Tetrabiblos* the art of astrology receives sanction and exposition from perhaps the ablest mathematician and closest scientific observer of the day or at least from one who seemed so to succeeding generations. Hence from that time on astrology was able to take shelter from any criticism under the aegis of his authority. Not that it lacked

[1] Digby 51, 13th Century, fols. 79-114, "Liber iiii tractatuum Batolomei Alfalisobi in sciencia judiciorum astrorum. . . . Et perfectus est eius translatio de Arabico in Latinum a Tiburtino Platone cui Deus parcat die Veneris hora tertia XXa die mensis Octobris anno Domini MCXXVIII (*sic*) XV die mensis Saphar anno Arabum DXXXIII (*sic*) in civitate Barchinona. . . ." The date of translation is given as October 2, 1138, in CUL 1767, 1276 A.D., fols. 240-76, "Liber 4 Partium Ptholomei Auburtino Palatone."

[2] It is found in an edition printed at Venice in 1493, "per Bonetum locatellum impensis nobilis viri Octaviani scoti civis Modoetiensis."

[3] In the British Museum are editions of Venice, 1484, 1493, 1519; Paris, 1519; Basel, 1533; Louvain, 1548; it was also printed in 1551, 1555, 1578.

[4] In the British Museum are but three editions of the Greek text, all with an accompanying Latin translation: Nürnberg, 1535; Basel, 1553; and 1583.

other exponents and defenders of great name and ability. Naturally the authenticity of the *Tetrabiblos* has been questioned by modern admirers of Hellenic philosophy and science who would keep the reputations of the great men of the past free from all smudge of superstition. But Franz Boll has shown that it is by Ptolemy by a close comparison of it with his other works.[1] The astrological *Centiloquium* or *Karpos,* and other treatises on divination and astrological images ascribed to Ptolemy in medieval Latin manuscripts are probably spurious, but there is no doubt of his belief in astrology. German research as usual regards its favorite Posidonius as the ultimate source of much of the *Tetrabiblos,* but this is not a matter of much consequence for our present investigation.

Validity of astrology.

In the *Tetrabiblos* Ptolemy first engages in argument as to the validity of the art of judicial astrology. If his remarks in this connection were not already trite contentions, they soon came to be regarded as truisms. The laws of astronomy are beyond dispute, says Ptolemy, but the art of prediction of human affairs from the courses of the stars may be assailed with more show of reason. Opponents of astrology object that the art is uncertain, and that it is useless since the events decreed by the force of the stars are inevitable. Ptolemy opens his argument in favor of the art by assuming as evident that a certain force is diffused from the heavens over all things on earth. If ignorant sailors are able to judge the future weather from the sky, a highly trained astronomer should be able to predict concerning its influence on man. The art itself should not be rejected because impostors frequently abuse it, and Ptolemy admits that it has not yet been brought to the point of perfection and that even the skilful investigator often makes mistakes owing to the incomplete state of human science. For one thing, Ptolemy regards the doctrine of the nature of matter held in his time as hypothetical rather than certain. Another difficulty is that old configurations of the stars can-

[1] *Studien über Claudius Ptolemäus,* 1894.

not safely be used as the basis of present day predictions. Indeed, so manifold are the different possible positions of the stars and the different possible arrangements of terrestrial matter in relation to the stars that it is difficult to collect enough observations on which to base rules of general judgment. Moreover, such considerations as diversity of place, of custom, and of education must be taken into account in foretelling the future of different persons born under the same stars. But although for these reasons predictions frequently fail, yet the art is not to be condemned any more than one rejects the art of navigation because of frequent shipwrecks.

Influence of the stars not inevitable.

Nor it is true that the art is useless because the decrees of the stars are inevitable. It is often an advantage to have previous knowledge even of what cannot be avoided. Even the prediction of disaster serves to break the news gently. But not all predictions are inevitable and immutable; this is true only of the motion of the sky itself and events in which it is exclusively concerned. "But other events which do not arise solely from the sky's motion, are easily altered by application of opposite remedies," just as we can in part remedy the hurt of wounds and diseases or counteract the heat of summer by use of cooling things. The Egyptians have always found astrology useful in the practice of medicine.

Astrology as natural science.

Ptolemy next proceeds to set forth the natures and powers of the stars "according to the observations of the ancients and conformably to natural science." Later, when he comes to the prediction of particulars, he still professes "to follow everywhere the law of natural causation," and in a third passage he states that he "will omit all those things which do not have a probable natural cause, which many nevertheless scrutinize curiously and to excess: nor will I pile up divinations by lot-castings or from numbers, which are unscientific, but I will treat of those which have an investigated certainty based on the positions of the stars and the properties of places." Connecting the positions of the stars with earthly regions,—it is an art that fits in well with Ptolemy's other occupations of astronomer and geographer! The *Tetrabiblos* has been called "Science's surrender," [1] but was it not more truly divination purified and made scientific?

Properties of the planets.

Taking up first the properties of the seven planets, Ptolemy associates with each one or more of the four elemental qualities, hot, cold, dry, and moist. Thus the sun warms and to some extent dries, for the nearer it comes to our pole the more heat and drought it produces. The moon is moist, since it is close to the earth and is affected by the vapors from the latter, while its influence renders other bodies soft and causes putrefaction. But it also warms a little owing to the rays it receives from the sun. Saturn chills and to some extent dries, for it is remote from the sun's heat and earth's damp vapors. Mars emits a parching heat, as its color and proximity to the sun indicate. Jupiter, situated between cold Saturn and burning Mars, is of a rather lukewarm nature but tends more to warmth and moisture than to their opposites. So does Venus, but conversely, for it warms less than Jupiter does but moistens more, its large surface catching many vapors from the neighboring earth. In Mercury, situated near sun, moon, and earth alike, neither drought nor dampness predominates, but the velocity of that planet makes it a potent cause of sudden changes. In general, the planets exert a good or evil influence as they abound in the two rich and vivifying qualities, heat and moisture, or in the detrimental ones, cold and drought. Wet stars like the moon and Venus, are feminine; Mercury is neuter; the other planets are masculine. The sex of a planet may also, however, be reckoned according to its position in relation to the sun and the horizon; and changes in the influences exerted by the planets are noted according to their position or relation to the sun. This discussion of the properties of the planets is neither convinc-

[1] "C'était la capitulation de la science." Bouché-Leclercq in *Rev. Hist.*, LXV, 257, note 3.

ing nor scientific. It seems arguing in a circle to make their effects upon the earth depend to such an extent upon themselves being affected by vapors from the earth. Indeed we are rather surprised that an astronomer like Ptolemy should represent vapors from the earth as affecting the planets at all. But his discussion is at least an effort, albeit a feeble one, to express the potencies of the planets in physical terms.

Remaining contents of Book One.

Ptolemy goes on to discuss the powers of the fixed stars which seem to depend upon their positions in constellations and their relations to the planets. Then he treats of the influence of the four seasons of the year and four cardinal points, each of which he relates to one of the four qualities, hot, cold, dry, and moist. With a discussion of the signs of the zodiac and their division into Houses and relation in *Trigones* or *Triplicitates* or groups of three connected with the four qualities, of the exaltation of the planets in the signs and of other divisions of the signs and relations of the planets to them, the first book ends.

Book Two: Regions.

The second book begins by distinguishing prediction of events for whole regions or countries, such as wars, pestilences, famines, earthquakes, winds, drought, and weather, from the prediction of events in the lives of individuals. Ptolemy holds that events which affect large areas or whole peoples and cities are produced by greater and more valid causes than are the acts of individual men, and also that in order to predict aright concerning the individual it is necessary to know his region and nationality. He characterizes the inhabitants of the three great climatic zones,[1] quarters the inhabited world into Europe, Libya, and two parts for Asia in the style of the T maps, and subdivides these into different countries whose peoples are described, including such races as the Amazons. The effects of the stars vary according to time as well as place, so that the period in which any individual lives is as important to take into

[1] In the medieval Latin translation the Slavs replace the Scythians of Ptolemy's text.

account as his nationality. Ptolemy also discusses how the heavenly bodies influence the *genus* of events, a matter which depends largely upon the signs of the zodiac, and also how they determine their quality, good or bad, and species, which depends on the dominant stars and their conjunctions. Consequently he gives a list of the things which belong under the rule of each planet. The remainder of the second book is concerned chiefly with prediction of wind and weather through the year and with other meteorological phenomena such as comets.

Nativities.

The last two books take up the prediction of events in the lives of individuals from the stars, in other words the science of nativities or genethlialogy. The third book discusses conception and birth, how to take the horoscope—Ptolemy insists that the astrolabe is the only reliable instrument for determining the exact time; sun-dials or water-clocks will not do—and how to predict concerning parents, brothers and sisters, sex, twins, monstrous births, length of life, the physical constitution of the child born and what accidents and diseases may befall it, and finally concerning mental traits and defects. The fourth book deals less with the nature of the individual and more with the prediction of external events which befall the individual: honors, office, marriage, offspring, slaves, travel, and the sort of death that he will die. Ptolemy in opening the fourth book makes the distinction that, while in the third book he treated of matters antecedent to birth or immediately related to birth or which concern the temperament of the individual, now he will deal with those external to the body and which happen to the individual from without. But of course it is difficult to maintain such a distinction with entire consistency.

Future influence of the *Tetrabiblos*.

The great influence of the *Tetrabiblos* is shown not only in medieval Arabic commentaries and Latin translations, but more immediately in the astrological writings of the declining Roman Empire, when such astrologers as Hephaestion of Thebes,[1] Paul of Alexandria, and Julius Firmicus Maternus cite it as a leading authoritative work. Only the opponents of astrology appear to have remained ignorant of the *Tetrabiblos,* continuing to make criticisms of the art which do not apply to Ptolemy's presentation of it or which had been specifically answered by him. Thus Sextus Empiricus, attacking astrology about 200 A. D., does not mention the *Tetrabiblos* and some of the Christian critics of astrology apparently had not read it. Whether the Neo-Platonists, Porphyry and Proclus, wrote an introduction to and commentary upon it is disputed.

[1] Indeed, Hephaestion's first two books are nothing but Ptolemy repeated. About contemporary with Ptolemy seems to have been Vettius Valens whose astrological work is extant: Vettius Valens, *Anthologiarum libri primum edidit* Guilelmus Kroll, Berlin, 1908. See also CCAG *passim* concerning both Hephaestion and Vettius Valens, and Engelbrecht, *Hephästion von Theben und sein astrologisches Compendium,* Vienna, 1887.

CHAPTER IV

GALEN

I. *The Man and His Times*

Recent ignorance of Galen—His voluminous works—The manuscript tradition of his works—His vivid personality—Birth and parentage—Education in philosophy and medicine—First visit to Rome—Relations with the emperors; later life—His unfavorable picture of the learned world—Corruption of the medical profession—Lack of real search for truth—Poor doctors and medical students—Medical discovery in his time—The drug trade—The imperial stores—Galen's private supply of drugs—Mediterranean commerce—Frauds of dealers in wild beasts—Galen's ideal of anonymity—The ancient book trade—Falsification and mistakes in manuscripts—Galen as a historical source—Ancient slavery—Social life; food and wine—Allusions to Judaism and Christianity—Galen's monotheism—Christian readers of Galen.

II. *His Medicine and Experimental Science*

Four elements and four qualities—His criticism of atomism—Application of the theory of four qualities in medicine—His therapeutics obsolete—Some of his medical notions—Two of his cases—His power of rapid observation and inference—His happy guesses—Tendency toward scientific measurement—Psychological tests with the pulse—Galen's anatomy and physiology—Experiments in dissection—Did he ever dissect human bodies?—Dissection of animals—Surgical operations—Galen's argument from design—Queries concerning the soul—No supernatural force in medicine—Galen's experimental instinct—His attitude toward authorities—Adverse criticism of past writers—His estimate of Dioscorides—Galen's dogmatism; logic and experience—His account of the Empirics—How the Empirics might have criticized Galen—Galen's standard of reason and experience—Simples knowable only through experience—Experience and food science—Experience and compounds—Suggestions of experimental method—Difficulty of medical experiment—Empirical remedies—Galen's influence upon medieval experiment—His more general medieval influence.

III. *His Attitude Toward Magic*

Accusations of magic against Galen—His charges of magic against others—Charms and wonder-workers—Animal substances inadmissible in medicine—Nastiness of ancient medicine—Parts of animals—Some scepticism—Doctrine of occult virtue—Virtue of the flesh of vipers—Theriac—Magical compounds—Amulets—Incantations and characters—Belief in magic dies hard—*On Easily Procurable Remedies*—Specimens of its superstitious contents—External signs of the temperaments of internal organs—Marvelous statements repeated by Maimonides—Dreams—Absence of astrology in most of Galen's medicine—*The Prognostication of Disease by Astrology*—Critical days—*On the History of Philosophy*—Divination and demons—Celestial bodies.

ἀλλ' εἴ τις καταγνῷ μου τόδε, ὁμολογῶ τὸ πάθος τοὐμὸν ὃ παρ' ὅλον ἐμαυτοῦ τὸν βίον ἔπαθον, οὐδενὶ πιστεύσας τῶν διηγουμένων τὰ τοιαῦτα, πρὶν πειραθῆναι καὶ αὐτὸς ὧν δυνατὸν ἦν εἰς πεῖραν ἐλθεῖν με.

Kühn, IV, 513.

διὸ κἂν μετ' ἐμέ τις ὁμοίως ἐμοὶ φιλόπονός τε καὶ ζηλωτικὸς ἀληθείας γένηται, μὴ προπετῶς ἐκ δυοῖν ἢ τριῶν χρήσεων ἀποφαινέσθω. πολλάκις γὰρ αὐτῷ φανεῖται διὰ τῆς μακρᾶς πείρας ὥσπερ ἐφάνη κἀμοὶ . . .

Kühn, XIII, 96-1.

χρὴ γὰρ τὸν μέλλοντα γνώσεσθαί τι τῶν πολλῶν ἄμεινον εὐθὺς μὲν καὶ τῇ φύσει καὶ τῇ πρώτῃ διδασκαλίᾳ πολὺ τῶν ἄλλων διενεγκεῖν ἐπειδὰν δὲ γένηται μειράκιον ἀληθείας τινὸς ἔχειν ἐρωτικὴν μανίαν ὥσπερ ἐνθουσιῶντα, καὶ μήθ' ἡμέρας μήτε νυκτὸς διαλείπειν σπεύδοντα τε καὶ συντεταμένον ἐκμαθεῖν, ὅσα τοῖς ἐνδοξοτάτοις εἴρηται τῶν παλαιῶν· ἐπειδὰν δ' ἐκμάθῃ, κρίνειν αὐτὰ καὶ βασανίζειν χρόνῳ παμπόλλῳ καὶ σκοπεῖν πόσα μὲν ὁμολογεῖ τοῖς ἐναργῶς φαινομένοις πόσα δὲ διαφέρεται καὶ οὕτως τὰ μὲν αἱρεῖσθαι τὰ δ' ἀποστρέφεσθαι.

Kühn, II, 179.

"But if anyone charges me therewith, I confess my disease from which I have suffered all my life long, to trust none of those who make such statements until I have tested them for myself in so far as it has been possible for me to put them to the test."

"So if anyone after me becomes like me fond of work and zealous for truth, let him not conclude hastily from two or three cases. For often he will be enlightened through long experience, just as I have been." (It is remarkable that Ptolemy spoke similarly of his predecessor, Hipparchus, as a "lover of toil and truth"—φιλόπονον καὶ φιλαλήθεα, quoted by Orr (1913), 122.)

"For one who is to understand any matter better than most men do must straightway differ much from other persons in his nature and earliest education. And when he becomes a lad he must be madly in love with the truth and carried away by enthusiasm for it, and not let up by day or by night but press on and stretch every nerve to learn whatever the ancients of most repute have said. But having learned it, he must judge the same and put it to the test for a long, long time and observe what agrees with visible phenomena and what disagrees, and so accept the one and reject the other."

I. *The Man and His Times*

Recent ignorance of Galen.

At the close of the nineteenth century one English student of the history of medicine said, "Galen is so inaccessible to English readers that it is difficult to learn about him at first hand." [1] Another wrote, "There is, perhaps, no other instance of a man of equal intellectual rank who has been so persistently misunderstood and even misinterpreted." [2] A third obstacle to the ready comprehension of Galen has been that while more critical editions of some single works have been published by Helmreich and others in recent times,[3] no complete edition of his works has appeared since that of Kühn a century ago,[4] which is now regarded as very faulty.[5] A fourth reason for neglect or

[1] James Finlayson, *Galen: Two Bibliographical Demonstrations in the Library of the Faculty of Physicians and Surgeons of Glasgow*, 1895. Since then I believe that the only work of Galen to be translated into English is *On the Natural Faculties*, ed. A. J. Brock, 1916 (Loeb Library).

[2] J. F. Payne, *The Relation of Harvey to his Predecessors and especially to Galen:* Harveian Oration of 1896, in *The Lancet*, Oct. 24, 1896, p. 1136.

[3] In the Teubner texts: *Scriptora minora*, 1-3, ed. I. Marquardt, I. Mueller, G. Helmreich, 1884-1893; *De victu*, ed. Helmreich, 1898; *De temperamentis*, ed. Helmreich, 1904; *De usu partium*, ed. Helmreich, 1907, 1909.

In *Corpus Medicorum Graecorum*, V, 9, 1-2, 1914-1915, *The Hippocratic Commentaries*, ed. Mewaldt, Helmreich, Westenberger, Diels, Hieg.

[4] Carolus Gottlob Kühn, *Claudii Galeni Opera Omnia*, Leipzig, 1821-1833, 21 vols. My citations will be to this edition, unless otherwise specified. An older edition which is often cited is that of Renatus Charterius, Paris, 1679, 13 vols.

[5] The article on Galen in PW regards some of the treatises as printed in Kühn as almost unreadable.

misunderstanding of Galen is probably that there is so much by him to be read.

His voluminous works.

Athenaeus stated that Galen wrote more treatises than any other Greek, and although many are now lost, more particularly of his logical and philosophical writings, his collected extant works in Greek text and Latin translation fill some twenty volumes averaging a thousand pages each. When we add that often there are no chapter headings or other brief clues to the contents,[1] which must be ploughed through slowly and thoroughly, since some of the most valuable bits of information come in quite incidentally or by way of unlooked-for digression; that errors in the printed text, and the technical vocabulary with numerous words not found in most classical dictionaries increase the reader's difficulties;[2] and that little if any of the text possesses any present medical value, while much of it is dreary enough reading even for one animated by historical interest, especially if one has no technical knowledge of medicine and surgery:—when we consider all these deterrents, we are not surprised that Galen is little known. "Few physicians or even scholars in the present day," continues the English historian of medicine quoted above, "can claim to have read through this vast collection; I certainly least of all. I can only pretend to have touched the fringe, especially of the anatomical and physiological works."[3]

[1] Although Kühn's Index fills a volume, it is far from dependable.

[2] Liddell and Scott often fail to allude to germane passages in Galen's works, even when they include, with citation of some other author, the word he uses.

[3] Perhaps at this point a similarly candid confession by the present writer is in order. I have tried to do a little more than Dr. Payne in his modesty seems ready to admit of himself, and to look over carefully enough not to miss anything of importance those works which seemed at all likely to bear upon my particular interest, the history of science and magic. In consequence I have examined long stretches of text from which I have got nothing. For the most part, I thought it better not to take time to read the Hippocratic commentaries. At first I was inclined to depend upon others for Galen's treatises on anatomy and physiology, but finally I read most of them in order to learn at first hand of his argument from design and his attitude towards dissection. Further than this the reader can probably judge for himself from my citations as to the extent and depth of my reading. My first draft was completed before I discovered that Puschmann had made considerable use of Galen for

The manuscript tradition of Galen's works.

Although the works of Galen are so voluminous, they have reached us for the most part in comparatively late manuscripts,[1] and to some extent perhaps only in their medieval form. The extant manuscripts of the Greek text are mostly of the fifteenth century and represent the enthusiasm of humanists who hoped by reviving the study of Galen in the original to get something new and better out of him than the schoolmen had. In this expectation they seem to have been for the most part disappointed; the middle ages had already absorbed Galen too thoroughly. If it be true, as Dr. Payne contends,[2] that the chief original contributions to medical science of the Renaissance period were the work of men trained in Greek scholarship, this was because, when they failed to get any new ideas from the Greek texts, they turned to the more promising path of experimental research which both Galen and the middle ages had already advocated. The bulky medieval Latin translations [3] of Galen are older than most of the extant Greek texts; there are also versions in Arabic and Syriac.[4] For the last five books of the *Anatomical Exercises* the only extant text is an Arabic manuscript not yet published.[5]

medical conditions in the Roman Empire in his *History of Medical Education,* English translation, London, 1891, pp. 93-113. For the sake of a complete and well-rounded survey I have thought it best to retain those passages where I cover about the same ground. I have been unable to procure T. Meyer-Steineg, *Ein Tag im Leben des Galen,* Jena, 1913, 63 pp.

[1] For an account of the MSS see H. Diels, *Berl. Akad. Abh.* (1905), 58ff. Some fragments of Galen's work on medicinal simples exist in a fifth century MS of Dioscorides at Constantinople and have been reproduced by M. Wellmann in *Hermes,* XXXVIII (1903), 292ff. The first two books of his περὶ τῶν ἐν ταῖς τροφαῖς δυνάμεων were discovered in a Wolfenbüttel palimpsest of the fifth or sixth century by K. Koch; see *Berl. Akad. Sitzb.* (1907), 103ff.

[2] *Lancet* (1896), p. 1135.

[3] For these see V. Rose, *Analecta Graeca et Latina,* Berlin, 1864. As a specimen of these medieval Latin translations may be mentioned a collection of some twenty-six treatises in one huge volume which I have seen in the library of Balliol College, Oxford: Balliol 231, a large folio, early 14th century (a note of ownership was added in 1334 at Canterbury) fols. 437, double columned pages. For the titles and *incipits* of the individual treatises see Coxe (1852).

[4] A. Merx, "Proben der syrischen Uebersetzung von Galenus' Schrift über die einfachen Heilmittel," *Zeitsch. d. Deutsch. Morgendl. Gesell.* XXXIX (1885), 237-305.

[5] Payne, *Lancet* (1896), p. 1136.

Galen's vivid personality.

If so comparatively little is generally known about Galen, it is not because he had an unattractive personality. Nor is it difficult to make out the main events of his life. His works supply an unusual amount of personal information, and throughout his writings, unless he is merely transcribing past prescriptions, he talks like a living man, detailing incidents of daily life and making upon the reader a vivid and unaffected impression of reality. Daremberg asserts [1] that the exuberance of his imagination and his vanity frequently make us smile. It is true that his pharmacology and therapeutics often strike us as ridiculous, but he did not imagine them, they were the medicine of his age. It is true that he mentions cases which he has cured and those in which other physicians have been at fault, but official war despatches do the same with their own victories and the enemy's defeats. *Vae victis!* In Galen's case, at least, posterity long confirmed his own verdict. And dull or obsolete as his medicine now is, his scholarly and intellectual ideals and love of hard work at his art are still a living force, while the reader of his pages often feels himself carried back to the Roman world of the second century. Thus "the magic of literature," to quote a fine sentence by Payne, "brings together thinkers widely separated in space and time." [2]

Birth and parentage.

Galen—he does not seem to have been called Claudius until the time of the Renaissance—was born about 129 A.D.[3] at Pergamum in Asia Minor. His father, Nikon, was an architect and mathematician, trained in arithmetic, geometry, and astronomy. Much of this education he transmitted to his son, but even more valuable, in Galen's opinion, were his precepts to follow no one sect or party but to hear and judge them all, to despise honor and glory, and to magnify truth alone. To this teaching Galen attributes his own peaceful and painless passage through life. He has never

[1] Ch. V. Daremberg, *Exposition des connaissances de Galien sur l'anatomie, la physiologie, et la pathologie du système nerveux*, Paris, 1841.

[2] *Lancet* (1896), p. 1140.

[3] Brock (1916), p. xvi, says in 131 A.D. Clinton, *Fasti Romani*, placed it in 130.

grieved over losses of property but managed to get along somehow. He has not minded much when some have vituperated him, thinking instead of those who praise him. In later life Galen looked back with great affection upon his father and spoke of his own great good fortune in having as a parent that gentlest, justest, most honest and humane of men. On the other hand, the chief thing that he learned from his mother was to avoid her failings of a sharp temper and tongue, with which she made life miserable for their household slaves and scolded his father worse than Xanthippe ever did Socrates.[1]

Education in philosophy and medicine.

In one of his works Galen speaks of the passionate love and enthusiasm for truth which has possessed him since boyhood, so that he has not stopped either by day or by night from quest of it.[2] He realized that to become a true scholar required both high natural qualifications and a superior type of education from the start. After his fourteenth year he heard the lectures of various philosophers, Platonist and Peripatetic, Stoic and Epicurean; but when about seventeen, warned by a dream of his father,[3] he turned to the study of medicine. This incident of the dream shows that neither Galen nor his father, despite their education and intellectual standards, were free from the current belief in occult influences, of which we shall find many more instances in Galen's works. Galen first studied medicine for four years under Satyrus in his native city of Pergamum, then under Pelops at Smyrna, later under Numisianus at Corinth and Alexandria.[4] This was about the time that the great mathematician and astronomer, Ptolemy, was completing his observations [5] in the neighborhood of Alexandria, but Galen does not mention him, despite his own belief that a first-rate physician should also know such subjects as

[1] These details are from the *De cognoscendis curandisque animi morbis*, cap. 8, Kühn, V, 40-44.

[2] *De naturalibus facultatibus*, III, 10, Kühn, II, 179.

[3] Kühn, X, 609 (*De methodo medendi*); also XVI, 223; and XIX, 59.

[4] *De anatom. administ.*, Kühn, II, 217, 224-25, 660. See also XV, 136; XIX, 57.

[5] His recorded astronomical observations extend from 127 to 151 A.D.

geometry and astronomy, music and rhetoric.[1] Galen's interest in philosophy continued, however, and he wrote many logical and philosophical treatises, most of which are lost.[2] His father died when he was twenty, and it was after this that he went to other cities to study.

First visit to Rome.

Galen returned to Pergamum to practice and was, when but twenty-nine, made the doctor for the gladiators by five successive pontiffs.[3] During his thirties came his first residence at Rome.[4] The article on Galen in Pauly-Wissowa states that he was driven away from Rome by the plague, and in *De libris propriis* he does say that, "when the great plague broke out there, I hurriedly departed from the city for my native land." [5] But in *De prognosticatione ad Epigenem* his explanation is that he became disgusted with the malice of the envious physicians of the capital, and determined to return home as soon as the sedition there was over.[6] Meanwhile he stayed on and gained great fame by his cures but their jealousy and opposition multiplied, so that presently, when he learned that the sedition was over, he went back to Pergamum.

Relations with the emperors: later life.

His fame, however, had come to the imperial ears and he was soon summoned to Aquileia to meet the emperors on their way north against the invading Germans. An outbreak of the plague there prevented their proceeding with the campaign immediately,[7] and Galen states that the emperors fled for Rome with a few troops, leaving the rest to suffer from the plague and cold winter. On the way Lucius Verus died, and when Marcus Aurelius finally returned to the front, he allowed Galen to go back to Rome as court

[1] Kühn, X, 16.

[2] *Fragments du commentaire de Galien sur le Timée de Platon,* were published for the first time, both in Greek and a French translation, together with an *Essai sur Galien considéré comme philosophe,* by Ch. Daremberg, Paris, 1848.

[3] Kühn, XIII, 599-600.

[4] Clinton, *Fasti Romani,* I, 151 and 155, speaks of a first visit of Galen to Rome in 162 and a second in 164, but he has misinterpreted Galen's statements. When Galen speaks of his second visit to Rome, he means his return after the plague.

[5] Kühn, XIX, 15.

[6] Kühn, XIV, 622, 625, 648; see also I, 54-57, and XII, 263.

[7] Kühn, XIV, 649-50.

physician to Commodus.[1] The prevalence of the plague at this time is illustrated by a third encounter which Galen had with it in Asia, when he claims to have saved himself and others by thorough venesection.[2] The war lasted much longer than had been anticipated and meanwhile Galen was occupied chiefly in literary labors, completing a number of works. In 192 some of his writings and other treasures were lost in a fire which destroyed the Temple of Peace on the Sacred Way. Of some of the works which thus perished he had no other copy himself. In one of his works on compound medicines he explains that some persons may possess the first two books which had already been published, but that these had perished with others in a shop on the Sacra Via when the whole shrine of peace and the great libraries on the Palatine hill were consumed, and that his friends, none of whom possessed copies, had besought him to begin the work all over again.[3] Galen was still alive and writing during the early years of the dynasty of the Severi, and probably died about 200.

His unfavorable picture of the learned world.

Although the envy of other physicians at Rome and their accusing him of resort to magic arts and divination in his marvelous prognostications and cures were perhaps neither the sole nor the true reason for Galen's temporary withdrawal from the capital, there probably is a great deal of truth in the picture he paints of the medical profession and learned world of his day. There are too many other ancient witnesses, from the encyclopedist Pliny and the satirist Juvenal to the fourth century lawyer and astrologer, Firmicus, who substantiate his charges to permit us to explain them away as the product of personal bitterness or

[1] R. M. Briau, *L'Archiatrie Romaine,* Paris, 1877, however, held that Galen never received the official title, *archiater;* see p. 24, "il est difficile de comprendre pourquoi le médecin de Pergame qui donnait des soins à l'empereur Marc Aurèle, ne fut jamais honoré de ce titre." But he is given the title in at least one medieval MS—Merton 219, early 14th century, fol. 36*v*—"Incipit liber Galieni archistratos medicorum de malitia complexionis diversae."

[2] *De venae sectione,* Kühn, XIX, 524.

[3] Kühn, XIII, 362-63; for another allusion to this fire see XIV, 66. Also II, 216; XIX, 19 and 41.

pessimism. We feel that these men lived in an intellectual society where faction and villainy, superstition and petty-mindedness and personal enmity, were more manifest than in the quieter and, let us hope, more tolerant learned world of our time. Selfishness and pretense, personal likes and dislikes, undoubtedly still play their part, but there is not passionate animosity and open war to the knife on every hand. The *status belli* may still be characteristic of politics and the business world, but scholars seem able to live in substantial peace. Perhaps it is because there is less prospect of worldly gain for members of the learned professions than in Galen's day. Perhaps it is due to the growth of the impartial scientific spirit, of unwritten codes of courtesy and ethics within the leading learned professions, and of state laws concerning such matters as patents, copyright, professional degrees, pure food, and pure drugs. Perhaps, in the unsatisfactory relations between those who should have been the best educated and most enlightened men of that time we may see an important symptom of the intellectual and ethical decline of the ancient world.

Corruption of the medical profession.

Galen states that many tire of the long struggle with crafty and wicked men which they have tried to carry on, relying upon their erudition and honest toil alone, and withdraw disgusted from the madding crowd to save themselves in dignified retirement. He especially marvels at the evil-mindedness of physicians of reputation at Rome. Though they live in the city, they are a band of robbers as truly as the brigands of the mountains. He is inclined to account for the roguery of Roman physicians compared to those of a smaller city by the facts that elsewhere men are not so tempted by the magnitude of possible gain and that in a smaller town everyone is known by everyone else and questionable practices cannot escape general notice. The rich men of Rome fall easy prey to these unscrupulous practitioners who are ready to flatter them and play up to their weaknesses. These rich men can see the use of arithmetic and geometry, which enable them to keep their books straight and to build houses for their domestic comfort, and of divination and astrology, from which they seek to learn whose heirs they will be, but they have no appreciation of pure philosophy apart from rhetorical sophistry.[1]

Lack of real search for truth.

Galen more than once complains that there are no real seekers after truth in his time, but that all are intent upon money, political power, or pleasure. You know very well, he says to one of his friends in the *De methodo medendi,* that not five men of all those whom we have met prefer to be rather than to seem wise.[2] Many make a great outward display and pretense in medicine and other arts who have no real knowledge.[3] Galen several times expresses his scorn for those who spend their mornings in going about saluting their friends, and their evenings in drinking bouts or in dining with the rich and powerful. Yet even his friends have reproached him for studying too much and not going out more. But while they have wasted their hours thus, he has spent his, first in learning all that the ancients have discovered that is of value, then in testing and practicing the same.[4] Moreover, now-a-days many are trying to teach others what they have never accomplished themselves.[5] Thessalus not only toadied to the rich but secured many pupils by offering to teach them medicine in six months.[6] Hence it is that tailors and dyers and smiths are abandoning their arts to become physicians. Thessalus himself, Galen ungenerously taunts, was educated by a father who plucked wool badly in the women's apartments.[7] Indeed, Galen himself, by the violence of his invective and the occasional passionateness of his animosity in his controversies with other individuals or schools of medicine, illustrates that state of war in the intellectual world of his age to which we have adverted.

[1] For the statements of this paragraph see Kühn, XIV, 603-5, 620-23.

[2] Kühn, X, 114.

[3] Kühn, XIV, 599-600.

[4] Kühn, X, 1, 76.

[5] Kühn, X, 609.

[6] Kühn, X, 4-5.

[7] Kühn, X, 10.

Poor doctors and medical students.

We suggested the possibility that learning compared to other occupations was more remunerative in Galen's day than in our own, but there were poor physicians and medical students then, as well as those greedy for gain or who associated with the rich. Many doctors could not afford to use the rarer or stronger simples and limited themselves to easily procured, inexpensive, and homely medicaments.[1] Many of his fellow-students regarded as a counsel of perfection unattainable by them Galen's plan of hearing all the different medical sects and comparing their merits and testing their validity.[2] They said tearfully that this course was all very well for him with his acute genius and his wealthy father behind him, but that they lacked the money to pursue an advanced education, perhaps had already lost valuable time under unsatisfactory teachers, or felt that they did not possess the discrimination to select for themselves what was profitable from several conflicting schools.

Medical discovery in Galen's time.

Galen was, it has already been made apparent, an intellectual aristocrat, and possessed little patience with those stupid men who never learn anything for themselves, though they see a myriad cures worked before their eyes. But that, apart from his own work, the medical profession was not entirely stagnant in his time, he admits when he asserts that many things are known to-day which had not been discovered before, and when he mentions some curative methods recently invented at Rome.[3]

The drug trade.

Galen supplies considerable information concerning the drug trade in Rome itself and throughout the empire. He often complains of adulteration and fraud. The physician must know the medicinal simples and their properties himself and be able to detect adulterated medicines, or the merchants, perfumers, and *herbarii* will deceive him.[4] Galen refuses to reveal the methods employed in adulterating opobalsam, which he had investigated personally, lest the

[1] Kühn, XII, 909, 916, and in vol. XIV the entire treatise *De remediis parabilibus*.

[2] Kühn, X, 560.

[3] Kühn, X, 1010-11.

[4] Kühn, XIII, 571-72.

evil practice spread further.[1] At Rome at least there were dealers in unguents who corresponded roughly to our druggists. Galen says there is not an unguent-dealer in Rome who is unacquainted with herbs from Crete, but he asserts that there are equally good medicinal plants growing in the very suburbs of Rome of which they are totally ignorant, and he taxes even those who prepare drugs for the emperors with the same oversight. He tells how the herbs from Crete come wrapped in cartons with the name of the herb written on the outside and sometimes the further statement that it is *campestris*.[2] These Roman drug stores seem not to have kept open at night, for Galen in describing a case speaks of the impossibility of procuring the medicines needed at once because "the lamps were already lighted." [3]

The imperial stores.

The emperors kept a special store of drugs of their own and had botanists in Sicily, Crete, and Africa who supplied not only them with medicinal herbs, but also the city of Rome as well, Galen says. However, the emperors appear to have reserved a large supply of the finest and rarest simples for their own use. Galen mentions a large amount of Hymettus honey in the imperial stores—*ἐν ταῖς αὐτοκρατορικαῖς ἀποθήκαις*,[4] whence our word "apothecary." [5] He proves that cinnamon [6] loses its potency with time by his own ex-

[1] Kühn, XIV, 62, and see Puschmann, *History of Medical Education* (1891), p. 108.

[2] Kühn, XIV, 10, 30, 79; and see Puschmann (1891), 109-11, where there is bibliography of the subject.

[3] Kühn, X, 792.

[4] Kühn, XIV, 26.

[5] The meaning of the word "apothecary" is explained as follows in a fourteenth century manuscript at Chartres which is a miscellany of religious treatises with a bestiary and lapidary and bears the title, "Apothecarius moralis monasterii S. Petri Carnotensis."

"Apothecarius est, secundum Hugucium, qui nonnullas diversarum rerum species in apothecis suis aggregat. . . . Apothecarius dicitur is qui species aromaticas et res quacunque arti medicine et cirurgie necessarias habet penes se et venales exponit," fol. 3. "According to Hugutius an apothecary is one who collects samples of various commodities in his stores. An apothecary is called one who has at hand and exposes for sale aromatic species and all sorts of things needful in medicine and surgery."

[6] The nest of the fabled cinnamon bird was supposed to contain supplies of the spice, which Herodotus (III, 111) tells us the Arabian merchants procured by leaving heavy pieces of flesh for the birds to carry to their nests, which then broke down under the excessive weight. In Aristotle's *History of Animals* (IX, 13) the

perience as imperial physician. An assignment of the spice sent to Marcus Aurelius from the land of the barbarians (ἐκ τῆς βαρβάρου) was superior to what had stood stored in wooden jars from the reigns of Trajan, Hadrian, and Antoninus Pius. Commodus exhausted all the recent supply, and when Galen was forced to turn to what had been on hand in preparing an antidote for Severus, he found it much weaker than before, although not thirty years had elapsed. That cinnamon was a commodity little known to the populace is indicated by Galen's mentioning his loss in the fire of 192 of a few precious bits of bark he had stored away in a chest with other treasures.[1] He praises the Severi, however, for permitting others to use theriac, a noted medicine and antidote of which we shall have more to say presently. Thus, he says, not only have they as emperors received power from the gods, but in sharing their goods freely they are like the gods, who rejoice the more, the more people they save.[2]

Galen's private supply of drugs: *terra sigillata.*

Galen himself, and apparently other physicians, were not content to rely for medicines either upon the unguent-sellers or the bounty of the imperial stores. Galen stored away oil and fat and left them to age until he had enough to last for a hundred years, including some from his father's lifetime. He used some forty years old in one prescription.[3] He also traveled to many parts of the Roman Empire and procured rare drugs in the places where they were produced. Very interesting is his account of going out of his way in journeying back and forth between Rome and Pergamum in order to stop at Lemnos and procure a supply of the famous *terra sigillata,* a reddish clay stamped into pellets with the sacred seal of Diana.[4] On the way to Rome, instead of journeying on foot through Thrace and Macedonia, he took ship from the Troad to Thessalonica; but the vessel stopped

nests are shot down with arrows tipped with lead. For other allusions to the cinnamon bird in classical literature see D'Arcy W. Thompson, *A Glossary of Greek Birds,* Oxford, 1895, p. 82.

[1] Kühn, XIV, 64-66.

[2] *Ad Pisonem de theriaca, Kühn,* XIV, 217.

[3] Kühn, XIII, 704.

[4] Kühn, XII, 168-78.

in Lemnos at Myrine on the wrong side of the island, which Galen had not realized possessed more than one port, and the captain would not delay the voyage long enough to enable him to cross the island to the spot where the *terra sigillata* was to be found. Upon his return from Rome through Macedonia, however, he took pains to visit the right port, and for the benefit of future travelers gives careful instructions concerning the route to follow and the distances between stated points. He describes the solemn procedure by which the priestess from the neighboring city gathered the red earth from the hill where it was found, sacrificing no animals, but wheat and barley to the earth. He brought away with him some twenty thousand of the little discs or seals which were supposed to cure even lethal poisons and the bite of mad dogs. The inhabitants laughed, however, at the assertion which Galen had read in Dioscorides that the seals were made by mixing the blood of a goat with the earth. Berthelot, the historian of chemistry, believed that this earth was "an oxide of iron more or less hydrated and impure."[1] In another passage Galen advises his readers,

[1] M. Berthelot, "Sur les voyages de Galien et de Zosime dans l'Archipel et en Asie, et sur la matière médicale dans l'antiquité," in *Journal des Savants* (1895), pp. 382-7. The article is chiefly devoted to showing that an alchemistic treatise attributed to Zosimus copies Galen's account of his trips to Lemnos and Cyprus. Of such future copying of Galen we shall encounter many more instances.

As for the *terra sigillata,* C. J. S. Thompson, in a paper on "Terra Sigillata, a famous medicament of ancient times," published in the *Proceedings of the Seventeenth International Congress of Medical Sciences,* London, 1913, Section XXIII, pp. 433-44, tells of various medieval substitutes for the Lemnian earth from other places, and of the interesting religious ceremony, performed in the presence of the Turkish officials on only one day in the year by Greek monks who had replaced the priestess of Diana. Pierre Belon witnessed it on August 6th, 1533. By that time there were many varieties of the tablets, "because each lord of Lemnos had a distinct seal." When Tozer visited Lemnos in 1890 the ceremony was still performed annually on August sixth and must be completed before sunrise or the earth would lose its efficacy. Mohammedan khodjas now shared in the religious ceremony, sacrificing a lamb. But in the twentieth century the entire ceremony was abandoned. Through the early modern centuries the *terra sigillata* continued to be held in high esteem in western Europe also, and was included in pharmacopeias as late as 1833 and 1848. Thompson gives a chemical analysis of a sixteenth century tablet of the Lemnian earth and finds no evidence therein of its possessing any medicinal property.

if they are ever in Pamphylia, to lay in a good supply of the drug *carpesium*.[1] In the ninth book of his work on medicinal simples he tells of three strata of sory, chalcite, and misy, which he had seen in a mine in Cyprus thirty years before and from which he had brought away a supply, and of the surprising chemical change which the misy underwent in the course of these years.[2]

Mediterranean commerce.

Galen speaks of receiving other drugs from Great Syria, Palestine, Egypt, Cappadocia, Pontus, Macedonia, Gaul, Spain, and Mauretania, from the Celts, and even from India.[3] He names other places in Greece and Asia Minor than Mount Hymettus where good honey may be had, and states that much so-called Attic honey is really from the Cyclades, although it is brought to Athens and there sold or reshipped. Similarly, genuine Falernian wine is produced only in a small part of Italy, but other wines like it are prepared by those who are skilled in such knavery. As the best iris is that of Illyricum and the best asphalt is from Judea, so the best *petroselinon* is that of Macedonia, and merchants export it to almost the entire world just as they do Attic honey and Falernian wine. But the *petroselinon* crop of Epirus is sent to Thessalonica and there passed off for Macedonian. The best turpentine is that of Chios but a good variety may be obtained from Libya or Pontus. The manufacture of drugs has spread recently as well as the commerce in them. The

Agricola in the sixteenth century wrote in his work on mining (*De re metal.*, ed. Hoover, 1912, II, 31), "It is, however, very little to be wondered at that the hill in the Island of Lemnos was excavated, for the whole is of a reddish-yellow color which furnishes for the inhabitants that valuable clay so especially beneficial to mankind."

[1] Kühn, XIV, 72.

[2] Kühn, XII, 226-9. See the article of Berthelot just cited in a preceding note for an explanation of the three names and of Galen's experience. Mr. Hoover, in his translation of Agricola's work on metallurgy (1912), pp. 573-4, says, "It is desirable here to enquire into the nature of the substances given by all of the old mineralogists under the Latinized Greek terms, chalcitis, misy, sory, and melanteria." He cites Dioscorides (V, 75-77) and Pliny (NH, XXXIV, 29-31) on the subject, but not Galen. Yule (1903) I, 126, notes that Marco Polo's account of *Tutia* and *Spodium* "reads almost like a condensed translation of Galen's account of *Pompholyx* and *Spodos*."

[3] Kühn, XIV, 7-8; XIII, 411-2; XII, 215-6.

best form of unguent was formerly made only in Laodicea, but now it is similarly compounded in many other cities of Asia Minor.[1]

Frauds of dealers in wild beasts.

We are reminded that parts of animals as well as herbs and minerals were important constituents in ancient pharmacy by Galen's invective against the frauds of hunters and dealers in wild beasts as well as of unguent-sellers. They do not hunt them at the proper season for securing their medicinal virtues, but when they are no longer in their prime or just after their long period of hibernation, when they are emaciated. Then they fatten them upon improper food, feed them barley cakes to stuff up and dull their teeth, or force them to bite frequently so that virus will run out of their mouths.[2]

Galen's ideal of anonymity.

Besides the ancient drug trade, Galen gives us some interesting glimpses of the publishing trade, if we may so term it, of his time. Writing in old age in the *De methodo medendi,*[3] he says that he has never attached his name to one of his works, never written for the popular ear or for fame, but fired by zeal for science and truth, or at the urgent request of friends, or as a useful exercise for himself, or, as now, in order to forget his old age. Popular fame is only an impediment to those who desire to live tranquilly and enjoy the fruits of philosophy. He asks Eugenianus, whom he addresses in this passage, not to praise him immoderately before men, as he has been wont to do, and not to inscribe his name in his works. His friends nevertheless prevailed upon him to write two treatises listing his works,[4] and he also is free enough in many of his books in mentioning others which are essential to read before perusing the present volume.[5] Perhaps he felt differently at different times on the question of fame and anonymity. He also objected

[1] Kühn, XIV, 22-23, 77-78; XIII, 119.

[2] Kühn, XIV, 255-56. The beasts of course were also in demand for the arena.

[3] Kühn, X, 456-57, opening passage of the seventh book.

[4] *περὶ τῶν ἰδίων βιβλίων*, Kühn, XIX, 8ff.; and *περὶ τῆς τάξεως τῶν ἰδίων βιβλίων*, XIX, 49 ff.

[5] See, for instance, in the *De methodo medendi* itself, X, 895-96 and 955.

to those who read his works, not to learn anything from them, but only in order to calumniate them.[1]

The ancient book trade.

It was in a shop on the Sacra Via that most of the copies of some of Galen's works were stored when they, together with the great libraries upon the Palatine, were consumed in the fire of 192. But in another passage Galen states that the street of the Sandal-makers is where most of the book-stores in Rome are located.[2] There he saw some men disputing whether a certain treatise was his. It was duly inscribed *Galenus medicus* and one man, because the title was unfamiliar to him, bought it as a new work by Galen. But another man who was something of a philologer asked to see the introduction, and, after reading a few lines, declared that the book was not one of Galen's works. When Galen was still young, he wrote three commentaries on the throat and lungs for a fellow student who wished to have something to pass off as his own work upon his return home. This friend died, however, and the books got into circulation.[3] Galen also complains that notes of his lectures which he has not intended for publication have got abroad,[4] that his servants have stolen and published some of his manuscripts, and that others have been altered, corrupted, and mutilated by those into whose possession they have come, or have been passed off by them in other lands as their own productions.[5] On the other hand, some of his pupils keep his teachings to themselves and are unwilling to give others the benefit of them, so that if they should die suddenly, his doctrines would be lost.[6] But his own ideal has always been to share his knowledge freely with those who sought it, and if possible with all mankind. At least one of Galen's works was taken down from his dictation by short-hand writers, when, after his convincing demonstration by dissection concerning respiration and the voice, Boëthus asked him for commentaries on the subject and

[1] Kühn, XIV, 651: henceforth this text will generally be cited without name.
[2] XIX, 8.
[3] II, 217.
[4] XIX, 9.
[5] XIX, 41.
[6] II, 283.

sent for stenographers.[1] Although Galen in his travels often purchased and carried home with him large quantities of drugs, when he made his first trip to Rome he left all his books in Asia.[2]

Falsification and mistakes in manuscripts.

Galen dates the falsification of title pages and contents of books back to the time when kings Ptolemy of Egypt and Attalus of Pergamum were bidding against each other for volumes for their respective libraries.[3] Works were often interpolated then in order to make them larger and so bring a better price. Galen speaks more than once of the deplorable ease with which numbers, signs, and other abbreviations are altered in manuscripts.[4] A single stroke of the pen or slight erasure will completely change the meaning of a medical prescription. He thinks that such alterations are sometimes malicious and not mere mistakes. So common were they that Menecrates composed a medical work written out entirely in complete words and entitled *Autocrator Hologrammatos* because it was also dedicated to the emperor. Another writer, Damocrates, from whom Galen often quotes long passages, composed his book of medicaments in metrical form so that there might be no mistake made even in complete words.

Galen as a historical source.

Galen's works contain occasional historical information concerning many other matters than books and drugs. Clinton in his *Fasti Romani* made much use of Galen for the chronology of the period in which he lived. His allusions to several of the emperors with whom he had personal relations are valuable bits of source-material. Trajan was, of course, before his time, but he testifies to the great improvement of the roads in Italy which that emperor had effected.[5] Galen sheds a little light on the vexed question

[1] XIV, 630.
[2] XIX, 34.
[3] XV, 109.
[4] XIII, 995-96; XIV, 31-32.
[5] X, 633. Duruy refers to the passage in his *History of Rome* (ed. J. P. Mahaffy, Boston, 1886, V, i, 273), but says, "Extensive sanitary works were undertaken throughout all Italy, and the celebrated Galen, who was almost a contemporary, extols their happy effects upon the public health." But Galen does not have sanitary considerations especially in mind, since he mentions Trajan's road-building only by way of illustration, comparing his own systematic

of the population of the empire, if Pergamum is the place he refers to in his estimate of forty thousand citizens or one hundred and twenty thousand inhabitants, including women and slaves but perhaps not children.[1]

Ancient slavery.

Galen illustrates for us the evils of ancient slavery in an incident which he relates to show the inadvisability of giving way to one's passions, especially anger.[2] Returning from Rome, Galen fell in with a traveler from Gortyna in Crete. When they reached Corinth, the Cretan sent his baggage and slaves from Cenchrea[3] to Athens by boat, but himself with a hired vehicle and two slaves went by land with Galen through Megara, Eleusis, and Thriasa. On the way the Cretan became so angry at the two slaves that he hit them with his sheathed sword so hard that the sheath broke and they were badly wounded. Fearing that they would die, he then made off to escape the consequences of his act, leaving Galen to look after the wounded. But later he rejoined Galen in penitent mood and insisted that Galen administer a beating to him for his cruelty. Galen adds that he himself, like his father, had never struck a slave with his own hand and had reproved friends who had broken their slaves' teeth with blows of their fists. Others go farther and kick their slaves or gouge their eyes out. The emperor Hadrian in a moment of anger is said to have blinded a slave with a stylus which he had in his hand. He, too, was sorry afterwards and offered the slave money, but the latter refused it, telling the emperor that nothing could compensate him for the loss of an eye. In another passage Galen discusses how many slaves and "clothes" one really needs.[4]

treatment of medicine to the emperor's great work in repairing and improving the roads, straightening them by cut-offs that saved distance, but sometimes abandoning an old road that went straight over hills for an easier route that avoided them, filling in wet and marshy spots with stone or crossing them by causeways, bridging impassable rivers, and altering routes that led through places now deserted and beset by wild beasts so that they would pass through populous towns and more frequented areas. The passage thus bears witness to a shifting of population.

[1] V, 49.

[2] V, 17-19.

[3] Mentioned in *Acts*, xviii, 18, ". . . having shorn his head in Cenchrea: for he had a vow."

[4] V, 46-47.

Social life: food and wine.

Galen also depicts the easy-going, sociable, and pleasure-loving society of his time. Not only physicians but men generally begin the day with salutations and calls, then separate again, some to the market-place and law courts, others to watch the dancers or charioteers.[1] Others play at dice or pursue love affairs, or pass the hours at the baths or in eating and drinking or some other bodily pleasure. In the evening they all come together again at symposia which bear no resemblance to the intellectual feasts of Socrates and Plato but are mere drinking bouts. Galen had no objection, however, to the use of wine in moderation and mentions the varieties from different parts of the Mediterranean world which were especially noted for their medicinal properties.[2] He believed that drinking wine discreetly relieved the mind from all worry and melancholy and refreshed it. "For we use it every day." [3] He affirmed that taken in moderation wine aided digestion and the blood.[4] He classed wine with such boons to humanity as medicines, "a sober and decent mode of life," and "the study of literature and liberal disciplines." [5] Galen's treatise in three books on food values (*De alimentorum facultatibus*) supplies information concerning the ancient table and dietary science.

Allusions to Judaism and Christianity.

Galen's allusions to Judaism and Christianity are of considerable interest. He scarcely seems to have distinguished between them. In two passages in his treatise on differences in the pulse he makes incidental allusion to the followers of Moses and Christ, in both cases speaking of them rather lightly, not to say contemptuously. In criticizing Archigenes for using vague and unintelligible language and not giving a sufficient explanation of the point in question, Galen says that it is "as if one had come to a school of

[1] X, 3-4.

[2] X, 831-36; XIII, 513; XIV, 27-29, and 14-19 on the heating and storage of wine.

[3] IV, 777-79.

[4] Similarly Milward (1733), p. 102, wrote of Alexander of Tralles, "He has in most distempers a separate article concerning wine and I much doubt whether there be in all nature a more excellent medicine than this in the hands of a skillful and judicious practitioner."

[5] IV, 821.

Moses and Christ and had heard undemonstrated laws."[1] And in criticizing opposing sects for their obstinacy he remarks that it would be easier to win over the followers of Moses and Christ.[2] Later we shall speak more fully of a third passage in *De usu partium*[3] where Galen criticizes the Mosaic view of the relation of God to nature, representing it as the opposite extreme to the Epicurean doctrine of a purely mechanistic and materialistic universe. This suggests that Galen had read some of the Old Testament, but he might have learned from other sources of the Dead Sea and of salts of Sodom, of which he speaks in yet another context.[4] According to a thirteenth century Arabian biographer of Galen, he spoke more favorably of Christians in a lost commentary upon Plato's *Republic,* admiring their morals and admitting their miracles.[5] This last, as we shall see, is unlikely, since Galen believed in a supreme Being who worked only through natural law. "A confection of Ioachos, the martyr or metropolitan," and "A remedy for headache of the monk Barlama" occur in the third book of the *De remediis parabilibus* ascribed to Galen, but this third book is greatly interpolated or entirely spurious, citing Galen himself as well as Alexander of Tralles, the sixth century writer, and mentioning the Saracens. Wellmann regards it as composed between the seventh and eleventh centuries of our era.[6]

Galen's monotheism.

Like most thoughtful men of his time, Galen tended to believe in one supreme deity, but he appears to have derived

[1] Kühn, VIII, 579, *ὡς εἰς Μωϋσοῦ καὶ Χριστοῦ διατριβὴν ἀφιγμένος νόμων ἀναποδείκτων ἀκούῃ.*

[2] *Ibid.*, p. 657, *θᾶττον γὰρ ἄν τις τοὺς ἀπὸ Μωϋσοῦ καὶ Χριστοῦ μεταδιδάξειεν..*' I have been unable to find a passage in which, according to Moses Maimonides of the twelfth century in his *Aphorisms* from Galen, Galen said that the wealthy physicians and philosophers of his time were not prepared for discipline as were the followers of Moses and Christ. Perhaps it is a mistranslation of one of the above passages. Particula 24 (56), "medici et philosophi cum aere augmentati non sunt preparati ad disciplinam sicut parati fuerunt ad disciplinam moysis et christi socii predictorum. decimotercio megapulsus."

[3] Kühn, III, 905-7.

[4] Kühn, XI, 690-4; XII, 372-5.

[5] Finlayson (1895); pp. 8-9; Harnack, *Medicinisches aus der ältesten Kirchengeschichte,* Leipzig, 1892.

[6] Wellmann (1914), p. 16 note.

this conception from Greek rather than Hebraic sources. It was to philosophy and the Greek mysteries that he turned for revelation of the deity, as we shall see. Hopeless criminals were for him those whom neither the Muses nor Socrates could reform.[1] It is Plato, not Christ, whom in another treatise he cites as describing the first and greatest God as ungenerated and good. "And we all naturally love Him, being such as He is from eternity." [2]

Galen's Christian readers.

But while Galen's monotheism cannot be regarded as of Christian or Jewish origin, it is possible that his argument from design and supporting theology by anatomy made him more acceptable to both Mohammedan and Christian readers. At any rate he had Christian readers at Rome at the opening of the third century, when a hostile controversialist complains that some of them even worship Galen.[3] These early Christian enthusiasts for natural science, who also devoted much time to Aristotle and Euclid, were finally excommunicated; but Aristotle, Euclid, and Galen were to return in triumph in medieval learning.

II. *His Medicine and Experimental Science*

Four elements and four qualities.

Galen held as his fundamental theory of nature the view which was to prevail through the middle ages, that all natural objects upon this globe are composed of four elements, earth, air, fire, and water,[4] and the cognate view, which he says Hippocrates first introduced and Aristotle later demonstrated, that all natural objects are characterized by four qualities, hot, cold, dry, and moist. From the combinations of these four are produced various secondary qualities.[5] Neither hypothesis was as yet universally accepted, however, and Galen felt it incumbent upon him to argue against those

[1] Kühn, IV, 816.

[2] Kühn, IV, 815.

[3] Quoted by Eusebius, V, 28, and reproduced by Harnack, *Medicinisches aus der ältesten Kirchengeschichte*, 1892, p. 41, and by Finlayson (1895), pp. 9-10.

[4] Kühn, X, 16-17. J. Leminne, *Les quatre éléments*, in *Mémoires couronnés par l'Académie de Belgique*, vol. 65, Brussels, 1903, traces the influence of the theory in medieval thought.

[5] Kuhn, XIII, 763-4.

who contended that the human body and world of nature were made from but one element.[1] There were others who ridiculed the four quality hypothesis, saying that hot and cold were words for bath-keepers, not for physicians to deal with.[2] Galen explains that philosophers do not regard any particular variety of earth or any other mineral substance as representing the pure element earth, which in the philosophical sense is an extremely cold and dry substance to which adamant and rocks make perhaps the closest approach. But the earths that we see are all compound bodies.[3]

Criticism of atomism.

Galen rejected the atomism of Democritus and Epicurus, in which the atoms were indivisible particles differing in shape and size, but not differing in quality as chemical atoms are supposed to do. He credits Democritus with the view that such qualities as color and taste are sensed by us from the concourse of atoms, but do not reside in the atoms themselves.[4] Galen also makes the criticism that the mere regrouping of "impassive and immutable" atoms is not enough to account for the new properties of the compound, which are often very different from those of the constituents, as when "we alter the qualities of medicines in artificial mixtures." [5] Thus he virtually says that the purely physical atomism of Democritus will not account for what today we call chemical change. He also, as we shall see, rejected Epicurus' theory of a world of nature ruled by blind chance.

Application of the theory of four qualities in medicine.

Galen of course thought that a dry medicine was good for a moist disease, and that in a compound medicine, by mixing a very cold with a slightly cold drug in varying proportions a medicine of any desired degree of coldness might be obtained.[6] In general he regarded solids like stones and metals as dry and cold, while he thought that hot and moist objects tended to evaporate rapidly into air.[7] So he declared that dryness of solid bodies was incurable, while he believed that children's bodies were more easily dissolved

[1] Kühn, I, 428.
[2] Kühn, X, 111.
[3] Kühn, XII, 166.
[4] I, 417.
[5] XIV, 250-53.
[6] XIII, 948.
[7] X, 657.

than adults' because moister and warmer.[1] The Stoics and many physicians believed that heat prolonged life, but Asclepiades pointed out that the Ethiopians are old at thirty because the hot sun dries up their bodies so, while the inhabitants of Britain sometimes live to be one hundred and twenty years old. This last, however, was regarded as probably due to the fact that their thicker skins conserved their innate heat longer.[2]

Galen's therapeutics obsolete.

As an offset to the evidence which will be presented later of the traces of occult virtues, magic, and astrology in Galen's therapeutics I should like to be able to indicate the good points in it. But his entire system, like the four quality theory upon which it is largely based, seems now obsolete, and what evidenced his superiority to other physicians in his own day would probably strike the modern reader only as a token of his distinct inferiority to present practice. Eighty odd years of modern medical progress since have added further emphasis to Daremberg's declaration that we have had to throw overboard "much of his physiology, nearly all of his pathology and general therapeutics." [3]

Some of his medical notions.

Nevertheless, we may note a few specimens which perhaps represent his ordinary theory and practice as distinguished from passages in which the influence of magic enters. He holds that bleeding and cold drink are the two chief remedies for fever.[4] He notes that children occasionally resemble their grandparents rather than their parents.[5] He disputes the assertion of Epicurus—one by which some of his followers failed to be guided—that there is no benefit to health in Aphrodite, and contends that at certain intervals and in certain individuals and circumstances sexual intercourse is beneficial.[6] His discussion of anodynes and stupor or sleep-producing medicines shows that the ancients had anaesthetics of a sort.[7] He recognized the importance

[1] X, 872.

[2] XIX, 344-45.

[3] More recently Galen's *Materia medica* has been treated of in a German doctoral dissertation by L. Israelson, *Die materia medica des Klaudios Galenos*, 1894, 204 pp.

[4] X, 624.

[5] XIV, 253-54.

[6] V, 911.

[7] X, 817-19.

of breathing plenty of fresh, invigorating, and unpolluted air, free from any intermixture of impurity from mines, pits, or ovens, or of putridity from decaying vegetable or animal matter, or of noxious vapors from stagnant water, swamps, and rivers.[1] As was usual in ancient and medieval times, he attributes plagues to the corruption of the air, which poisons men breathing it, and tells how Hippocrates tried to allay a plague at Athens by purifying the air by fumigation with fires, odors, and unguents.[2]

Two of Galen's cases.

Two specimens may be given of Galen's accounts of his own cases. In the first, some cheese, which he had told his servants to take away as too sharp, when mixed with boiled salt pork and applied to the joints, proved very helpful to a gouty patient and to several others whom he induced to try it.[3] In the second case Galen administered the following heroic treatment to a woman at Rome who was afflicted with catarrh to the point of throwing up blood.[4] He did not deem it wise to bleed her, since for four days past she had gone almost without food. Instead he ordered a sharp clyster, rubbed and bound her hands and feet with a hot drug, shaved her head and put on it a medicament made of doves' dung. After three hours she was bathed, care being taken that nothing oily touched her head, which was then covered up. At first he fed her only gruel, afterwards some bitter autumn fruit, and as she was about to go to sleep he administered a medicament made from vipers four months before. On the second day came more rubbing and binding except the head, and at evening a somewhat smaller dose of the viper remedy. Again she slept well and in the morning he gave her a large dose of cooked honey. Again her body was well rubbed and she was given barley water and a little bread to eat. On the fourth day an older and therefore stronger variety of viper-remedy was administered and her head was covered with the same medicament as before. Its properties, Galen explains, are vehemently drying and heat-

[1] X, 843.
[2] XIV, 281.
[3] XII, 270-71.
[4] X, 368-71.

ing. Again she was given a bath and a little food. On the fifth day Galen ventured to purge her lungs, but he returned at intervals to the imposition upon her head. Meanwhile he continued the process of rubbing, bathing, and dieting, until finally the patient was well again,—a truly remarkable cure!

His power of rapid observation and inference.

These two cases, however, do not give us a just comprehension of Galen's abilities at their best. In his medical practice he could be as quick and comprehensive an observer and as shrewd in drawing inferences from what he observed as the famous Sherlock Holmes, so that some of his slower-witted contemporaries accused him of possessing the gift of divination. His immediate diagnosis of the case of the Sicilian physician by noting as he entered the house the excrements in a vessel which a servant was carrying out to the dungheap, and as he entered the sick-room a medicine set on the window-sill which the patient-physician had been preparing for himself, amazed the patient and the philosopher Glaucon[1] more than, let us hope in this case in view of his profession, they would have amazed the estimable Dr. Watson.

His happy guesses.

Puschmann has pointed out that Galen employs certain expressions which seem happy guesses at later discoveries. He writes: "Galen was supported in his researches by an extremely happy imaginative faculty which put the proper word in his mouth even in cases where he could not possibly arrive at a full understanding of the matter,—where he could only conjecture the truth. When, for instance, he declares that sound is carried 'like a wave' (Kühn, III, 644), or expresses the conjecture that the constituent of the atmosphere which is important for breathing also acts by burning (IV, 687), he expresses thoughts which startle us, for it was only possible nearly two thousand years later to understand their full significance."[2]

[1] Kühn, VIII, 363. Finlayson (1895), pp. 39-40, gives an English translation of Galen's full account of the case.

[2] Puschmann (1891), pp. 105-6. Vitruvius, too, however (V, iii), states that sound spreads in waves like eddies in a pond.

Tendency towards scientific measurement.

Galen was keenly alive to the need of exactness in weights and measurements. He often criticizes past writers for not stating precisely what ailment the medicament recommended is good for, and in what proportions the ingredients are to be mixed. He also frequently complains because they do not specify whether they are using the Greek or Roman system of weights, or the Attic, Alexandrine, or Ephesian variety of a certain measure.[1] Moreover, he saw the desirability of more accurate means of measuring the passage of time.[2] When he states that even some illustrious physicians of his acquaintance mistake the speed of the pulse and are unable to tell whether it is slow, fast, or normal, we begin to realize something of the difficulties under which medical practice and any sort of experimentation labored before watches were invented, and how much depended upon the accuracy of human machinery and judgment. Yet Galen estimates that the chief progress made in medical prognostication since Hippocrates is the gradual development of the art of inferring from the pulse.[3] Galen tried to improve the time-pieces in use in his age. He states that in any city the inhabitants want to know the time of day accurately, not merely conjecturally; and he gives directions how to divide the day into twelve hours by a combination of a sun-dial and a *clepsydra,* and how on the water clock to mark the duration of the longest, shortest, and equinoctial days of the year.[4]

Psychological tests with the pulse.

Delicate and difficult as was the task of measuring the pulse in Galen's time, he was clever enough to anticipate by seventeen centuries some of the tests which modern psychologists have urged should be applied in criminal trials. He detected the fact that a female patient was not ill but in love by the quickening of her pulse when someone came in from the theater and announced that he had just seen Py-

[1] XIII, 435, 893, are two instances.

[2] V, 80; XIV, 670.

[3] Various treatises on the pulse by Galen will be found in vols. V, IX, and X of Kühn's edition.

[4] Galen's contributions to the arts of clock-making and timekeeping have been dealt with in an article which I have not had access to and of which I cannot now find even the author and title.

lades dance. When she came again the next day, Galen had purposely arranged that someone should enter and say that he had seen Morphus dancing. This and a similar test on the third day produced no perceptible quickening in the woman's pulse. But it bounded again when on the fourth day Pylades' name was again spoken. After recounting another analogous incident where he had been able to read the patient's mind, Galen asks why former physicians have never availed themselves of these methods. He thinks that they must have had no conception of how the bodily health in general and the pulse in particular can be affected by the "psyche's" suffering.[1] We might then call Galen the first experimental psychologist as well as the first to elaborate the physiology of the nervous system.

Galen's anatomy and physiology.

It would scarcely be fair to discuss Galen's science at all without saying something of his remarkable work in anatomy and physiology. Daremberg went so far as to hold that all there is good or bad in his writings comes from good or bad physiology, and regarded his discussion of the bones and muscles as especially good.[2] He is generally considered the greatest anatomist of antiquity, but it is barely possible that he may have owed more to predecessors and contemporaries and less to personal research than is apparent from his own writings, which are the most complete anatomical treatises that have reached us from antiquity. Herophilus, for example, who was born at Chalcedon in the closing fourth century B. C. and flourished at Alexandria under the first Ptolemy, discovered the nerves and distinguished them from the sinews, and thought the brain the center of the nervous system, so that it is perhaps questionable whether Payne is justified in calling Galen "the founder of the physiology of the nervous system," and in declaring that

[1] XIV, 631-34.

[2] C. V. Daremberg, *Exposition des connaissances de Galien sur l'anatomie, la physiologie, et la pathologie du système nerveux*, Paris, 1841. J. S. Milne discussed "Galen's Knowledge of Muscular Anatomy" at the International Congress of Medical Sciences held at London in 1913; see pp. 389-400 of the volume devoted to the history of medicine, Section XXIII.

"in physiological diagnosis he stands alone among the ancients."[1] However, if Galen owed something to Herophilus, we owe much of our knowledge of the earlier physiologist to Galen.[2]

Experiments in dissection.

Aristotle had held that the heart was the seat of the sensitive soul[3] and the source of nervous action, "while the brain was of secondary importance, being the coldest part of the body, devoid of blood, and having for its chief or only function to cool the heart." Galen attacked this theory by showing experimentally that "all the nerves originated in the brain, either directly or by means of the spinal cord, which he thought to be a conducting organ merely, not a center." "A thousand times," he says, "I have demonstrated by dissection that the cords in the heart called nerves by Aristotle are not nerves and have no connection with nerves." He found that sensation and movement were stopped and even the voice and breathing were affected by injuries to the brain, and that an injury to one side of the brain affected the opposite side of the body. His public demonstration by dissection, performed in the presence of various philosophers and medical men, of the connection between the brain and voice and respiration and the commentaries which he immediately afterwards dictated on this point were so convincing, he tells us fifteen years later, that no one has ventured openly to dispute them.[4] His "experimental investigation of the spinal cord by sections at different levels and by half sections was still more remarkable."[5] Galen opposed these experimental proofs to such unscientific arguments on the part of the Stoic philosopher, Chrysippus, and others, as that the heart must be the chief organ because it is in the center of the body, or because one lays

[1] *Lancet* (1896), p. 1139.

[2] I have failed to obtain K. F. H. Mark, *Herophilus, ein Beitrag zur Geschichte der Medizin*, Carlsruhe, 1838.

[3] D'Arcy W. Thompson (1913), 22-23, thinks that the precedence of the heart over all other organs in appearing in the embryo of the chick led Aristotle to locate in it the central seat of the soul.

[4] XIV, 626-30.

[5] II, 683, 696. This and the other quotations in this paragraph are from Dr. Payne's Harveian Oration as printed in *The Lancet* (1896), pp. 1137-39.

one's hand on one's heart to indicate oneself, or because the lips are moved in a certain way in saying "I"(ἐγώ).[1] Another noteworthy experiment by Galen was that in which, by binding up a section of the femoral artery he proved that the arteries contain blood and not air or *spiritus* as had been generally supposed.[2] He failed, however, to perform any experiments with the pulmonary veins, and so the notion persisted that these conveyed "spirit" and not blood from the lungs to the heart.[3]

Did Galen ever dissect human bodies?

It has usually been stated that Galen never dissected the human body and that his inferences by analogy from his dissection of animals involved him in serious error concerning human anatomy and physiology. Certainly he speaks as if opportunities to secure human cadavers or even skeletons were rare.[4] He mentions, however, the possibility of obtaining the bodies of criminals condemned to death or cast to beasts in the arena, or the corpses of robbers which lie unburied in the mountains, or the bodies of infants exposed by their parents.[5] It is not sufficient, he states in another passage,[6] to read books about human bones; one should have them before one's eyes. Alexandria is the best place for the student to go to see actual exhibitions of this sort made by the teachers.[7] But even if one cannot go there, one may be able to procure human bones for oneself, as Galen did from a skeleton which had

[1] Kühn, V, 216, cited by Payne.

[2] Kühn, II, 642-49; IV, 703-36, "An in arteriis natura sanguis contineatur." J. Kidd, *A Cursory Analysis of the Works of Galen so far as they relate to Anatomy and Physiology,* in *Transactions of the Provincial Medical and Surgical Association,* VI (1837), 299-336.

[3] *Lancet* (1896), p. 1137, where Payne states that Colombo (*De re anatomica,* Venet. 1559, XIV, 261) was the first to prove by experiment on the living heart that these veins conveyed blood from the lungs.

[4] II, 146-47.

[5] II, 384-86.

[6] II, 220-21.

[7] Augustine testifies in two passages of his *De anima et eius origine* (Migne PL 44, 475-548), that vivisection of human beings was practiced as late as his time, the early fifth century: IV, 3, "Medici tamen qui appellantur anatomici per membra per venas per nervos per ossa per medullas per interiora vitalia etiam vivos homines quamdiu inter manus rimantium vivere potuerunt dissiciendo scrutati sunt ut naturam corporis nossent"; and IV, 6 (Migne, PL 44, 528-9).

been washed out of a grave by a flooded stream and from the corpse of a robber slain in the mountains. If one cannot get to see a human skeleton by these means or some other, he should dissect monkeys and apes.

Dissection of animals.

Indeed Galen advises the student to dissect apes in any case, in order to prepare himself for intelligent dissection of the human body, should he ever have the opportunity. From lack of such previous experience the doctors with the army of Marcus Aurelius, who dissected the body of a dead German, learned nothing except the position of the entrails. Galen at any rate dissected a great many animals. Tiny animals and insects he let alone, for the microscope was not yet discovered, but besides apes and quadrupeds he cut up many reptiles, mice, weasels, birds, and fish.[1] He also gives an amusing account of the medical men at Rome gathering to observe the dissection of an elephant in order to discover whether the heart had one or two vertices and two or three ventricles. Galen assured them beforehand that it would be found similar to the heart of any other breathing animal. This particular dissection was not, however, performed exclusively in the interests of science, since it was scarcely accomplished when the heart was carried off, not to a scientific museum, but by the imperial cooks to their master's table.[2] Galen sometimes dissected animals the moment he killed them. Thus he observed that the lungs always sensibly shrank from the diaphragm in a dying animal, whether he killed it by suffocation in water, or strangling with a noose, or severing the spinal medulla near the first vertebrae, or cutting the large arteries or veins.[3]

Surgical operations.

Surgical operations and medical practice were a third way of learning the human anatomy, and Galen complains of the carelessness of those physicians and surgeons who do not take pains to observe it before performing an operation or cure. He himself had had one case where the

[1] II, 537. [2] II, 619-20. [3] II, 701.

human heart was laid bare and yet the patient recovered.[1] As a young practitioner before he came to Rome Galen worked out so successful a method of treating wounds of the sinews that the care of the health of the gladiators in his native city of Pergamum was entrusted to him by several successive pontifices [2] and he hardly lost a life. In the same passage he again speaks contemptuously of the doctors in the war with the Germans who were allowed to cut open the bodies of the barbarians but learned no more thereby than a cook would. When Galen came from Pergamum to Rome he found the professions of physicians and surgeons distinct and left cases to the latter which he before had attended to himself.[3] We may note finally that he invented a new form of surgical knife.[4]

Galen's argument from design.

In Galen's opinion the study of anatomy was important for the philosopher as well as for the physician. An understanding of the use of the parts of the body is helpful to the doctor, he says, but much more so to "the philosopher of medicine who strives to obtain knowledge of all nature." [5] In the *De usu partium* [6] he came to the conclusion that in the structure of any animal we have the mark of a wise workman or demiurge, and of a celestial mind; and that "the investigation of the use of the parts of the body lays the foundation of a truly scientific theology which is much greater and more precious than all medicine," and which reveals the divinity more clearly than even the Eleusinian mysteries or Samothracian orgies. Thus Galen adopts the argument from design for the existence of God. The modern doctrine of evolution is of course subversive of his premise that the parts of the body are so well constructed for and marvelously adapted to their functions that nothing better is possible, and consequently of his conclusion that this necessitates a divine maker and planner.

[1] II, 631ff.

[2] XIII, 599-600. Galen states that the pontifex's term of office was seven months, a fact which perhaps had some astrological bearing.

[3] X, 454-55.

[4] II, 682.

[5] II, 291.

[6] IV, 360, *et passim.*

In the treatise *De foetuum formatione* Galen displays a similar inclination but more tentatively and timidly. He thinks that the human body attests the wisdom and power of its maker,[1] whom he wishes the philosophers would reveal to him more clearly and tell him "whether he is some wise and powerful god."[2] The process of the formation of the child in the womb, the complex human muscular system, the human tongue alone, seem to him so wonderful that he will not subscribe to the Epicurean denial of any all-ruling providence.[3] He thinks that nature alone cannot show such wisdom. He has, however, sought vainly from philosopher after philosopher for a satisfactory demonstration of the existence of God, and is by no means certain himself.[4]

Queries concerning the soul.

Galen is also at a loss concerning the existence and substance of the soul. He points out that puppies try to bite before their teeth come and that calves try to hook before their horns grow, as if the soul knew the use of these parts beforehand. It might be argued that the soul itself causes the parts to grow,[5] but Galen questions this, nor is he ready to accept the Platonic world-soul theory of a divine force permeating all nature.[6] It offends his instinctive piety and sense of fitness to think of the world-soul in such things as reptiles, vermin, and putrefying corpses. On the other hand, he disagrees with those who deny any innate knowledge or standards to the soul and attribute everything to sense perception and certain imaginations and memories based thereon. Some even deny the existence of the reasoning faculty, he says, and affirm that we are led by the affections of the senses like cattle. For these men courage, prudence, temperance, continence are mere names.[7]

No supernatural force in medicine.

In commenting upon the works of Hippocrates, Galen insists that in speaking of "something divine" in diseases

[1] IV, 687.
[2] IV, 694, 696.
[3] IV, 688.
[4] IV, 700.
[5] IV, 692; II, 537. Others contend, he says (IV, 693), that one soul constructs the parts and another soul incites them to voluntary motion.
[6] IV, 701.
[7] II, 28.

Hippocrates could not have meant supernatural influence, which he never admits into medicine in other passages. Galen tries to explain away the expression as having reference to the effect of the surrounding air.[1] Thus while Galen might look upon nature or certain things in nature as a divine work, he would not admit any supernatural force in science or medicine, or anything bordering upon special providence. In the *De usu partium* Galen states that he agrees with Moses that "the beginning of genesis in all things generated" was "from the demiurge," but that he does not agree with him that anything is possible with God and that God can suddenly turn a stone into a man or make a horse or cow from ashes. "In this matter our opinion and that of Plato and of others among the Greeks who have written correctly concerning natural science differs from the view of Moses." In Galen's view God attempts nothing contrary to nature but of all possible natural courses invariably chooses the best. Thus Galen expresses his admiration at nature's providence in keeping the eyebrows and eyelashes of the same length and not letting them grow long like the beard or hair, but this is because a harder cartilaginous flesh is provided for them to grow in, and the mere will of God would not keep hairs from growing in soft flesh. If God had not provided the cartilaginous substance for the eyelashes, "he would have been more careless, not merely than Moses but than a worthless general who builds a wall in a swamp."[2] As between the views on God of Moses and Epicurus, Galen prefers to steer a middle course.

Galen's experimental instinct

Already in describing Galen's dissections and tests with the pulse we have seen evidence of the accurate observation and experimental instincts which accompanied his zest for hard work and zeal for truth. In one of his treatises he

[1] XVIII B, 17ff.

[2] *De usu partium,* XI, 14 (Kühn, III, 905-7). The passage seems to me an integral part of the work and not a later interpolation. Moses Maimonides in the twelfth century took exception at some length, in the 25th *Particula* of his *Aphorisms* from Galen, to this criticism of his national lawgiver.

confesses that it was a passion of his always to test everything for himself. "And if anyone accuses me of this, I will confess my disease, from which I have suffered all my life long, that I have trusted no one of those who narrate such things until I have tested it myself, if it was possible for me to have experience of it." [1] Galen also recognized that general theories were not sufficient for exact knowledge and that specific examples seen with one's own eyes were indispensable.[2] He maintains that, if all teachers and writers would realize and observe this, they would make comparatively few false statements. He saw the danger of making absolute assertions and the need of noting the particular circumstances of each individual case.[3] Galen more than once declared that things, not names, were important and refused to waste time in disputing about terminology and definitions which might be spent in "pursuing the knowledge of things themselves." [4] Thus we see in Galen a pragmatic scientist intent upon concrete facts and exact knowledge; but at the same time it must be recognized that he accepted some universal theorems and general views.

Attitude towards authorities.

Galen did not believe in merely repeating in new books the statements of previous authorities. Ever since boyhood, he writes in his *Anatomical Administrations,* it has seemed to him that one should record in writing only one's new discoveries and not repeat what has been said already.[5] Nevertheless in some of his writings he collects the prescriptions of past physicians at great length, and a previous treatise by Archigenes is practically embodied in one of Galen's works on compound medicines. On another occasion, however, after stating that Crito had combined previous treatises upon cosmetics, including the work of Cleopatra, into four books of his own which constitute a well-nigh exhaustive treatment of the subject, Galen says that

[1] IV, 513; see also II, 55, ὡς ἔγωγε πρῶτον μὲν ἀκούσας τὸ γινόμενον, ἐθαύμασα καὶ αὐτὸς ἐβουλήθην αὐτόπτης αὐτοῦ καταστῆναι.

[2] X, 608; XIII, 887-88.

[3] XIII, 964.

[4] II, 136; X, 385; XII, 311; he credited Plato with the same attitude, see II, 581.

[5] II, 659-60.

he sees no profit in copying Crito's work again and merely reproduces its table of contents.[1] On the other hand, as this passage shows, Galen thought that the ancients had stated many things admirably and he had little patience with contemporaries who would learn nothing from them but were always ambitiously weaving new and complicated dogmas, or misinterpreting and perverting the teachings of the ancients.[2] His method was rather first to "make haste and stretch every nerve to learn what the most celebrated of the ancients have said;" [3] then, having mastered this teaching, to judge it and put it to the test for a long time and determine by observation how much of it agrees and how much disagrees with actual phenomena, and then embrace the former portion and reject the latter.

Adverse criticism of past writers.

This critical employment of past authorities is frequently illustrated in Galen's works. He mentions a great many names of past physicians and writers, thereby shedding some light upon the history of Greek medicine; but at times he criticizes his predecessors, not sparing even Empedocles and Aristotle. Although he cites Aristotle a great deal, he declares that it is not surprising that Aristotle made many errors in the anatomy of animals, since he thought that the heart in large animals had a third ventricle.[4] As we have already seen in discussing the topic of weights and measurements, Galen especially objects to the vagueness and inaccuracy of many past medical writers,[5] or praises individuals like Heras who give specific information.[6] He also shows a preference for writers who give first-hand information, commending Heraclides of Tarentum as a trustworthy man, if there ever was one, who set down only those things proved by his own experience.[7] Galen declares that one could spend a life-time in reading the books that have already been written upon medicinal simples. He urges his readers, however, to abstain from Andreas and

[1] XII, 446.
[2] II, 141, 179.
[3] II, 179; X, 609.
[4] II, 621.
[5] XIII, 891.
[6] XIII, 430-31.
[7] XIII, 717.

other liars of that stamp, and above all to eschew Pamphilus who never saw even in a dream the herbs which he describes.

Galen's estimate of Dioscorides.

Of all previous writers upon *materia medica* Galen preferred Dioscorides. He writes, "But Anazarbensis Dioscorides in five books discussed all useful material not only of herbs but of trees and fruits and juices and liquors, treating besides both all metals and the parts of animals." [1] Yet he does not hesitate to criticize certain statements of Dioscorides, such as the story of mixing goat's blood with the *terra sigillata* of Lemnos. Dioscorides had also attributed marvelous virtues to the stone Gagates which he said came from a river of that name in Lycia; Galen's comment is that he has skirted the entire coast of Lycia in a small boat and found no such stream.[2] He also wonders that Dioscorides described butter as made of the milk of sheep and goats, and correctly states that "this drug" is made from cows' milk.[3] Galen does not mention its use as a food in his work on medicinal simples, and in his treatise upon food values he alludes to butter rather incidentally in the chapter on milk, stating that it is a fatty substance and easily recognized by tasting it, that it has many of the properties of oil, and in cold countries is sometimes used in baths in place of oil.[4] Galen further criticizes Dioscorides for his unfamiliarity with the Greek language and consequent failure to grasp the significance of many Greek names.

Galen's dogmatism: logic and experience.

Daremberg said of Galen that he represented at the same time the most exaggerated dogmatism and the most advanced experimental school. There is some justification for the paradox, though the latter part seems to me the truer. But Galen was proud of his training in philosophy and logic and mathematics; he stood fast by many Hippocratic dogmas such as the four qualities theory, he thought [5] that in medicine as in geometry there were a certain num-

[1] XI, 794; also XIII, 658; XIV, 61-62, and many other passages of the *Antidotes*.

[2] XII, 203. Pliny, NH XXXVI, 34, makes the same statement as Dioscorides.

[3] XII, 272.

[4] Pliny, NH XXVIII, 35, however, both tells how butter is made and of its use as food among the barbarians.

[5] X, 40-41.

ber of self-evident maxims upon which reason, conforming to the rules of logic, might build up a scientific structure. In the *De methodo medendi*[1] he makes a distinction between the discovery of drugs and medicines, simple or compound, by experience and the methodical treatment of disease which he now sets forth and which should proceed logically and independently of mere empiricism, and he wishes that other medical writers would make it clear when they are relying merely on experience and when exclusively upon reason.[2] At the same time he expresses his dislike for mere dogmatizers who shout their *ipse dixits* like tyrants without the support either of reason or experience.[3] He also grants that the ordinary man, taught by nature alone, often instinctively pursues a better course of action for his health than "the sophists" are able to advise.[4] Indeed, he is of the opinion that some doctors would do well to stick to experience alone and not try to mix in reasoning, since they are not trained in logic, and when they endeavor to divide or analyze a theme, perform like unskilled carvers who fail to find the joints and mutilate the roast.[5] Later on in the same work[6] he again affirms that persons who will not read and profit by the books of medical authorities and whose own reasoning is defective, should limit themselves to experience.

Galen's account of the Empirics.

Normally, however, Galen upholds both reason and experience as criteria of truth against the opposing schools of Dogmatics and Empirics. The former attacked experience as uncertain and impossible to regulate, slow and unmethodical. The latter replied that experience was consistent, adaptable to art, and proof enough.[7] Galen's chief objection to the Empirics is that they reject reason as a criterion of truth and wish the medical art to be irrational.[8] "The Empirics say that all things are discovered by experi-

[1] X, 127, 962.
[2] X, 31.
[3] X, 29.
[4] X, 668.
[5] X, 123.
[6] X, 915-16.
[7] I, 75-76; XIV, 367.
[8] I, 145; II, 41-43; X, 30-31, 782-83; XIII, 188, 366, 375, 463, 579, 594, 892; XIV, 245, 679.

ence, but we say that some are found by experience and some by reason."[1] Galen also objects to Herodotus's explanation of the medical art as originating in the conversation of patients exposed at crossroads who told one another of their complaints and recoveries and thus evolved a fund of common experience.[2] Galen criticizes such experience as irrational and not yet put into scientific form (οὔπω λογική). Of the Empirics he tells us further that they regard phenomena only and ignore causes and put no trust in reasoning. They hold that there is no system or necessary order in medical discovery or doctrine, and that some remedies have been discovered by dreams, others by chance. They also accepted written accounts of past experiences and thus to a certain extent trusted in tradition. Galen argues that they should test these statements of past authorities by reason.[3] His further contention that, if they test them by experience, they might as well reject all writings and trust only to present experience from the start, is a sophistical quibble unworthy of him. He adds, however, that the Empirics themselves say that past tradition or "history" (ἱστορία) should not be judged by experience, but it is unlikely that he represents their view correctly in this particular. In another passage[4] he says that they distinguish three kinds of experience, chance or accidental, offhand or impromptu, and imitative or the repetition of the same thing. In a third passage[5] he repeats that they held that observation of one or two instances was not enough, but that oft-repeated observation was needed with all conditions the same each time. In yet another place[6] he says that the Empirics observe coincidences in things joined by experience. He himself defines experience as the comprehending and remembering of something seen often and in the same condition,[7] and makes the good point that one cannot observe satisfactorily without use of reason.[8] He also admits

[1] X, 159.
[2] XIV, 675-76.
[3] I, 144-55.
[4] XVI, 82.
[5] I, 135.
[6] XIV, 680.
[7] I, 131.
[8] I, 134.

in one place that some Empirics are ready to employ reason as well as experience.[1]

How the Empirics might have criticized Galen.

Having noted Galen's criticism of the Empirics, we may imagine what their attitude would be towards his medicine. They would probably reject all his theories—which we, too, have finally discarded—of four elements and four qualities and the like, and would accept only his specific recommendations for the cure of disease based upon his medical experience; except that they would also be credulous concerning anything which he assured them was based upon his own or another's experience, whether it truly was or not. They would, however, have probably questioned much of his anatomical inference from the dissection of the lower animals, since he tells us that they "have written whole books against anatomy." [2] Considering the state of knowledge in their time, their refusal to attempt any large generalizations or to hazard any scientific hypotheses or to build any risky medical system was in a way commendable, but their credulity as to particulars was a weakness.

Galen's standard of reason and experience.

On the whole Galen's attitude towards experience seems an improvement upon theirs. He was apparently more critical towards the "experiences" of past writers than the average Empiric, and in his combination of reason and experience he came a little nearer to modern experimental method. Reason alone, he says, discovers some things, experience alone discovers some, but to find others requires use of both experience and reason.[3] In his treatise upon critical days he keeps reiterating that their existence is proved both by reason and experience. These two instruments in judging things given us by nature supplement each other.[4] "Logical methods have force in finding what is sought, but in believing what has been well found there are two criteria for all men, reason and experience." [5] "What can you do with men who cannot be persuaded either by reason or by

[1] XVI, 82.
[2] II, 288.
[3] IX, 842; XIII, 887.
[4] XIII, 116-17.
[5] X, 28-29.

practice?"[1] Galen also speaks of discovering a truth by logic and being thereby encouraged to try it in practice and of then verifying it by experience.[2] This, however, is not quite the same thing as saying that the scientist should aim to discover new truth by purposive experiments, or that from a number of experiences reason may infer some general law of nature.

Simples knowable only from experience.

It is perhaps in his work on medicinal simples that Galen lays most stress upon the importance of experience. Indeed he sees no other way to learn the properties of natural objects than through the experience of the senses.[3] "For by the gods," he exclaims, "how is it that we know that fire is hot? Are we taught it by some syllogism or persuaded of it by some demonstration? And how do we learn that ice is cold except from the senses?"[4] And Galen sees no advantage in spending further time in arguments and hair-splitting where one can learn the truth at once from the senses. This thought he keeps repeating through the treatise, saying, for example, "The surest judge of all will be experience alone, and those who abandon it and reason on any other basis not only are deceived but destroy the value of the treatise."[5] Moreover, he restricts his account of medicinal simples to those with which he is personally acquainted. In the three books treating of plants he does not mention all those found in all parts of the world, but only as many as it has been his privilege to know by experience.[6] He proposes to follow the same rule in the ensuing discussion of animals and to say nothing of virtues which he has not tested or of substances mentioned in the writings of past physicians but unknown to him. He dares not trust their statements when he reflects how some have lied in such matters. In the middle ages Albertus Magnus talks in much the same strain in his works on animals, plants, and minerals, and perhaps he was stimulated to such ideals, consciously or un-

[1] X, 684.
[2] X, 454-55.
[3] XI, 420.
[4] XI, 434-35.
[5] XI, 456.
[6] XII, 246.

consciously, directly by reading Galen or indirectly through Arabic works, by Galen's earlier expression of them. Galen mentions some virtues ascribed to substances which he has tested by experience and found false, such as the medicinal properties attributed to the belly of a seagull[1] and some of those claimed for the marine animal called torpedo.[2] Anointing the place with frog's blood or dog's milk will not prevent eyebrows that have been plucked out from growing again, nor will bat's blood and viper's fat remove hair from the arm-pits.[3] Also the brain of a hare is only fairly good for boys' teeth.[4]

Experience and food science.

In beginning his work on food values [5] Galen states that many have discussed the properties of aliments, some on the basis of reason alone, some on the basis of experience alone, but that their statements do not agree. On the whole, since reasoning is not easy for everyone, requiring natural sagacity and training from childhood, he thinks it better to start from experience, especially since not a few physicians are of the opinion that only thus can the properties of foods be learned.

Experience and compounds.

The Empirics contended that most compound medicines had been hit upon by chance, and Galen grants that the Dogmatics usually are unable to give reasons for the ingredients of their doses and find difficulty in reproducing a lost prescription.[6] But he holds that reasons can be given for the constituents of the compound and that the logical discovery of such remedies differs from the empirical.[7] His own method was to learn the nature of each disease and the varied properties of simples, and then prepare a compound suited to the disease and to the patient.[8] On the other hand, we see how much depends upon experience from his confession that sometimes he has hastily prepared a compound from a few simples, sometimes from more, sometimes from a great variety. If the compound worked well, he would

[1] XII, 336.
[2] XII, 365.
[3] XII, 258, 262, 269, 331.
[4] XII, 334.
[5] VI, 453-55.
[6] XIII, 463.
[7] XII, 895.
[8] XIV, 222.

continue to use it, sometimes making it stronger and sometimes weaker.[1] For as you cannot put together compounds without rational method, so you cannot tell their strength certainly and accurately without experience.[2] He admits that no one can tell the exact quantity of each ingredient to employ without the aid of experience,[3] and says, "The proper proportions in the mixture we shall find conjecturally before experience, scientifically after experience." [4] In these treatises upon compound medicines, unlike that on medicinal simples, Galen gives the prescriptions of former physicians as well as some tested by his own experience.[5] Sometimes, however, he expresses a preference for the medicines of those writers who were "most experienced"; and once says that he will give some compounds of the more recent writers, who in their turn had selected the best from older writers of long experience and added later discoveries.[6] We suspect, however, that some of these prescriptions had not been tested for centuries.

Suggestions of experimental method.

Galen gives a few directions how to regulate medical observation and experience, although they cannot be said to carry us very far on the road to modern laboratory research. He saw the value of "long experience," a phrase which he often employs.[7] He states that one experience is enough to learn how to prepare a drug, but to learn to know the best medicines in each kind and in different places many experiences are required.[8] Medicinal simples should be frequently inspected, "since the knowledge of things perceived by the senses is strengthened by careful examination." [9] Galen advises the student of medicine to study herbs, trees, and fruit as they grow, to find out when it is best to pluck them, how to preserve them, and so on. But elsewhere he states that it is possible to estimate the general virtue of the simple

[1] XIII, 700-701.
[2] XIII, 706-707.
[3] XIII, 467.
[4] XIII, 867.
[5] XII, 392-93, 884; XIII, 116-17, 123, 125, 128-29, 354, 485, 502-503, 582, 656.
[6] XII, 968, 988.
[7] See XII, 988; XIII, 960-61; XIV, 12, 60, 341.
[8] XIV, 82.
[9] XIII, 570.

from one or two experiences.[1] However, he suggests that their effect be noted in the three cases of a perfectly healthy person, a slightly ailing patient, and a really sick man.[2] In the last case one should further note their varying effects as the disease is marked by any excess of heat, cold, dryness, or moisture. Care should be taken that the simples themselves are pure and free from any admixture of a foreign substance.[3] "It is also essential to test the relation to the nature of the patient of all those things of which great use is made in the medical art." [4] One condition to be observed in experimental investigation of critical days is to count no cases where any slip has been made by physician or patient or bystanders or where any other foreign factor has done harm.[5] Galen was acquainted with physical experiments in siphoning, for he says that, if one withdraws the air from a vessel containing sand and water, the sand will follow before the water, which is the heavier (*sic?*).[6]

Difficulty of medical experiment.

Galen also points out some of the difficulties of medical experimentation. One is the extreme unlikelihood of ever being able to observe in even two cases the same combination of symptoms and circumstances.[7] The other is the danger to the life of the patient from rash experimenting.[8] Thus Galen more than once tells us of abstaining from testing some remedy because he had others of whose effects he was surer.

Empirical remedies.

In the treatise on easily procurable remedies ascribed to Galen,[9] in which we have already seen evidence of later interpolation or authorship, some recipes are concluded by

[1] XII, 350.
[2] XVI, 86-87; XI, 518.
[3] XI, 485.
[4] XVI, 85.
[5] IX, 842.
[6] II, 206.
[7] I, 138.
[8] XVI, 80.
[9] There would seem to be something wrong, at least with its arrangement as it now stands, for the first book ends (XIV, 389) with the words, "This my fourth book, O Glaucon, ends thus. If it has been useful to you, you will readily follow what I've written to Salomon the archiater." But then the present second book opens with the words (XIV, 390), "Since you've asked me to write you about easily procurable remedies, O dearest Solon," and goes on to say that the author will state what he has learned from experience beginning with the hair and closing with the feet.

such expressions as, "This has been experienced; it works unceasingly," [1] or "Another remedy tested by us in many cases." [2] This became a custom in many subsequent medical works, including those of the middle ages. One recipe is introduced by the caution, "But don't cure anybody unless you have been paid first, for this has been tested in many cases." [3] But we are left in some doubt whether we should infer that remedies tested by experience are so superior that they call for cash payment rather than credit, or so uncertain that it is advisable that the physician secure his fee before the outcome is known. In the middle ages the word *experimentum* was used a great deal as a synonym for any medical treatment, recipe, or prescription. Galen approaches this usage, which we have already noticed in Pliny's *Natural History,* when he describes "a very important experiment" in bleeding performed by certain doctors at Rome.[4]

Galen's influence upon medieval experiment.

Indeed Galen appears to have exerted a great influence in the middle ages by his passages concerning experience in particular as well as by his medicine in general. Medieval writers cite him as an authority for the recognition of experience and reason as criteria of truth.[5] Gilbert of England cites "experiences from the book of experiments experienced by Galen," [6] and we shall find more than one such apocryphal work ascribed to Galen in the middle ages. John of St. Amand seems to have developed seven rules [7] which he gives for discovering experimentally the properties of medicinal simples from what we have heard Galen say on the subject, and in another work, the *Concordances,* John collects a number of passages about experience from

[1] XIV, 378.

[2] XIV, 462.

[3] XIV, 534.

[4] XI, 205.

[5] John of St. Amand, *Expositio in Antidotarium Nicolai,* fol. 231, in *Mesuae medici clarissimi opera,* Venice, 1568. Pietro d'Abano, *Conciliator,* Venice, 1526, Diff. X, fol. 15; Diff. LX, fol. 83. Arnald of Villanova, *Repetitio super Canon "Vita brevis,"* fol. 276, in his *Opera,* Lyons, 1532.

[6] Gilbertus Anglicus, *Compendium medicinae,* Lyons, 1510, fol. 328v., "Experimenta ex libro experimentorum Gal. experta."

[7] In his *Expositio in Antidotarium Nicolai,* as cited above (note 5).

the works of Galen.[1] Peter of Spain, who died as Pope John XXI in 1277, cites Galen in his discussion of "the way of experience" and "the way of reason" in his *Commentaries on Isaac on Diets.*[2] We have already suggested Galen's possible influence upon Albertus Magnus, and we might add Roger Bacon who wrote some treatises on medicine. But it is hardly possible to tell whether such ideas were in the air, or were due to Galen individually either in their origin or their transmission. But he made a rather close approach to the medieval attitude in his equal regard for logic and for experimentation.

His more general medieval influence.

The more general influence of Galen upon all sides of the medicine of the following fifteen centuries has often been stated in sweeping terms, but is difficult to exaggerate. His general theories, his particular cures, his occasional marvelous stories, were often repeated or paraphrased. Oribasius has been called "the ape of Galen," and we shall see that the epithet might with equal reason be applied to Aëtius of Amida. Indeed, as in the case of Pliny, we shall find plenty of instances of Galen's influence in our later chapters. Perhaps as good a single instance of medieval study of Galen as could be given is from the *Concordances* of John of St. Amand already mentioned, which bear the alternative title, "Recalled to Mind" (*Revocativum memoriae*), since they were written to "relieve from toil and worry scholars who often spend sleepless nights in searching for points in the books of Galen." [3] Or we may note how the associates of the twelfth century translator from the Arabic, Gerard of Cremona, added a list of his works at the close of his translation of Galen's *Tegni,* "imitating Galen in the commemoration of his books at the end of the same treatise," as they themselves state.[4]

Not that medieval men did not make additions of their

[1] J. L. Pagel, *Die Concordanciae des Johannes de Sancto Amando,* Berlin, 1894, pp. 102-104. John also wrote commentaries on Galen, (*Histoire Littéraire de la France,* XXI, 263-65).

[2] ed. Lyons, 1515, fols. 19v-20v.

[3] Berlin, 902, 14th century, fol. 175; Berlin 903, 1342 A.D., fol. 2.

[4] Boncompagni (1851), pp. 3-4.

own to Galen. For instance, the noted Jewish philosopher, Moses Maimonides, in adding his collection of medical *Aphorisms* to the many previous compilations of this sort by Hippocrates, Rasis (Muhammad ibn Zakariya), Mesuë (Yuhanna ibn Masawaih), and others, states that he has drawn them mainly from the works of Galen, but that he supplements these with some in his own name and some by other "moderns."[1] Not that Galen was not sometimes criticized or questioned. A later Greek writer, Symeon Seth, ventured to devote a special treatise to a refutation of some of Galen's physiological views. In it, addressing himself to those "persons who regard you, O Galen, as a god," he endeavored to make them realize that no human being is infallible.[2] Among the medical treatises of Gentile da Foligno, who was papal physician and performed a public dissection at Padua in 1341,[3] is found a brief argument against Galen's fifth aphorism.[4] But such criticism or opposition

[1] Moses ben Maimon, *Aphorisms,* 1489. "Incipiunt aphorismi excellentissimi Raby Moyses secundum doctrinam Galieni medicorum principis . . . collegi eos ex verbis Galieni de omnibus libris suis. . . . Et ego protuli super his afforismis quedam dicta que circumspexi et ea meo nomine nominavi et similiter protuli aliquos aphorismos aliquorum modernorum quos denominavi eorum nomine."

[2] Ed. C. V. Daremberg, *Notices et Extraits des manuscrits médicaux,* 1853, pp. 44-47, Greek text; pp. 229-33, French translation.

[3] Garrison, *History of Medicine,* 2nd edition, 1917, p. 141. But at p. 151 Garrison would seem mistaken in stating that Gentile died in 1348, for in the MS of which I shall speak in the next footnote his treatise on critical days is dated back in the year 1362: "Tractatus de enumeratione dierum creticorum m'i Gentilis anni 1362," at fol. 125; while at fol. 162 we read, "Explicit questio . . . m'i Zentilis anno Domini 1359 de mense marcii, et scripta Pisis de mense octobris 1359." It is possible but rather unlikely that the dates later than 1348 refer to the labors of copyists. Venetian MSS contain not only a *De reductione medicinarum ad actum* by Gentile, written at Perugia in April, 1342 (S. Marco, XIV, 7, 14th century, fols. 44-48); but also "Suggestions concerning the pestilence which was at Genoa in 1348," by him (S. Marco, XIV, 26, 15th century, fols. 99-100, consilia de peste quae fuit Ianuae anno 1348). Valentinelli's catalogue of the MSS in the Library of St. Mark's does not help, however, to clear up the question when Gentile died, since in one place (IV, 235) Valentinelli assures us that he died at Bologna in 1310, and in another place (V, 19) says that he died at Perugia in 1348.

[4] Cortona 110, early years of 15th century, fol. 128, Rationes Gentilis contra Galenum in quinto aphorismi. This MS contains several other works by Gentile da Foligno.

only shows how generally Galen was accepted as an authority.

III. *His Attitude Towards Magic*

From Galen's habits of critical estimation rather than blind acceptation of authority, of scientific observation, careful measurement, and personal experiment, from his brilliant demonstrations by dissection, and his medical prognostication and therapeutics, sane and shrewd for his time,—from these we have now to turn to the other side of the picture, and examine what information his works afford us concerning the magic and astrology in ancient medicine, concerning the belief in occult virtues, suspensions, characters, incantations, and the like. We may first consider what he has to say concerning magic and divination as he understands those words, and then take up his attitude to those other matters which we look upon as almost equally deserving classification under those heads.

Accusations of magic against Galen.

Apollonius of Tyana and Apuleius of Madaura were not the only celebrated men of learning in the early Roman Empire to be accused of magic; we have already alluded to the charges of magic made against Galen by the envious physicians of Rome during his first residence in that city. It is hard to escape the conviction that at that time learned men were very liable to be suspected or accused of magic. Indeed, Galen makes the general assertion that when a physician prognosticates aright concerning the future course of a malady, this seems so marvelous to most men that they would receive him with great affection, if they did not often regard him as a wizard.[1] Soon after saying this, Galen begins the story of the prognostications he made and the cure he wrought, when all the other doctors took an opposite view of the case.[2] One of them then jealously suggested that Galen's diagnosis was due to divination.[3] When asked by what kind of divination, he gave different answers

[1] XIV, 601. [2] XIV, 605. [3] XIV, 615.

at different times and to different persons, sometimes saying by dreams, sometimes by sacrificing, again by symbols, or by astrology. Afterwards such charges against Galen kept multiplying.[1] As a result, Galen says that since then he has not gone about advertising his prognostications like a herald, lest the physicians and philosophers hate him the more and slander him as a wizard and diviner, but that he now reveals his discoveries only to his friends.[2] In another treatise he represents Hippocrates as saying that a proficient doctor should be able to prognosticate the course of diseases, but adds that contemporary physicians call such a doctor a sorcerer and wonder-worker (γόητά τε καὶ παραδοξολόγον).[3] Again in his work on medicinal simples [4] he states that he abstained from testing the supposed virtue of crocodile's blood in sharpening the vision, and the blood of house mice in removing warts, partly because he had other reliable eye-medicines and cures for warts—such as *myrmecia,* a gem with wart-like lumps, partly because by employing such substances he feared to incur the reputation of a sorcerer, since jealous physicians were already slandering his medical prognostications as divination. This last passage affords a good illustration of the close connection with magic of certain natural substances supposed to possess marvelous virtues, while Galen's wart stone also seems magical to the modern reader.

His charges of magic against others.

Galen himself sometimes calls other physicians magicians. Certain men with whom he does not agree are called by him "liars or wizards or I don't know what to say," [5] and another man who used mouse dung to excess he calls superstitious and a sorcerer.[6] In the same work on simples [7] he says that he will list herbs in alphabetical order as Pamphilus did, but that he will not like him descend to old wives' tales, Egyptian sorceries and incantations, amulets and other magical devices, which not only do not belong in the medical art

[1] XIV, 625.
[2] XIV, 655.
[3] I, 54-55.
[4] XII, 263.
[5] XII, 306.
[6] XII, 307.
[7] XI, 792-93

but are utterly false. Pamphilus never saw most of the herbs he mentioned, much less tested their virtues, but copied anything he found, piling up names, incantations, and wizardry. Galen accuses Xenocrates Aphrodisiensis also of not having eschewed sorcery, and he notes that medical writers have either said nothing about sweat or what is superstitious and bordering upon magic.[1]

Charms and wonder-workers.

Philters, love-charms, dream-draughts, and imprecations Galen regards as impossible or injurious, and intends to have nothing to do with them. He thinks it ridiculous to believe that by such spells one can bewitch one's adversaries so that they cannot plead in court, or conceive or bear children. He considers it worse to advertise and perpetuate such false or criminal notions in writings than to practice such a crime but once.[2] In one passage,[3] however, to illustrate his theory that the gods prepare the sperms of plants and animals, and set them going as it were, and afterwards leave them to themselves, Galen compares them to the wonder-workers—who were perhaps not magicians but men similar to our sidewalk fakirs who exhibit mechanical toys—who start things moving and then go away themselves while what they have prepared moves on artificially for a time.

Animal substances inadmissible in medicine.

Galen's own works are not entirely free from the magical devices of which he accuses others. We may begin with animal substances, since he himself has testified that the use of sweat, crocodile's blood, and mouse's dung is suggestive of magic. Moreover, he attributes more bizarre virtues to the parts of animals than to herbs or stones. In a passage somewhat similar to that in which Pliny[4] expressed his horror at the use of human blood, entrails, and skulls as medicines, Galen declares that he will not mention the abominable and detestable, as Xenocrates and some others have done. The Roman law has long forbidden eating human flesh, while Galen regards even the mention of certain secretions and excrements of the human body as

[1] XII, 283.
[2] XII, 251-53.
[3] IV, 688.
[4] *Natural History*, XXVIII, 2.

offensive to modest ears.[1] Nevertheless, before long he offends against his own standard and describes how he administered to patients the very substance which he had before characterized as most unmentionable.[2] It may also be noted that he repeats unquestioningly such a tale as that the cubs of the bear are born unformed and licked into shape by their mother.[3]

Nastiness of ancient medicine.

Further milder illustrations of the fact that such nasty substances were then not merely recommended in books but freely employed in actual medical practice, are seen in the frequent use by one of Galen's teachers of the dung of dogs who for two days before had eaten nothing but bones,[4] in Galen's own wonderfully successful treatment of a tumor on a rustic's knee with goat dung—which is, however, too sharp for the skins of children or city ladies,[5] and in his discovery by repeated experience that the dung of doves who take little exercise is less potent than that of those who take much,[6] Galen also says that he has known of doctors who have cured many persons by giving them burnt human bones in drink without their knowledge.[7]

Parts of animals.

Galen's medicinal simples include the bile of bulls, hyenas, cocks, partridges, and other animals.[8] A digestive oil can be manufactured by cooking foxes and hyenas, some alive and some dead, whole in oil.[9] Galen discusses with perfect seriousness the relative strength of various animal fats, those of the goose, hen, hyena, goat, pig, and so forth.[10] He decides that lion's fat is by far the most potent, with that of the pard next. Among his simples are also found the slough of a snake, a sheepskin, the lichens of horses, a spider's web,[11] and burnt young swallows, for whose introduction into medicine he gives Asclepiades credit.[12] Of

[1] XII, 248, 284-85, 290.
[2] XII, 293.
[3] XIV, 255. (*To Piso on theriac.*)
[4] XII, 291-92.
[5] XII, 298.
[6] XII, 304.
[7] XII, 342.
[8] XII, 276-77.
[9] XII, 367-69.
[10] XIII, 949-50, 954-55.
[11] XII, 343. These form the titles of four successive chapters, *De simplic.*, XI, i, caps. 19-22.
[12] XII, 359, 942-43, 977.

Archigenes' prescriptions for toothache he repeats that which recommended holding for some time in the mouth a frog boiled in water and vinegar, or a dog's tooth, burnt, pulverized, and boiled in vinegar.[1] Cavities may be filled with toasted earth-worms or spiders' eggs diluted with unguent of nard. Teething infants are benefited, if their gums are moistened with dog's milk or anointed with hare's brains.[2] For colic he recommends dried cicadas with three, five, or seven grains of pepper.[3]

Some scepticism.

Galen is less confident as to the efficacy for earache of the multipedes which roll themselves up into a ball, and which, cooked in oil, are employed especially by rural doctors.[4] He is still more sceptical whether the liver of a mad dog will cure its bite.[5] Many say so, and he knows of some who have tried it and survived, but they took other remedies too.[6] Galen has heard that some who trusted to it alone died. In one treatise [7] Galen discusses the strange virtues of the basilisk in much the usual way, but in his work on simples [8] he remarks drily that it is obviously impossible to employ it in pharmacy, since, if the tales about it be true, men cannot see it and live or even approach it without danger. He therefore will not include it or elephants or Nile horses (hippopotamuses?) or any other animals of which he has had no personal experience.

Doctrine of occult virtue.

Galen tries to find some satisfactory explanation of the strange properties which he believes exist in so many things. The attractive power of the magnet and of drugs suggests to him that nature in us is divine, as Homer says, and leads like to like and thus shows its divine virtues.[9] Galen rejects Epicurus's explanation of the magnet's attractive power.[10] It was that the atoms flowing off from both the magnet and iron fit one another so closely that the two sub-

[1] XII, 856.
[2] XII, 860.
[3] XII, 360.
[4] XII, 366-67.
[5] XII, 335.
[6] A fact which—one cannot help remarking—considering the character of most ancient remedies for hydrophobia, only tends to make their recovery seem the more marvelous.
[7] XIV, 233.
[8] XII, 250-51.
[9] XIV, 224-25.
[10] II, 45-48.

stances are drawn together. Galen objects that this does not explain how a whole series of rings can be suspended in a row from a magnet. Galen's teacher Pelops, who claimed to be able to tell the cause of everything, explained why ashes of river crabs are used for the bite of a mad dog as follows.[1] The crab is efficacious against hydrophobia because it is an aquatic animal. River crabs are better for this purpose than salt water crabs because salt dries up moisture. He also thought the ashes of crabs very potent in absorbing the venom. But this type of reasoning is unsatisfactory to Galen, who finds the best explanation of all such action in the peculiar property, or occult virtue, of the substance as a whole. Upon this subject [2] he proposes to write a separate treatise, and in the fragment *De substantia facultatum naturalium* (περὶ οὐσίας τῶν φυσικῶν δυνάμεων) he again discusses the matter.[3]

Virtue of the flesh of vipers.

Among parts of animals Galen regarded the flesh of vipers as especially medicinal, particularly as an antidote to poisons. Of the following cures wrought by vipers' flesh which Galen narrates [4] two were repeated without giving him credit by Aëtius of Amida in the sixth, and Bartholomew of England in the thirteenth century, and doubtless by other writers. When Galen was a youth in Asia, some reapers found a dead viper in their jug of wine and so were afraid to drink any of it. Instead they gave it to a man near by who suffered from the terrible skin disease elephantiasis and whom they thought it would be a mercy to put quietly out of his misery. He drank the wine but instead of dying recovered from his disease. A similarily unexpected cure was effected when a slave wife in Mysia tried to kill her hus-

[1] XII, 358-59. Concerning the virtue of river crabs we may also quote from a story told in Nias Island, west of Sumatra: "for had he only eaten river crabs, men would have cast their skin like crabs, and so, renewing their youth perpetually, would never have died."—From J. G. Frazer (1918), I, 67. The belief that the serpent annually changes its skin and renews its youth may account for the virtues ascribed to the flesh of vipers and to theriac in the following paragraphs.

[2] περὶ τῶν ἰδιότητι τῆς ὅλης οὐσίας ἐνεργοῦντων.

[3] IV, 760-61, ἐνεργεῖν τὰς οὐσίας κατ' ἰδίαν ἑκάστην φύσιν.

[4] XII, 311-15.

band by offering him a like drink. A third case was that of a patient whom Galen told of these two previous cures. After resorting to augury to learn if he too should try it and receiving a favorable response, the patient drank wine infected by venom with the result that his elephantiasis changed into leprosy, which Galen cured a little later with the usual drugs. A fourth man, while hunting vipers, was stung by one. Galen bled him, extracted black bile with a drug, and then made him eat the vipers which he had caught and which were prepared in oil like eels. A fifth man, warned by a dream, came from Thrace to Pergamum. Another dream instructed him both to drink, and to anoint himself with, a concoction of vipers. This changed his disease into leprosy which in its turn was cured by drugs which the god prescribed.

Theriac.

The flesh of vipers was an important ingredient in the famous antidote and remedy called theriac, concerning which Galen wrote two special treatises [1] besides discussing it in his works on simples and antidotes. Mithridates, like King Attalus in Galen's native land, had tested the effects of various drugs upon condemned criminals, and had thus discovered antidotes against spiders, scorpions, sea-hares, aconite, and other poisons. He then combined the results of his research into one grand compound which should be an antidote against any and every poison. But he did not include the flesh of the viper, which was added with some other changes by Andromachus, chief physician to Nero.[2] The divine Marcus Aurelius used to take a dose of theriac daily and it had since come into general use.[3] Galen gives a long list of ills which it will cure, including the plague and hydrophobia,[4] and adds that it is beneficial in keeping a man in good health.[5] He advises its use when traveling or in wintry weather, and tells Piso that it will prolong his life.[6] He explains more than once[7] how to prepare the

[1] *Ad Pisonem de theriaca; De theriaca ad Pamphilianum.*
[2] XIV, 2-3.
[3] XIV, 217.
[4] XIV, 271-80.
[5] XIV, 283.
[6] XIV, 294.
[7] XII, 317-18; XIV, 45-46, 238.

viper's flesh, why the head and tail must be cut off, how it is cleaned and boiled until the flesh falls from the backbone, how it is mixed with pounded bread into pills, how the flesh of the viper is best in early summer. Galen also accepts the legend,[1] quoting six lines of verse from Nicander to that effect, that the viper conceives in the mouth and then bites off the male's head, and that the young viper avenges its father's death by gnawing its way out of its mother's vitals. The *Marsi* at Rome denied the existence of the *dipsas* or snake whose bite causes one to die of thirst, but Galen is not quite sure whether to agree with them.

Magical compounds.

Already we have had occasion to refer to Galen's two works on compound medicines which occupy the better part of two bulky volumes in Kühn's edition and contain a vast number of prescriptions. It is not uncommon for one of these to contain as many as twenty-five ingredients. It seems unlikely that such elaborate concoctions would have been discovered by chance, as the Empirics held, but the modern reader is ready to agree that it was chance, if anyone was ever cured of anything by one of them. Yet Galen, as we have seen, believes that reasons can be given for the ingredients and would not for a moment admit that they are no better than the messes of witches' cauldrons. He argues that, if all diseases could be cured by simples, no one would use compounds, but that they are essential for some diseases, especially such as require the simultaneous application of contrary virtues.[2] Also where a simple is too strong or weak, it can be toned up or down to just the right strength in a compound. Plasters and poultices seem always to be compounds. Of panaceas Galen is somewhat more chary, except in the case of theriac; he opines that a medicine which is good for a number of ills cannot be very good for any one of them.[3]

Amulets.

Procedure as well as substances suggestive of magic is found to some extent in Galen's works. He instructs, for

[1] XIV, 238-39. [2] XIII, 371, 374. [3] XIII, 134.

example, to pluck an herb with the left hand before sunrise.[1] He also recommends the suspension of a peony to cure epilepsy.[2] He saw a boy who wore this root remain free from that disease for eight months, when the root happened to drop off and the boy soon fell in a fit. When another peony root was hung about his neck, he remained in good health until Galen for the sake of experiment removed it a second time, whereupon another epileptic fit ensued as before. In this case Galen suggests that perhaps some particles from the root were drawn in by the patient's breathing or altered the surrounding air. In another passage he holds that there is no medical reason to account for the virtues of amulets, but that those who have tested them by experience say that they act by some marvelous antipathy unknown to man.[3] A ligature recommended by Galen is to bind about the neck of the patient a viper which has been suffocated by tying several strings, preferably of marine purple, about its neck.[4] Galen marvels that *stercus lupinum,* even when simply suspended from the neck, "sometimes evidently is beneficial."[5] It should not have touched the ground but should have been taken from trees or bushes. It also works better, as Galen has found in his own practice, if suspended by the wool of a sheep who has been torn by a wolf.

Incantations and characters.

While Galen thus employs ligatures and suspensions and sanctions magic logic, he draws the line at use of images, characters, and incantations. In the passage just cited he goes on to say that he has found other suspended substances efficacious, but not the barbarous names such as wizards use. Some say that the gem jasper comforts the stomach if bound about the abdomen,[6] and some wear it in a ring engraved with a dragon and rays,[7] as King Nechepso directs in his fourteenth book. Galen has employed it suspended about the neck without any engraving upon it and

[1] XIII, 242.
[2] XI, 859.
[3] XII, 573; see also XIII, 256.
[4] XI, 860.
[5] XII, 295-96.
[6] XII, 207.
[7] A representation of the Agathodaemon; see C. W. King, *The Gnostics and their Remains,* London, 1887, p. 220.

found it equally beneficial. In illustrating the virtue of human saliva, especially that of a fasting man, Galen tells of a man who promised him to kill a scorpion by means of an incantation which he repeated thrice. But at each repetition he spat on the scorpion and Galen afterwards killed one by the same procedure without any incantation, and more quickly with the spittle of a fasting than of a full man.[1]

Belief in magic dies hard.

The preceding paragraph gives a good illustration of the slow progress of human thought away from magic and towards science. Men are discovering that marvels can be worked as well without characters and incantations. Similar passages may be found in Arabic and Latin medieval writers. But while Galen questions images and incantations, he still clings to the notions of marvelous virtue in a fasting man's spittle or in a gem suspended about the neck. And these and other passages in which he clung to old superstitions were unfortunately equally influential upon succeeding writers, who sometimes, we fear, took them as an excuse for further indulgence in magic. Indeed, we shall find Alexander of Tralles in the sixth century arguing that Galen finally became a believer in the efficacy of incantations. Thus the old notions and practices die hard.

On easily procurable remedies.

In the treatise on easily procurable remedies, where popular and rustic remedies enter rather more largely than in Galen's other writings, superstitious recipes are also met with more frequently, and, if that be possible, the doses become even more calculated to make one's gorge rise, it being felt that the unfastidious tastes and crude constitutions of peasants and the poorer classes can stand more than daintier city patients. Another reason for separate consideration of the contents of this treatise is the possibility, already mentioned, that it is interpolated and misarranged, and the fact that it is in part of much later date than Galen.

[1] XII, 288-89. At II, 163, Galen again accepts the notion that human saliva is fatal to scorpions.

Specimens of its superstitious contents.

We must limit ourselves to a hasty survey of a few specimens of its prescriptions. Following Archigenes, ligatures and crowns are employed for headaches.[1] In contrast to Galen's previous scepticism concerning depilatories for eyebrows we now find a number mentioned, including the blood of a bed-bug.[2] To cure lumbago,[3] if the pain is in the right foot, reduce to powder with your right hand the wings of a swallow. Then make an incision in the swallow's leg and draw off all its blood. Skin it and roast it and eat it entire. Then anoint yourself all over with the oil for three days and you will marvel at the result. "This has been often proved by experience." To prevent hair from falling out take many bees and burn them and mix with oil and use as an ointment.[4] For a sty in the eye catch flies, cut off their heads, and rub the sty with the rest of their bodies.[5] A cooked black chameleon performs the double duty of curing toothache and killing mice.[6] To extract a tooth in the upper jaw surround it with the worms found in the tops of cabbages; for a lower tooth use the worms on the lower parts of the leaves.[7] Pain in the intestines will vanish, if the patient drinks water in which his feet have been washed.[8] A net transferred from a woman's hair to the patient's head acts as a laxative, especially if the net is first heated.[9] Various superstitious devices are suggested to insure the birth of a child of the sex desired.[10] Bituminous trefoil,[11] boiled and applied hot, cures snake or spider bite, but let no one use it who is not so afflicted or it will make him feel as if he was.[12] For cataract is recommended a mixture of equal parts of mouse's blood, cock's gall, and woman's milk,

[1] XIV, 321.
[2] XIV, 349.
[3] XIV, 386-87.
[4] XIV, 343.
[5] XIV, 413.
[6] XIV, 427.
[7] XIV, 430.
[8] XIV, 471.
[9] XIV, 472.
[10] XIV, 476. And others, "Ut ne cui penis arrigi possit," and "Ad arrectionem pudendi."
[11] "The *Psoranthea bituminosa* of Linnaeus. It is found on declivities near the sea-coast in the south of Europe," says a note in Bostock and Riley's *The Natural History of Pliny* (Bohn Library), IV, 330. Pliny, too (XXI, 88), states that trefoil is poisonous itself and to be used only as a counter-poison.
[12] XIV, 491; a good example of the power of suggestion.

dried.[1] For pain on one side of the head or face smear with fifteen earthworms and fifteen grains of pepper powdered in vinegar.[2] To stop a cough wear the tongue of an eagle as an amulet.[3] Wearing a root of rhododendron makes one fearless of dogs and would cure a mad dog itself, if it could be tied on the animal.[4] A "confection" covering three pages is said to prolong life, to have been used by the emperors, and to have enabled Pythagoras, its inventor, who began to make use of it at the age of fifty, to live to be one hundred and seventeen without disease. "And he was a philosopher and unable to lie about it." [5]

External signs of the temperaments of internal organs.

It remains to note what there is in Galen's works in the way of divination and astrology. We are not entirely surprised that contemporary doctors confused his medical prognostic with divination, when we read what he has to say concerning the outward signs of hot or cold internal organs. In the treatise, entitled *The Healing Art* (τέχνη ἰατρική),[6] which Mewaldt says was the most studied of Galen's works and spread in a vast number of medieval Latin manuscript translations,[7] he devotes a number of chapters to such subjects as signs of a hot and dry heart, signs of a hot liver, and signs of a cold lung. Among the signs of a cold brain are excessive excrements from the head, stiff straight red hair, a late birth, mal-nutrition, susceptibility to injury from cold causes and to catarrh, and somnolence.[8]

Marvelous statements repeated by Maimonides.

In his commentary on the *Aphorisms* of Hippocrates Galen adds other signs by which it may be foretold whether the child will be a boy or girl to those signs already mentioned by Hippocrates.[9] Some of these seem superstitious enough to us. And it was a case of the evil that men do living after them, for Moses Maimonides, the noted Jewish physician of Cordova in the twelfth century, in his collection

[1] XIV, 498.
[2] XIV, 502.
[3] XIV, 505.
[4] XIV, 517.
[5] XIV, 567ff.
[6] I, 305-412.
[7] *Galen* in PW.
[8] I, 325-6.
[9] XVII B, 212 and 834.

of *Aphorisms,* drawn chiefly from the works of Galen, repeats the following method of prognostication: *Puerum cum primo spermatizat perscrutare, quem si invenis habere testiculum dextrum maiorem sinistro,* you will know that his first child will be a male, otherwise female. The same may be determined in the case of a girl by a comparison of the size of her breasts. Maimonides also repeats, from Galen's work to Caesar on theriac,[1] the story of the ugly man who secured a beautiful son by having a beautiful boy painted on the wall and making his wife keep her eyes fixed upon it. Maimonides also repeats from Galen[2] the story of the bear's licking its unformed cubs into shape.[3]

Dreams.

In another treatise on *Diagnosis from Dreams* Galen makes a closer approach to the arts of divination.[4] He states that dreams are affected by our daily life and thought, and describes a few corresponding to bodily states or caused by them. He thinks that if you dream you see fire, you are troubled by yellow bile, and if you dream of vapor or darkness, by black bile. In diagnosing dreams one should note when they occurred and what had been eaten. But Galen also believes that to some extent the future can be predicted from dreams, as has been testified, he says, by experience.[5] We have already mentioned the effect of his father's dream upon Galen's career. In the Hippocratic commentaries[6] he says that some scorn dreams and omens and signs, but that he has often learned from dreams how to prognosticate or cure diseases. Once a dream instructed him to let blood between the index and great fingers of the right hand until the flow of blood stopped of its own accord. "It is necessary," he concludes, "to observe dreams accurately both as to what is seen and what is done in sleep in order that you

[1] Partic. 6, Kühn, XIV, 253.

[2] Kühn, XIV, 255.

[3] These passages all come from the 24th *Particula* of Maimonides' *Aphorisms,* which is devoted especially to marvels:—"Incipit particula xxiiii continens aphorismos dependentes a miraculis repertis in libris medicorum," from an edition of the *Aphorisms* dated 1489 and numbered IA.28878 in the British Museum. The same section contains still other marvels from the works of Galen.

[4] Kühn, VI, 832-5.

[5] VI, 833.

[6] XVI, 222-23.

may prognosticate and heal satisfactorily." Perhaps he had a dim idea along Freudian lines.

Lack of astrology in most of Galen's medicine.

In the ordinary run of Galen's pharmacy and therapeutics there is very little mention or observance of astrological conditions, although Hippocrates is cited as having said that a study of geometry and astronomy—which may well mean astrology—is essential in medicine.[1] In the *De methodo medendi* he often urges the importance of the time of year, the region, and the state of the sky.[2] But this expression seems to refer to the weather rather than to the position of the constellations. The dog-star is also occasionally mentioned,[3] and one passage [4] tells how "Aeschrion the Empiric, . . . an old man most experienced in drugs and our fellow citizen and teacher," burned live river crabs on a plate of red bronze after the rise of the dog-star when the sun entered Leo and on the eighteenth day of the moon. We are also informed that many Romans are in the habit of taking theriac on the first or fourth day of the moon.[5] But Galen ridicules Pamphilus for his thirty-six sacred herbs of the horoscope—or decans, taken from an Egyptian Hermes book.[6] On the other hand, one of his objections to the atomists is that "they despise augury, dreams, portents, and all astrology," as well as that they deny a divine artificer of the world and an innate moral law to the soul.[7] Thus atheism and disbelief in astrology are put on much the same plane.

The Prognostication of Disease by Astrology.

Whereas there is so little to suggest a belief in astrology in most of Galen's works, we find among them two devoted especially to astrological medicine, namely, a treatise on critical days in which the influence of the moon upon disease is assumed, and the *Prognostication of Disease by Astrology*. In the latter he states that the Stoics favored astrology, that Diocles Carystius represented the ancients

[1] I, 53.

[2] *Coeli status*, or ἡ κατάστασις. X, 593-96, 625, 634, 645, 647-48, 658, 662, 685, 737, 759-60, 778, 829, etc.

[3] X, 688; XIII, 544; XIV, 285.

[4] XII, 356.

[5] XIV, 298.

[6] XI, 798.

[7] II, 26-28.

as employing the course of the moon in prognostications, and that, if Hippocrates said that physicians should know physiognomy, they ought much more to learn astrology, of which physiognomy is but a part.[1] There follows a statement of the influence of the moon in each sign of the zodiac and in its relations to the other planets.[2] On this basis is foretold what diseases a man will have, what medical treatment to apply, whether the patient will die or not, and if so in how many days. This treatise is the same as that ascribed in many medieval manuscripts to Hippocrates and translated into Latin by both William of Moerbeke and Peter of Abano.

Critical days.

The treatise on critical days discusses them not by reason or dogma, lest sophists befog the plain facts, but solely, we are told, upon the basis of clear experience.[3] Having premised that "we receive the force of all the stars above," [4] the author presents indications of the especially great influence of sun and moon. The latter he regards not as superior to the other planets in power, but as especially governing the earth because of its nearness.[5] He then discusses the moon's phases, holding that it causes great changes in the air, rules conceptions and birth, and "all beginnings of actions." [6] Its relations to the other planets and to the signs of the zodiac are also considered and much astrological technical detail is introduced.[7] But the Pythagorean theory that the numbers of the critical days are themselves the cause of their significance in medicine is ridiculed, as is the doctrine that odd numbers are masculine and even numbers feminine.[8] Later the author also ridicules those who talk of seven Pleiades and seven stars in either Bear and the seven gates of Thebes or seven mouths of the Nile.[9] Thus he will not accept the doctrine of perfect or magic numbers along with his astrological theory. Much of this rather

[1] XIX, 529-30.
[2] XIX, 534-73.
[3] IX, 794.
[4] IX, 901-2.
[5] IX, 904.
[6] IX, 908-10.
[7] IX, 913.
[8] IX, 922.
[9] IX, 935.

long treatise is devoted to a discussion of the duration of a moon, and it is shown that one of the moon's quarters is not exactly seven days in length and that the fractions affect the incidence of the critical days.

On the history of philosophy.

A treatise on the history of philosophy, which is marked "spurious" in Kühn's edition, I have also discovered among the essays of Plutarch where, too, it is classed as spurious.[1] In some ways it is suggestive of the middle ages. After an account of the history of Greek philosophy somewhat in the style of the brief reviews of the same to be found in the church fathers, it adds a sketch of the universe and natural phenomena not dissimilar to some medieval treatises of like scope. There are chapters on the universe, God, the sky, the stars, the sun, the moon, the *magnus annus,* the earth, the sea, the Nile, the senses, vision and mirrors, hearing, smell and taste, the voice, the soul, breathing, the processes of generation, and so on.

Divination and demons.

In discussing divination [2] the treatise states that Plato and the Stoics attributed it to God and to divinity of the spirit in ecstasy, or to interpretation of dreams or astrology or augury. Xenophanes and Epicurus denied it entirely. Pythagoras admitted only divination by *haruspices* or by sacrifice. Aristotle and Dicaearchus admit only divination by enthusiasm and by dreams. For although they deny that the human soul is immortal, they think that there is something divine about it. Herophilus said that dreams sent by God must come true. Other dreams are natural, when the mind forms images of things useful to it or about to happen to it. Still others are fortuitous or mere reflections of our desires. The treatise also takes up the subject of heroes and demons.[3] Epicurus denied the existence of

[1] Kühn, XIX, 22-345. Plutarch, *Opera,* ed. Didot, *De placitis philosophorum,* pp. 1065-1114; in *Plutarch's Miscellanies and Essays,* English translation, 1889, III, 104-92. The wording of the two versions differs somewhat and in Galen's works it is divided simply into 37 chapters, whereas in Plutarch's works it is divided into five books and many more chapters.

[2] XIX, 320-21; *De plac. philos.,* V, 1-2.

[3] XIX, 253; *De plac. philos.,* I, 8.

either, but Thales, Plato, Pythagoras, and the Stoics agree that demons are natural substances, while heroes are souls separate from bodies, and are good or bad according to the lives of the men who lived in those bodies.

Celestial bodies.

The treatise also gives the opinions of various Greek philosophers on the question whether the universe or its component spheres are either animals or animated. Fate is defined on the authority of Heracleitus as "the heavenly body, the seed of the genesis of all things." [1] The question is asked why babies born after seven months live, while those born after eight months die.[2] On the other hand, a very brief discussion of how the stars prognosticate does not go into particulars beyond their indication of seasons and weather, and even this Anaximenes ascribed to the effect of the sun alone.[3] Philolaus the Pythagorean is quoted concerning some lunar water about the stars[4] which reminds one of the waters above the firmament in the first chapter of Genesis.

[1] Kühn, XIX, 261-62; *De placitis philosophorum*, I, 28; " ἡ δὲ εἱμαρμένη ἐστὶν αἰθέριον σῶμα. σπέρμα τῆς-τῶν πάντων γενέσεως."

[2] XIX, 333.

[3] XIX, 274; *De plac. philos.*, II, 19.

[4] XIX, 265; *De plac. philos.*, II, 5.

CHAPTER V

ANCIENT APPLIED SCIENCE AND MAGIC: VITRUVIUS, HERO, AND THE GREEK ALCHEMISTS

The sources—Vitruvius depicts architecture as free from magic—But himself believes in occult virtues and perfect numbers—Also in astrology—Divergence between theory and practice, learning and art—Evils in contemporary learning—Authorities and inventions—Machines and Ctesibius—Hero of Alexandria—Medieval working over of the texts—Hero's thaumaturgy—Instances of experimental proof—Magic jugs and drinking animals—Various automatons and devices—Magic mirrors—Astrology and occult virtue—Date of extant Greek alchemy—Legend that Diocletian burned the books of the alchemists—Alchemists' own accounts of the history of their art—Close association of Greek alchemy with magic—Mystery and allegory—Experiment: relation to science and philosophy.

"doctum ex omnibus solum neque in alienis locis peregrinum . . . sed in omni civitate esse civem."

—*Vitruvius, VI, Introd.* 2.

The sources.

THIS chapter will examine what may be called ancient applied science and its relations to magic, taking observations at three different points, the ten books of Vitruvius on architecture, the collection of writings which pass under the name of Hero of Alexandria, and the compositions of the Greek alchemists. The remains of Greek and Roman literature in the field of applied science are scanty, not because they were not treasured, and even added to, by the periods following, but apparently because there had thus far been so little development in the way of machinery or of power other than manual and animal. So we must make the best of what we have. The writings to be considered are none of them earlier than the period of the Roman Empire but like other writings of that time they more or less reflect the scientific achievements or the occult lore of the preceding Hellenistic period.

Vitruvius depicts architecture as free from magic.

Vitruvius lived just at the beginning of the Empire under Julius and Augustus Caesar. He is not much of a writer, but architecture as set forth in his book appears sane, straightforward, and solid. The architect is represented as going about his business with scarcely any admixture of magical procedure or striving after marvelous results. The combined guidance of practical utility and of high standards of art—Vitruvius stresses reality and propriety now and again, and has little patience with mere show—perhaps accounts for this high degree of freedom from superstition. Perhaps permanent building is an honest, downright, open, constructive art where error is at once apparent and superstition finds little hold. If so, one wonders how there came to be so much mystery enveloping Free-Masonry. At any rate, not only in his building directions, but even in his instructions for the preparation of lime, stucco, and bricks, or his discussion of colors, natural and artificial, Vitruvius seldom or never embodies anything that can be called magical.[1]

Occult virtue and number.

This is the more noteworthy because passages in the very same work show him to have accepted some of the theories which we have associated with magic. Thus he appears to believe in occult virtues and marvelous properties of things in nature, since he affirms that, while Africa in general abounds in serpents, no snake can live within the boundaries of the African city of Ismuc, and that this is a property of the soil of that locality which it retains when exported.[2] Vitruvius also mentions some marvelous waters. One

[1] As much can hardly be said of our present day architects, whose fantastic tin cornices projecting far out from the roofs of high buildings and rows of stones poised horizontally in midair, with no other visible support than a plate glass window beneath, remind one forcibly and painfully of the deceits and levitations of magicians.

[2] *De architectura,* ed. F. Krohn, Leipzig, Teubner, 1912, VIII, iii, 24. A recent English translation of Vitruvius is by M. H. Morgan, Harvard University Press, 1914.

breaks every metallic receptacle and can be retained only in a mule's hoof. Some springs intoxicate; others take away the taste for wine. Others produce fine singing voices.[1] Vitruvius furthermore speaks of six and ten as perfect numbers and contends that the human body is symmetrical in the sense that the distances between the different parts are exact fractions of the whole.[2] He also tells how the Pythagoreans composed books on the analogy of the cube, allowing in any one treatise no more than three books of 216 lines each.[3]

Astrology. Vitruvius also more than once implies his confidence in the art of astrology. In mapping out the ground-plan of his theater he advises inscribing four equilateral triangles within the circumference of a circle, "as the astrologers do in a figure of the twelve signs of the zodiac, when they are making computations from the musical harmony of the stars." [4] I cannot make out that there is any astrological significance or magical virtue in this so far as the arrangement of the theater is concerned, but it shows that Vitruvius and his readers are familiar with the technique of astrology and the *trigona* of the signs. In another passage, comparing the physical characteristics and temperaments of northern and southern races, which astrologers generally interpreted as evidence of the influence of the constellations upon mankind, Vitruvius patriotically contends that the inhabitants of Italy, and especially the Romans, represent a happy medium between north and south, combining the greater courage of the northerners with the keener intellects of the southerners, just as the planet Jupiter is a golden mean between the extreme influences of Mars and Saturn. So the Romans are fitted for world rule, overcoming barbarian valor by their superior intelligence and the devices of the southerners by their valor.[5] In a third passage Vitruvius says more expressly of the art of astrology: "As for the branch of

[1] VIII, iii, 16, 20-21, 24-5.
[2] III, i.
[3] V, Introduction, 3-4.
[4] V, vi, 1. The wording is that of Morgan's translation.
[5] VI, i, 3-4, 9-10.

astronomy which concerns the influences of the twelve signs, the five stars, the sun, and the moon upon human life, we must leave all this to the calculations of the Chaldeans, to whom belongs the art of casting nativities, which enables them to declare the past and the future by means of calculations based on the stars. These discoveries have been transmitted by men of genius and great acuteness who sprang directly from the nations of the Chaldeans; first of all, by Berosus, who settled in the island state of Cos, and there opened a school. Afterwards Antipater pursued the subject; then there was Archinapolus, who also left rules for casting nativities, based not on the moment of birth but on that of conception." After listing a number of natural philosophers and other astronomers and astrologers, Vitruvius concludes: "Their learning deserves the admiration of mankind; for they were so solicitous as even to be able to predict, long beforehand, with divining mind, the signs of the weather which was to follow in the future."[1]

Divergence between theory and practice, learning and art.

Such a passage demonstrates plainly enough Vitruvius' full confidence in the art of casting nativities and of weather prediction, but it has no integral connection with his practical architecture or even any necessary connection with the construction of a sun-dial, which is what he is actually driving at. But Vitruvius believed that an architect should not be a mere craftsman but broadly educated in history, medicine, and philosophy, geometry, music, and astronomy, in order to understand the origin and significance of details inherited from the art of the past, to assure a healthy building, proper acoustics, and the like. It is in an attempt to air his learning and in the theoretical portions of his work that he is prone to occult science. But the practical processes of architecture and military engineering are free from it.

Evils in contemporary learning.

The attitude of Vitruvius towards other architects of his own age, to past authorities, and to personal experimentation is of interest to note, and roughly parallels the attitude of Galen in the field of medicine. Like Galen he com-

[1] IX, vi, 2-3, Morgan's translation.

plains that the artist must plunge into the social life of the day in order to gain professional success and recognition.[1] "And since I observe that the unlearned rather than the learned are held in high favor, deeming it beneath me to struggle for honors with the unlearned, I will rather demonstrate the virtue of our science by this publication." [2] He also objects to the self-assertion and advertising of themselves in which many architects of his time indulge.[3] He recognizes, however, that the state of affairs was much the same in time past, since he tells a story how the Macedonian architect, Dinocrates, forced himself upon the attention of Alexander the Great solely by his handsome and stately appearance,[4] and since he asserts that the most famous artists of the past owe their celebrity to their good fortune in working for great states or men, while other artists of equal merit are seldom heard of.[5] He also speaks of those who plagiarize the writings of others, especially of the men of the past.[6] But all this does not lead him to despair of art and learning; rather it confirms him in the conviction that they alone are really worth while, and he quotes several philosophers to that effect, including the saying of Theophrastus that "the learned man alone of all others is no stranger even in foreign lands . . . but is a citizen in every city." [7]

Authorities and inventions.

In contradistinction to the plagiarists Vitruvius expresses his deep gratitude to the men of the past who have written books, and gives lists of his authorities,[8] and declares that "the opinions of learned authors . . . gain strength as time

[1] III, Introduction, 3, ". . . There should be the greatest indignation when, as often, good judges are flattered by the charm of social entertainments into an approbation which is a mere pretence."

[2] *Idem.*

[3] VI, Introduction, 5.

[4] II, Introduction. Vitruvius continues, "But as for me, Emperor, nature has not given me stature, age has marred my face, and my strength is impaired by ill health. Therefore, since these advantages fail me, I shall win your approval, as I hope, by the help of my knowledge and my writings."

[5] III, Introduction, 2.

[6] VII, Introduction, 1-10.

[7] VI, Introduction, 2. Also IX, Introduction, where authors are declared superior to the victorious athletes in the Olympian, Pythian, Isthmian, and Nemean games.

[8] VII, Introd., 11-14; IX, Introd.

goes on."[1] "Relying upon such authorities, we venture to produce new systems of instruction."[2] Or, as he says in discussing the properties of waters, "Some of these things I have seen for myself, others I have found written in Greek books."[3] But in describing sun-dials he frankly remarks, "I will state by whom the different classes and designs of dials have been invented. For I cannot invent new kinds myself at this late day, nor do I think that I ought to display the inventions of others as my own."[4] He also gives an account of a number of notable miscellaneous discoveries and experiments by past mathematicians and physicists.[5] Also he sometimes repeats the instruction which he had received from his teachers. Like Pliny a little later he thinks that in some respects artistic standards have been lowered in his own time, notably in fresco-painting.[6] But also, like Galen, he once admits that there are still good men in his own profession besides himself, affirming that "our architects in the old days, and a good many even in our own times, have been as great as those of the Greeks."[7] He describes a basilica which he himself had built at Fano.[8]

Machines and Ctesibius.

Vitruvius's last book is devoted to machines and military engines. Here he makes a feeble effort to introduce the factor of astrological influence, asserting that "all machinery is derived from nature, and is founded on the teaching and instruction of the revolution of the firmament."[9] Among the devices described is the pump of Ctesibius of Alexandria, the son of a barber.[10] He had already been mentioned in the preceding book[11] for the improvements which he introduced in water-clocks, especially regulating their flow according to the changing length of the hours of the day in summer and winter. Vitruvius also asserts that he constructed the first water organs, that he "discovered

[1] IX, Introd., 17.
[2] VII, Introd., 10.
[3] VIII, iii, 27.
[4] IX, vii, 7.
[5] IX, Introd.
[6] VII, v.
[7] VII, Introd., 18.
[8] V, i, 6-10.
[9] X, i, 4.
[10] X, vii.
[11] IX, viii.

the natural pressure of the air and pneumatic principles, . . . devised methods of raising water, automatic contrivances, and amusing things of many kinds, . . . blackbirds singing by means of waterworks, and *angobatae,* and figures that drink and move, and other things that have been found to be pleasing to the eye and the ear." [1] Vitruvius states that of these he has selected those that seemed most useful and necessary and that the reader may turn to Ctesibius's own works for those which are merely amusing. Pliny more briefly mentions the invention of pneumatics and water organs by Ctesibius.[2]

Hero of Alexandria.

This characterization by Vitruvius of the writings of Ctesibius also applies with astonishing fitness to some of the works current under the name of Hero of Alexandria," [3] who is indeed in a Vienna manuscript of the *Belopoiika* spoken of as the disciple or follower of Ctesibius.[4] Hero, however, is not mentioned either by Vitruvius or Pliny, and it is now generally agreed as a result of recent studies that he belongs to the second century of our era.[5] His writings are objective and impersonal and tell us much less about himself than Vitruvius's introductions to the ten books of *De architectura.*

[1] IX, viii, 2 and 4; X, vii, 4.

[2] NH, VII, 38.

[3] The work of Martin, *Recherches sur la vie et les ouvrages d'Héron d'Alexandrie,* Paris, 1854, and the accounts of Hero in histories of physics and mathematics such as those of Heller and Cajori, must now be supplemented by the long article in Pauly and Wissowa, *Realencyclopädie der classischen Altertums-wissenschaft,* (1912), cols. 992-1080. A recent briefer summary in English is the article by T. L. Heath, EB, 11th edition, XIII, 378. See also Hammer-Jensen, *Ptolemaios und Heron,* in *Hermes,* XLVIII (1913), p. 224, *et seq.*

The writings ascribed to Hero, hitherto scattered about in various for the most part inaccessible editions and MSS, are now appearing in a single Teubner edition, of which five vols. have appeared, 1899, 1900, 1903, 1912, 1914, including respectively, the *Pneumatics* and *Automatic Theater,* the *Mechanics* and *Mirrors,* the *Metrics* and *Dioptra,* the *Definitions* and geometrical remains, *Stereometrica* and *De mensuris* and *De geodaesia.* For the *Belopoiika* or work on military engines see C. Wescher, *Poliorcétique des Grecs,* Paris, 1867. In English we have *The Pneumatics of Hero of Alexandria,* translated for Bennet Woodcroft by J. G. Greenwood, London, 1851. A number of articles on Hero by Heiberg, Carra de Vaux, Schmidt, and others will be found in *Bibliotheca Mathematica* and Sudhoff's *Archiv f. d. Gesch. d. Naturwiss. u. d. Technik.*

[4] *παρὰ Ἥρωνος Κτησιβίου.*

[5] Heath in EB, XIII, 378; Heiberg (1914), V, ix.

The similarity in content of his writings to those of the much earlier Ctesibius as well as the character of his terminology suggest that he stands at the end of a long development. He speaks of his own discoveries, but perhaps in the main simply continues and works over the previous principles and mechanisms of men like Ctesibius. As things stand, however, his works constitute our most important, and often our only, source for the history of exact science and of technology in antiquity.[1]

Medieval working over of the texts.

Not only does Hero seem to have been in large measure a compiler and continuer of previous science, his works also have evidently been worked over and added to in subsequent periods and bear marks of the Byzantine, Arabian, and medieval Latin periods as well as of the Hellenistic and Roman. Indeed Heiberg regards the *Geometry* and *De stereometricis* and *De mensuris* as later Byzantine collections which have perhaps made some use of the works of Hero, while the *De geodaesia* is an epitome of, or extract from, a pseudo-Heronic collection. The *Catoptrica* is known only from the Latin translation of 1269, probably by William of Moerbeke, and long known as *Ptolemy on Mirrors.* It appears, however, to be directly translated from the Greek and not from the Arabic. The *Mechanics,* on the other hand, is known only from the Arabic translation by Costa ben Luca. Of the *Pneumatics* we have Greek, Arabic, and Latin versions. It was apparently known to the author of the thirteenth century *Summa philosophiae* ascribed to Robert Grosseteste, since he speaks of the investigations of vacuums made by "Hero, that eminent philosopher, with the aid of water-clocks, siphons, and other instruments." [2] Scholars are of the opinion that the Arabic adaptation, which is of popular character and limited to the entertaining side, comes closer to the original Greek version of Hero's time than does the Latin version which devotes more attention to experimental physics. The *Automatic Theater,* for which there is the same

[1] PW, *Heron.*

[2] Baur (1912), p. 417.

chief manuscript as for the *Pneumatics,* also seems to have been worked over and added to a great deal.

Hero's thaumaturgy.

From Vitruvius's allusions to the works of Ctesibius and from a survey of those works current under Hero's name which are chiefly concerned with mechanical contrivances and devices, the modern reader gets the impression that, aside from military engines and lifting appliances, the science of antiquity was applied largely to purposes of entertainment rather than practical usefulness. However, in Hero's case at least there is something more than this. His apparatus and experiments are not intended so much to divert as to deceive the spectator, and not so much to amuse as to astound him. The mechanism is usually concealed; the cause acts indirectly, intermediately, or from a distance to produce an apparently marvelous result. It is a case of thaumaturgy, as Hero himself says,[1] of apparent magic. In fine, the experimental and applied scientist is largely interested in vying with the feats of the magicians or supplying the temples and altars of religion with pseudo-miracles.

Instances of experimental proof.

The introduction or proemium to the *Pneumatics* is rather more truly scientific and has been called an unusual instance in antiquity of the use as proof of purposive observation of nature and experiment. Thus the existence of air is demonstrated by the experiment of pressing an inverted vessel, kept carefully upright, into water, which will not enter the vessel because of the resistance offered by the air already within the vessel. Or the elasticity of air and the existence of empty spaces between its particles is shown by the experiment of blowing more air into a globe through a siphon, and then holding one's finger over the orifice. As soon as the finger is removed the surplus air rushes out with a loud report. Along with such admirable experimental proof, however, the introduction contains some astonishingly erroneous assertions, such as that "slime and mud are transformations of water into earth," and that air released from

[1] In the first chapter of the *Automatic Theater* he says, "The ancients called those who constructed such things thaumaturges because of the astounding character of the spectacle."

a vessel under water "is transformed so as to become water." Hero believes that heat and light rays are particles of matter which penetrate the interstices between the particles composing air and water.

Magic jugs and drinking animals.

The *Pneumatics* consist of some seventy-eight theorems or experiments or tricks, call them what you will, which in different manuscripts and editions are variously grouped in a single book or two books. The same idea or method, however, is often repeated in the different chapters. Thus we encounter over half a dozen times the magic water-jar or drinking horn from which either wine or water or a mixture of both can be poured, or a choice of other liquids. And in all these cases the explanation of the trick is the same. When the air-hole in the top of the vessel is closed so that no air can enter, the liquid will not flow out through the narrow orifice in the bottom. Changes are rung on this principle by means of inner compartments and connecting tubes. Different kinds of siphons, the bent, the enclosed, and the uniform discharge, are described in the opening chapters and are utilized in working the ensuing wonders, such as statues of animals which drink water offered to them, inexhaustible goblets or those that will not overflow, and harmonious jars. By this last expression is meant pairs of vessels, secretly connected by tubes and so arranged that nothing will flow from one until the other is filled, when wine will pour from one jar and water from the other. Or when water is poured into one jar, wine or mixed wine and water flows from the other. Or, when water is drawn off from one jar, wine flows from the other. Other vessels are made to commence or cease to pour out wine or water, when a little water is poured in. Others will receive no more water once you have ceased pouring it in, no matter how little may have been poured in, or, when you cease for a moment to pour water in and then begin again, will not resume their outpour until half full. In another case the water will not flow out of a hole in the bottom of the vessel at all until the vessel is entirely filled. Others are made to flow by dropping a coin in a slot or working a lever, or turning a wheel. In the last case the vessel of water is concealed behind the entrance column of a temple. In one magic drinking horn the flow of water from the bottom is checked by putting a cover over the open top. When another pitcher is tipped up, the same amount of liquid will always flow out.

Various automatons and devices.

In half a dozen chapters mechanical birds are made to sing by driving air through a pipe by the pressure of flowing water. In other chapters a dragon is made to hiss and a thyrsus to whistle by similar methods. By the force of compressed air water is made to spurt forth and automatons to sound trumpets. The heat of the sun's rays is used to warm air which expands and causes water to trickle out. In a number of cases as long as a fire burns on an altar the expansion of enclosed air caused thereby opens temple doors by the aid of pulleys, or causes statues to pour libations, dancing figures to revolve, and a serpent to hiss. The force of steam is used to support a ball in mid-air, revolve a sphere, and make a bird sing or a statue blow a horn. Inexhaustible lamps are described as well as inexhaustible goblets, and a self-trimmed lamp in which a float resting on the oil turns a cog-wheel which pushes up the wick as it and the oil are consumed. Floats and cog-wheels are also used in some of the tricks already mentioned. In another the flow of a liquid from a vessel is regulated by a float and a lever. Cog-wheels are also employed in constructing the neck of an automaton so that it can be cut completely through with a knife and yet the head not be severed from the body. A cupping glass, a syringe, a fire engine pump with valves and pistons, a hydraulic organ and one worked by wind pretty much exhaust the contents of the *Pneumatics*. In its introduction Hero alludes to his treatise in four books on water-clocks, but this is not extant. Hero's water-organ is regarded as more primitive than that described by Vitruvius.[1]

Magic mirrors.

If magic jugs and marvelous automatons make up most of the contents of the *Pneumatics* and *Automatic Theater*,

[1] PW, 1045.

comic and magic mirrors play a prominent part in the *Catoptrics*. The spectator sees himself upside down, with three eyes, two noses, or an otherwise distorted countenance. By means of two rectangular mirrors which open and close on a common axis Pallas is made to spring from the head of Zeus. Instructions are given how to place mirrors so that the person approaching will see no reflection of himself but only whatever apparition you select for him to see. Thus a divinity can be made suddenly to appear in a temple. Clocks are also described where figures appear to announce the hours.

Astrology and occult virtue.

Hero displays a slight tendency in the direction of astrology, discussing the music of the spheres in the first chapters of the *Catoptrics,* and in the *Pneumatics* describing an absurdly simple representation of the cosmos by means of a small sphere placed in a circular hole in the partition between two halves of a transparent sphere of glass. One hemisphere is to be filled with water, probably in order to support the ball in the center.[1] The marvelous virtues of animals other than automatons are rather out of his line, but he alludes to the virtue of the marine torpedo which can penetrate bronze, iron, and other bodies.

Date of extant Greek alchemy.

Although we have seen some indications of its earlier existence in Egypt, alchemy seems to have made its appearance in the ancient Greek-speaking and Latin world only at a late date. There seems to be no allusion to the subject in classical literature before the Christian era, the first mention being Pliny's statement that Caligula made gold from orpiment.[2] The papyri containing alchemistic texts are of

[1] But perhaps this is a medieval interpolation in the nature of a crude Christian attempt to depict "the firmament in the midst of the waters" (Genesis, I, 6). However, it also somewhat resembles the universe of the Greek philosopher, Leucippus, who "made the earth a hemisphere with a hemisphere of air above, the whole surrounded by the supporting crystal sphere which held the moon. Above this came the planets, then the sun"—Orr (1913), p. 63 and Fig. 13. See also K. Tittel, "Das Weltbild bei Heron," in *Bibl. Math.* (1907-1908), pp. 113-7.

[2] Berthelot (1885), pp. 68-9. For the following account of Greek alchemy I have followed Berthelot's three works, *Les Origines de l'Alchimie,* 1885; *Collection des anciens Alchimistes Grecs,* 3 vols., 1887-1888; *Introduction à l'Étude*

the third century, and the manuscripts containing Greek works of alchemy, of which the oldest is one of the eleventh century in the Library of St. Mark's, seem to consist of works or remnants of works written in the third century and later, many being Byzantine compilations, excerpts, or additions. Also Syncellus, the polygraph of the eighth century, gives some extracts from the alchemists.

Legend that Diocletian burned the books of the alchemists.

Syncellus and other late writers [1] are our only extant sources for the statement that Diocletian burned the books of the alchemists in Egypt, so that they might not finance future revolts against him. If the report be true, one would fancy that the imperial edict would be more effective as a testimonial to the truth of transmutation in encouraging the art than it would be in discouraging it by destroying a certain amount of its literature. Thus the edict would resemble the occasional laws of earlier emperors banishing the astrologers—except their own—from Rome or Italy because they had been too free in predicting the death of the emperor, which only serve to show what a hold astrology had both on emperors and people. But the report concerning Diocletian sounds improbable on the face of it and must be doubted for want of contemporary evidence. Certainly we are not justified in explaining the air of secrecy so often assumed by writers on alchemy as due to the fear of persecution which this action of Diocletian [2] or the fear of being accused of magic aroused in them. Persons who wish to keep matters secret do not rush into publication, and the air of secrecy of the alchemists is too often evidently assumed for purposes of

de la Chimie, 1889. Berthelot made a good many books from too few MSS; went over the same ground repeatedly; and sometimes had to correct his previous statements; but still remains the fullest account of the subject. E. O. v. Lippmann, *Entstehung und Ausbreitung der Alchemie,* 1919, is still based largely on Berthelot's publications. In English see C. A. Browne, "The Poem of the Philosopher Theophrastos upon the Sacred Art: A Metrical Translation with Comments upon the History of Alchemy," in *The Scientific Monthly,* September, 1920, pp. 193-214.

[1] The earliest of them is John of Antioch of the reign of Heraclius, about 620 A.D., although they seem to use Panodorus, an Egyptian monk of the reign of Arcadius. Even he would be a century removed from the event.

[2] Berthelot (1885), pp. 26, 72, etc., took this story about Diocletian far too seriously.

show and to impress the reader with the idea that they really have something to hide. Sometimes the alchemists themselves realize that this adoption of an air of secrecy has been overdone. Thus Olympiodorus wrote in the early fifth century, "The ancients were accustomed to hide the truth, to veil or obscure by allegories what is clear and evident to everybody."[1] Nor can we accept the story of Diocletian's burning the books of alchemy as the reason why none have reached us which can be certainly dated as earlier than the third century.

Alchemists' own accounts of the history of their art.

The alchemists themselves, of course, claimed for their art the highest antiquity. Zosimus of Panopolis, who seems to have written in the third century, says that the fallen angels instructed men in alchemy as well as in the other arts, and that it was the divine and sacred art of the priests and kings of Egypt, who kept it secret. We also have an address of Isis to her son Horus repeating the revelation made by Amnael, the first of the angels and prophets. To Moses are ascribed treatises on domestic chemistry and doubling the weight of gold.[2] The manuscripts of the Byzantine period discuss what "the ancients" meant by this or that, or purport to repeat what someone else said of some other person. Zosimus seems fond of citing himself in the texts reproduced by Berthelot, so that it may be questioned how much of his original works has been preserved. Hermes is often cited by the alchemists, although no work of alchemy ascribed to him has reached us from this early period. To Agathodaemon is ascribed a commentary on the oracle of Orpheus addressed to Osiris, dealing with the whitening and

[1] Berthelot (1885), 192-3.

[2] But the *Labyrinth of Solomon*, which Berthelot (1885), p. 16, had cited as an example of the sort of ancient magic figures which had been largely obliterated by Christians, and of the antiquity of alchemy among the Jews (*ibid.*, p. 54), although he granted (*ibid.*, p. 171) that it might not be as old as the Papyrus of Leyden of the third century, later when he had secured the collaboration of Ruelle (1888), I, 156-7, and III, 41, he had to admit was not even as old as the eleventh century MS in which it occurred but was an addition in writing of the fourteenth century and "a cabalistic work of the middle ages which does not belong to the old tradition of the Greek alchemists."

yellowing of metals and other alchemical recipes. Other favorite authorities are Ostanes, whom we have elsewhere heard represented as the introducer of magic into the Greek world, and the philosopher Democritus, whom the alchemists represent as the pupil of Ostanes and whom we have already heard Pliny charge with devotion to magic. Seneca says in one of his letters that Democritus discovered a process to soften ivory, that he prepared artificial emerald, and colored vitrified substances. Diogenes Laertius ascribes to him a work on the juices of plants, on stones, minerals, metals, colors, and coloring glass. This was possibly the same as the four books on coloring gold, silver, stones, and purple ascribed to Democritus by Synesius in the fifth, and Syncellus in the eighth, century. More recent presumably than Ostanes and Democritus are the female alchemists, Cleopatra and Mary the Jewess, although one text represents Ostanes and his companions as conversing with Cleopatra. A few of the spurious works ascribed to these authors may have come into existence as early as the Hellenistic period, but those which have reached us, at least in their present form, seem to bear the marks of the Christian era and later centuries of the Roman Empire, if not of the early medieval and Byzantine periods. And those authors whose names seem genuine: Zosimus, Synesius, Olympiodorus, Stephanus, are of the third, fourth and fifth centuries, at the earliest.

Close association of Greek alchemy with magic.

The associations of the names above cited and the fact that pseudo-literature forms so large a part of the early literature of alchemy suggest its close connection at that time with magic. Whereas Vitruvius, although not personally inhospitable to occult theory, showed us the art of architecture free from magic, and Hero told how to perform apparent magic by means of mechanical devices and deceits, the Greek alchemists display entire faith in magic procedure with which their art is indissolubly intermingled. Indeed the papyri in which works of alchemy occur are primarily magic papyri, so that alchemy may be said to spring from the brow of magic. The same is only somewhat less true of the manuscripts. In the earliest one of the eleventh century the alchemy is in the company of a treatise on the interpretation of dreams, a sphere of divination of life or death, and magic alphabets. The treatises of alchemy themselves are equally impregnated with magic detail. Cleopatra's art of making gold employs concentric circles, a serpent, an eight-rayed star, and other magic figures. *Physica et mystica,* ascribed to Democritus, after a purely technical fragment on purple dye, invokes his master Ostanes from Hades, and then plunges into alchemical recipes. There are also frequent bits of astrology and suggestions of Gnostic influence. Often the encircling serpent Ouroboros, who bites or swallows his tail, is referred to.[1] Sometimes the alchemist puts a little gold into his mixture to act as a sort of nest egg, or mother of gold, and encourage the remaining substance to become gold too.[2] Or we read in a work ascribed to Ostanes of "a divine water" which "revives the dead and kills the living, enlightens obscurity and obscures what is clear, calms the sea and quenches fire. A few drops of it give lead the appearance of gold with the aid of God, the invisible and all-powerful. . . ."[3]

Mystery and allegory.

These early alchemists are also greatly given to mystery and allegory. "Touch not the philosopher's stone with your hands," warns Mary the Jewess, "you are not of our race, you are not of the race of Abraham."[4] In a tract concerning the serpent Ouroboros we read, "A serpent is stretched out guarding the temple. Let his conqueror begin by sacrifice, then skin him, and after having removed his flesh to the very bones, make a stepping-stone of it to enter the temple. Mount upon it and you will find the object sought. For the priest, at first a man of copper, has changed his color and nature and become a man of silver; a few days later, if you wish, you will find him changed into a man of gold."[5] Or in the preparation of the aforesaid divine

[1] Berthelot (1885), p. 59.
[2] *Ibid.*, p. 53.
[3] Berthelot (1888), III, 251.
[4] Berthelot (1885), p. 56.
[5] Berthelot (1888), III, 23.

water Ostanes tells us to take the eggs of the serpent of oak who dwells in the month of August in the mountains of Olympus, Libya, and the Taurus.[1] Synesius tells that Democritus was initiated in Egypt at the temple of Memphis by Ostanes, and Zosimus cites the instruction of Ostanes, "Go towards the stream of the Nile; you'll find there a stone; cut it in two, put in your hand, and take out its heart, for its soul is in its heart." [2] Zosimus himself often resorts to symbolic jargon to obscure his meaning, as in the description of the vision of a priest who was torn to pieces and who mutilated himself.[3] He, too, personifies the metals and talks of a man of gold, a tin man, and so on.[4] A brief example of his style will have to suffice, as these allegories of the alchemists are insufferably tedious reading. "Finally I had the longing to mount the seven steps and see the seven chastisements, and one day, as it chanced, I hit upon the path up. After several attempts I traversed the path, but on my return I lost my way and, profoundly discouraged, seeing no way out, I fell asleep. In my dream I saw a little man, a barber, clothed in purple robe and royal raiment, standing outside the place of punishment, and he said to me. . . ." [5] When Zosimus was not dreaming dreams and seeing visions, he was usually citing ancient authorities.

Experimentation in alchemy: relation to science and philosophy.

At the same time even these early alchemists cannot be denied a certain scientific character, or at least a connection with natural science. Behind alchemy existed a constant experimental progress. "Alchemy," said Berthelot, "rested upon a certain mass of practical facts that were known in antiquity and that had to do with the preparation of metals, their alloys, and that of artificial precious stones; it had there an experimental side which did not cease to progress during the entire medieval period until positive modern chemistry emerged from it." [6] The various treatises of the Greek alchemists describe apparatus and experiments which are real

[1] Berthelot (1888), III, 251.

[2] Berthelot (1885), p. 164.

[3] *Ibid.*, pp. 179-80.

[4] *Ibid.*, p. 60.

[5] Berthelot (1888), II, 115-6; III, 125.

[6] Berthelot (1885), pp. 211-2.

but with which they associated results which were impossible and visionary. Their theories of matter seem indebted to the earlier Greek philosophers, while in the description of nature Berthelot noted a "direct and intimate" relation between them and the works of Dioscorides, Vitruvius, and Pliny.[1]

[1] Berthelot (1889), p. vi.

CHAPTER VI

PLUTARCH'S ESSAYS

Themes of ensuing chapters—Life of Plutarch—Superstition in Plutarch's *Lives*—His *Morals* or *Essays*—Question of their authenticity—Magic in Plutarch—*Essay on Superstition*—Plutarch hospitable toward some superstitions—The oracles of Delphi and of Trophonius—Divination justified—Demons as mediators between gods and men—Demons in the moon: migration of the soul—Demons mortal: some evil—Men and demons—Relation of Plutarch's to other conceptions of demons—The astrologer Tarrutius—*De fato*—Other bits of astrology—Cosmic mysticism—Number mysticism—Occult virtues in nature—Asbestos—*On Rivers and Mountains*—Magic herbs—Stones found in plants and fish—Virtues of other stones—Fascination—Animal sagacity and remedies—Theories and queries about nature—The Antipodes.

Themes of ensuing chapters

HAVING noted the presence of magic in works so especially devoted to natural science as those of Pliny, Galen, and Ptolemy, we have now to illustrate the prominence both of natural science and of magic in the life and thought of the Roman Empire by a consideration of some writers of a more miscellaneous character, who should reflect for us something of the interests of the average cultured reader of that time. Of this type are Plutarch, Apuleius and Philostratus, whom we shall consider in the coming chapters in the order named, which also roughly corresponds to their chronological sequence.

Life of Plutarch.

Plutarch flourished during the reigns of Trajan and Hadrian at the turn of the first and second centuries, but *The Letter on the Education of a Prince to Trajan* [1] probably is not by him, and the legend that Hadrian was his pupil is a medieval invention. He was born in Boeotia about 46-48 A. D. and was educated in rhetoric and philosophy, science and mathematics, at Athens, where he was a student

[1] *De institutione principis epistola ad Traianum,* a treatise extant only in Latin form.

when Nero visited Greece in 66 A. D. He also made several visits to Rome and resided there for some time. He held various public positions in the province of Achaea and in his small native town of Chaeronea, and had official connections with the Delphic oracle and amphictyony. Artemidorus in the *Oneirocriticon* states that Plutarch's death was foreshadowed in a dream.[1]

Superstition in Plutarch's *Lives.*

With Plutarch's celebrated *Lives of Illustrious Men,* as with narrative histories in general, we shall not be much concerned, although they of course abound in omens and portents, in bits of pseudo-science which details in his narrative bring to the mind of the biographer, and in cases of divination and magic. Thus theories are advanced to explain why birds dropped dead from mid-air at the shout set up by the Greeks at the Isthmian games when Flamininus proclaimed their freedom. Or we are told how Sulla received from the Chaldeans predictions of his future greatness, how in the dedication to his *Memoirs* he admonished Lucullus to trust in dreams, and how Lucullus's mind was deranged by a love philter administered by his freedman in the hope of increasing his master's affection towards him.[2] Such allusions and incidents abound also of course in Dio Cassius, Tacitus, and other Roman historians.

His *Morals* or *Essays.*

But we shall be concerned rather with Plutarch's other writings, which are usually grouped together under the title of *Morals,* or, more appropriately, *Miscellanies and Essays.* Not only is there great variety in their titles, but in any given essay the attention is usually not strictly held to one theme or problem but the discussion diverges to other points. Some are by their very titles and form rambling dialogues, symposiacs, and table-talk, where the conversation lightly flits from one topic to other entirely different ones, never dwelling for long upon any one point and never re-

[1] IV, 72. On the biography and bibliography of Plutarch consult Christ, *Gesch. d. Griechischen Litteratur,* 5th ed., Munich, 1913, II, 2, "Die nachklassische Periode," pp. 367ff.

[2] See also the essay, "Whether an old man should engage in politics," cap. 16.

turning to its starting-point. This dinner-table and drinking-bout type of cultured and semi-learned discourse has other extant ancient examples such as the *Attic Nights* of Aulus Gellius and the *Deipnosophists* of Athenaeus, but Plutarch will have to serve as our main illustration of it. His *Essays* reflect in motley guise and disordered array the fruits of extensive reading and a retentive memory in ancient philosophy, science, history, and literature.

Question of their authenticity.

The authenticity of some of the essays attributed to him has been questioned, and very likely with propriety, but for our purpose it is not important that they should all be by the same author so long as they represent approximately the same period and type of literature. The spurious treatise, *De placitis philosophorum,* we have already considered in the chapter on Galen, to whom it has also been ascribed. The essay *On Rivers and Mountains* we shall treat by itself in the present chapter. The *De fato* has also been called spurious.[1] Superstitious content is not a sufficient reason for denying that a treatise is by Plutarch,[2] since he is superstitious in writings of undoubted genuineness and since we have found the leading scientists of the time unable to exclude superstition from their works entirely. Moreover, many of the essays are in the form of conversations expressing the divergent views of different speakers, and it is not always possible to tell which shade of opinion Plutarch himself favors. Suffice it that the views expressed are those of men of education.

Magic in Plutarch.

Plutarch does not specifically discuss magic under that name at any length in any of his essays, but does treat of

[1] See R. Schmertosch, in *Philol.-Hist. Beitr. z. Ehren Wachsmuths,* 1897, pp. 28ff.

[2] Language and literary form are surer guides and have been applied by B. Weissenberger, *Die Sprache Plutarchs von Chäronea und die pseudoplutarchischen Schriften,* II Progr. Straubing, 1896, pp. 15ff. In 1876 W. W. Goodwin, editing a revised edition of the seventeenth century English translation of the *Morals,* declared that no critical translation was possible until a thorough revision of the text had been undertaken with the help of the best MSS. Since then an edition of the text by G. N. Bernadakes, 1888-1896, has appeared, but it has not escaped criticism.

such subjects as superstition in general, dreams, oracles, demons, number, fate, the craftiness of animals, and other "natural questions." Certain vulgar forms of magic, at least, were regarded by him with disapproval or incredulity.[1] He rejects as a fiction the statement that the women of Thessaly can draw down the moon by their spells, but thinks that the notion perhaps originated in the fact or story that Aglaonice, daughter of Hegetor, was so skilful in astrology or astronomy as to be able to foresee the occurrence of lunar eclipses, and that she deluded the people into believing that at such times she brought down the moon from heaven by charms and enchantments.[2] Thus we have one more instance of the union of magic and science, this time of pseudo-magic with real science as at other times of magic with pseudo-science.

Essay on superstition.

The essay entitled *περὶ δεισιδαιμονίας* deals with superstition in the usual Greek sense of dread or excessive fear of demons and gods. We are accustomed to think of Hellenic paganism as a cheerful faith, full of naturalism, in which the gods were humanized and made familiar. Plutarch apparently regards normal religion as of this sort, and attacks the superstitious dread of the supernatural. He contends that such fear is worse, if anything, than atheism, for it makes men more unhappy and is an equal offense against the divinity, since it is at least as bad to believe ill of the gods as not to believe in them at all. Nothing indeed encourages the growth of atheism so much as the absurd practices and beliefs of such superstitious persons, "their words and

[1] The English translation of Plutarch's *Morals* "by several hands," first published in 1684-1694, sixth edition corrected and revised by W. W. Goodwin, 5 vols., 1870-1878, IV, 10, renders a passage in the seventh chapter of *De defectu oraculorum,* in which complaint is made of the "base and villainous questions" which are now put to the oracle of Apollo, as follows: "some coming to him as a mere paltry astrologer to try his skill and impose upon him with subtle questions." But the corresponding clause in the Greek text is merely *οἱ μὲν ὡς σοφιστοῦ διάπειραν λαμβάνοντες*, and there seems to be no reason for taking the word "sophist" in any other than its usual meaning. The passage therefore cannot be interpreted as an attack upon even vulgar astrologers.

[2] *De defectu oraculorum,* 13.

motions, their sorceries and magics, their runnings to and fro and beatings of drums, their impure rites and their purifications, their filthiness and chastity, their barbarian and illegal chastisements and abuse." [1] Plutarch seems to be in part animated by the common prejudice against all other religions than one's own, and speaks twice with distaste of Jewish Sabbaths. He also, however, as the passage just quoted shows, is opposed to the more extreme and debasing forms of magic, and declares that the superstitious man becomes a mere peg or post upon which all the old-wives hang any amulets and ligatures upon which they may chance.[2] He further condemns such historic instances of superstition as Nicias's suspension of military operations during a lunar eclipse on the Sicilian expedition.[3] There was nothing terrible, says Plutarch, with his usual felicity of antithesis, in the periodic recurrence of the earth's shadow upon the moon; but it was a terrible calamity that the shadow of superstition should thus darken the mind of a general at the very moment when a great crisis required the fullest use of his reason.

Plutarch hospitable toward some superstitions.

In the essay upon the demon of Socrates one of the speakers, attacking faith in dreams and apparitions, commends Socrates as one who did not reject the worship of the gods but who did purify philosophy, which he had received from Pythagoras and Empedocles full of phantasms and myths and the dread of demons, and reeling like a Bacchanal, and reduced it to facts and reason and truth.[4] Another of the company, however, objects that the demon of Socrates outdid the divination of Pythagoras.[5] These conflicting opinions may be applied in some measure to Plutarch himself. His censure of dread of demons and excessive superstition is not to be taken as a sign of scepticism on his part in oracles, dreams, or the demons themselves. To these matters we next turn.

[1] Cap. 12.
[2] Cap. 7.
[3] Cap. 8.
[4] Cap. 9.
[5] Cap. 10.

The oracles of Delphi and of Trophonius.

Plutarch's faith and interest in oracles in general and in the Delphian oracle of Apollo in particular are attested by three of his essays, the *De defectu oraculorum, De Pythiae oraculis* and *De Ei apud Delphos*. At the same time these essays attest the decline of the oracles from their earlier popularity and greatness. The oracular cave of Trophonius, of which we shall hear again in the *Life of Apollonius of Tyana,* also comes into Plutarch's works, and the prophetic and apocalyptic vision is described of a youth who spent two nights and a day there in an endeavor to learn the nature of the demon of Socrates.[1]

Divination justified.

Plutarch further had faith in divination in general, whether by dreams, sneezes or other omens; but he attempted to give a dignified philosophical and theological explanation of it. Few men receive direct divine revelation, in his opinion, but to many signs are given on which divination may be based.[2] He held that the human soul had a natural faculty of divination which might be exercised at favorable times and when the bodily state was not unfavorable.[3] A speaker in one of his dialogues justifies divination even from sneezes and like trivial occurrences upon the ground that as the faint beat of the pulse has meaning for the physician and a small cloud in the sky is for a skilful pilot a sign of impending storm, so the least thing may be a clue to the truly prophetic soul.[4] The extent of Plutarch's faith in dreams may be inferred from his discussion of the problem, Why are dreams in autumn the least reliable?[5] First there is Aristotle's suggestion that eating autumn fruit so disturbs the digestion that the soul is left little opportunity to exercise its prophetic faculty undistracted. If we accept the doctrine of Democritus that dreams are caused by images from other bodies and even minds or souls, which enter the body of the sleeper through the open pores and affect the mind, revealing to it the present passions and future de-

[1] *De genio Socratis,* 21-22.
[2] *Ibid.,* 24.
[3] *De defectu oraculorum,* 40.
[4] *De genio Socratis,* 12.
[5] *Sympos.* VIII, 10.

signs of others,—if we accept this theory, it may be that the falling leaves in autumn disturb the air and ruffle these extremely thin and film-like emanations. A third explanation offered is that in the declining months of the year all our faculties, including that of natural divination, are in a state of decline. In the case of oracles like that at Delphi it is suggested that the Pythia's natural faculty of divination is stimulated by "the prophetical exhalations from the earth" which induce a bodily state favorable to divination.[1] The god or demon, however, is the underlying and directing cause of the oracle.[2]

Demons as mediators between gods and men.

To the demons and their relations to the gods and to men we therefore next come. Plutarch's view is that they are essential mediators between the gods and men. Just as one who should remove the air from between the earth and moon would destroy the continuity of the universe, so those who deny that there is a race of demons break off all intercourse between gods and men.[3] On the other hand, the theory of demons solves many doubts and difficulties.[4] When and where this doctrine originated is uncertain, whether among the *magi* about Zoroaster, or in Thrace with Orpheus, or in Egypt or Phrygia. Plutarch likens the gods to an equilateral, the demons to an isosceles, and human beings to a scalene triangle; and again compares the gods to sun and stars, the demons to the moon, and men to comets and meteors.[5] In the youth's vision in the cave of Trophonius the moon appeared to belong to earthly demons, while those stars which have a regular motion were the demons of sages, and the wandering and falling stars the demons of men who have yielded to irrational passions.[6]

Demons in the moon: migration of the soul.

These suggestions that the moon and the air between earth and moon are the abode of the demons and this reminiscence of the Platonic doctrine of the soul and its migrations receive further confirmation in a discussion whether

[1] *De defectu oraculorum*, 44.
[2] *Ibid.*, 48.
[3] *Ibid.*, 13.
[4] *Ibid.*, 10.
[5] *Ibid.*, 13.
[6] *De genio Socratis*, 22.

the moon is inhabited in the essay, *On the Face in the Moon.* A story is there told [1] of a man who visited islands five days' sail west of Britain, where Saturn is imprisoned and where there are demons serving him. This man who acquired great skill in astrology during his stay there stated upon his return to Europe that every soul after leaving the human body wanders for a time between earth and moon, but finally reaches the latter planet, where the Elysian fields are located, and there becomes a demon.[2] The demons do not always remain in the moon, however, but may come to earth to care for oracles or be imprisoned in a human body again for some crime.[3] The man who repeats the stranger's story leaves it to his hearers, however, to believe it or not. But the struggle upward of human souls to the estate of demons is again described in the essay on the demon of Socrates,[4] where it is explained that those souls which have succeeded in freeing themselves from all union with the flesh become guardian demons and help those of their fellows whom they can reach, just as men on shore wade out as far as they can into the waves to rescue those sea-tossed, shipwrecked mariners who have succeeded in struggling almost to land. The soul is plunged into the body, the uncorrupted mind or demon remains without.[5]

The demons differ from the gods in that they are mortal, though much longer-lived than men. Hesiod said that crows live nine times as long as men, stags four times as long as crows, ravens three times as long as stags, a phoenix nine times as long as a raven, and the nymphs ten times as long as the phoenix.[6] There are storms in the isles off Britain whenever one of the demons residing there dies.[7] Some demons are good spirits and others are evil; some are more passive and irrational than others; some delight in gloomy festivals, foul words, and even human sacrifice.[8]

Demons mortal: some evil.

[1] Cap. 26.
[2] Cap. 29.
[3] Cap. 30.
[4] Cap. 24.
[5] Cap. 22.
[6] *De defectu oraculorum,* 10.
[7] *Ibid.,* 18.
[8] *Ibid.,* 13-14.

Men and demons.

Once a year in the neighborhood of the Red Sea a man is seen who spends the remainder of his time among "nymphs, nomads and demons." [1] At his annual appearance many princes and great men come to consult him concerning the future. He also has the gift of tongues to the extent of understanding several languages perfectly. His speech is like sweetest music, his breath sweet and fragrant, his person the most graceful that his interlocutor had ever seen. He also was never afflicted with any disease, for once a month he ate the bitter fruit of a medicinal herb. As to the exact nature of Socrates' demon there is some diversity of opinion. One man suggests that it was merely the sneezing of himself or others, sneezes on the left hand warning him to desist from his intended course of action, while a sneeze in any other quarter was interpreted by him as a favorable sign.[2] The weight of opinion, however, inclines towards the view that his demon did not appear to him as an apparition or phantasm, or even communicate with him as an audible voice, but by immediate impression upon his mind.[3]

Relation of Plutarch's to other conceptions of demons.

Plutarch's account of demons is the first of a number which we shall have occasion to note. As the discussion of them by Apuleius in the next chapter and the rather crude representation of them given in Philostratus's *Life of Apollonius of Tyana* will show, there was as yet among non-Christian writers no unanimity of opinion concerning demons. On the other hand there are several conceptions in Plutarch's essays which were to be continued later by Christians and Neo-Platonists: namely, the conception of a mediate class of beings between God and men, the hypothesis of a world of spirits in close touch with human life, the association of divination and oracles with demons, and the location of spirits in the sphere of the moon or the air between earth and moon,—although Plutarch sometimes connected demons with the stars above the moon. This occasional association of stars with spirits and of sinning souls with falling stars

[1] *De defectu oraculorum*, 21.
[2] *De genio Socratis*, 11.
[3] *Ibid.*, 20.

bears some resemblance to the depiction of certain stars as sinners in the Hebraic *Book of Enoch,* which was written before Plutarch's time and which we shall consider in our next book as an influence upon the development of early Christian thought.

The astrologer Tarrutius.

As for the stars apart from demons, Plutarch discusses the art of astrology as little as he does "magic" by that name. Mentions of individuals as skilled in "astrology" may simply mean that they were trained astronomers. When a veritable astrologer in our sense of the word is mentioned in one of Plutarch's *Lives,*[1] he is described as a μαθηματικός —a word often used for a caster of horoscopes and predicter of the future. Here, however, it carries no reproach of charlatanism, since in the same phrase he is called a philosopher. This Tarrutius was a friend of Varro, who asked him to work out the horoscope of Romulus backward from what was known of the later life and character of the founder of Rome. "For it was possible for the same science which predicted man's life from the time of his birth to infer the time of his birth from the events of his life." Tarrutius set to work and from the data at his disposal figured out that Romulus was conceived in the first year of the second Olympiad, on the twenty-third day of the Egyptian month Khoeak at the third hour when there was a total eclipse of the sun; and that he was born on the twenty-first day of the month Thoth about sunrise. He further estimated that Rome was founded by him on the ninth day of the month Pharmuthi between the second and third hour. For, adds Plutarch, they think that the fortunes of cities are also controlled by the hour of their genesis. Plutarch, however, seems to look upon such doctrines as rather strange and fabulous.[2] Varro, on the other hand, may have regarded it as the most scientific method possible of settling disputed questions of historical chronology.

[1] *Romulus,* cap. 12.

[2] 'Αλλὰ ταῦτα μὲν ἴσως καὶ τὰ τοιαῦτα τῷ ξένῳ καὶ περιττῷ προσάξεται μᾶλλον ἢ διὰ τὸ μυθῶδες ἐνοχλήσει τοὺς ἐντυγχάνοντας αὐτοῖς.

The *De fato.*

A favorable attitude towards astrology is found mainly in those essays by Plutarch which are suspected of being spurious, the *De fato* and *De placitis philosophorum.* Of the latter we have already treated under Galen. In the former fate is described as "the soul of the universe," and the three main divisions of the universe, namely, the immovable heaven, the moving spheres and heavenly bodies, and the region about the earth, are associated with the three Fates, Clotho, Atropos, and Lachesis.[1] It is similarly stated in the essay on the demon of Socrates [2] that of the four principles of all things, life, motion, genesis or generation, and corruption, the first two are joined by the One indivisibly, the second and third Mind unites through the sun; the third and fourth Nature joins through the moon. And over each of these three bonds presides one of the three Fates, Atropos, Clotho, and Lachesis. In other words, the one God or first cause, invisible and unmoved, in whom is life, sets in motion the heavenly spheres and bodies, through whose instrumentality generation and corruption upon earth are produced and regulated,—which is substantially the Aristotelian view of the universe. Returning to the *De fato* we may note that it repeats the Stoic theory of the *magnus annus* when the heavenly bodies resume their rounds and all history repeats itself.[3] Despite this apparent admission that human life is subject to the movements of the stars, the author of the *De fato* seems to think that accident, fortune or chance, the contingent, and "what is in us" or free-will, can all co-exist with fate, which he practically identifies with the motion of the heavenly bodies,[4] Fate is also comprehended by divine Providence but this fact does not militate against astrology, since Providence itself divides into that of the first God, that of the secondary gods or stars "who move through the heavens regulating mortal affairs, and that of the demons who act as guardians of men.[5]

[1] Cap. 2.
[2] Cap. 22.
[3] Cap. 3.
[4] Caps. 5-8.
[5] Cap. 9

One or two bits of astrology may be noted in Plutarch's other essays. The man who learned "astrology" among demons in the isle beyond Britain affirmed that in human generation earth supplies the body, the moon furnishes the soul, and the sun provides the intellect.[1] In the *Symposiacs* [2] the opinion of the mythographers is repeated that monstrous animals were produced during the war with the giants because the moon turned from its course then and rose in unaccustomed quarters. Plutarch was, by the way, inclined to distinguish the moon from other heavenly bodies as passive and imperfect, a sort of celestial earth or terrestrial star. Such a separation of the moon from the other stars and planets would have, however, no necessary contrariety with astrological theory, which usually ascribed a peculiar place to the moon and represented it as the medium through which the more distant planets exerted their effects upon the earth.

Other bits of astrology.

Sometimes Plutarch's cosmology carries Platonism to the verge of Gnosticism, a subject of which we shall treat in a later chapter. The diviner who had communed with demons, nomads, and nymphs in the desert asserted that there was not one world, but one hundred and eighty-three worlds arranged in the form of a triangle with sixty to each side and one at each angle. Within this triangle of worlds lay the plain of truth where were the ideas and models of all things that had been or were to be, and about these was eternity from which time flowed off like a river to the one hundred and eighty-three worlds. The vision delectable of those ideas is granted to men only once in a myriad of years, if they live well, and is the goal toward which all philosophy strives. The stranger, we are informed, told this tale artlessly, like one in the mysteries, and produced no demonstration or proof of what he said. We have already heard Plutarch liken gods, demons, and men to different kinds of triangles; he also repeats Plato's association of the

Cosmic mysticism.

[1] *De facie in orbe lunae*, 28. [2] VIII, 9.

five regular solids with the elements, earth, air, fire, water, and ether.[1] He states that the nature of fire is quite apparent in the pyramid from "the slenderness of its decreasing sides and the sharpness of its angles," [2] and that fire is engendered from air when the octahedron is dissolved into pyramids, and air produced from fire when the pyramids are compressed into an octahedron.[3]

Number mysticism.

These geometrical fancies are naturally accompanied by considerable number mysticism. In this particular passage the merits of the number five are enlarged upon and a long list is given of things that are five in number.[4] Five is again extolled in the essay on *The Ei at Delphi,*[5] but there one of the company remarks with much reason that it is possible to praise any number in many ways, but that he prefers to five "the sacred seven of Apollo." [6] Platonic geometrical reveries and Pythagorean number mysticism are indulged in even more extensively in the essay *On the Procreation of the Soul in Timaeus.* The number and proportion existing in planets, stars and spheres are touched on,[7] and it is stated that the divine demiurge produced the marvelous virtues of drugs and organs by employing harmonies and numbers.[8] Thus in the potency of number and numerical relations is suggested a possible explanation of astrology and magic force in nature.

Occult virtues in nature.

Plutarch, indeed, shows the same faith in the existence of occult virtues in natural objects and in what may be called natural magic as most of his contemporaries. At his symposium when one man avers that he saw the tiny fish *echeneïs* stop the ship upon which he was sailing until the look-out man picked it off,[9] some laugh at his credulity but

[1] *De defectu oraculorum,* 31-32. The resemblance of the stranger's tale to the vision of Er in Plato's *Republic* is also evident.

[2] *Ibid.,* 34.

[3] *Ibid.,* 37.

[4] *Ibid.,* 36; and see 11-12.

[5] Caps. 8-16.

[6] Cap. 17.

[7] Cap. 31.

[8] Cap. 33.

[9] *Symposiacs,* II, 7. D'Arcy W. Thompson in his translation of Aristotle's *History of Animals* comments on II, 14, "The myth of the 'ship-holder' has been elegantly explained by V. W. Elkman, 'On Dead Water,' in the Reports of Nansen's North Polar Expedition, Christiania, 1904."

others narrate other cases of strange antipathies in nature. Mad elephants are quieted by the sight of a ram; vipers will not move if touched with a leaf from a beech tree; wild bulls become tame when tied to a fig tree;[1] if light objects are oiled, amber fails to attract them as usual; and iron rubbed with garlic does not respond to the magnet. "These things are proved by experience but it is difficult if not quite impossible to learn their cause." At the Symposium[2] the question also is raised why salt is called divine, and it is suggested that it may be because it preserves bodies from decay after the soul has left them, or because mice conceive without sexual intercourse by merely licking salt. In *The Delay of the Deity* Plutarch again treats of occult virtues.[3] They pass from body to body with incredible swiftness or to an incredible distance. He wonders why it is that if a goat takes a piece of sea-holly in her mouth, the entire herd will stand still until the goatherd removes it. We see once more how closely such notions are associated with magical practices, when in the same paragraph he mentions the custom of making the children of those who have died of consumption or dropsy sit soaking their feet in water until the corpse has been buried so that they may not catch their parent's disease.

On the other hand, how difficult it must have been with the limited scientific knowledge of that time to distinguish true from false marvelous properties may be inferred from Plutarch's description[4] of a certain soft and pliable stone that used to be produced at Carystus and from which handkerchiefs and hair-nets were made which could not be burnt and were cleaned by exposure to fire,—a description, it would seem, of our asbestos, although Plutarch does not give the stone any name. Strabo also ascribes similar properties to a stone from Carystus without naming it.[5] Dioscorides and Asbestos.

[1] See above p. 77 for the somewhat different statement of Pliny (NH, XXIII, 64).

[2] *Symposiacs*, V, 10.

[3] *De sera numinis vindicta*, 14.

[4] *De defectu oraculorum*, 43.

[5] X, 1 (Casaub., 446); for this and some other source citations and a brief bibliography of modern discussions on the subject see the article, "Amiantus" (3) in Pauly-Wissowa.

other Greek authors, we are told,[1] apply the word "asbestos" to quick-lime, but Pliny in the *Natural History*[2] describes what he says the Greeks call ἀσβέστινον much as Plutarch does. He adds that it is employed in making shrouds for royal funerals to separate the ashes of the corpse from those of the pyre.[3] But he seems to regard it as a plant, not a stone, listing it as a variety of linen in one of his books on vegetation. He also states incorrectly that it is found but rarely and in desert and arid regions of India where there is no rain and a hot sun and amid terrible serpents.[4] Probably Pliny or his source argued that anything which resisted the action of fire must have been inured by growth under fiery suns and among serpents. Furthermore it obviously should possess other marvelous properties, so we are not surprised to find Anaxilaus cited to the effect that if this "linen" is tied around a tree trunk, the blows with which the tree is felled cannot be heard. It was thus that imaginations inured to magic enlarged upon unusual natural properties.

[1] Article on "Asbestos" in the *Encyclopedia Britannica,* 11th edition, which further states that Charlemagne was said to own a tablecloth which was cleaned by throwing it into the fire, and that in 1676 a merchant from China exhibited to the Royal Society a handkerchief of "salamander's wool" or *linum asbesti* (asbestos linen). See also Marco Polo, I, 42, and Cordier's note in Yule (1903), I, 216.

[2] XIX, 4. In Bostock and Riley's English translation, note 44 states that "the wicks of the inextinguishable lamps of the middle ages, the existence of which was an article of general belief, were said to be made of asbestus." On its use in lamp-wicks see also Pausanias, I, 26, 7.

[3] "In the year 1702 there was found near the Naevian Gate at Rome a funeral urn, in which there was a skull, calcined bones, and other ashes, enclosed in a cloth of asbestus of a marvelous length. It is still preserved in the Vatican," (Bostock and Riley, note 45).

[4] "On the contrary, it is found in the Higher Alps in the vicinity of glaciers, in Scotland, and in Siberia even" (Bostock and Riley, note 46). The article on "Amiantus (3)" in Pauly-Wissowa incorrectly assumes that in XIX, 4, Pliny has it in mind. In XXXVI, 31, however, Pliny briefly describes the stone amianthus, which Bostock and Riley (note 52) call "the most delicate variety of asbestus," as "losing nothing in fire" and "resisting all potions (or, spells) even of the *magi,*"—"Amiantus alumini similis nihil igni deperdit. Hic veneficis resistit omnibus privatim magorum." In XXXVII, 54, in an alphabetical list of stones, he briefly states that asbestos is iron-colored and found in the mountains of Arcadia,—"Asbestos in Arcadiae montibus nascitur coloris ferrei."

On rivers and mountains.

A treatise upon rivers and mountains in which the marvelous virtues of herbs and stones figure very prominently has sometimes been included among the works of Plutarch, but also has been omitted entirely from some editions.[1] Some have ascribed it to Parthenius of the time of Nero. It is made up of some thirty-five chapters in each of which a river and a mountain are mentioned. Usually some myth or tragic history is recounted, from which the river took its name or with which it was otherwise intimately connected. A similar procedure is followed in the case of the mountain. The writer, whoever he may be, makes a show of extensive reading, citing over forty authorities, most of whom are Greek and not mentioned in the full bibliographies of Pliny's *Natural History.* The titles cited have to do largely with stones, rivers, and different countries. It has been questioned, however, whether these citations are not bogus.[2]

Magic herbs.

The properties attributed to herbs and stones in this treatise are to a large extent magical. A white reed found in the river Phasis while one is sacrificing at dawn to Hecate, if strewn in a wife's bedroom, drives mad any adulterer who enters and makes him confess his sin.[3] Another herb mentioned in the same chapter was used by Medea to protect Jason from her father. In a later chapter [4] we are told how Hera called upon Selene to aid her in securing her revenge upon Heracles, and how the moon goddess filled a large chest with froth and foam by her magic spells until presently a huge lion leaped out of the chest. Returning from such sorceresses as Hecate, Medea, and Selene to herbs alone, in other rivers are plants which test the purity of gold, aid dim sight or blind one, wither at the mention of the word "step-mother" or burst into flames whenever a step-mother has evil designs against her step-son, free their bearers from fear of apparitions, operate as charms in love-making and

[1] Ed. by R. Hercher, Lipsiae, 1851; and by C. Müller in *Geograph. Graeci Minores,* II, 637ff.

[2] In Christ's *Gesch. d. Griech. Litt.,* not only is the *On Rivers and Mountains* itself called a "Schwindelbuch," but these citations are rejected as fraudulent.

[3] Cap. 5.

[4] Cap. 18.

childbirth, cure madmen of their frenzy, check quartan agues if applied to the breasts, protect virginity or wither at a virgin's touch, turn wine into water except that it retains its bouquet, or preserve persons anointed with their juice from sickness to their dying day.

Stones found in plants and fish.

An easy transition from the theme of magic herbs to that of stones is afforded by a sort of poppy which grows in a river of Mysia and bears black, harp-shaped stones which the natives gather and scatter over their ploughed fields.[1] If these stones then lie still where they have fallen, it is taken as a sign of a barren year; but if they fly away like locusts, this prognosticates a plentiful harvest. Other marvelous stones are found in the head of a fish in the river Arar, a tributary of the Rhone. The fish is itself quite wonderful since it is white while the moon waxes and black when it wanes.[2] Presumably for this reason the stone cures quartan agues, if applied to the left side of the body while the moon is waning. There is another stone which must be sought after under a waxing moon with pipers playing continually.[3]

Virtues of other stones.

Other stones guard treasuries by sounding a trumpet-like alarm at the approach of thieves; or change color four times a day and are ordinarily visible only to young girls. But if a virgin of marriageable age chances to see this stone, she is safe from attempts upon her chastity henceforth.[4] Some stones drive men mad and are connected with the Mother of the Gods or are found only during the celebration of the mysteries.[5] Others stop dogs from barking, expel demons, grow black in the hands of false witnesses, protect from wild beasts, and have varied medicinal powers or other effects similar to those already mentioned in the case of herbs.[6] In a river where the Spartans were defeated is a stone which leaps towards the bank, if it hears a trumpet,

[1] Cap. 21.
[2] Cap. 6.
[3] Cap. 1.
[4] Cap. 7.
[5] Caps. 9, 10, 12.
[6] Caps. 16, 18, 24.

but sinks at the mention of the Athenians.[1] Certainly a marvelous stone, capable of both hearing and motion!

Leaving the treatise on rivers and mountains, for the occult virtue of human beings we may turn to a discussion of fascination in the *Symposiacs*.[2] Some of the company ridiculed the idea, but their host asserted that a myriad of events went to prove it and that if you reject a thing simply because you cannot give a reason for it, you "take away the marvelous from all things." He pointed out that some men hurt little and tender children by looking at them, and argued that, as the plumes of other birds are ruined when mixed with those of the eagle, so men may injure by their touch or mere glance. Plutarch, who was of the company, suggested effluvia or emanations from the body as a possible explanation, pointing out that love begins with glances, that no disease is more contagious than sore eyes, and that gazing upon the curlew cures jaundice. The bird appears to attract the disease to itself, and averts its head and closes its eyes, not, as some think, because it is jealous of the remedy sought from it, but because it feels wounded as if from a blow. Others of the company contended that the passions and affections of the soul may have a powerful effect through the eyes and glance upon other persons, and argued that the sufferings of the soul strengthen the powers of the body, and that the same counter-charms are efficacious against envy as against fascination. The emanations which Democritus believed that envious and malicious persons sent forth are also mentioned; fathers have fascinated their own children, and it is even possible that one might injure oneself by reflection of one's gaze. It is suggested that young children may sometimes be fascinated in this manner rather than by the glance of others.

Fascination.

Plutarch devotes two essays to the familiar theme of the craftiness and sagacity of animals and the remedies used by them. In one essay [3] a companion of Odysseus refuses to

Animal sagacity and remedies.

[1] Cap. 17.
[2] V, 7.
[3] *Bruta animalia ratione uti*, cap. 9; also *Quaest. Nat.*, cap. 26, "Why certain brutes seek certain remedies."

allow Circe to turn him back from a pig to human form. He boasts among other things that beasts know how to cure themselves. Without ever having been taught swine when sick run to rivers to search for craw-fish; tortoises physic themselves with origanum after eating vipers; and Cretan goats devour dittany to extract arrows and darts which have been shot into their bodies. In the other essay [1] on the cleverness of animals we find many familiar stories repeated, including some of the inevitable excerpts from Juba on elephants. We meet again the dolphins with their love for mankind,[2] the bird who picks the crocodile's teeth and warns him of the ichneumon,[3] the fish who rescue one another by biting the line or dragging one another by the tail out of nets,[4] the trained elephant who was slow to learn and was beaten for it and was afterwards seen practicing his exercises by himself in the moonlight,[5] the sentinel cranes who stand on one foot and hold a stone in the other to awaken them if they let it drop.[6] More novel perhaps is the story how herons open oysters by first swallowing them, shells and all, until they are relaxed by the internal heat of the bird, which then vomits them up and eats them out of the shells. Or the account of the tunny fish who needs no astrological canons and is familiar with arithmetic, "Yes, by Zeus, and with optics, too." [7]

Theories and queries about nature.

Plutarch's essays bring out yet other interests and defects of the science of the time. One on *The Principle of Cold* is a good illustration of the failings of the ancient hypothesis of four elements and four qualities and of the silly, limited arguing which usually and almost of necessity accompanied it. He denies that cold is mere privation of heat, since it seems to act positively upon fluids and solids and exists in different degrees. After considering various assertions such as that air becomes cold when it becomes

[1] *De solertia animalium.*

[2] *Ibid.*, 36-37; also the closing chapters of *The Banquet of the Seven Sages.*

[3] Cap. 31.

[4] Cap. 25.

[5] Cap. 12.

[6] Cap. 10.

[7] Cap. 29.

dark; that air whitens things and water blackens them; that cold objects are always heavy; he finally associates the element earth especially with the quality cold. In another essay [1] he states that there are no females of a certain type of beetle which was engraved as a charm upon the rings warriors wore to battle, but that the males begat offspring by rolling up balls of earth. He declares that "diseases do not have distinct germs" in a discussion in the *Symposiacs* whether there can be new diseases.[2] Other natural questions discussed in the treatise of that name and the *Symposiacs* are: Why a man who often passes near dewy trees contracts leprosy in those limbs which touch the wood? Why the Dorians pray for bad hay-making? Why bears' paws are the sweetest and most palatable food? Why the tracks of wild beasts smell worse at the full of the moon? Why bees are more apt to sting fornicators than other persons? [3] Why the flesh of sheep bitten by wolves is sweeter than that of other sheep? Why mushrooms are thought to be produced by thunder? Why flesh decays sooner in moonlight than sunlight? Whether Jews abstain from pork because they worship the pig or because they have an antipathy towards it? [4]

The Antipodes.

Plutarch sometimes shows evidence of considerable astronomical knowledge. For instance, he knows that the mathematicians figure that the distance from sun to earth is immense, and that Aristarchus demonstrated the sun to be eighteen or twenty times as far off as the moon, which is distant fifty-six times the earth's radius at the lowest estimate.[5] Yet in the same essay [6] Plutarch has scoffed at the idea of a spherical earth and of antipodes, and at the assertion that bars weighing a thousand talents would stop falling at the earth's center, if a hole were opened up through the earth, or that two men with their feet in opposite directions

[1] *Isis and Osiris*, 10.

[2] VIII, 9, ἴδια δὲ σπέρματα νόσων οὐκ ἔστιν.

[3] *Nat. Quaest.*, caps. 6, 14, 22, 24, 36.

[4] *Symposiacs*, II, 9; IV, 2; III, 10; IV, 5.

[5] *De facie in orbe lunae*, 9-10; also the opening chapters of *De defectu oraculorum*.

[6] Cap. 7.

at the center of the earth might nevertheless both be right side up, or that one man whose middle was at the center might be half right side up and half upside down. He admits, however, that the philosophers think so. Thus we see that Christian fathers like Lactantius were not the first to ridicule the notion of the Antipodes; apparently as well educated and omnivorous a pagan reader as Plutarch could do the same.

CHAPTER VII

APULEIUS OF MADAURA

I. *Life and Works*

Magic and the man—Stylistic reasons for regarding the *Metamorphoses* as his first work—Biographical reasons—No mention of the *Metamorphoses* in the *Apology*.

II. *Magic in the Metamorphoses*

Powers claimed for magic—Its actual performances—Its limitations —The crimes of witches—Male magicians—Magic as an art and discipline—Materials employed—Incantations and rites—Quacks and charlatans—Various superstitions—Bits of science and religion—Magic in other Greek romances.

III. *Magic in the Apology*

Form of the *Apologia*—Philosophy and magic—Magic defined—Good and bad magic—Magic and religion—Magic and science—Medical and scientific knowledge of Apuleius—He repeats familiar errors—Apparent ignorance of magic and occult virtue—Despite an assumption of knowledge—Attitude toward astronomy—His theory of demons—Apuleius in the middle ages.

I. *His Life and Works*

Magic and the man as reflected in his works.

ONE of the fullest and most vivid pictures of magic in the ancient Mediterranean world which has reached us is provided by the writings of Apuleius. He lived in the second century of our era and was not merely a rhetorician of great note in his day and the writer of a romance which has ever since fascinated men, but also a Platonic philosopher, an initiate into many religious cults and mysteries, and a student of natural science and medicine. To him has been ascribed the Latin version of *Asclepius,* a supposititious dialogue of Hermes Trismegistus. No author perhaps ever more readily and complacently talked of himself than

Apuleius, yet it is no easy task to make out the precise facts of his life, partly because in his romance, *The Metamorphoses,* or *The Golden Ass,* he has hopelessly confused himself with the hero Lucius and introduced an autobiographical element of uncertain extent into what is in the main a work of fiction; partly because his *Apology,* or defense when tried on the charge of magic at Oea in Africa, is more in the nature of special pleading intended to refute and confound his accusers than of a frank confession or accurate history of his career. However, he appears to have been born at Madaura in North Africa, to have studied first at Carthage and then at Athens, to have visited Rome and wandered rather widely about the Mediterranean world, but to have spent more time altogether at Carthage than at any other one place.

Stylistic reasons for regarding the *Metamorphoses* as his first work.

Besides the *Metamorphoses* and *Apologia,* with which we shall be chiefly concerned, four other works are extant which are regarded as genuine, *The God of Socrates, The Dogma of Plato, Florida,* and *On the Universe.* The order in which these works were written is uncertain, but it seems almost sure that the *Metamorphoses* was the first. In it Apuleius not only more or less identifies himself with the hero Lucius, who is represented as quite a young man, he also apologizes for his Latin and speaks of the difficulty with which he had acquired that language at Rome. But in the *Florida* [1] we find him repeating a hymn and a dialogue in both Latin and Greek, or, after delivering half an address in Greek, finishing it in Latin, or boasting that he writes poems, satires, riddles, histories, scientific treatises, orations, and philosophical dialogues with equal facility in either language.[2] Instead now of craving pardon if he offends by his rude, exotic, and forensic speech, he feels that his reputation for literary refinement and elegance has become such that his audience will not pardon him a solitary solecism or a single syllable pronounced with a barbarous accent.[3] It

[1] Cap. 18.

[2] "Tam graece quam latine, gemino voto, pari studio, simili studio."

[3] *Florida,* cap. 9.

therefore looks as if the *Metamorphoses* was his first published effort in Latin and as if his peculiar style had proved so popular that he did not find it necessary to apologize for it again. In the *Apology* he seems supremely confident of his rhetorical powers in the Latin language, and even the accusers describe him as a philosopher of great eloquence both in Greek and Latin.[1] Three years before in the same town his first public discourse had been greeted with shouts of "Insigniter," and many in the audience at the time of his trial can still repeat a passage from it on the greatness of Aesculapius.[2] In the *Apology,* too, he displays a more extensive learning than in the *Metamorphoses* and has written already poems and scientific treatises as well as orations. Indeed, practically all the doctrines set forth in his other philosophical works may be found in brief in the *Apology.*

Biographical reasons.

Moreover, while in the *Metamorphoses* Apuleius ends the narrative with what seems to be his own comparatively recent initiation into the mysteries of Isis in Greece and of Osiris at Rome, in the *Apology*[3] he speaks of having been initiated in the past into all sorts of sacred rites, although he does not mention Rome or Isis and Osiris specifically. It is implied, however, that he has been at Rome in more than one passage of the *Apology.* Pontianus, his future step-son, with whom Apuleius had become acquainted at Athens "not so many years ago," was "an adult at Rome" before Apuleius came to Oea. After they had met again at Oea and had both married there, Apuleius gave Pontianus a letter of introduction to the proconsul Lollianus Avitus at Carthage, of whom he says, "I have known intimately many cultured men of Roman name in the course of my life, but have never admired anyone as much as him." Perhaps Apuleius may have met Lollianus at Carthage, but in the *Florida,*[4] in a panegyric on Scipio Orfitus, proconsul of Africa in 163-164 A. D., he alludes to the time "when I moved among your friends in Rome." All this fits in nicely

[1] *Apologia,* cap. 4.
[2] Caps. 73 and 55.
[3] Caps. 55-56.
[4] Cap. 17.

with the statements in the closing chapters of the *Metamorphoses* concerning his rising fame as an orator in the courts of law and "the laborious doctrine of my studies" at Rome. We may therefore reconstruct the course of events as follows. After meeting Pontianus at Athens and concluding his studies in Greece, Apuleius came to Rome, where he remained for some time, perfecting his Latin style, engaging in forensic oratory, and publishing the *Metamorphoses.* Pontianus, who was younger than Apuleius, either accompanied or followed his friend to Rome, in which city he was still residing after Apuleius had returned to Africa. But Pontianus, too, had left Rome and come back to his African city of Oea to settle the question of his mother's proposed second marriage, before Apuleius, who had probably revisited Carthage in the meantime and was now traveling east again with the intention of visiting Alexandria, arrived at Oea and was induced to wed the widow, who was considerably older than he. On the delicate question of this lady's exact age depends our dating of the birth of Apuleius and the chronology of his entire career. At the trial of Apuleius for magic Aemilianus, the accuser, declared that she was sixty when she married Apuleius, and he had previously proposed to marry her to his brother, Clarus, whom Apuleius calls "a decrepit old man." [1] On the other hand, Apuleius asserts that the records, which he produces in court, of her being accepted in infancy by her father as his child show that she is "not much over forty," [2]—a tactful ambiguity which, inasmuch as we no longer have the records, it would probably be idle to attempt to fathom.

No mention of the *Metamorphoses* in the *Apology.*

The chief, if not the only, objection to dating the *Metamorphoses* before the *Apology* is that nothing is said of it in the latter.[3] But obviously Apuleius, when on trial for magic, would not mention the *Metamorphoses* unless his

[1] *Apologia,* cap. 70.

[2] Cap. 89.

[3] To Professor Butler (*Apulei Apologia,* ed. H. E. Butler and A. S. Owen, Oxford, 1914) this difficulty seems so insurmountable that he places the *Apology* earlier. But for the reasons already given I agree with the article on Apuleius in Pauly and Wissowa and its citations that the *Metamorphoses* is Apuleius's first work.

accusers forced him to do so. They may not have yet heard of it or it may at first have been published anonymously, although the probability is that Apuleius would not have spent three years at Oea without bringing it to his admirers' attention. Or they may know of it, but the judge may not have admitted it as evidence on the ground that they must prove that Apuleius has practiced magic. The *Metamorphoses* does not recount any personal participation of Apuleius himself in magic arts, unless one identifies him throughout with the hero Lucius; it purports to be a Latin rendition of Milesian tales [1] and does not seem to have been taken very seriously until the church fathers began to cite it. Or the accusers may have dwelt upon it and Apuleius simply have failed to take notice of their charge. All these suppositions may not seem very plausible, but on the other hand we may ask, how would Apuleius dare to write a work like the *Metamorphoses* after he had been accused and tried of magic? One would expect him then to drop the subject rather than to display an increasing interest in it. But let us turn to his treatment of that theme in both those works, and first consider the *Metamorphoses.*

II. *Magic in the Metamorphoses*

Powers claimed for magic.

Vast power over nature and spirits is attributed to magic and its practitioners in the opening chapters of the *Metamorphoses.* "By magic's mutterings swift streams are reversed, the sea is calmed, the sun stopped, foam drawn from the moon, the stars torn from the sky, and day turned into night." [2] While such assertions are received with some scepticism by one listener, they are largely borne out by the subsequent experiences of the characters in the story and by the feats which witches are made to perform. These are sometimes humorously and extravagantly presented, but as crime and ferocious cruelty are treated in the same spirit,

[1] The work opens with the statement that the author "will stitch together varied stories in the so-called Milesian manner," and that "we begin with a Grecian story."

[2] I, 3.

this light vein cannot be regarded as an admission of magic's unreality. On the contrary, the magic of Thessaly is celebrated with one accord the world over.[1] Meroë the witch can "displace the sky, elevate the earth, freeze fountains, melt mountains, raise ghosts, bring down the gods, extinguish the stars, and illuminate the bottomless pit."[2] Submerging the light of starry heaven to the lowest depths of hell is a power also attributed to the witch Pamphile.[3] "By her marvelous secrets she makes ghosts and elements obey and serve her, disturbs the stars and coerces the divinities."[4]

Its actual performances.

In none of the episodes recorded in *The Golden Ass,* however, do the witches find it necessary or advisable to go to quite so great lengths as these, although Pamphile once threatens the sun with eternal darkness because he is so slow in yielding to night when she may ply her sorcery and amours.[5] The witches content themselves with such accomplishments as carrying on love affairs with inhabitants of distant India, Ethopia, and even the Antipodes,—"trifles of the art these and mere bagatelles";[6] with transforming their enemies into animal forms or imprisoning them helpless in their homes, or transporting them house and all to a spot a hundred miles off;[7] and, on the other hand, with breaking down bolted doors to murder their victims,[8] or assuming themselves the shape of weasels, birds, dogs, mice, and even insects in order to work their mischief unobserved;[9] they then cast their victims into a deep sleep and cut their throats or hang them or mutilate them.[10] They often know what is being said about them when apparently absent, and they sometimes indulge in divination of the future.[11] But to whatever fields of activity they may extend or confine them-

[1] II, 1.
[2] I, 8.
[3] II, 5.
[4] III, 15. The wording of the translated passages throughout this chapter is mainly my own, but I have made some use of existing English translations.
[5] III, 16.
[6] I, 8.
[7] I, 9-10.
[8] I, 11-13.
[9] II, 22 and 25.
[10] II, 20 and 30; IX, 29.
[11] I, 11; II, 11.

selves, their violent power is irresistible, and we are given to understand that it is useless to try to fight against it or to escape it. Its secret and occult character is also emphasized, and the adjective *caeca* or noun *latebrae* are more than once employed to describe it.[1]

Its limitations.

Yet there are also suggested certain limitations to the power of magic. The witches seem to break down the bolted doors, but these resume their former place when the hags have departed, and are to all appearances as intact as before. The man, too, whose throat they have cut, whose blood they have drained off, and whose heart they have removed, awakes apparently alive the next morning and resumes his journey. All the events of the preceding night seem to have been merely an unpleasant dream. The witches had stuffed a sponge into the wound of his throat [2] with the adjuration, "Oh you sponge, born in the sea, beware of crossing running water." In the morning his traveling companion can see no sign of wound or sponge on his friend's throat. But when he stoops to drink from a brook, out falls the sponge and he drops dead. The inference, although Apuleius draws none, is obvious; witches can make a corpse seem alive for a while but not for long, and magic ceases to work when you cross running water. We also get the impression that there is something deceptive and illusive about the magic of the witches, and that only the lusts and crimes are real which their magic enables them or their employers to commit and gratify. They may seem to draw down the sun, but it is found shining next day as usual. When Lucius is transformed into an ass, he retains his human appetite and tenderness of skin,[3]—a deplorable state of mind and body which must be attributed to the imper-

[1] II, 20, 22; III, 18.

[2] Very similar practices are recounted by A. W. Howitt, *Native Tribes of South-East Australia*, pp. 355-96; "the medicine-men of hostile tribes sneak into the camp in the night, and with a net of a peculiar construction garotte one of the tribe, drag him a hundred yards or so from the camp, cut up his abdomen obliquely, take out the kidney and caul-fat, and then stuff a handful of grass and sand into the wound."

[3] VI, 26.

fections of the magic art as well as to the humorous cruelty of the author.

The crimes of witches.

In *The Golden Ass* the practitioners of magic are usually witches and old and repulsive. We have to deal with wonders worked by old-wives and not by *Magi* of Persia or Babylon. As we have seen and shall see yet further, their deeds are regarded as illicit and criminal. They are "most wicked women" (*nequissimae mulieres*),[1] intent upon lust and crime. They practice *devotiones,* injurious imprecations and ceremonies.[2]

Male magicians.

Male practitioners of magic are represented in a less unfavorable light. An Egyptian, who in return for a large sum of money engages to invoke the spirit of a dead man and restore the corpse momentarily to life, is called a prophet and a priest, though he seems a manifest necromancer and is himself adjured to lend his aid and to "have pity by the stars of heaven, by the infernal deities, by the elements of nature, and by the silence of night," [3]—expressions which are certainly suggestive of the magic powers elsewhere ascribed to witches. The hero of the story, Lucius, is animated in his dabblings in the magic art by idle curiosity combined with thirst for learning, but not by any criminal motive.[4] Yet after he has been transformed into an ass by magic, he fears to resume his human form suddenly in public, lest he be put to death on suspicion of practicing the magic art.[5]

Magic as an art and discipline.

Magic is depicted not merely as irresistible or occult or criminal or fallacious; it is also regularly called an art and a discipline. Even the practices of the witches are so dignified. Pamphile has nothing less than a laboratory on the roof of her house,—a wooden shelter, concealed from view but open to the winds of heaven and to the four points of the compass,—where she may ply her secret arts and where she spreads out her "customary apparatus." [6] This consists

[1] II, 22.
[2] I, 10; VII, 14; IX, 23, 29.
[3] II, 28.
[4] II, 6; III, 19.
[5] III, 29.
[6] III, 17.

of all sorts of aromatic herbs, of metal plates inscribed with cryptic characters, a chest filled with little boxes containing various ointments,[1] and portions of human corpses obtained from sepulchers, shipwrecks (or birds of prey, according as the reading is *navium* or *avium*), public executions, and the victims of wild beasts.[2] It will be recalled that Galen represented medical students as most likely to secure human skeletons or bodies to dissect from somewhat similar sources; and possibly they might incur suspicion of magic thereby.

Materials employed.

All this makes it clear that to work magic one must have materials. The witches seem especially avid for parts of the human body. Pamphile sends her maid, Fotis, to the barber's shop to try to steal some cuttings of the hair of a youth of whom she is enamoured; [3] and another story is told of witches who by mistake cut off and replaced with wax the nose and ears of a man guarding the corpse instead of those of the dead body.[4] Other witches who murdered a man carefully collected his blood in a bladder and took it away with them.[5] But parts of other animals are also employed in their magic, and stones as well as varied herbs and twigs.[6] In trying to entice the beloved Boeotian youth Pamphile used still quivering entrails and poured libations of spring water, milk, and honey, as well as placing the hairs—which she supposed were his—with many kinds of incense upon live coals.[7] To turn herself into an owl she anointed herself from top to toe with ointment from one of her little boxes, and also made much use of a lamp.[8] To regain her human form she has only to drink, and bathe in, spring water mixed with anise and laurel leaf,—"See how great a result is attained by such small and insignificant herbs!" [9]—while Lucius is told that eating roses will re-

[1] III, 21.
[2] I, 10; II, 20-21.
[3] III, 16.
[4] II, 23-30.
[5] I, 13.
[6] II, 5. "Surculis et lapillis et id genus frivolis inhalatis."
[7] III, 18.
[8] III, 21.
[9] III, 23.

store him from asinine to human form.[1] The Egyptian prophet makes use of herbs in his necromancy, placing one on the face and another on the breast of the corpse; and he himself wears linen robes and sandals of palm leaves.[2]

Incantations and rites.

Besides materials, incantations are much employed,[3] while the Egyptian prophet turns towards the east and "silently imprecates" the rising sun. As this last suggests, careful observance of rite and ceremony also play their part, and Pamphile's painstaking procedure is described in precise detail. Divine aid is once mentioned [4] and is perhaps another essential for success. More than one witch is called *divina*,[5] and magic is termed a divine discipline.[6] But we have also heard the witches spoken of as coercing the gods rather than depending upon them for assistance. Their magic seems to be performed mainly by using things and words in the right ways.

Quacks and charlatans.

Besides the witches (*magae* or *sagae*) and what Apuleius calls magic by name, a number of other charlatans and superstitions of a kindred nature are mentioned in *The Golden Ass*. Such a one is the Egyptian "prophet" already described. Such was the Chaldean who for a time astounded Corinth by his wonderful predictions, but had been unable to foresee his own shipwreck.[7] On learning this last fact, a business man who was about to pay him one hundred *denarii* for a prognostication snatched up his money again and made off. Such were the painted disreputable crew of the Syrian goddess who went about answering all inquiries concerning the future with the same ambiguous couplet.[8] Such were the jugglers whom Lucius saw at Athens swallowing swords or balancing a spear in the

[1] III, 25.

[2] II, 28.

[3] Examples are: I, 3, magico susurramine; II, 1, artis magicae nativa cantamina; II, 5, omnis carminis sepulchralis magistra creditur; II, 22, diris cantaminibus somno custodes obruunt; III, 18, tunc decantatis spirantibus fibris; III, 21, multumque cum lucerna secreta collocuta.

[4] I, 11, quo numinis ministerio.

[5] I, 8, saga, inquit, et divina; IX, 29, saga illa et divini potens.

[6] III, 19.

[7] II, 12-14.

[8] VIII, 26-27; IX, 8.

throat while a boy climbed to the top of it.[1] Such were the physicians who turned poisoners.[2]

Various superstitions.

Other passages allude to astrology [3] besides that already cited concerning the Chaldean. Divination from dreams is also discussed. In the fourth book the old female servant tells the captive maiden not to be terrified "by the idle figments of dreams" and explains that they often go by contraries; but in the last book the hero is several times guided or forewarned by dreams. Omens are believed in. Starting left foot first loses a man a business opportunity,[4] and another is kicked out of a house for his ill-omened words.[5] The violent deaths of all three sons of the owner of another house are presaged by the following remarkable conglomeration of untoward portents: a hen lays a chick instead of an egg; blood spurts up from under the table; a servant rushes in to announce that the wine is boiling in all the jars in the cellar; a weasel is seen dragging a dead snake out-of-doors; a green frog leaps from the sheep-dog's mouth and then a ram tears open the dog's throat at one bite.[6]

Some bits of science and religion.

Of scientific discussion or information there is little in the *Metamorphoses.* When Pamphile foretells the weather for the next day by inspection of her lamp, Lucius suggests that this artificial flame may retain some properties from its heavenly original.[7] The herb mandragora is described as inducing a sleep similar to death, but as not fatal; and the beaver is said to emasculate itself in order to escape its hunters.[8] We should feel lost without mention of a dragon in a book of this sort, and one is introduced who is large enough to devour a man.[9] It is interesting to note for purposes of comparison,—inasmuch as we shall presently take up the *Life of Apollonius of Tyana,* a Neo-Pythagorean, and later shall learn from the *Recognitions of Clement* that the apostle Peter was accustomed to bathe at dawn in the

[1] I, 4.
[2] X, 11, 25.
[3] VIII, 24; XI, 22, 25.
[4] I, 5.
[5] II, 26.
[6] IX, 33-34.
[7] II, 11-12.
[8] X, 11. For bibliography on the mandragora see Frazer (1918) I, 377, note 2, in his chapter, "Jacob and the Mandrakes."
[9] VIII, 21.

sea,—that Lucius, while still in the form of an ass, in his zeal for purification plunged into the sea and submerged his head beneath the wave seven times, because the divine Pythagoras had proclaimed that number as especially appropriate to religious rites.[1] "It has been said that *The Golden Ass* is the first book in European literature showing piety in the modern sense, and the most disreputable adventures of Lucius lead, it is true, in the end to a religious climax." But, adds Professor Duncan B. Macdonald, "Few books, in spite of fantastic gleams of color and light, move under such leaden-weighted skies as *The Golden Ass.* There is no real God in that world; all things are in the hands of enchanters; man is without hope for here and hereafter; full of yearnings he struggles and takes refuge in strange cults." [2]

Magic in other Greek romances.

While magic plays a larger part in *The Golden Ass* than in any other extant Greek romance, it is not unusual in the others to find the hero and heroine exposed to perils from magicians, or themselves falsely charged with magic, as in the *Aethiopica* of Heliodorus, where Charicles is "condemned to be burned on a charge of poisoning." [3] In the Christian romances, too, as the *Recognitions* will show us later, there are plenty of allusions to magic and demons. Meanwhile we are reminded that in the Roman Empire accusations of magic were made not merely in story books but in real life by the trial for magic of the author of the *Metamorphoses* himself, and we next turn to the *Apology* which he delivered upon that occasion.

III. *Magic in the Apology*

Form of the *Apologia.*

The *Apologia* has every appearance of being preserved just as it was delivered and perhaps as it was taken down by shorthand writers; it does not seem to have undergone the subsequent revision to which Cicero subjected some of his orations. It must have been hastily composed, since

[1] XI, 1.
[2] Macdonald (1909), p. 128.
[3] VIII, 9.

Apuleius states that it has been only five or six days since the charges were suddenly brought against him, while he was occupied in defending another lawsuit brought against his wife.[1] There also are numerous apparently extempore passages in the oration, notably those where Apuleius alludes to the effect which his statements produce, now upon his accusers, now upon the proconsul sitting in judgment. From the *Florida* we know that Apuleius was accustomed to improvise.[2] Moreover, in the *Apology* certain statements are made by Apuleius which might be turned against him with damaging effect and which he probably would have omitted, had he had the leisure to go over his speech carefully before the trial. For instance, in denying the charge that he had caused to be made for himself secretly out of the finest wood a horrible magic figure in the form of a ghost or skeleton, he declares that it is only a little image of Mercury made openly by a well-known artisan of the town.[3] But he has earlier stated that "Mercury, carrier of incantations," is one of the deities invoked in magic rites;[4] and in another passage[5] has recounted how the outcome of the Mithridatic war was investigated at Tralles by magic, and how a boy, gazing at an image of Mercury in water, had predicted the future in one hundred and sixty verses. But this is not all. In a third passage[6] he actually quotes Pythagoras to the effect that Mercury ought not to be carved out of every kind of wood.

[1] Cap. 1.

[2] *Florida*, caps. 24-26.

[3] Caps. 61-63. The following passages from E. A. W. Budge, *Egyptian Magic* (1899), perhaps furnish an explanation of the true purpose and character of Apuleius's wooden figure: p. 84, "Under the heading of 'Magical Figures' must certainly be included the so-called Ptah-Seker-Ausar figure, which is usually made of wood; it is often solid, but is sometimes made hollow, and is usually let into a rectangular wooden stand which may be either solid or hollow." To get the protection of Ptah, Seker, and Osiris, says Budge at p. 85, "a figure was fashioned in such a way as to include the chief characteristics of the forms of these gods, and was inserted in a rectangular wooden stand which was intended to represent the coffin or chest out of which the trinity Ptah-Seker-Ausar came forth. On the figure itself and on the sides of the stand were inscribed prayers. . . ." Such a figure in a coffin might well be described by the accusers as the horrible form of a ghost or skeleton.

[4] Cap. 31.

[5] Cap. 42.

[6] Cap. 43.

Philosophy and magic.

If in the *Metamorphoses* the practice of magic is imputed chiefly to old-wives, in the *Apology* a main concern of Apuleius is to defend philosophers in general [1] and himself in particular from "the calumny of magic." [2] Epimenides, Orpheus, Pythagoras, Ostanes, Empedocles, Socrates, and Plato have been so suspected, and it consoles Apuleius in his own trial to reflect that he is but sharing the undeserved fate of "so many and such great men." [3] In this connection he states that those philosophers who have taken an especial interest in theology, "who investigate the providence of the universe too curiously and celebrate the gods too enthusiastically," are the ones to be suspected of magic; while those who devote themselves to natural science pure and simple are more liable to be called irreligious atheists.

Magic defined.

But what is it to be a magician, Apuleius asks the accusers,[4] and therewith we face again the question of the definition of magic, and Apuleius gradually answers his own query in the course of the oration. Magic, in the ordinary use of the word, is described in much the same way as in the *Metamorphoses*. It has been proscribed by Roman law since the Twelve Tables; it is hideous and horrible; it is secret and solitary; it murmurs its incantations in the darkness of the night.[5] It is an art of ill repute, of illicit evil deeds, of crimes and enormities.[6] Instead of simply calling it *magia,* Apuleius often applies to it the double expression, *magica maleficia.*[7] Perhaps he does this intentionally. In one passage he states that he will refute certain charges which the accusers have brought against him, first, by showing that the things he has been charged with have nothing to do with magic; and second, by proving that, even if he were a magician, there was no cause or occasion for his having committed any *maleficium* in this connection.[8]

[1] Caps. 1-3.

[2] Cap. 2.

[3] Caps. 27 and 31. For the same thought applied in the case of medieval men see Gabriel Naudé, *Apologie pour tous les grands personages qui ont esté faussement soupçonnez de Magie,* Paris, 1625.

[4] Cap. 25.

[5] Cap. 47.

[6] Cap. 25.

[7] Caps. 9, 42, 61, 63.

[8] Cap. 28.

That is to say, *maleficium,* literally "an evil deed," means an injury done another by means of magic art. The proconsul sitting in judgment takes a similar view and has asked the accusers, Apuleius tells us,[1] when they asserted that a woman had fallen into an epileptic fit in his presence and that this was due to his having bewitched her, whether the woman died or what good her having a fit did Apuleius. This is significant as hinting that Roman law did not condemn a man for magic unless he were proved to have committed some crime or made some unjust gain thereby.

Good and bad magic.

Does Apuleius for his part mean to suggest a distinction between *magia* and *magica maleficia,* and to hint, as he did not do in the *Metamorphoses,* that there is a good as well as a bad magic? He cannot be said to maintain any such distinction consistently; often in the *Apology magia* alone as well as *maleficium* is used in a bad sense. But he does suggest such a thought and once voices it quite explicitly.[2] "If," he says, "as I have read in many authors, *magus* in the Persian language corresponds to the word *sacerdos* in ours, what crime, pray, is it to be a priest and duly know and understand and cherish the rules of ceremonial, the sacred customs, the laws of religion?" Plato describes magic as part of the education of the young Persian prince by the four wisest and best men of the realm, one of whom instructs him in the magic of Zoroaster which is the worship of the gods. "Do you hear, you who rashly charge me with magic, that this art is acceptable to the immortal gods, consists in celebrating and reverencing them, is pious and prophetic, and long since was held by Zoroaster and Oromazes, its authors, to be noble and divine?" [3] In common speech, however, Apuleius recognizes that a magician is one "who by his power of addressing the immortal gods is able to accomplish whatever he will by an almost incredible force of incantations." But anyone who believes that another man possesses such a power as this should be afraid to accuse him,

[1] Cap. 48. [2] Cap. 25. [3] Cap. 26.

says Apuleius, who thinks by this ingenious dilemma to prove the insincerity of his accusers. Nevertheless he presently mentions that Mercury, Venus, Luna, and Trivia are the deities usually summoned in the ceremonies of the magicians.[1]

Magic and religion.

It will be noted that Apuleius connects magic with the gods and religion more in the *Apology* than in the *Metamorphoses.* There his emphasis was on the natural materials employed by the witches and their almost scientific laboratories. But in the *Apology* both Persian *Magi* and common magicians are associated with the worship or invocation of the gods, and it is theologians rather than natural philosophers who incur suspicion of magic.

Magic and science.

But it may be that the reason why Apuleius abstains in the *Apology* from suggesting any connection or confusion between magic and natural science is that the accusers have already laid far too much stress upon this point for his liking. He has been charged with the composition of a tooth-powder,[2] with use of a mirror,[3] with the purchase of a sea-hare, a poisonous mollusc, and two other fish appropriate from their obscene shapes and names for use as love-charms.[4] He is said to have had a horrible wooden image or seal constructed secretly for use in his magic,[5] to keep other instruments of his art mysteriously wrapped in a handkerchief in the house,[6] and to have left in the vestibule of another house where he lodged "many feathers of birds" and much soot on the walls.[7] All these charges make it evident that natural and artificial objects are, as in the *Metamorphoses,* considered essential or at least usual in performing magic. Moreover, so ready have the accusers shown themselves to interpret the interest of Apuleius in natural science as an evidence of the practice of magic by him, that he sarcastically remarks [8] that he is glad that they were unaware that he had read Theophrastus *On beasts that bite and sting* and Ni-

[1] Cap. 31.
[2] Cap. 6.
[3] Cap. 13.
[4] Caps. 30, 33.
[5] Cap. 61.
[6] Cap. 53.
[7] Cap. 58.
[8] Cap. 41.

cander *On the bites of wild beasts* (usually called *Theriaca*),[1] or they would have accused him of being a poisoner as well as a magician.

Medical and scientific knowledge of Apuleius.

Apuleius shows that he really is a student, if not an authority, in medicine and natural science. The gift of the tooth-powder and the falling of the woman in a fit were incidents of his occasional practice of medicine, and he also sees no harm in his seeking certain remedies from fish.[2] He repeats Plato's theory of disease from the *Timaeus* and cites Theophrastus's admirable work *On Epileptics*.[3] Mention of the mirror starts him off upon an optical disquisition in which he remarks upon theories of vision and reflection, upon liquid and solid, flat and convex and concave mirrors, and cites the *Catoptrica* of Archimedes.[4] He also regards himself as an experimental zoologist and has conducted all his researches publicly.[5] He procures fish in order to study them scientifically as Aristotle, Theophrastus, Eudemus, Lycon, and other pupils of Plato did.[6] He has read innumerable books of this sort and sees no harm in testing by experience what has been written. Indeed he is himself writing in both Greek and Latin a work on *Natural Questions* in which he hopes to add what has been omitted in earlier books and to remedy some of their defects and to arrange all in a handier and more systematic fashion. He has passages from the section on fishes in this work read aloud in court.

He repeats familiar errors.

Throughout the *Apology* Apuleius occasionally airs his scientific attainments by specific statements and illustrations from the zoological and other scientific fields. Indeed the

[1] Nicander lived in the second century B.C. under Attalus III of Pergamum. Of his works there are extant the *Theriaca* in 958 hexameters and another poem, the *Alexipharmaca,* of 630 lines; ed. J. G. Schneider, 1792 and 1816; by O. Schneider, 1856. There is an illuminated eleventh century manuscript of the *Theriaca* in the Bibliothèque Nationale at Paris, which O. M. Dalton (*Byzantine Art and Archaeology,* p. 483) says "is evidently a painstaking copy of a very early original, perhaps almost contemporary with Nicander himself."

[2] Cap. 40.

[3] Caps. 49-51.

[4] Caps. 15-16.

[5] Cap. 40.

[6] Cap. 36.

presence of such allusions is as noticeable in the *Apology* as was their absence from the *Metamorphoses*. But they go to show that his knowledge was greater than his discretion, since for the most part they repeat familiar errors of contemporary science. We are told—the story is also in Aristotle, Pliny, and Aelian—how the crocodile opens its jaws to have its teeth picked by a friendly bird,[1] that the viper gnaws its way out of its mother's womb,[2] that fish are spontaneously generated from slime,[3] and that burning the stone *gagates* will cause an epileptic to have a fit.[4] On the other hand, the skin shed by a spotted lizard is a remedy for epilepsy, but you must snatch it up speedily or the lizard will turn and devour it, either from natural appetite or just because he knows that you want it.[5] This tale, so characteristic of the virtues attributed to parts of animals and the human motives ascribed to the animals themselves, is taken by Apuleius from a treatise by Theophrastus entitled *Jealous Animals*.

Apparent ignorance of magic and occult virtue.

In defending what he terms his scientific investigations from the aspersion of magic Apuleius is at times either a trifle disingenuous and inclined to trade upon the ignorance of his judge and accusers, or else not as well informed himself as he might be in matters of natural science and of occult science. He contends that fish are not employed in magic arts, asks mockingly if fish alone possess some property hidden from other men and known to magicians, and affirms that if the accuser knows of any such he must be a magician rather than Apuleius.[6] He insists that he did not make use of a sea-hare and describes the "fish" in question in detail,[7] but this description, as is pointed out in Butler and Owen's edition of the *Apology*,[8] tends to convince us that it really was a sea-hare. In the case of the two fish with obscene names, he ridicules the arguing from similarity of names to similarity of powers in the things so designated, as

[1] Cap. 8.
[2] Cap. 85.
[3] Cap. 38.
[4] Cap. 45.
[5] Cap. 51.
[6] Caps. 30, 42.
[7] Cap. 40.
[8] P. 98.

if that were not what magicians and astrologers and believers in sympathy and antipathy were always doing. You might as well say, he declares, that a pebble is good for the stone and a crab for an ulcer,[1] as if precisely these remedies for those diseases were not found in the Pseudo-Dioscorides and in Pliny's *Natural History*.[2]

Despite an assumption of knowledge.

It is hardly probable that in the passages just cited Apuleius was pretending to be ignorant of matters with which he was really acquainted, since as a rule he is eager to show off his knowledge even of magic itself. Thus the accusers affirmed that he had bewitched a boy by incantations in a secret place with an altar and a lamp; Apuleius criticizes their story by saying that they should have added that he employed the boy for purposes of divination, citing tales which he has read to this effect in Varro and many other authors.[3] And he himself is ready to believe that the human soul, especially in one who is still young and innocent, may, if soothed and distracted by incantations and odors, forget the present, return to its divine and immortal nature, and predict the future. When he reads some technical Greek names from his treatise on fishes, he suspects that the accuser will protest that he is uttering magic names in some Egyptian or Babylonian rite.[4] And as a matter of fact, when later he mentioned the names of a number of celebrated magicians,[5] the accusers appear to have raised such a tumult that Apuleius deemed it prudent to assure the judge that he had simply read them in reputable books in public libraries, and that to know such names was one thing, to practice the magic art quite another matter.

Attitude toward astrology.

Apuleius affirms that one of his accusers had consulted he knows not what Chaldeans how he might profitably marry off his daughter, and that they had prophesied truthfully that her first husband would die within a few months. "As for what she would inherit from him, they fixed that up, as

[1] Cap. 35.
[2] So Abt has pointed out: *Die Apologie des Apuleius von Madaura und die antike Zauberei*, Giessen, 1908, p. 224.
[3] Caps. 42-43.
[4] Cap. 38.
[5] Cap. 90.

they usually do, to suit the person consulting them." [1] But in this respect their prediction turned out to be quite incorrect. We are left in some doubt, however, whether their failure in the second case is not regarded as due merely to their knavery, and their first successful prediction to the rule of the stars. Elsewhere, however, Apuleius does state that belief in fate and in magic are incompatible, since there is no place left for the force of spells and incantations, if everything is ruled by fate.[2] But in other extant works [3] he speaks of the heavenly bodies as visible gods, and Laurentius Lydus attributes astrological treatises to him.[4]

His theory of demons.

In one passage of the *Apology* Apuleius affirms his belief with Plato in the existence of certain intermediate beings or powers between gods and men, who govern all divinations and the miracles of the magicians.[5] In the treatise on the god or demon of Socrates [6] he repeats this thought and tells us more of these mediators or demons. Their native element is the air, which Apuleius thought extended as far as the moon,[7] just as Aristotle [8] tells of animals who live in fire and are extinguished with it, and just as the fifth element, that "divine and inviolable" ether, contains the divine bodies of the stars. With the superior gods the demons have immortality in common, but like mortals they are subject to passions and to feeling and capable of reason.[9] But their bodies are very light and like clouds, a point peculiar to themselves.[10] Since both Plutarch and Apuleius wrote essays on the demon of Socrates and both derived, or thought that they derived, their theories concerning demons from Plato, it is interesting to note some divergences between their accounts. Apuleius confines them to the atmosphere beneath the moon more exclusively than Plutarch does; unlike Plutarch he represents them as immortal, not merely long-lived; and he has more to say about the sub-

[1] Cap. 97.
[2] Cap. 84.
[3] *De mundo*, cap. 1; *De deo Socratis*, cap. 4.
[4] *De mens.*, IV., 7, 73; *De ostent.*, 3, 4, 7, 10, 44, 54.
[5] Cap. 43.
[6] Cap. 6.
[7] *De deo Socratis*, cap. 8.
[8] *Hist. Anim.*, V, 19.
[9] *De deo Socratis*, cap. 13.
[10] *Ibid.*, caps. 9-10.

stance of their bodies and less concerning their relations with disembodied souls.

Apuleius in the middle ages.

Apuleius would have been a well-known name in the middle ages, if only indirectly through the use made by Augustine in *The City of God*[1] of the *Metamorphoses* in describing magic and of the *De deo Socratis* in discussing demons.[2] He also speaks of Apuleius in three of his letters,[3] declaring that for all his magic arts he could win neither a throne nor judicial power. Augustine was not quite sure whether Apuleius had actually been transformed into an ass or not. A century earlier Lactantius[4] spoke of the many marvels remembered of Apuleius. That manuscripts of the *Metamorphoses, Apology* and *Florida* were not numerous until after the twelfth and thirteenth centuries may be inferred from the fact that all the extant manuscripts seem to be derived from a single one of the later eleventh century, written in a Lombard hand and perhaps from Monte Cassino.[5] The article on Apuleius in Pauly and Wissowa states that the best manuscripts of his other works are an eleventh century codex at Brussels and a twelfth century manuscript at Munich,[6] but does not mention a twelfth century manuscript of the *De deo Socratis* in the British Museum.[7] Another indication that in the twelfth century there were manuscripts of Apuleius in England or at Chartres and Paris is that John of Salisbury borrows from the *De dogmate Platonis* in his *De nugis curialium*.[8] In the earlier middle ages there was ascribed to Apuleius a work on herbs of which we shall treat later.

[1] XVIII, 18.

[2] VIII, 14-22.

[3] Epistles 102, 136, 138, in Migne, PL, vol. 33.

[4] *Divin. Instit.*, V, 3.

[5] Codex Laurentianus, plut. 68, 2. The same MS contains the *Histories* and *Annals* (XI-XVI) of Tacitus. A subscription to the ninth book of the *Metamorphoses* indicates that the original manuscript from which this was derived or copied was produced in 395 A.D. and 397 A.D. G. Huet, "Le roman d'Apulée était-il connu au moyen âge," *Le Moyen Age* (1917), 44-52, holds that the *Metamorphoses* was not known directly to the medieval vernacular romancers. See also B. Stumfall, *Das Märchen von Amor und Psyche in Seinem Fortleben*, Leipzig, 1907.

[6] CLM 621.

[7] Harleian 3969.

[8] VII, 5.

CHAPTER VIII

PHILOSTRATUS'S LIFE OF APOLLONIUS OF TYANA

Compared with Apuleius—Philostratus's sources—Time and space covered—Philostratus's audience—Object of the *Life*—Apollonius charged with magic—A confusion of terms—The *Magi* and magic—Apollonius and the *Magi*—Philostratus on wizards—Apollonius and wizards—Quacks and old-wives—The Brahmans—Marvels of the Brahmans—Magical methods of the Brahmans—Medicine of the Brahmans—Some signs of astrology—Interest in natural science—Natural law or special providence?—Cases of scepticism—Anecdotes of animals—Dragons of India—Occult virtues of gems—Absence of number mysticism—*Mantike* or the art of divination—Divining power of Apollonius—Dreams—Interpretation of omens—Animals and divination—Divination by fire—Other so-called predictions—Apollonius and the demons—Not all demons are evil—Philostratus's faith in demons—The ghost of Achilles—Healing the sick and raising the dead—Other marvels—Golden wrynecks and the *iunx*—Why named *iunx?*—Apollonius in the middle ages.

Compared with Apuleius.

SOME fifty years after the birth of Apuleius occurred that of Philostratus, whose career and interests were somewhat similar, although he came from the Aegean island of Lemnos instead of the neighborhood of Carthage and wrote in Greek rather than Latin. But like Apuleius he was a student of rhetoric and went first to Athens and then to Rome. The resemblance is perhaps closer between Apuleius and Apollonius of Tyana, whose life Philostratus wrote and of whom we know more than of his biographer. Like Apuleius Apollonius had to defend himself in court against the accusation of magic, and Philostratus gives us what purports to be his apology on that occasion. Two centuries afterwards Augustine in one of his letters [1] names Apollonius and Apuleius as examples of men who were addicted to the magic art and who, the pagans said, performed greater

[1] Ep. 136.

miracles than Christ did. A century before Augustine Lactantius states [1] that a certain philosopher who had "vomited forth" three books "against the Christian religion and name" had compared the miracles of Apollonius favorably with those of Christ; Lactantius marvels that he did not mention Apuleius as well. Like Apuleius, Apollonius was a man of broad learning who traveled widely and sought initiation into mysteries and cults. Apuleius was a Platonist; Apollonius, a Pythagorean. We may also note a resemblance between the *Metamorphoses* and the *Life of Apollonius.* Both seem to elaborate earlier writings and both have much to say of transformations, wizards, demons, and the occult. The *Life of Apollonius of Tyana,* however, must be taken more seriously than the *Metamorphoses.* If the African's work is a rhetorical romance embodying a certain autobiographical element, a Milesian tale to which personal religious experiences are annexed, then the work by Philostratus is a rhetorical biography with a tinge of romance and a good deal of sermonizing.

Philostratus's sources.

Philostratus [2] composed the *Life of Apollonius* about 217 A. D. at the request of the learned wife of the emperor Septimius Severus, to whose literary circle he belonged. The empress had come into possession of some hitherto unknown memoirs of Apollonius by a certain Damis of Nineveh, who had been his disciple and had accompanied him upon many of his travels. Some member of Damis's family had brought these documents to the empress's attention. Some scholars incline to the view that she was deceived by an impostor, but it hardly seems that there would be sufficient profit in the venture to induce anyone to take the pains to forge such memoirs. Also I can see no reason why a contemporary of Apollonius should not have said and believed everything which Philostratus represents Damis as saying; on the contrary it seems to me just what would be

[1] *Divin. Instit.,* V, 2-3.

[2] Concerning other writers named Philostratus and which works should be assigned to each, see Schmid (1913) 608-20.

said by a naïf, gullible, and devoted disciple, who was inclined to exaggerate the abilities and achievements of his master and to take literally everything that Apollonius uttered ironically or figuratively. Other accounts of Apollonius were already in existence by a Maximus of Aegae, where Apollonius had spent part of his life, and by Moeragenes, but the memoirs of Damis seem to have offered much new material. Philostratus accordingly wrote a new life based largely upon Damis, but also making use of the will and epistles of Apollonius, many of which the emperor Hadrian had earlier collected, and of the traditions still current in the cities and temples which Apollonius had frequented and which Philostratus now took the trouble to visit. It has sometimes been suggested, chiefly by Christian writers intent upon discrediting the career of Apollonius, that Philostratus invented Damis and his memoirs. But Philostratus seems straightforward in describing the pains he has been to in preparing the *Life,* and certainly is more explicit and systematic in stating his sources than other ancient biographers like Plutarch and Suetonius are. He appears to follow his sources rather closely and not to invent new incidents, although he may, like Thucydides and other ancient historians, have taken liberties with the speeches and arguments put into his characters' mouths. And through the work, despite his belief in demons and marvels, he now and then gives evidence of a moderate and sceptical mind, at least for his times.

Time and space covered.

Apollonius lived in the first century of our era and died during the reign of Nerva well advanced in years. It is therefore of a period over a century before his own that Philostratus writes. He is said to commit a number of errors in history and geography,[1] but we must remember that mistakes in geography were a failing of the best ancient his-

[1] See article on Apollonius of Tyana in Pauly-Wissowa. Priaulx, *The Indian Travels of Apollonius of Tyana,* London, 1873, p. 62, found the geography of Apollonius's Indian travels so erroneous that he came to the conclusion that either Apollonius never visited India, or, if he did, that Damis "never accompanied him but fabricated the journal Philostratus speaks of."

torians such as Polybius, and the general picture drawn of the emperors and politics of Apollonius's time is not far wrong. It is true that Philostratus also makes use of tradition which has gradually formed since the death of Apollonius, and introduces explanations or comments of his own on various matters. It is, however, not the facts either of Apollonius's career or of his times that concern us but the beliefs and superstitions which we find in Philostratus's *Life* of him. Whether these are of the first, second, or early third century is scarcely necessary or possible for us to distinguish. If Damis records them, Philostratus accepts them, and the probability is that they apply not only to all three centuries but to a long period before and after. The territory covered in the *Life* is almost as extensive; it ranges all over the Roman Empire, alludes occasionally to the Celts and Scythians, and opens up Ethiopia and India [1] to our gaze. Apollonius was a great traveler and there are many interesting and informing passages concerning ships, sailing, pilots, merchants and sea-trade.[2]

Philostratus's audience

If we ask further, for what class of readers was the *Life* intended, the answer is, for the intellectual and learned. Apollonius himself was distinctly a Hellene. Philostratus represents him as often quoting Homer and other bygone Greek authors, or mentioning names from early Greek history such as Lycurgus and Aristides. One of his aims was to restore the degenerate Greek cities of his own day to their ancient morality. Furthermore, Apollonius never cared for many disciples, and neither required them to observe all the rules of life which he himself followed, nor admitted them to all his interviews with other sages and his initiations into sacred mysteries. This aloofness of the sage is somewhat reflected in his biographer. The *Life* is an attempt not to

[1] Priaulx, however, regarded its statements concerning India as such as might have been "easily collected at that great mart for Indian commodities and resort for Indian merchants — Alexandria," or from earlier authors.

[2] III, 23, 35; IV, 9, 32; V, 20; VI, 12, 16; VII, 10, 12, 15-16.

popularize the teachings of Apollonius but to justify him before the learned world.

Object of the *Life.*

The charge had been frequently made that Apollonius came illegitimately by his wisdom and acquired it violently by magic. Philostratus would restore him to the ranks of true philosophers who gained wisdom by worthy and licit methods. He declares that he was not a wizard, as many suppose, but a notable Pythagorean, a man of broad culture, an intellectual and moral teacher, a religious ascetic and reformer, probably even a prophet of divine and superhuman nature. It is not now so generally held by Christian writers as it used to be that Philostratus wrote the *Life* with the Gospel story of Christ in mind, and that his purpose was to imitate or to parody or to oppose a rival narrative to the Christian story and teaching. At no point in the *Life* does Philostratus betray unmistakably even a passing acquaintance with the Gospels, much less display any sign of animus against them. Moreover, the Christian historian and apologist, Eusebius, who lived in the century following Philostratus and was familiar with his *Life* of Apollonius, in writing a reply to a treatise in which Hierocles, a provincial governor under Diocletian, had compared Apollonius with Jesus, distinctly states that Hierocles was the first to suggest such an idea.[1] Such similarities then as may exist between the *Life* and the Gospels must be taken as examples of beliefs common to that age.

Apollonius charged with magic.

Apollonius was accused of sorcery or magic during his lifetime by the rival philosopher Euphrates. The four books on Apollonius written by Moeragenes also portrayed him as a wizard;[2] and Eusebius in his reply to Hierocles ascribed the miracles wrought by Apollonius to sorcery and the aid of evil demons.[3] Earlier the satirist Lucian de-

[1] See the treatise of Eusebius *Against Apollonius.* Lactantius (*Divin. Inst.,* V, 2-3) probably had reference to Hierocles in speaking of a philosopher who had written three books against Christianity and declared the miracles of Apollonius as wonderful as those of Christ.

[2] So Origen says (*Against Celsus,* VI, 41) and Philostratus implies (I, 3).

[3] See the *Against Apollonius,* caps. 31, 35.

scribed Alexander the pseudo-prophet as having been in his youth an apprentice to "one of the charlatans who deal in magic and mystic incantations, . . . a native of Tyana, an associate of the great Apollonius, and acquainted with all his heroics." [1]

A confusion of terms

In defending his hero against these charges Philostratus is guilty himself both of some ambiguous use of terms and of some loose thinking. The same ambiguous terminology, however, will be found in other discussions of magic. In a few passages Philostratus denies that Apollonius was a *μάγος* but much oftener exculpates him from the charge of being a *γόης* or *γοήτης*. With the latter word or words there is no difficulty. It means a wizard, sorcerer, or enchanter, and is always employed in a sinister or disreputable sense. With the term *μάγος* the case is different, as with the Latin *magus*. It may signify an evil magician, or it may refer to one of the Magi of the East, who are generally regarded as wise and good men. This delicate distinction, however, is not easy to maintain and Philostratus fails to do so, while Mr. Conybeare in his English translation [2] makes confusion worse confounded not only by translating *μάγος* as "wizard" instead of "magician," but by sometimes doing this when it really should be rendered as "one of the Magi." It may also be noted that Philostratus locates the Magi in Babylonia as well as in Persia.

The Magi and magic

To begin with, in his second chapter Philostratus says that some consider Apollonius a magician "because he consorted with the Magi of the Babylonians, and the Brahmans of the Indians, and the Gymnosophists in Egypt." But they are wrong in this. "For Empedocles and Pythagoras himself and Democritus, although they associated with the Magi and spake many divine utterances, yet did not stoop to the art" (of magic). Plato, too, he goes on to say, although

[1] *Ἀλέξανδρος, ἢ ψευδόμαντις*, cap. 5. In the passage quoted I have used Fowler's translation.

[2] In other respects, however, I have usually found this translation, which accompanies the Greek text in the recent Loeb Classical Library edition, both racy and accurate, and have employed it in a number of the quotations which follow.

he visited Egypt and its priests and prophets, was never regarded as a magician. In this passage, then, Philostratus closely associates the Magi with the magic art, and I am not sure whether the last "Magi" should not be "magicians." On the other hand his acquittal of Democritus and Pythagoras from the charge of magic does not agree with Pliny, who ascribed a large amount of magic to them both.

Apollonius and the Magi.

Apollonius himself evidently did not regard the Magi whom he met in Babylon and Susa as evil magicians. One of the chief aims of his scheme of oriental travel "was to acquaint himself thoroughly with their lore." He wished to discover whether they were wise in divine things, as they were said to be.[1] Sacrifices and religious rites were performed under their supervision.[2] Apollonius did not permit Damis to accompany him when he visited the Magi at noon and again about midnight and conversed with them.[3] But Apollonius himself said that he learned some things from them and taught them some things; he told Damis that they were "wise men, but not in all respects"; on leaving their country he asked the king to give the presents which the monarch had intended for Apollonius himself to the Magi, whom he described then as "men who both are wise and wholly devoted to you." [4]

Philostratus on wizards.

Quite different is the attitude towards witchcraft and wizards of both Apollonius and his biographer. In the opinion of Philostratus wizards are of all men most wretched.[5] They try to violate nature and to overcome fate by such methods as inquisition of spirits, barbaric sacrifices, incantations and besmearings. Simple-minded folk attribute great powers to them; and athletes desirous of winning victories, shopkeepers intent upon success in business ventures, and lovers in especial are continually resorting to them and apparently never lose faith in them despite repeated failures, despite occasional exposure or ridicule of their methods in

[1] I, 32.
[2] I, 29.
[3] I, 26.
[4] I, 40.
[5] V, 12.

books and writing, and despite the condemnation of witchcraft both by law and nature.[1] Apollonius was certainly no wizard, argues Philostratus, for he never opposed the Fates but only predicted what they would bring to pass, and he acquired this foreknowledge not by sorcery but by divine revelation.[2]

Apollonius and wizards.

Nevertheless Apollonius is frequently accused of being a wizard by others in the pages of Philostratus. At Athens he was refused initiation into the mysteries on this ground,[3] and at Lebadea the priests wished to exclude him from the oracular cave of Trophonius for the same reason.[4] When the dogs guarding the temple of Dictynna in Crete fawned upon him instead of barking at his approach, the guardians of the shrine arrested him as a wizard and would-be temple robber who had bewitched the dogs by something that he had given them to eat.[5] Apollonius also had to defend himself against the accusation of witchcraft in his hearing or trial before Domitian.[6] He then denied that one is a wizard merely because one has prescience, or that wearing linen garments proves one a sorcerer. Wizards shun the shrines and temples of the gods; they make use of trenches dug in the earth and invoke the gods of the lower world. They are greedy for gain and pseudo-philosophers. They possess no true science, depending for success in their art upon the stupidity of their dupes and devotees. They imagine what does not exist and disbelieve the truth. They work their sorcery by night and in darkness when those employing them cannot see or hear well. Apollonius himself was accused to Domitian of having sacrificed an Arcadian boy at night and consulted his entrails with Nerva in order to determine the latter's prospects of becoming emperor.[7] When before his trial Domitian was about to put Apollonius in fetters, the sage proposed the dilemma that if he were a wizard he could not be kept in bonds, or that if Domitian were able

[1] VII, 39.
[2] V, 12.
[3] IV, 18.
[4] VIII, 19.
[5] VIII, 30.
[6] VIII, 7.
[7] VII, 20.

to fetter him, he was obviously no wizard.[1] This need not imply, however, that Apollonius believed that wizards really could free themselves, for he was at times ironical. If so, Domitian replied in kind by assuring him that he would at least keep him in fetters until he transformed himself into water or a wild beast or a tree.

Quacks and old-wives.

Closely akin to the *goëtes* or wizards are the old hags and quack-doctors who offer one Indian spices or boxes supposed to contain bits of stone taken from the moon, stars, or depths of earth.[2] Likewise the divining old-wives who go about with sieves in their hands and pretend by means of their divination to heal sick animals for shepherds and cowherds.[3] We also read that Apollonius expelled from the cities along the Hellespont various Egyptians and Chaldeans who were collecting money on the pretense of offering sacrifices to avert the earthquakes which were then occurring.[4]

The Brahmans.

We have heard Philostratus mention the Brahmans of India in the same breath with the Magi of Persia and imply that Apollonius's association with them contributed to his reputation as a magician.[5] In another passage [6] Philostratus places *goëtes* and Brahmans in unfortunate juxtaposition, and, immediately after condemning the wizards and defending Apollonius from the charge of sorcery, goes on to say that when he saw the automatic tripods and cup-bearers of the Indians, he did not ask how they were operated. "He applauded them, it is true, but did not think fit to imitate them." But of course Apollonius should not even have applauded these automatons, which set food and poured wine before the guests of the Brahmans, if they were the contrivances of wizards. And in another passage,[7] where he defends the signs and wonders wrought by the Brahmans against the aspersions cast upon them by the Gymnosophists of Ethiopia, Apollonius explains their practice of levitation

[1] VII, 34.
[2] VII, 39.
[3] VI, 11; III, 43.
[4] VI, 41.
[5] I, 2.
[6] V, 12.
[7] VI, 11.

as an act of worship and communion with the sun god, and hence far removed from the rites performed in deep trenches and hollows of the earth to the gods of the lower world which we have heard him mention before as a practice characteristic of wizards.

Marvels of the Brahmans.

Nevertheless the feats ascribed to the Brahmans are certainly sufficiently akin to magic to excuse Philostratus for mentioning them along with the Magi and wizards and to justify us in considering them. Indeed, modern scholarship informs us that in the Vedic texts the word "bráhman" in the neuter means a "charm, rite, formulary, prayer," and "that the caste of the Brahmans is nothing but the men who have *bráhman* or magic power.[1] In marked contrast to the taciturnity of Apollonius as to his interviews with the Magi of Babylon and Susa is the long account repeated by Philostratus from Damis of the sayings and doings of the sages of India. As for Apollonius himself, "he was always recounting to everyone what the Indians said and did." [2] They knew that he was approaching when he was yet afar off and sent a messenger who greeted him by name.[3] Iarchas, their chief, also knew that Apollonius had a letter for him and that a delta was missing in it, and he told Apollonius many events of his past life. "We see, O Apollonius," he said, "the signs of the soul, tracing them by a myriad symbols." [4] The Brahmans lived in a castle concealed by clouds, where they rendered themselves invisible at will. The rocks along the path up to their abode were still marked by the cloven feet, beards, faces, and backs of the Pans who had tried to scale the height under the leadership of Dionysus and Heracles, but had been hurled down headlong.[5] Here too was a well for testing oaths, a purify-

[1] J. E. Harrison, *Themis,* Cambridge, 1912, p. 72. "The Buddha himself condemned as worthless the whole system of Vedic sacrifices, including in his ban astrology, divination, spells, omens, and witchcraft; but in the earliest Buddhist stupas known to us, the symbolism is entirely borrowed from the sacrificial lore of the Vedas: "E. B. Havell, *A Handbook of Indian Art,* 1920, p. 6, and see p. 32 for the birth of Buddha under the sign Taurus.

[2] VI, 10.

[3] III, 12.

[4] III, 16.

[5] III, 13.

ing fire, and the jars in which the winds and rain were bottled up.

Magical methods of the Brahmans.

When the messenger of the Brahmans greeted Apollonius by name, the latter remarked to the astounded Damis, "We have come to men who are wise without art (ἀτεχνῶς), for they seem to have the gift of foreknowledge." [1] As a matter of fact, however, most of the subsequent wonders wrought by the Brahmans were not performed without the use of paraphernalia and rites very similar to those of magic. Each Brahman carries a staff—or magic wand—and wears a ring, which are both prized for their occult virtue by which the Brahmans can accomplish anything they wish.[2] They clothe themselves in sacred garments made of "a wool that springs wild from the ground" (cotton?) and which the earth will not permit anyone else to pluck. Iarchas also showed Apollonius and Damis a marvelous stone called *Pantarbe,* which attracted and bound other stones to itself and which, although only the size of his finger-nail and formed in earth four fathoms deep, had such virtue that it broke the earth open.[3] But it required great skill to secure this gem. "We only," said the Brahman, "can obtain this *pantarbe,* partly by doing things and partly by saying things," in other words by incantations and magical operations. Before performing their rite of levitation they bathed and anointed themselves with a certain drug. "Then they stood like a chorus with Iarchas as leader and with their rods uplifted struck the earth, which heaving like the sea-wave raised them up in the air two cubits high." [4] The metallic tripods and cup-bearers which served the king of the country when he came to visit the Brahmans appeared from nowhere laden with food and wine exactly as if by magic.[5]

Medicine of the Brahmans.

The medical practice, if we may so call it, of the Brahmans was tinged, to say the least, with magic. A dislocated hip, indeed, they appear to have cured by massage, and a

[1] III, 12. But perhaps the translation should be, "men who are exceedingly wise."

[2] III, 15.

[3] III, 46-47.

[4] III, 17.

[5] III, 27.

blind man and a paralytic are healed by unspecified methods.[1] But a boy is cured of inherited alcoholism by chewing owl's eggs that have been boiled; a woman who complains that her sixteen-year-old son has for two years been vexed by a demon is sent away with a letter full of threats or incantations to employ against the spirit; and another woman's sufferings in childbirth are prevented by directing her husband to enter her chamber with a live hare concealed in his bosom and to release the hare after he has walked around his wife once. Iarchas, indeed, attributed the origin of medicine to divination or divine revelation.[2] His theory was that Asclepius, as the son of Apollo, learned by oracles what drugs to employ for the different diseases, in what amounts to mix the drugs, what the antidotes for poisons were, and how to use even poisons as remedies. This last especially he affirmed that no one would dare attempt without foreknowledge.

Some signs of astrology.

The Brahmans seem to have made some use of astrology in working their feats of magic. Damis at any rate said that when Apollonius bade farewell to the sages, Iarchas made him a present of seven rings named after the planets, which he wore in turn upon the appropriate days of the week.[3] Perhaps, too, the seven swords of adamant which Iarchas had rediscovered as a child had some connection with the planets.[4] Moeragenes ascribed four books on foretelling the future by the stars to Apollonius himself, but Philostratus was unable to find any such work by Apollonius extant in his day.[5] And unless it be an allusion to Chaldeans which we have already noted, there is no further mention of astrology in Philostratus's *Life*—a rather remarkable fact considering that he wrote for the court of Septimius Severus, the builder of the Septizonium.

Interest in natural science.

The philosopher Euphrates, who is represented by Philostratus as jealous of Apollonius, once advised the emperor Vespasian, when Apollonius was present, to embrace natural

[1] III, 38-40.
[2] III, 44.
[3] III, 41.
[4] III, 21.
[5] III, 41.

philosophy—or a philosophy in accordance with natural law—but to beware of philosophers who pretended to have secret intercourse with the gods.[1] There was justification in the latter charge against Apollonius, but it should not be assumed that his mysticism rendered him unfavorable to natural science. On the contrary he is frequently represented by Philostratus as whiling away the time along the road by discussing with Damis such natural problems as the delta of the Nile or the tides at the mouth of the Guadalquivir. He was especially interested in the habits of animals and the properties of gems. Vespasian was fond of listening to "his graphic stories of the rivers of India and the animals" of that country, as well as to "his statements of what the gods revealed concerning the empire."[2] Some of the questions which Apollonius put to the Brahmans concerned nature.[3] He asked of what the world was composed, and when they said, "Of elements," he asked if there were four. They believed, however, in a fifth element, ether, from which the gods had been generated and which they breathe as men breathe air. They also regarded the universe as a living animal. He further inquired of them whether land or sea predominated on the earth's surface,[4] and this same attitude of scientific inquiry and of curiosity about natural forces and objects is frequently met in the *Life*.

Natural law or special providence?

Apollonius believed, as we shall see, in omens and portents, and interpreted an earthquake at Antioch as a divine warning to the inhabitants.[5] The Brahman sages, moreover, regarded prolonged drought as a punishment visited by the world soul upon human sinfulness.[6] On the other hand, Apollonius gave a natural explanation of volcanoes and denied the myths concerning Enceladus being imprisoned under Mount Aetna and the battle of the gods and giants.[7] And in the case of the earthquake the people had already accepted it as a portent and were praying in terror, when

[1] V, 37.
[2] V, 37.
[3] III, 34.
[4] III, 37.
[5] VI, 38.
[6] III, 34.
[7] V, 17.

Apollonius took the opportunity to warn them to cease from their civil factions. As a matter of fact, both Apollonius and Philostratus appear to regard portents as an extraordinary sort of natural phenomena. A knowledge of natural science helps in recognizing them and in interpreting them. When a lioness of enormous size with eight whelps in her is slain by hunters, Apollonius at once recognizes the event as portentous because as a rule lionesses have whelps only thrice and only three of them on the first occasion, two in the second litter, and finally but a single whelp, "but I believe a very big one and preternaturally fierce." [1] Here Apollonius is not in strict agreement with Pliny and Aristotle [2] who say that the lioness produces five whelps at the first birth and one less every succeeding year.

Cases of scepticism

The scepticism of Apollonius concerning the Aetna myth is not an isolated instance. At Sardis he ridiculed the notion that trees could be older than earth,[3] and he was one of the few ancients to question the swan's song.[4] He denied "the silly story that the young of vipers are brought into the world without mothers" as "consistent neither with nature nor experience," [5] and also the tale that the whelps of the lioness claw their way out into the world.[6] In India Apollonius saw a wild ass or unicorn from whose single horn a magic drinking horn was made.[7] A draught from this horn was supposed to protect one for that day from disease, wounds, fire, or poison, and on that account the king

[1] I, 22.

[2] NH, VIII, 17; *Hist. Anim.*, VI, 31.

[3] VI, 37.

[4] The ancient authorities, pro and con, will be found listed in D. W. Thompson, *Glossary of Greek Birds*, 106-107. He adds: "Modern naturalists accept the story of the singing swans, asserting that though the common swan cannot sing, yet the Whooper or whistling swan does so. It is certain that the Whooper sings, for many ornithologists state the fact, but I do not think that it can sing very well; at the very best, *dant sonitum rauci per stagna loquacia cygni*. This concrete explanation is quite inadequate; it is beyond a doubt that the swan's song (like the halcyon's) veiled, and still hides, some mystical allusion."

[5] II, 14.

[6] I, 22. Pliny, NH, VIII, 17, repeats a slightly different popular notion that the lioness tears her womb with her claws and so can bear but once; against this view he cites Aristotle's statement that the lioness bears five times, as described above.

[7] III, 2.

alone was permitted to hunt the animal and to drink from the horn. When Damis asked Apollonius if he credited this story, the sage ironically replied that he would believe it if he found the king of the country to be immortal. Either, however, the scepticism of Apollonius, as was the case with so many other ancients and medieval men, was sporadic and inconsistent, or it came to be overlaid with the credulity of Damis and Philostratus, as the following example suggests. Iarchas told Damis and Apollonius flatly that the races described by Scylax of men with long heads or huge feet with which they were said to shade themselves did not exist in India or anywhere else; yet in a later book Philostratus states that the shadow-footed people are a tribe in Ethiopia.[1]

Anecdotes of animals.

At any rate the marvels of India are more frequently credited than criticized in the *Life* by Philostratus, and the same holds true of the extraordinary conduct and well-nigh human intelligence attributed to animals. Especially delightful reading are six chapters on the remarkable sagacity of elephants and their love for mankind.[2] On this point, as by Pliny, use is made of the work of Juba. We read again of sick lions eating apes, of the lioness's love affair with the panther, of the fondness of leopards for the fragrant gum of a certain tree and of goats for the cinnamon tree; of apes who are made to collect pepper for men by appealing to their instinct towards mimicry;[3] and of the tiger, whose loins alone are eaten by the Indians. "For they decline to eat the other parts of this animal, because they say that as soon as it is born it lifts up its front paws to the rising sun."[4] In the river Hyphasis is a creature like a white worm which yields when melted down a fat or oil that once set afire cannot be extinguished and which the king uses to burn walls

[1] III, 47; VI, 25. Scylax was a Persian admiral under Darius who traveled to India and wrote an account of his voyages. The work extant under his name is of doubtful authorship (Isaac Vossius, *Periplus Scylacis Caryandensis*, 1639), but some date it as early as the fourth century B.C.

[2] II, 11-16.

[3] II, 2; III, 4.

[4] II, 28.

and capture cities.[1] In India are griffins who quarry gold with their powerful beaks, and the luminous phoenix with its nest of spices and swan-like funeral song.[2]

Dragons of India.

Especially remarkable are the snakes or dragons with which all India is filled and which often are of enormous size, thirty or even seventy cubits long.[3] Those found in the marshes are sluggish and have no crests; but those on the hills and ridges move faster than the swiftest rivers and have both beards and crests.[4] Those in the plain engage in combats with elephants which terminate fatally for both parties as we have already learned from Pliny.[5] The mountain dragons have bushy beards, fiery crests, golden scales, and a ferocious glance.[6] They burrow into the earth, making a noise like clashing brass, or go hissing down to the shore and swim far out to sea. Terrifying as they are, the Indians charm them by showing them golden characters embroidered on a cloak of scarlet and by incantations of a secret wisdom. They eat the dragon's heart and liver in order to be able to understand the language and thoughts of animals.[7]

Occult virtues of gems.

The dragons, however, are prized more for the precious stones in their heads, which the Indians quickly cut off as soon as they have bewitched them. The pupils of the eyes of the hill dragons are a fiery stone possessing irresistible virtue for many occult purposes,[8] while in the heads of the mountain dragons are many brilliant stones of flashing colors which exert occult virtue if set in a ring, "and they say that Gyges had such a ring." [9] But there are many marvelous stones outside the heads of dragons. "Who does not know the habits of birds," says Apollonius to Damis in one of his disquisitions upon natural phenomena,[10] "and that eagles and storks will not build their nests without placing in them, the one the stone *aetites,* and the other the *lychnites,*

[1] III, 1. Greek fire?
[2] III, 48-9.
[3] III, 6; II, 17.
[4] III, 7.
[5] NH, VIII, 11.
[6] III, 8.
[7] III, 9.
[8] III, 7.
[9] III, 8.
[10] II, 14.

as aids in hatching and to drive snakes away?" On parting from the Indian king Phraotes, Apollonius as usual refused to accept money presents but picked up one of the gems that were offered him with the exclamation, "O rare stone, how opportunely and providentially have I found you!" [1] Philostratus supposes that he detected some occult and divine power in this particular stone. The Brahmans had gems so huge that from one of them a goblet could be carved large enough to slake the thirst of four men in midsummer, but in this case nothing is said of occult virtue.[2] The Brahman Iarchas felt sure that he was the reincarnation of the hero Ganges, son of the river Ganges, because as a mere child he knew where to dig for the seven swords of adamant which Ganges had fixed in the earth.[3] Presumably these were magic swords and their virtue in part due to the stone adamant of which they were made. Less is said in the *Life* of the virtues of herbs than of gems, but the Indians made a nuptial ointment or love-charm from balm distilled from trees,[4] and drugs and poisons are mentioned more than once, mandragora being described as a soporific drug rather than a deadly poison.[5]

Absence of number mysticism.

Considering that Apollonius was a Pythagorean, there is surprisingly little said concerning perfect numbers and their mystic significance. Aside from the seven rings and seven swords already mentioned, about the only instance is the question asked by Apollonius whether eighteen, the number of the Brahman sages at the time of his visit, had any especial importance.[6] He remarked that eighteen was not a square, nor a number usually held in esteem and honor like ten, twelve, and sixteen. The Brahmans agreed that there was no particular significance in eighteen, and further informed him that they maintained no fixed number of members but had varied from only one to as many as seventy according to the available supply of worthy men.

[1] II, 40.
[2] III, 27.
[3] III, 21.
[4] III, 1.
[5] VIII, 7.
[6] III, 30.

If Philostratus denies that Apollonius was a magician, he does depict him as endowed with prophetic gifts, with power over demons, and with "secret wisdom." He rather likes to give the impression that the sage foretold things by innate prophetic gift or divine inspiration, but even μαντική or the art of divination is not condemned as γοητεία or witchcraft was. Iarchas the Brahman says that those who delight in *mantike* become divine thereby and contribute to the safety of mankind.[1] Apollonius himself, when condemning wizards as pseudo-wise, made the reservation that *mantike,* if true in its predictions, was not a pseudo-science, although he professed ignorance whether it could be called an art or not.[2] He denied that he practiced it, when he was examined by Tigellinus, the favorite of Nero, who was persecuting philosophers on the ground that they were addicted to *mantike.*[3] His accusers before Domitian again adduced his alleged practice of divination as evidence that he was a wizard.[4]

Mantike or the art of divination.

If Apollonius practiced neither wizardry nor *mantike,* the question arises how he was able to foretell the future. In his trial before Domitian he did not attempt to deny that he had predicted the plague at Ephesus, but attributed his "sense of the coming disaster" to his abstemious diet, which kept his senses clear and enabled him to see as in an unclouded mirror "all that is happening or about to occur." [5] For he was credited with knowledge of distant events the moment they occurred as well as with foreknowledge of the future. Thus at Ephesus he was aware of the assassination of Domitian at Rome; and at Tarsus, although he arrived after the incident had occurred, he was able to describe and to find the mad dog by whom a boy had been bitten.[6] Iarchas told Apollonius that health and purity were requisite for

Divining power of Apollonius.

[1] III, 42.
[2] VIII, 7.
[3] IV, 44.
[4] VIII, 7.
[5] VIII, 7.
[6] VIII, 26; VI, 43. The historian, Dio Cassius, a contemporary of Philostratus, also states that Apollonius announced the assassination of Domitian and even named the assassin in Ephesus on the very day that the event occurred at Rome. His account differs too much from that by

divination;[1] and Apollonius in turn, in recounting his life story to the naked sages of Egypt, represented the Pythagorean philosophy as appearing before him and promising, "And when you are pure, I will grant you the faculty of foreknowledge."[2]

Dreams.

Apollonius often was warned by dreams. When he dreamt of fish who were cast gasping upon dry land and who appealed for succour to a dolphin swimming by, he knew that he ought to visit and restore the graves and assist the descendants of the Eretrians whom Darius had taken captive to the Persian kingdom over five centuries before.[3] Another dream he interpreted as a command to visit Crete.[4] In defending his linen apparel before Domitian he declared, "It is a pure substance under which to sleep at night, for to those who live as I do dreams bring the truest of their revelations."[5] He was not the only dreamer of the time, however, and when some of his followers were afraid to accompany him to Rome in Nero's reign, they made warning dreams their excuse for deserting him.[6]

Interpretation of omens.

It has been seen that Apollonius not only had prophetic dreams but was skilful in interpreting them. He was equally adept in explaining the meaning of omens. The dead lion with her eight unborn whelps he took as a sign that Damis and he would remain a year and eight months in that land.[7] When Damis objected that Homer interpreted the sparrow and her eight nestlings whom the snake devoured as nine years' duration of the Trojan war, Apollonius retorted that the birds had been hatched but that the whelps, being yet unborn, could not signify complete years. On another occasion he interpreted the birth of a three-headed child as a sign of the year of the three emperors.[8]

Philostratus to have been copied from it. He concludes it with the positive assertion, "This is really what took place, though there should be ten thousand doubters." (LXVII, 18.)

[1] III, 42.

[2] VI, 11.

[3] I, 23.

[4] IV, 34.

[5] VIII, 7.

[6] IV, 37.

[7] I, 22.

[8] V, 13.

Animals and divination.

Such interpretation of dreams and omens suggests an art or arts of divination rather than foreknowledge by direct divine inspiration. So does the passage in which Apollonius informs Domitian, when accused before him of having divined the future by sacrificing a boy, that human entrails are inferior to those of animals for purposes of divination, since the beasts are less perturbed by knowledge of their approaching death.[1] Apollonius himself would not sacrifice even animal victims, but he enlarged his powers of divination during his sojourn among the Arab tribes by learning to understand the language of animals and to listen to the birds as these predict the future.[2] The Arabs acquire this power by eating, some say the heart, others the liver, of dragons,—a fact which gave the church historian Eusebius an opportunity to charge Apollonius with having broken his taboo of animal flesh.

Divination by fire.

Although he did not sacrifice animals and divine from their entrails, Apollonius appears to have employed practices akin to those of the art of pyromancy when he threw a handful of frankincense into the sacrificial fire with a prayer to the sun, "and watched to see how the smoke of it curled upwards, and how it grew turbid, and in how many points it shot up; and in a manner he caught the meaning of the fire, and observed how it appeared of good omen and pure." [3] Again he visited an Egyptian temple and sacrificed an image of a bull made of frankincense and told the priest that if he really understood the science of divination by fire (*ἐμπύρου σοφίας*), he would see many things revealed in the circle of the rising sun.[4]

Other so-called predictions.

It should be added that only a very ardent admirer of Apollonius or an equally ardent seeker after prophecies would see anything prophetic in some of the apparently chance remarks of the sage which have been perverted into predictions. At Ephesus he did not actually predict the plague, which had already begun to spread judging from the

[1] VIII, 7.
[2] I, 20.
[3] I, 31.
[4] V, 25.

account of Philostratus, but rather warned the heedless population to take measures to prevent its becoming general.[1] When visiting the isthmus of Corinth he began to say that it would be cut through, an idea which had doubtless occurred again and again to many; but then said that it would not be cut through.[2] This sane, if somewhat vacillating, state of mind received confirmation soon afterwards when Nero attempted an Isthmian canal but left it uncompleted. Another similarly ambiguous utterance was elicited from Apollonius by an eclipse of the sun accompanied by thunder: "There shall be some great event and there shall not be."[3] This was believed to receive miraculous fulfillment three days later when a thunderbolt dashed the cup out of which Nero was drinking from his hands but left him unharmed. Once Apollonius saved his life by changing from a ship which sank soon afterwards to another vessel.[4] An instance of more specific prophecy is the case of the consul Aelian, who testified that when he was but a tribune under Vespasian, Apollonius took him aside and told him his name and country and parentage, "and you foretold to me that I should hold this high office which is accounted by the multitude the highest of all."[5] But Aelian may have exaggerated the accuracy of Apollonius's prediction, or the latter may have made a shrewd guess that Aelian was likely to rise to high office.

Apollonius and the demons.

The divining faculty of Apollonius enabled him to detect the presence and influence of demons, phantoms, and goblins, whose ways he understood as well as the language of the birds. At Ephesus he detected the true cause of the plague in a ragged old beggar whom he ordered the people to stone to death.[6] At this command the blinking eyes of the aged mendicant suddenly shot forth malevolent and fiery gleams and revealed his demon character. Afterwards, when the people removed the stones, they found underneath, pounded to a pulp, an enormous hound still vomiting foam

[1] IV, 4.
[2] IV, 24.
[3] IV, 43.
[4] V, 18.
[5] VII, 18.
[6] IV, 10.

as mad dogs do. Later, when accused of magic before Domitian, Apollonius requested that the emperor question him in private about the causes of this pestilence at Ephesus, which he said were too deep to be discussed publicly.[1] And earlier in the reign of Nero, when asked by Tigellinus how he got the better of demons and phantasms, he evaded the question by a saucy retort.[2] On one occasion, however, we are told that he got rid of a ghostly apparition by heaping abuse upon it; [3] and a satyr, who remained invisible but created annoyance by running amuck through the camp, he disposed of by the expedient of filling a trough with wine and letting the spirit get drunk on it. When the wine had all disappeared, Apollonius led his companions to the cave of the nymphs where the satyr was now visible in a drunken sleep.[4] He also reformed the character of a licentious youth by expelling a demon from him,[5] and at Corinth exposed a lamia who, under the disguise of a dainty and wealthy lady, was fattening up a beautiful youth named Menippus with the intention of eventually devouring his blood.[6] On his return by sea from India Apollonius passed a sacred island where lived a sea nymph or female demon who was as destructive to mariners as Scylla or the Sirens were of old.

Not all demons are evil

But the word "demon" is not always employed by Philostratus in the sense of an evil spirit. The annunciation ot the birth of Apollonius was made to his mother by Proteus in the form of an Egyptian demon.[7] Damis looked upon Apollonius himself as a demon and worshiped him as such, when he heard him say that he comprehended not only all human languages but also those things concerning which men maintain silence.[8] In a letter to Euphrates [9] Apollonius affirms that the all-wise Pythagoras should be classed among demons. But when Domitian, on first meeting Apollonius

[1] VIII, 7.
[2] IV, 44.
[3] II, 4.
[4] VI, 27.
[5] IV, 20.
[6] IV, 25.
[7] I, 4.
[8] I, 19.
[9] Epist. 50.

said that he looked like a demon, the sage replied that the emperor was confusing demons and human beings.[1]

Philostratus's faith in demons.

Philostratus adds his own bit of personal testimony to the existence of demons, although it cannot be said to be very convincing. After telling the satyr story he warns his readers not to be incredulous as to the existence of satyrs or to doubt that they make love. For they should not mistrust what is supported by experience and by Philostratus's own word. For he knew in Lemnos a youth of his own age whose mother was said to be visited by a satyr, and such he probably was, since he wore a fawn skin tied around his neck by the two front paws.[2]

The ghost of Achilles.

Apollonius had an interview with the ghost of Achilles which strongly suggests necromancy. He sent his companions on board ship and passed the night alone at the hero's tomb. Nor did he allude to what had happened until questioned by the curious Damis. He then averred that his method of invoking the dead had not been that of Odysseus, but that he had prayed to Achilles much as the Indians do to their heroes. A slight earthquake then occurred and Achilles appeared. At first he was five cubits tall but gradually increased to some twelve cubits in height. At cockcrow he vanished in a flash of summer lightning.[3]

Healing the sick and raising the dead.

Apollonius, as well as the Brahmans, wrought some cures. One was of a boy who had been bitten by a mad dog and consequently "behaved exactly like a dog, for he barked and howled and went on all fours." [4] Apollonius first found and quieted the dog, and then made it lick the wound, a homeopathic treatment which cured the boy. It now only remained to cure the dog, too, and this the philosopher effected by praying to the river which was near by and then making the dog swim across it. "For," concludes Philostratus, "a drink of water will cure a mad dog if he only can be induced to take it." The modern reader will suspect that the dog was not mad to begin with and that Apollonius

[1] VII, 32.
[2] VI, 27.
[3] IV, 11, 15-16.
[4] VI, 43.

cleverly cured the boy's complaint by the same force that had induced it—suggestion. Apollonius once revived a maiden who was being borne to the grave by touching her and saying something to her, but Philostratus honestly admits that he is not sure whether he restored her to life or detected signs of life in the body which had escaped the notice of everyone else.[1]

Other marvels.

When Apollonius was brought before Tigellinus, the scroll on which the charges against him had been written was found to have become quite blank when Tigellinus unrolled it.[2] Upon that occasion and again before Domitian he intimated that his body could not be bound or slain against his will.[3] The former contention he proved to the satisfaction of Damis, who visited him in prison, by suddenly removing his leg from the fetters and then inserting it again.[4] Damis regarded this exhibition as a divine miracle, since Apollonius performed it without magical ceremony or incantations. He is also represented as escaping from his bonds at about midnight when imprisoned later in life in Crete.[5] Philostratus, too, implies that he vanished miraculously from the courtroom of Domitian and that he sometimes passed from one place to another in an incredibly short time, and is somewhat doubtful whether he ever died. But we have seen that even on the testimony of Damis and Philostratus themselves many of the marvels and predictions of Apollonius were not "artless" but involved a knowledge of contemporary natural science and medicine, or of arts of divination, or the employment, in a way not unlike the procedure of magic, of forces and materials outside himself, namely, the occult virtues of things in nature or incantations, rites, and ceremonies.

Golden wrynecks and the *iunx*.

So much for Apollonius and his magic, but the *Life* contains some interesting allusions to the ἴυγξ or wryneck, which throw light upon the use of that bird in Greek magic, but which have seldom been noted and then not correctly

[1] IV, 45.
[2] IV, 44.
[3] VIII, 8.
[4] VII, 38.
[5] VIII, 30.

interpreted.[1] The wryneck was so much employed in Greek magic, as references to it from Pindar to Theocritus show, that the word *iunx* was sometimes used as a synonym or figurative expression for spells or charms in general. Philostratus, too, employs it in this sense, representing the Gymnosophists as accusing the Brahmans of "appealing to the crowd with varied enchantments (or *iunges*)." [2] But in other passages he makes it clear that the wryneck is still employed as a magic bird. Describing the royal palace at Babylon [3] he states that the Magi have hung four golden wrynecks, which they themselves attune and which they call the tongues of the gods, from the ceiling of the judgment hall to remind the king of divine judgment and not to set himself above mankind. Golden wrynecks were also suspended in the Pythian temple at Delphi, and in this connection they are said to possess some of the virtue of the Sirens,[4] or, as Mr. Cook translates it, "to echo the persuasive note of siren voices." These two passages seem to point clearly to the employment of mechanical metal birds which sang and moved as if by magic. The Greek mathematician Hero in his explanation of mechanical devices employed in temples tells how to make a bird turn itself about and whistle by turning a wheel.[5]

Why named *iunx?*

Now this is precisely what the wryneck does in its "wonderful way of writhing its head and neck" and emitting hissing sounds. The bird's "unmistakable note" is "que, que,

[1] The passages are not listed in Liddell and Scott, nor mentioned by Professor Bury in his note on "The ἴυγξ in Greek Magic," *Journal of Hellenic Studies* (1886), pp. 157-60. Hubert's article on "Magia" in Daremberg-Saglio cites only one passage and seems to regard the *iunx* solely as a magic wheel. D'Arcy W. Thompson, *A Glossary of Greek Birds,* Oxford, 1895, also cites but one passage from Philostratus. A. B. Cook, *Zeus,* Cambridge, 1914, I, 253-65, notes both main passages but tries to interpret the *iunges* as solar wheels rather than birds. But the *iunx* is found as a bird on several Greek vases of the latest period; see *British Museum Catalogue of Vases,* vol. IV, figs. 94, 98, 342, 163, 331b; magic wheels are also represented on the vases, but are not described as *iunges* in the catalogue; see vol. IV, figs. 331a, 373, 385, 399, 409, 436, 450, 458, and vol. III, E 774, F 223, F 279.

[2] VI, 10; see also VIII, 7.

[3] I, 25.

[4] VI, 11.

[5] Cited by Cook, *Zeus,* I, 266, who, however, fails to connect it with the *iunx.*

que, repeated many times in succession, at first rapidly, but gradually slowing and in a continually falling key." [1] I would therefore suggest that as the English name for the bird is derived from its writhing its neck, so the Greek name comes from its cry, for "que" and the root ἰυγ, if repeated rapidly many times in succession, sound much alike.[2]

Apollonius in the middle ages.

The name, Apollonius, continued to be associated with magic in the middle ages, when the *Golden Flowers* of Apollonius, a work on the notory art or theurgy,[3] is found in the manuscripts. And we shall find Cecco d'Ascoli [4] in the early fourteenth century citing a "book of magic art" by Apollonius and also a treatise on spirits, *De angelica factione.* In 1412 Amplonius listed in the catalogue of his manuscripts a "book of Apollonius the magician or philosopher which is called Elizinus." [5] Works on the causes and properties of things are also ascribed to Apollonius in medieval manuscripts,[6] and a Balenus or Belenus to whom works on astrological images and seals are ascribed in the manuscripts [7] is perhaps a corruption for Apollonius.[8]

[1] Newton's *Dictionary of Birds;* a reference supplied me by the kindness of my colleague, Professor F. H. Herrick.

[2] Professor Bury's theory that "the bird was called ἴυγξ from its call which sounded like ἰώ ἰώ; and it was used in lunar enchantments because it was supposed to be calling on Io, the moon": and that "ἴυγξ originally meant a moon-song independently of the wryneck," which came to be employed in magic moon-worship on account of its cry, has already been refuted by Professor Thompson, who pointed out that "the bird does not cry ἰώ,, ἰώ, and the suggested derivation of its name and sanctity from such a cry cannot hold."

[3] See Chapter 49 for a fuller account of it.

[4] See Chapter 71.

[5] Math. 54, Liber Appollonii magi vel philosophi qui dicitur Elizinus.

[6] BN 13951, 12th century, Liber Apollonii de principalibus rerum causis. Vienna 3124, 15th century, fols. 57v-58v, "Verba de proprietatibus rerum quomodo virtus unius frangitur per alium. Adamas nec ferro nec igne domatur . . . / . . . cito medetur."

[7] Royal 12-C-XVIII, Baleni de imaginibus; Sloane 3826, fols. 100v-101, Beleemus de imaginibus; Sloane 3848, fols. 52-8, Liber Balamini sapientis de sigillis planetarum, fols. 59-62, liber sapientis Baleym de ymaginibus septem planetarum. But these forms might suggest Balaam. We also hear of Flacius Affricus, a disciple of Belenus.

[8] M. Steinschneider, "Apollonius von Thyana (oder Balinas) bei den Arabern," in *Zeitschrift der Deutschen Morgenländischen Gesellschaft,* XLV (1891), 439-46.

CHAPTER IX

LITERARY AND PHILOSOPHICAL ATTACKS UPON SUPERSTITION: CICERO, FAVORINUS, SEXTUS EMPIRICUS, AND LUCIAN

Authors to be considered—Their standpoint—*De divinatione;* argument of Quintus—Cicero attacks past authority—Divination distinct from natural science—Unreasonable in method—Requires violation of natural law—Cicero and astrology—His crude historical criticism—Favorinus against astrologers—Sextus Empiricus—*Lucius,* or *The Ass:* is it by Lucian?—Career of Lucian—*Alexander the pseudo-prophet*—Magical procedure in medicine satirized—Snake-charming—A Hyperborean magician—Some ghost stories—Pancrates, the magician—Credulity and scepticism—*Menippus,* or *Necromancy*—Astrological interpretation of Greek myth—History and defense of astrology—Lucian not always sceptical—Lucian and medicine—Inevitable intermingling of scepticism and superstition—Lucian on writing history.

Authors to be considered.

HAVING noted the large amount of magic that still existed both in the leading works of natural science of the early Roman empire and in the more general literature of that period, it is only fair that we should note such extremes of scepticism towards the superstitions then current as can be found during the same period. They are, however, few and far between, and we shall have to go back to the close of the Republican period for the best instance in the *De divinatione* of Cicero. As Pliny's *Natural History* was mainly a compilation of earlier Greek science, so Cicero's arguments against divination were not entirely original with him. As his other philosophical writings are largely indebted to the Greeks, so his attack upon divination is supposed to be under considerable obligations to Clitomachus and Panaetius,[1] philosophers of the New Academy and the

[1] T. Schiche, *De fontibus librorum Ciceronis qui sunt de divinatione,* Jena, 1875; K. Hartfelder, *Die Quellen von Ciceros zwei Büchern de Divinatione,* Freiburg, 1878.

Stoic school who flourished respectively at Carthage and Athens and at Rhodes and Rome in the second century before our era. We shall next briefly note the criticisms of astrologers and astrology made by Favorinus, a rhetorician from Gaul who resided at Rome under Hadrian and was a friend of Plutarch but whose argument against the astrologers has been preserved only in the *Attic Nights* of Aulus Gellius,[1] and by Sextus Empiricus,[2] a sceptical philosopher who wrote about 200. Finally we shall consider Lucian's satirical depiction of various superstitions of his time.

Their standpoint.

It will be noticed that no one of these critics of magic, if we may so designate them, is primarily a natural scientist. Cicero and Lucian and Favorinus are primarily men of letters and rhetoricians. And all four of our critics write to a greater or less extent from the professed standpoint of a general sceptical attitude in all matters of philosophy and not merely in the matter of superstition. Thus the attack of Sextus Empiricus upon astrology occurs in a work which is directed against learning in general, and in which he assails grammarians, rhetoricians, geometricians, arithmeticians, students of music, logicians, physicists, and students of ethics, as well as the casters of horoscopes. Aulus Gellius did not know whether to take the arguments of Favorinus against the astrologers seriously or not. He says that he heard Favorinus make the speech the substance of which he repeats, but that he is unable to state whether the philosopher really meant what he said or argued merely in order to exercise and to display his genius. There was reason for this perplexity of Aulus Gellius, since Favorinus was inclined to such *tours de force* as eulogies of Thersites or of Quartan Fever.

De divinatione: argument of Quintus.

De divinatione takes the form of a supposititious conversation, or better, informal debate, between the author and his brother Quintus. In the first book Quintus, in a rather rambling and leisurely fashion and with occasional repetition

[1] Aulus Gellius, *Noctes Atticae*, XIV, 1.

[2] *Adv. astrol.*, in *Opera*, ed. Johannes Albertus Fabricius, Leipzig, 1718.

of ideas, upholds divination to the best of his ability, citing many reported instances of successful recourse to it in antiquity. In the second book Tully proceeds with a somewhat patronizing air to pull entirely to pieces the arguments of his brother who assents with cheerful readiness to their demolition. On the whole the appeal to the past is the main point in the argument of Quintus. What race or state, he asks, has not believed in some form of divination? "For before the revelation of philosophy, which was discovered but recently, public opinion had no doubt of the truth of this art; and after philosophy emerged no philosopher of authority thought otherwise. I have mentioned Pythagoras, Democritus, Socrates. I have left out no one of the ancients save Xenophanes. I have added the Old Academy, the Peripatetics, the Stoics. Epicurus alone dissented." [1] Quintus closes his long argument in favor of the truth of divination by solemnly asserting that he does not approve of sorcerers, nor of those who prophesy for the sake of gain, nor of the practice of questioning the spirits of the dead—which nevertheless, he says, was a custom of his brother's friend Appius.[2]

Cicero attacks past authority.

When Tully's turn to speak comes, he rudely disturbs his brother's reliance upon tradition. "I think it not the part of a philosopher to employ witnesses, who are only haply true and often purposely false and deceiving. He ought to show why a thing is so by arguments and reasons, not by events, especially those I cannot credit." [3] "Antiquity," Cicero declares later, "has erred in many respects." [4] The existence of the art of divination in every age and nation has little effect upon him. There is nothing, he asserts, so widespread as ignorance.[5]

Divination distinct from natural science.

Both brothers distinguish divination as a separate subject from the natural or even the applied sciences. Quintus says that medical men, pilots, and farmers foresee many things, yet their arts are not divination. "Not even Phere-

[1] *De divinatione,* I, 39.
[2] *Ibid.,* I, 58.
[3] *Ibid.,* II, 11.
[4] *Ibid.,* II, 33.
[5] *Ibid.,* II, 36.

cydes, that famous Pythagorean master, who predicted an earthquake when he saw that the water had disappeared from a well which usually was well filled, should be regarded as a diviner rather than a physicist."[1] Tully carries the distinction a step further and asserts that the sick seek a doctor, not a soothsayer; that diviners cannot instruct us in astronomy; that no one consults them concerning philosophic problems or ethical questions; that they can give us no light on the problems of the natural universe; and that they are of no service in logic, dialectic, or political science.[2] An admirable declaration of independence of natural science and medicine and other arts and constructive forms of thought from the methods of divination! But also one more easy to state in general terms of theory than to enforce in details of practice, as Pliny, Galen, and Ptolemy have already shown us. None the less it is indeed a noteworthy restriction of the field of divination when Cicero remarks to his brother, "For those things which can be perceived beforehand either by art or reason or experience or conjecture you regard as not the affair of diviners but of scientists."[3] But the question remains whether too large powers of prediction may not be claimed by "science."

Unreasonable in method.

Cicero proceeds to attack the methods and assumptions of divination as neither reasonable nor scientific. Why, he asks, did Calchas deduce from the devoured sparrows that the Trojan war would last ten years rather than ten weeks or ten months?[4] He points out that the art is conducted in different places according to quite different rules of procedure, even to the extent that a favorable omen in one locality is a sinister warning elsewhere.[5] He refuses to believe in any extraordinary bonds of sympathy between things which, in so far as our daily experience and our

[1] I, 50.

[2] II, 3-4.

[3] II, 5. "Quae enim praesentiri aut arte aut ratione aut usu aut coniectura possunt, ea non divinis tribuenda putas sed peritis."

[4] II, 30.

[5] II, 12. An astrologer, however, would probably say that seeming contradiction could be accounted for by the varying influence of the constellations upon different regions.

knowledge of the workings of nature can inform us, have no causal connection. What intimate connection, he asks, what bond of natural causality can there be between the liver or heart or lung of a fat bull and the divine eternal cause of all which rules the universe?[1] "That anything certain is signified by uncertain things, is not this the last thing a scientist should admit?"[2] He refuses to accept dreams as fit channels either of natural divination or divine revelation.[3] The Sibylline Books, like most oracles, are vague and the evident product of labored ingenuity.[4]

Requires violation of natural law.

Moreover, divination asserts the existence of phenomena which science denies. Such a figment, Cicero scornfully affirms, as that the heart will vanish from the carcass of a victim is not believed even by old-wives now-a-days. How can the heart vanish from the body? Surely it must be there as long as life lasts, and how can it disappear in an instant? "Believe me, you are abandoning the citadel of philosophy while you defend its outposts. For in your effort to prove soothsaying true you utterly pervert physiology. . . . For there will be something which either springs from nothing or suddenly vanishes into nothingness. What scientist ever said that? The soothsayers say so? Are they then, do you think, to be trusted rather than scientists?"[5] Cicero makes other arguments against divination such as the stock contentions that it is useless to know predetermined events beforehand since they cannot be avoided, and that even if we can learn the future, we shall be happier not to do it, but his outstanding argument is that it is unscientific.

Cicero and astrology.

Cicero's attack upon divination is mainly directed against liver divination and analogous methods of predicting the future, but he devotes a few chapters[6] to the doctrines of the Chaldeans. They postulate a certain force in the constellations called the zodiac and hold that between

[1] II, 12.

[2] II, 19. "Quid igitur minus a physicis dici debet quam quidquam certi significari rebus incertis?"

[3] II, 60-71.

[4] II, 54.

[5] II, 16.

[6] II, 42-47.

man and the position of the stars and planets at the moment of his birth there exists a relation of sympathy so that his personality and all the events of his life are thereby determined. Diogenes the Stoic limited this influence to the determination of one's aptitude and vocation, but Cicero regards even this much as going too far. The immense spaces intervening between the different planets seem to him a reason for rejecting the contentions of the Chaldeans. His further criticism that they insist that all men born at the same moment are alike in character regardless of horizons and different aspects of the sky in different places is one that at least did not hold good permanently against astrology and is not true of Ptolemy. He asks if all the men who perished at Cannae were born beneath the same star and how it came about that there was only one Homer if several men are born every instant. He also adduces the stock argument from twins. He attacks the practice, which we shall find continued in the middle ages, of astrological prediction of the fate of cities. He says that if all animals are to be subjected to the stars, then inanimate things must be, too, than which nothing can be more absurd. This suggests that he hardly conceives of the fundamental hypothesis of medieval science that all inferior nature is under the influence of the celestial bodies and their motion and light. At any rate his arguments are directed against the casting of horoscopes or genethlialogy. And in the matter of the influence of the planets upon man he was not entirely antagonistic, at least in other writings than the *De divinatione,* for in the *Dream of Scipio* he speaks of Jupiter as a star wholesome and favorable to the human race, of Mars as most unfavorable. He further calls seven and eight perfect numbers and speaks of their product, fifty-six, as signifying the fatal year in Scipio's life. Incidentally, as another instance that Cicero was not always sceptical, it may be recalled that it was in Cicero that Pliny read of a man who could see one hundred and thirty-five miles.[1]

[1] NH, VII, 21.

His crude historical criticism.

Such apparent inconsistency is perhaps a sign of somewhat indiscriminating eclecticism on Cicero's part. We experience something of a shock, although perhaps we should not be surprised, to find him in his *Republic* [1] arguing as seriously in favor of the ascension or apotheosis of Romulus as a historic fact as a professor of natural science in a denominational college might argue in favor of the historicity of the resurrection of Christ. Although in the *De divinatione* he impatiently brushed aside the testimony of so great a cloud of witnesses and of most philosophers in favor of divination, he now argues that the opinion that Romulus had become a god "could not have prevailed so universally unless there had been some extraordinary manifestation of power," and that "this is the more remarkable because other men, said to have become gods, lived in less learned times when the mind was prone to invent and the inexperienced were easily led to believe," whereas Romulus lived only six centuries ago when literature and learning had already made great progress in removing error, when "Greece was already full of poets and musicians, and little faith was placed in legends unless they concerned remote antiquity." Yet a few chapters later Cicero notes that Numa could not have been a pupil of Pythagoras, since the latter did not come to Italy until 140 years after his death; [2] and in a third chapter [3] when Laelius remarks, "That king is indeed praised but Roman History is obscure, for although we know the mother of this king, we are ignorant of his father," Scipio replies, "That is so; but in those times it was almost enough if only the names of the kings were recorded." We can only add, "Consistency, thou art a jewel!"

Favorinus against astrologers.

Favorinus denied that the doctrine of nativities was the work of the Chaldeans and regarded it as the more recent invention of marvel-mongers, tricksters, and mountebanks. He regards the inference from the effect of the moon on tides to that of the stars on every incident of our daily life

[1] *Republic,* II, 10. [2] *Ibid.,* II, 15. [3] *Ibid.,* II, 18.

as unwarranted. He further objects that if the Chaldeans did record astronomical observations these would apply only to their own region and that observations extended over a vast lapse of time would be necessary to establish any system of astrology, since it requires ages before the stars return to their previous positions. Like Cicero, Favorinus probably manifests his ignorance of the technique of astrology in complaining that astrologers do not allow for the different influence of different constellations in different parts of the earth. More cogent is his suggestion that there may be other stars equal in power to the planets which men cannot see either for their excess of splendor or because of their position. He also objects that the position of the stars is not the same at the time of conception and the time of birth, and that, if the different fate of twins may be explained by the fact that after all they are not born at precisely the same moment, the time of birth and the position of the stars must be measured with an exactness practically impossible. He also contends that it is not for human beings to predict the future and that the subjection of man not merely in matters of external fortune but in his own acts of will to the stars is not to be borne. These two arguments of the divine prerogative and of human free will became Christian favorites. He complains that the astrologers predict great events like battles but cannot predict small ones, and declares that they may congratulate themselves that he does not propose such a question to them as that of astral influence on minute animals. This and his further question why, out of all the grand works of nature, the astrologers limit their attention to petty human fortune, suggest that like Cicero he did not realize that astrology was or would become a theory of all nature and not mere genethlialogy.

Sextus Empiricus.

To the arguments against nativities that men die the same death who were not born at the same time and that men who are born at the same time are not identical in character or fortune Sextus Empiricus adds the derisive question whether a man and an ass born in the same instant would suffer exactly the same destiny. Ptolemy would of course reply that while the influence of the stars is constant in both cases it is variably received by men and donkeys; and Sextus's query does not show him very well versed in astrology. He mentions the obstacle of free will to astrological theory but does not make very much of it. The chief point which he makes is that even if the stars do rule human destiny, their effect cannot be accurately measured. He lays stress on the difficulty of exactly determining the date of birth or of conception, or the precise moment when a star passes into a new sign of the zodiac. He notes the variability and unreliability of water-clocks. He calls attention to the fact that observers at varying altitudes as well as in different localities would arrive at different conclusions. Differences in eyesight would also affect results, and it is difficult to tell just when the sun sets or any sign of the zodiac drops below the horizon owing to reflection and refraction of rays. Sextus thus leaves us somewhat in doubt whether his objections are to be taken as indicative of a spirit of captious criticism towards an art, the fundamental principles of which he tacitly recognizes as well-nigh incontestible, or whether he is simply trying to make his case doubly sure by showing astrology to be impracticable as well as unreasonable. In any case we shall find his argument that the influence of the stars cannot be measured accurately repeated by Christian writers.

Lucius or *The Ass:* is it by Lucian?

The main plot of the *Metamorphoses* of Apuleius appears, shorn of the many additional stories, the religious mysticism, and the autobiographical element which characterize his narrative, in a brief and perhaps epitomized Greek version, entitled *Lucius* or *The Ass,* among the works of Lucian of Samosata, the contemporary of Apuleius and noted satirist. The work is now commonly regarded as spurious, since the style seems different from that of Lucian and the Attic Greek less pure. The narrative, too, is bare, at least compared with the exuberant fancy of Apuleius, and seems to avoid the marvelous and romantic details in which he abounds. Photius, patriarch of Constantinople in the ninth century, who regarded the work as Lucian's, said that he wrote in it as one deriding the extravagance of superstition. Whether this be true of *The Ass* or not, it is true of other satires by Lucian of undisputed genuineness, in which he ridicules the impostures of the magic and pseudoscience of his day. In place of the genial humor and fantastic imagination with which his African contemporary credulously welcomed the magic and occult science of his time, the Syrian satirist probes the same with the cool mockery of his keen and sceptical wit.

Career of Lucian.

Lucian was born at Samosata near Antioch about 120 or 125 A. D. and after an unsuccessful beginning as a sculptor's apprentice turned to literature and philosophy. He practiced in the law courts at Antioch for some time and also wrote speeches for others. For a considerable period of his life he roamed about the Mediterranean world from Paphla gonia to Gaul as a rhetorician, and like Apuleius resided both at Athens and Rome. After forty he ceased teaching rhetoric and devoted himself to literary production, living at Athens. Towards the close of his life, "when he already had one foot in Charon's boat," [1] he was holding a well paid and important legal position in Egypt. His death occurred perhaps about 200 A. D. Some ascribe it to gout, probably because he wrote two satires on that disease. Suidas states that Lucian was torn to pieces by dogs as a punishment for his attacks upon Christianity, which again is probably a perversion of Lucian's own statement in *Peregrinus* that he narrowly escaped being torn to pieces by the Cynics.

Alexander the pseudo-prophet.

It was at the request of that same adversary of Christianity against whom Origen composed the *Reply to Celsus* that Lucian wrote his account of the impostor, Alexander of Abonutichus, a pseudo-prophet of Paphlagonia. This Alexander pretended to discover the god Asclepius in the form of a small viper which he had sealed up in a goose egg.

[1] *Apologia pro mercede conductis.* Most of Lucian's Essays have been translated into English by H. W. Fowler and F. G. Fowler, 1905, 4 vols.

He then replaced the tiny viper by a huge tame serpent which he had purchased at Pella in Macedon and which was trained to hide its head in Alexander's armpit, while to the crowd, who were also permitted to touch the tail and body of the real snake, was shown a false serpent's head made of linen with human features and a mouth that opened and shut and a tongue that could be made to dart in and out. Having thus convinced the people that the viper had really been a god and had miraculously increased in size, Alexander proceeded to sell oracular responses as from the god. Inquirers submitted their questions in sealed packages which were later returned to them with appropriate answers and with the seals unbroken and apparently untouched. Similarly Plutarch tells of a sceptical opponent of oracles who became converted into their ardent supporter by receiving such an answer to a sealed letter.[1] Lucian, however, explains that Alexander sometimes used a hot needle to melt the seal and then restore it to practically its original shape, or employed other methods by which he took exact impressions of the seal, then boldly broke it, read the question, and afterwards replaced the seal by an exact replica of the original made in the mould. Lucian adds that there are plenty of other devices of this sort which he does not need to repeat to Celsus who has already made a sufficient collection of them in his "excellent treatises against the magicians." Lucian tells later, however, how Alexander made his god seem to speak by attaching a tube made of the windpipes of cranes to the artificial head and having an assistant outside speak through this concealed tube. In our later discussion of the church father Hippolytus we shall find that he apparently made use of this exposé of magic by Lucian as well as of the arguments of Sextus Empiricus against astrology. Lucian's personal experiences with this Alexander were quite interesting but are less germane to our investigation.

[1] *De defectu oraculorum*, 45.

We must not fail, however, to note another essay, *Philopseudes* or *Apiston,* in which the superstition and pseudo-science of antiquity are sharply satirized in what purports to be a conversation of several philosophers, including a Stoic, a Peripatetic, and a Platonist, and a representative of ancient medicine in the person of Antigonus, a doctor. Some of the magical procedure then employed in curing diseases is first satirized. Cleodemus the Peripatetic advises as a remedy for gout to take in the left hand the tooth of a field mouse which has been killed in a prescribed manner, to wrap it in the skin of a lion freshly-flayed, and thus to bind it about the ailing foot. He affirms that it will give instant relief. Dinomachus the Stoic admits that the occult virtue of the lion is very great and that its fat or right fore-paw or the bristles of its beard, if combined with the proper incantations, have wonderful efficacy. But he holds that for the cure of gout the skin of a virgin hind would be superior on the ground that the hind is speedier than the lion and so more beneficial to the feet. Cleodemus retorts that he used to think the same, but that a Libyan has convinced him that the lion can run faster than the hind or it would never catch one. The sceptical reporter of this conversation states that he vainly attempted to convince them that an internal disease could not be cured by external attachments or by incantations, methods which he regards as the veriest sorcery (*goetia*).

Magical procedure in medicine satirized.

His protests, however, merely lead Ion the Platonist to recount how a Magus, a Chaldean of Babylonia, cured his father's gardener who had been stung by an adder on the great toe and was already all swollen up and nearly dead. The magician's method was to apply a splinter of stone from the statue of a virgin to the toe, uttering at the same time an incantation. He then led the way to the field where the gardener had been stung; pronounced seven sacred names from an ancient volume, and fumigated the place thrice with torches and sulphur. All the snakes in the field then came forth from their holes with the exception of one very aged

Snake-charming.

and decrepit serpent, whom the magician sent a young snake back to fetch. Having thus assembled every last serpent, he blew upon them, and they all vanished into thin air.

A Hyperborean magician.

This tale reminds the Stoic of another magician, a barbarian and Hyperborean, who could walk through fire or upon water and even fly through the air. He could also "make people fall in love, call up spirits, resuscitate corpses, bring down the moon, and show you Hecate herself as large as life."[1] More specific illustration of the exercise of these powers is given in an account of a love spell which he performed for a young man for a big fee. Digging a trench, he raised the ghost of the youth's father and also summoned Hecate, Cerberus, and the Moon. The last named appeared in three successive forms of a woman, an ox, and a puppy. The sorcerer then constructed a clay image of the god of love and sent it to fetch the girl, who came and stayed until cock-crow, when all the apparitions vanished with her. In vain the sceptic argues that the girl in question would have come willingly enough without any magic. The Platonist matches the previous story with one of a Syrian from Palestine who cast out demons.

Some ghost stories.

The discussion then further degenerates into ghost stories and tales of statuettes that leave their pedestals after the household has retired for the night. One speaker says that he no longer has any fear of ghosts since an Arab gave him a magic ring made of nails from crosses and taught him an incantation to use against spooks. At this juncture a Pythagorean philosopher of great repute enters and adds his testimony in the form of an account of how he laid a ghost at Corinth by employing an Egyptian incantation.

Pancrates, the magician.

Eucrates, the host, then tells of Pancrates, whom he had met in Egypt and who "had spent twenty-three years underground learning magic from Isis," and whom crocodiles would allow to ride on their backs. They traveled a time together without a servant, since Pancrates was able to dress up the door-bar or a broom or pestle, turn it into human

[1] Fowler's translation.

form, and make it wait upon them. There follows the familiar story of Eucrates' overhearing the incantation of three syllables which Pancrates employed and of trying it out himself when the magician was absent. The pestle turned into human form all right enough and obeyed his order to bring in water, but then he discovered that he could not make it stop, and when he seized an axe and chopped it in two, the only effect was to produce two water-carriers in place of one.

Credulity and scepticism.

The conversation is turning to the subject of oracles when the sceptic can stand it no longer and retires in disgust. As he tells what he has heard to a friend, he remarks upon the childish credulity of "these admired teachers from whom our youth are to learn wisdom." At the same time, the stories seem to have made a considerable impression even upon him, and he wishes that he had some lethal drug to make him forget all these monsters, demons, and Hecates that he seems still to see before him. His friend, too, declares that he has filled him with demons. Their dialogue then concludes with the consoling reflection that truth and sound reason are the best drugs for the cure of such empty lies.

Menippus, or *Necromancy.*

The *Menippus* or *Necromancy,* while an obvious imitation and parody of Odysseus' mode of descent to the underworld to consult Teiresias, also throws some light on the magic of Lucian's time. In order to reach the other world Menippus went to Babylon and consulted Mithrobarzanes, one of the Magi and followers of Zoroaster. He is also called one of the Chaldeans. Besides a final sacrifice similar to that of Odysseus, the procedure by which the magician procured their passage to the other world included on his part muttered incantations and invocations, for the most part unintelligible to Menippus, spitting thrice in the latter's face, waving torches about, drawing a magic circle, and wearing a magic robe. As for Menippus, he had to bathe in the Euphrates at sunrise every morning for the full twenty-nine days of a moon, after which he was purified at midnight in the Tigris and by fumigation. He had to sleep out-of-doors and observe a special diet, not look anyone in the eye on his way home, walk backwards, and so on. The ultimate result of all these preparations was that the earth was burst asunder by the final incantation and the way to the underworld laid open. When it came time to return Menippus crawled up with difficulty, like Dante going from the Inferno to Purgatory, through a narrow tunnel which opened on the shrine of Trophonius.

Astrological interpretation of Greek myth.

An essay on astrology ascribed to Lucian is usually regarded as spurious.[1] Denial of its authenticity, however, should rest on such grounds as its literary style and the manuscript history of the work rather than upon its—to modern eyes—superstitious character. In antiquity a man might be sceptical about most superstitions and yet believe in astrology as a science. Lucian's sceptical friend Celsus, for example, as we shall see in our chapter on Origen's *Reply to Celsus*, believed that the future could be foretold from the stars. And whether the present essay is genuine or spurious, it is certainly noteworthy that for all his mockery of other superstition Lucian does not attack astrology in any of his essays. Moreover, this essay on astrology is very sceptical in one way, since it denies the literal truth of various Greek myths and gives an astrological interpretation of them, as in the case of Zeus and Kronos and the so-called adultery of Mars. This is not inconsistent with Lucian's ridicule elsewhere of the anthropomorphic Olympian divinities. What Orpheus taught the Greeks was astrology, and the planets were signified by the seven strings of his lyre. Teiresias taught them further to distinguish which stars were masculine and which feminine in character and influence. A proper interpretation of the myth of Atreus and Thyestes also shows the Greeks at an early date acquainted with astrological doctrine. Bellerophon soared to the sky, not on a

[1] Fowler omits it. It appears in the Teubner edition, *Luciani Samosatensis opera*, ed. C. Jacobitz, II (1887), 187-95, but both Jacobitz and Dindorf mark it as spurious. Croiset, *Essai sur la vie et les œuvres de Lucien*, Paris, 1882, p. 43, also rejects it.

horse but by the scientific power of his mind. Daedalus taught Icarus astrology and the fable of Phaëthon is to be similarly interpreted. Aeneas was not really the son of the goddess Venus, nor Minos of Jupiter, nor Aesculapius of Mars, nor Autolycus of Mercury. These are to be taken simply as the planets under whose rule they were born. The author also connects Egyptian animal worship with the signs of the zodiac.

History and defense of astrology.

The author of the essay also delves into the history of astrology, to which he assigns a high antiquity. The Ethiopians were the first to cultivate it and handed it on in a still imperfect stage to the Egyptians who developed it. The Babylonians claim to have studied it before other peoples, but our author thinks that they did so long after the Ethiopians and Egyptians. The Greeks were instructed in the art neither by the Ethiopians nor the Egyptians, but, as we have seen, by Orpheus. Our author not only states that the ancient Greeks never built towns or walls or got married without first resorting to divination, but even asserts that astrology was their sole method of divination, that the Pythia at Delphi was the type of celestial purity and that the snake under the tripod represented the dragon among the constellations. Lycurgus taught his Lacedaemonians to observe the moon, and only the uncultured Arcadians held themselves aloof from astrology. Yet at the present day some oppose the art, declaring either that the stars have naught to do with human affairs or that astrology is useless since what is fated cannot be avoided. To the latter objection our author makes the usual retort that forewarned is forearmed; as for the former denial, if a horse stirs the stones in the road as it runs, if a passing breath of wind moves straws to and fro, if a tiny flame burns the finger, will not the courses and deflexions of the brilliant celestial bodies have their influence upon earth and mankind?

Lucian not always sceptical.

The manner of the essay does not seem like Lucian's usual style, and the astrological interpretation of religious myth was characteristic of the Stoic philosophy, whereas

Lucian's philosophical affinities, if he can be said to have any, are perhaps rather with the Epicureans. But Celsus was an Epicurean and yet believed in astrology. It must not be thought, however, that Lucian in his other essays is always sceptical in regard to what we should classify as superstition. He tells us how his career was determined by a dream in the autobiographical essay of that title. In the *Dialogues of the Gods* magic is mentioned as a matter-of-course, Zeus complaining that he has to resort to magic in order to win women and Athene warning Paris to have Aphrodite remove her girdle, since it is drugged or enchanted and may bewitch him.

Lucian and medicine.

The writings of Lucian contain many allusions to the doctors, diseases, and medicines of his time.[1] On the whole he confirms Galen's picture. Numerous passages show that the medical profession was held in high esteem, and Lucian himself first went to Rome in order to consult an oculist. At the same time Lucian satirizes the quacks and medical superstition of the time, as we have already seen, and describes several statues which were believed to possess healing powers. In the burlesque tragedy on gout, *Tragodopodagra*, whose authenticity, however, is questioned, the disease personified is triumphant, and the moral seems to be that all the remedies which men have tried are of no avail. On the other hand, Lucian wrote seriously of the African snake whose bite causes one to die of thirst (*De dipsadibus*). He admits that he has never seen anyone in this condition and has not even been in Libya where these snakes are found, but a friend has assured him that he has seen the tombstone epitaph of a man who had died thus, a rather indirect mode of proof which we are surprised should satisfy the author of *How to Write History*. Lucian also repeats the common notion that persons bitten by a mad dog can be cured only by a hair or other portion of the same animal.[2]

[1] See the interesting paper of J. D. Rolleston, "Lucian and Medicine," 1915, 23 pp., reprinted from *Proceedings of the Royal Society of Medicine*, VIII, 49-58, 72-84.

[2] See the close of *Nigrinus*.

Inevitable intermingling of scepticism and superstition.

Our chapter which set out to note cases of scepticism in regard to superstition has ended by including a great deal of such superstition. The sceptics themselves seem credulous on some points, and Lucian's satire perhaps more reveals than refutes the prevalence of superstition among even the highly educated. The same is true of other literary satirists of the Roman Empire whose jibes against the astrologers and their devotees only attest the popularity of the art and who themselves very probably meant only to ridicule its more extreme pretensions and were perhaps at bottom themselves believers in the fundamentals of the art. Our authors to some extent, as we have pointed out, provided an arsenal of arguments from which later Christian writers took weapons for their assaults upon pagan magic and astrology. But sometimes subsequent writers confused scepticism with credulity, and the influence of our authors upon them became just the opposite of what they intended. Thus Ammianus Marcellinus, the soldier-historian of the falling Roman Empire upon whom Gibbon placed so much reliance, was so attached to divination that he even quoted its arch-opponent, Cicero, in support of it. For he actually concludes his discussion of the subject in these words: "Wherefore in this as in other matters Tully says most admirably, 'Signs of future events are shown by the gods.' "[1]

Lucian on writing history.

But in order to conclude our chapter on scepticism with a less obscurantist passage, let us return to Lucian. His essay, *How to Write History,* gives serious expression to those ideals of truth and impartiality which also lie behind his mockery of impostors and the over-credulous. "The historian's one task," in his estimation, "is to tell the thing as it happened." He should be "fearless, incorruptible, independent, a believer in frankness, . . . an impartial judge, kind to all but too kind to none." "He has to make of his brain a mirror, unclouded, bright, and true of surface." "Facts are not to be collected at haphazard but with careful,

[1] *Rerum gestarum libri qui supersunt,* XXI, i, 14.

laborious, repeated investigation." "Prefer the disinterested account." [1] Such sentences and phrases as these reveal a scientific and critical spirit of high order and seem a vast improvement upon the frailty of Cicero's historical criticism. But how far Lucian would have been able to follow his own advice is perhaps another matter.

[1] The wording of these excerpts is that of Fowler's translation.

CHAPTER X

THE SPURIOUS MYSTIC WRITINGS OF HERMES, ORPHEUS, AND ZOROASTER

Mystic works of revelation—The Hermetic books—*Poimandres* and the Hermetic *Corpus*—Astrological treatises ascribed to Hermes—Hermetic works of alchemy—Nechepso and Petosiris—Manetho—The *Lithica* of Orpheus—Argument of the poem—Magic powers of stones—Magic rites to gain powers of divination—Power of gems compared with herbs—Magic herbs and demons in Orphic rites—Books ascribed to Zoroaster—*The Chaldean Oracles.*

Mystic works of revelation.

THERE were in circulation in the Roman Empire many writings which purported to be of divine origin and authorship, or at least the work of ancient culture-heroes and founders of religions who were of divine descent and divinely inspired. These oracular and mystic compositions usually pretend to great antiquity and often claim as their home such hoary lands as Egypt and Chaldea, although in the Hellenic past Apollo and in the Roman past the Sibylline books [1] also afford convenient centers about which forgeries cluster. Assuming as these writings do to disclose the secrets of ancient priesthoods and to publish what should not be revealed to the vulgar crowd, they may be confidently expected to embody a great deal of superstition and magic along with their expositions of mystic theologies. Also the authors, editors, or publishers of astrological, alchemistic, and other pseudo-scientific treatises could not be expected to resist the temptation of claiming a venerable and cryptic origin for some of their books. Moreover, such pseudo-literature was not entirely unjustified in its affirmation of high antiquity. Few things in intellectual history antedate magic, and these spurious compositions are not especially

[1] See Sackur, *Sibyllinische Texte und Forschungen,* Halle, 1898; Alexandre, *Oracula Sibyllina,* 2nd ed., Paris, 1869; Charles (1913) II, 368 ff.

distinguished by new ideas, although they to some extent reflect the progress made in learning, occult as well as scientific, in the Hellenistic age. It must be added that much of their contents depends for its effect entirely upon its claim to eminent authorship and great antiquity and upon the impressionability of its public. To-day most of it seems trivial commonplace or marked by the empty vagueness characteristic of oracular utterances. I shall attempt no complete exposition or exhaustive treatment of such writings [1] but touch upon a few examples which bear upon the relations of science and magic.

The Hermetic books.

Chief among these are the Hermetic books or writings attributed to Hermes the Egyptian or Trismegistus. "Under this name," wrote Steinschneider in 1906, "there exists in many languages a literature, for the most part superstitious, which seems to have not yet been treated in its totality." [2] The Egyptian god Thoth or Tehuti, known in Greek as Θωύθ, Θώθ, and Τάτ, was identified with Hermes, and the epithet "thrice-great" is also derived from the Egyptian *aā aā*, "the great Great." Citations of works ascribed to this Hermes Trismegistus can be traced back as early as the first century of our era.[3] He is also mentioned or quoted by various church fathers from Athenagoras to Augustine and often figures in the magical papyri. The historian Ammianus Marcellinus [4] in the fourth century ranks him with the great sages of the past such as Pythagoras, Socrates, and Apollonius of Tyana. Our two chief descriptions of the Hermetic books from the period of the Roman Empire are found in the *Stromata* [5] of the Christian Clement of Alexandria (c.150-c.220 A.D.) and in the *De mysteriis* [6] ascribed to the Neo-Platonist Iamblichus (died about 330

[1] Besides the works to be cited later in this chapter, the reader may consult: A. Dieterich, *Abraxas* (*Studien z. relig. gesch. d. spät. alt.*), Leipzig, 1891, especially chapter II (pp. 136ff.), "Jüdisch-orphisch-gnostiche Kulte und die Zauberbücher"; and G. A. Lobeck, *Aglaophamus,* 1829, 2 vols.

[2] Steinschneider (1906), 24. He mentions the dissertation of R. Pietschmann, *Hermes Trismegistus,* Leipzig, 1875.

[3] See Galen, citing Pamphilus, Kühn, XI, 798.

[4] XXI, 14, 15.

[5] VI, 4.

[6] I, 1; VIII, 1-4.

A. D.). Clement speaks of forty-two books by Hermes which are regarded as "indispensable." Of these ten are called "Hieratic" and deal with the laws, the gods, and the training of the priests. Ten others detail the sacrifices, prayers, processions, festivals, and other rites of Egyptian worship. Two contain hymns to the gods and rules for the king. Six are medical, "treating of the structure of the body and of diseases and instruments and medicines and about the eyes and the last about women." Four are astronomical or astrological, and the remaining ten deal with cosmography and geography or with the equipment of the priests and the paraphernalia of the sacred rites. Clement does not say so, but from his brief summary one can imagine how full these volumes probably were of occult virtues of natural substances, of magical procedure, and of intimate relations and interactions between nature, stars, and spirits. Iamblichus repeats the statement of Seleucus that Hermes wrote twenty thousand volumes and the assertion of Manetho that there were 36,525 books, a number doubtless connected with the supposed length of the year, three hundred and sixty-five and one-quarter days.[1] Iamblichus adds that Hermes wrote one hundred treatises on the ethereal gods and one thousand concerning the celestial gods.[2] He is aware, however, that most books attributed to Hermes were not really composed by him, since in other passages he speaks of "the books which are circulated under the name of Hermes," [3] and explains that "our ancestors . . . inscribed all their own writings with the name of Hermes," [4] thus dedicating them to him as the patron deity of language and theology. By the time of Iamblichus these books had been translated from the Egyptian tongue into Greek.

Poimandres and the Hermetic *Corpus*.

There has come down to us under the name of Hermes a collection of seventeen or eighteen fragments which is generally known as the Hermetic *Corpus*. Of the frag-

[1] VIII, 1.
[2] VIII, 2.
[3] VIII, 4.
[4] I, 1.

ments the first and chief is entitled *Poimandres* (Ποιμάνδρης), a name which is sometimes applied to the entire *Corpus*. Another fragment entitled *Asclepius,* since it is in the form of a dialogue between him and "Mercurius Trismegistus," exists in a Latin form which has been ascribed probably incorrectly to Apuleius of Madaura as translator (*Asclepius . . . Mercurii trismegisti dialogus Lucio Apuleio Madaurensi philosopho Platonico interprete*). None of the Greek manuscripts of the *Corpus* seems older than the fourteenth century, although Reitzenstein thinks that they may all be derived from the version which Michael Psellus had before him in the eleventh century.[1] But the concluding prayer of the *Poimandres* exists in a third century papyrus, and the alchemist Zosimus in the fourth century seems acquainted with the entire collection. The treatises in this *Corpus* are concerned primarily with religious philosophy or theosophy, with doctrines similar to those of Plato concerning the soul and to the teachings of the Gnostics. The moral and religious instruction is associated, however, with a physics and cosmology very favorable to astrology and magic. Of magic in the narrow sense there is little in the *Corpus,* but a Hermetic fragment preserved by Stobaeus affirms that "philosophy and magic nourish the soul." Astrology plays a much more prominent part, and the stars are ranked as visible gods, of whom the sun is by far the greatest. All seven planets nevertheless control the changes in the world of nature; there are seven human types corresponding to them; and the twelve signs of the zodiac also govern the human body. Only the chosen few who possess *gnosis* or are capable of receiving *nous* can escape the decrees of fate as administered by the stars and ultimately return to the spiritual world, passing through "choruses of demons" and "courses of stars" and reaching the Ogdoad or eighth heaven above and beyond the spheres of the seven planets.[2] Such

[1] R. Reitzenstein, *Poimandres,* Leipzig, 1904, p. 319. This work is the fullest scientific treatment of the subject.

[2] Citations supporting this and the preceding sentences may be found in Kroll's article on Hermes Trismegistus in Pauly-

Gnostic cosmology and demonology, especially the location of demons amid the planetary spheres, provides favorable ground for the development of astrological necromancy.

Astro-logical treatises ascribed to Hermes.

Not only is a belief in astrology implied throughout the *Poimandres,* but a number of separate astrological treatises are extant in whole or part under the name of Hermes Trismegistus,[1] and he is frequently cited as an authority in other Greek astrological manuscripts.[2] The treatises attributed to him comprise one upon general method,[3] one on the names and powers of the twelve signs, one on astrological medicine addressed to Ammon the Egyptian,[4] one on thunder and lightning, and some hexameters on the relation of earthquakes to the signs of the zodiac. This last is also ascribed to Orpheus.[5] There are various allusions to and versions of tracts concerning the relation of herbs to the planets or signs of the zodiac or thirty-six decans.[6] These treatises attribute magic virtues to plants, include a prayer to be repeated when plucking each herb, and tell how to use the

Wissowa, 809-820. The *Poimandres* was translated into English by John Everard, D.D., a mystic but also a popular preacher whose outspoken sermons caused his frequent arrest and imprisonment during the reigns of James I and Charles I. James is reported to have said of him, "What is this Dr. Ever-out? His name shall be Dr. Never-out," (*Dict. Nat. Biog,*). Dr. Everard's translation was printed in 1650 and again in 1657 when the "Asclepius" was added to it. In 1884 it appeared again in the Bath Occult Reprint Series with an introduction by Hargrave Jennings, and the second volume in the same series was Hermes' *The Virgin of the World,* published at London. Kroll mentions only the more recent translation by Mead, *Thrice Greatest Hermes.* London, 1906.

[1] Consult the bibliography in Kroll's article in Pauly-Wissowa.

[2] See the various volumes of *Catalogus codicum astrologorum Graecorum, passim.*

[3] Unprinted.

[4] An English translation by John Harvey was printed in London, 1657, 12mo. It also exists in manuscript form in the British Museum; Sloane 1734, fols. 283-98, "The learned work of Hermes Trismegistus intituled hys Phisicke Mathematycke or Mathematicall Physickes, direct to Hammon Kinge of Egypte."

[5] *Orphica,* ed. Abel (1885), p. 141.

[6] It was to a work on this last subject that Pamphilus, cited by Galen, referred in mentioning the herb ἀετοῦ, but this plant is not named in the extant treatise on the decans. Such treatises are more or less addressed to Asclepius: printed in J. B. Pitra, *Analecta Sacra,* V, ii, 279-90; *Cat. cod. astrol. Graec.,* IV, 134; VI, 83; VII, 231; VIII, ii, 159; VIII, iii, 151; and by Ruelle, *Rev. Phil.,* XXXII, 247.

astrological figures of the decans, engraved on stones, as healing amulets.

Hermetic works of alchemy.

Works under the name of Hermes Trismegistus are cited by Greek alchemists of the closing Roman Empire, such as Zosimus, Stephanus, and Olympiodorus, but those Hermetic treatises of alchemy which are extant are of late date and much altered.[1] Some treatises are preserved only in Arabic; others are medieval Latin fabrications. The Greek alchemists, however, seem to have recited the mystic hymn of Hermes from the *Poimandres.*[2]

Nechepso and Petosiris.

Hellenistic and Roman astrology sought to extend its roots far back into Egyptian antiquity by putting forth spurious treatises under the names, not only of Hermes Trismegistus, but also of Nechepso and Petosiris,[3] who were regarded respectively as an Egyptian king and an Egyptian priest who had lived at least seven centuries before Christ. Indeed, they were held to be the recipients of divine revelation from Hermes and Asclepius. A lengthy astrological treatise, which Pliny[4] is the first to cite and from a fourteenth book of which Galen[5] mentions a magic ring of jasper engraved with a dragon and rays, seems to have appeared in their names probably at Alexandria in the Hellenistic period. Only fragments and citations ascribed to Nechepso and Petosiris are now extant.[6]

Manetho.

Yet another astrological work which claims to be drawn from the secret sacred books and cryptic monuments of ancient Egypt is ascribed to Manetho. It is a compilation

[1] Berthelot (1885), pp. 133-6, and his article on Hermes Trismegistus in *La Grande Encyclopédie;* also Kroll on Hermes in Pauly-Wissowa, 799.

[2] Berthelot (1885), p. 134.

[3] Bouché-Leclercq, *L'Astrologie grecque,* 1899, pp. xi, 519-20, 563-4.

[4] NH, II, 21; VII, 50.

[5] Kühn, XII, 207.

[6] They have been collected and edited by E. Riess, *Nechepsonis et Petosiridis fragmenta magica,* in *Philologus,* Supplbd. VI, Göttingen (1891-93), pp. 323-394. See also F. Boll, *Die Erforschung der antiken Astrologie,* in *Neue Jahrb. für das klass. Altert.,* XI (1908), p. 106, and his dissertation of the same title published at Bonn, 1890. I have found that Riess, while including some of the passages attributed to Nechepso by the sixth century medical writer, Aetius, seems to have overlooked the "Emplastrum Nechepsonis e cupresso," Aetius, *Tetrabibl.,* IV, Sermo III, cap. 19 (p. 771 in the edition of Stephanus, 1567).

in verse of prognostications from the various constellations and is regarded as the work of several writers, of whom the oldest is placed in the reign of Alexander Severus in the third century.[1]

The *Lithica* of Orpheus.

Orpheus is another author more cited than preserved by classical antiquity. Pliny called him the first writer on herbs and suspected him of magic. Ernest Riess affirms that Rohde (*Psyche,* p. 398) "has abundantly proved that Orpheus' followers were among the chief promulgators of purifications and charms against evil spirits."[2] Among poems of some length extant under Orpheus' name the one of most interest to us is the *Lithica,* where in 770 lines the virtues of some thirty gems are set forth with considerable allusion to magic.[3] The authorship is uncertain, but the verse is supposed to follow the prose treatise by Damigeron who lived in the second century B. C. The date of the poem is now generally fixed in the fourth century of our era, although King[4] argued for an earlier date. I agree with him that the allusion in lines 71-74 to decapitation on the charge of magic is, taken alone, too vague and blind to be associated with any particular event or time; editors since Tyrwhitt have connected it with the law of Constantius against magic and the persecution of magicians in 371 A. D. But King's contention that the *Lithica* is by the same author as the *Argonautica,* also ascribed to Orpheus, and is therefore of early date, falls to the ground since the *Argonautica,* too, is now dated in the fourth century.

[1] Bouché-Leclercq, *L'Astrologie grecque,* 1898, p. xiii. Axt and Riegler, *Manethonis Apotelesmaticorum libri sex,* Cologne, 1832. Also edited by Koechly.

[2] E. Riess, On Ancient Superstition, in *Transactions American Philological Association* (1895), XXVI, 40-55. Grenfell (1921), p. 151, announces that J. G. Smyly is about to publish "a remarkable fragment of an Orphic ritual" among some thirty papyrus texts in the *Cunningham Memoirs of the Royal Irish Academy.*

[3] The Greek text of the Lithica is contained in *Orphica,* ed. E. Abel, Lipsiae et Pragae, 1885. A rather too free English verse translation, *Orpheus on Gems,* is given in C. W. King, *The Natural History, Ancient and Modern, of Precious Stones and Gems and of Precious Metals,* London, 1865.

[4] Pp. 397-98.

Argument of the poem.

The *Lithica* opens by representing Hermes as bestowing upon mankind the precious lore of the marvelous virtues of gems. In his cave are stored stones which banish ghosts, robbers, and snakes, which bring health, happiness, victory in war and games, honor at courts and success in love, and which insure safety on journeys, the favor of the gods, and enable one to read the hidden thoughts of others and to understand the language of the birds as they predict the future. Few persons, however, avail themselves of this mystic lore, and those who do so are liable to be executed on the charge of magic. After this introduction, which may be regarded as a piquant appetizer to whet the reader's taste for further details, the virtues of individual stones are described, first in the words of Theodamas, a wise and divine man [1] whom the author meets on his way to perform annual sacrifice at an altar of the Sun, where as a child he narrowly escaped from a deadly snake, and then in a speech of the seer Helenus to Philoctetes which Theodamas quotes. Greek gods are often mentioned; as the poem proceeds the virtues of a number of gems are attributed to Apollo rather than Hermes; and there are allusions to Greek mythology and the Trojan war. Some gems are found in animals, for instance, in the viper or the brain of the stag.

Magic powers of stones.

Let us turn to some examples of the marvelous virtues of particular stones. The crystal wins favorable answers from the gods to prayers; kindles fire, if held over sticks, yet itself remains cold; as a ligature benefits kidney trouble. Sacrifices in which the adamant is employed win the favor of the gods; it is also called Lethaean because it makes one forget worries, or the milk-stone (*galactis*) because it renews the milk of sheep or goats when powdered in brine and sprinkled over them. Worn as an amulet it counteracts the evil eye and gains royal favor for its bearer. The agate is an agricultural amulet and should be attached to the plowman's arm and the horns of the oxen. Other stones help vineyards, bring rain or avert hail and pests from the crops.

[1] Line 94, *περίφρονι Θειοδάμαντι*; line 165, *δαιμόνιος φώς*.

Lychnis prevents a pot from boiling on a fire and makes it boil when the fire is dead. The magnet was used by the witches Circe and Medea in their spells; an unchaste wife is unable to remain in the bed where this stone has been placed with an incantation. Other stones cure snake-bite and various diseases, serve as love-charms or aids in child-birth, or counteract incantations and enchantments.

Magic rites to gain powers of divination.

To make the gem *sideritis* or *oreites* utter vocal oracles the operator must abstain for three weeks from animal food, the public baths, and the marriage bed; he is then to wash and clothe the gem like an infant and employ various sacrifices, incantations, and illuminations. The gem *Liparaios,* known to the learned Magi of Assyria, when burnt on a bloodless altar with hymns to the Sun and Earth attracts snakes from their holes to the flame. Three youths robed in white and carrying two-edged swords should cut up the snake who comes nearest the fire into nine pieces, three for the Sun, three for the earth, three for the wise and prophetic maiden. These pieces are then to be cooked with wine, salt, and spices and eaten by those who wish to learn the language of birds and beasts. But further the gods must be invoked by their secret names and libations poured of milk, wine, oil, and honey. What is not eaten must be buried, and the participants in the feast are then to return home wearing chaplets but otherwise naked and speaking to no one whom they may meet. On their arrival home they are to sacrifice mixed spices. It will be recalled that Apollonius of Tyana and the Arabs also learned the language of the birds by eating snake-flesh.

Powers of gems compared with herbs.

Thus gems are potent in religion and divination, love-charms and child-birth, medicine and agriculture. The poem fails, however, to touch upon their uses in alchemy or relations to the stars, nor does it contain much of anything that can be called necromancy. But the author ranks the virtues of stones above those of herbs, whose powers disappear with age. Moreover, some plants are injurious, whereas the marvelous virtues of stones are almost all beneficial as well as permanent. "There is great force in herbs," he says, "but far greater in stones," [1] an observation often repeated in the middle ages.

Magic herbs and demons in Orphic rites.

More stress is laid upon the power of demons and herbs in a description which has been left us by Saint Cyprian,[2] bishop of Antioch in the third century, of some pagan mysteries upon Mount Olympus into which he was initiated when a boy of fifteen and which have been explained as Orphic rites. His initiation was under the charge of seven hierophants, lasted for forty days, and included instruction in the virtues of magic herbs and visions of the operations of demons. He was also taught the meaning of musical notes and harmonies, and saw how times and seasons were governed by good and evil spirits. In short, magic, pseudo-science, occult virtue, and perhaps astrology formed an important part of Orphic lore.

Books ascribed to Zoroaster.

Cumont states in his *Oriental Religions in Roman Paganism* that "towards the end of the Alexandrine period the books ascribed to the half-mythical masters of the Persian science, Zoroaster, Hosthanes and Hystaspes, were translated into Greek, and until the end of paganism those names enjoyed a prodigious authority." [3] Pliny regarded Zoroaster as the founder of magic and we have met other examples of his reputation as a magician. Later we shall find him cited several times in the Byzantine *Geoponica* which seems to use a book ascribed to him on the sympathy and antipathy existing between natural objects.[4] Naturally a number of pseudo-Zoroastrian books were in circulation, some of which Porphyry, the Neo-Platonist, is said to have suppressed. At least he tells us in his *Life of Plotinus* [5] that certain Christians and other men

[1] Lines 410-411.

[2] *Confessio S. Cypriani*, in *Acta Sanctorum*, ed. Bollandists, Sept., VII, 222; L. Preller, *Philologus* (1846), I, 349ff.; cited by A. B. Cook, *Zeus*, Cambridge, 1914, I, 110-111. The work is treated more fully below in Chapter 18.

[3] Franz Cumont, *op. cit.*, Chicago, 1911, p. 189. See also Windischmann, *Zoroastrische Studien*, Berlin, 1863.

[4] See below, Chapter 26.

[5] Cap. 16.

claimed to possess certain revelations of Zoroaster, but that he advanced many arguments to show that their book was not written by Zoroaster but was a recent composition.

The Chaldean Oracles.

There has been preserved, however, in the writings of the Neo-Platonists a collection of passages known as the Zoroastrian Logia or Chaldean Oracles [1] and which "present . . . a heterogeneous mass, now obscure and again bombastic, of commingled Platonic, Pythagorean, Stoic, Gnostic, and Persian tenets." [2] Not only are these often cited by the Neo-Platonists, but Porphyry, Iamblichus, and Proclus composed commentaries upon them.[3] Some think that these citations and commentaries have reference to a single work put together by Julian the Chaldean in the period of the Antonines. This "mass of oriental superstitions, a medley of magic, theurgy, and delirious metaphysics," [4] was reverenced by the Neo-Platonists of the following centuries as a sacred authority equal to the *Timaeus* of Plato. Our next chapter will therefore deal with the writings of the Neo-Platonists upon whom this spurious mystic literature had so much influence.

[1] Edited by Kroll, *De oraculis Chaldaicis*, in *Breslau Philolog. Abhandl.*, VII (1894), 1-76. Cory, *Ancient Fragments*, London, 1832.

[2] L. A. Gray in A. V. W. Jackson, *Zoroaster*, 1901, pp. 259-60.

[3] G. Wolff, *Porphyrii de philosophia ex oraculis hauriendis*, Berlin, 1886. Pitra, *Analecta Sacra*, V, 2, pp. 192-95, Πρόκλου ἐκ τῆς Χαλδαικῆς φιλοσοφίας. Many quotations of oracles from Porphyry's *De philosophia ex oraculis hausta* are made by Eusebius, *Praeparatio evangelica*, in PG, XXI.

[4] Bouché-Leclercq, *L'Astrologie grecque*, p. 599.

CHAPTER XI

NEO-PLATONISM AND ITS RELATIONS TO ASTROLOGY AND THEURGY

Neo-Platonism and the occult—Plotinus on magic—The life of reason is alone free from magic—Plotinus unharmed by magic—Invoking the demon of Plotinus—Rite of strangling birds—Plotinus and astrology—The stars as signs—The divine star-souls—How do the stars cause and signify?—Other causes and signs than the stars—Stars not the cause of evil—Against the astrology of the Gnostics—Fate and free-will—Summary of the attitude of Plotinus to astrology—Porphyry's *Letter to Anebo*—Its main argument—Questions concerning divine natures—Orders of spiritual beings—Nature of demons—The art of theurgy—Invocations and the power of words—Magic a human art: theurgy divine—Magic's abuse of nature's forces—Its evil character—Its deceit and unreality—Porphyry on modes of divination—Iamblichus on divination—Are the stars gods?—Is there an art of astrology?—Porphyry and astrology—Astrological images—Number mysticism—Porphyry as reported by Eusebius—The emperor Julian on theurgy and astrology—Julian and divination—Scientific divination according to Ammianus Marcellinus—Proclus on theurgy—Neo-Platonic account of magic borrowed by Christians—Neo-Platonists and alchemy.

Neo-Platonism and the occult.

THAT the Neo-Platonists were much given to the occult has been a common impression among those who have written upon the period of the decline of the Roman Empire, of the end of paganism, and the passing of classical philosophy. This is perhaps in some measure the result of Christian viewpoint and hostility; probably the Christians of the period would seem equally superstitious to a modern Neo-Platonist. If the lives of the philosophers by Eunapius sound like fairy tales,[1] what do the lives of the saints of the same period sound like? If the Neo-Platonists were like our mediums,

[1] Paul Allard, *La transformation du Paganisme romain au IVe siècle*, pp. 113-33, in *Compte Rendu du Congrès Scientifique International des Catholiques. Deuxième Section, Sciences religieuses.* Paris, 1891.

what were the Christian exorcists like? But let us turn to the writings of the leading Neo-Platonists themselves, the only accurate mirror of their views.

Plotinus on magic.

Plotinus,[1] who lived from about 204 to 270 A. D. and is generally regarded as the founder of Neo-Platonism, was apparently less given to occult sciences than some of his successors.[2] One of his charges against the Gnostics[3] is that they believe that they can move the higher and incorporeal powers by writing incantations and by spoken words and various other vocal utterances, all which he censures as mere magic and sorcery. He also attacks their belief that diseases are demons and can be expelled by words. This wins them a following among the crowd who are wont to marvel at the powers of magicians, but Plotinus insists that diseases are due to natural causes.[4] Even he, however, accepted incantations and the charms of sorcerers and magicians as valid, and accounted for their potency by the sympathy or love and hatred which he said existed between different objects in nature, which operates even at a dis-

[1] *Plotini opera omnia, Porphyrii liber de vita Plotini, cum Marsilii Ficini commentariis* . . . ed D. Wyttenbach, G. H. Moser, and F. Creuzer, Oxford, 1835, 3 vols. Page references in my citations are to this edition, but I have also employed: *Plotini Enneades,* ed. R. Volkmann, Leipzig, 1883; *Select Works of Plotinus translated from the Greek with an Introduction containing the substance of Porphyry's Life of Plotinus,* by Thomas Taylor, new edition with preface and bibliography by G. R. S. Mead, London, 1909; K. S. Guthrie, *The Philosophy of Plotinus,* Philadelphia, 1896, and *Plotinos, Complete Works,* 4 vols., 1918, English Translation. Where my citations give the number of the chapter in addition to the *Ennead* and Book, these agree with Volkmann's text and Guthrie's translation,—which, however, are not quite identical in this respect. A noteworthy recent publication is W. R. Inge, *The Philosophy of Plotinus,* 1918, 2 vols.

[2] H. F. Müller, *Plotinische Studien II,* in *Hermes,* XLIX, 70-89, argues that the philosophy of Plotinus was genuinely Hellenic and free from oriental influence, that all theurgy was hateful to him, and that he opposed Gnosticism and astrology. Müller seems to me to overstate his case and to be too ready to exculpate Plotinus, or perhaps rather Hellenism, from concurrence in the superstition of the time.

[3] For Gnosticism see Chapter 15.

[4] *Ennead,* II, 9, 14. Πλωτίνου πρὸς τοὺς Γνωστικούς, ed. G. A. Heigl, 1832; and *Plotini De Virtutibus et Adversus Gnosticos libellos,* ed. A. Kirchhoff, 1847; are simply extracts from the *Enneads.* See also C. Schmidt, *Plotin's Stellung zum Gnosticismus u. kirchl. Christentum,* 1900; in TU, X, 90 pp.

tance, and which is an expression of one world-soul animating the universe.[1]

The life of reason is alone free from magic.

Plotinus held further, however, that only the physical and irrational side of man's nature was affected by drugs and sorcery, just as "even demons are not impassive in their irrational part," [2] and so are to some extent subject to magic. But the rational soul may free itself from all influence of magic.[3] Moreover, remorselessly adds the clear-headed Plotinus with a burst of insight that may well be attributed to Hellenic genius, he who yields to the charms of love and family affection or seeks political power or aught else than Truth and true beauty, or even he who searches for beauty in inferior things; he who is deceived by appearances, he who follows irrational inclinations, is as truly bewitched as if he were the victim of magic and *goetia* so-called. The life of reason is alone free from magic.[4] Whereat one is tempted to paraphrase a remark of Aelian [5] and exclaim, "What do you think of that definition of magic, my dear anthropologists and sociologists and modern students of folk-lore?"

Plotinus unharmed by magic.

This immunity of the true philosopher and sincere follower of truth from magic received illustration, according to Porphyry,[6] in the case of Plotinus himself, who suffered no harm from the magic arts which his enemy, Alexandrinus Olympius, directed against him. Instead the baleful defluxions from the stars which Olympius had tried to draw down upon Plotinus were turned upon himself. Porphyry also states [7] that Plotinus was aware at the time of the "sidereal enchantments" of Olympius against him. Incidentally the episode provides one more proof of the essential unity of astrology and magic.

[1] *Ennead*, IV, 4, 40 (II, 805 or 434). Τὰς δὲ γοητείας πῶς; ἢ τῇ συμπαθείᾳ, καὶ τῷ πεφυκέναι συμφωνίαν εἶναι ὁμοίων καὶ ἐναντίωσιν ἀνομοίων, καὶ τῇ τῶν δυνάμεων τῶν πολλῶν ποικιλίᾳ εἰς ἓν ζῷον συντελούντων. *Ibid.* 42 (II, 808 or 436) . . . καὶ τέχναις καὶ ἰατρῶν καὶ ἐπαοιδῶν ἄλλο ἄλλῳ ἠναγκάσθη παρασχεῖν τι τῆς δυνάμεως τῆς αὐτοῦ. *Ennead*, IV, 9 (II, 891 or 479). εἰ δὲ καὶ ἐπῳδαὶ καὶ ὅλως μαγεῖαι συνάγουσι καὶ συμπαθεῖς πόῤῥωθεν ποιοῦσι, πάντως τοι διὰ ψυχῆς μιᾶς.

[2] *Ennead*, IV, 4 (II, 810 or 437).

[3] *Ennead*, IV, 4, 43-44.

[4] *Ennead*, IV, 4, 44.

[5] See Chapter XII, pp. 323-4.

[6] *Vita Plotini*, cap. 10.

[7] *Vita*, cap. 10.

Invoking the demon of Plotinus.

Plotinus, indeed, was regarded by his admirers as divinely inspired, as another incident from the *Life* by Porphyry will illustrate.[1] An Egyptian priest had little difficulty in persuading Plotinus, who although of Roman parentage had been born in Egypt, to allow him to try to invoke his familiar demon. Plotinus was then teaching in Rome where he resided for twenty-six years, and the temple of Isis was the only pure place in the city which the priest could find for the ceremony. When the invocation had been duly performed, there appeared not a mere demon but a god. The apparition was not long enduring, however, nor would the priest permit them to question it, on the ground that one of the friends of Plotinus present had marred the success of the operation. This man had feared he might suffer some injury when the demon appeared and as a counter-charm had brought some birds which he held in his hands, apparently by the necks, for at the critical moment when the apparition appeared he suffocated them, whether from fright or from envy of Plotinus Porphyry declares himself unable to state.

The rite of strangling birds.

This practice of grasping birds by the necks in both hands is shown by a number of works of art to have been a custom of great antiquity. We may see a winged Gorgon strangling a goose in either hand upon a plate of the seventh century B.C. from Rhodes now in the British Museum.[2] A gold pendant of the ninth century B.C. from Aegina, now also in the British Museum, consists of a figure holding a water-bird by the neck in either hand, while from its thighs pairs of serpents issue on whose folds the birds stand with their bills touching the fangs of the snakes.[3] There also is a figure of a winged goddess grasping two water-birds by the necks upon an ivory fibula excavated at Sparta.[4]

[1] Cap. 10.

[2] A748.

[3] Shown in the article on "Jewelry" in the eleventh edition of the *Encyclopedia Britannica*, Plate I, Figure 50. The article says of the pendant, "Here we find the themes of archaic Greek art, such as a figure holding up two water-birds, in immediate connexion with Mycenaean gold patterns." See further A. J. Evans in *Journal of Hellenic Studies*, 1893, p. 197.

[4] J. E. Harrison, *Themis*, Cambridge, 1912. p. 114, Fig. 20.

Plotinus and astrology.

Porphyry also tells us in the *Life* that Plotinus devoted considerable attention to the stars and refuted in his writings the unwarrantable claims of the casters of horoscopes.[1] Such passages are found in the treatises on fate and on the soul, while one of his treatises is devoted entirely to the question, "Whether the stars effect anything?"[2] This was one of four treatises which Plotinus a little before his death sent to Porphyry, and which are regarded as rather inferior to those composed by him when in the prime of life. In the next century the astrologer, Julius Firmicus Maternus, regards Plotinus as an enemy of astrology and represents him as dying a horrible and loathsome death from gangrene.[3]

The stars as signs.

As a matter of fact the criticisms made by Plotinus were not necessarily destructive to the art of astrology, but rather suggested a series of amendments by which it might be made more compatible with a Platonic view of the universe, deity, and human soul. These amendments also tended to meet Christian objections to the art. His criticisms were not new; Philo Judaeus had made similar ones over two centuries before.[4] But the great influence of Plotinus gave added emphasis to these criticisms. For instance, the point made by him several times that the motion of the stars "does not cause everything but signifies the future concerning each"[5] man and thing, is noted by Macrobius both in the *Saturnalia*[6] and the *Dream of Scipio;*[7] while in the twelfth century John of Salisbury, arguing against astrology, fears that its devotees will take refuge in the authority of Plotinus and say that they detract

[1] *Vita*, cap. 15. It will be noted that like some of the church fathers Plotinus attacked genethlialogy rather than astrology. Προσεῖχε δὲ τοῖς μὲν περὶ τῶν ἀστέρων κανόσιν οὐ πάνυ τι μαθηματικῶς, τοῖς δὲ τῶν γενεθλιαλόγων ἀποτελεστικοῖς ἀκριβέστερον. καὶ φωράσας τῆς ἐπαγγελίας τὸ ἀνεχέγγυον ἐλέγχειν πολλαχοῦ καὶ (τῶν) ἐν τοῖς συγγράμμασιν οὐκ ὤκνησε.

[2] *Ennead* II, 3, Περὶ τοῦ εἰ ποιεῖ τὰ ἄστρα. Porphyry arranged his master's treatises in the form of six enneads of nine each and perhaps somewhat revised them at the same time.

[3] *Matheseos libri VIII*, ed. Kroll et Skutsch, Lipsiae, 1897. I, 7, 14-22.

[4] See below, pp. 353-4.

[5] *Ennead* II, 3 (p. 242), Ὅτι ἡ τῶν ἄστρων φορὰ σημαίνει περὶ ἕκαστον τὰ ἐσόμενα ἀλλ' οὐκ αὐτὴ πάντα ποιεῖ, ὡς τοῖς πολλοῖς δοξάζεται, εἴρηται μὲν πρότερον ἐν ἄλλοις. See also *Ennead* III, 1, and IV, 3-4.

[6] I, 18.

[7] Cap. 19.

nothing from the Creator's power, since He established once for all an unalterable natural law and disposed all future events as He foresaw them. Thus the stars are merely His instruments.[1]

The divine star-souls.

But let us see what Plotinus says himself rather than what others took to be his meaning. Like Plato, who regarded the stars as happy, divine, and eternal animals, Plotinus not only believes that the stars have souls but that their intellectual processes are far above the frailties of the human mind and nearer the omniscience of the world-soul. Memory, for example, is of no use to them,[2] nor do they hear the prayers which men address to them.[3] Plotinus often calls them gods. They are, however, parts of the universe, subordinate to the world-soul, and they cannot alter the fundamental principles of the universe, nor deprive other beings of their individuality, although they are able to make other beings better or worse.[4]

How do the stars cause and signify?

In his discussion of problems concerning the soul Plotinus says that "it is abundantly evident . . . that the motion of the heavens affects things on earth and not only in bodies but also the dispositions of the soul," [5] and that each part of the heavens affects terrestrial and inferior objects. He does not, however, think that all this influence can be accounted for "exclusively by heat or cold,"—perhaps a dig at Ptolemy's *Tetrabiblos*.[6] He also objects to ascribing the crimes of men to the will of the stars or every human act

[1] *Polycraticus,* II, 19, (ed. C. C. I. Webb, 1909, I, 112). Mr. Webb (I, xxviii) holds that John of Salisbury "certainly did not have Plotinus," and derived some passages from his works through Macrobius and Augustine; but he is unable to state in what intermediate source John could have found the passage now in question. It does not seem to reflect Plotinus' doctrine very accurately.

[2] *Ennead* IV, iv, 6 and 8.

[3] *Ibid.,* 30. Guthrie's translation, "We have shown that memory is useless to the stars: we have agreed that they have senses, namely, sight and hearing," is quite misleading, as caps. 40-42 make evident.

[4] *Ennead* II, iii, 6 and 13 (249-50).

[5] *Ennead* IV, iv, 31. *ὅτι μὲν οὖν ἡ φορὰ ποιεῖ . . . ἀναμφισβητήτως μὲν τὰ ἐπίγεια οὐ μόνον τοῖς σώμασιν ἀλλὰ καὶ ταῖς τῆς ψυχῆς διαθέσεσι, καὶ τῶν μερῶν ἕκαστον εἰς τὰ ἐπίγεια καὶ ὅλως τὰ κάτω ποιεῖ, πολλαχῇ δῆλον.*

[6] *Idem.* Guthrie heads the passage, "Absurdity of Ptolemean Astrology." See also *Ennead,* II, iii, 1-5.

to a sidereal decision,[1] and to speaking of friendships and enmities as existing between the planets according as they are in this or that aspect towards one another.[2] If then the admittedly vast influence of the stars cannot be satisfactorily accounted for either as material effects caused by them as bodies or as voluntary action taken by them, how is it to be explained? Plotinus accounts for it by the relation of sympathy which exists between all parts of the universe, that single living animal, and by the fact that the universe expresses itself in the figures formed by the movements of the celestial bodies, which "exert what influence they do exert on things here below through contemplation of the intelligible world." [3] These figures, or constellations in the astrological sense, have other powers than those of the bodies which participate in them, just as many plants and stones "among us" have marvelous occult powers for which heat and cold will not account.[4] They both exert influence effectively and are signs of the future through their relation to the universal whole. In many things they are both causes and signs, in others they are signs only.[5]

Other causes and signs than the stars.

For Plotinus, however, the universe is not a mechanical one where but one force prevails, namely, that produced by or represented by the constellations. The universe is full of variety with countless different powers, and the whole would not be a living animal unless each living thing in it lived its own life, and unless life were latent even in inanimate objects. It is true that some powers are more effective than others, and that those of the sky are more so than those of earth, and that many things lie under their power. Nevertheless Plotinus sees in the reproduction of life and species in the universe a force independent of the stars. In

[1] *Ennead* II, iii, 6.

[2] *Ennead* II, iii, 4.

[3] Guthrie's translation, *Ennead* IV, iv, 35. εἰ δὴ δρᾷ τι ὁ ἥλιος καὶ τὰ ἄλλα ἄστρα εἰς τὰ τῇδε, χρὴ νομίζειν αὐτὸν μὲν ἄνω βλέποντα εἶναι.

[4] *Idem.* καὶ ἐν τοῖς παρ' ἡμῖν εἰσι πολλαί, ἃς οὐ θερμὰ ἢ ψυχρὰ παρέχεται, ἀλλὰ γενόμενα ποιότησι διαφόροις καὶ λόγοις εἰδοποιηθέντα καὶ φύσεως δυνάμεως μεταλαβόντα, οἷον καὶ λίθων φύσεις καὶ βοτανῶν ἐνέργειαι θαυμαστὰ πολλὰ παρέχονται.

[5] *Ennead* IV, iv, 34. καὶ ποιήσεις καὶ σημασίας ἐν πολλοῖς ἀλλαχοῦ δὲ σημασίας μόνον.

the generation of any animal, for example, the stars contribute something, but the species must follow that of its forebears.[1] And after they have been produced or begotten, terrestrial beings add something of their own. Nor are the stars the sole signs of the future. Plotinus holds that "all things are full of signs," and that the sage can not merely predict from stars or birds, but infer one thing from another by virtue of the harmony and sympathy existing between all parts of the universe.[2]

Stars not the cause of evil.

Nor can the gods or stars be said to cause evil on earth, since their influence is affected by other forces which mingle with it. Like the earlier Jewish Platonist, Philo, Plotinus denies that the planets are the cause of evil or change their own natures from good to evil as they enter new signs of the zodiac or take up different positions in relation to one another. He argues that they are not changeable beings, that they would not willingly injure men, or, if it is contended that they are mere bodies and have no wills, he replies that then they can produce only corporeal effects. He then solves the problem of evil in the usual manner by ascribing it to matter, in which reason and the celestial force are received unevenly, as light is broken and refracted in passing through water.[3]

Against the astrology of the Gnostics.

Plotinus repeats much the same line of argument in his book against the Gnostics, where he protests against "the tragedy of terrors which they think exists in the spheres of the universe," [4] and the tyranny they ascribe to the heavenly bodies. His belief is that the celestial spheres are in perfect harmony both with the universe as a whole and with our globe, completing the whole and constituting a great part of it, supplying beauty and order. And often they are to be regarded as signs rather than causes of the future. Their natures are constant, but the sequence of events may be varied by chance circumstances, such as different hours of

[1] *Ennead* II, iii (p. 256).
[2] *Ibid.* (pp. 250-1).
[3] *Ibid.*, II, iii (pp. 243-6, 254-5, 263-5).
[4] *Ennead*, II, ix, 13. τῆς τραγῳδίας τῶν φοβερῶν, ὡς οἴονται, ἐν ταῖς τοῦ κόσμου σφαίραις.

nativities, place of residence, and the dispositions of individual souls. Amid all this diversity one must also expect both good and evil, but not on that account call nature or the stars either evil themselves or the cause of evil.

Fate and free-will.

As the allusion just made in the preceding paragraph to "the dispositions of individual souls" shows, Plotinus made a distinction between the extent of the control exercised by the stars over inanimate, animate, and rational beings. The stars signify all things in the sensible world but the soul is free unless it slips and is stained by the body and so comes under their control. Fate or the force of the stars is like a wind which shakes and tosses the ship of the body in which the soul makes its passage. Man as a part of the world does some things and suffers many things in accordance with destiny. Some men become slaves to this world and to external influences, as if they were bewitched. Others look to their inner souls and strive to free themselves from the sensible world and to rise above demonic nature and all fate of nativities and all necessity of this world, and to live in the intelligible world above.[1]

Summary of the attitude of Plotinus to astrology.

Thus Plotinus arrives at practically what was to be the usual Christian position in the middle ages regarding the influence of the stars, maintaining the freedom of the human will and yet allowing a large field to astrological prediction. He is evidently more concerned to combat the notion that the stars cause evil or are to be feared as evil powers than he is to combat the belief in their influence and significations. His speaking of the stars both as signs and causes in a way doubles the possibility of prediction from them. If he attacked the language used by astrologers of the planets, and perhaps to a certain extent the technique of their art, he supported astrology by reconciling the existence of evil and of human freedom with a great influence of the stars and by his emphasis upon the importance of the figures made by the

[1] The references for the statements in this paragraph are in the order of their occurrence: *Ennead,* II, iii (pp. 257, 251-2); III, iv (p. 521); IV, iv (p. 813); II, iii (p. 260); III, iv (p. 520); IV, 3 (p. 711): in these cases the higher page-numbering is used.

movements of the heavenly bodies above any purely physical effects of their bodies as such. Thus he reinforced the conception of occult virtue, always one of the chief pillars, if not the chief support, of occult science and magic. On the other hand, men were not likely to reform a language and technique sanctioned by as great an astronomer as Ptolemy merely because a Neo-Platonist questioned its propriety.

Porphyry's *Letter to Anebo.*

Although Plotinus denied that diseases were due to demons, we once heard him speak of "demonic nature," and one of the *Enneads* discusses *Each man's own demon.* Here, however, the discussion is limited to the power presiding in each human soul, and nothing is said of magic. For the connection of demons with magic and for the art of theurgy we must turn to the writings of Porphyry and Iamblichus, and especially to *The Letter to Anebo* of Porphyry, who lived from about 233 to 305, and the reply thereto of the master Abammon, a work which is otherwise known as *Liber de mysteriis.*[1] The attribution of the latter work to Iamblichus, who died about 330, is based upon an anonymous assertion prefixed to an ancient manuscript of Proclus and upon the fact that Proclus himself quotes a passage from the *De mysteriis* as the words of Iamblichus. This attribution has been questioned, but if not by Iamblichus, the work seems to be at least by some disciple of his with similar views.[2] Other works of Iamblichus are largely philosophical and mathematical; among the chief works of Porphyry, apart from his literary work in connection with Plotinus, were his commentaries on Aristotle and fifteen books against the Christians.

Its main argument.

The Letter to Anebo inquires concerning the nature of the gods, the demons, and the stars; asks for an explanation of divination and astrology, of the power of names and incantations; and questions the employment of invocations

[1] Edited Venice, Aldine Press, 1497 and 1516; Oxford, 1678; by G. Parthey, Berlin, 1857. In the following quotations from it I have usually adhered to T. Taylor's English translation, London, 1821.

[2] Carl Rasche, *De Iamblicho libri qui inscribitur de mysteriis auctore,* Aschendorff, 1911, 82 pp.

and sacrifice. Other topics brought up are the rule of spirits over the world of nature, partitioned out among them for this purpose; the divine inspiration or demoniacal possession of human beings; and the occult sympathy between different things in the material universe. In especial the art of theurgy, a word said to be used now for the first time by Porphyry,[1] is discussed. It may be roughly defined for the moment as a sort of pious necromancy or magical cult of the gods. Porphyry raises various objections to the procedure and logic of the theurgists, diviners, enchanters, and astrologers, which Iamblichus, as we shall henceforth call the author of the *De mysteriis* as a matter of convenience if not of certainty, endeavors to answer, and to justify the art of theurgy.

Questions concerning divine natures.

We may first note the theory of demons which is elicited from Iamblichus in response to Porphyry's trenchant and searching questions. The latter, declaring that ignorance and disingenuousness concerning divine natures are no less reprehensible than impiety and impurity, demands a scientific discussion of the gods as a holy and beneficial act. He asks why, if the divine power is infinite, indivisible, and incomprehensible, different places and different parts of the body are allotted to different gods. Why, if the gods are pure intellects, they are represented as having passions, are worshiped with phallic ritual, and are tempted with invocations and sacred offerings? Why boastful speech and fantastic action are taken as indications of the divine presence; and why, if the gods dwell in the heavens, theurgists invoke only terrestrial and subterranean deities? How superior beings can be invoked with commands by their inferiors, why the Sun and Moon are threatened, why the man must be just and chaste who invokes spirits in order to secure unjust ends or gratify lust, and why the worshiper must abstain from animal food and not touch a corpse when sacrifices to the gods consist of the bodies of dead victims? Porphyry

[1] Bouché-Leclercq, *L'Astrologie grecque* (1898), p. 599, citing Kroll, *De oraculis Chaldaicis.*

wishes further an explanation of the various *genera* of gods, visible and invisible, corporeal and incorporeal, beneficent and malicious, aquatic and aerial. He wants to know whether the stars are not gods, how gods differ from demons, and what the distinction is between souls and heroes.

Orders of spiritual beings.

Iamblichus in reply states that as heroes are elevated above souls, so demons are inferior and subservient to the gods and translate the infinite, ineffable, and invisible divine transcendent goodness into terms of visible forms, energy, and reason.[1] He further distinguishes "the etherial, empyrean, and celestial gods," and angels, archangels, and archons.[2] As for corporeal, visible, aerial, and aquatic gods, he affirms that the gods have no bodies and no particular allotments of space, but that natural objects participate in or are related to the gods etherially or aerially or aquatically, each according to its nature.[3] "The celestial divinities," for example, "are not comprehended by bodies but contain bodies in their divine lives and energies. They are not themselves converted to body, but they have a body which is converted to its divine cause, and that body does not impede their intellectual and incorporeal perfection." [4] Iamblichus denies that there are any maleficent gods, saying that "it is much better to acknowledge our inability to explain the occurrence of evil than to admit anything impossible and false concernings the gods." [5] But he admits the existence of both good and evil demons and makes of the latter a convenient scapegoat upon whom to saddle any inconsistencies or impurities in religious rites and magical ceremony.

Nature of demons.

Iamblichus does not, however, hold the view of Apuleius that demons are subject to passions. They are impassive and incapable of suffering.[6] He scorns the notion that even the worst demons can be allured by the vapors of animal sacrifice or that petty mortals can supply such beings with anything;[7] it is rather in the consumption of foul matter

[1] *De mysteriis,* I, 5.
[2] VIII, 2.
[3] I, 9.
[4] I, 17 (Taylor's translation).
[5] IV, 6.
[6] I, 10.
[7] V, 10-12.

by pure fire in the act of sacrifice that they take delight. Demons are not, however, like the gods entirely separated from bodies. The world is divided up into prefectures among them and they are more or less inseparable from and identified with the natural objects which they govern.[1] Thus they may serve to enmesh the soul in the bonds of matter and of fate, and to afflict the body with disease.[2] Also the evil demons "are surrounded by certain noxious, blood-devouring, and fierce wild beasts," probably of the type of vampires and *empousas.*[3] Iamblichus further holds that there is a class of demons who are without judgment and reason, each of whom has some one function to perform and is not adapted to do anything else.[4] Such demons or forces in nature men may well address as superiors in invoking them, since they are superior to men in their one special function; but when they have once been invoked, man as a rational being may also well issue commands to them as his irrational inferiors.[5]

The art of theurgy.

Iamblichus also undertakes the defense of theurgy and carefully distinguishes it from magic, as we shall soon see. It is also different from science, since it does not merely employ the physical forces of the natural universe,[6] and from philosophy, since its ineffable works are beyond the reach of mere intelligence, and those who merely philosophize theoretically cannot hope for a theurgic union or communion with the gods.[7] Even theurgists cannot as a rule endure the light of spiritual beings higher than heroes, demons, and angels,[8] and it is an exceedingly rare occurrence for one of them to be united with the supramundane gods.[9] This theurgy, or "the art of divine works," operates by means of "arcane signatures" and "the power of inexplicable symbols."[10] It is thus that Iamblichus explains away most of the details in sacred rites and sacrifices to which Porphyry

[1] I, 20.
[2] II, 6.
[3] II, 7.
[4] IV, 1.
[5] IV, 2.
[6] IV, 10.
[7] II, 11.
[8] II, 3.
[9] V, 20.
[10] I, 9; VI, 6; II, 11.

had objected as obscene or material and as implying that the gods themselves were passive and passionate. They are mystic symbols, "consecrated from eternity" for some hidden reason "which is more excellent than reason." [1] Occult virtues indeed! We have already heard Iamblichus state that natural objects participate in or are related to the gods etherially or aerially or aquatically; theurgists therefore quite properly employ in their art certain stones, herbs, aromatics, and sacred animals.[2] By employing such potent symbols mere man takes on such a sacred character himself that he is able to command many spiritual powers.[3]

Invocations and the power of words.

Invocations and prayers are also much used in theurgical operations. But such invocations do not draw down the impassive and pure gods to this world; rather they purify those who employ them from their passions and impurity and exalt them to union with the pure and the divine.[4] These prayers are symbolic, too. They do not appeal to human passions or reason, "for they are perfectly unknown and arcane and are alone known to the God whom they invoke." [5] In another passage [6] Iamblichus replies to Porphyry's objection that such prayers are often composed of meaningless words and names without signification by declaring—somewhat inconsistently with his previous assertion that these invocations are "perfectly unknown"—that some of the names "which we can scientifically analyze" comprehend "the whole divine essence, power and order." Moreover, if translated into another language, they do not have exactly the same meaning, and even if they do, they no longer retain the same power as in the original tongue. We shall meet a similar passage concerning the power of words and divine names in the church father Origen who lived earlier in the third century than Porphyry and Iamblichus. Iamblichus concludes that "it is necessary that

[1] I, 11.

[2] V, 23.

[3] IV, 2.

[4] I, 12.

[5] I, 15; III, 24 (Taylor's translation).

[6] VII, 4.

ancient prayers . . . should be preserved invariably the same." [1]

Magic a human art: theurgy divine.

Neither Porphyry nor Iamblichus, I believe, employs the word, "magic," but they both often allude to its practitioners and methods by such expressions as "jugglers" and "enchanters" or by contrasting what is done "artificially" or by means of art with theurgical operations. In the last case the distinction is between what on the one hand is regarded as a divine mystery or revelation and what on the other hand is looked upon as a mere human art and contrivance. And "nothing . . . which is fashioned by human art is genuine and pure." [2] Christian writers drew a like distinction between prophecy or miracle and divination or magic. Sometimes, however, Iamblichus speaks of theurgy itself as an art, an involuntary admission of the close resemblance between its methods and those of magic. We are also told that if the theurgist makes a slip in his procedure, he thereby reduces it to the level of magic.[3]

Magic's abuse of nature's forces.

Another distinction is that theurgy aims at communion with the gods while magic has to do rather with "the physical or corporeal powers of the universe." [4] Both Porphyry and Iamblichus believed that harmony, sympathy, and mutual attraction existed between the various objects in the universe, which Iamblichus asserted was one animal.[5] Thus it is possible for man to draw distant things to himself or to unite them to, or separate them from, one another.[6] But art may also use this force of sympathy between objects in an extreme and unseemly manner, and this disorderly forcing of nature, we are left to infer, constitutes an essential feature of magic, whose procedure is not truly natural or scientific.

Its evil character.

Magic not only disorders the law and harmony, and makes a perverse and contrary use of natural forces. Its practitioners are also represented as aiming at evil ends and as

[1] VII, 5.
[2] III, 29.
[3] II, 10.
[4] IV, 10.
[5] IV, 12.
[6] IV, 3.

themselves of evil character.[1] They may try by their illicit and impure procedure to have intercourse with the gods or with pure spirits, but they are unable to accomplish this. All that they succeed in doing is to secure the alliance of evil demons by associating with whom they become more depraved than ever. Such wicked demons may pose as angels of light by requiring that those who invoke them should be just or chaste, but afterwards they show their true colors by assisting in crimes and the gratification of lusts.[2] It is they, too, who assuming the guise of superior spirits are responsible for the boastful and arrogant utterances of which Porphyry complained in persons supposed to be divinely inspired.[3]

Its deceit and unreality.

Finally magic is unstable and fantastic. "The imaginations artificially produced by enchantment" are not real objects. Those who foretell the future by "standing on characters" are no theurgists, but employ a superficial, false, and deceptive procedure which can attract only evil demons.[4] These demons are themselves deceitful and produce "fictitious images." [5] Porphyry in the *Letter to Anebo* also alluded to the frauds of "jugglers." Although the attitude both of Porphyry and Iamblichus is thus professedly unfavorable to the magic arts, we find that one of Iamblichus's disciples, named Sopater, was executed under Constantine on a charge of having charmed the winds.[6]

Porphyry on modes of divination.

How is divination to be placed in reference to magic and theurgy? Porphyry had inquired concerning various methods of divination: in sleep, in trances, and when fully conscious; in ecstasy, in disease, and in states of mental aberration or enchantment. He mentioned divination on hearing drums and cymbals, by drinking water and other potions, by inhaling vapor; divination in darkness, in a wall, in the open air or in the sunlight; by observing entrails or the flight of birds or the motion of the stars, or even by means

[1] IV, 10; III, 31.
[2] IV, 7.
[3] II, 10.
[4] VI, 5; III, 25; III, 13.
[5] II, 10.
[6] E. S. Bouchier, *Syria as a Roman Province*, Oxford, 1916, p. 231.

of meal. Yet other modes of determining the future which he lists are by characters, images, incantations, and invocations, with which the use of stones and herbs is often combined. These details make it evident how impossible it is to draw any dividing line between the methods of magic and divination, and Porphyry himself states that those who invoke the gods concerning the future not only "have about them stones and herbs," but are able to bind and to free from bonds, to open closed doors, and to change men's intentions. Among the virtues of parts of animals mentioned in his treatise upon abstinence from animal food are the powers of divination which may be obtained by eating the heart of a hawk or crow.[1]

Iamblichus on divination.

Porphyry states that all diviners attribute their predictions to gods or demons, but that he wonders if foreknowledge may not be a power of the human soul or perhaps accountable for by the sympathy which exists between different parts of the universe. Iamblichus holds, however, that divination is neither a human art nor the work of nature but of divine origin.[2] He perhaps regards it as little more than a branch of theurgy. He distinguishes between human dreams which are sometimes true, sometimes false, and dreams and visions divinely sent.[3] If one is able to predict the future by drinking water, it is because the water has been divinely illuminated.[4] That we can predict when the mind is diseased and disordered, and that stupid or simple-minded men are often better able to prophesy than the wise and learned, are for him but further proofs that foreknowledge is a divine gift and not a human science, while divination by such means as rods, pebbles, grains of corn and wheat simply excites the more his pious admiration at the greatness of divine power.[5] He disapproves of divination by standing on characters,[6] but sees no reason why divination in darkness, in a wall, or in sunlight, or by potions and incantations, may not be divinely directed. He will not, how-

[1] *De abstinentia,* II, 48.
[2] III, 1, 10.
[3] III, 2-3.
[4] III, 11.
[5] III, 24; III, 17.
[6] III, 14.

ever, connect the disordered imaginations excited by disease with divine presentiments.[1] From true divination he also separates the "natural prescience" of certain animals whose acuteness of sense or occult sympathy with other parts and forces of nature enables them to perceive some coming events before men do. Their power resembles prophecy, "yet falls short of it in stability and truth." [2] Augury is an art whose conjectures have great probability, but they are based upon divine signs or portents effected in nature by the agency of demons.[3]

Are the stars gods?

The stars are on a totally different plane from the other substances employed in divination. To Porphyry's question whether they are not gods Iamblichus is not content to reply that the celestial divinities comprehend these heavenly bodies and that the bodies in no way impede "their intellectual and incorporeal perfection." [4] He must needs go on to argue that the stars themselves, as simple indivisible bodies, unchanging in quality and uniform in movement, closely approach to "the incorporeal essence of the gods." He then triumphantly if illogically concludes, "Thus therefore the visible celestials are all of them gods and after a certain manner incorporeal." We may add the opinion of Chaeremon and others, noted by Porphyry, that the only gods were the physical ones of the Egyptians and the planets, signs of the zodiac, decans, and horoscope; all religious myths were explained by Chaeremon as astrological allegories.

Is there an art of astrology?

Porphyry objected that those who thus reduce religion to astrology submit everything to fate and leave the human soul no freedom, and furthermore that in any case astrology is an unattainable science. Iamblichus defends it against these objections, insisting that the universe is divided under the rule of planets, signs, and decans; [5] that the Egyptians

[1] III, 25. Although, as stated above, one may be divinely inspired while diseased. But there is no causal connection between the two.

[2] III, 26.

[3] III, 15.

[4] I, 17.

[5] VIII, 4.

do not make everything physical but ascribe two souls to man, one of which obeys the revolutions of the stars, while the other is intellectual and free;[1] and that there is a systematic art of astrology based on divine revelation and the long observations of the Chaldeans, although like any other science it may at times degenerate and become contaminated by error.[2] Iamblichus further regards as ridiculous the contention of those "who ascribe depravity to the celestial bodies because their participants sometimes produce evil."[3] In the brief separate treatise, *De fato*,[4] he again holds that all things are bound by the indissoluble chain of necessity which men call fate, but that the gods can loose the bonds of fate, and that the human mind, too, has power to rise above nature, unite with the gods, and enjoy eternal life.

Porphyry and astrology.

Whether Porphyry in his other extant works evidences a belief in astrology or not, and whether he wrote an *Introduction to the Tetrabiblos* or astrological handbook of Ptolemy, has been disputed.[5] This *Introduction* ascribed to Porphyry was much cited by subsequent astrologers[6] and was printed in 1559 together with a much longer anonymous commentary on the *Tetrabiblos* which some ascribe to Proclus.[7]

Astrological images.

Towards astrological images at least, Porphyry shows himself in the *Letter to Anebo* more favorable than Iamblichus, saying, "Nor are the artificers of efficacious images to be despised, for they observe the motion of celestial bodies." Iamblichus, on the other hand, rather grudgingly admits that "the image-making art attracts a certain very obscure genesiurgic portion from the celestial effluxions."[8] He seems to have the same feeling against images as against

[1] VIII, 6.

[2] IX, 3-4.

[3] I, 18.

[4] Iamblichus, *In Nicomachi Geraseni arithmeticam introductionem et De fato*, published by Tennulius, Deventer and Arnheim, 1668.

[5] Zeller, *Philos. d. Gr.*, III, 2, 2, p. 608, cites passages to show Porphyry's leanings towards astrology; but F. Boll, *Studien über Claudius Ptolemaeus*, 115-17, and Bouché-Leclercq, *L'Astrologie grecque*, 601-602, are inclined to the opposite view.

[6] CCAG, *passim*.

[7] Ed. Hieronymus Wolf, Basel, 1559, Greek and Latin.

[8] III, 28.

characters, perhaps regarding both as bordering upon idolatry.[1]

Number mysticism.

Plotinus, Porphyry, and Iamblichus were all given to number mysticism. The sixth book of the sixth *Ennead* is entirely devoted to this subject, while Porphyry and Iamblichus both wrote *Lives* of Pythagoras and treatises upon his doctrine of number.

Porphyry as reported by Eusebius.

Other works by Porphyry than the *Letter to Anebo* are cited or quoted a good deal by Eusebius in *Praeparatio evangelica,* especially his Περὶ τῆς ἐκ λογίων φιλοσοφίας, but the extracts are made for Eusebius's own purposes, which are to discredit pagan religion, and neither express Porphyry's complete thought nor probably even tend to prove his original point. Besides showing that Porphyry was inconsistent in distinguishing the different victims to be sacrificed to terrestrial and subterranean, aerial, celestial, and sea gods in the above-mentioned work, when in his *De abstinentia a rebus animatis* he held that beings who delighted in animal sacrifice were no gods but mere demons, Eusebius quotes him a good deal to show that the pagan gods were nothing but demons, that they themselves might be called magicians and astrologers, that they loved characters, and that they made their predictions of the future not from their own foreknowledge but from the stars by the art of astrology, and that like men they could not even always read the decrees of the stars aright. The belief is also mentioned that the fate foretold from the stars may be avoided by resort to magic.[2]

The Emperor Julian on theurgy and astrology.

The Emperor Julian was an enthusiastic follower of Iamblichus whom he praises [3] in his *Hymn to the Sovereign Sun* delivered at the Saturnalia of 361 A. D. He also describes "the blessed theurgists" as able to comprehend unspeakable mysteries which are hidden from the crowd, such as Julian the Chaldean prophesied concerning the god

[1] III, 29.

[2] Eusebius, *Praep. evang.*, IV, 6-15, 23; V, 6, 11, 14-15; VI, 1, 4-5; etc., in Migne, PG, XXI.

[3] Loeb Library edition of Julian's works, I, 398, 412, 433.

of the seven rays.[1] The emperor tells us that from his youth he was regarded as over-curious (περιεργότερον, a word which almost implies the practice of magic) and as a diviner by the stars (ἀστρόμαντιν). His *Hymn to the Sun* contains a good deal of astrological detail, speaks of the universe as eternal and divine, and regards planets, signs, and decans as "the visible gods." In short, "there is in the heavens a great multitude of gods." [2] The Sun, however, is superior to the other planets, and as Aristotle has pointed out "makes the simplest movement of all the heavenly bodies that travel in a direction opposite to the whole." [3] The Sun is also the link between the visible universe and the intelligible world, and Julian infers from his middle station among the planets that he is also king among the intellectual gods.[4] For behind his visible self is the great Invisible. He frees our souls entirely from the power of "Genesis," or the force of the stars exercised at nativity, and lifts them to the world of the pure intellect.[5]

Julian and divination.

Julian believed in almost every form of pagan divination as well as in astrology. To the oracles of Apollo he ascribed the civilizing of the greater part of the world through the foundation of Greek colonies and the revelation of religious and political law.[6] The historian Ammianus Marcellinus [7] tells us that Julian was continually inspecting entrails of victims and interpreting dreams and omens, and that he even proposed to re-open a prophetic fountain whose predictions were supposed to have enabled Hadrian to become emperor, after which that emperor blocked it up from fear that someone else might supplant him through its instrumentality. In another passage [8] he defends Julian from the charge of magic, saying, "Inasmuch as malicious persons have attributed the use of evil arts to learn the future to this ruler who was a learned inquirer into all branches of knowledge, we shall briefly indicate how a wise man is able

[1] I, 482, 498.
[2] I, 405.
[3] I, 374-75.
[4] I, 366-67.
[5] I, 368.
[6] I, 419.
[7] XXII, xii, 8.
[8] XXI, i, 7.

to acquire this by no means trivial variety of learning. The spirit behind all the elements, seeing that it is incessantly and everywhere active in the prophetic movement of perennial bodies, bestows upon us the gift of divination by the different arts which we employ; and the forces of nature, propitiated by varied rites, as from exhaustless springs provide mankind with prophetic utterances."

Scientific divination.

Ammianus thus regards the arts of divination as serious sciences based upon natural forces, although of course in the characteristic Neo-Platonic way of thinking he confuses the spiritual and physical and substitutes propitiatory rites for scientific experiments. His phrase, "the prophetic movement of perennial bodies" almost certainly means the stars and shows his belief in astrology. In another passage [1] he indicates the widespread trust in astrology among the Roman nobles of his time, the later fourth century, by saying that even those "who deny that there are superior powers in the sky," nevertheless think it imprudent to appear in public or dine or bathe without having first consulted an almanac as to the whereabouts of Mercury or the exact position of the moon in Cancer. The passage is satirical, no doubt, but Ammianus probably objects quite as much to their disbelief in superior powers in the sky as he does to the excess of their superstition. That astrology and divination may be studied scientifically he again indicates in a description of learning at Alexandria. Besides praising the medical training to be had there, and mentioning the study of geometry, music, astronomy, and arithmetic, he says, "In addition to these subjects they cultivate the science which reveals the ways of the fates." [2]

Proclus on theurgy.

Iamblichus's account of theurgy is repeated in more condensed form by Proclus (412-485) in a brief treatise or fragment which is extant only in its Latin translation by the Florentine humanist Ficinus, entitled *De sacrificio et magia.*[3] Neither magic nor theurgy, however, is mentioned

[1] XXVIII, iv, 24.

[2] XXII, xvi, 17-18.

[3] Published at Venice (Aldine), 1497, along with the *De mysteriis*, and other works edited or composed by Marsilius Ficinus. See

by name in the Latin text. Proclus states that the priests of old built up their sacred science by observing the sympathy existing between natural objects and by arguing from manifest to occult powers. They saw how things on earth were associated with things in the heavens and further discovered how to bring down divine virtue to this lower world by the force of likeness which binds things together. Proclus gives several examples of plants, stones, and animals which evidence such association. The cock, for instance, is reverenced by the lion because both are under the same planet, the sun, but the cock even more so than the lion. Therefore demons who appear with the heads of lions (*leonina fronte*) vanish suddenly at the sight of a cock unless they chance to be demons of the solar order. After thus indicating the importance of astrology as well as occult virtue in theurgy or magic, Proclus tells how demons are invoked. Sometimes a single herb or stone "suffices for the divine work"; sometimes several substances and rites must be combined "to summon that divinity." When they had secured the presence of the demons, the priests proceeded, partly under the instruction of the demons and partly by their own industrious interpretation of symbols, to a study of the gods. "Finally, leaving behind natural objects and forces and even to a great extent the demons, they won communion with the gods."

Neo Platonic account of magic borrowed by Christians.

Despite the writings of Porphyry and other Neo-Platonists against Christianity, much use was made by Christian theologians of the fourth and fifth centuries of the Neo-Platonic accounts of magic, astrology, and divination, especially of Porphyry's *Letter to Anebo*. Eusebius in his *Praeparatio Evangelica*[1] made large extracts from it on these themes and also from Porphyry's work on the Chaldean oracles. Augustine in *The City of God*[2] accepted Por-

also *Procli Opera*, ed. Cousin, Paris, 1820-1827, III, 278; and Kroll, *Analecta Graeca*, Greisswald, 1901, where a Greek translation accompanies the Latin text.

[1] *Eusebii Caesariensis Opera, Pars II, Apologetica, Praep. Evang.*, IV, 22; V, 6, 8, 10, 12, 14; VI, 1, 4; XIV, 10 (Migne, *Patrologia Graeca*, vol. 21).

[2] X, 9-10.

phyry as an authority on the subjects of theurgy and magic. On the other hand, we do not find the Christian writers repeating the attitude of Plotinus that the life of reason is alone free from magic, except as they substitute the word "Christianity" for "the life of reason."

Neo-Platonists and alchemy.

The Neo-Platonists showed some interest in alchemy as well as in theurgy and astrology. Berthelot published in his *Collection des Alchimistes Grecs* "a little tract of positive chemistry" which is extant under the name of Iamblichus; and Proclus treated of the relations between the metals and planets and the generation of the metals under the influence of the stars.[1] Of Synesius, who was both a Neo-Platonist and a Christian bishop, and who seems to have written works of alchemy, we shall treat in a later chapter.

[1] Berthelot (1889), p. ix.

CHAPTER XII

AELIAN, SOLINUS AND HORAPOLLO

Aelian *On the Nature of Animals*—General character of the work—Its hodge-podge of unclassified detail—Solinus in the middle ages—His date—General character of his work; its relation to Pliny—Animals and gems—Occult medicine—Democritus and Zoroaster not regarded as magicians—Some bits of astrology—Alexander the Great—The *Hieroglyphics* of Horapollo—Marvels of animals—Animals and astrology—The cynocephalus—Horapollo the cosmopolitan.

Aelian *On the Nature of Animals.*

FROM mystic and theurgic compositions we return to works of the declining Roman Empire which deal more directly with nature but, it must be confessed, in a manner somewhat fantastic. About the beginning of the third century, Aelian of Praeneste, who is included by Philostratus in his *Lives of the Sophists,* wrote *On the Nature of Animals.*[1] Its seventeen books, written in Greek, which Aelian used fluently despite his Latin birth, are believed to have reached us partly in interpolated form through two families of manuscripts, of which the older and less interpolated text is found in a thirteenth century manuscript at Paris and a somewhat earlier Vatican codex.[2] A number of its chapters are similar to and perhaps borrowed from Pliny's *Natural History;* at any rate they are commonplaces of ancient science; but the work also has a marked individuality. Parallels have also been noted between this work and the later *Hexaemeron* of the church father Basil. Aelian was much cited in Byzantine literature and learning, and if he was not directly used in the Latin west, at least the attitude

[1] Περὶ ζώων ἰδιότητος. I have used both the *editio princeps* by Gesner, Zurich, 1556, and the critical edition by R. Hercher, Paris, 1858, and Teubner, 1864. The work will henceforth be cited without title in the notes.

[2] See PW, and Christ, *Gesch. d. griech. Litt.,* for further details.

toward animals which he displays and his selection of material concerning them are as apt precursors of medieval Latin as of medieval Greek scientific literature.

General character of the work.

In preface and epilogue Aelian himself adequately indicates the character of his work. He is impressed by the customs and characteristics of animals, and marvels at their wisdom and native shrewdness, their justice and modesty, their affection and piety, which should put human beings to blush. Thus Aelian's work is marked by that tendency which runs through ancient and medieval literature to admire actions in the irrational brutes which seem to indicate almost human intelligence and virtue on their part, and to moralize therefrom at the expense of human beings. Another striking feature of his work is its utterly whimsical and haphazard order. He mentions things simply as they happen to occur to him. This fact, too, he recognizes, but refuses to apologize for, stating that it suits him, if it does not suit anyone else, and that he regards a mixed-up order as more motley, variegated, and pleasing. Not only does he attempt no classification whatever of his animals and mention snakes and quadrupeds and birds in the same breath; he also does not complete the treatment of a given animal in one passage but may scatter detached items about it throughout his work. There is, for instance, probably at least one chapter concerning elephants in each of his seventeen books.

Its hodgepodge of unclassified detail.

It would therefore be absurd for us to attempt any logical arrangement in discussing his contents; we may do justice to him most adequately by adopting his own lack of method and noting a few items and topics taken more or less at random from his work. Ants never go out in the new moon. Yet they neither gaze at the sky, nor count the number of days on their fingers, like the learned Babylonians and Chaldeans, but have this marvelous gift from nature.[1] In sexual intercourse the female viper conceives through the mouth and bites off the head of the male; afterwards her young gnaw their way out of her vitals. "What have your

[1] I, 22.

Oresteses and Alcmaeons to say to that, my dear tragedians?" [1] Doves put laurel boughs in their nests to guard against fascination and the evil eye, and the hoopoe similarly employs ἀδίαυτον or καλλίτριχον as an amulet; [2] and other unreasoning animals guard against sorcery by some mystic and marvelous natural power. Another chapter treats of divinations from the crow and how hairs are dyed black with its eggs.[3] Others tell us of the generation of serpents from the marrow of a dead man's spine,[4] and of venomous women like Medea and Circe who are worse than the asp with its incurable sting, since they kill by mere touch.[5]

We go on to read of swift little beasts called *Pyrigoni* who are generated from fire and live in it, of salamanders who extinguish flames, of the remedies used by the tortoise against snakes, of the chastity of doves whose marriages never result in divorce, and of the incontinence of the partridge.[6] Also of the jealousies of certain animals like the stag which hides its right horn, the lizard who devours its cast-off skin, and the mare who eats the hippomanes from its colt, lest men obtain these precious substances.[7] Of the care taken by storks, herons, and pelicans of their aged parents.[8] How the swallow by the virtue of an herb gives sight to its young who are born blind, and how a hoopoe found an herb whose virtue dissolved the mud with which the caretaker of a building had plugged up the hole in the wall which it used for its nest.[9] How the lion and basilisk fear the cock, and of a lake without fish in a place where the cocks do not crow.[10]

How elephants venerate the waxing moon; how the weasel eats rue when about to fight the snake; and of the jeal-

[1] I, 24.

[2] I, 35. D. W. Thompson, *Glossary of Greek Birds,* p. 57, notes that in the *Birds* of Aristophanes, where the hoopoe appears, "the mysterious root in verse 654 is the magical ἀδίαυτον."

[3] I, 48.

[4] I, 52.

[5] I, 54.

[6] II, 2 and 31; III, 5.

[7] III, 17.

[8] III, 23 and 25.

[9] III, 26; in I, 45, the woodpecker similarly employs the virtue of an herb to remove a stone blocking the entrance to its nest.

[10] III, 32 and 38.

ousy of the hedge-hog and lynx, the latter concealing his precious urine, the other watering his own hide when he is captured in order to spoil it.[1] How the Indians fight griffins when collecting gold.[2] How the presence of a cock aids a woman's delivery.[3] Of unnamed beasts in Libya who know how to count and leave an eleventh part of their prey untouched.[4] That the sea dragon is easily captured with the left hand but not with the right.[5] Dragons know the force of herbs and cure themselves with some and increase their venom with others.[6] How dogs, cows, and other animals sense a famine or plague beforehand.[7] How the Egyptians by their magic charm birds from the sky and snakes from their holes.[8] When it rains in Egypt, mice are born from the small drops and plague the country. Traps and fences and ditches are of no avail against them, as they can leap over trenches and walls. Consequently the Egyptians are forced to pray God to end the calamity,[9]—an interesting variant on the Old Testament account of the plagues of Egypt.

In dogs there exists a certain dialectical faculty of ratiocination.[10] The weather may be predicted from birds, quadrupeds, and flies.[11] The she-goat can cure suffusion of its eyes.[12] Eagles drop tortoises on rocks to break their shells and the bald-headed poet Aeschylus met his death by having his pate mistaken thus for a smooth round stone.[13] Some predict the future by birds, others by entrails, or by grains, sieves, and cheeses; the Lycians practice divination by fish.[14] A stork whom a widow of Tarentum helped when it was too young to fly brought her a luminous precious stone the following year.[15] Solon did not have to enact a law ordering

[1] IV, 10, 14, 17.
[2] IV, 27.
[3] IV, 29.
[4] IV, 53.
[5] V, 37.
[6] VI, 4.
[7] VI, 16.
[8] VI, 33.
[9] VI, 41.
[10] VI, 59.
[11] VII, 7-8.
[12] VII, 14.
[13] VII, 16. The story is also found in Pliny NH, X, 3, where it is added that Aeschylus remained out-doors that day, because an oracle predicted that he would be killed by the fall of a (tortoise's) house.
[14] VIII, 5.
[15] VIII, 22.

children to support their aged parents in the case of lions, whose cubs are taught by nature filial piety toward their elders.[1] Only the horn of the Scythian ass can hold the water of the Arcadian river Styx; Alexander the Great sent a sample of it to Delphi with some accompanying verses which Aelian quotes.[2] In Epirus dragons sacred to Apollo are employed in divination, and in the Lavinian Grove dragons spit out again the frumenty offered them by unchaste virgins.[3] By flying beneath it an eagle saved the life of its young one who had been thrown down from a tower.[4] Different fish eat different sea herbs.[5] There are fish who live in boiling water.[6] There are scattered mentions of the marvels of India throughout Aelian's work, and in his sixteenth book the first fourteen chapters are almost exclusively concerned with the animals of that land.

Solinus in the middle ages.

A well-known work in the middle ages dating from the period of the Roman Empire was the *Collectanea rerum memorabilium* or *Polyhistor* of Solinus. Mommsen's edition lists 153 manuscripts from 32 places,[7] and we shall find many citations of Solinus in our later medieval authors. Martianus Capella and Isidore were the first to make extensive use of his work. In the thirteenth century Albertus Magnus had little respect for Solinus as an authority and expressed more than once the quite accurate opinion that his work was full of lies. Nevertheless copies of it continued to abound in the fourteenth and fifteenth centuries, and by 1554 five printed editions had appeared. "From it directly come most of the fables in works of object so different as those of Dicuil, Isidore, Capella, and Priscian."[8]

His date.

The first extant author to make use of Solinus is Augustine in *The City of God*, while he is first named in the *Genealogus* of 455 A. D. None of the manuscripts of the work

[1] IX, 1.
[2] X, 40.
[3] XI, 2 and 16.
[4] XII, 21.
[5] XIII, 3.
[6] XIV, 19.
[7] *C. Iulii Solini Collectanea rerum memorabilium iterum recensuit* Th. Mommsen, Berlin, 1895, pp. xxxi-li. Beazley, *Dawn of Modern Geography*, I, 520-2, lists 152 MSS.
[8] Beazley, *Dawn of Modern Geography*. I, 247.

antedate the ninth century, but many of them have copied an earlier subscription from a manuscript written "by the zeal and diligence of our lord Theodosius, the unconquered prince." This is taken to refer to the emperor Theodosius II, 401-450. The work itself, however, has no Christian characteristics; on the contrary it is very fond of mentioning places famed in pagan religion and Greek mythology and of recounting miracles and marvels connected with heathen shrines and rites. Indeed, Solinus seldom, if ever, mentions anything later than the first century of our era. He speaks of Byzantium, not of Constantinople, and makes no mention of the Roman provinces as divided in the system of Diocletian. His book, however, is a compilation from earlier writings so that we need not expect allusions to his own age. The Latin style and general literary make-up of the work are characteristic of the declining empire and early medieval period. Mommsen was inclined to date Solinus in the third rather than the fourth century, but the work seems to have been revised about the sixth century, after which date it became customary to call it the *Polyhistor* rather than the *Collectanea rerum memorabilium.* It is also referred to, however, as *De mirabilibus mundi,* or *Wonders of the World.*

General character of his work: its relation to Pliny.

The work is primarily a geography and is arranged by countries and places, beginning with Rome and Italy. As each locality is considered, Solinus sometimes tells a little of its history, but is especially inclined to recount miraculous religious events or natural marvels associated with that particular region. Thus in describing two lakes he rather apologizes for mentioning the first at all because it can scarcely be called miraculous, but assures us that the second "is regarded as very extraordinary." [1] Sometimes he digresses to other topics such as calendar reform.[2] Solinus draws both his geographical data and further details very largely from Pliny's *Natural History;* but inasmuch as Pliny treated of these matters in separate books, Solinus has

[1] Mommsen (1895), p. 48. [2] *Ibid.,* p. 7.

to re-organize the material. He also selects simply a few particulars from Pliny's wealth of detail on any given subject, and furthermore considerably alters Pliny's wording, sometimes condensing the thought, sometimes amplifying the phraseology—apparently in an effort to make the point clearer and easier reading. Of Pliny's thirty-seven books only those from the third to the thirteenth inclusive and the last book are used to any extent by Solinus. That is to say, he either was acquainted with only, or confined himself to, those books dealing with geography, man and other animals, and gems, omitting almost entirely, except for the twelfth and thirteenth books, Pliny's elaborate treatment of vegetation and of medicinal simples [1] and discussion of metals and the fine arts. Solinus does not acknowledge his great debt to Pliny in particular, although he keeps alluding to the fulness with which everything has already been discussed by past authors, and although he cites other writers who are almost unknown to us. Of his known sources Pomponius Mela is the chief after Pliny but is used much less. On the other hand, the number of passages for which Mommsen was unable to give any source is not inconsiderable. As may have been already inferred, the work of Solinus is brief; the text alone would scarcely fill one hundred pages.[2]

Animals and gems.

It would perhaps be rash to conjecture which quality commended the book most to the following period: its handy size, or its easy style and fairly systematic arrangement, or its emphasis upon marvels. The last characteristic is at least the most germane to our investigation. Solinus rendered the service, if we may so term it, of reducing Pliny's treatment of animals and precious stones in particular to a few common examples, which either were already the best known or became so as a result of his selection. Indeed, King was of the opinion that the descriptions of gems in Solinus were more precise, technical, and systematic than

[1] Yet one medieval MS of Solinus is described as *De variarum herbarum et radicum qualitate et virtute medica;* Vienna 3959, 15th century, fols. 156-74.

[2] In Mommsen's edition critical apparatus occupies more than one-half of the 216 pages.

those in Pliny, and found his notices "often extremely useful." [1] Solinus describes such animals as the wolf, lynx, bear, lion, hyena, *onager* or wild ass, basilisk, crocodile, hippopotamus, phoenix, dolphin, and chameleon; and recounts the marvelous properties of such gems as *achates* or agate, *galactites, catochites,* crystal, *gagates,* adamant, heliotrope, hyacinth, and *paeanites.* The dragons of India and Ethiopia also occupy his attention, as they did that of Philostratus in the *Life of Apollonius of Tyana;* indeed, he repeats in different words the statement found in Philostratus that they swim far out to sea.[2] In Sardinia, on the contrary, there are no snakes, but a poisonous ant exists there. Fortunately there are also healing waters there with which to counteract its venom, but there is also native to Sardinia an herb called *Sardonia* which causes those who eat it to die of laughter.[3]

Occult medicine.

Although Solinus makes no use of Pliny's medical books, he shows considerable interest in the healing properties of simples and in medicine. He tells us that those who slept in the shrine of Aesculapius at Epidaurus were warned in dreams how to heal their diseases,[4] and that the third daughter of Aeetes, named Angitia, devoted herself "to resisting disease by the salubrious science" of medicine.[5] According to Solinus Circe as well as Medea was a daughter of Aeetes, but usually in Greek mythology she is represented as his sister.

[1] C. W. King, *The Natural History, Ancient and Modern, of Precious Stones and Gems,* London, 1865, p. 6.

[2] Mommsen (1895), pp. 132, 188.

[3] *Ibid.,* 46-7. Mommsen could give no source for these statements concerning Sardinia, and they do not appear to be in Pliny. But it is from a footnote in the English translation of the *Natural History* by Bostock and Riley (II, 208, citing Dalechamps, and Lemaire, III, 201) that I learn that the laughter which Pliny (NH, VII, 52) speaks of as a premonitory sign of death in cases of madness, "is not the indication of mirth, but what has been termed the *risus Sardonicus,* the 'Sardonic laugh,' produced by a convulsive action of the muscles of the face." This form of death may be what Solinus has in mind. Agricola in his work on metallurgy and mines still believes in the poisonous ants of Sardinia; *De re metallica,* VI, near close, pp. 216-7, in Hoover's translation, 1912.

[4] Mommsen (1895), p. 57.

[5] *Ibid.,* p. 39.

Democritus and Zoroaster not regarded as magicians.

This allusion to Circe and Medea shows that magic, to which medicine and pharmacy are apparently akin, does not pass unnoticed in Solinus's page. He copies from Mela the account of the periodical transformation of the *Neuri* into wolves.[1] But instead of accusing Democritus of having employed magic, as Pliny does, Solinus represents him as engaging in contests with the *Magi,* in which he made frequent use of the stone *catochites* in order to demonstrate the occult power of nature.[2] That is to say, Democritus was apparently opposing science to magic and showing that all the latter's feats could be duplicated or improved upon by employing natural forces. In two other passages[3] Solinus calls Democritus *physicus,* or scientist, and affirms that his birth in Abdera did more to make that town famous than any other thing connected with it, despite the fact that it was founded by and named after the sister of Diomedes. Zoroaster, too, whom Pliny called the founder of the magic art, is not spoken of as a magician by Solinus, although he is mentioned three times and is described as "most skilled in the best arts," and is cited concerning the power of coral and of the gem *aetites.*[4]

Some bits of astrology.

It is not part of Solinus's plan to describe the heavens, but he occasionally alludes to "the discipline of the stars,"[5] as he calls astronomy or astrology. On the authority of L. Tarrutius, "most renowned of astrologers,"[6] he tells us that the foundations of the walls of Rome were laid by Romulus in his twenty-second year on the eleventh day of the kalends of May between the second and third hours, when Jupiter was in Pisces, the sun in Taurus, the moon in Libra, and the other four planets in the sign of the scorpion. He also

[1] Mommsen (1895), p. 82.

[2] *Ibid.*, pp. 45-46.

[3] *Ioid.*, pp. 13, 68.

[4] *Ibid.*, pp. 18, 41, 159.

[5] *Ibid.*, p. 50, and elsewhere, "siderum disciplinam."

[6] *Ibid.*, p. 5, "mathematicorum nobilissimus." Solinus probably takes this from Varro, who, as Plutarch informs us in his *Life of Romulus,* asked "Tarrutius, his familiar acquaintance, a good philosopher and mathematician," to calculate the horoscope of Romulus. See above, p. 209.

speaks of the star Arcturus destroying the Argive fleet off Euboea on its return from Ilium.[1]

Alexander the Great.

Alexander the Great figures prominently in the pages of Solinus, being mentioned a score of times, and this too corresponds to the medieval interest in the Macedonian conqueror. Stories concerning him are repeated from Pliny, but Solinus also displays further information. He insists that Philip was truly his father, although he adds that Olympias strove to acquire a nobler father for him, when she affirmed that she had had intercourse with a dragon, and that Alexander tried to have himself considered of divine descent.[2] The statement concerning Olympias suggests the story of Nectanebus, of which a later chapter will treat, but that individual is not mentioned, although Aristotle and Callisthenes are spoken of as Alexander's tutors, so that it is doubtful if Solinus was acquainted with the *Pseudo-Callisthenes.* He describes Alexander's line of march with fair accuracy and not in the totally incorrect manner of the *Pseudo-Callisthenes.*

The *Hieroglyphics* of Horapollo.

In seeking a third text and author of the same type as Aelian and Solinus to round out the present chapter, our choice unhesitatingly falls upon the *Hieroglyphics* of Horapollo, a work which pretends to explain the meaning of the written symbols employed by the ancient Egyptian priests, but which is really principally concerned with the same marvelous habits and properties of animals of which Aelian treated. In brief the idea is that these characteristics of animals must be known in order to comprehend the significance of the animal figures in the ancient hieroglyphic writing. Horapollo is supposed to have written in the Egyptian language in perhaps the fourth or fifth century of our era,[3] but his work is extant only in the Greek translation of it made by a Philip who lived a century or two later and who seems to have made some additions of his own.[4]

[1] Mommsen (1905), pp. 75-6.

[2] *Ibid.*, p. 66.

[3] PW, for the problem of his identity and further bibliography.

[4] I have used the text and English translation of A. T. Cory, *The Hieroglyphics of Horapollo Nilous*, 1840. Philip's Greek is so

Marvels of animals.

The zoology of Horapollo is for the most part not novel, but repeats the same erroneous notions that may be found in Aristotle's *History of Animals,* Pliny's *Natural History,* Aelian, and other ancient authors. Again we hear of the basilisk's fatal breath, of the beaver's discarded testicles, of the unnatural methods of conception of the weasel and viper, of the bear's licking its cubs into shape, of the kindness of storks to their parents, of wasps generated from a dead horse, of the phoenix, of the swan's song, of the sick lion's eating an ape to cure himself, of the bull tamed by tying it to the branch of a wild fig tree, of the elephant's fear of a ram or a dog and how it buries its tusks.[1] Less familiar perhaps are the assertions that the mare miscarries, if she merely treads on a wolf's tracks;[2] that the pigeon cures itself by placing laurel in its nest;[3] that putting the wings of a bat on an ant-hill will prevent the ants from coming out.[4] The statement that if the hyena, when hunted, turns to the right, it will slay its pursuer, while if it turns to the left, it will be slain by him, is also found in Pliny.[5] But his long enumeration of virtues ascribed to parts of the hyena by the *Magi* does not include the assertion in Horapollo's next chapter[6] that a man girded with a hyena skin can pass through the ranks of his enemies without injury, although it ascribes somewhat similar virtues to the animal's skin. In Horapollo it is the hawk rather than the eagle which surpasses other winged creatures in its ability to gaze at the sun; hence physicians use the hawk-weed in eye-cures.[7]

bad that some would date it in the fourteenth or fifteenth century. The oldest extant Greek codex was purchased in Andros in 1419. The work was translated into Latin by the fifteenth century at latest; see Vienna 3255, 15th century, 82 fols., Horapollo, Hieroglyphicon latine versorum liber I et libri II introductio cum figuris calamo exaratis et coloratis.

[1] I, 1; II, 61; II, 65; II, 36 and 59; II, 57; II, 83; I, 34-5; II, 57; II, 44 and 39 and 76-7 and 85-6 and 88.

[2] II, 45.

[3] II, 46; Aelian says the same, however, as we stated above.

[4] II, 64.

[5] NH, XXVIII, 27.

[6] II, 72.

[7] I, 6. According to Pliny (NH, XX, 26), the hawk sprinkles its eyes with the juice of this herb; Apuleius (*Metamorphoses,* cap. 30) says that the eagle does so.

Animals and astrology.

Animals also serve as astronomical or astrological symbols in the system of hieroglyphic writing as interpreted by Horapollo. Not only does a palm tree represent the year because it puts forth a new branch every new moon,[1] but the phoenix denotes the *magnus annus* in the course of which the heavenly bodies complete their revolutions.[2] The scarab rolls his ball of dung from east to west and gives it the shape of the universe.[3] He buries it for twenty-eight days conformably to the course of the moon through the zodiac, but he has thirty toes to correspond to the days of the month. As there is no female scarab, so there is no male vulture. The female vulture symbolizes the Egyptian year by spending five days in conceiving by the wind, one hundred and twenty in pregnancy, the same period in rearing its young, and the remaining one hundred and twenty days in preparing itself to repeat the process.[4] The vulture also visits battlefields seven days in advance and by the direction of its glance indicates which army will be defeated.

The cynocephalus.

The cynocephalus, dog-headed ape, or baboon, was mentioned several times by Pliny, but Horapollo gives more specific information concerning it, chiefly of an astrological character. It is born circumcised and is reared in temples in order to learn from it the exact hour of lunar eclipses, at which times it neither sees nor eats, while the female *ex genitalibus sanguinem emittit.* The cynocephalus represents the inhabitable world which has seventy-two primitive parts, because the animal dies and is buried piecemeal by the priests during a period of as many days, until at the end of the seventy-second day life has entirely departed from the last remnant of its carcass.[5] The cynocephalus not only marks the time of eclipses but at the equinoxes makes water twelve times by day and by night, marking off the hours; hence a figure of it is carved by the Egyptians on their water-clocks.[6] Horapollo associates together the god of the universe and fate and the stars which are five in number, for he believes

[1] I, 3.
[2] II, 57.
[3] I, 10.
[4] I, 11.
[5] I, 14.
[6] I, 16.

that five planets carry out the economy of the universe and that they are subject to God's government.[1]

Horapollo the cosmopolitan.

Horapollo cannot be given high rank either as a zoologist and astronomer, or a philologer and archaeologist; but at least he was no narrow nationalist and had some respect for history. The Egyptians, he says, "denote a man who has never left his own country by a human figure with the head of an ass, because he neither hears any history nor knows of what is going on abroad." [2]

[1] I, 13.

[2] I, 23.

BOOK II. EARLY CHRISTIAN THOUGHT

BOOK II. EARLY CHRISTIAN THOUGHT

FOREWORD

WE now turn back chronologically to the point from which we started in our survey of classical science and magic in order to trace the development of Christian thought in regard to the same subjects. How far did Christianity break with ancient science and superstition? To what extent did it borrow from them?

Magic and religion.

It has often been remarked that, as a new religion comes to prevail in a society, the old rites are discredited and prohibited as magic. The faith and ceremonies of the majority, performed publicly, are called religion: the discarded cult, now practiced only privately and covertly by a minority, is stigmatized as magic and contrary to the general good. Thus we shall hear Christian writers condemn the pagan oracles and auguries as arts of divination, and classify the ancient gods as demons of the same sort as those invoked in the magic arts. Conversely, when a new religion is being introduced, is as yet regarded as a foreign faith, and is still only the private worship of a minority, the majority regard it as outlandish magic. And this we shall find illustrated by the accusations of sorcery and magic heaped upon Jesus by the Jews, and upon the Jews and the early Christians by a world long accustomed to pagan rites. The same bandying back and forth of the charge of magic occurred between Mohammed and the Meccans.[1]

Relation between early Christian and medieval literature.

It is perhaps generally assumed that the men of the middle ages were widely read in and deeply influenced by the fathers of the early church, but at least for our subject this influence has hardly been treated either broadly or

[1] Sir William Muir, "Ancient Arabic Poetry, its Genuineness and Authenticity," in *Royal Asiatic Society's Journal* (1882), p. 30.

in detail. Indeed, the predilection of the humanists of the fifteenth and sixteenth centuries for anything written in Greek and their aversion to medieval Latin has too long operated as a bar to the study of medieval literature in general. And scholars who have edited or studied the Greek, Syriac, and other ancient texts connected with early Christianity have perhaps too often neglected the Latin versions preserved in medieval manuscripts, or, while treasuring up every hint that Photius lets fall, have failed to note the citations and allusions in medieval Latin encyclopedists. Yet it is often the case that the manuscripts containing the Latin versions are of earlier date than those which seem to preserve the Greek original text.

Method of presenting early Christian thought.

There is so much repetition and resemblance between the numerous Christian writers in Greek and Latin of the Roman Empire that I have even less than in the case of their classical contemporaries attempted a complete presentation of them, but, while not intending to omit any account of the first importance in the history of magic or experimental science, have aimed to make a selection of representative persons and typical passages. At the same time, in the case of those authors and works which are discussed, the aim is to present their thought in sufficiently specific detail to enable the reader to estimate for himself their scientific or superstitious character and their relations to classical thought on the one hand and medieval thought on the other.

Before we treat of Christian writings themselves it is essential to notice some related lines of thought and groups of writings which either preceded or accompanied the development of Christian thought and literature, and which either influenced even orthodox thought powerfully, or illustrate foreign elements, aberrations, side-currents, and undertows which none the less cannot be disregarded in tracing the main current of Christian belief. We therefore shall successively treat of the literature extant under the name of Enoch, of the works of Philo Judaeus, of the doctrines of the Gnostics, of the Christian *Apocrypha,* of the *Pseudo-*

Clementines and Simon Magus, and of the *Confession* of Cyprian and some similar stories. We shall then make Origen's *Reply to Celsus,* in which the conflict of classical and Christian conceptions is well illustrated, our point of departure in an examination of the attitude of the early fathers towards magic and science. Succeeding chapters will treat of the attitude toward magic of other fathers before Augustine, of Christianity and natural science as shown in Basil's *Hexaemeron,* Epiphanius' *Panarion,* and the *Physiologus,* and of Augustine himself. A final chapter on the fusion of paganism and Christianity in the fourth and fifth centuries will terminate this second division of our investigation and also serve as a supplement to the preceding division and an introduction to the third book on the early middle ages. Our arrangement is thus in part topical rather than strictly chronological. The dates of many authors and works are too dubious, there is too much of the apocryphal and interpolated, and we have to rely too much upon later writers for the views of earlier ones, to make a strictly or even primarily chronological arrangement either advisable or feasible.

CHAPTER XIII

THE BOOK OF ENOCH

Enoch's reputation as an astrologer in the middle ages—Date and influence of the literature ascribed to Enoch—Angels governing the universe; stars and angels—The fallen angels teach men magic and other arts—The stars as sinners—Effect of sin upon nature—Celestial phenomena—Mountains and metals—Strange animals.

Enoch's reputation as an astrologer in the middle ages.

IN collections of medieval manuscripts there often is found a treatise on fifteen stars, fifteen herbs, fifteen stones, and fifteen figures engraved upon them, which is attributed sometimes to Hermes, presumably Trismegistus, and sometimes to Enoch, the patriarch, who "walked with God and was not."[1] Indeed in the prologue to a Hermetic work on astrology in a medieval manuscript we are told that Enoch and the first of the three Hermeses or Mercuries are identical.[2] This

[1] Ascribed to Enoch in Harleian MS 1612, fol. 15r, Incipit: "Enoch tanquam unus ex philosophis super res quartum librum edidit, in quo voluit determinare ista quatuor: videlicet de xv stellis, de xv herbis, de xv lapidibus preciosis et de xv figuris ipsis lapidibus sculpendis," and Wolfenbüttel 2725, 14th century, fols. 83-94v; BN 13014, 14th century, fol. 174v; Amplon, Quarto 381 (Erfurt), 14th century, fols. 42-45: for "Enoch's prayer" see Sloane MS 3821, 17th century, fols. 190v-193.

Ascribed to Hermes in Harleian 80, Sloane 3847, Royal 12-C-XVIII; Berlin 963, fol. 105; Vienna 5216, 15th century, fols. 63r-66v; "Dixit Enoch quod 15 sunt stelle / ex tractatu Heremeth (i. e. Hermes) et enoch compilatum"; and in the Catalogue of Amplonius (1412 A.D.), Math. 53. See below, II, 220-21.

The stars are probably fifteen in number because Ptolemy distinguished that many stars of first magnitude. Dante, *Paradiso,* XIII, 4, also speaks of "quindici stelle." See Orr (1913), pp. 154-6, where Ptolemy's descriptions of the fifteen stars of first magnitude and their modern names are given.

[2] Digby 67, late 12th century, fol. 69r, "Prologus de tribus Mercuriis." They are also identified by other medieval writers. Some would further identify with Enoch Nannacus or Annacus, king of Phrygia, who foresaw Deucalion's flood and lamented. See J. G. Frazer (1918), I, 155-6, and P. Buttmann, *Mythologus,* Berlin, 1828-1829, and E. Babelon, *La tradition phrygienne du déluge,* in *Rev. d. l'hist. d. religs.,* XXIII (1891), which he cites.

Roger Bacon stated that some would identify Enoch with "the

treatise probably has no direct relation to the *Book of Enoch,* which we shall discuss in this chapter and which was composed in the pre-Christian period. But it is interesting to observe that the same reputation for astrology, which led the middle ages sometimes to ascribe this treatise to Enoch, is likewise found in "the first notice of a book of Enoch," which "appears to be due to a Jewish or Samaritan Hellenist," which "has come down to us successively through Alexander Polyhistor and Eusebius," and which states that Enoch was the founder of astrology.[1] The statement in Genesis that Enoch lived three hundred and sixty-five years would also lead men to associate him with the solar year and stars.

Date and influence of the literature ascribed to Enoch.

The *Book of Enoch* is "the precipitate of a literature, once very active, which revolved . . . round Enoch," and in the form which has come down to us is a patchwork from "several originally independent books." [2] It is extant in the form of Greek fragments preserved in the *Chronography* of G. Syncellus,[3] or but lately discovered in (Upper) Egypt, and in more complete but also more recent manuscripts giving an Ethiopic and a Slavonic version.[4] These last two versions are quite different both in language and content, while some of the citations of Enoch in ancient writers apply to neither of these versions. While "Ethiopic did not exist as a literary language before 350 A. D.," [5] and none

great Hermogenes, whom the Greeks much commend and laud, and they ascribe to him all secret and celestial science." Steele (1920) 99.

[1] R. H. Charles, *The Book of Enoch,* Oxford, 1893, p. 33, citing Euseb. *Praep. Evan.,* ix, 17, 8 (Gaisford).

[2] Charles (1893), p. 10, citing Ewald.

[3] ed. Dindorf, 1829.

[4] Lods, Ad. *Le Livre d'Hénoch, Fragments grecs découverts à Akhmin,* Paris, 1892.

Charles, R. H., *The Book of Enoch,* Oxford, 1893, "translated from Professor Dillman's Ethiopic text, amended and revised in accordance with hitherto uncollated Ethiopic manuscripts and with the Gizeh and other Greek and Latin fragments, which are here published in full." *The Book of Enoch, translated anew,* etc., Oxford, 1912. Also translated in Charles (1913) II, 163-281. There are twenty-nine Ethiopic MSS of Enoch.

Charles, R. H., and Morfill, W. R., *The Book of the Secrets of Enoch,* translated from the Slavonic, Oxford, 1896. Also by Forbes and Charles in Charles (1913) II, 425-69.

[5] Charles (1893), p. 22.

of the extant manuscripts of the Ethiopic version is earlier than the fifteenth century,[1] Charles believes that they are based upon a Greek translation of the Hebrew and Aramaic original, and that even the interpolations in this were made by an editor living before the Christian era. He asserts that "nearly all the writers of the New Testament were familiar with it," and influenced by it,—in fact that its influence on the New Testament was greater than that of all the other apocrypha together, and that it "had all the weight of a canonical book" with the early church fathers.[2] After 300 A. D., however, it became discredited, except as we have seen among Ethiopic and Slavonic Christians. Before 300 Origen in his *Reply to Celsus*[3] accuses his opponent of quoting the *Book of Enoch* as a Christian authority concerning the fallen angels. Origen objects that "the books which bear the name Enoch do not at all circulate in the Churches as divine." Augustine, in the *City of God*,[4] written between 413 and 426, admits that Enoch "left some divine writings, for this is asserted by the Apostle Jude in his canonical epistle." But he doubts if any of the writings current in his own day are genuine and thinks that they have been wisely excluded from the course of Scripture. Lods writes that after the ninth century in the east and from a much earlier date in the west, the *Book of Enoch* is not mentioned, "At the most some medieval rabbis seem still to know of it."[5] Yet Alexander Neckam, in the twelfth century, speaks as if Latin Christendom of that date had some acquaintance with the Enoch literature. We shall note some passages in Saint Hildegard which seem parallel to others in the *Book of Enoch*, while Vincent of Beauvais in his *Speculum naturale* in the thirteenth century, in justifying a certain discriminating use of the apocryphal books, points out that Jude quotes Enoch whose book is now called apocryphal.[6]

[1] Charles (1913), II, 165-6.
[2] Charles (1893), pp. 2 and 41.
[3] V., 54.
[4] XV, 23.
[5] Introd., vi.
[6] *Spec. Nat.*, I, 9. A Latin fragment, found in the British Museum in 1893 by Dr. M. R. James and

Angels governing the universe: stars and angels.

The Enoch literature has much to say concerning angels, and implies their control of nature, man, and the future. We hear of Raphael, "who is set over all the diseases and wounds of the children of men"; Gabriel, "who is set over all the powers"; Phanuel, "who is set over the repentance and hope of those who inherit eternal life."[1] The revolution of the stars is described as "according to the number of the angels," and in the Slavonic version the number of those angels is stated as two hundred.[2] Indeed the stars themselves are often personified and we read "how they keep faith with each other" and even of "all the stars whose privy members are like those of horses."[3] The Ethiopic version also speaks of the angels or spirits of hoar-frost, dew, hail, snow and so forth.[4] In the Slavonic version Enoch finds in the sixth heaven the angels who attend to the phases of the moon and the revolutions of stars and sun and who superintend the good or evil condition of the world. He finds angels set over the years and seasons, the rivers and sea, the fruits of the earth, and even an angel over every herb.[5]

The fallen angels teach men magic and other arts.

The fallen angels in particular are mentioned in the *Book of Enoch.* Two hundred angels lusted after the comely daughters of men and bound themselves by oaths to marry them.[6] After having thus taken unto themselves wives, they instructed the human race in the art of magic and the science of botany—or to be more exact, "charms and enchantments" and "the cutting of roots and of woods." In another chapter various individual angels are named who taught respectively the enchanters and botanists, the breaking of charms, astrology, and various branches thereof.[7] In the Greek fragment preserved by Syncellus there are further mentioned pharmacy, and what probably denote geomancy ("sign of

published in the Cambridge *Texts and Studies,* II, 3, *Apocrypha Anecdota,* pp. 146-50, "seems to point to a Latin translation of Enoch"—Charles (1913) II, 167.

[1] *Book of Enoch,* XL, 9.

[2] *Ibid.,* XLIII; *Secrets of Enoch,* IV.

[3] *Book of Enoch,* XLIII; XC, 21.

[4] *Ibid.,* LX, 17-18.

[5] *Secrets of Enoch,* XIX.

[6] Caps. VI-XI in both Lods and Charles.

[7] *Book of Enoch,* VIII, 3, in both Charles and Lods.

the earth") and aeromancy (*aeroskopia*). Through this revelation of mysteries which should have been kept hid we are told that men "know all the secrets of the angels, and all the violence of the Satans, and all their occult power, and all the power of those who practice sorcery, and the power of witchcraft, and the power of those who make molten images for the whole earth." [1] The revelation included, moreover, not only magic arts, witchcraft, divination, and astrology, but also natural sciences, such as botany and pharmacy—which, however, are apparently regarded as closely akin to magic—and useful arts such as mining metals, manufacturing armor and weapons, and "writing with ink and paper"—"and thereby many sinned from eternity to eternity and until this day." [2] As the preceding remark indicates, the author is decidedly of the opinion that men were not created to the end that they should write with pen and ink. "For man was created exactly like the angels to the intent that he should continue righteous and pure, . . . but through this their knowledge men are perishing." [3] Perhaps the writer means to censure writing as magical and thinks of it only as mystic signs and characters. Magic is always regarded as evil in the Enoch literature, and witchcraft, enchantments, and "devilish magic" are given a prominent place in a list in the Slavonic version [4] of evil deeds done upon earth.

The stars as sinners.

In connection with the fallen angels we find the stars regarded as capable of sin as well as personified. In the Ethiopic version there is more than one mention of seven stars that transgressed the command of God and are bound against the day of judgment or for the space of ten thousand years.[5] One passage tells how "judgment was held first over the stars, and they were judged and found guilty, and went to the place of condemnation, and they were cast into an abyss." [6] A similar identification of the stars with the fallen angels is found in one of the visions of Saint

[1] *Book of Enoch,* LXV, 6.
[2] *Ibid.,* LXV, 7-8; LXIX, 6-9.
[3] *Ibid.,* LXIX, 10-11.
[4] *Secrets of Enoch,* X.
[5] *Book of Enoch,* XVIII, XXI.
[6] *Ibid.,* XC, 24.

Hildegard in the twelfth century. She writes, "I saw a great star most splendid and beautiful, and with it an exceeding multitude of falling sparks which with the star followed southward. And they examined Him upon His throne almost as something hostile, and turning from Him, they sought rather the north. And suddenly they were all annihilated, being turned into black coals . . . and cast into the abyss that I could see them no more."[1] She then interprets the vision as signifying the fall of the angels.

Effect of sin upon nature.

An idea which we shall find a number of times in other ancient and medieval writers appears also in the *Book of Enoch*. It is that human sin upsets the world of nature, and in this particular case, even the period of the moon and the orbits of the stars.[2] Hildegard again roughly parallels the Enoch literature by holding that the original harmony of the four elements upon this earth was changed into a confused and disorderly mixture after the fall of man.[3]

Celestial phenomena

The natural world, although intimately associated with the spiritual world and hardly distinguished from it in the Enoch literature, receives considerable attention, and much of the discussion in both the Ethiopic and Slavonic versions is of a scientific rather than ethical or apocalyptic character. One section of the Ethiopic version is described by Charles[4] as the *Book of Celestial Physics* and upholds a calendar based upon the lunar year. The Slavonic version, on the other hand, while mentioning the lunar year of 354 days and the solar year of 365 and ¼ days, seems to prefer the latter, since the years of Enoch's life are given as 365, and he writes 366 books concerning what he has seen in his visions and voyages.[5] The *Book of Enoch* supposes a plurality of heavens.[6] In the Slavonic version Enoch is

[1] Singer's translation. *Studies in the History and Method of Science*, Vol. I, p. 53, of *Scivias*, III, 1, in Migne, PL, 197, 565. See also the Koran XV, 18.

[2] Charles, p. 32 and cap. LXXX.

[3] Singer, 25-26.

[4] Pp. 187-219.

[5] *Secrets of Enoch*, I and XXX.

[6] See Morfill-Charles, pp. xxxiv-xxxv, for mention of three and seven heavens in the apocryphal *Testaments of the Twelve Patriarchs*, "written about or before the beginning of the Christian era," and for "the probability of an Old Testament belief in the plurality of the heavens." For the

taken through the seven heavens, or ten heavens in one manuscript, with the signs of the zodiac in the eighth and ninth. An account is also given of the creation, and the waters above the firmament, which were to give the early Christian apologists and medieval clerical scientists so much difficulty, are described as follows: "And thus I made firm the waters, that is, the depths, and I surrounded the waters with light, and I created seven circles, and I fashioned them like crystal, moist and dry, that is to say, like glass and ice, and as for the waters and also the other elements I showed each of them their paths, (viz.) to the seven stars, each of them in their heaven, how they should go." [1] The order of the seven planets in their circles is given as follows: in the first and highest circle the star Kruno, then Aphrodite or Venus, Ares (Mars), the sun, Zeus (Jupiter), Hermes (Mercury), and the moon.[2] God also tells Enoch that the duration of the world will be for a week of years, that is, seven thousand, after which "let there be at the beginning of the eighth thousand a time when there is no computation and no end; neither years nor months nor weeks nor days nor hours." [3]

Mountains and metals.

Turning from celestial physics to terrestrial phenomena, we may note a few allusions to minerals, vegetation, and animals. "Seven mountains of magnificent stones" are more than once mentioned in the Ethiopic version and are described as each different from the other.[4] Another passage speaks of "seven mountains full of choice nard and aromatic trees and cinnamon and pepper." [5] But whether

seven heavens in the apocryphal *Ascension of Isaiah* see Charles' edition of that work (1900), xlix.

[1] *Secrets of Enoch,* XXVII. Charles prefaces this passage by the remark, "I do not pretend to understand what follows": but it seems clear that the waters above the firmament are referred to from what the author goes on to say, "And thus I made firm the circles of the heavens, and caused the waters below which are under the heavens to be gathered into one place." It would also seem that each of the seven planets is represented as moving in a sphere of crystal. In the Ethiopic version, LIV, 8, we are told that the water above the heavens is masculine, and that the water beneath the earth is feminine; also LX, 7-8, that Leviathan is female and Behemoth male.

[2] *Secrets of Enoch,* XXX.

[3] *Ibid.,* 45-46, see also the Ethiopic *Book of Enoch,* XCIII, for "seven weeks."

[4] *Book of Enoch,* XVIII, XXIV.

[5] *Ibid.,* XXXII.

these groups of seven mountains are to be astrologically related to the seven planets is not definitely stated. We are also left in doubt whether the following passage may have some astrological or even alchemical significance, or whether it is merely a figurative prophecy like that in the Book of Daniel concerning the image seen by Nebuchadnezzar in his dream. "There mine eyes saw all the hidden things of heaven that shall be, an iron mountain, and one of copper, and one of silver, and one of gold, and one of soft metal, and one of lead." [1] At any rate Enoch has come very near to listing the seven metals usually associated with the seven planets. In another passage we are informed that while silver and "soft metal" come from the earth, lead and tin are produced by a fountain in which an eminent angel stands.[2]

Strange animals.

As for animals we are informed that Behemoth is male and Leviathan female.[3] When Enoch went to the ends of the earth he saw there great beasts and birds who differed in appearance, beauty, and voice.[4] In the Slavonic version we hear a good deal of phoenixes and *chalkydri,* who seem to be flying dragons. These creatures are described as "strange in appearance with the feet and tails of lions and the heads of crocodiles. Their appearance was of a purple color like the rainbow; their size, nine hundred measures. Their wings were like those of angels, each with twelve, and they attend the chariot of the sun, and go with him, bringing heat and dew as they are ordered by God." [5]

[1] *Book of Enoch,* LII, 2.
[2] *Ibid.,* LXV, 7-8.
[3] *Ibid.,* LX, 7.
[4] *Ibid.,* XXXIII.
[5] *Secrets of Enoch,* XII, XV, XIX.

CHAPTER XIV

PHILO JUDAEUS

Bibliographical note—Philo the mediator between Hellenistic and Jewish-Christian thought—His influence upon the middle ages was indirect—Good and bad magic—Stars not gods nor first causes—But rational and virtuous animals, and God's viceroys over inferiors—They do not cause evil; but it is possible to predict the future from their motions—Jewish astrology—Perfection of the number seven—And of fifty—Also of four and six—Spirits of the air—Interpretation of dreams—Politics are akin to magic—A thought repeated by Moses Maimonides and Albertus Magnus.

"But since every city in which laws are properly established has a regular constitution, it became necessary for this citizen of the world to adopt the same constitution as that which prevailed in the universal world. And this constitution is the right reason of nature."

—*On Creation,* cap. 50.

Philo the mediator between Hellenistic and Jewish-Christian thought.

THERE probably is no other man who marks so well the fusion of Hellenic and Hebrew ideas and the transition from them to Christian thought as Philo Judaeus.[1] He flourished at Alexandria in the first years of our era—the exact dates both of his birth and of his death are uncertain—and speaks of himself as an old man at the time of

[1] The literature dealing in general with Philo and his philosophy is too extensive to indicate here, while there has been no study primarily devoted to our interest in him. It may be useful to note, however, the most recent editions of his works and studies concerning him, from which the reader can learn of earlier researches. See also Leopold Cohn, *The Latest Researches on Philo of Alexandria* (Reprinted from *The Jewish Quarterly Review*), London, 1892. The most recent edition of the Greek text of Philo's works is by L. Cohn and P. Wendland, *Philonis Alexandrini opera quae supersunt,* Berlin, 1896-1915, in six vols. The earlier edition was by Mangey. Recent editions of single works are: F. C. Conybeare, *Philo about the Contemplative Life,* critically edited with a defence of its genuineness, 1895. E. Bréhier, *Commentaire allégorique des Saintes Lois après l'œuvre des six jours,* Greek and

his participation in the embassy of Jews to the Emperor Gaius or Caligula in 40 A. D. He repeats the doctrines of the Greek philosophers and anticipates much that the church fathers discuss. Before the Neo-Platonists he regards matter as the source of all evil and feels the necessity of mediators, angels or demons, between God and man. Before the medieval revival of Aristotle and natural philosophy he tries to reconcile the Mosaic account of creation with belief in a world soul, and monotheism with astrology. Before the rise of Christian monasticism he describes in his treatise *On the Contemplative Life* an ascetic community of *Therapeutae* at Lake Maerotis.[1] After Pythagoras he enlarges upon the mystic significance of numbers. After Plato he repeats the conception of an ideal city of God

French, 1909. In the passages from Philo quoted in this chapter I have often availed myself of the wording of the English translation by C. D. Yonge in four vols., 1854-1855. The Latin translation of Philo's works made from the Greek by Lilius Tifernates for Popes Sixtus IV and Innocent VIII is preserved at the Vatican in a series of six MSS written during the years 1479-1484: Vatic. Lat., 180-185.

J. d'Alma, *Philon d'Alexandrie et le quatrième Évangile*, 1910.

N. Bentwich, *Philo-Judaeus of Alexandria*, 1910 (a small general book).

T. H. Billings, *The Platonism of Philo Judaeus*, 1919.

W. Bousset, *Jüdisch-Christlicher Schulbetrieb in Alexandria und Rom*, 1915.

E. Bréhier, *Les Idées philosophiques et religieuses de Philon d'Alexandrie*, 1908, a scholarly work with a ten-page bibliography.

M. Caraccio, *Filone d'Alessandria e le sue opere*, 1911, a brief indication of the contents of each work.

K. S. Guthrie, *The Message of Philo Judaeus*, 1910, popular.

H. Guyot, *Les Réminiscences de Philon le Juif chez Plotin*, 1906.

P. Heinsch, *Der Einfluss Philos auf die älteste christliche Exegese*, 1908, 296 pp.

H. A. A. Kennedy, *Philo's contribution to Religion*, 1919.

J. Martin, *Philon*, 1907, with a five-page bibliography.

L. H. Mills, *Zarathustra, Philo, the Achaemenids and Israel*, 1905, 460 pp.

L. Treitel, *Philonische Studien*, 1915, is of limited scope.

H. Windisch, *Die Frömmigkeit Philos und ihre Bedeutung für das Christentum*, 1909.

[1] The genuineness of this treatise, denied by Graetz and Lucius in the mid-nineteenth century, was amply demonstrated by L. Massebieau, *Revue de l'Histoire des Religions*, XVI (1887), 170-98, 284-319; Conybeare, *Philo about the Contemplative Life*, Oxford, 1895; and P. Wendland, *Die Therapeuten und die Philonische Schrift vom Beschaulichen Leben*, in *Jahrb. f. Class. Philologie*, Band 22 (1896), 693-770. In St. John's College Library, Oxford, in a manuscript of the early eleventh century (MS 128, fol. 215 ff) with Dionysius the Areopagite on the ecclesiastical hierarchy, is, Philonis de excircumcisione credentibus in Aegypto Christianis simul et monachis ex suprascripto ab eo sermone de vita theorica aut de orantibus.

which was to gain such a hold upon Christian imagination.[1] After the Stoics he proclaims the doctrine of the law of nature, holds that the institution of human slavery is absolutely contrary to it, and writes "a treatise to prove that every virtuous man is free" and that to be virtuous is to live in conformity to nature.[2] He had previously written another treatise designed to show that "every wicked man was a slave," [3] and he held a theory which we met in the Enoch literature and shall meet again in a number of subsequent writers that sin was punished naturally by forces of nature such as floods and thunderbolts. He did not originate the practice of allegorical interpretation of the Bible but he is our first great extant example thereof. He even went so far as to regard the tree of life and the story of the serpent tempting Eve as purely symbolical, an attitude which found little favor with Christian writers.[4] His effort by means of the allegorical method to find in the books of the Pentateuch all the attractive concepts and theories which he had learned from the Greeks became later in the Christian apologists an assertion that Plato and Pythagoras had borrowed their doctrines from Abraham and Moses. His doctrine of the *logos* had a powerful influence upon the writers of the New Testament and the theology of the early church.[5] Yet Philo affirms that no more perfect good than philosophy exists in human life and in both literary style and erudition he is a Hellene to his very finger tips. The recent tendency, seen especially in German scholarship, to deny the writers of the Roman Empire any capacity for original thought and to trace back their ideas to unextant authors of a supposedly much more productive Hellenistic age has perhaps been carried too far. But if we may not regard Philo as a great originator, and it is evident that he borrowed many of his ideas, he was at any rate a great

[1] *De mundi opificio,* caps. 49 and 50.

[2] *On the Contemplative Life,* Chapter 9.

[3] So he states in the opening sentences of the other treatise; it is not extant.

[4] *De mundi opificio,* caps. 54 and 55.

[5] Réville, J., *Le logos, d'après Philon d'Alexandrie,* Genève, 1877.

transmitter of thought, a mediator after his own heart between Jews and Greeks, and between them both and the Christian writers to come. Standing at the close of the Hellenistic age and at the opening of the Roman period, he occupies in the history of speculative and theological thought an analogous position to that of Pliny the Elder in the history of natural science, gathering up the lore of the past, perhaps improving it with some additions of his own, and exercising a profound influence upon the age to come.

His influence upon the middle ages was indirect.

Philo's medieval influence, however, was probably more indirect than Pliny's and passed itself on through yet other mediators to the more remote times. Comparatively speaking, the *Natural History* of Pliny probably was more important in the middle ages than in the early Roman Empire when other authorities prevailed in the Greek-speaking world. Philo's influence on the other hand must soon be transmitted through Christian, and then again through Latin, mediums. This is indicated by the fact that to-day many of his works are wholly lost or extant only in fragments [1] or in Armenian versions,[2] and that we have no sure information as to the order in which they were composed.[3] But his initial force is none the less of the greatest moment, and seems amply sufficient to justify us in selecting his writings as one of our starting points. The extent to which one is apt to find in the writings of Philo passages which are forerunners of the statements of subsequent writers, may be illustrated by the familiar story of King Canute and the tide. Philo in his work *On Dreams* [4] speaks of the custom of the Germans of charging the incoming tide with their drawn swords. But what especially concern us are Philo's

[1] Lincoln College, Oxford, has a 12th century MS in Greek of the *De vita Mosis* and *De virtutibus*, —MS 34.

[2] The *Alexander sive de animalibus* and the complete text of the *De providentia* exist only in Armenian translation,—see Cohn (1892), p. 16. *The Biblical Antiquities*, extant only in an imperfect Latin version, is not regarded as a genuine work,—see W. O. E. Oesterley and G. H. Box, *The Biblical Antiquities of Philo*, now first translated from the old Latin version by M. R. James (1917), p. 7.

[3] Cohn (1892), 11.

[4] II, 17.

statements concerning magic, astrology, the stars, the perfection and power of numbers, demons, and the interpretation of dreams.

Good and bad magic.

Philo draws a distinction between magic in the good and bad sense. The former and true magical art is the lore of learned Persians called *Magi* who investigate nature more minutely and deeply than is usual and explain divine virtues clearly.[1] The latter magic is a spurious imitation of the other, practised by quacks and impostors, old-wives and slaves, who by means of incantations and the like procedure profess to change men from love to hatred or vice versa and who "deceive unsuspecting persons and waste whole families away by degrees and without making any noise." It is to this adulterated and evil magic that Philo again refers when he likens political life to Joseph's coat of many colors, stained with the blood of wars, and in which a very little truth is mixed up with a great deal of sophistry akin to that of the augurs, ventriloquists, sorcerers, jugglers and enchanters, "from whose treacherous arts it is very difficult to escape."[2] This distinction between a magic of the wise and of nature and that of vulgar impostors is one which we shall find in many subsequent writers, although it was not recognized by Pliny. Philo also antecedes numerous Christian commentators upon the Book of Numbers[3] in considering the vexed question whether Balaam was an evil enchanter and diviner, or a divine prophet, or whether he combined magic and prophecy, and thus indicated that the former art is not evil but has divine approval. Philo's conclusion is the more usual one that Balaam was a celebrated diviner and magician, and that it is impossible that "holy inspiration should be combined with magic," but that in the particular case of his blessing Israel the spirit of divine

[1] (*Quod omnis probus liber sit,* cap. xi); also *The Law Concerning Murderers,* cap. 4.

[2] *On Dreams,* I, 38.

[3] Numbers XXII-XXV. Balaam is, of course, referred to in a number of other passages of the Bible: Deut., XXIII, 3-6; Joshua, XIII, 22; XXIV, 9-10; Nehemiah, XIII, 1ff; Micah, VI, 5; Second Peter, II, 15-16; Jude, 11; Revelation, II, 14.

prophecy took possession of him and "drove all his artificial system of cunning divination out of his soul."[1]

Stars not gods nor first causes.

Philo has considerably more to say upon the subject of astrology than upon that of magic. He was especially concerned to deny that the stars were first causes or independent gods. He chided the Chaldean adepts in genethlialogy for recognizing no other god than the universe and no other causes than those apparent to the senses, and for regarding fate and necessity as gods and the periodical revolutions of the heavenly bodies as the cause of all good and evil.[2] Philo more than once exhorts the reader to follow Abraham's example in leaving Chaldea and the science of genethlialogy and coming to Charran to a comprehension of the true nature of God.[3] He agreed with Moses that the stars should not be worshiped and that they had been created by God, and more than that, not created until the fourth day, in order that it might be perfectly clear to men that they were not the primary causes of things.[4]

But rational and virtuous animals: and God's viceroys over inferiors.

Philo, nevertheless, despite his attack on the Chaldeans, believed in much which we should call astrological. The stars, although not independent gods, are nevertheless divine images of surpassing beauty and possess divine natures, although they are not incorporeal beings. Philo distinguishes between the stars, men, and other animals as follows. The beasts are capable of neither virtue nor vice; human beings are capable of both; the stars are intelligent animals, but incapable of any evil and wholly virtuous.[5] They were native-born citizens of the world long before its first human citizen had been naturalized.[6] God, moreover, did not post-

[1] *Vita Mosis,* I, 48-50. Besides discussion of Balaam in various Biblical commentaries, dictionaries, and encyclopedias, see Hengstenberg, *Die Geschichte Bileams und seine Weissagungen,* 1842.

[2] *De migrat. Abrahami,* cap. 32.

[3] *Idem,* and *De somniis,* cap. 10.

[4] *De monarchia,* I, 1. *De mundi opificio,* cap. 14.

[5] *De mundi opificio,* caps. 18, 50 and 24. See also his *De gigantibus* and Περὶ τοῦ θεοπέμπτους εἶναι τοὺς ὀνείρους.

[6] *Ibid.,* Cap. 50. Huet, the noted French scholar of the 17th century, states in his edition of Origen that "Philo after his custom repeats an opinion of Plato's and almost his very words for . . . he asserts that the stars are not only animals but also the purest intellects." Migne PG, XVII, col. 978.

pone their creation until the fourth day because superiors are subject to inferiors. On the contrary they are the viceroys of the Father of all and in the vast city of this universe the ruling class is made up of the planets and fixed stars, and the subject class consists of all the natures beneath the moon.[1] A relation of natural sympathy exists between the different parts of the universe, and all things upon the earth are dependent upon the stars.[2]

They do not cause evil: but it is possible to predict the future from their motions.

Philo of course will not admit that evil is caused either by the virtuous stars or by God working through them. As has been said, he attributed evil to matter or to "the natural changes of the elements," [3] drawing a line between God and nature in much the fashion of the church fathers later. But he granted that "before now some men have conjecturally predicted disturbances and commotions of the earth from the revolutions of the heavenly bodies, and innumerable other events which have turned out most exactly true." [4] Philo's interest in astronomy and astrology is further suggested by his interpretation of the eleven stars of Joseph's dream as referring to the signs of the zodiac,[5] Joseph himself making the twelfth; and by his interpreting the ladder in Jacob's dream which stretched between earth and heaven as referring to the air,[6] into which earth's evaporations dissolve, while the moon is not pure ether like the other stars but itself contains some air. This accounts, Philo thinks, for the spots upon the moon—an explanation which I do not remember having met in subsequent writers.

Jewish astrology.

Josephus [7] and the Jews in general of Philo's time were equally devoted to astrology according to Münter, who says: "Only their astrology was subordinated to theism. The one God always appeared as the master of the host of heaven. But they regarded the stars as living divine beings and

[1] *De monarchia,* I, 1; *De mundi opificio,* cap. 14.

[2] *De monarchia,* I, 1; *De migratione Abrahami,* cap. 32; *De mundi opificio,* cap. 40.

[3] Eusebius, *De praep. Evang.,* cap. 13.

[4] *De mundi opificio,* cap. 19.

[5] *De somniis,* II, 16.

[6] *Ibid.,* I, 22.

[7] *De bello Jud.,* V, 5, 5; *Antiq.,* III, 7, 7-8.

powers of heaven."[1] In the Talmud later we read that the hour of Abraham's birth was announced by the stars and that he feared from his observations of the constellations that he would go childless. Münter also gives examples of the belief of the rabbis in the influence of the stars upon the destiny of the Jewish people and upon the fate of individual men, and of their belief that a star would announce the coming of the Messiah.[2]

Perfection of the number seven.

From Philo's astrology it is an easy step to his frequent reveries concerning the perfection and mystic significance of certain numbers,—a train of thought which was continued by many of the church fathers, and is also found in various pagan writers of the Roman Empire.[3] Thomas Browne in his enquiry into "Vulgar Errors"[4] was inclined to hold Philo even more responsible than Pythagoras or Plato for the dissemination of such doctrines. Philo himself recognizes the close connection between astrology and number mysticism, when, after affirming the dependence of all earthly things upon the heavenly bodies, he adds: "It is in heaven, too, that the ratio of the number seven began."[5] Philo doubts if it is possible to express adequately the glories of the number seven, but he feels that he ought at least to attempt it and devotes a dozen chapters of his treatise on the creation of the world to it,[6] to say nothing of other passages. He notes that there are seven planets, seven circles of heaven, four quarters of the moon of seven days each, that such constellations as the Pleiades and Ursa Major consist of seven stars, and that children born at the end of

[1] *Der Stern der Weisen* (1827), p. 36. "Nur war ihre Astrologie dem Theismus untergeordnet. Der Eine Gott erschien immer als der Herrscher des Himmelsheeres. Sie betrachteten aber die Sterne als lebende göttliche Wesen und Mächte des Himmels."

[2] Münter (1827), pp. 38-39, 43, 45, etc. On the subject of Jewish astrology see also: D. Nielsen, *Die altarabische Mondreligion und die mosaische Überlieferung,* Strasburg, 1904; F. Hommel, *Der Gestirndienst der alten Araber und die altisraelitische Überlieferung,* Munich, 1901.

[3] Such as Aulus Gellius, Macrobius, and Censorinus. These writers seem to have taken it from Varro. We have also noted number mysticism in Plutarch's *Essays.*

[4] Browne (1650) IV, 12.

[5] *De mundi opificio,* cap. 40.

[6] *Ibid.,* caps. 30-42.

seven months live, while those who see the light in the eighth month die. In diseases the seventh is a critical day. Also there are either seven ages of man's life, as Hippocrates says, or, in accordance with Solon's lines, man's three-score years and ten may be subdivided into ten periods of seven years each. The lyre of seven strings corresponds to the seven planets, and in speech there are seven vowels. There are seven divisions of the head—eyes, ears, nostrils, and mouth, seven divisions of the body, seven kinds of motion, seven things seen, and even the senses are seven rather than five if we add the vocal and generative organs.[1]

And of fifty.

Philo's ideal sect, the *Therapeutae,* are wont to assemble as a prelude to their greatest feast at the end of seven weeks, "venerating not only the simple week of seven days but also its multiplied power," [2] but the chief festival itself occurs on the fiftieth day, "the most holy and natural of numbers, being compounded of the power of the right-angled triangle, which is the principle of the origination and condition of the whole." [3]

Also of four and six.

The numbers four and six, however, yield little to seven and fifty in the matter of perfection. It was the fourth day that God chose for the creation of the heavenly bodies, and He did not need six days for the entire work of creation, but it was fitting that that perfect work should be accomplished in a perfect number of days. Six is the product of the first female number, two, and the first male number, three. Indeed, the first three numbers, one, two, and three, whether added or multiplied, give six.[4] As for four, there are that many elements and seasons; it is the only number produced by the same number—two—whether added to

[1] For the later influence of such doctrines in the Mohammedan world see D. B. Macdonald, *Muslim Theology, Jurisprudence, and Constitutional Theory,* 1903, pp. 42-3, concerning the "Seveners" and the "Twelvers" and the doctrine of the hidden Iman.

[2] *Ibid.,* "Thus we have a series of seven times seven Imans, the first, and thereafter each seventh, having the superior dignity of Prophet. The last of the forty-nine Imans, this Muhammad ibn Isma'il, is the greatest and last of the Prophets."

[3] *De vita contemplativa,* cap. 8. It will be recalled that the fifty books of the *Digest* of Justinian are similarly divided.

[4] *De mundi opificio,* cap. 3.

itself or multiplied by itself; it is the first square and as such the emblem of justice and equality; it also represents the cube or solid, as the number one stands for a point, two for a line, and three for a surface.[1] Furthermore four is the source of "the all-perfect decade," since one and two and three and four make ten. At this we begin to suspect, and with considerable justification, as the writings of other devotees of the philosophy of numbers would show, that the number of perfect numbers is legion. We may not, however, follow Philo much farther on this topic. Suffice it to add that he finds the fifth day fitting for the creation of animals possessed of five senses,[2] while he divides the ten plagues of Egypt into three dealing with the more solid elements, earth and water, and performed by Aaron; three dealing with air and fire which were entrusted to Moses; the seventh was committed to both Aaron and Moses; while the other three God reserved for Himself.[3]

Spirits of the air.

Philo believed in a world of spirits, both the angels of the Jews and the demons of the Greeks. When God said: "Let us make man," Philo believed that He was addressing those assistant spirits who should be held responsible for the viciousness to which man alone of all creation is liable.[4] Of the divine rational natures Philo regarded some as incorporeal, others like the stars as possessed of bodies.[5] He also believed that there were spirits in the air as well as afar off in heaven. He could not see why the air should not be inhabited when there were stars in the ether and fish in the sea as well as other animals upon land.[6] Indeed he argued that it would be absurd that the element which was essential for the vitality even of land and aquatic animals should have no living beings of its own. That these spirits of the air must be invisible did not trouble him, since the human soul is also invisible.

[1] *De mundi opificio,* caps. 15-16. See also on perfect numbers *On the Allegories of the Sacred Laws.*

[2] *Ibid.,* cap. 20.

[3] *Vita Mosis,* I, 17.

[4] *De mundi opificio,* cap. 24.

[5] *Ibid.,* cap. 50.

[6] *De somniis,* II, 21-22.

Interpretation of dreams.

Of Philo's five books on dreams only two are extant. They suffice to show, however, that he accepted the art of divination from dreams. Of dreams he distinguished three varieties: those direct from God which require no interpretation; those in which the dreamer's mind moves in unison with the world soul, and which are neither entirely clear nor yet very obscure—an instance is Jacob's vision of the ladder; and third, those in which the mind is moved by a prophetic frenzy of its own, and which require the science of interpretation—such dreams were Joseph's concerning his brothers, and those of the butler and the baker at Pharaoh's court.[1]

Politics akin to magic.

The recent war and its accompaniments and sequels have brought home to some the conviction that our modern civilization is after all not vastly superior to that of some preceding ages. To those who still imagine that because modern science has freed us from much past superstition concerning nature, we are therefore free from political fakirs, from social absurdities, and from fallacious procedure and reasoning in many departments of life, the reading may be recommended of a passage in Philo's treatise on dreams,[2] in which he classifies the art of politics along with that of magic. He compares Joseph's coat of many colors to "the much-variegated web of political affairs" where along with "the smallest possible portion of truth" falsehoods of every shade of plausibility are interwoven; and he compares politicians and statesmen to augurs, ventriloquists, and sorcerers, "men skilful in juggling and in incantations and in tricks of all kinds, from whose treacherous arts it is very difficult to escape." He adds that Moses very naturally represented Joseph's coat as blood-stained, since all statecraft is tainted with wars and bloodshed.

Twelve centuries later we find Philo's association of politicians with magicians repeated by his compatriot Moses Maimonides in the *More Nevochim* or *Guide for the Per-*

[1] *De somniis,* II, 1.

[2] Cap. 38.

A thought repeated by Moses Maimonides and Albertus Magnus.

plexed,[1] a work which appeared almost immediately in Latin translation and from which this very passage is cited by Albertus Magnus in his discussion of divination by dreams.[2] There are some men, says Albert, in whom the intellect is abundant and active and clear. Such men are akin to the superior substances, that is, to the angels and stars, and therefore Moses of Egypt, *i.e.,* Maimonides, calls them sages. But there are others who, according to Albert, confound true wisdom with sophistry and are content with mere probabilities and imaginations and are at home in "rhetorical and civil matters." Maimonides, however, described this class a little differently, saying that in them the imaginative faculty is preponderant and the rational faculty imperfect. "Whence arises the sect of politicians, of legislators, of diviners, of enchanters, of dreamers, . . . and of prestidigiteurs who work marvels by strange cunning and occult arts." [3]

[1] II, 37.

[2] Cap. 5.

[3] Since I finished this chapter, I have noted that the "folk-lore in the Old Testament" has led Sir James Frazer to write a passage on "the harlequins of history" somewhat similar to that of Philo on Joseph's coat of many colors. After remarking that friends and foes behold these politicians of the present and historical figures of the future from opposite sides and see only that particular hue of the coat which happens to be turned toward them, Sir James concludes (1918), II, 502, "It is for the impartial historian to contemplate these harlequins from every side and to paint them in their coats of many colors, neither altogether so white as they appeared to their friends nor altogether so black as they seemed to their enemies." But who can paint out the blood-stains?

CHAPTER XV

THE GNOSTICS

Difficulty in defining Gnosticism—Magic and astrology in Gnosticism—Simon Magus as a Gnostic—Simon's Helen—The number thirty and the moon—Ophites and Sethians—A magical diagram—Employment of names and formulae—Seven metals and planets—Magic of Simon's followers—Magic of Marcus in the Eucharist—Other magic and occult lore of Marcus—Name and number magic—The magic vowels—Magic of Carpocrates—The Abraxas and the number 365—Astrology of Basilides—*The Book of Helxai*—Epiphanius on the Elchasaites—*The Book of the Laws of Countries*—Personality of Bardesanes—Sin possible for men, angels, and stars—Does fate in the astrological sense prevail?—National laws and customs as a proof of free will—*Pistis-Sophia;* attitude to astrology—"Magic" condemned—Power of names and rites—Interest in natural science—"Gnostic gems" and astrology—The planets in early Christian art—Gnostic amulets in Spain—Syriac Christian charms—Priscillian executed for magic—Manichean manuscripts—The Mandaeans.

Difficulty in defining Gnosticism.

GNOSTICISM [1] is not easy to define and the term Gnostic appears to have been applied to a great variety of sects with a confusing diversity of beliefs. Many of the constituents and roots at least of Gnosticism were older than Christianity, and it is now the custom to associate the Gnosis or superior knowledge and revelation, which gives the movement its name, not with Greek philosophy or mysteries but with oriental speculation and religions. Anz [2] has been impressed by its connection with Babylonian star-worship; Amélineau [3] has urged its debt to Egyptian magic and

[1] A good account of the Gnostic sources and bibliography of secondary works on Gnosticism will be found in CE, "Gnosticism" (1909) by J. P. Arendzen.

[2] Anz, *Zur Frage nach dem Ursprung des Gnostizismus*, 1897, 112 pp., in TU, XV, 4.

[3] Amélineau, *Essai sur le gnosticisme égyptien, ses développements et son origine égyptienne*, 1887, 330 pp., in *Musée Guimet*, tom. 14; and various other publications by the same author.

religion; Bousset [1] has argued for Persian origins. The main features of the great oriental religions which swept westward over the Roman Empire were shared by Gnosticism: the redeemer god, even the great mother goddess conception to some extent, the divinely revealed mysteries, the secret symbols, the dualism, and the cosmic theory. Gnosticism as it is known to us, however, is more closely connected with Christianity than with any other oriental religion or body of thought, for the extant sources consist almost entirely either of Gnostic treatises which pretend to be Christian Scriptures and were almost entirely written in Coptic in the second or third century of our era,[2] or of hostile descriptions of Gnostic heresies by the early church fathers. However, the philosopher Plotinus also criticized the Gnostics, as we have seen.

Magic and astrology in Gnosticism.

What especially concerns our investigation is the great use made, or said to be made, by the Gnostics of sacred formulae, symbols, and names of demons, and the prevalence among them of astrological theory as shown by their widespread notion of the seven planets as the powers who have created our inferior and material world and who rule over its affairs. Gnosticism was deeply influenced by, albeit it to some extent represents a reaction against, the Babylonian star-worship and incantation of spirits. The seven planets and the demons occupy an important place in Gnostic myth because they intervene between our world and the world of supreme light, and their spheres must be traversed —much as in the *Book of Enoch* and Dante's *Paradiso*— both by the redeeming god in his descent and return and by any human soul that would escape from this world of fate, darkness, and matter. What encouragement there is for such views in the canonical Scriptures themselves may be

[1] Bousset, *Hauptprobleme der Gnosis,* 1911; and "Gnosticism" in EB, 11th edition.

[2] The dating is somewhat disputed. Some of the Gnostic writings discovered in 1896 have, I believe, not yet been published, although announced to be edited by C. Schmidt in TU. Grenfell and Hunt will soon publish "a small group of 21 papyri . . . among which is a gnostic magical text of some interest": Grenfell (1921), p. 151.

inferred from the following passage in which Christ foretells His second coming: "Immediately after the tribulation of those days shall the sun be darkened, and the moon shall not give her light, and the stars shall fall from heaven, *and the powers of the heavens shall be shaken.* And then shall appear *the sign* of the Son of man *in heaven;* and then shall all the tribes of the earth mourn, and they shall see the Son of man coming in the clouds of heaven with power and great glory. And He shall send His angels with a great sound of a trumpet, and they shall gather together His elect from the four winds, from one end of heaven to the other." [1] But in order to pass the demons and the spheres of the planets, who are usually represented as opposed to this, one must, as in the Egyptian *Book of the Dead,* know the passwords, the names of the spirits, the sacred formulae, the appropriate symbols, and all the other apparatus suggestive of magic and necromancy which forms so large a part of the *gnosis* that gives its name to the system. This will become the more apparent from the following particular accounts of Gnostic sects and doctrines found in the works of the Christian fathers and in the scanty remains of the Gnostics themselves. The philosopher Plotinus we have already heard charge the Gnostics with resort to magic and sorcery, and with ascribing evil and fatal influence to the stars. At the same time we shrewdly suspect that Gnosticism has been made a scapegoat for the sins in these regards of both early Christianity and pagan philosophy.

Simon Magus as a Gnostic.

Simon Magus, of whose magical exploits as recorded by many a Christian writer we shall treat in another chapter, is also represented by the fathers as holding Gnostic doctrine, although some writers have contended that Simon the magician named in *Acts* was an entirely different person from Simon the heretic and author of *The Great Declaration.*[2] Simon declared himself the Great Power of God, or

[1] The Gospel of Matthew, XXIV, 29-31. Not to mention Paul's "angels and principalities and powers."

[2] St. George Stock, "Simon Magus," in EB, 11th edition. See also George Salmon in *Dict. Chris. Biog.*, IV, 681.

the Being who was over all, who had appeared in Samaria as the Father, in Judea as the Son, and to other nations as the Holy Spirit.[1] In the *Pseudo-Clementines* Simon is represented as arguing against Peter in characteristically Gnostic style that "he who framed the world is not the highest God, but that the highest God is another who alone is good and who has remained unknown up to this time." [2] According to Epiphanius Simon claimed to have descended from heaven through the planetary spheres and spirits in the manner of the Gnostic redeemer. He is quoted as saying, "But in each heaven I changed my form in accordance with the form of those who were in each heaven, that I might escape the notice of my angelic powers and come down to the Thought, who is none other than she who is likewise called Prounikon and the Holy Spirit." Epiphanius further informs us that Simon believed in a plurality of heavens, assigned certain powers to each firmament and heaven, and applied barbaric names to these spirits or cosmic forces. "Nor," adds Epiphanius, "can anyone be saved unless he learns this mystic lore and offers such sacrifices to the Father of all through these archons and authorities." [3]

Simon's Helen.

The fathers tell us that Simon went about with a woman called Helena or Helen, who Justin Martyr says had formerly been a prostitute.[4] Simon is said to have called her the mother of all, through whom God had created the angels and aeons, who in their turn had formed the world and men. These cosmic powers had then, however, cast her down to earth, where she had been confined in various successive human and animal bodies. She seems to have obtained her name of Helen from the fact that it was for her that the Trojan war had been fought, an event which Simon seems to have subjected to much allegorical interpretation. He also spoke of Helen as "the lost sheep," whom he, the Great

[1] Irenaeus, *Against Heresies,* I, 23.

[2] *Homilies,* XVIII, 1-.

[3] Epiphanius, *Panarion,* A-B-XXI; Petavius, 55-60; Dindorf, II, 6-12.

[4] *First Apology,* cap. 26.

Power, had descended from heaven to release from the bonds of the flesh. She was that Thought or Holy Spirit which we have heard him say he came down to recover. Simon's Helen also corresponds to Pistis-Sophia, who in the extant Gnostic work named after her descends through the twelve aeons, deceived by a lion-faced power whom they have formed to mislead her, and then reascends by the aid of Jesus or the true light. It seems fairly evident that the fathers [1] have taken literally and travestied by a scandalous application to an actual woman a beautiful Gnostic myth or allegory concerning the human soul. At the same time Simon's Helen reminds us of Jesus's relations with the woman taken in adultery, the woman of Samaria, and Mary Magdalene. Mary Magdalene, it may be noted, in the Gnostic writing, *Pistis-Sophia,* takes a rôle superior to the twelve disciples, a fact of which Peter complains to his Lord more than once. But Simon's Helen was that spirit of truth which lies latent in the human mind and which he endeavored to release by means of the philosophy, astrology, and magic of his time. May modern scientific method prove more successful in setting the prisoner free!

The number thirty and the moon.

We find in the *Pseudo-Clementines* other details concerning Simon and Helen which bring out the astrological side of Gnosticism. We are told that John the Baptist had thirty disciples, a number suggestive of the days of the moon and also of the thirty aeons of the Gnostics of whom we elsewhere hear a great deal.[2] But the revolution of the moon does not occupy thirty full days, so that we are not surprised to learn that one of these disciples was a woman and furthermore that she was the very Helen of whom we have been speaking. At least, she is so called in the *Homilies* of the Pseudo-Clement; in the *Recognitions* she is actually

[1] Irenaeus and Epiphanius as cited above; also Hippolytus, *Philosophumena,* VI, 2-15; X, 8.

[2] See, for example, Irenaeus, *Against Heresies,* I, i, 3, where we are told among other things that the disciples of the Gnostic Valentinus affirm that the number of these aeons is signified by the thirty years of Christ's life which elapsed before He began His public ministry.

called *Luna* or the Moon.[1] After the death of John the Baptist Simon by his magic power supplanted Dositheus as leader of the thirty, and then fell in love with Luna and went about with her, proclaiming that she was Wisdom or Truth, "brought down . . . from the highest heavens to this world." [2] The number thirty is again associated with Simon and Dositheus in a curiously insistent, although apparently unconscious, manner by Origen, who in one passage of his *Reply to Celsus,* written in the first half of the third century, expresses doubt whether thirty followers of Simon, the Samaritan magician, can be found in all the world, and in a second passage, while asserting that "Simonians are found nowhere throughout the world," adds that of the followers of Dositheus there are now not more than thirty in all.[3]

Ophites and Sethians.

Similar to Simon's account of the heavens and of his descent through them were the teachings of the Ophites and Sethians who, according to Irenaeus,[4] held that Christ "descended through the seven heavens, having assumed the likeness of their sons, and gradually emptied them of their power." These heretics also represented the "heavens, potentates, powers, angels, and creators as sitting in their proper order in heaven, according to their generation, and as invisibly ruling over things celestial and terrestrial." All ruling spirits were not invisible, however, since the Ophites and Sethians identified with the seven planets their Holy Hebdomad, consisting of Ialdabaoth, Iao, Sabaoth, Adonaus (or, Adonai), Eloeus, Oreus, and Astanphaeus,—names often employed in the Greek magical papyri,[5] in medieval incantations, and in the Jewish Cabbala. The Ophites and Sethians further asserted that when the serpent was cast down into the lower world by the Father, he begat six sons

[1] *Homilies,* II, 23-25; *Recognitions,* II, 8-9.

[2] *Homilies,* II, 25.

[3] *Reply to Celsus,* I, 57, and VI, 11.

[4] Irenaeus, *Against Heresies,* I, 30.

[5] G. Parthey, *Zwei griech. Zauberpapyri des Berliner Museums,* 1860, p. 128; C. Wessely, *Griech. Zauberpapyrus von Paris und London,* 1888, p. 115; F. G. Kenyon, *Greek Papyri in the British Museum,* 1893, p. 469ff.

who, with himself, constitute a group of seven corresponding and in contrast to the Holy Hebdomad which surround the Father. They are the seven mundane demons who are ever hostile to humanity. The Sethians of course took their name from Seth, son of Adam, who in the middle ages was regarded sometimes, like Enoch, as the especial recipient of divine revelation and as the author of sacred books. The historian Josephus states in his *Jewish Antiquities* that Seth and his descendants discovered the art of astronomy and that one of the two pillars on which they recorded their findings was still extant in his time, the first century.[1] Under the caption, *Sethian Tablets of Curses,* Wünsch has published some magical imprecations scratched on lead tablets between 390 and 420 A. D. at Rome.[2] Eight revelations ascribed to Adam and Seth are also extant in Armenian.[3]

A magical diagram.

In Origen's *Reply to Celsus* is described a mystic diagram with details redolent of magic and astrological necromancy,[4] which Celsus had laid to the charge of Christians generally but which Origen declares is probably the product of the "very insignificant sect called Ophites." Origen himself has seen this diagram or one something like it, and assures his readers that "we know the depth of these unhallowed mysteries," but he declares that he has never met anybody anywhere who put any faith in this diagram. Obviously, however, such a diagram would not have been in existence if no one had ever had faith in it. Furthermore, its survival into Origen's time, when he asserts that men had ceased to use it, is evidence of the antiquity of the sect and the superstition. In this diagram ten distinct circles were united by a single circle representing the soul of all

[1] Josephus, *Antiquities,* I, ii, 3.

[2] R. Wünsch, *Sethianische Verfluchungstafeln aus Rom,* Leipzig, 1898.

[3] E. Preuschen, *Die apocryph. gnost. Adamschrift,* 1900. *Mechitarist collection of Old Testament Apocrypha,* Venice, 1896.

[4] The diagram is described in the *Reply to Celsus,* VI, 24-38; in the following description I have somewhat altered the order. An attempt to reproduce this diagram will be found in CE, "Gnosticism," p. 597.

things and called Leviathan. Celsus spoke of the upper circles, of which at least some were in colors, as "those that are above the heavens." On these were inscribed such words and phrases as "Father and Son," "Love," "Life," "Knowledge," and "Understanding." Then there were "the seven circles of archontic demons," who are probably to be connected with the spheres of the seven planets. These seven ruling demons were represented by animal heads or figures, somewhat resembling the symbols of the four evangelists to be seen in the mosaics at Ravenna and elsewhere in Christian art. The angel Michael was depicted by a sort of chimaera, the words of Celsus being, "The goat was shaped like a lion"; Suriel, by a bull; Raphael, by a dragon; Gabriel, by an eagle; Thautabaoth, by a bear; Erataoth, by a dog; and Thaphabaoth or Onoel, by an ass. The diagram was divided by a thick black line called Gehenna and beneath the lowest circle was placed "the being named Behemoth." There was also "a square pattern" with inscriptions concerning the gates of paradise, a flaming circle with a flaming sword as its diameter guarding the tree of knowledge and of life, "a barrier inscribed in the shape of a hatchet," and a rhomboid with the words, "The foresight of wisdom." Celsus further mentioned a seal with which the Father impresses the Son, who says, "I have been anointed with white ointment from the tree of life," and seven angels who contend with the seven ruling demons for the soul of the dying body.

Employment of names and formulae.

Origen further informs us of the forms of salutation to each ruling spirit employed by "those sorcerers," as they pass through "the fence of wickedness" or the gate to the realm of each spirit. The names of the spirits are now given as Ialdabaoth, who is the lion-like archon and with whom the planet Saturn is in sympathy, Iao or Jah, Sabaoth, Adonaeus, Astaphaeus, Aloaeus or Eloaeus, and Horaeus. The following is an example of the salutations or invocations addressed to these spirits: "Thou, O second Iao, who shinest by night, who art the ruler of the secret mysteries of Son and Father, first prince of death, and portion of the innocent, bearing now thine own beard as symbol, I am ready to pass through thy realm, having strengthened him who is born of thee by the living word. Grace be with me; Father, let it be with me!" Origen also states that the makers of this diagram have borrowed from magic the names Ialdabaoth, Astaphaeus, and Horaeus, while the other four are names of God drawn from the Hebrew Scriptures.

Seven metals and planets.

It is worth noting that immediately before this account of the diagram Celsus had described similar Persian mysteries of Mithras, in which seven heavens through which the soul has to pass were arranged in an ascending scale like a ladder.[1] Each successive heaven was entered by a gate of a metal corresponding to the planet in question, lead for Saturn, tin for Venus, copper for Jupiter, iron for Mercury, a mixed metal for Mars, silver for the moon, and gold for the sun. This association of metals and planets became a common feature of medieval alchemy. At the same time the passage is said to be our chief literary source for the mysteries of Mithras.[2]

Magic of Simon's followers.

The Simonians, according to Irenaeus, were as addicted to magic as their founder had been, employing exorcisms and incantations, love-philters and enchantments, familiar spirits and "dream-senders." "And whatever other curious arts may be resorted to are eagerly employed by them." Menander, the immediate successor of Simon in Samaria, was "a perfect adept in the practice of magic" and taught that by means of it one could overcome the angels who had created this world.[3] In a treatise on rebaptism, falsely ascribed to Cyprian but very likely contemporary with him, it is stated that the Simonians regard their baptism as superior to that of orthodox Christians, because when they descend into the water fire appears upon its surface. The writer thinks that this is done by some trick, or that there is some natural explanation of it, or that they merely imag-

[1] *Reply to Celsus*, VI, 22.
[2] Anz. (1897), p. 78.
[3] *Adv. haer.*, I, 23.

ine that they see a flame on the water, or that it is the work of some evil one and of magic power.[1] Epiphanius states that Simon employed such obscene substances as *semen* and *menstruum* in his magic,[2] but this seems to be a slander, at least against Gnosticism, since in a passage of the Gnostic *Book of the Saviour,* adjoined to the *Pistis-Sophia,* Thomas asks Jesus what shall be the punishment of men who eat "*semen maris et menstruum feminae*" mixed with lentils, saying as they do so, "We believe in Esau and Jacob," and is told that this is the worst of sins and that the souls of those committing it will be absolutely blotted out.[3]

Magic of Marcus in the Eucharist.

Next to Simon Magus, Marcus was the Gnostic and heretic most notorious as a practitioner of the magic arts, as Irenaeus states at the close of the second century, and Hippolytus and Epiphanius repeat in the third and fourth centuries respectively.[4] In performing the Eucharist he would change white wine placed in three wine cups into three different colors, one blood-red, one purple, and one dark blue, according to Epiphanius, while Irenaeus and Hippolytus more vaguely state, although they lived closer to Marcus's time, that he gave the wine a purple or reddish hue as if it had been changed into blood, an alteration which Marcus himself regarded as a manifestation of divine grace. Epiphanius attributes the change to an incantation muttered by Marcus while pretending to perform the Eucharist.

[1] Wm. Hartel, *S. Thasci Caecili Cypriani Opera Omnia,* Pars III, *Opera Spuria* (1870), p. 90, *De rebaptismate,* cap. 16, "quod si aliquo lusu perpetrari potest, sicut adfirmantur plerique huiusmodi lusus Anaxilai esse, sive naturale quid est quo pacto possit hoc contingere, sive illi putant hoc se conspicere, sive maligni opus et magicum virus ignem potest in aqua exprimere."

[2] *Contra haereses,* II, 2.

[3] *Pistis-Sophia,* ed. Schwartze and Petermann (1851), pp. 386-7; ed. Mead (1896), p. 390.

[4] Irenaeus, *Against Heresies,* I, 13, *et seq.;* Hippolytus, *Philosophumena,* VI, 34, *et seq.;* Epiphanius, *Panarion,* ed. Dindorf, II, 217, *et seq.* (ed. Petav., 232, *et seq.*). Concerning Marcus see further Tertullian, *De praescript.,* L; Theodoret, *Haeret. Fab.,* I, 9; Jerome, *Epist.,* 29; Augustine, *Haer.,* xiv. "D'après Reuvens," says Berthelot (1885), p. 57, "le papyrus n° 75 de Leide renferme un mélange de recettes magiques, alchimiques, et d'idées gnostiques; ces dernières empruntées aux doctrines de Marcus."

Hippolytus, who ascribes Marcus's feats partly to sleight-of-hand and partly to demons, in this case charges that he furtively dropped some drug into the wine. Marcus was also accustomed to fill a large cup from a smaller one so that it would overflow, a marvel which Hippolytus again tries to account for by stating that "very many drugs, when mingled in this way with liquid substances" temporarily increase their volume, "especially when diluted in wine."

Other magic and occult lore of Marcus.

Irenaeus, who is quoted verbatim by Epiphanius, further states that Marcus had a familiar demon by whose aid he was able to prophesy, and that he pretended to confer this gift upon others. He also accuses Marcus of seducing women by means of philters and love potions which he compounded. Hippolytus does not make these charges, but unites with the others in describing at length Marcus's theory of mystic names and his symbolical and mystical interpretation of the letters of the alphabet and of numbers. Marcus made various calculations based upon the number of letters in a name, the number of letters in the name of each letter, and so on. When Christ, whose ineffable name has thirty letters, said, "I am Alpha and Omega," He was believed by Marcus to have displayed the dove, whose number is 801. These reveries "are mere bits," as Hippolytus says, of astrological theory and Pythagorean philosophy. We shall find them perpetuated in the middle ages in the method of divination known as the Sphere of Pythagoras.

Name and number magic.

Such symbolism and mysticism concerning numbers and letters seldom indeed remain a matter of mere theory but readily lend themselves to operative magic. Thus Hippolytus can speak in the same breath of "magical arts and Pythagorean numbers" or tell that Pythagoras himself "also touched on magic, as they say, and himself discovered an art of physiognomy, laying down as a basis certain numbers and measures." Or note a third passage where Hippolytus is discussing Egyptian theology based on the theory of numbers.[1] After treating of the monad, duad, and enneads,

[1] Hippolytus, *Philosophumena,* VI, preface; I, 2; and IV, 43-4.

of the four elements in pairs, of the 360 parts of the circle, of "ascending and beneficent and masculine names" which end in odd numbers, and of feminine and malicious and descending names which terminate in even numbers, Hippolytus continues, "Moreover, they assert that they have calculated the word, 'Deity.' Now this name is an even number, and they write it down and attach it to the body and accomplish cures by it. In the same way an herb which terminates in this number is bound around the body and operates by reason of a similar calculation of the number. Nay, even a doctor cures the sick by such calculations." Similarly Censorinus states that the number seven is ascribed to Apollo and used in the cure of bodily ills, while nine is associated with the Muses and heals mental diseases.[1] But to return to Gnosticism.

The magic vowels.

The seven vowels were much employed by the Gnostics, undoubtedly as symbols for the seven planets and the spirits associated with them, but as symbols possessed of magic power as well as of mystic significance. "The Saviour and His disciples are supposed in the midst of their sentences to have broken out in an interminable gibberish of only vowels; magic spells have come down to us consisting of vowels by the fourscore; on amulets the seven vowels, repeated according to all sorts of artifices, form a very common inscription." [2] As the seven planets made the music of the spheres, so the seven vowels seem to have represented the musical scale, "and many a Gnostic sheet of vowels is in fact a sheet of music." [3]

Magic of Carpocrates.

Other heretics with Gnostic views who were accused of magic by the fathers were the followers of Carpocrates, who employed incantations and spells, philters and potions, who attracted spirits to themselves and made light of the cosmic angels, and who pretended to have great power over all

[1] Censorinus, *De die natali,* caps. 7 and 14.

[2] Arendzen, *Gnosticism,* in CE.

[3] Ruelle et Poirée, *Le chant gnostico-magique,* Solesmes, 1901.

things so that they were able by their magic to satisfy every desire.[1]

The Abraxas and the number 365.

Saturninus and Basilides were charged with "practicing magic, and employing images, incantations, invocations, and every other kind of curious art." They also believed in a supreme power named Abrasax or Abraxas, whose number was 365; and they contended that there were 365 heavens and as many bones in the human body; "and they strive to set forth the names, principles, angels, and powers of the 365 imagined heavens." [2]

Astrology of Basilides.

Hippolytus gives further indication of the astrological leanings of Basilides, who held that each thing had its own particular time, and supported his view by citing the *Magi* gazing wistfully at the star of Bethlehem and the remark of Christ Himself, "Mine hour is not yet come." [3] I suppose that by this Hippolytus means to suggest that Basilides held the astrological doctrine of elections; Basilides further affirmed, according to Hippolytus, that Jesus was "mentally preconceived at the time of the generation of the stars; and of the complete return to their starting point of all the seasons in the vast conglomeration," that is, at the end of the astronomical *magnus annus,* variously reckoned as of 36,000 or 15,000 years in duration.

The Book of Helxai.

In his *Refutation of all Heresies* [4] Hippolytus tells of an Alcibiades from Apamea in Syria who in his time brought to Rome a book supposed to contain revelations made to a holy man, Elchasai or Helxai, by an angel ninety-six miles in height and from sixteen to twenty-four miles in breadth and leaving a footprint fourteen miles long. This angel was the Son of God, and was accompanied by a female of corresponding size who was the Holy Spirit. This apparition and revelation was accompanied by a preaching of a new remission of sins in the third year of Trajan's reign, at which time we are led to suppose that the *Book of Helxai*

[1] Irenaeus, I, 25; Hippolytus, VII, 20; Epiphanius, ed. Dindorf, II, 64.

[2] Irenaeus, I, 24; Epiphanius, ed. Dindorf, II, 27-8.

[3] Hippolytus, VII, 14-15.

[4] The more correct title for the *Philosophumena,* see IX, 8-12.

came into existence. It imposed secrecy upon those initiated into its mysteries. The sect, according to Hippolytus, were much given to magic, astrology, and the number mysticism of Pythagoras. The Elchasaites employed incantations and formulae to cure persons bitten by mad dogs or afflicted with disease. In such cases and also in the case of rebaptism for the remission of sins it was customary with them to invoke or adjure "seven witnesses," not however in this case the planets, but "the heaven, and the water, and the holy spirits, and the angels of prayer, and the oil (or, the olive), and the salt, and the earth." Hippolytus declares that their formulae of this sort were "very numerous and very ridiculous." They dipped consumptives and persons possessed by demons in cold water forty times in seven days. They believed in the astrological doctrine of elections, since their sacred book warned them not to baptize or begin other important undertakings upon those days which were governed by the evil stars. They also seem to have predicted political events from the stars, foretelling that three years after Trajan's subjugation of the Parthians "war rages between the impious angels of the northern (constellations), and on this account all kingdoms of impiety are in confusion."

Epiphanius on the Elchasaites.

In the next century Epiphanius adds one or two further details to Hippolytus' account of the Elchasaites. Besides the list of seven witnesses already given he mentions another slightly different one: salt, water, earth, wheat, heaven, ether, and wind. He also tells of two sisters in the time of Constantine who were supposed to be descendants of Helxai. One of them was still alive the last Epiphanius knew, and crowds followed "this witch" to collect the dust of her footprints or her spittle to use in curing diseases.[1]

The Book of the Laws of Countries.

We possess an important document for the attitude of early Christianity and Gnosticism towards astrology in *The Dialogue concerning Fate* or *The Book of the Laws of Countries* of Bardesanes or Bardaisan.[2] The complete

[1] Dindorf, II, 109-10, 507-9.

[2] A. Merx, *Bardesanes der letzte Gnostiker*, Jena, 1864. F. Haase, *Zur bardesanischen Gnosis*, Leipzig, 1910, in TU, XXIV, 4.

Syriac text is extant;[1] there is a long and somewhat modified extract adopted from it in the Latin *Recognitions* of Clement,[2] and briefer fragments in the Greek fathers. Strictly speaking, the text seems to be written by some follower of Bardesanes named Philip who represents his master as discussing the problem of human free will with Avida, himself, and other disciples. The bulk of the treatise is in any case put in Bardesanes' mouth and it probably reflects his views with fair accuracy. Eusebius ascribed it to Bardesanes himself.

Personality of Bardesanes

Bardesanes (154-222 A. D.) was born in Edessa. He spent most of his life in Mesopotamia but for a time went to Armenia as a missionary. His many works in Syriac included apologies for Christianity, attacks upon heresies, and numerous hymns, but the only work extant is the treatise we are about to examine, with the possible exception of *The Hymn of the Soul*[3] ascribed to him and contained in the Syriac *Acts of St. Thomas*. His doctrines were regarded by Ephraem Syrus and others as tainted with Gnostic heresy. He is often represented as a follower of Valentinus, but the ancient authorities, such as Epiphanius and Eusebius, disagree as to whether he degenerated from orthodoxy to Valentinianism or reformed in the opposite direction. In the dialogue which we consider he is represented as a Christian, but his remarks have often been thought to have a Gnostic flavor. F. Nau, however, has argued that he was not a Gnostic and that the statements in question in the dialogue can be explained as purely astrological.[4]

Sin possible for men, angels, and stars.

The treatise opens with the query, why did not God make men so that they could not sin? The reply of course is that moral freedom for good or evil is a greater gift of God than compulsory morality. By virtue of his individual freedom of action man is equal to the angels, some of whom,

[1] English translation in AN, VIII, 723-34.

[2] *Recognitions*, IX, 17 and 19-29.

[3] English translations by A. A. Bevan, 1897; F. C. Burkett, 1899; G. R. S. Mead, 1906.

[4] F. Nau, *Une biographie inédite de Bardesane l'astrologue*, 1897.

too, have sinned with the daughters of men and fallen, and is superior even to the sun, moon, and signs of the zodiac which are fixed in their courses. The stars, however, as in *The Book of Enoch,* "are not absolutely destitute of all freedom" and will be held responsible at the day of judgment. Presently some of them are called evil.

Does fate in the astrological sense prevail?

After some discussion whether man does wrong from his nature, the treatise turns to the question, how far are men controlled by fate, that is, by the power of the seven planets in accordance with the doctrine of the Chaldeans, which is the term here usually employed for astrologers. Some men attack astrology as "a lying invention" and hold that the human will is free and that such evils as man cannot avoid are due to chance or to divine punishment but not to the stars. Between these extremes Bardesanes takes middle ground. He believes that there is such a force in the stars, whom he refers to as Potentates and Governors, as the fate of which the astrologers speak, but that this fate evidently does not rule everything, since it is itself established by the one God who imposed upon the stars and elements that motion in conformity with which "intelligences undergo change when they descend to the soul, and souls undergo change when they descend to bodies," a statement which appears to have a Gnostic flavor. This fate furthermore is limited by nature on the one hand and human free will on the other hand. The vital processes and periods which are common to all men, such as birth, generation, childbearing, eating, drinking, old age, and death, Bardesanes regards as governed by nature. "The body," he says, "is neither hindered nor helped by fate in the several acts it performs," a view which most astrologers would probably not accept. On the contrary, in Bardesanes' opinion wealth and honors, power and subjection, sickness and health, are controlled by fate which often disturbs the regular course of nature. This is because in genesis or the nativity the stars, some of which work with and some against nature, are in conflict. In short, some stars are good and some are evil.

National laws and customs as a proof of free will.

If nature is thus often upset by the stars, fate in its turn may be resisted and overpowered by man's exercise of will. This assertion Bardesanes proceeds to prove by the argument which has given to the dialogue the title, *The Book of the Laws of the Countries,* and which we find much repeated in subsequent writers. Briefly it is that in various nations certain laws are enforced upon, or customs observed by all the people alike regardless of their diverse individual horoscopes. In illustration of this are listed various prohibitions and practices fondly supposed by Bardesanes and his audience to characterize the Seres, Brahmans, Persians, Geli, Bactrians, Arabs, Britons, Parthians, Amazons, and other peoples. Savage tribes are mentioned among whom there are no artists, bankers, perfumers, musicians, and poets to fit the nativities decreed by the constellations for certain times. Bardesanes is aware of the astrological theory of seven zones or climes, by which the science of individual horoscopes is corrected and modified, but he contends that there are many different laws in each of these zones, and would be, even if the number were raised to twelve according to the number of the signs or to thirty-six after the decans. He also contends that men retain their laws or customs when they migrate to other climes, and adduces the fidelity of Jews and Christians to the commandments of their respective religions as a further illustration of the triumph of free will over the stars. He concedes, however, as before that "in every country and in every nation there are rich and poor, and rulers and subjects, and people in health and those who are sick, each one according as fate and his nativity have affected him." Incidentally to the foregoing discussion it is affirmed that the astrology of Egypt and that of the Chaldeans in Babylon are identical. At the close of the treatise is appended a note stating that Bardesanes estimated the duration of the world at six thousand years on the basis of sixty as the least number of years in which the seven planets complete an even number of revolutions.

The *Pistis-Sophia*: attitude to astrology.

If the work ascribed to Bardesanes is not certainly Gnostic, the *Pistis-Sophia* is, and we turn next to it and first of all to its attitude towards astrology. This treatise is extant in a Coptic codex of the fifth or sixth century;[1] the Greek original text was probably written in the second half of the third century. It gives the revelations made by Jesus to his disciples after He had ascended to heaven and returned again to them. When He ascended through the heavens, He changed the fatal influence of the lords of the spheres and made the planets turn to the right for six months of the year, whereas before they had faced the left continually.[2] In a long passage near the close of the *Pistis-Sophia* proper [3] Jesus asserts the absolute control of human destiny hitherto by "the rulers of the fate" and describes how they fashion the new soul, control the process of generation and of the formation of the child in the womb, and decree every event of life down to the day and manner of death. Only by the Gnostic key to the mysteries can one escape their control.[4] In the following *Book of the Saviour*, moreover, even the finding of this key is subjected to astral control, since a constellation is described under which all souls descending to this world will be just and good and will discover the mysteries of light.[5]

"Magic" condemned.

The *Pistis-Sophia* assumes the usual attitude of condemnation of magic so-called. Among the evils which Jesus warns his followers to renounce are superstition and invocations and drugs or magic potions.[6] One object of his reducing by one-third the power of the lords of the spheres when He ascended through the heavens was that men might not henceforth invoke them by magic rites for evil pur-

[1] ed. Coptic and Latin by M. G. Schwartze and J. H. Petermann, 1851; French translation by E. Amélineau, 1895; English by G. R. S. Mead, 1896; German by C. Schmidt, 1905. The Coptic text is thickly interspersed with Greek words and phrases. In the same manuscript occurs the *Book of the Saviour* of which we shall also treat.

[2] *Pistis-Sophia*, 25-6.

[3] *Ibid.*, 336-50.

[4] *Ibid.*, 355, *et seq.*

[5] *Ibid.*, 389-90.

[6] *Ibid.*, 255 and 258.

poses. Marvels may still, however, be accomplished by "those who know the mysteries of the magic of the thirteenth aeon" or power above the spheres.[1]

Power of names and rites.

But while magic is renounced, great faith is shown in the power of names and rites. Thus after a description of the dragon of outer darkness and the twelve main dungeons into which it divides and the animal faces and names of the twelve rulers thereof, who evidently represent in an inaccurate fashion the signs of the zodiac, it is added that even unrepentant sinners, if they know the mystery of any one of these twelve names, can escape from these dungeons.[2] In the *Book of the Saviour* Jesus not only utters several long lists of strange and presumably magic words by way of invocation to the Power or powers above, but these are accompanied by careful observance of ceremonial. On both occasions Jesus and the disciples are clad in linen.[3] In the first case the disciples are carefully grouped with reference to the points of the compass, towards which Jesus turns successively as He utters the magic words standing at a sacrificial altar. The result of this ceremony and invocation was that the heavens were displaced and the earth left behind and that Jesus and the disciples found themselves in the region of mid-air. Before uttering the other invocation Jesus commanded that fire and vine branches be brought, placed an offering on the flame, and carefully arranged two vessels of wine, two cups of water, and as many pieces of bread as there were disciples. In this case the object was to remit the sins of the disciples. In the *Book of Jeû* in the Bruce Papyrus there is a perfect riot of such magic names and invocations, seals and diagrams, and accompanying ceremonial.[4]

Interest in natural science.

The interest of the Gnostics in natural science is seen in the list of things that will be known by one who has pene-

[1] *Pistis-Sophia*, 29-30.

[2] *Ibid.*, 319-35.

[3] *Ibid.*, 357-8, 375-6.

[4] Carl Schmidt, *Gnostische Schrifte in koptischer Sprache aus dem codex Brucianus*, 1892, 692 pp., in TU, VIII, 2, with German translation of the Coptic text at pp. 142-223. Portions have been translated into English by G. R. S. Mead, *Fragments of a Faith Forgotten*, 1900.

trated all the mysteries and fully entered upon the inheritance of the kingdom of light. Not only will he understand why there is light and darkness, and why sin and vice exist and life and death, but also why there are reptiles and wild beasts and why they shall be destroyed, why there are birds and beasts of burden, why there are gems and precious metals, why there are brass, iron and steel, lead, glass, wax, herbs, waters, "and why the wild denizens of the sea." Why there are four points of the compass, why demons and men, why heat and cold, stars, winds, and clouds, frost, snow, planets, aeons, decans, and so on and so forth.[1]

"Gnostic gems" and astrology.

King has shown that many of the so-called "Gnostic gems" are purely astrological talismans and that "only a very small minority amidst their multitude present any traces of the influence of Christian doctrines." [2] Many are for medicinal or magical purposes rather than of a religious character. Some nevertheless are engraved with the truly Gnostic figure of Pantheus Abraxas which King regards as "the actual invention of Basilides." Another common symbol, borrowed from Egypt, is the Agathodaemon, which by the third century had become the popular designation of the hooded snake of Egypt, or Chnuphis or Chneph, a great serpent with a lion's head encircled by a crown of seven or twelve rays, representing the planets or signs. Often the seven Greek vowels are placed at the tips of the seven rays. On the obverse of the gem the letter "s" is engraved thrice and traversed by a straight rod, a design probably meant to depict a snake twisting about a wand. We are reminded, not only with King of the club of Aesculapius, but of Aaron's rod, the magicians of Pharaoh, and the serpent lifted up in the wilderness; also of Lucian's tale of the pretended discovery of the god Asclepius by the pseudo-prophet, Alexander. At least one "Gnostic amulet" has on the back the legend "Iao Sabao" (th).[3]

[1] *Pistis-Sophia,* 205-15.

[2] C. W. King, *The Gnostics and their Remains,* 1887, pp. xvi-xviii, 215-8. Also his *The Natural History, Ancient and Modern, of Precious Stones and Gems,* London, 1865.

[3] A. B. Cook, *Zeus,* p. 235, citing J. Spon, *Miscellanea eruditae antiquitatis,* Lyons, 1685, p. 297.

The planets in early Christian art.

The influence of astrology may be seen in other and more certainly genuine works of early Christian art than many of the so-called Gnostic gems. On a lamp in the catacombs Christ is depicted as the good shepherd with a lamb on His shoulder. Above His head are the seven planets, although the sun and moon are shown again at either side, and about His feet press seven lambs, perhaps an indication that He is freeing the peoples of the seven climes from the fatal influence of the stars. In the *Poemander* attributed to Hermes it is stated that there are seven peoples from the seven planets. On a gem of perhaps the third century a similar scene is engraved except that the sun and moon are not shown apart from the seven planets, and that the lamb on Christ's shoulders is counted as one of the seven, so that there are but six at His feet.[1]

Gnostic amulets in Spain.

"Gnostic amulets and other works of art" are occasionally found in Spain, especially the Asturian northwest which remained Christian at the time of the Mohammedan conquest of the rest of the peninsula. One ring is inscribed with the sentence, "Zeus, Serapis, and Iao are one." On another octagonal ring are Greek letters signifying the Gnostic *Anthropos* or father of wisdom. A stone is carved with a candelabrum and the seven planets, "the sacred hebdomad of the Chaldeans." [2]

Syriac Christian charms.

Gollancz in his *Selection of Charms from Syriac Manuscripts* presents a number of spells and incantations which, whether any of them are Gnostic or not, certainly seem to be Christian, since they mention the divine persons of Christianity, Mary, and various Biblical characters.[3]

At the close of the fourth century the views of the Gnostics were revived in Gaul and Spain by Priscillian, who

Reitzenstein, *Poimandres,* pp. 111-3. On the planets in later medieval art see Fuchs, *Die Ikonographie der 7 Planeten in der Kunst Italiens bis zum Ausgange des Mittelalters,* Munich, 1909.

[2] E. S. Bouchier, *Spain under the Roman Empire,* p. 125.

[3] Hermann Gollancz, *Selection of Charms from Syriac Manuscripts,* 1898; also pp. 77-97 in *Acts of International Congress of Orientalists,* Sept., 1897; Syriac text and English translation.

seems to have been much influenced by astrology and who was put to death at Treves in 385 A. D. on a charge of magic. He confessed under torture, but was afterwards thought innocent. We are not told, however, what the magical practices were of which he was accused.[1] Both Sulpicius Severus and Isidore of Seville [2] state that he was accused of *maleficium,* which should mean witchcraft, sorcery, or magical operations with the intent to injure someone. But further details are wanting, except that Sulpicius calls Priscillian a man "more puffed up than was right with the knowledge of profane things, and who was further believed to have practiced magic arts since adolescence," while Isidore states that Bishop Itacius (Ithaicus), who was largely responsible for pushing the charges against Priscillian, showed in a book which he wrote against Priscillian's heresy that "a certain Marcus of Memphis, most learned in magic art, was a disciple of Mani and master of Priscillian." Priscillian himself states in his extant works that Itacius had accused him of magic in 380. As the final trial proceeded, Itacius gave way as accuser to a public prosecutor (*fisci patronus*) who continued the case on behalf of the emperor Maximus who seems to have had his eye upon Priscillian's large fortune. St. Martin of Tours in vain obtained from Maximus a promise that Priscillian should not be put to death.[3] But his execution brought his persecutor Itacius into such bad odor that he was excommunicated and condemned to exile for the rest of his life.

Priscillian executed for magic.

We have just heard that Priscillian was taught by a disciple of Mani, while Ephraem Syrus states that Bardesanes

Manichean Manuscripts

[1] In 1885-1886 eleven tracts by Priscillian were discovered by G. Schepss in a Würzburg MS. They shed, however, little light upon the question whether he was addicted to magic. They have been published in *Priscilliani quae supersunt.*, etc., ed. G. Schepss, 1889, in CSEL, XVIII.

See also E. Ch. Babut, *Priscillien et la Priscillienisme*, Paris, 1909 (*Bibl. d. l'École d. Hautes Études,* Fasc. 169), which supersedes the earlier works of Paret, 1891; Dierich, 1897; and Edling, 1902.

[2] *Sulpicii Severi Historia Sacra,* II, 46-51 (Migne, PL, XX, 155, *et seq.*) S. Isidori Hispalensis Episcopi, *De viris illustribus,* Cap. 15 (Migne, PL, LXXXIII, 1092).

[3] *Realencyklopädie für protestantische Theologie,* XVI, 63.

was the teacher of Mani. Augustine in his youth, when a follower of the Manicheans, had been devoted to astrology. This connection between Gnosticism and astrology and Manicheism has been further attested by the fragments of Manichean manuscripts recently discovered in central Asia.[1] In them the sun-god and moon-god and five other planets play a prominent part. Besides the five planets we have five elements—ether, wind, light, fire, and water—five plants, five trees, and five beings with souls—man, quadrupeds, reptiles, aquatic, and flying animals. The five gods or luminous bodies are represented as good forces who imprisoned five kinds of demons; but the devil had his revenge by imprisoning luminous forces in man, whom he made a microcosm of the universe. And whereas the good spirit had created sun and moon, the devil formed male and female. The great sage of beneficent light then appeared in the world and brought forth from his own five members five liberators—pity, contentment, patience, wisdom, and good faith—corresponding to the five elements just as among the Christians we shall find four virtues and four elements. Then ensued the struggle of the old man with the new man. Although we are commonly told that idolatry and magic were strictly prohibited by the Manicheans, the envoy of light is in one text represented as "employing great magic prayers" in his effort to deliver living beings. When men eat living beings, they offend against the five gods, the earth dry and moist, the five orders of animate beings, the five different herbs and five trees. Other numbers than five appear in these Manichean fragments: four seals of light and four praises, four courts with iron barriers; three vestments and three wheels and three calamities; ten vows and ten layers of heavens above, and eight layers of earth beneath; twelve

[1] My following statements in the text are based upon E. Chavannes et P. Pelliot, *Un traité manichéen retrouvé en Chine*, 1913,—they date the Chinese translation about 900 A.D. and the MS of it within a century later; W. Radloff, *Chuastuanift, Das Bussgebet der Manichäer*, Petrograd, 1909; A. v. Le Coq, *Chuastuanift, ein Sündenbekenntnis der Manichäischen Auditores*, Berlin, 1911. There are further publications on the subject.

great kings and twelve evil natures; thirteen great luminous forces and thirteen parts of the carnal body and thirteen vices,—elsewhere fourteen parts; fifteen enumerations of sins for which forgiveness is sought; fifty days in the year to be observed; and so on.

The Mandaeans.

A sect derived either from Gnosticism or from common sources seems still to exist in the case of the Mandaeans of southern Babylonia.[1] They believe that the earth and man were formed by a Demiurge, who corresponds to the Ialdabaoth of the Ophites, and who was aided by the spirits of the seven planets. They divide the history of the world into seven ages and represent Jesus Christ as a false prophet and magician produced by the planet Mercury. The lower world consists of four vestibules and three hells proper and has seven iron and seven golden walls. A dying Mandaean is clothed in a holy dress of seven pieces. The spirits of the planets, however, are represented as evil beings, and the first two of three sets of progeny borne by the spirit of hell fire were the seven planets and the twelve signs of the zodiac. The influence of these two numbers, seven and twelve, may be further seen in the regulation that a candidate for the priesthood should be at least nineteen years old and have had twelve years of previous training, which we infer would normally begin when he reached his seventh year and not before. Other prominent numbers in Mandaean lore are five,[2] perhaps indicative of the planets other than sun and moon, and three hundred and sixty, suggestive of the number of degrees in the circle of the zodiac. Thus the main manifestations of the primal light are five, and the third generation produced by the spirit of hell fire was of like number. The number of aeons is often stated as three hundred and sixty, and the delivering deity or Messiah of the

[1] The following details are drawn from the articles on the Mandaeans in EB, 11th edition, by K. Kessler and G. W. Thatcher, and in ERE by W. Brandt, author of *Mandäische Religion*, 1889, and *Mandäische Schriften*, 1893, and from Anz (1897), pp. 70-8. Further bibliography will be found in these references.

[2] The number five also appears in the *Pistis-Sophia* and other Gnostic literature.

Mandaeans is said to have sent forth that number of disciples before his return to the realm of light. We hear of yet other numbers, such as 480,000 years for the duration of the world, 60,000, and 240, but these too are commensurate, if not identical, with astrological periods such as those of conjunctions and the *magnus annus*. A peculiarity of Mandaean astronomy and astrology is that the other heavenly bodies are all believed to rotate about the polar star. Mandaeans always face it when praying; their sanctuaries are built so that persons entering face it; and even the dying man is placed so that his feet point and eyes gaze in its direction. Like the Gnostics, the Mandaeans invoke by many strange names their spirits and aeons who are divided into numerous orders. Their names for the planets seem to be of Babylonian origin. Passages from their sacred books are recited like incantations and are considered more effective in danger and distress than prayer in the ordinary sense of the word. Such recitations are also employed to aid the souls of the dead to ascend through various stages or prisons to the world of light. Earthenware vessels have recently been brought to light with Mandaean inscriptions and incantations to avert evil.[1]

[1] H. Pognon, *Une Incantation contre les génies malfaisants en Mandäite*, 1893; *Inscriptions mandäites des coupes de Khonabir*, 1897-1899. M. Lidzbarski, *Mandäische Zaubertexte*, in *Ephemeris f. semit. Epig.*, I (1902), 89-106. J. A. Montgomery, *Aramaic Incantation Texts from Nippur*, 1913.

CHAPTER XVI

THE CHRISTIAN APOCRYPHA

Magic in the Bible—Apocryphal Gospels of the Infancy—Question of their date—Their medieval influence—Resemblances to Apuleius and Apollonius in the Arabic *Gospel of the Infancy*—Counteracting magic and demons—Other miracles and magic by the Christ child—Sometimes with injurious results—Further marvels from the *Pseudo-Matthew*—Learning of the Christ child—Other charges of magic against Christ and the apostles—The *Magi* and the star—Allegorical zoology of Barnabas—Traces of Gnosticism in the apocryphal Acts—Legend of St. John—Legend of St. Sousnyos—Old Testament Apocrypha of the Christian era.

Magic in the Bible.

IT is hardly necessary to rehearse here in detail the numerous allusions to, prohibitions of, and descriptions of the practice of magic, witchcraft, and astrology, enchantments and exorcisms, divination and interpretation of dreams, which are to be found scattered through the pages of the Old and New Testaments. Such passages had a profound influence upon Christian thought on such themes in the early church and during the middle ages, and we shall have occasion to mention many, if not most, of such scriptural passages, in connection with this later discussion of them by the church fathers and others. For instance, Pharaoh's magicians and their contests with Moses and Aaron; Balaam and his imprecations and enchantments and prediction that a star would come out of Jacob and a scepter out of Israel; the witch of Endor or ventriloquist and her invocation of what seemed to be the ghost of Samuel; the repeated use of the numbers seven and twelve, suggestive of the planets and signs of the zodiac, as in the twelve cakes of showbread and candlestick with seven branches; the dreams and interpretation of dreams of Joseph and Daniel, not to mention the former's silver divining cup;[1] the wise men who saw Christ's star in the east; Christ's own allusion to the shaking of "the powers of the heavens" and the gathering of His elect from the four winds at His second coming; the accusation against Christ that He cast out demons by the aid of the prince of demons; the eclipse of the sun at the time of the crucifixion; the adventures of the apostles with Simon Magus, with Elymas the sorcerer, and with the damsel possessed with a spirit of divination who brought her master much gain by soothsaying; the burning of their books of magic by the vagabond Jewish exorcists; the prohibitions of heathen divination and witchcraft by the Mosaic law and by the prophets; the penalties prescribed for sorcerers in the Book of Revelation; at the same time the legalized practice of similar superstitions, such as the ordeal to test a wife's faithfulness by making her drink "the bitter water that causeth the curse,"[2] the engraved gold plate upon the high priest's forehead,[3] or the use of Paul's handkerchief and underwear to cure the sick and dispel demons; the promise to believers in the closing verses or appendix of *The Gospel according to St. Mark* that they shall cast out devils, speak with new tongues, handle serpents and drink poison without injury, and cure the sick by laying on of hands. The foregoing scarcely exhaust the obvious allusions or analogies to astrology and other magic arts in the Bible, to say nothing of less explicit passages[4] which were later taken to justify certain occult arts, as Exodus XIII, 9, to support chiromancy, and the Gospel of John XI, 9, to support the astrological doctrine of elections. Suffice it for the present to say that the prevailing atmosphere of the Bible is one of

[1] Genesis XLIV, 5, and J. G. Frazer (1918), II, 426-34.

[2] In the apocryphal *Protevangelium of James,* cap. 16, both Joseph and Mary undergo the test.

[3] Joachim consults the plate in the *Protevangelium,* cap. 5.

[4] See J. G. Frazer, *Folk-Lore in the Old Testament,* 1918, 3 vols., and also his other works; for instance, *The Magic Art,* 1911, I, 258, for the contest in magic rain-making between Elijah and the priests of Baal in First Kings, Chapter XVIII, while I do not understand why Joshua is not mentioned in connection with "The magical control of the sun," *Ibid.,* I, 311-19.

prophecy, vision, and miracle, and that with these go, like the obverse face of a coin or medal, their inevitable accompaniments of divination, demons, and magic.

Apocryphal gospels of the infancy.

This is also the case in apocryphal literature of the New Testament which is now so much less familiar and accessible especially to English readers,[1] but which had wide currency in the early Christian and medieval periods. We may begin with the apocryphal gospels and more particularly those dealing with the infancy and childhood of Christ. Of these two are believed to date from the second century, namely, the Gospel of James or "Gospel of the Infancy" (*Protoevangelium Iacobi*) [2] and the Gospel of St. Thomas, which is mentioned by Hippolytus. However, he cites a sentence which is not in the present text—of which the manuscripts are scanty and for the most part of late date [3]—and the gospel as we have it is not Gnostic, as he says it is, so that our version has probably been altered by some Catholic.[4] Later in date is the Latin gospel of the Pseudo-Matthew—perhaps of the fourth or fifth century—and the Arabic Gospel of the Infancy, which is believed to be a translation from a lost Syriac original. We are the worst off of all for manuscripts of its text and apparently there is no Latin manuscript of it now extant, although a Latin

[1] However, the *Apocrypha of the New Testament* may be read in English translation by Alexander Walker in *The Ante-Nicene Fathers* (American edition), VIII, 357-598, and in that by Hone in 1820, which has since been reprinted without change. It includes only a part of the apocrypha now known and presents these in a blind fashion without explanation. It differs from Tischendorf's text of the apocryphal gospels (*Evangelia Apocrypha*, ed. Tischendorf, Lipsiae, 1876) both in the titles of the gospels, the distribution of the texts under the respective titles, and the division into chapters. I have, however, sometimes used Hone's wording in making quotations. Older than Tischendorf is Thilo, *Codex apocryphus Novi Testamenti*, Leipzig, 1832; Fabricius, etc.

[2] It is ascribed to the second century both by Tischendorf and *The Catholic Encyclopedia* ("Apocrypha," 607). There are plenty of fairly early Greek MSS for it.

[3] The Greek MSS are of the 15th and 16th centuries; Tischendorf examined only partially a Latin palimpsest of it which is probably of the fifth century.

[4] So argues *The Catholic Encyclopedia*, 608; Tischendorf seems inclined to date the Gospel of Thomas a little later than that of James, and to hold that we possess only a fragment of it.

text has reached us through the printed editions. Tischendorf was, however, "unwilling to omit in this new collection of the apocryphal gospels that ancient and memorable monument of the superstition of oriental Christians," and for the same reason we shall survey its medley of miracle and magic in the present chapter. Speaking of the flight into Egypt this gospel says, "And the Lord Jesus performed a great many miracles in Egypt which are not found recorded either in the Gospel of the Infancy or in the Perfect Gospel." [1] Tischendorf noted the close resemblance of its first nine chapters to the Gospel of James and of chapters 36-55 to the Gospel of Thomas, while the intervening chapters "contain especially fables of the sort you may fittingly call oriental, filled with allusions to Satan and demons and sorceries and magic arts." [2] We find, however, the same sort of fables in the other three apocryphal gospels; there are simply more of them in the Arabic Gospel of the Infancy. It appears to be a compilation and may embody other earlier sources no longer extant as well as passages from the pseudo-James and pseudo-Thomas.

Question of their date.

There is a tendency on the part of orthodox Christian scholars to defer the writing of apocryphal works to as late a date as possible, and they seem to have a notion that they can save the credibility or purity of the miracles of the New Testament [3] by representing such miracles as those recorded of the infancy of Christ as the inventions of a later age. And it is probably true that all these marvels were not the invention of a single century but of a succession of

[1] *Evang. Inf. Arab.*, cap. 25, "fecitque dominus Iesus plurima in Egypto miracula quae neque in evangelio infantiae neque in evangelio perfecto scripta reperiuntur."

[2] Tischendorf (1876), p. xlviii. As I have already intimated on other occasions, it seems to me no explanation to call such stories "oriental." Christianity was an oriental religion to begin with. Moreover, as our whole investigation goes to show, both classical antiquity and the medieval west were ready enough both to repeat and to invent similar tales.

[3] It may be noted, however, that the chief miracles of the Gospels were attacked as "absurd or unworthy of the performer" nearly two centuries ago by Thomas Woolston in his *Discourses on the Miracles of our Saviour*, 1727-1730. The words in quotation marks are from J. B. Bury's *History of Freedom of Thought*, 1913, p. 142.

centuries. On the other hand, I know of no reason for thinking Christians of the first century any less credulous than Christians of the fifth century; it was not until the latter century that Pope Gelasius' condemnation of apocryphal books was drawn up, but apocryphal books had long been in existence before that time; nor for thinking the Christians of the thirteenth century any more credulous than those of the other two centuries. It is only in our own age that Christians have become really critical of such matters. Moreover, these unacceptable miracles, whenever they were invented, were presumably invented by and accepted by Christians, who must bear the discredit for them. Whatever the century was, the same men believed in them who believed in the miracles recorded in the New Testament. If the plant has flowered into such rank superstition, can the original seed escape responsibility? The Arabic Gospel of the Infancy is no doubt an extreme instance of Christian credence in magic, but it is an instance that cannot be overlooked, whatever its date, place, or language.

Their medieval influence.

These apocryphal gospels of the Infancy, which are in part extant only in Latin, continued to be influential in the medieval period. At the beginning of it we find included in Pope Gelasius' list of apocryphal works, published at a synod at Rome in 494,[1] besides apocryphal gospels of Matthew and of Thomas—which last we are told, "the Manicheans use"—a *Liber de infantia Salvatoris* and a *Liber de nativitate Salvatoris et de Maria et obstetrice.* There are numerous manuscripts of such gospels in the later medieval centuries but it would not be safe to attempt to identify or classify them without examining each in detail. As Tischendorf said, the Latins do not seem to have long remained content with mere translations of the Greek pseudo-gospel of James but combined the stories told there with others from the Pseudo-Thomas or other sources into new

[1] Migne, PL, 59, 162 ff. The list was reproduced with slight variations by Hugh of St. Victor in the twelfth century in his *Didascalicon* (IV, 15), and in the thirteenth century by Vincent of Beauvais in the *Speculum Naturale* (I, 14).

apocryphal treatises. Thus the extant Latin apocrypha in no case reproduce the Gospel of James accurately but rather are imitated after it, and include some of it, omit some of it, embellish some of its tales, and add to it.[1] Mâle states in his work on religious art in France in the thirteenth century that *The Gospel of the Pseudo-Matthew* and *The Gospel of Nicodemus* or *Acts of Pilate* were the two apocryphal gospels especially used in the twelfth and thirteenth centuries.[2]

Resemblances to Apuleius and Apollonius in the Arabic Gospel of the Infancy.

That the fables of the Arabic Gospel of the Infancy were at least not fresh from the orient is indicated by the way in which some of the incidents in the stories of Apuleius and Apollonius of Tyana are closely paralleled.[3] In the parlor of a well furnished house where lived two sisters with their widowed mother stood a mule caparisoned in silk and with an ebony collar about his neck, "whom they kissed and were feeding."[4] He was their brother, transformed into a mule by the sorcery of a jealous woman one night a little before daybreak, although all the doors of the house were locked at the time. "And we," they tell a girl who had been instantly cured of leprosy by use of perfumed water in which the Christ child had been washed and who had then become the maid-servant of the virgin Mary,[5] "have applied to all the wise men, magicians, and diviners in the world, but they have been of no service to us."[6] The girl recommends them to consult Mary, who restores their brother to human form by placing the Christ child upon his back. This romantic episode is then brought to a fitting conclusion by the marriage of the brother to the girl who had assisted in his restoration to his right body. As the demon, who

[1] Tischendorf (1876), pp. xxiii-xxiv.

[2] Mâle (1913), pp. 207-8.

[3] Since writing this, I find that Mâle has been impressed by the same resemblance. He writes (1913), p. 207, "Some chapters in the apocryphal gospels are like the *Life of Apollonius of Tyana* or even like *The Golden Ass*, permeated with the belief in witchcraft and magic." The resemblance to Apuleius is also noted in AN, VIII, 353.

[4] Tischendorf, *Evang. Infantiae Arabicum*, caps. 20-21.

[5] *Ibid.*, cap. 17.

[6] *Ibid.*, cap. 20, "nullum in mundo doctum aut magum aut incantatorem omisimus quin illum accerseremus; sed nihil nobis profuit."

in the form of an artful beggar was causing the plague at Ephesus and whom Apollonius had stoned to death, turned at the last moment into a mad dog, so Satan, when forced by the presence of the Christ child to leave the boy Judas, ran away like a mad dog.[1] The reviving of a corpse by an Egyptian prophet in the *Metamorphoses* in order that the dead man may tell who murdered him is paralleled in both the Arabic Infancy and the gospels of Thomas and the Pseudo-Matthew by the conduct of Jesus when accused of throwing another boy down from a house-top. The text reads: "Then the Lord Jesus going down stood over the dead boy and said with a loud voice, 'Zeno, Zeno, who threw you down from the house-top?' Then the dead boy answered, 'Lord, thou didst not throw me down, but so-and-so did.'"[2]

Counter-acting magic and demons.

Many were the occasions upon which the Christ child or his mother counteracted the operations of magic or relieved persons who were possessed by demons. Kissing him cured a bride whom sorcerers had made dumb at her wedding,[3] and a bridegroom who was kept by sorcery from enjoying his wife was cured of his impotence by the mere presence of the holy family who lodged in his house for the night.[4] Mary's pitying glance was sufficient to expel Satan from a woman possessed by demons.[5] Another upright woman who was often vexed by Satan in the form of a serpent when she went to bathe in the river,[6] which reminds one somewhat of Olympias and Nectanebus,[7] was permanently cured by kissing the Christ child. And a girl, whose blood Satan used to suck, miraculously discomfited him when he

[1] *Evang. Inf. Arab.*, cap. 35, "Extemplo exivit ex puero illo satanas fugiens cani rabido similis." The apocryphal gospel adds, "This same boy who struck Jesus," i.e., while he was still possessed by the demon, "and out of whom Satan went in the form of a dog, was Judas Iscariot, who betrayed Him to the Jews. And that same side, on which Judas struck him, the Jews pierced with a lance."

[2] *Ibid.*, cap. 44; *Evang. Thomae Lat.*, cap. 7; *Ps. Matth.*, cap. 32.

[3] *Evang. Inf. Arab.*, cap. 15.

[4] *Ibid.*, cap. 19, "qui veneficio tactus uxore frui non poterat."

[5] *Ibid.*, cap. 14.

[6] *Ibid.*, cap. 16.

[7] See below, chapter 24.

appeared in the shape of a huge dragon by putting upon her head and about her eyes a swaddling cloth of Jesus which Mary had given to her. Fire then went forth and was scattered upon the dragon's head and eyes, as from the blinking eyes of the artful beggar who caused the plague in the *Life of Apollonius of Tyana,* and he fled in a panic.[1] A priest's three-year-old son who was possessed by a great multitude of devils, who uttered many strange things, and who threw stones at everybody, was likewise cured by placing on his head one of Christ's swaddling clothes which Mary had hung out to dry. In this case the devils made their escape through his mouth "in the shape of crows and serpents." [2] Such marvels may offend modern taste but have their probable prototype in the miracles wrought by use of Paul's handkerchief and underwear in the New Testament and illustrate, like the placing of spittle on the eyes of the blind man, the great healing virtue then ascribed to the perspiration and other secretions and excretions of the human body.

Other miracles and magic by the Christ child.

Sick children as well as lepers were cured by the water in which Jesus had bathed or by wearing coats made of his swaddling clothes,[3] while the child Bartholomew was snatched from the very jaws of death by the mere smell of the Christ child's garments the moment he was placed on Jesus' bed.[4] On the road to Egypt is a balsam which was produced "from the sweat which ran down there from the Lord Jesus." [5] The Christ child cured snake-bite, in the case of his brother James by blowing on it, in the case of his playfellow, Simon the Canaanite, by forcing the serpent who had stung him to come out of its hole and suck all the poison from the wound, after which he cursed the snake "so that it immediately burst asunder and died." [6] When the boy Jesus took all the cloths waiting to be dyed with different colors in a dyer's shop and threw them into the furnace, the dyer began to scold him for this mischief, but the cloths all

[1] *Evang. Inf. Arab.*, caps. 33-34.
[2] *Ibid.*, caps. 10-11.
[3] *Ibid.*, caps. 27-32.
[4] *Ibid.*, cap. 30.
[5] *Ibid.*, cap. 24.
[6] *Ibid.*, caps. 42-43; *Ps. Matth.*, 41; *Evang. Thom. Lat.*, 14. Compare pp. 279-80 above.

came out of the desired colors.[1] Jesus also miraculously remedied the defective carpentry of Joseph, who had worked for two years on a throne for the king of Jerusalem and made it too short. Jesus and Joseph took hold of the opposite sides and pulled the throne out to the required dimensions.[2]

Sometimes with injurious results.

The usual result of the Christ child's miracles was that all the bystanders united in praising God. But when his little playmates went home and told their parents how he had made his clay animals walk and his clay birds fly, eat, and drink, their elders said, "Take heed, children, for the future of his company, for he is a sorcerer; shun and avoid him, and from henceforth never play with him." [3] Indeed, if the theory of the fathers is correct that the surest hall-mark by which divine miracles may be distinguished from feats of magic is that the former are never wrought for any evil end while the latter are, it must be admitted that his contemporaries were sometimes justified in suspecting the Christ child of resort to magic. After his playmates had been thus forbidden to associate with Jesus, they hid from him in a furnace, and some women at a house near by told him that there were not boys but kids in the furnace. Jesus then actually transformed them into kids who came skipping forth at his command.[4] It is true that he soon changed them back into human form, and that the women worshiped Christ and asserted their conviction that he was "come to save and not to destroy." But on several subsequent occasions Jesus is represented in the apocryphal gospels of the infancy as causing the death of his playmates. When another boy broke a little fish-pool which Jesus had constructed on the Sabbath day, he said to him, "In like manner as this water has vanished, so shall thy life vanish," and the boy pres-

[1] *Evang. Inf. Arab.*, cap. 37.

[2] *Ibid.*, 38-39; *Ps. Matth.*, 37; *Evang. Thom. Lat.*, 11.

[3] *Evang. Inf. Arab.*, cap. 36; *Ps. Matth.*, 27; *Evang. Thom. Lat.*, 4.

[4] *Evang. Inf. Arab.*, cap. 40. See Ad-Damîrî, translated by A. S. G. Jayakar, 1906, I, 703, for a Moslem tale of Jews who called Jesus "the enchanter the son of the enchantress," and were transformed into pigs.

ently died.[1] When a third boy ran into Jesus and knocked him down, he said, "As thou hast thrown me down, so shalt thou fall, nor ever rise;" and that instant the boy fell down and died.[2] When Jesus' teacher started to whip him, his hand withered and he died. After which we are not surprised to hear Joseph say to Mary, "Henceforth we will not allow him to go out of the house; for everyone who displeases him is killed." [3]

Further marvels from the *Pseudo-Matthew.*

As has been indicated in the foot-notes many of the foregoing marvels are recounted in the Pseudo-Matthew and Latin Gospel of Thomas as well as in the Arabic Gospel of the Infancy. The Pseudo-Matthew also tells how lions adored the Christ child and were bade by him to go in peace.[4] And how he "took a dead child by the ear and suspended him from the earth in the sight of all. And they saw Jesus speaking with him like a father with his son. And his spirit returned unto him and he lived again. And all marveled thereat." [5] When a rich man named Joseph died and was lamented, Jesus asked his father Joseph why he did not help his dead namesake. When Joseph asked what there was that he could do, Jesus replied, "Take the handkerchief which is on your head and go and put it over the face of the corpse and say to him, 'May Christ save you.'" Joseph followed these instructions except that he said, *"Salvet te Iesus,"* instead of *"Salvet te Christus,"* which was possibly the reason why the dead man upon reviving asked, "Who is Jesus?" [6]

Learning of the Christ child.

While no very elaborate paraphernalia or ceremonial were involved in the miracles ascribed to the Christ child in the Arabic Gospel of the Infancy, it is perhaps worth noting that he was already possessed of all learning and nonplussed his masters, when they tried to teach him the alpha-

[1] *Evang. Inf. Arab.*, 46; *Evang. Thom. Lat.*, 4; *Ps. Matth.*, 26, where Mary afterwards induces Jesus to restore him to life, and 28.

[2] *Evang. Inf. Arab.*, cap. 47; *Evang. Thom. Lat.*, 5; *Ps. Matth.*, 29.

[3] *Evang. Inf. Arab.*, cap. 49; *Evang. Thom. Lat.*, 12; *Ps. Matth.*, 38.

[4] *Ps. Matth.*, caps. 35-36.

[5] *Ibid.*, cap. 29.

[6] *Ibid.*, cap. 40.

bet, by asking the most abstruse questions. And when he appeared before the doctors in the temple, he expounded to them not only the books of the law,[1] but natural philosophy, astronomy, physics and metaphysics, physiology, anatomy, and psychology. He is represented as telling them "the number of the spheres and heavenly bodies, as also their triangular, square, and sextile aspect; their progressive and retrograde motion; their twenty-fourths and sixtieths of twenty-fourths" (perhaps corresponding to our hours and minutes!) "and other things which the reason of man had never discovered." Furthermore, "the powers also of the body, its humors and their effects; also the number of its members, and bones, veins, arteries, and nerves; the several constitutions of the body, hot and dry, cold and moist, and the tendencies of them; how the soul operates upon the body; what its various sensations and faculties are; the faculty of speaking, anger, desire; and lastly, the manner of the body's composition and dissolution, and other things which the understanding of no creature had ever reached." [2] It may be added that in the apocryphal epistles supposed to have been interchanged between Christ and Abgarus, king of Edessa, that monarch writes to Christ, "I have been informed about you and your cures, which are performed without the use of herbs and medicines." [3]

Other charges of magic against Christ and the apostles.

Jesus is again accused of magic in *The Gospel of Nicodemus* or *Acts of Pontius Pilate,* where the Jews tell Pilate that he is a conjurer. After Pilate has been warned by his wife, the Jews repeat, "Did we not say unto thee, He is a magician? Behold, he hath caused thy wife to dream." [4] In the *Acts of Paul and Thecla,* to which Tertullian refers and which are now seen to be an excerpt from the apocry-

[1] Later the same gospel (cap. 54) rather inconsistently represents Jesus as engaged in the study of law until his thirtieth year.

[2] *Evang. Inf. Arab.*, caps. 51-52.

[3] Eusebius states that he discovered these letters written in Syriac in the public records of Edessa. Hone says that it used to be a common practice among English people to have the epistle ascribed to Christ framed and place a picture of the Saviour before it.

[4] *Gospel of Nicodemus,* I, 1-2.

phal *Acts of Paul,* discovered in 1899 in a Coptic papyrus,[1] the mob similarly cries out against Paul, "He is a magician; away with him." In the *Acts of Peter and Andrew*[2] they are both accused of being sorcerers by Onesiphorus, who also, however, denies that Peter can make a camel go through the eye of a needle. Nor is he satisfied when the feat is successfully performed with a needle and camel of Peter's selection, but insists upon its being repeated with an animal and instrument of his own selection. Onesiphorus also has "a polluted woman" ride upon his camel's back, apparently with the idea that this will break the magic spell. But Peter sends the camel through the eye of the needle, "which opened up like a gate," as successfully as before, and also back again through it once more from the opposite direction.

The *Magi* and the star.

Some details are added by the apocrypha to the account of the star at Christ's birth. The Arabic Gospel states that Zoroaster (Zeraduscht) had predicted the coming of the *Magi,* that Mary gave the *Magi* one of Christ's swaddling clothes, that they were guided on their homeward journey by an angel in the form of the star which had led them to Bethlehem, and that after their return they found that the swaddling cloth would not burn in fire.[3] The *Epistle of Ignatius to the Ephesians* states that this star shone with a brightness far exceeding all others, filling men with fear, and that with its coming the power of magic was destroyed and the new kingdom of God ushered in.[4]

Allegorical zoology of Barnabas.

In the apocryphal *Epistle of Barnabas* occurs some of that allegorical zoology which we are apt to associate especially with the Physiologus. In its ninth chapter the hyena and weasel are adduced as examples of its contention that the Mosaic distinction between clean and unclean animals has a spiritual meaning. Thus the command not to eat the hyena means not to be an adulterer or corrupter of

[1] CE, *Apocrypha,* p. 611.

[2] Greek text in Tischendorf, *Apocalypses Apocryph.,* pp. 161-7; English translation, *The Ante-Nicene Fathers,* VIII, 526-7.

[3] *Evang. Inf. Arab.,* 7-8.

[4] Cap. 19 (AN, I, 57).

others, for the hyena changes its sex annually. The weasel which conceives with its mouth signifies persons with unclean mouths. In the *Acts of Barnabas* he cures the sick of Cyprus by laying a copy of the *Gospel of Matthew* upon their bodies.[1]

Traces of Gnosticism in the apocryphal Acts.

If we turn again to the various apocryphal Acts, where we have already noted charges of magic made against the apostles, we may find traces of gnosticism which have already been noted by Anz.[2] In the *Acts of Thomas* the Holy Ghost is called the pitying mother of seven houses whose rest is the eighth house of heaven. In the *Acts of Philip* that apostle prays, "Come now, Jesus, and give me the eternal crown of victory over every hostile power . . . Lord Jesus Christ . . . lead me on . . . until I overcome all the cosmic powers and the evil dragon who opposes us. Now therefore Lord Jesus Christ make me to come to Thee in the air." *The Acts of John,* too, speak of overcoming fire and darkness and angels and demons and archons and powers of darkness who separate man from God.

Legend of John.

We deal in another chapter with the struggle of the apostles with Simon Magus as recounted in the apocryphal *Acts of Peter and Paul,* and with similar legends of the contests of other apostles with magicians. Here, however, we may mention some of the marvels in the apocryphal legend of St. John, supposed to have been written by his disciple Procharus and "which deluded the Greek Church by its air of sincerity and its extreme precision of detail," [3] although it does not seem to have reached the west until the sixteenth century. John is represented as drinking without injury a poison which had killed two criminals, and as reviving two corpses without going near them by directing an incredulous pagan to lay his cloak over them. A Stoic philosopher had

[1] *Ante-Nicene Fathers,* VIII, 494.

[2] W. Anz, *Zur Frage nach dem Ursprung des Gnostizisnus* (1897), pp. 36-41. Lipsius et Bonnet, *Acta apostolorum apocrypha,* 1891-.

[3] Mâle (1913), 299. For the text of this apocryphal work see Migne, *Dictionnaire des Apocryphes,* II, 759, *et seq.*, or more recently, Bonnet, *Acta apostolorum apocrypha,* 1898, II, 151-216.

persuaded some young men to embrace the life of poverty by converting their property into gems and then pounding the gems to pieces. John made the criticism that this wealth might have better been distributed among the poor, and when challenged to do so by the Stoic, prayed to God and had the gems made whole again. Later when the young men longed for their departed wealth, he turned the pebbles on the seashore into gold and precious stones, a miracle which is said to have persuaded the medieval alchemists that he possessed the secret of the philosopher's stone.[1] At any rate Adam of St. Victor in the twelfth century wrote the following lines concerning St. John in a chant to be used in the church service:

Cum gemmarum partes fractas
Solidasset, has distractas
  Tribuit pauperibus;
Inexhaustum fert thesaurum
Qui de virgis fecit aurum,
  Gemmas de lapidibus.[2]

Legend of St. Sousnyos.

The brief legend of St. Sousnyos, which Basset has included in his edition of Ethiopian Apocrypha,[3] is all magic, beginning with an incantation or magic prayer against disease and demons. There is also a Slavonic version. This Sousnyos is presumably the same as the Sisinnios who is said by the author of the apocryphal *Acts of Archelaus,*[4] forged about 330-340 A. D., to have abandoned Mani, embraced Christianity, and revealed to Archelaus secret teachings which enabled him to triumph over his adversary.

[1] Mâle (1913), 300. But one would think that they must needs be Byzantine alchemists, if the legend did not reach the west until the sixteenth century.

[2] HL, XV, 42.
When the gems, all smashed to pieces,
He had mended, then their prices
  To the poor he handed;
Quite exhaustless was his treasure
Who from sticks made gold at pleasure,
  Gems from stones commanded.

[3] René Basset, *Les apocryphes Éthiopiens,* Paris, 1893-1894, vol. iv.

[4] See Migne, PG, X (1857), for the old Latin version; the Greek text is extant only in fragments; the tradition, going back to Jerome, that there was a Syriac original is unfounded; the work is first cited by Cyril.

Old Testament apocrypha of the Christian era.

While on the subject, mention may be made of two works which properly belong to the apocrypha of the Old Testament, but which first appear during the Christian era and so fall within our period. *The Ascension of Isaiah,*[1] of which the old Latin version was printed at Venice in 1522, and which dates back to the second century, is something like the *Book of Enoch,* describing Isaiah's ascent through the seven heavens and vision of the mission of Christ. In the *Book of Baruch,* of which the original version was written in Greek by a Christian of the third or fourth century,[2] the most interesting episode is the magic sleep into which, like Rip Van Winkle, Abimelech falls during the destruction of Jerusalem by the Chaldeans. In the legend of Jeremiah the prophet's soul is absent from his body on one occasion for three days, while on another occasion he dresses up a stone to impersonate himself before the populace who are trying to stone him to death, in order that he may gain time to make certain revelations to Abimelech and Baruch. When he has had his say, the stone asks the people why they persist in stoning it instead of Jeremiah, against whom they then turn their missiles.[3]

Such is no exhaustive listing but rather a few examples of the encouragement given to belief in magic by the Christian Apocrypha.

[1] The Ethiopic version, made from the Greek between the fifth and seventh centuries, is translated by Basset (1894), vol. iii; and was printed before him by Dillmann, *Ascensio Isaiae aethiopice et latine,* Leipzig, 1877, and by Laurence, *Ascensio Isaiae vatis, opusculum pseudepigraphus,* Oxford, 1819. See also R. H. Charles, *Ascension of Isaiah,* 1900; reprinted 1917 in Oesterley and Box, *Translations of Early Documents,* Series I, vol. 7.

[2] The fragments of the *Book of Baruch* by Justin, preserved in the *Philosophumena* of Hippolytus, are from an entirely different Gnostic work.

[3] R. Basset, *Les apocryphes Éthiopiens,* Paris, 1893-1894, vol. i, *Le Livre de Baruch et la légende de Jérémie.*

CHAPTER XVII

THE RECOGNITIONS OF CLEMENT AND SIMON MAGUS

The Pseudo-Clementines—Was Rufinus the sole medieval version?—Previous Greek versions—Date of the original version—Internal evidence—Resemblances to Apuleius and Philostratus—Science and religion—Interest in natural science—God and nature—Sin and nature—Attitude to astrology—Arguments against genethlialogy—The virtuous Seres—Theory of demons—Origin of magic—Frequent accusations of magic—Marvels of magic—How distinguish miracle from magic?—Deceit in magic—Murder of a boy—Magic is evil—Magic is an art—Other accounts of Simon Magus: Justin Martyr to Hippolytus—Peter's account in the *Didascalia et Constitutiones Apostolorum*—Arnobius, Cyril, and Philastrius—Apocryphal *Acts of Peter and Paul*—An account ascribed to Marcellus—Hegesippus—A sermon on Simon's fall—Simon Magus in medieval art.

"The Truth herself shall receive thee a wanderer and a stranger, and enroll thee a citizen of her own city."

—*Recognitions* I, 13.

The Pseudo-Clementines.

THE starting-point and chief source for this chapter will be the writings known as the *Pseudo-Clementines* and more particularly the Latin version commonly called *The Recognitions.* We shall then note other accounts of its villain-hero, Simon Magus, in patristic literature.[1] The *Pseudo-*

[1] Text of *The Recognitions* in Migne, PG, I; of *The Homilies* in PG, II, or P. de Lagarde, *Clementina,* 1865. E. C. Richardson had an edition of *The Recognitions* in preparation in 1893, when a list of some seventy MSS communicated by him was published in A. Harnack's *Gesch. d. altchr. Lit.,* I, 229-30, but it has not yet appeared. In quoting *The Recognitions* I often avail myself of the language of the English translation in the *Ante-Nicene Fathers.*

Since A. Hilgenfeld, *Die klement. Rekogn. u. Homilien,* 1848, the Pseudo-Clementines have provided a much frequented field of research and controversy, of which the articles in CE, EB, and *Realencyklopädie* (1913), XXIII, 312-6, provide fairly recent summaries from varying ecclesiastical standpoints. For bibliography see pp. 4-5 in the recent monograph of W. Heintze, *Der Klemensroman und seine griechischen Quellen,* 1914, in TU, XL, 2. In the same series, TU, XXV, 4, H.

Clementines, as the name implies, are works or different versions of one work ascribed to Clement of Rome, who is represented as writing to James, the brother of the Lord, an account of events and discussions in which he and the apostle Peter had participated not long after the crucifixion. This Pseudo-Clementine literature has a double character, combining romantic narrative concerning Peter, Simon Magus, and the family of Clement with long, argumentative, didactic, and doctrinal discussions and dialogues in which the same persons participate but Peter takes the leading and most authoritative part. Not only the authorship, origin, and date, but even the title or titles and the make-up and arrangement of the various versions and their original are doubtful or disputed matters. The versions now extant and published seem by no means to have been the only ones, but we will describe them first. In Greek we have the version known as *The Homilies* in twenty books, in which the didactic element preponderates. It is extant in only two manuscripts of the twelfth and fourteenth centuries at Paris and Rome,[1] but is also preserved in part in epitomes. Different from it is the Latin version in which the narrative element plays a greater part.

Was Rufinus the sole medieval version?

This Latin version, now usually referred to as *The Recognitions,* because the main point in its plot is the successive bringing together again of, and recognition of one another by, the members of a family long separated, is the translation made by Rufinus, who is last heard from in 410. It is usually divided into ten books. Numerous manuscripts of this version attest its popularity and influence in the middle ages, when we early find Isidore of Seville quoting

Waitz, *Die Pseudo-Klementinen,* 1904.

Concerning Simon Magus may be mentioned: H. Schlurick, *De Simonis Magi fatis Romanis;* A. Hilgenfeld, *Der Magier Simon,* in *Zeitschr. f. wiss. Theol.,* XII (1869), 353 ff.; G. Frommberger, *De Simone Mago,* Pars I, *De origine Pseudo-Clementinorum,* Diss. inaug., Warsaw, 1866; G. R. S. Mead (Fellow of the Theosophical Society), *Simon Magus,* 1892; H. Waitz, *Simon Magus in d. altchr. Lit.,* in *Zeitschr. f. d. neutest. Wiss.,* V (1904), 121-43.

[1] BN, Greek, 930; Ottobon, 443.

Clement several times as an authority on natural science.[1] Arevalus, however, thought that Isidore used some other version of the Pseudo-Clementines than that of Rufinus,[2] and in the medieval period another title was common, namely, *The Itinerary of Clement,* or *The Itinerary of Peter.*[3] William of Auvergne, for instance, in the first half of the thirteenth century cites the *Itinerarium Clementis* or "Book of the disputations of Peter against Simon Magus." [4] This *Itinerary of Clement* also heads the list of works condemned as apocryphal by Pope Gelasius at a synod at Rome in 494,[5] a list reproduced by Vincent of Beauvais in his *Speculum naturale* in the thirteenth century [6] and in the previous century rather more accurately by Hugh of St. Victor in his *Didascalicon.*[7] In all three cases the full title is given in practically the same words, "The Itinerary by the name of the Apostle Peter which is called Saint Clement's, an apocryphal work in eight books." [8] Here we encounter a difficulty, since as we have said *The Recognitions* are in ten books. We find, however, that in another passage [9] Vincent correctly cites the ninth book of *The Recognitions* as Clement's ninth book, and that the number of books into which *The Recognitions* is divided varies in the manuscripts, and that they, too, more often call it *The Itinerary of Clement* or even apply other designations. Rabanus Maurus in the ninth century quotes an utterance of the apostle Peter from *The History of Saint Clement,* but the passage is found in *The Recognitions.*[10] Vincent of Beauvais also

[1] Isidore, *De natura rerum,* caps. xxxi, xxxvi, xxxix-xli (PL, 83, 1003-12).

[2] PL, 83, 1003, note, "Sunt haec lib. VIII Recognitionum sed apparet Isidorum alia interpretatione usum ac dubitare posse an ea quae circumfertur Rufini sit."

[3] See CU, Trinity 1041, 14th century, fols. 7-105, "Inc. prologus in librum quem moderni itinerarium beati Petri vocant."

[4] Valois (1880), p. 204.

[5] PL, 59, 162, "Notitia librorum apocryphorum qui non recipiuntur."

[6] Vincent of Beauvais, *Speculum naturale,* 1485, I, 14.

[7] PL, 176, 787-8, *Erudit. Didasc.,* IV, 15.

[8] "Itinerarium nomine Petri apostoli quod appellatur sancti Clementis libri octo apocryphum (or, apocryphi)."

[9] *Speculum naturale,* XXXII, 129, concerning the morality of the Seres.

[10] Compare *Recognitions,* I, 27 (PG, I, 122) with Rabanus, *Comment. in Genesim,* I, 2 (PL, 107, 450).

quotes "the blessed apostle Peter in a certain letter attached to *The Itinerary of Clement.*" No letter by Peter is prefaced to the printed text of *The Recognitions,* nor does Rufinus mention such a letter, although he does speak in his preface of a letter by Clement which he has already translated elsewhere. Prefixed to the printed *Homilies,* however, and in the manuscripts found also with *The Recognitions,* are letters of Peter and Clement respectively to James. But the passage quoted by Vincent does not occur in either, but comes from the tenth book of *The Recognitions.*[1] It would seem, therefore, despite variations in the number of books and in the arrangement of material, that the Latin version by Rufinus was the only one current in the middle ages, but we cannot be sure of this until all the extant manuscripts have been more carefully examined.[2]

Previous Greek versions.

The version by Rufinus differed from previous ones not only in being in Latin but also in various omissions which he admits he made and perhaps other changes to suit it to his Latin audience. That there was already more than one version in Greek he shows in his preface by describing another text than that upon which his translation or adaptation was based. Neither of these two Greek texts appears to have been the same as the present *Homilies.*[3] Yet *The Homilies* were apparently in existence at that time, since a Syriac manuscript of 411 A. D. contains four books of *The Homilies* and three of *The Recognitions,*[4] thus in itself

[1] *Speculum naturale,* I, 7. Peter is represented as saying, "When anyone has derived from divine Scripture a sound and firm rule of truth, it will not be absurd if to the assertion of true dogma he joins something from the education and liberal studies which he may have pursued from boyhood. Yet so that in all points he teaches what is true and shuns what is false and pretense." This corresponds to the close of the 42nd chapter of the tenth book of *The Recognitions.*

[2] Since writing this I learn that Professor E. C. Richardson has examined most of the known MSS of *The Recognitions* and has found them all to be the version by Rufinus, except for a few additional chapters which someone has added in the French group of MSS, — chapters which Rufinus seems to have omitted because they were difficult to translate.

[3] Heintze (1914), 23, however, argues that the conclusion of *The Recognitions* is dependent upon *The Homilies.*

[4] Professor E. C. Richardson, after kindly reading this chapter in manuscript, writes me (Sept. 5, 1921) that he doubts if this Syriac

furnishing an illustration of the ease with which new versions might be compounded from old. Both *The Homilies* and *The Recognitions* as they have reached us would seem to be confusions and perversions of this sort, as their incidents are obviously not arranged in correct order. For instance, when the story of *The Recognitions* begins Christ is still alive and reports of His miracles are reaching Rome; the same year Barnabas pays a visit to Rome and Clement almost immediately follows him back to Syria, making the passage from Rome to Caesarea in fifteen days;[1] but on his arrival there he meets Peter who tells him that "a week of years" have elapsed since the crucifixion and of other intervening events involving a considerable lapse of time. Or again, in the third book of *The Recognitions* Simon is said to have sunk his magical paraphernalia in the sea and gone to Rome, but as late as the tenth and last book we find him still in Antioch and with enough paraphernalia left to transform the countenance of Faustus.

Date of the original version.

Yet this late and misarranged version on which Rufinus bases his text must have been already in existence for some time, since he confesses that he has been a long while about his translation. The virgin Sylvia who "once enjoined it upon" him to "render Clement into our language" is now spoken of as "of venerable memory," and it is to Bishop Gaudentius that Rufinus "after many delays" in his old age "at length" presents the work. We might thus infer that the original and presumably more self-consistent Pseudo-Clementine narrative, which Rufinus evidently does not use, must date back to a much earlier period. We hear from other sources of *The Circuits* or *Periodoi of Peter* by Clement, but this may have been the version translated by Ru-

MS is correctly described as three books of *The Recognitions* and four books of *The Homilies,* and that he thinks it may represent an earlier form in the evolution than either of them. He writes further, "I have a strong notion that a study of Greek MSS of the Epitomes will reveal still more variant forms in Greek, and there are certainly other oriental compilations not yet brought into comparison with the Greek, Latin, and Syriac forms."

[1] In *The Homilies* it is a trip only from Alexandria to Caesarea that consumes this number of days.

finus.[1] Conservative Christian scholars regard as the oldest unmistakable allusion to the Pseudo-Clementines that by Eusebius early in the fourth century, who, without giving any specific titles, speaks of certain "verbose and lengthy writings, containing dialogues of Peter forsooth and Apion," which are ascribed to Clement but are really of recent origin. As for the date of the original work from which *Homilies* and *Recognitions* are derived,[2] from 200 to 280 A. D. is suggested by Harnack and his school, who take middle ground between the extreme contentions of Hilgenfeld and Chapman. But the original Pseudo-Clement is supposed to have utilized *The Teachings of Peter* and *The Acts of Peter*, which Waitz would date between 135 and 210 A. D.[3]

Internal evidence.

The work itself, even in the perverted form preserved by Rufinus, makes pretensions to the highest Christian antiquity. Not only is it addressed to James and put into the mouth of Clement, but Paul is never mentioned, and no book of the New Testament is cited by name, while sayings of Jesus are cited which are not found in the Bible. Christ is often alluded to in a veiled and mystic fashion as "the true prophet," who had appeared aforetime to Abraham and Moses, and interesting and vivid incidental glimpses are given of what purports to be the life of an early Christian community and perhaps is that of the Ebionites, Essenes, or some Gnostic sect. Emphasis is laid upon the purifying power of baptism, upon Peter's practice of bathing early every morning, preferably in the sea or running water, upon secret prayers and meetings, a separate table for the initiated, esoteric discussions of religion at cock-crow and in the night, and upon power over demons. All this may be mere clever invention, but there certainly is an atmosphere of verisimilitude about it; and it is rather odd that a later

[1] About 375 A.D. Epiphanius (Dindorf, II, 107-9) describes *The Circuits* in such a way that he might have either *The Homilies* or *The Recognitions* in mind. On the other hand, the *Philocalia*, composed about 358 by Basil and Gregory, cites a passage on astrology from the fourteenth book of *The Circuits* which is in the tenth book of *The Recognitions* and not in *The Homilies* at all.

[2] Heintze (1914), p. 113.

[3] Waitz (1904), pp. 151 and 243.

writer should be "very careful to avoid anachronisms," in whose account as it now stands are such glaring chronological confusions as those already noted concerning Clement's voyage to Caesarea and Simon's departure for Rome. But, as in the case of the New Testament Apocrypha, the exact date of composition makes little difference for our purpose, for which it is enough that the *Pseudo-Clementines* played an important part in the first thirteen centuries of Christian thought viewed as a whole. Eusebius and Epiphanius may find them unpalatable in certain respects and reject them as heretical, but Basil and Gregory utilize their arguments against astrology. Gelasius may classify them as apocryphal, but Vincent of Beauvais justifies a discriminating use of the apocryphal books in general and cites this one in particular more than once as an authority, and the incidents of its story were embodied, as we shall see, in medieval art.

Resemblances to Apuleius and Philostratus.

The same resemblance to the works of Apuleius and Philostratus that we noted in the case of an apocryphal gospel is observable in the *Pseudo-Clementines*. We see in *The Recognitions* the same mixed interest in natural science and in magic combined with religion and romantic incident that characterized the variegated and motley page of the author of the *Metamorphoses* and the biographer of Apollonius of Tyana. It is probably only a coincidence that two of the works of Apuleius are dedicated to a Faustinus whom he calls "my son," while Clement's father is named Faustus or Faustinianus, and the legend of Faust is believed to originate with him and the episodes in which he is concerned.[1] Less accidental may be the connection between Peter's religious sea-bathing and that purification in the sea by which the hero of the *Metamorphoses* began the process by which he succeeded in regaining his lost human form. More considerable are the detailed parallels to the work of Philostratus.[2] Peter corresponds roughly to Apollonius and Clem-

[1] See E. C. Richardson in *Papers of the American Society of Church History,* VI (1894).

[2] Neither Philostratus nor Apollonius of Tyana is mentioned, however. in the index of W.

ent to Damis, while the wizards and *magi* are ably personified by the famous Simon Magus. If Apollonius abstained from all meat and wine and wore linen garments, Peter lives upon "bread alone, with olives, and seldom even with potherbs; and my dress," he says, "is what you see, a tunic with a pallium: and having these, I require nothing more." [1] Like Philostratus the Pseudo-Clement speaks of bones of enormous size which are still to be seen as proof of the existence of giants in former ages; [2] and the accounts of the Brahmans and allusions to the Scythians in the *Life of Apollonius of Tyana* are paralleled in *The Recognitions* by a series of brief chapters on these and other strange races.[3] Peter is, of course, a Jew, not a Hellene like Apollonius, but in his train are men who are thoroughly trained in Greek philosophy and capable of discussing its problems at length. They also are not without appreciation of pagan art and turn aside, with Peter's consent, to visit a temple upon an island and "to gaze earnestly" upon "the wonderful columns" and "very magnificent works of Phidias." [4] Just as Apollonius knew all languages without having ever studied them, so Peter is so filled with the Spirit of God that he is "full of all knowledge" and "not ignorant even of Greek learning"; but to descend from his usual divine themes to discuss it is considered to be rather beneath him. Clement, however, felt the need of coaching Peter up a little in Greek mythology.[5] This mingled attitude of contempt for "the babblings of the Greeks" when compared to divine revelation, and of respect for Greek philosophy when compared with anything else is, it is hardly necessary to say, a very common one with Christian writers throughout the Roman Empire.

Science and religion.

The same attitude prevails toward natural science. At the very beginning of the Clementines the curiosity of the

Heintze's *Der Klemensroman und seine griechischen Quellen* (1914), 144 pp.

[1] *Recogs.*, VII, 6.

[2] *Recogs.*, I, 29; not mentioned in the corresponding chapter of *The Homilies*, VIII, 15.

[3] *Recogs.*, IX, 19-29.

[4] *Recogs.*, VII, 12.

[5] *Recogs.*, X, 15, *et seq.*

ancient world in regard to things of nature is shown by the question which someone propounded to Barnabas when he began to preach, at Rome according to *The Recognitions,* at Alexandria according to *The Homilies,* of the Son of God. The heckler wanted to know why so small a creature as a fly has not only six feet but wings in addition, while the elephant, despite its enormous bulk, has only four feet and no wings at all. Barnabas did not answer the question, although he asserted that he could if he wished to, making the excuse that it was not fitting to speak of mere creatures to those who were still ignorant of their Creator.[1]

Interest in natural science.

This unwillingness to discuss natural questions by no means continues characteristic of the Clementines, however. Not only does Peter explain to Clement the creation of the world and propound the extraordinary[2] doctrine that after completing the process of creation God "set an angel as chief over the angels, a spirit over the spirits, a star over the stars, a demon over the demons, a bird over the birds, a beast over the beasts, a serpent over the serpents, a fish over the fishes," and "over men a man who is Christ Jesus.[3] Not only does he later in public defend baptism with water on the ground that "all things are produced from waters" and that waters were first created.[4] We also find Niceta accepting the Greek hypothesis of four elements, of the sphericity of the universe, and of the motions of the heavenly bodies "assigned to them by fixed laws and periods," citing Plato's *Timaeus,* mentioning Aristotle's introduction of a fifth element,[5] disputing the atomic theory of Epicurus,[6] and alluding to "mechanical science."[7] He further discusses the generation of plants, animals, and human beings as evidences of divine design and providence,[8] in which connection he collects a number of examples of marvelous gen-

[1] *Recogs.,* I, 8; *Homilies,* I, 10.

[2] Extraordinary, of course, only in that single animals instead of angels, as in the Enoch literature, are set over birds, beasts, serpents, etc.

[3] *Recogs.,* I, 27 and 45.

[4] *Recogs.,* VI, 8.

[5] *Recogs.,* VIII, 9, 20-22.

[6] *Recogs.,* VIII, 15-17.

[7] *Recogs.,* VIII, 21.

[8] *Recogs.,* VIII, 25-32.

eration of animals such as moles from earth and vipers from ashes, and affirms that "the crow conceives through the mouth and the weasel generates through the ear." [1] Simon Magus declared himself immortal on the theory, which we shall find cropping out again in the thirteenth century in Roger Bacon and Peter of Abano, that his flesh was "so compacted by the power of his divinity that it can endure to eternity." [2] On the other hand, Niceta describes the action of the intestines in a fairly intelligent manner,[3] and tells how the blood flows like water from a fountain, "and first borne along in one channel, and then spreading through innumerable veins as through canals, irrigates the entire territory of the human body with vital streams." [4] A little later on Aquila gives a natural explanation of rainbows.[5]

God and nature.

There is noticeable, it is true, a tendency, common in patristic literature and found even among those fathers who hold the dualism of the Manichees in the deepest detestation, to make a distinction between God and nature and to attribute any flaws in the universe to the latter.[6] Niceta cannot agree with "those who speak of nature instead of God and declare that all things were made by nature"; he holds that God created the universe. But Aquila, who supports his brother in the discussion, seems to think that God's responsibility for the universe ceased, at least in part, after it was once created. At any rate he admits that "in this world some things are done in an orderly and some in a disorderly fashion. Those things therefore," he continues, "that are done rationally, believe that they are done by Providence; but those that are done irrationally and inordinately, believe that they befall naturally and happen accidentally." [7]

Sin and nature.

But even nature sometimes rises up against the sins of mankind according to Peter and his associates. Aquila be-

[1] On the other hand, in the apocryphal *Epistle of Barnabas*, IX, 9, it is stated that the weasel conceives with its mouth and hence typifies persons with unclean mouths.

[2] *Recogs.*, II, 7.

[3] *Recogs.*, VIII, 31.

[4] *Recogs.*, VIII, 30.

[5] *Recogs.*, VIII, 42.

[6] *Recogs.*, VIII, 34.

[7] *Recogs.*, VIII, 44.

lieves that the sins of men are the cause of pestilences;[1] that "when chastisement is inflicted upon men according to the will of God, he" (i. e. the Sun, already called "that good servant" and whom the early Christians found it difficult to cease to personify) "glows more fiercely and burns up the world with more vehement fires";[2] and that "those who have become acquainted with prophetic discourse know when and for what reason blight, hail, pestilence, and such like have occurred in every generation, and for what sins these have been sent as a punishment."[3] Peter gives the impression that nature sometimes acts rather independently of God in thus punishing the wicked. He says: "But this also I would have you know, that upon such souls God does not take vengeance directly, but His whole creation rises up and inflicts punishments upon the impious. And although in the present world the goodness of God bestows the light of the world and the services of the earth alike upon the pious and the impious, yet not without grief does the Sun afford his light and the other elements perform their services to the impious. And, in short, sometimes even in opposition to the goodness of the Creator, the elements are worn out by the crimes of the wicked; and hence it is that either the fruit of the earth is blighted, or the composition of the air is vitiated, or the heat of the sun is increased beyond measure, or there is an excess of rain or cold."[4] This is a close approach to the notion of *The Book of Enoch* that human sin upsets the world of nature, and an even closer approach to the theory of the Brahmans in *The Life of Apollonius of Tyana* that prolonged drought is a punishment visited by the world-soul upon human sinfulness.

Attitude to astrology.

Such vestiges of the world-soul doctrine, such a tendency to ascribe emotion and will to the elements and planets, to personify them, and to think of God as ruling the world indirectly through them, prepare us to find an attitude rather favorable to astrological theory. Indeed, in the first book

[1] *Recogs.*, VIII, 45.
[2] *Recogs.*, VIII, 46.
[3] *Recogs.*, VIII, 47.
[4] *Recogs.*, V, 27.

of *The Recognitions* [1] we are told in so many words that the Creator adorned the visible heaven with stars, sun, and moon in order that "they might be for an indication of things past, present, and future," and that these celestial signs, while seen by all, are "understood only by the learned and intelligent." Astrology is respectfully described as "the science of mathesis," [2] and, as was common in the Roman Empire, astrologers are called *mathematici*.[3] A defender even of the most extreme pretensions of the art is not abused as a charlatan but is courteously greeted as "so learned a man," [4] and all admire his eloquence, grave manners, and calm speech, and accord him a respectful hearing.[5] Astrology, far from being regarded as necessarily contrary to religion, is thought to furnish arguments for the existence of God, and it is said that Abraham, "being an astrologer, was able from the rational system of the stars to recognize the Creator, while all other men were in error, and understood that all things are regulated by His Providence."[6] The number seven is somewhat emphasized [7] and the twelve apostles are called the twelve months of Christ who is the acceptable year of the Lord.[8] Somewhat similarly the Gnostic followers of the heretic Valentinus made much of the Duodecad, a group of twelve aeons, and believed, according to Irenaeus, "that Christ suffered in the twelfth month. For their opinion is that He continued to preach for one year only after His baptism." [9] Peter, too, has a group of twelve disciples.[10] Niceta speaks of "man who is a microcosm in the great world." [11] It is admitted that the stars exert evil as well as good influence,[12] and that the astrologer "can indicate the evil desire which malign

[1] *Recogs.*, I, 28.

[2] *Recogs.*, VIII, 57, "frater meus Clemens tibi diligentius respondebit qui plenius scientiam mathesis attigit; IX, 18, "quoniam quidem scientia mihi mathesis nota est."

[3] *Recogs.*, X, 11-12.

[4] *Recogs.*, IX, 18.

[5] *Recogs.*, VIII, 2.

[6] *Recogs.*, I, 32.

[7] *Recogs.*, I, 21, 43, 72.

[8] *Recogs.*, IV, 35.

[9] Irenaeus, I, 3.

[10] *Recogs.*, III, 68.

[11] *Recogs.*, VIII, 28, "qui est parvus in alio mundus."

[12] *Recogs.*, VIII, 45.

virtue produces." [1] But it is contended that, "possessing freedom of the will, we sometimes resist our desires and sometimes yield to them," and that no astrologer can predict beforehand which course we will take.

Arguments against genethlialogy.

In fine, astrology is criticized adversely only when it goes to the length of contending that "there is neither any God, nor any worship, neither is there any Providence in the world, but all things are done by fortuitous chance and *genesis*"; that "whatever your *genesis* contains, that shall befall you"; [2] and that the constellations force men to commit murder, adultery, and other crimes.[3] On this point Niceta and Aquila, and finally Clement himself, have long discussions with an aged adept in genethlialogy which fill a large portion of the last three books of *The Recognitions,* and include a dozen chapters which are little more than an extract from *The Laws of Countries* of Bardesanes. Divine Providence and human free will are defended, and genethlialogy is represented as an error which has received confirmation through the operations of demons.[4] It is asserted that men can be kept from committing crimes by fear of punishment and by law, even if they are naturally so inclined, and races like the Seres (Chinese) and Brahmans are adduced as examples of entire races of men who never commit the crimes into which men are supposed to be forced by the constellations. The argument is also advanced, "Since God is righteous and since He Himself made human nature, how could it be that He should place *genesis* in opposition to us, which should compel us to sin, and then that He should punish us when we do sin?" [5] It is further charged that the constellations are so complicated,

[1] *Recogs.,* X, 12. In *Homilies,* XIV, 5, the existence of astrological medicine is implied when Peter promises to cure by prayer to God any bodily ill, even "if it is utterly incurable and entirely beyond the range of the medical profession—a case, indeed, which not even the astrologers profess to cure."

[2] *Recogs.,* VIII, 2. In *The Homilies,* however, Peter argues that, even if Genesis prevails, which he does not admit, still he can "worship Him who is also Lord of the stars," and that the doctrine of genesis is far more destructive to polytheism and pagan worship.

[3] *Recogs.,* IX, 16-17.

[4] *Recogs.,* IX, 6 and 12.

[5] *Recogs.,* IX, 30.

that for any given moment one astrologer may infer a favorable and another a disastrous influence,[1] and that most successful explanations of the effects of the stars are made after the event, like dreams of which men can make nothing at the time, but "when any event occurs, then they adapt what they saw in the dream to what has occurred." [2] Finally the aged defender of *genesis*, who believed that his own fate and that of his wife had been accurately prescribed by their horoscopes, turns out to be Faustinianus (called Faustus in *The Homilies*), the long-lost father of Clement, Niceta, and Aquila; is also restored to his wife; and learns that his previous interpretation of events from the stars was quite erroneous.[3]

The virtuous Seres.

The ideal picture of the Seres or Chinese, "who dwell at the beginning of the world," which *The Recognitions* apparently borrows from Bardesanes, is perhaps worth repeating here as an odd admission that a non-Christian people can attain a state of moral perfection and sinlessness, as well as an interesting bit of ancient ethnology. "In all that country which is very large there is neither temple nor image nor harlot nor adulteress, nor is any thief brought to trial. But neither is any man ever slain there. . . . For this reason they are not chastened with those plagues of which we have spoken; they live to extreme old age, and die without sickness." [4] Perhaps these virtuous Seres are the blameless Hyperboreans in another guise.

Theory of demons.

Demons and angels abound in *The Recognitions*. One may be rebuked and scourged at night by an angel of God.[5] Peter says that every nation has an angel, since God has divided the earth into seventy-two sections and appointed an angel as governor and prince of each.[6] Once, before beginning to preach, Peter expelled demons from a number of persons in the audience.[7] In another passage is described the cure of a girl of twenty-seven who for twenty years

[1] *Recogs.*, X, 11.
[2] *Recogs.*, X, 12.
[3] *Recogs.*, IX, 32-7.
[4] *Recogs.*, IX, 19, and VIII, 48.
[5] *Recogs.*, X, 66.
[6] *Recogs.*, II, 42.
[7] *Recogs.*, IV, 7.

had been vexed by an unclean spirit and had been shut up in a closet in chains because of her violence and superhuman strength. The mere presence of Peter put this demon to rout and the chains fell off the girl of their own accord.[1] Besides these personal encounters with demons, the theory of demoniacal possession is discussed more than once, and anything of which the author does not approve, such as the art of horoscopes, heathen oracles, the excesses of pagan rites and festivals, and the animal gods of the Egyptians, is attributed to the influence of demons.[2] One becomes susceptible to demoniacal possession who eats meat sacrificed to idols or who merely eats and drinks immoderately.[3] Demons are apt to get into the very bowels of those who frequent drunken banquets.[4] Incontinence, too, is accompanied by demons whose "noxious breath" produces "an intemperate and vicious progeny. . . . And therefore parents are responsible for their children's defects of this sort, because they have not observed the law of intercourse." [5] As much care should be taken in human generation as in the sowing of crops. But while demons abound, God has given every Christian power over them, since they may be driven out by uttering "the threefold name of blessedness." [6] Moreover, "what is spoken by the true God, whether by prophets or varied visions, is always true; but what is foretold by demons is not always true." [7]

Origin of magic.

With demons is associated the origin of the magic art. "Certain angels . . . taught men that demons could be made to obey man by certain arts, that is, by magical invocations." [8] The first magicians were Ham and his son Mesraim, from whom the Egyptians, Babylonians, and Assyrians are descended, and who tried to draw sparks from the stars [9] but set himself on fire "and was consumed by the demon

[1] *Recogs.*, IX, 38.

[2] *Recogs.*, IX, 6 and 12; IV, 21; V, 20 and 31.

[3] *Recogs.*, II, 71; IV, 16.

[4] *Recogs.*, IV, 30.

[5] *Recogs.*, IX, 9.

[6] *Recogs.*, IV, 32-33.

[7] *Recogs.*, IV, 21.

[8] *Recogs.*, IV, 26.

[9] Reminding one of Benjamin Franklin's more successful attempt to "snatch the thunderbolt from heaven."

whom he had accosted with too great importunity."[1] But on this account he was called Zoroaster or "living star" after his death. Moreover, the magic art did not perish but was transmitted to Nimrod "as by a flash."[2] With this may be compared the slightly different account of the origin of magic given by Epiphanius in the *Panarion,* written about 374-375 A. D. Magic is older than heresy and was already in existence before the time of Ham or Mesraim in the antediluvian days of Jared, when it coexisted with "pharmacy," a term here used to cover sorcery and poisoning, licentiousness, adultery, and injustice. After the flood Epiphanius mentions Nimrod (Νεβρώδ) as the first tyrant and the inventor of the evil disciplines of astrology and magic. He states that the Greeks incorrectly confuse him with Zoroaster whom they regard as the founder of magic and astrology. According to Epiphanius, "pharmacy" and magic passed from Egypt to Greece in the time of Cecrops.[3]

Frequent accusations of magic.

In *The Recognitions* everyone, Christian, heretic, pagan, and philosopher, condemns or professes to condemn magic, and reference is made to the laws of the Roman emperors against it.[4] But Christians, pagans, and heretics, while claiming divine power and protection for themselves, freely accuse one another of the practice of magic. An unnamed person, by whom Paul is perhaps meant, stirs up the people of Jerusalem to persecute the apostolic community there as "most miserable men, who are deceived by Simon, a magician."[5] The guards at the sepulcher, unable to prevent the resurrection, said that Jesus was a magician, a charge which is repeated by one of the scribes and by Simon Magus. Simon also calls Peter a magician on more than one occasion.[6] Peter, of course, makes similar charges against Simon; he had been especially sent by James to Caesarea in order to refute this magician who was giving himself out to be the *Stans* or Christ.[7] The gods of Greek

[1] *Recogs.,* IV, 27, and I, 30.
[2] *Recogs.,* IV, 29.
[3] Dindorf, I, 282, 286-7.
[4] *Recogs.,* X. 55; III, 64.
[5] *Recogs.,* I, 70.
[6] *Recogs.,* I, 42 and 58; III, 12, 47, and 73; X, 54.
[7] *Recogs.,* I, 72.

mythology, too, are accused of having resorted to magic transformations and sorcery.[1] Philosophy, however, escapes the accusation of magic in *The Recognitions,*[2] and it was a philosopher who deterred Clement, before the latter had become a Christian, from his plan of investigating the problem of the immortality of the soul by hiring an Egyptian magician to evoke a soul from the infernal regions by the art of necromancy.[3] The philosopher condemned such an attempt as unlawful, impious, and "hateful to the Divinity." [4]

Marvels of magic.

But while magic is condemned, its great powers are admitted. Simon Magus makes great boasts of the marvels which he can perform. These include becoming invisible, boring through rocks and mountains as if they were clay, passing through fire without being burned, flying through the air, loosing bonds and barriers, transformation into animal shapes, animation of statues, production of new plants or trees in a moment, and growing beards upon little boys.[5] He also asserted that he had formed a boy by turning air into water and the water into blood, and then solidifying this into flesh, a feat which he regarded as superior to the creation of Adam from earth. Later Simon unmade him and restored him to the air, "but not until I had placed his image and picture in my bedchamber as a proof and memorial of my work.[6] Not only does Simon himself make such boasts; Niceta and Aquila, who had been his disciples before their conversion by Zaccheus, also bear witness to

[1] *Recogs.,* X, 22 and 25.

[2] But by no means always in early Christian writings: thus Clement of Alexandria (c150-c220) in the *Stromata,* II, 1, asserts that the Greeks eulogize "astrology and mathematics and magic and sorcery" as the highest sciences.

[3] In contrast to Lucian's *Menippus* or *Necromancy,* in which the Cynic philosopher Menippus resorts to a *Magus* at Babylon in order to gain entrance to the lower world and question Teiresias.

Necromancy is given as a proof of the immortality of the soul in Justin's *First Apology,* cap. 18, where we read, "For let even necromancy, and the divinations you practise by means of immaculate children, and the evoking of departed human souls . . . let these persuade you that even after death souls are in a state of sensation."

[4] *Recogs.,* I, 5.

[5] *Recogs.,* II, 9.

[6] *Recogs.,* II, 15.

his amazing feats. "Who would not be astonished at the wonderful things which he does? Who would not think that he was a god come down from heaven for the salvation of men?" [1] He can fly through the air, or so mingle himself with fire as to become one body with it, he can make statues walk and dogs of brass bark. "Yea, he has also been seen to make bread of stones." [2] When Dositheus tried to beat Simon, the rod passed through his body as if it had been smoke.[3] The woman called Luna who goes about with Simon was seen by a crowd to look out of all the windows of a tower at the same time,[4] an illusion possibly produced by mirrors. When Simon fears arrest, he transforms the face of Faustinianus into the likeness of his own, in order that Faustinianus may be arrested in his place.[5]

How distinguish miracle from magic?

So great, indeed, are the marvels wrought by Simon and by magicians generally that Niceta asks Peter how they may be distinguished from divine signs and Christian miracles, and in what respect anyone sins who infers from the similarity of these signs and wonders either that Simon Magus is divine or that Christ was a magician. Speaking first of Pharaoh's magicians, Niceta asks, "For if I had been there, should I not have thought, from the fact that the magicians did like things (to those which Moses did), either that Moses was a magician, or that the feats displayed by the magicians were divinely wrought? . . . But if he sins who believes those who work signs, how shall it appear that he also does not sin who has believed on our Lord for His signs and occult virtues?" Peter's reply is that Simon's magic does not benefit anyone, while the Christian miracles of healing the sick and expelling demons are performed for the good of humanity. To Antichrist alone among workers of magic will it be permitted at the end of the world to mix in some beneficial acts with his evil marvels. Moreover, "by this means going beyond his bounds, and

[1] *Recogs.*, II, 6.
[2] *Recogs.*, III, 57.
[3] *Recogs.*, II, 11.
[4] *Recogs.*, II, 12.
[5] *Recogs.*, X, 53, *et seq.*

being divided against himself, and fighting against himself, he shall be destroyed."[1] Later in *The Recognitions,* however, Aquila states that even the magic of the present has found ways of imitating by contraries the expulsion of demons by the word of God, that it can counteract the poisons of serpents by incantations, and can effect cures "contrary to the word and power of God." He adds, "The magic art has also discovered ministries contrary to the angels of God, placing the evocation of souls and the figments of demons in opposition to these."[2]

Deceit in magic.

But while the marvels of magic are admitted, there is a feeling that there is something deceitful and unreal about them. The teachings of the true prophet, we are told, "contain nothing subtle, nothing composed by magic art to deceive,"[3] while Simon is "a deceiver and magician."[4] Nor is he deceitful merely in his religious teaching and his opposition to Peter; even his boasts of magic power are partly false. Aquila, his former disciple, says, "But when he spoke thus of the production of sprouts and the perforation of the mountain, I was confounded on this account, because he wished to deceive even us, in whom he seemed to place confidence; for we knew that those things had been from the days of our fathers, which he represented as having been done by himself lately."[5] Moreover, not only does Simon deceive others; he is himself deceived by demons as Peter twice asserts:[6] "He is deluded by demons, yet he thinks that he sees the very substance of the soul." "Although in this he is deluded by demons, yet he has persuaded himself that he has the soul of a murdered boy ministering to him in whatever he pleases to employ it."

Murder of a boy.

This story of having sacrificed a pure boy for purposes of magic or divination was a stock charge, which we have previously heard made against Apollonius of Tyana and which was also told of the early Christians by their

[1] *Recogs.,* III, 57-60; X, 66.
[2] *Recogs.,* VIII, 53.
[3] *Recogs.,* VIII, 60.
[4] *Recogs.,* II, 5.
[5] *Recogs.,* II, 10.
[6] *Recogs.,* II, 16, and III, 49.

pagan enemies and of the Jews and heretics in the middle ages. Simon is said to have confessed to Niceta and Aquila, when they asked how he worked his magic, that he received assistance from "the soul of a boy, unsullied and violently slain, and invoked by unutterable adjurations." He went on to explain that "the soul of man holds the next place after God, when once it is set free from the darkness of the body. And immediately it acquires prescience, wherefore it is invoked in necromancy." When Aquila asked why the soul did not take vengeance upon its slayer instead of performing the behests of magicians, Simon answered that the soul now had the last judgment too vividly before it to indulge in vengeance, and that the angels presiding over such souls do not permit them to return to earth unless "adjured by someone greater than themselves."[1] Niceta then indignantly interposed, "And do you not fear the day of judgment, who do violence to angels and invoke souls?" As a matter of fact, the charge that Simon had murdered or violently slain a boy is rather overdrawn, since the boy in question was the one whom he had made from air in the first place and whom he simply turned back into air again, claiming, however, to have thereby produced an unsullied human soul. According to *The Homilies,* however, he presently confided to Niceta and Aquila that the human soul did not survive the death of the body and that a demon really responded to his invocations.[2]

Nevertheless, the charge of murder thus made against Simon illustrates the criminal character here as usually ascribed to magic. Simon is said to be "wicked above measure," and to depend upon "magic arts and wicked devices," and Peter accuses him of "acting by nefarious arts."[3] Magic is evil.

[1] Similarly, in a passage contained only in *The Homilies,* V, 5, Appion, recommending to Clement a love incantation which he had learned from an Egyptian who was well versed in magic, explains that demons obey the magician when invoked by the names of superior angels, who in their turn may be adjured by the name of God.

[2] Concerning this boy see *Recogs.,* II, 13-15; III, 44-45; *Homilies,* II, 25-30.

[3] *Recogs.,* II, 6; III, 13.

Simon in his turn calls Peter "a magician, a godless man, injurious, cunning, ignorant, and professing impossibilities," and again "a magician, a sorcerer, a murderer." [1]

Magic is an art.

A further characteristic of magic which comes out clearly in *The Recognitions* is that it is an art. Demons and souls of the dead may have a great deal to do with it, but it also requires a human operator and makes use of materials drawn from the world of nature. It was by anointing his face with an ointment which the magician had compounded that the countenance of Faustinianus was transformed into the likeness of Simon, while Appion and Anubion, who anointed their faces with the juice of a certain herb, were thereby enabled still to recognize Faustinianus as himself.[2] In another passage one of Simon's disciples who has deserted him and come to Peter tells how Simon had made him carry on his back to the seashore a bundle "of his polluted and accursed secret things." Simon took the bundle out to sea in a boat and later returned without it.[3] Simon not only employed natural materials in his magic, but was regarded as a learned man, even by his enemies. He is "by profession a magician, yet exceedingly well trained in Greek literature." [4] He is "a most vehement orator, trained in the dialectic art, and in the meshes of syllogisms; and what is most serious of all, he is greatly skilled in the magic art." [5] And he engages with Peter in theological debates. It is also interesting to note as an illustration of the connection between magic and experimental science that Simon, in boasting of his feats of magic, says, "For already I have achieved many things by way of experiment." [6]

In the Pseudo-Clementines we are told that Simon intended to go to Rome, but *The Recognitions* and *The Homilies* deal only with the conflicts between Peter and Simon in various Syrian cities and do not follow them to

[1] *Recogs.*, III, 73; X, 54.

[2] *Recogs.*, X, 58.

[3] *Recogs.*, III, 63.

[4] *Recogs.*, II, 7.

[5] *Recogs.*, II, 5.

[6] *Recogs.*, II, 9, "Multa etenim iam mihi experimenti causa consummata sunt."

Other accounts of Simon Magus: Justin Martyr to Hippolytus.

Rome, where, as other Christian writers tell us, they had yet other encounters in which Simon finally came to his bitter end. Justin Martyr, writing about the middle of the second century, states that Simon, a Samaritan of Gitto, came to Rome in the reign of Claudius and performed such feats of magic by demon aid that a statue was erected to him as a god. In this matter of the statue Justin is thought to have confused Semo Sancus, a Sabine deity, with Simon. Justin adds that almost all Samaritans and a few persons from other nations still believe in Simon as the first God, and that a disciple of his, named Menander, deceived many by magic at Antioch. Justin complains that the followers of these men are still called Christians and on the other hand that the emperors do not persecute them as they do other Christians, although Justin charges them with practicing promiscuous sexual intercourse as well as magic.[1] Irenaeus gives a very similar account.[2] Origen, as we have seen, denied that there were more than thirty of Simon's followers left,[3] but his contemporary Tertullian wrote, "At this very time even the heretical dupes of this same Simon are so much elated by the extravagant pretensions of their art, that they undertake to bring up from Hades the souls of the prophets themselves. And I suppose that they can do so under cover of a lying wonder." [4] But Origen and Tertullian add nothing to the story of Simon Magus himself. Hippolytus, too, implies that Simon still has followers, since he devotes a number of chapters to stating and refuting Simon's doctrines and to "teaching anew the parrots of Simon that Christ . . . was not Simon." [5] But Hippolytus also gives further details concerning Simon's visit to Rome, stating that he there encountered the apostles and was repeatedly opposed by Peter, until finally Simon declared that if he were buried alive he would rise again upon the third day.

[1] *First Apology,* caps. 26 and 56; *Dialogue with Trypho,* 120.
[2] *Adv. haer.*, I, 23.
[3] See above, chapter 15, p. 365.
[4] Tertullian, *De anima,* cap. 57, in PL, II, 794; *De idolatria,* cap. 9.
[5] *Philosophumena,* VI, 2-15.

His disciples buried him, as they were directed, but he never reappeared, "for he was not the Christ."

Peter's account in the *Didascalia et Constitutiones Apostolorum.*

Peter himself is represented as briefly recounting his struggle at Rome with Simon Magus in the *Didascalia Apostolorum,* an apocryphal work of probably the third century, extant in Syriac and Latin, and more fully in the parallel passage of the Greek *Constitutiones Apostolorum,* written perhaps about 400 A. D.[1] Peter found Simon at Rome drawing many away from the church as well as seducing the Gentiles by his "magic operation and virtues," or, in the Greek version, "magic experiments and the working of demons." [2] In the Syriac and Latin account Peter then states that one day he saw Simon flying through the air. "And standing beneath I said, 'In the virtue of the holy name, Jesus, I cut off your virtues.' And so falling he broke the arch (thigh?) of his foot (leg?)." [3] But he did not die, since Peter goes on to say that while "many then departed from him, others who were worthy of him remained with him." In the longer Greek version Simon announced his flight in the theater. While all eyes were turned on Simon, Peter prayed against him. Meanwhile Simon mounted aloft into mid-air, borne up, Peter says, by demons, and telling the people that he was ascending to heaven, whence he would return bringing them good tidings. The people applauded him as a god, but Peter stretched forth his hands to heaven, supplicating God through the Lord Jesus to dash down the corrupter and curtail the power of the demons. He asked further, however, that Simon might not be killed by his fall but merely bruised. Peter also addressed Simon and the evil powers who were supporting him, requiring that he might fall and become a laughing-stock to those who had been deceived by him. Thereupon Simon fell with a great commotion and bruised

[1] F. X. Funk, *Didascalia et Constitutiones Apostolorum,* 1905, I, 320-1.

[2] τὰ δὲ ἔθνη ἐξιστῶν μαγικῇ ἐμπειρίᾳ καὶ δαιμόνων ἐνεργείᾳ.

[3] ". . . in una die procedens vidi illum per aera volantem et ferebatur. Et subsistens dixi: In virtute sancti nominis Iesu excido virtutes tuas. Et sic ruens femur pedis sui fregit."

his bottom and the soles of his feet. It will be noted that here, as in the accounts by some other authors, Peter alone struggles with Simon Magus, lending color to the Tübingen theory once suggested in connection with the Pseudo-Clementines, that Simon Magus is meant to represent the apostle Paul.

Arnobius, Cyril, and Philastrius.

Arnobius, writing about 300 A. D., gives a somewhat different account of Simon's mode of flight and fall. He says that the people of Rome "saw the chariot of Simon Magus and his four fiery horses blown away by the mouth of Peter and vanish at the name of Christ. They saw, I say, him who had trusted false gods and been betrayed by them in their fright precipitated by his own weight and lying with broken legs. Then, after he had been carried to Brunda, worn out by his shame and sufferings, he again hurled himself down from the highest ridge of the roof." [1] Cyril of Jerusalem, 315-386 A. D., also speaks of Simon's being borne in air in the chariot of demons, "and is not surprised that the combined prayers of Peter and Paul brought him down, since in addition to Jesus's promise to answer the petition of two or three gathered together it is to be remembered that Peter carried the keys of heaven and that Paul had been rapt to the third heaven and heard secret words.[2] Philastrius, another writer of the fourth century, describes Simon's death more vaguely, stating that after Peter had driven him from Jerusalem he came to Rome where they engaged in another contest before Nero. Simon was worsted by Peter on every point of argument, and, "smitten by an angel died a merited death in order that the falsity of his magic might be evident to all men." [3] But it is hardly worth while to pile up such brief allusions to Simon in the writings of the fathers.[4]

[1] Arnobius, *Adversus gentes,* II, 12.

[2] Cyril, *Cathechesis,* VI, 15, in PG 33, 564.

[3] *Filastrii diversarum hereseon liber,* cap. 23, ed. F. Marx, 1898, in CSEL; also in PL, vol. 12.

[4] Sulpicius Severus, 363-420, *Chron.,* II, 28, and Theodoret, c386-456, *Haereticarum fabularum compendium,* I, 1 (PG 83, 344) have nothing new to say.

Apocryphal *Acts of Peter and Paul.*

Other fuller accounts of Simon's doings at Rome are contained in the Syriac *Teaching of Simon Cephas*[1] and in the apochryphal *Acts of Peter and Paul.*[2] In the former Peter urges the people of Rome not to allow the sorcerer Simon to delude them by semblances which are not realities, and he raises a dead man to life after Simon has failed to do so. In the latter work Simon opposes Peter and Paul in the presence of Nero and as usual they charge one another with being magicians. Simon also as usual affirms that he is Christ, and we are told that the chief priests had called Jesus a wizard. Simon had already made a great impression upon Nero by causing brazen serpents to move and stone statues to laugh, and by altering both his face and stature and changing first to a child and then to an old man. Nero also asserts that Simon has raised a dead man and that Simon himself rose on the third day after being beheaded. It is later explained, however, that Simon had arranged to have the beheading take place in a dark corner and through his magic had substituted a ram for himself. The ram appeared to be Simon until after it had been decapitated, when the executioner discovered that the head was that of a ram but did not dare report the fact to Nero. When Simon met the apostles in Nero's presence, he caused great dogs to rush suddenly at Peter, but Peter made them vanish into air by showing them some bread which he had been secretly blessing and breaking. As a final test Simon promised to ascend to heaven if Nero would build him a tower in the Campus Martius, where "my angels may find me in the air, for they cannot come to me upon earth among sinners." The tower was duly provided, and Simon, crowned with laurel, began to fly successfully until Peter, tearfully entreated by Paul to make haste, adjured the angels of Satan who were supporting Simon to let him drop. Simon then fell upon the *Sacra Via* and his body was broken into

[1] AN, VIII, 673-5.

[2] *Ibid.*, 477-85; Greek text in Tischendorf, *Acta Apostolorum Apocrypha*, 1851, pp. 1-39. The Greek scholar, Constantine Lascaris, translated part of the work into Latin in 1490.

four parts.[1] Nero, however, chose to regard the apostles as Simon's murderers and put them to death, after which a Marcellus, who had been Simon's disciple but left him to join Peter, secretly buried Peter's body.

An account ascribed to Marcellus.

To this Marcellus is ascribed a very similar narrative which is found in an early medieval manuscript and was perhaps written in the seventh or eighth century.[2] Fabricius and Florentinus give its title as, *Of the marvelous deeds and acts of the blessed Peter and Paul and of Simon's magic arts.*[3] I have read it in a Latin pamphlet printed at some time before 1500, where the full title runs: *The Passion of the Apostles Peter and Paul, and their disputation before the emperor Nero against Simon, a certain magician, who, when he saw that he could not resist the utterances of St. Peter, cast all his books of magic into the sea lest he be adjudged a magician. Then when the same Simon Magus presumed to ascend to heaven, overcome by St. Peter he fell to earth and perished most miserably.* At its close occurs the statement, "I, Marcellus, a disciple of my lord, the apostle Peter, have written what I saw." When this Marcellus began to desert his former master, Simon, to follow Peter, Simon procured a big dog to keep Peter away from Marcellus, but at Peter's order the dog turned upon Simon himself. Peter then humanely forbade the beast to do Simon any serious bodily injury, but the dog tore the magician's clothing off his back, and Simon was chased from town by the mob and did not venture to return until after a year's time.[4]

[1] Mead (1892), p. 37, notes that Dr. Salmon (article *Simon Magus* in *Dict. Chris. Biog.* IV, 686) "connects this with the story, told by Suetonius and Dio Chrysostom, that Nero caused a wooden theater to be erected in the Campus, and that a gymnast who tried to play the part of Icarus fell so near the emperor as to bespatter him with blood." Hegesippus (*De bello judaico,* III, 2), Abdias (*Hist.* 1), and Maximus Taurinensis (*Patr.* VI, *Synodi ad Imp. Const. Act.* 18) compare Simon's flight with that of Icarus.

[2] Tischendorf (1851), p. xix.

[3] "De mirificis rebus et actibus beatorum Petri et Pauli, et de magicis artibus Simonis:" Fabricius, *Cod. apocr.,* III, 632; Florentinus, *Martyrologium Hieronymi,* 103.

[4] A slightly different version of the dog incident is found in the *Acts of Nereus and Achilles* (AS, May III, 9).

Hegesippus.

A chapter is devoted to Simon Magus in the *History of the Jewish War* of the so-called Hegesippus, a name which is thought to be a corruption of Josephus, since the work in large measure reproduces that historian. At any rate it was not written until the fourth century and is probably a translation or adaptation by Ambrose. Its account of Simon Magus combines the story of his competition with Peter in raising the dead, "for in such works Peter was held most celebrated," with that of his flight and fall. He is represented as launching his flight from the Capitoline Hill and leaping off the Tarpeian rock. The people marveled at his flight, some remarking that Christ had never performed such a feat as this. But when Peter prayed against him, "straightway his propeller was tangled up in Peter's voice, and he fell, nor was he killed, but, weakened by a broken leg, withdrew to Aricia and died there." [1]

A sermon on Simon's fall.

Finally, passing over other Latin accounts of the contest between the apostles and Simon Magus to be found in the *Apostolic Histories* of the Pseudo-Abdias [2] and in a work ascribed to Pope Linus,[3] we may note a sermon which has been variously ascribed in the manuscripts and printed editions to Augustine, Ambrose, and Maximus.[4] This sermon, intended for the anniversary of the day of martyrdom of Peter and Paul, proceeds to inquire the cause of their death and finds it in the fact that among other marvels they "prostrated by their prayers that magician Simon in a headlong fall from the empty air. For when the same Simon called himself Christ and asserted that as the Son he could ascend unto the Father by flying, and, suddenly

[1] *Hegesippus,* III, 2 ed. C. F. Weber and J. Caesar, Marburg, 1864, "et statim in voce Petri implicatiis remigiis alarum quas sumserat corruit, nec exanimatus est, sed fracto debilitatus crure Ariciam concessit atque ibi mortuus est." I earnestly recommend this passage to those who delight in finding ancient precursors of modern inventions as an example of remarkable insight into the effect of air-waves upon delicate mechanisms.

[2] ed. Fabricius, *Cod. apocr.,* I, 411; AS, June V, 424.

[3] *Biblioth. Patrum,* Cologne, 1618, I, 70.

[4] Printed PL, 39, 2121-2, among the works of Augustine, *Sermones Supposititi,* CCII. The greater number of MSS assign it to Maximus.

raised up by magic arts, began to fly, then Peter on his knees prayed the Lord, and by sacred prayer overcame the magical levitation. For the prayer ascended to the Lord before the flier, and the just petition arrived ere the iniquitous presumption. Peter, I say, though placed on the ground, obtained what he sought before Simon reached the heaven towards which he was tending. So then Peter brought him down like a captive from high in air, and, falling precipitately upon a rock, he broke his legs. And this in contumely of his feat, so that he who just before had tried to fly, of a sudden could not even walk, and he who had assumed wings lost even his feet. But lest it appear strange that, while the apostle was present, that magician should fly through the air even for a while, let it be explained that this was due to Peter's patience. For he let him soar the higher in order that he might fall the farther; for he wished him to be carried aloft where everyone could see him, in order that all might see him when he fell from on high." The preacher then draws the moral that pride goes before a fall.

Simon Magus in medieval art.

The struggle of Peter and Paul with Simon Magus at Rome appears in *The Golden Legend,* compiled by Jacopo de Voragine in the thirteenth century, and was likewise a favorite theme of Gothic stained glass. At Chartres and Angers Peter may be seen routing Simon's dogs by blessing bread; at Bourges and Lyons Simon and Peter compete in raising the dead; while windows at Chartres, Bourges, Tours, Reims, and Poitiers show the apostles praying and Simon falling and breaking his neck.[1] This last scene and also the disputation before Nero are represented in the earlier mosaics of the eleventh or twelfth century which the Norman rulers of Sicily had executed in the cathedral of Monreale and the royal chapel of their castle at Palermo.[2]

[1] Mâle, *Religious Art in France,* 1913, p. 297, notes 3 and 4; p. 298, note 1.

[2] The two representations are essentially identical. Simon falls head first, and the accompanying legend reads, *"Hic praecepto Petri oratione Pauli Simon Magus cecidit in terram,"*—"Here at Peter's command and Paul's prayer Simon Magus falls to earth."

CHAPTER XVIII

THE CONFESSION OF CYPRIAN AND SOME SIMILAR STORIES

The *Confession* of Cyprian—His initiation into mysteries—His thorough study of nature, divination, and magic—The lore of Egypt—And of Chaldea—Cyprian's practice of magic at Antioch—A Christian virgin defeats the magic of the demons—Summary of Cyprian's picture of magic—Christians accused of magic—A story from Epiphanius—Joseph's experience of miracle and magic—Legend of St. James and Hermogenes the magician—Other contests of apostles and magicians in *The Golden Legend.*

The *Confession* of Cyprian.

To the accounts of the contests of Peter and Paul with Simon Magus which were recorded in our last chapter we shall add in this some other encounters of early Christians with magicians, and to the picture of magic contained in the Pseudo-Clementines that presented by Cyprian in his *Confession.* If Simon Magus died impenitent in the midst of his magic, very different was the end of Cyprian, a magician by profession in the third century, who, after being educated from childhood in heathen mysteries and the magic art, repented and was baptized, became bishop of Antioch, and finally achieved a martyr's crown. In the *Confession* [1] current under his name and which most critics agree was composed before the time of Constantine [2] is described his

[1] Greek and Latin text in parallel columns in AS, Sept. VII (1867), pp. 204ff. For an account of previous editions see *Ibid.*, p. 182. Bishop John Fell published a Latin text from three Oxford MSS. In Digby 30, 15th century, fol. 29-, which I have examined, the wording differed considerably from that of the Latin text in AS. The brief *Martyrium* of Cyprian and Justina follows in the same volume of AS at pp. 224-6. *Sahidische Bruchstücke der Legende von Cyprian von Antiochen*, ed. O. v. Lamm, 1899, Ethiopic, Greek, and German, in *Petrograd Acad. Scient. Imper. Mémoires, VIII série, Cl. hist. philol.*, IV, 6. Πρᾶξις τῶν ἁγίων μαρτύρων Κυπριανοῦ καὶ Ἰουστίνης, with an Arabic version, ed. Margaret D. Gibson, 1901, in *Studia Sinaitica*, No. 8.

[2] *Ibid.*, p. 180, "ipsa S. Cypriana nomine vulgata Confessio quam ante Constantini aetatem scriptam esse critici plurimi etiam rigidiores fatentur."

education in and subsequent practice of magic. For us perhaps the most interesting feature of his account of his education is the association of magic, not only with pagan mysteries and the operations of demons, but also with natural science.

His initiation into mysteries.

"I am Cyprian," says the author, "who from a tender age was consecrated a gift to Apollo and while yet a child was initiated into the arts of the dragon." When not yet seven years old, he entered the mysteries of Mithra, and at ten his parents enrolled him a citizen at Athens, and he carried a torch in the mysteries of Demeter and "ministered to the dragon on the citadel of Pallas." When not yet fifteen, he also visited Mount Olympus for forty days, and "was initiated into sonorous speeches and noisy narrations."[1] There he saw in phantasy trees and herbs which seemed to be moved by the presence of the gods, spirits who regulated the passage of time, and choruses of demons who sang, while others waged war or plotted, deceived, and permeated.[2] He saw the phalanx of each god and goddess, and how from Mount Olympus as from a palace spirits were despatched to every nation of the earth. He was fed only after sunset and upon fruits, and was taught the efficacy of each of them by seven hierophants.

His thorough study of nature, divination, and magic.

Cyprian's parents were determined that he should learn whatever there was in earth and air and sea, and not merely the natural generation and corruption of herbs and trees and bodies, but also the virtues implanted in all these, which the prince of this world impressed upon them in order that he might oppose the divine constitution. Cyprian also participated at Argos in the sacred rites of Hera, and saw the union of air with ether and of ether with air, also of earth with water, and water with air. He penetrated the Troad and to Artemis Tauropolos who is at Lacedaemon to learn

[1] *Ibid.*, p. 205, "et initiatus sum sonis sermonum ac strepitum narrationibus." L. Preller in *Philologus*, I (1846), 349ff., and A. B. Cook, *Zeus*, 110-1, suggest that these rites on Mount Olympus were Orphic.

[2] "Et aliorum insidiantium decipientium permiscentium. . . ."

how matter was confused and divided "and the profundities of sinister and cruel legends." From the Phrygians he learned liver divination; among the barbarians he studied auspices and the significance of the movements of quadrupeds, and how to interpret omens and the language of birds, and the sounds made by every kind of wood and stone, or by the dead in tombs and the creaking of doors. He became acquainted with the palpitations of the limbs, the movement of the blood and pulse in bodies, all the extensions and corollaries of ratios and numbers, diseases simulated as well as natural, "and oaths which are heard yet are not audible, and pacts for discord." There was, in fine, nothing whatever in earth or sea or air that he did not know, whether it was a matter of science or phantasy, of mechanics or artifice, "even down to the magic translation of writings and other things of that sort."

The lore of Egypt.

At twenty Cyprian was admitted to the shrines at ancient Memphis in Egypt and learned what communication and relationship existed between demons and earthly things and "in what stars and laws and objects they delight." He witnessed imitations of earthquakes, rain, and storms at sea. He saw the souls of giants held in darkness and fancied that they sustained the earth as a load on their shoulders. He saw the communications of serpents with demons, ideas of transfigurations, impious piety, science without reason, iniquitous justice, and things topsy-turvy generally. Besides the forms of various sins and vices, such as fornication and avarice, which suggest the medieval personification of the seven deadly sins, he saw the three hundred and sixty-five varieties of ailments, "and the empty glory and the empty virtue" with which the priests of Egypt had deceived the Greek philosophers.

And of Chaldea.

At thirty Cyprian left Egypt for Chaldea in order to acquire its lore concerning air, fire, and light. Here he was instructed in the qualities of stars as well as of herbs, and their "choruses like drawn-up battle lines." He was taught the house and relationships of each star and its appropriate food and drink. Also the meetings of spirits with men in light, the three hundred and sixty-five demons who divide as many parts of the ether between them, and the sacrifices, libations, and words appropriate to each. Cyprian's education had now advanced to such a point that the devil himself hailed him, mere youth as he was, as a new Jambres, a skilful and reliable practitioner, and worthy of communication with himself. Cyprian again explains at this point that in all the stars and plants and other works of God the devil has bound to himself likenesses in preparation to wage war with God and His angels, but these likenesses are shadowy images, not solid substances. The devil's rain is not water, his fire does not burn, his fish are not food, and his gold is not genuine. The devil obtains the material for his products from the vapors of sacrifices.

Cyprian's practice of magic at Antioch.

Cyprian now returned from Chaldea and wrought marvels at Antioch "like one of the ancients," and "made many experiments of magic and became celebrated as a magician and philosopher endowed with vast knowledge of things invisible." Men came to him to be taught magic or to secure their ends by his assistance. And he easily helped them all, some to the gratification of pleasure, others to triumph over their adversaries or even to slay their rivals. His conscience sometimes pricked him at the evil deeds which he thus wrought with the aid of demons, but as yet he did not doubt that the devil was all powerful.

A Christian virgin defeats the magic of the demons.

But then the case of the Christian girl Justina revealed to him the weakness and fraud of the devil. Determined to dedicate herself to a life of virginity, Justina repulsed the love of the youth Aglaïdes, who sought Cyprian's assistance. But in vain: the demon failed to alter Justina's determination and was not even able to give another girl the form of Justina and so deceive Aglaïdes. Justina was shown the form of her lover, but she called upon the Virgin, and the devil was forced to vanish in smoke. Nor did disease and other plagues and torments affect her resolution. Her parents, however, were similarly afflicted until they besought her to marry Aglaïdes, but instead she cured them of their ailments by the sign of the cross. The devil then inflicted a plague on the entire community and delivered an oracle to the effect that the pest could be stayed only by the marriage of Justina and Aglaïdes, but her prayers turned the wrath of the public from herself against Cyprian. When the magician in disgust cursed the demon for the evil pass to which he had thus brought him, the demon made a ferocious attack upon him, from which Cyprian saved himself just in the nick of time by calling upon God for aid and making the sign of the cross. He then publicly confessed his crimes as a magician, burned his books of magic, and was baptized into the Christian faith.[1]

Summary of Cyprian's picture of magic.

Cyprian's *Confession* thus represents magic as a very elaborate art, requiring long study and a thorough knowledge of natural objects and processes. The magician has his books, and he must also be able to read the book of nature. Astrology and other arts of divination are integral parts of magic. But magic is also represented as the work of evil spirits. This involves not merely a Neo-Platonic sort of association of demons with natural forces and regions of earth or sky, but also the specific association of the devil for evil purposes with objects in nature, a doctrine which we shall find again in the works of a medieval saint, Hildegard of Bingen. Furthermore, magic aids in the commission of crime and is dangerous even to the magician against whom the devil may turn. While magic involves study of nature and use of natural forces and associations, and we also hear of "many experiments of magic," it is scarcely represented as operating scientifically in the *Confession*. It is mystic, confused, shadowy, imitative, imaginary, lacking in solidity and reality, fraudulent and deceptive. Finally, this complex art, this universal system of knowledge, is easily balked and overthrown by the far simpler

[1] Shelley, it may be recalled, in 1822 translated some scenes, published in 1824, from Calderón's *Magico Prodigioso*, in which Cyprian, Justina, and the demon figure.

counter-magic of Christianity, by such methods as a prayer to the Virgin, calling on the name of God, or merely making the sign of the cross.

Christians accused of magic.

Such counter-magic was apt to be regarded as magic by the pagans, and the account of the martyrdom of Cyprian states that the devil, that "very bad serpent," suggested to the Count of the Orient that Cyprian, together with a certain virgin who is assumed to be Justina, was destroying the ancient worship of the gods by his magic tricks as well as stirring up the orient and the whole world by his epistles. He was accordingly arrested and finally beheaded. According to one account he and Justina were first placed together in a cauldron of tallow and pitch over a fire. But when they sang a hymn, the flames left them uninjured and instead shot out and caused the death of an unreformed magician who happened to be standing near by.[1] Another case of Christian martyrs who were probably accused of magic is found in Spain about 287 A. D. Two Christian sisters who were dealers in pottery refused to sell their earthenware for purposes of pagan worship. One day, as a pagan religious procession passed by their shop, the crowd trampled upon their wares which were exposed for sale. But thereupon the idol which was being borne in the procession fell and broke in pieces. "Being probably suspected of magical practices," the two sisters were arrested; one died in prison and the other was strangled; whereupon the bishop rescued their bones, and these were cherished as the remains of martyrs.[2]

A story from Epiphanius.

Epiphanius in the next century tells a story similar to that of Cyprian, Aglaïdes, and Justina, of a youth who was led astray by evil companions who employed magic arts, love philters, and incantations to force free women to gratify their licentious desires. By means of magic the youth went through the air to a very beautiful woman in

[1] Bouchier, *Syria as a Roman Province*, p. 237.

[2] Bouchier, *Spain under the Roman Empire*, p. 123, citing AS, July 19.

the public bath, but she repelled him by making the sign of the cross. His companions then tried to devise some more powerful magic for his benefit, and took him at sunset to a cemetery full of caves where for three successive nights the wizards vainly plied their arts in the attempt to gratify his lust. But in every instance they were foiled by the name of Christ and the sign of the cross.[1]

Joseph's experience of miracle and magic.

Joseph, the guardian of this same young man, finally became converted to Christianity after Christ had appeared repeatedly to him in dreams and cured him of diseases and after he himself, by employing the name of Jesus, had cured a man of a demoniacal possession which made him go shamelessly about the town in a nude state. After his conversion, Joseph started to complete as a Christian church an unfinished structure in Tiberias called the Adrianaion, which the citizens previously had tried to convert into a public bath. When the Jews endeavored to ruin his undertaking by bewitching the furnaces which he had erected for the preparation of quick-lime, he counteracted their magic by making the sign of the cross, sprinkling his furnaces with holy water, and saying in the name of Jesus of Nazareth, "Let there be power in this water to counteract all pharmacy and magic employed by these men and to instill sufficient energy into the fire to complete the house of the Lord." With that his fires blazed up violently.[2]

Legend of St. James and Hermogenes the magician.

Very similar both to the *Confession* of Cyprian and the story of Simon Magus is the legend of St. James the Great

[1] Epiphanius, *Panarion,* ed. Dindorf, II, 97-104; ed. Petavius, 131A-137C.

[2] *Idem.* The attempt to bewitch the furnaces reminds one of the fourteenth Homeric epigram, in which the bard threatens to curse the potters' furnaces if they do not pay him for his song, and to summon "the destroyers of furnaces,"—Σύντριβ' ὁμῶς Σμάραγόν τε καὶ Ἄσβετον ἠδὲ Σαβάκτην,—words usually interpreted as names for mischievous Pucks and brawling goblins who smash pottery. But the two middle names suggest the stones, smaragdus or emerald, and asbestos. The poet also invokes "Circe of many drugs" to cast injurious spells, and appeals to Chiron to complete the work of destruction. He further prays that the face of any potter who peers into the furnace may be burned. This epigram is probably of late date. See A. Abel, *Homeri Hymni, Epigrammata, Batrachomyomachia,* Lipsiae, 1886, pp. 123-4.

and Hermogenes the magician, which is found in *The Golden Legend* and which was often reproduced in medieval stained glass windows.[1] James converted to Christianity a disciple of Hermogenes whom the magician had sent against him when he was preaching in Judea. When the angry wizard cast a spell over his erstwhile disciple, the latter was freed by means of St. James's cloak. When the magician sent demons to fetch both the convert and the saint, James made them bring Hermogenes to him instead, but then set him free, telling him that Christians returned good for evil. Hermogenes now feared the vengeance that the demons would take upon himself, and so James gave his staff to him to protect himself with. Soon afterwards Hermogenes threw all his books of magic into the sea and was baptized.

Other contests of apostles and magicians in *The Golden Legend*.

"In *The Golden Legend,*" in fact, as Mâle says, "almost all the apostles have to contend with magicians. But it is St. Simon and St. Jude who strive with the most formidable of sorcerers, and they challenge him even in the very sanctuary of magic art, the temple of the Sun at Suanir, near Babylon. Undismayed by the science of Zoroaster and Aphaxad, they foretell the future, they cause a new-born babe to speak, they subdue tigers and serpents, and from a statue they cast out a demon, which shows itself in the shape of a black Ethiopian and flees uttering raucous cries."[2] If this last exorcism reminds us somewhat of the exploits of Apollonius of Tyana, still more do the performances of St. Andrew, who "must surpass all the marvels of the magicians before he can convert Asia and Greece. He drives away seven demons who in the shape of seven great dogs desolate the town of Nicaea, and he exorcises a spirit which dwells in the *thermae* and is wont to strangle the bathers."[3]

[1] Mâle, *Religious Art in France,* 1913, pp. 304-6.

[2] Mâle (1913), p. 306.

[3] *Ibid.,* p. 307.

CHAPTER XIX

ORIGEN AND CELSUS

Celsus' charges of magic against Christianity—Hebrew magic as depicted by Celsus—Various recriminations of magic—Origen's distinction between miracles and magic—Origen frees Jews as well as Christains from the charge of magic—Celsus' sceptical description of magic—Celsus suggests a connection between magic and occult virtues in nature—Celsus on magicians and demons—Origen ascribes magic to demons—Magic is an elaborate art—The Magi of Scripture were not different from other magicians—Origen's Biblical commentaries—Balaam and the power of words—Limitations to the power of Pharaoh's magicians—Was Balaam a prophet of God or a magician?—Balaam's magic experiments—Limitations to his magic power—Divine prophecy distinct from magic and divination—The ventriloquist really invoked Samuel for Saul—Christians less affected by magic than philosophers are—Their superstitious methods against magic—Incantations—The power of words—Origen admits a connection between the power of words and magic—Jewish and Christian employment of powerful names is really magic—Celsus' theory of demons—Origen calls demons wicked—But believes in presiding angels—A law of spiritual gravitation—Attitude of Celsus toward astrology—Attitude of Origen toward astrology—Further discussion in his *Commentary on Genesis*—Problems of the waters above the firmament and of one or more heavens—Augury, dreams, and prophecy—Animals and gems—Origen later accused of countenancing magic.

Celsus' charges of magic against Christianity.

IN the celebrated work of Origen *Against Celsus*,[1] written in the first half of the third century, the subject of magic is often touched upon, largely because Celsus in his *True Discourse* had so frequently brought charges of magic against Jesus, His Christian followers, and the Jewish people from whom they had sprung. Celsus had called Jesus

[1] Greek text in Migne PG, Vol. XI. English translation in the *Ante-Nicene Fathers*, of which I generally make use in quotations from the work. On the MSS of the *Against Celsus* see Paul Koetschau, *Die Textüberlieferung der Bücher des Origenes gegen Celsus in den Handschriften dieses Werkes und der Philokalia. Prolegomena zu einer kritischen Ausgabe*, 1889, 157 pp., (TU, VI, 1).

"a wicked and God-hated sorcerer"; [1] had contended that His miracles were wrought by magic, not by divine power; [2] and had compared them unfavorably, as less wonderful, to the tricks performed by jugglers and Egyptians in the middle of market-places.[3] It was the opinion of Celsus that Jesus in warning His disciples that "there shall arise false Christs and false prophets, and shall show great signs and wonders," had tacitly convicted Himself of the same magical practices.[4] Celsus, for his part, warned the Christians that they "must shun all deceivers and jugglers who will introduce you to phantoms"; [5] he accused them of employing incantations and the names of certain demons; [6] he asserted that he had seen in the hands of Christian presbyters "barbarous books containing the names and marvelous operations of demons," and that these presbyters "professed to do no good, but all that was calculated to injure human beings." [7]

Hebrew magic as depicted by Celsus.

Celsus regarded Moses equally with Jesus as a wizard,[8] and he evidently, like Juvenal and other classical writers, considered the Jews and Syrians as a race of charlatans, especially given to superstition, sorcery, incantations, ambiguous oracles and conjuration of spirits. "They worship angels," he declared, "and are addicted to sorcery, in which Moses was their instructor." [9] He stated that the Jews traced back their origin to "the first generation of lying wizards," by which phrase Origen thinks he referred to Abraham, Isaac, and Jacob, whose names Origen admits are much employed in the magic arts.[10] Celsus further characterized the Jews as "blinded by some crooked sorcery, or dreaming dreams through the influence of shadowy specters," [11] and as "induced to bow down to the angels in heaven by the incantations employed by jugglery and

[1] I, 71; also II, 32.
[2] I, 38; also VIII, 9; II, 48.
[3] I, 68; III, 52.
[4] II, 49.
[5] VII, 36.
[6] I. 6.
[7] VI, 40.
[8] V, 51.
[9] I, 26.
[10] IV, 33.
[11] V, 6.

sorcery, in consequence of which certain phantoms appear in obedience to the spells employed by the magicians." [1] Celsus, also, in describing the many self-styled prophets, Redeemers, and Sons of God in the Phoenicia and Palestine of his own time, states that they make use of "strange, fanatical, and quite unintelligible words, of which no rational person can find any meaning," [2] and that those prophets whom he himself had heard had afterwards confessed to him that these words "really meant nothing." [3] Yet even the Christians—Celsus complains—who condemn all other oracles, regard as marvelous and accept unquestioningly "those sayings which were uttered or were not uttered in Judea after the manner of that country, as indeed they are still delivered among the peoples of Phoenicia and Palestine." [4]

Various recriminations of magic.

To these accusations of Celsus Origen himself adds that the Jews affirm that Jesus passed Himself off as Christ by means of sorcery,[5] while the Egyptians charge Moses and the Hebrews with the practice of sorcery during their stay in Egypt.[6] Origen, on the other hand, speaks of "the magical arts and rites of the Egyptians" and holds that it was by divine aid and not by superior magic that Moses prevailed over Pharaoh's magicians.[7] Celsus for his part had accused Jesus during His residence in Egypt of "having there acquired some miraculous powers, on which the Egyptians greatly pride themselves." [8]

Origen's distinction between miracles and magic.

Origen repudiates the charges of magic made against Christ and His followers as slanders. He asserts that Christianity on the contrary strictly forbids the practice of magic arts,[9] and that these lost much of their force at the birth of Christ.[10] He contends that no magician would teach such noble doctrines as those of Christianity.[11] Origen goes so far as to deny that even the "false Christs and false

[1] V, 9.
[2] VII, 9.
[3] VII, 11.
[4] VII, 3.
[5] III, 1.
[6] III. 5.
[7] III, 46; IV, 51.
[8] I, 28.
[9] I, 38.
[10] I, 60.
[11] I, 38.

prophets," who "shall show great signs and wonders," will be sorcerers, and he states that no sorcerer has ever claimed to be Christ[1]—an amazing assertion in view of his own allusions to Simon Magus. Works of magic and miracles, Origen affirms, are no more alike than are a wolf and a dog or a wood-pigeon and a dove. They are, however, so closely related that if one admits the reality of magic he must also believe in divine miracles, just as the existence of sophistry proves that there is such a thing as sound argument and an art of dialectic.[2] Moreover, in one passage Origen admits that "there would indeed be a resemblance" between miracles and magic, "if Jesus, like the dealers in magic arts, had performed His works only for show; but now there is not a single juggler who, by means of his proceedings, invites his spectators to reform their manners, or trains those to the fear of God who are amazed at what they see, nor who tries to persuade them so to live as men who are to be justified by God."[3] On the contrary, Origen asserts that the magicians' "own lives are full of the grossest and most notorious sins."

Origen frees Jews as well as Christians from the charge of magic.

Since it is one of Origen's chief concerns to uphold Hebrew prophecy as a proof of Christ's divinity, although Celsus subjects the argument from prophecy to ridicule; to defend the Old Testament against Celsus' attacks as an inspired record of greater antiquity than Greek philosophy, history, and literature, which he asserts have stolen truths from it; and to maintain that "there is no discrepancy between the God of the Gospel and the God of the Law":[4]—since this is so, it is incumbent upon him to rebut also the accusations of magic laid by Celsus at the door of the Jews. Origen therefore asserts that the Jews "despised all kinds of divination as that which bewitches men to no purpose," and cites the prohibition of *Leviticus* (XIX, 31) against wizards and familiar spirits.[5]

[1] II, 49.
[2] II, 51.
[3] I, 68.
[4] VII, 25.
[5] V, 42.

Celsus' sceptical description of magic.

The *Reply to Celsus* is of especial interest to us because it presents as it were in parallel columns for our inspection the classical and the Christian conceptions of and attitudes towards magic. Before proceeding, therefore, to inquire how far justified Origen seems to be in thus acquitting, or Celsus, on the other hand, in condemning Christians and Jews on the charge of magic, it is essential to note what magic means for either author. Both evidently regard it as a term of reproach and as usually evil in character.[1] Celsus lists as feats of magic the expelling of demons and diseases from men, or the sudden production of tables, dishes, and food as for an expensive banquet, or of animals who move about as if alive. Celsus, however, seems to speak with a sneer of "their most venerated arts" and describes the banquet dishes as "dainties having no real existence" and the animals as "not really living but having only the appearance of life." Therefore the ensuing comment of Origen seems unusually stupid or unfair, when he tries to convict Celsus of inconsistency on the ground that "by these expressions he allows as it were the existence of magic," whereas Origen hints that it was he "who wrote several books against it." "These expressions" are, on the contrary, precisely those which a man who had attacked magic as deceptive would use. Celsus further stated that an Egyptian named Dionysius had told him that magic arts had power "only over the uneducated and men of corrupt morals," but had no effect upon philosophers, "because they were careful to observe a healthy manner of life." [2] Celsus himself observed that "those who in market-places perform most disreputable tricks and collect crowds around them . . . would never approach an assembly of wise men." [3] It was at the request of a Celsus, moreover, that the second century satirist Lucian wrote his *Alexander* or *Pseudomantis* [4] in which some of the tricks of a magician-impostor and oracle-monger are exposed, and in which allusion is

[1] I, 68.

[2] VI, 41.

[3] III, 52.

[4] See cap. 21.

made to the "excellent treatises against the magicians" written by Celsus himself. It seems reasonably certain that the Celsus of Lucian and the Celsus of Origen are identical, as there are no chronological difficulties and the same point of view is ascribed in either case to Celsus, whom both Lucian and Origen regard as an Epicurean or at least in sympathy with the Epicureans. Galen, in a treatise in which he lists his own writings, mentions an "Epistle to Celsus the Epicurean." [1] This, too, might be the same man.

Celsus suggests a connection between magic and occult virtues in nature.

Another passage in which Celsus, according to Origen at least, "mixed up together matters which belong to magic and sorcery" runs as follows: "What need to number up all those who have taught methods of purification, or expiatory hymns, or spells for averting evil, or images, or resemblances of demons, or the various sorts of antidotes against poison in clothing, or in numbers, or stones, or plants, or roots, or generally in all kinds of things?" [2] In another passage Celsus again closely connected sorcery with the knowledge of occult virtues in nature, arguing that men need not pride themselves upon their power of sorcery when serpents and eagles know of antidotes to poisons and amulets and the virtues of certain stones which help to preserve their young." [3] Origen objects that it is not customary to use the word sorcery (γοητεία) for such things, and suggests that Celsus is such an "Epicurean," i. e., so sceptical, that he wishes to discredit all those other beliefs and practices "as resting only on the professions of sorcerers." But we have already had proof enough in other chapters that Celsus was not unjustified in connecting the occult virtue of natural objects with magic, if not with sorcery.

Celsus on magicians and demons.

Celsus, as we shall see, believed in the existence of demons whom, however, he did not regard as necessarily evil spirits, and whom he probably regarded as above any connection with magic. Origen once says that if Celsus

[1] Kühn, XIX, 48 (*de libris propriis*). Μετροδώρου ἐπιστολὴ πρὸς Κέλσον Ἐπικούρειον.

[2] VI, 39.

[3] IV, 86.

"had been acquainted with the nature of demons" and their operations in the magic arts, he would not have blamed Christians for not worshiping them.[1] The natural inference from this statement is that Celsus did not associate demons with magic. Origen, however, depicts him as "speaking of those who employ the arts of magic and sorcery and who invoke the barbarous names of demons," [2] and we have already heard him censure certain Christian presbyters for their "barbarous books containing the names and marvelous doings of demons." [3] It therefore becomes evident that magicians attempt to avail themselves of the aid of demons, whether Celsus believes that they succeed in their attempt or not.

Origen ascribes magic to demons.

Origen at any rate believes that magicians are aided by evil spirits, and for him demons became the paramount factor in magic, just as it is they who are worshiped in pagan temples as gods and who inspire the pagan oracles.[4] Indeed, just as Celsus has kept calling the Christians sorcerers, so Origen is inclined to label all heathen religions, rites, and ceremonies as magic. He quotes the Psalmist as saying that "all the gods of the heathen are demons." [5] He states that the dedication of pagan temples, statues, and the like are accompanied by "curious magical incantations . . . performed by those who zealously serve the demons with magic arts." [6] Divination in general, he believes, "proceeds rather from wicked demons than from anything of a better nature." [7] He does not think of magic as a deception, he does not endeavor to expose its frauds, he accepts its marvels as facts, but declares that "magic and sorcery are produced by wicked spirits, held spellbound by elaborate incantations and yielding themselves to sorcerers." [8] Origen seems in doubt whether the demons are coerced by the spells and charms of magic or yield themselves willingly.[9]

[1] VII, 67.
[2] VI, 39.
[3] VI, 40.
[4] VII, 3 and 35.
[5] Ps. xcvi, 5.
[6] VII, 69.
[7] V, 42.
[8] II, 51. See also V, 38; VI, 45; VII, 69; VIII, 59; I, 60.
[9] See VII, 67, "demons . . . and their several operations, whether led on to them by the

As we shall see, Origen is at least ready to attribute great power to incantations, and he does not deny that magic is an elaborate art. With such various arts of magic he contrasts the simplicity of Christian prayers and adjurations "which the plainest person can use," or the Christian casting out of demons which is performed for the most part by "unlettered persons." [1] Origen also suggests that the natural properties of plants and animals are a factor in magic, when he cites Numenius the Pythagorean's description of the Egyptian deity Serapis. "He partakes of the essence of all the animals and plants that are under the control of nature, that he may appear to have been fashioned into a god, not only by the image-makers with the aid of profane mysteries and juggling tricks employed to invoke demons, but also by magicians and sorcerers (μάγων καὶ φαρμακῶν) and those demons who are bewitched by their incantations." [2] Another passage pointing in the same direction is Origen's description of "the man who is curiously inquisitive about the names of demons, their powers and agency, the incantations, the herbs proper to them, and the stones with the inscriptions graven on them, corresponding symbolically or otherwise to their traditional shapes." [3] Thus although Origen lays the emphasis upon demons, we see that he admits most of the other customary elements in magic.

Magic is an elaborate art.

Origen does not, like Philo Judaeus, Apuleius and some Christian writers, distinguish two uses of the word magic, one good and one evil. He does not differentiate between vulgar magic and malignant sorcery on the one hand and the lore of learned Magi of the east on the other hand. He

The Magi of Scripture were not different from other magicians.

conjurations of those who are skilled in the art, or urged on by their own inclinations. . . ."

Also VII, 5, "those spirits that are attached for entire ages, as I may say, to particular dwellings and places, whether by a sort of magical force or by their own natural inclinations."

Also VII, 64, ". . . the demons choose certain forms and places, whether because they are detained there by virtue of certain charms, or because for some other possible reason they have selected those haunts. . . ."

[1] VII, 4. ὡς ἐπίπαν γὰρ ἰδιῶται τὸ τοιοῦτον πράττουσι.

[2] V, 38.

[3] VIII, 61.

simply says that the art of magic gets its name from the Magi and that from them its evil influence has been transmitted to other nations.[1] Celsus had ranked the Magi among divinely inspired nations but Origen objects to this. Yet he recognizes that the wise men of the east who followed the star of Bethlehem and came to worship the infant Christ were Magi.[2] But he seems to regard them as ordinary magicians, who were accustomed to invoke evil spirits.[3] He thinks that the coming of Christ dispelled the demons and hindered the Magi's spells and charms from working as usual. Trying to find the reason for this, they would note the new star in the sky. Origen will not admit that they could do all this by means of astrology, nor even that they were astrologers at all; he accuses Celsus of blundering in calling them Chaldeans or astrologers.[4] Rather he thinks that they could find an explanation of the star in the prophecies of Balaam [5] which they possessed and which predicted, as Moses too records,[6] "There shall arise a star out of Jacob, and a man (or, as in the King James' version, a scepter) shall rise up out of Israel." [7] In another treatise than the *Reply to Celsus* Origen further explains that the Magi were descended from Balaam and so owned his written prophecies.[8] Balaam was perhaps alluding to these very Magi descended from him who came to adore Jesus when he prophesied that his seed should

[1] VI, 80.

[2] I, 58.

[3] I, 60.

[4] I, 58. The Magi had been confused with the Chaldeans several centuries before by Ctesias in his *Persica*, cap. 15; see D. F. Münter, *Der Stern der Weisen: Untersuchungen über das Geburtsjahr Christi*, Kopenhagen (1827), p. 14.

[5] Balaam himself was something of an astrologer according to Münter, *Der Stern der Weisen*, 1827, p. 31. "Die sieben Altäre die der moabitische Seher Bileam an verschiedenen Orten errichtete (IV B. Mose, XXIII) waren gewiss den sieben Planetfürsten gewidmet."

[6] Numbers, XXIV, 17.

[7] Similarly an English version (in an Oxford MS of the early 15th century, Laud Misc., 658) of *The History of the Three Kings of Cologne*, or medieval account of the translation of the relics of the Magi, in forty-one chapters with a preface, opens its first chapter with the words, "The mater of these three worshipful and blissid kingis token the begynnyng of the prophecye of Balaam."

[8] *In Numeros Homilia XIII*, in Migne, PG, XII, 675.

be as the seed of the just.[1] Origen seems to have been the first of the church fathers to state the number of these Magi as three, which he does in one of his homilies on the Book of Genesis.[2]

Origen's Biblical commentaries.

At this point indeed, we may well turn for a little while from the *Reply to Celsus* to those Biblical commentaries of Origen where he discusses such Old Testament passages connected with magic as the stories of Balaam and of the witch of Endor or ventriloquist. The commentary of Origen upon the Book of Numbers is extant only in the Latin translation by Rufinus, who literally snatched it for posterity as a brand from the burning, for he did not refrain from this learned and literary labor, although as he plied his pen in Messina in 410 A. D. he could see the invading barbarians ravaging the fields and burning Reggio just across the narrow strait which separates Sicily from Italy.[3]

Balaam and the power of words.

In commencing to speak of Balaam and his ass [4] Origen implies that much has already been written on this thorny theme and that he approaches it with considerable diffidence. He prays God again and again for grace to be able to explain it, not by means of fabulous Jewish narrations—by which expression he perhaps alludes to commentaries of the rabbis such as have reached us in the Talmud—but in a sense that shall be reasonable and worthy of the divine law. To begin with he admits the power of words, and not merely that of holy words or words of God, but of certain words used by men. That such words are in some respects more powerful than bodies is shown by the fact that Balaam's cursing could accomplish what armies and weapons could not effect. This calls to mind one of the Mohammedan tales concerning Balaam to the effect that by reading the books of Abraham he learned "the name

[1] *In Numeros Homilia XV,* col. 689.

[2] *In Genesim Homilia XIV,* 3, in PG, XII, 238.

[3] *Origenis in Numeros Homiliae, Prologus Rufini Interpretis ad Ursacium.* Migne, PG, XII, 583-86.

[4] *Origenis in Numeros Homilia XIII,* Migne, PG, XII, 670-677. In at least one medieval manuscript we find the homily upon Balaam preserved separately, BN 13350, 12th century, fol. 92v, et omeliae de Balaham et Balach.

Yahweh by virtue of which he predicted the future, and got from God whatever he wished." [1]

Limitations to the power of Pharaoh's magicians.

The magicians of Egypt, too, who withstood Moses and Aaron before Pharaoh, were able to turn rods into snakes and water into blood, feats which no man could accomplish by mere bodily strength. Indeed, because the king of Egypt knew that his magicians could do such things by a human art of words, he thought, at first at least, that Moses too was doing the same things not by the help of God but by the magic art. There was, however, a very serious limitation to the magicians' power. By the aid of demons they could turn good into evil but they could not repair the damage which they had done or restore the evil to good. The rod of Moses, on the other hand, not only devoured theirs but turned back from a snake into its original form,[2] and it was necessary for Moses to pray to God in order to stay the other plagues.

Was Balaam a prophet of God or a magician?

Origen classifies Balaam as a magician, not as a prophet. This seems to have been the prevalent patristic and medieval view, although the Biblical account in Numbers represents Balaam as in close and constant communication with God and the Second Epistle of Peter [3] calls him a prophet although it condemns his temporary madness in seeking "the wages of unrighteousness." Josephus too calls him the best prophet of his time but one who yielded to temptation.[4] A fifteenth century treatise on the translation of the relics of the three kings to Cologne tells us that "concerning this Balaam there is an altercation in the east between the Christians and the Jews"; the Jews holding that he was no prophet but a diviner who predicted by magic and diabolical arts, the Christians asserting that he was the first prophet of the Gentiles.[5] The problem continued to

[1] W. H. Bennett, *Balaam,* in EB, 11th edition.

[2] One cannot help wondering whether Pharaoh's magicians lost their rods for good as a result of this manœuvre, but it is a point upon which the Scriptural narrative fails to enlighten us.

[3] II, 15-16.

[4] *Antiq.*, IV, 6.

[5] Johannis Hildeshemensis, *Liber de trium regum translatione,* 1478, cap. 2.

exercise the ingenuity of Lutherans and theologians of the Reformed Churches, and in 1842 was the main theme of a treatise of 290 pages in which Hebrew words and quotations from Calvin abound.[1]

Balaam's magic experiments.

Origen remarks that magicians differ in the amount of power they possess. Balaam was a very famous and expert one, known throughout the whole orient. He had given many experimental proofs (*experimenta*) of his skill and Balak had frequently employed him. The translator Rufinus's repeated use of the words *experimenta* and *expertus* here is an interesting indication of the close connection between magic and experiment.[2]

Limitation to his magic power.

Great, however, as was Balaam's fame and power, he could only curse and not bless, an indication that he operated by the agency of demons who also only work evil and not good. It is true that King Balak said to him: "I know that whom you bless will be blessed," but Origen regards this as false flattery. Magicians employ the services of evil spirits, but cannot invoke such angels as Michael, Raphael, and Gabriel, much less God or Christ. Christians alone have the power to do this, and they must cease entirely from the invocation of demons or the Holy Spirit will flee from them.

Divine prophecy distinct from magic and divination.

It is true also that God in the end did speak through the mouth of Balaam and that he blessed instead of cursed Israel. Origen will not admit, however, that Balaam was worthy of this, or that a man can be both a magician and a prophet; if God spake through Balaam, it was only to prevent the demons from coming and helping Balaam to curse Israel. Origen also attempts to solve the difficulties

[1] E. W. Hengstenberg, *Die Geschichte Bileams und seine Weissagungen*, Berlin, 1842. Hengstenberg tried to take middle ground between Philo Judaeus, Ambrose, Augustine, Gregory of Nyssa, Theodoret, and others who regarded Balaam as a godless false prophet and magician, and the contrary opinion of Tertullian, Jerome, and some moderns who hold that Balaam was originally a devout man and true prophet who fell through his covetousness.

[2] "Et ideo quasi expertus in talibus in opinione erat omnibus qui erant in Oriente . . . Certus ergo Balach de hoc et frequenter expertus."

and inconsistencies involved in the repeated appearances and conflicting commands of God and the angel to Balaam. Finally we may note that Origen sees the similarity between the use of cauldron-shaped tripods in human arts of divination and the donning of the ephod by the prophets described in the Old Testament.[1] But he affirms that divine prophecy and divination are two different things and cites the Biblical prohibition of the latter.

The ventriloquist really invoked Samuel for Saul.

In his commentary upon the First Book of Samuel,[2] Origen takes the ground that when Saul consulted the witch or ventriloquist (ἐγγαστριμύθος), Samuel's ghost really appeared and spoke to Saul, for the Scriptural account plainly says that the woman saw Samuel [3] and that Samuel spoke to Saul. Consequently Origen cannot agree with those who have held that the woman deceived Saul or that both she and he were deluded by a demon who assumed the guise of Samuel. No demon, he thinks, could have prophesied that the kingdom would pass to David. It has been objected that the enchantress could not raise the spirit of Samuel from the infernal regions because he was a good man, but Origen holds that even Christ descended to hell and that all before Him had their abode there until He came to release them. From this position not even the parable of Dives and of Lazarus in Abraham's bosom with the great gulf fixed between them can shake Origen.

Christians less affected by magic than philosophers are.

Origen disputes the statement of Celsus that philosophers are not affected by the magic arts by pointing out that in Moiragenes's *Life of Apollonius of Tyana,* who was himself both a philosopher and magician, it is affirmed that other philosophers were won over by his magic power "and resorted to him as a sorcerer." [4] On the other hand Origen makes the counter-assertion that the followers of Christ "who live according to His gospel, using night and day con-

[1] In Homily XIV.

[2] Migne, PG, XII, 1011-28.

[3] J. G. Frazer (1918), II, 522, note, however, says of I. Samuel, XXVIII, 12: "It seems that we must read, 'And when the woman saw Saul,' with six manuscripts of the Septuagint and some modern critics, instead of, 'And when the woman saw Samuel.'"

[4] VI, 41.

tinuously and becomingly the prescribed prayers, are not carried away either by magic or demons."

Their superstitious methods against magic.

If these "prescribed prayers" were set forms of words, they would seem not far removed in character from the incantations of the magicians which they were supposed to counteract. An even clearer example of preventive magic is seen in Origen's explanation that the practice of circumcision was a safeguard against some angel (*sic*) hostile to the Jewish race.[1]

Incantations.

If demons are for Origen of primary importance in magic, incantations run a close second, since it is chiefly through them that men are able to utilize the power of the demons. Some of the barbarians, Origen tells us, "are admired for their marvelous powers of incantation."[2] And when he mentions the miraculous releases of Peter and Paul and Silas from prison, he adds that if Celsus had read of these events he "would probably say in reply that there are certain sorcerers who are able by incantations to unloose chains and to open doors."[3] But Celsus did not say this; we must therefore attribute the thought rather to Origen himself. Speaking elsewhere in his own person Origen more than once informs us that "almost all those who occupy themselves with incantations and magical rites" and "many who conjure evil spirits" employ in their spells and incantations such expressions as "God of Abraham."[4] Origen grants that these phrases are used by the Jews themselves in their prayers to God and exorcisms, and that the names of Abraham, Isaac, and Jacob possess great efficacy "when united with the word of God."[5] Yet he will not acknowledge that the Jews practice magic. He also denies the charge of Celsus that Christians use incantations and the names of

[1] V, 48.
[2] I, 30.
[3] II, 34.
[4] IV, 33, and I, 22.
[5] IV, 33. On the use of mystic names of God among the Jews of this period and "the new and greatly developed angelology that flourished at that time in Egypt and Palestine" see the Introduction to M. Gaster's edition of *The Sword of Moses*, 1896,—a book of magic found in a 13-14th century Hebrew MS, but which is mentioned in the 11th century and which he would trace back to ancient times.

certain demons, although he admits that Christians ward off magic by regular use of prescribed prayers and frequently expel demons by repetition of "the simple name of Jesus, and *certain other words* in which they repose faith, according to the holy Scriptures," or "the name of Jesus accompanied by the announcement of the narratives which relate to Him" (presumably a repetition of the names of the four Evangelists).[1] It is even possible for persons who are not true Christians to make use of the name of Jesus to work wonders just as magicians use the Hebrew names.[2]

The power of words.

Origen, however, does not try to justify these Hebrew and Christian formulae, adjurations, and exorcisms on the ground that they are simply prayers to God, who Himself then performs the cure or miracle without compulsion. Origen believes that there is power in the words themselves, as we have already heard him state in speaking of Balaam. This is seen from the fact that when translated into another language they lose their operative force, as those who are skilled in the use of incantations have noted.[3] Thus not what is signified by the words, but the qualities and peculiarities of the words themselves, are potent for this or that effect. It seems strange that Origen should thus cite enchanters, when in the sentence just preceding he had spoken of "our Jesus, whose name has been manifestly seen to have driven out demons from souls and bodies. . . ." Was the divine name alone and not God the cause of the miracle? It may be added, however, that Origen denied that languages were of human origin.[4] But he has already gone far along this line and in the previous chapter has stated that "the nature of powerful names" is a "deep and mysterious subject." [5] Some such names, he goes on to say, "are used by the learned amongst the Egyptians, or by the Magi among the Persians, and by the Indian philosophers called Brahmans."

[1] I, 6. It also, however, suggests the efficacy ascribed by the Mandaeans to the repetition of passages from their sacred books.

[2] II, 49.

[3] I, 25; V, 45.

[4] V, 45.

[5] I, 24.

Later on in the work, in a passage which we have already cited, Origen waxed indignant with Celsus for speaking favorably of the Magi, inventors of the destructive magic art. But now he speaks almost in a tone of respect of magic, stating that if "the so-called magic also is not, as followers of Epicurus" (i. e., men like Celsus whom Origen accuses of being an Epicurean) "and Aristotle think, an entirely chaotic affair but, as those skilled in such matters show, a connected system comprising words known to very few persons," then such names as Adonai and Sabaoth "pertain to some mystic theology," and, "when pronounced with that attendant train of circumstances which is appropriate to their nature, are possessed of great power."

Origen admits a connection between the power of words and magic.

These last clauses make it clear that Jews and Christians were guilty both of incantations and magic, however much Origen may protest to the contrary. It can hardly be argued that Origen means to distinguish this "so-called magic" from the magic art which he condemns in other passages, for not only is it evident that the followers of Epicurus and Aristotle make no such distinction, but Origen himself in other passages ascribes the employment of such Hebrew names to ordinary magicians and declares that such invocations of God are "found in treatises on magic in many countries." [1] Origen also states in his *Commentary upon Matthew* [2] that the Jews are regarded as adepts in adjuration of demons and that they employ adjurations in the Hebrew language drawn from the books of Solomon. Moreover, he continues in the present passage, "And other names, again, current in the Egyptian tongue, are efficacious against certain demons who can only do certain things; and others in the Persian language have corresponding power over other spirits; and so on in every different nation, for different purposes." ". . . And when one is able to philosophize about the mystery of names, he will find much to say respecting the titles of the angels of God, of whom one is

Jewish and Christian employment of powerful names is really magic.

[1] IV, 33; I, 22, etc.
[2] *In Math.* XXVI, 23 (Migne, PG, XIII, 1757).

called Michael, and another Gabriel, and another Raphael, appropriately to the duties which they discharge in the world. And a similar philosophy of names applies also to our Jesus." Between such mystic theology and philosophy of names, the Gnostic diagram of the Ophites,[1] and the downright incantations of the magicians, there is surely little to choose.

Celsus' theory of demons.

From the names of God and angels, by uttering which such wonders may be performed, we turn to the spirits themselves. Celsus seems to think of demons as spiritual beings who act as intermediaries between the supreme Deity and the world of nature and human society. He believes that "in all probability the various quarters of the earth were from the beginning allotted to different superintending spirits." [2] He warns the Christians that it is absurd for them to think that they can escape the demons by simply refusing to eat the meat that has been offered to idols; the demons are everywhere in nature, and one cannot eat bread or drink wine or taste fruit or breathe the very air without receiving these gifts of nature from the demons to whom the various provinces of nature have been assigned.[3] The Egyptians teach that even the most insignificant objects are committed to demon care, and they divide the human body into thirty-six parts, each in charge of a demon of the air who should be invoked in order to cure an ailment of that particular part.[4] Celsus mentions some of the names of these thirty-six demons: Chnoumen, Chnachoumen, Cnat, Sicat, Biou, Erou, and others. Celsus, however, does not accept this Egyptian doctrine without qualification. He suspects, Origen tells us, that it leads toward magic, and hence adds "the opinion of those wise men who say that most of the earth-demons are taken up with carnal indulgence, blood, odors, sweet sounds and other such sensual things; and therefore they are unable to do more than heal the body, or foretell the fortunes of men and cities, and do other such

[1] See p. 366 in Chapter XV on Gnosticism.

[2] V, 25.

[3] VIII, 28.

[4] VIII, 58.

things as relate to this mortal life."[1] Celsus himself, however, seems as unwilling to accept this Egyptian view as he is to condone magic, and concludes that "the more just opinion is that the demons desire nothing and need nothing, but that they take pleasure in those who discharge toward them offices of piety."[2] Celsus believes that divine providence regulates the acts of the demons and so asks: "Why are we not to serve demons?"[3]

Origen calls demons wicked.

Origen's reply to this question is that the demons are wicked spirits and concerned with magic and idolatry. He maintains that not only Christians "but almost all who acknowledge the existence of demons" regard them as evil spirits.[4] His own attitude toward them is invariably one of hostility. The thirty-six spirits who, as the Egyptians believe, have charge of different parts of the human body, Origen spurns as "thirty-six barbarous demons whom the Egyptian Magi alone call upon in some unknown way."[5] Really we probably have here to do with the astrological decans or sub-divisions of the signs of the zodiac into sections of ten degrees each.

But believes in presiding angels.

Yet Origen's notion of the spiritual world rather closely resembles that of Celsus, for he is ready to ascribe to angels or other good invisible beings much the same functions which Celsus attributed to demons. He does not, for example, dispute the theory that different parts of the earth and of nature are assigned to different spirits. Instead he "ventures to lay down some considerations of a profounder kind, conveying a mystical and secret view respecting the original distribution of the various quarters of the earth among different superintending spirits."[6] He quotes the Septuagint version of Deuteronomy, "When the most High divided the nations. . . . He set the bounds of the people according to the number of the angels of God."[7] He narrates how after Babel, men "were conducted by those angels

[1] VIII, 60.
[2] VIII, 63.
[3] VII, 68.
[4] VII, 69.
[5] VIII, 59.
[6] V, 28.
[7] V. 29; see *Deut.* xxxii, 8.

who imprinted on each his native language to the different parts of the earth according to their deserts." [1] He concludes by saying, "These remarks are to be understood as being made by us with a concealed meaning," [2] but there seems little doubt as to his substantial agreement with the view of Celsus. Indeed, later when Celsus asserts that Christians cannot eat, drink, or breathe without being indebted to demons, Origen responds, "We indeed also maintain . . . the agency and control of certain beings whom we may call invisible husbandmen and guardians; . . . but we deny that those invisible agents are demons." [3]

In his fourteenth homily on Numbers, as extant in Rufinus's translation,[4] Origen again speaks of presiding angels in these words. "And what is so pleasant, what is so magnificent as the work of the sun or moon by whom the world is illuminated? Yet there is work in the world itself too for angels who are over beasts and for angels who preside over earthly armies. There is work for angels who preside over the nativity of animals, of seedlings, of plantations, and many other growths. And again there is work for angels who preside over holy works, who teach the comprehension of eternal light and the knowledge of God's secrets and the science of divine things." How this passage might be used to encourage a belief in magic is made evident by the paraphrase of it in *The Occult Philosophy* of Henry Cornelius Agrippa,[5] written in 1510 at the close of the middle ages. He represents Origen as saying, "There is work in the world itself for angels who preside over earthly armies, kingdoms, provinces, men, beasts, the nativity and growth of animals, shoots, plants, and other things, giving that virtue which they say is in things from their occult property."

In the treatise *De Principiis*,[6] Origen states that particular offices are assigned to individual angels, as curing diseases to Raphael, and the conduct of wars to Gabriel. This notion he perhaps derived from the *Book of Enoch* which,

[1] V, 30.
[2] V, 32.
[3] VIII, 31.
[4] Migne, PG, XII, 680.
[5] III, 12.
[6] I, 8.

however, he states in his *Reply to Celsus* is not accepted by the churches as divinely inspired.[1] He further declares on the authority of passages in the New Testament that to one angel the Church of the Ephesians was entrusted; to another, that of Smyrna; that Peter had his angel and Paul his,—nay that "every one of the little ones of the Church" has his angel who daily beholds the face of God.[2]

A law of spiritual gravitation.

Origen advances a further theory concerning spirits, which may be described as a sort of law of spiritual gravitation. It is that when souls are pure and "not weighted down with sin as with a weight of lead," they ascend on high where other pure and ethereal bodies and spirits dwell, "leaving here below their grosser bodies along with their impurities." Polluted souls, on the contrary, have to stay close to earth where they wander about sepulchers as ghosts and apparitions.[3] Origen therefore infers that pagan gods "who are attached for entire ages to particular dwellings and places" on earth, are wicked and polluted spirits. Origen of course will not admit that Christians or Jews bow down even to angels; such worship they reserve for God alone.[4]

Attitude of Celsus toward astrology.

Both Celsus and Origen closely associate with the world of invisible spirits, whether these be angels or demons, the visible heavenly bodies, and thus lead us from magic, which Origen makes so dependent upon demons, to the kindred subject of astrology, the pseudo-science of the stars. Celsus had censured the Jews and by implication the Christians for worshiping heaven and the angels, and even apparitions produced by sorcery and enchantment, and yet at the same time neglecting what in his opinion formed the holiest and most powerful part of the heaven, namely, the fixed stars and the planets, "who prophesy to everyone so distinctly, through whom all productiveness results, the most conspicuous of supernal heralds, real heavenly angels." [5] This shows that Celsus was much more favorably inclined toward astrology

[1] V, 54; see *Book of Enoch,* XL, 9.

[2] Matthew, XVIII, 10.

[3] VII, 5.

[4] V, 6-9.

[5] V, 6.

than toward magic and less sceptical concerning its validity. Origen also represents Celsus—and furthermore the Stoics, Platonists, and Pythagoreans—as believing in the theory of the *magnus annus,* according to which, when the celestial bodies all return to their original positions after the lapse of some thousands of years, history will begin to repeat itself and the same events will occur and the same persons live over again.[1] Origen also complains that Celsus regards as a divinely-inspired nation the Chaldeans, who were the founders of "deceitful genethlialogy," [2] as well as the Magi whom Celsus elsewhere identified with the Chaldeans or astrologers, but whom Origen as we have seen regards rather as the founders of magic.

Attitude of Origen toward astrology.

Origen is opposed both to this art of casting horoscopes and determining the entire life of the individual from his nativity, and to the theory of the *magnus annus,*[3] because he is convinced that to admit their truth is to annihilate free-will. But he is far from having freed himself fundamentally from the astrological attitude toward the stars; indeed he still shows vestiges of the old pagan tendency to worship them as divinities. He is convinced that the celestial bodies are not mere fiery masses, as Anaxagoras teaches.[4] The body of a star is material, it is true, but also ethereal. But furthermore Origen is inclined to agree, both in the *De principiis* [5] and in the *Contra Celsum,*[6] that the stars are rational beings (λογικὰ καὶ σπουδαῖα—the latter word had already been applied to them by Philo Judaeus) possessed of free-will and "illuminated with the light of knowledge by that wisdom which is the reflection of everlasting light." He interprets a passage in Deuteronomy [7] to mean that the stars have in general been assigned by God to all the nations beneath the heaven, but asserts that from this system of astral satrapies God's chosen people were exempted. He

[1] IV, 67; V, 20-21.

[2] VI, 80.

[3] Duhem (1913-1917) II 447, treats of "Les Pères de l'Église et la Grande Année."

[4] V, 11.

[5] *De principiis,* I, 7.

[6] V, 10.

[7] *Deut.,* IV, 19-20.

is willing to admit that the stars foretell many things, and puts especial faith in comets as omens.[1] He states that they have appeared on the eve of dynastic changes, great wars, and other disasters, and inclines also to agree with Chaeremon the Stoic that they may come as signs of future good, as in the case of the star announcing the birth of Christ.[2] But while Origen will grant reasoning faculties and a certain amount of prophetic power to the stars, he refuses to permit worship of them. Rather he is persuaded "that the sun himself and moon and stars pray to the supreme God through his only begotten Son." [3]

Pierre Daniel Huet (1630-1721), the learned bishop of Avranches and editor of Origen, in his commentaries upon Origen [4] cites other works, commentaries on Matthew, the Psalms, the Epistle to the Romans, and Ezekiel, in which Origen again states that the stars are reasoning beings, honor God, praise and pray to Him, and even that they are capable of sin, a point upon which he agrees with the *Book of Enoch* and Bardesanes but not with Philo Judaeus. Nicephorus [5] states that Origen was condemned in the fifth synod for his error concerning the stars being animated. Sometimes, however, Huet points out, Origen leaves it an open question whether the heavenly bodies are animated or not.[6] Huet also asserts that in his own time such great men as Tycho Brahe and Kepler have defended the view that the stars are animated beings.

Further discussion in his *Commentary on Genesis.*

In a fragment from Origen's *Commentary on Genesis* preserved by Eusebius we have a further discussion of the stars and astrology.[7] Here he represents even Christians as troubled by the doctrine that the stars control human affairs absolutely. This theory he attacks as destructive to all morality, as rendering prayer to God of no avail, and as subjecting even such events as the birth of Christ and

[1] V, 12.

[2] I, 59.

[3] V, 11.

[4] P. D. Huet, *Origenianorum* Lib. II, Cap. II, Quaestio VIII, *De astris,* in Migne, *Patrologia Graeca,* XVII, 973, *et seq.*

[5] XVII, 28.

[6] "In prooemio libri prioris eiusdem Περὶ ἀρχῶν, num. 10."

[7] Eusebius, *Praep. Evang.,* VI, 11, in Migne, PG, XXI, 477-506.

the conversion of each individual to Christianity to fatal necessity. Like Philo Judaeus Origen holds that the stars are merely signs instituted by God, not causes of the future, and quotes passages from the Old Testament in support of his view; like the *Book of Enoch* he holds that men were instructed in the interpretation of the stars' significations by the fallen angels. He argues at length that divine foreknowledge does not impose necessity. While, however, God instituted the stars as signs of the future, He intended that only the angels should be able to read them, and deemed it best for mankind to remain in ignorance of the future. "For it is a much greater task than lies within human power to learn truly from the motion of the stars what each person will do and suffer." [1] The evil spirits have, however, taught men the art of astrology, but Origen believes that it is so difficult and requires such superhuman accuracy that the predictions of astrologers are more likely to be wrong than right. His tone toward astrology is thus distinctly more unfavorable here than in the *Reply to Celsus*. In arguing that the stars are merely signs, Origen asks why men admit that the flight of birds and condition of entrails in augury and liver-divination are only signs and yet insist that the stars are causes of future events.[2] The answer, of course, is simple enough: all nature is under the control of the stars which alike produce the events signified and the action of the birds or condition of the liver signifying them. But the question is notable because it was also put by Plotinus a little later in the same century.

Problems of the waters above the firmament and of one or more heavens.

In explaining the Book of Genesis Origen said that celestial and infernal virtues were represented by the waters above and below the firmament respectively. This figurative interpretation gave offence to many later Christian writers, although some of them were ready to interpret the waters above as celestial virtues, but not to take the waters below as signifying evil spirits.[3] Concerning the question of a

[1] PG, XXI, 489.

[2] *Ibid.*, 501-502.

[3] P. D. Huet, *Origenianorum* Lib., II, ii, v. 10, cites Basil, *Homil. 3 in Hexaem.;* Epiphanius, *Haer.*, LXIV, 4, and *Epist. ad*

plurality of heavens Origen says in the *Reply to Celsus,* "The Scriptures which are current in the Churches of God do not speak of seven heavens or of any definite number at all, but they do appear to teach the existence of heavens, whether that means the spheres of those bodies which the Greeks call planets or something more mysterious." [1]

Augury, dreams, and prophecy.

Of other pagan methods of divination than astrology Origen disapproved and classed them, as we have seen, as the work of demons. He was impressed by the weight of testimony to the validity of augury,[2] although he states that it has been disputed whether there is any such art, but he attributed the truth of the predictions to demons acting through the animals and pointed out that the Mosaic law forbade augury [3] and classified as unclean the animals commonly employed in divination. The true God, he held, would not employ irrational animals at all to reveal the future, nor even any chance human being, but only the purest of prophetic souls. Origen would appear for the moment to have forgotten Balaam's ass! Moreover, he himself accepted other channels of foreknowledge than holy prophecy, and believed that dreams often were of value in this respect. When Celsus, criticizing the Scriptural story of the flight into Egypt, stated that an angel descended from heaven to warn Joseph and Mary of the danger threatening the Christ child, Origen retorted that the angelic warning came rather in a dream—an occurrence which seemed in no way marvelous to him, since "in many other cases it has happened that a dream has shown persons the proper course of action." [4] Origen grants that all men desire to ascertain the future and argues that the Jews must have had divine prophets, or, since they were forbidden by the Mosaic law to consult "observers of times and diviners," they would have

Joan. Jerosolymit., cap. 3; Jerome, *Epist.* 61 *ad Pammach.*, cap. 3; Gregory Nyss., *lib. in Hexaem.;* Augustine, *Confess.*, XIII, 15; Isidore, *Origin.*, VII, 5.

See also Duhem (1913-1917) II, 487, "Les eaux supracélestes."

[1] VI, 21.

[2] IV, 90-95.

[3] Origen quotes, "Ye shall not practise augury nor observe the flight of birds," which is found in the Septuagint, *Levit.*, XIX, 26.

[4] I, 66.

had no means of satisfying this universal human craving. It was to slake this popular curiosity concerning the future, Origen thinks, that the Hebrew seers sometimes predicted things of no religious significance or other lasting importance.[1] Once Origen alludes to physiognomy, saying, "If there be any truth in the doctrine of the physiognomists, whether Zopyrus or Loxus or Polemon." [2]

Animals and gems.

The allusions to natural science in the *Reply to Celsus* are not numerous. There are a few passages where animals or gems are mentioned. The remarks concerning animals mention the usual favorites and embody familiar notions which we either have already met or shall meet again and again. Celsus speaks [3] of the knowledge of poisons and medicines possessed by animals, of predictions by birds, of assemblies held by other animals, of the fidelity with which elephants observe oaths, of the filial affection of the stork, and of the Arabian bird, the phoenix.[4] Origen implies the belief that the weasel conceives through its mouth when he says, "Observe, moreover, to what pitch of wickedness the demons proceed, so that they even assume the bodies of weasels in order to reveal the future." [5] Origen also adduces the marvelous methods of generation of several kinds of animals in support of the virgin birth of Jesus.[6] Origen's allusions to gems can scarcely be classified as natural science. He contends that Plato's statement that our precious stones are a reflection of gems in that better land is taken from Isaiah's description of the city of God.[7] In another passage Origen again quotes Isaiah regarding the walls, foundations, battlements, and gates of various precious stones, but states that he cannot stop to examine their spiritual meaning at present.[8] In one of his homilies on the Book of Numbers Origen displays a favorable attitude towards medical and pharmaceutical investigation, saying,

[1] I, 36.
[2] I, 33.
[3] IV, 86-88.
[4] IV, 98.
[5] IV, 93; it will be recalled that the witches in *The Golden Ass* of Apuleius assume the bodies of weasels in order to rob a corpse.
[6] I, 37.
[7] VII, 30.
[8] VIII, 19-20.

"For if there is any science from God, what will be more from Him than the science of health, in which too the virtues of herbs and the diverse properties of juices are determined." [1]

Origen later accused of countenancing magic.

Origen's belief that the stars were rational beings continued to be held by the sect called Origenists and also by the heretic Priscillian and his followers in the later fourth century. Priscillian, as we have seen, was accused of magic and executed in 385. But we are surprised to find Theophilus of Alexandria, who attacked some of Origen's views as heretical and persuaded Pope Anastasius to do the same, accusing Origen in a letter written in 405 and translated into Latin by Jerome, of having defended magic.[2] Theophilus states that Origen has written in one of his treatises, "The magic art seems to me a name for something which does not exist"—a bold and admirable assertion, but one which, as we have seen, the Epicurean Celsus would have been much more likely to make than the Christian Origen—"but if it does, it is not the name of an evil work." Theophilus cannot understand how Origen, who vaunts himself a Christian, can thus make himself a protector of Elymas the magician who opposed the apostles and of Jamnes and Mambres who resisted Moses. Huet, the learned seventeenth century editor of Origen, knew of no such passage in his extant works as that which Theophilus professes to quote.[3]

[1] Homily 18 on Numbers, Migne, PG, XII, 715.

[2] *Epistola* 96 in Migne, PL, XXII, 78.

[3] Migne, PG, XVII, 1091-92.

CHAPTER XX

OTHER CHRISTIAN DISCUSSION OF MAGIC BEFORE AUGUSTINE

Plan of this chapter—Tertullian on magic—Astrology attacked—Resemblance to Minucius Felix—Lactantius—Hippolytus on magic and astrology—Frauds of magicians in answering questions—Other tricks and illusions—Defects and merits of Hippolytus' exposure of magic and of magic itself—Hippolytus' sources—Justin Martyr and others on the witch of Endor—Gregory of Nyssa and Eustathius concerning the ventriloquist—Gregory of Nyssa *Against Fate*—Astrology and the birth of Christ—Chrysostom on the star of the Magi—*Sixth Homily on Matthew*—The spurious homily—Number, names, and home of the Magi—Liturgical drama of the Magi; *Three Kings of Cologne*—Another homily on the Magi—Priscillianists answered—Number and race of the Magi again.

Plan of this chapter.

IN this chapter we shall supplement the picture of the Christian attitude towards magic supplied us in preceding chapters by some accounts of magic in other Christian writers of the period before Augustine. After giving the opinions of a few Latin fathers, Minucius Felix, Tertullian, and Lactantius, we shall consider the exposure of magic devices in Hippolytus' *Refutation of All Heresies,* then compare the utterances of other fathers concerning the witch of Endor with those of Origen, and finally discuss the treatment of the Magi and the star of Bethlehem in both the genuine and the spurious homily of Chrysostom on that theme, adding some account of the medieval development of the legend of the three Magi, although leaving until later the statements of medieval theologians and astronomers concerning the star of the Magi. This makes a rather omnibus chapter, but its component parts are too brief to separate as distinct chapters and they all supplement the preceding chapter on Origen and Celsus.

Tertullian on magic.

Some important features of Origen's account of magic are duplicated in the writings of the western church father, Tertullian, who wrote at about the same time or perhaps a few years before Origen. Again the Jews are represented as calling Christ a magician,[1] and when Tertullian challenges the emperors to allow a Christian exorcist to appear before them and attempt to expel a demon from someone so possessed and force the spirit to confess its evil character, he expects that his Christian exorcist will be accused of employing magic.[2] Again divination and magic are attributed to the fallen angels; in fact, Tertullian follows the *Book of Enoch* in stating that men were instructed by the fallen angels in metallurgy and botany as well as in incantations and astrology.[3] The demons are represented as invisible and "everywhere in a moment." Living as they do in the air near the clouds and stars, they are enabled to predict the weather. They send diseases and then pretend to cure them by the recommendation of novel remedies or prescriptions quite contrary to accepted medical practice.[4] "There is hardly a human being who is unattended by a demon." [5] Magicians are described by Tertullian as producing phantasms, insulting the souls of the dead, injuring boys for purposes of divination, sending dreams, and performing many miraculous feats by their complicated jugglery.[6] "The science of magic" is well defined as "a multiform contagion of the human mind, an artificer of every error, a destroyer of safety and soul." As examples of well-known magicians Tertullian lists Ostanes and Typhon and Dardanus and Damigeron [7] and Nectabis [8] and Berenice. Ter-

[1] Tertullian, *Apology,* cap. 21; so also Cyprian, *Liber de idolorum vanitate,* cap. 13. Latin text of Tertullian in PL, vols. 1-2; English translation in AN, vol. 3.

[2] *Apology,* cap. 23.

[3] *De cultu feminarum,* I, 2.

[4] *Apology,* cap. 22.

[5] *De anima,* cap. 57.

[6] *Apology,* cap. 23.

[7] *De anima,* cap. 57. Damigeron is mentioned in the Orphic poem, *Lithica,* and in the *Apology* of Apuleius, cap. 45; is cited in the *Geoponica,* and was regarded by V. Rose as the Greek source of the Latin "Evax" and Marbod on stones. BN 7418, 14th century, *Amigeronis de lapidibus,* was printed by Pitra, *Spic. Solesm.,* III, 324-35, and Abel, *Orphei Lithica,* p. 157, *et seq.* See further PW, "Damigeron."

[8] Presumably Nectanebus.

tullian states that a literature is current which promises to evoke ghosts from the infernal regions, but that in such cases the dead are really impersonated by demons, as was the fact when the pythoness seemed to show Samuel to Saul, a point on which Tertullian disagrees with Origen. Magic is therefore fallacious, a point which Tertullian emphasizes more than Origen did, although Tertullian is not very explicit. He avers that "it is no great task to deceive the outer eye of him whose mental insight it is easy to blind." The rods of Pharaoh's magicians seemed to turn into snakes, "but Moses' [1] reality devoured their deceit."

Astrology attacked.

Tertullian further diverges from Origen in definitely classifying astrology as a species of magic along with that other variety of magic which works miracles. Astrology is an art which was invented by the fallen angels and with which Christians should have nothing to do. Tertullian would not mention it but for the fact that recently a certain person has defended his persistence in that profession, that is, presumably after he had become a Christian. Tertullian states, again unlike Origen, that the Magi who came from the east to the Christ child were astrologers—"We know the union existing between magic and astrology"—but that Christ's followers are under no obligation to astrology on their account, although he again implies the existence of Christian astrologers in the sarcastic remark, "Astrology now-a-days, forsooth, treats of Christ; is the science of the stars of Christ, not of Saturn and Mars." As Origen affirmed that the power of the demons and of magic was greatly weakened by the birth of Christ, so Tertullian affirms that the science of the stars was allowed to exist until the coming of the Gospel, but that since Christ's birth no one should cast nativities. "For since the Gospel you will never find sophist or Chaldean or enchanter or diviner or magician who has not been manifestly punished." [2] Tertullian rejoices that the *mathematici* or as-

[1] It is Aaron's rod in the King James version.

[2] *De idolatria,* cap. 9.

trologers are forbidden to enter Rome or Italy, the reason being, as he states in another passage,[1] that they are consulted so much in regard to the life of the emperor.

Resemblance to Minucius Felix.

Tertullian's account of magic is perhaps borrowed from the dialogue entitled *Octavius* by M. Minucius Felix,[2] which is generally regarded as the oldest extant work of Christian Latin literature and was probably written in the reign of Marcus Aurelius. Some of the words and phrases used by Tertullian and Minucius Felix in describing magic are almost identical,[3] and a third passage of the same sort appears in Cyprian of Carthage in the third century.[4] Ostanes, one of Tertullian's list of magicians, is also mentioned as the first prominent magician by both Minucius Felix and Cyprian. Minucius Felix ascribes magic to demons and seems to regard it as a deceptive and rather unreal art, saying, "The magicians not only are acquainted with demons, but whatever miraculous feats they perform, they do through demons; under their influence and inspiration they produce illusions, making things seem to be which are not, or making real things seem non-existent."

Lactantius.

A century after Tertullian Lactantius of Gaul treats of magic and demons in about the same way in his *Divine Institutes*,[5] written at the opening of the fourth century. He denies that Christ was a magician and declares that His miracles differed from those attributed to Apuleius and Apollonius of Tyana in that they were announced beforehand by the prophets. "He worked marvels," Lactantius says to his opponents, "and we should have thought Him a magician, as you think now and as the Jews thought at the time, had not all the prophets with one accord predicted that Christ would do these very things."[6] Lactantius believes

[1] *Apology*, cap. 35.

[2] PL, vol. 3; AN, vol. 4.

[3] Thus Minucius Felix says, *Octavius*, cap. 26, "Magi . . . quidquid miraculi ludunt . . . praestigias edunt," while Tertullian, *Apology*, cap. 23, writes, "Porro si et magi phantasmata edunt . . . si multa miracula circulatoriis praestigiis ludunt."

[4] Cyprian, *Liber de idolorum vanitate*, caps. 6-7.

[5] PL, vol. VI; AN, vol. VII; the following references are all to this work.

[6] V, 3.

that the offspring of the fallen angels and "the daughters of men" were a different variety of demon from their fathers and more terrestrial. Be that as it may, he affirms that the entire art and power of the magicians consist in invocations of demons who "deceive human vision by blinding illusions so that men do not see what does exist and think that they see what does not exist," [1] the very expression that we have just heard from Minucius Felix. More specifically Lactantius regards necromancy, oracles, liver-divination, augury, and astrology as all invented by the demons.[2] Like Origen he emphasizes the power of the sign of the cross and the name of Jesus against the evil spirits,[3] and he implies the power of the names of spirits when he states that, although demons may masquerade under other forms and names in pagan temples and worships, in magic and sorcery they are always summoned by their true names, those celestial ones which are read in sacred literature.[4]

Hippolytus on magic and astrology.

From these accounts of magic in Latin fathers, which do little more than reinforce the impressions which we had already gained concerning the Christian attitude, we come to a very different discussion by Hippolytus who wrote in Greek although he lived in Italy. Eusebius and Jerome state that Origen as a young man heard Hippolytus preach at Rome; in 235 he was exiled to Sardinia; the next year his body was brought back to Rome for burial. In Hippolytus, instead of attacks upon astrology as impious, immoral, and fatalistic, and upon magic as evil and the work of demons, we have an attempt to prove astrology irrational and impracticable, and to show that magic is based upon imposture and deceit. In the first four of the nine books of his *Philosophumena* or *Refutation of All Heresies* [5] Hippolytus set forth the tenets of the Greek philosophers, the system of the astrologers, and the practice of the magicians

[1] II, 15.
[2] II, 17.
[3] IV, 27.
[4] II, 17.
[5] The work was discovered in 1842 at Mount Athos and edited by E. Miller in 1851, Duncker and Schneidewin in 1859, and Abbé Cruice in 1860. Greek text in PG, vol. XVI, part 3; English translation in AN, vol. V.

in order later to be able to show how much the various heretics had borrowed from these sources. His second and third books are not extant; it is in the fourth book or what is left of it that we have portions of his discussion of astrology and magic.[1]

Frauds of magicians in answering questions.

In exposing the frauds of magicians Hippolytus uses the word μάγος, and not γόης, a sorcerer. He tells how the magicians pretend that the spirits give response through a medium to questions which those consulting them have written on papyrus, perhaps in invisible ink, and folded up, after which the papyrus is placed on coals and burned. The magician, however, operating in semi-darkness and making a great noise and diversion and pretending to invoke the demon, is really occupied in sprinkling the burnt papyrus with a mixture of water and copperas (vitriol?) or fumigating it with vapor of a gall nut or employing other methods to make the concealed letters visible. Having by some such method discovered the question, he instructs the medium, who is now supposed to be possessed of demons and is reclining upon a couch, what answer to give by whispering to him through a long hidden tube constructed out of the windpipe of a crane or ten brass pipes fitted together. It will be recalled that it was by such a tube made of the windpipes of cranes that Alexander the false prophet, according to Lucian, caused the artificial head of his god to give forth oracles. Hippolytus adds that at the same time the magician produces alarming flames and liquids by such chemical mixtures as fossil salts and Etruscan wax and a grain of salt. "And when this is consumed, the salts bound upward and give the impression of a strange vision." [2]

Other tricks and illusions.

Hippolytus also reveals how magicians secretly fill eggs with dyes, how they cause sheep to behead themselves against a sword by smearing their throats with a drug which makes them itch, how a ram dies if its head is merely bent back facing the sun, how they obstruct the ears of goats with

[1] R. Ganschinietz, *Hippolytos' Capitel gegen die Magier*, 1913, in TU, 39, 2, is a commentary on the text.

[2] *Refutation of All Heresies*, IV, 28.

wax so that they cannot breathe and presently die of suffocation, how out of sea foam they make a compound which, like alcohol, will itself burn but not consume the objects over which it is poured.[1] He tells how the magician produces stage thunder, how he is able to plunge his hand into a boiling cauldron or walk over hot coals without being burnt, and how he can set a seeming pyramid of stone on fire. He tells how the magicians loosen seals and seal them up again, just as Lucian did in his *Alexander* or *The Pseudo-Prophet;* how by means of trap-doors, mirrors, and the like devices they show demons in a cauldron; how they pretend to show flaming demons by igniting drawings which they have sketched on the wall with some inflammable substance or by loosing a bird which has been set on fire. They make the moon appear indoors and imitate the starry sky by attaching fish scales to the ceiling. They produce the sensation of an earthquake by burning the ordure of a weasel with the stone magnet upon an open fire. They construct a false skull from the caul of an ox, some wax, and some gum, make it speak by means of a hidden tube, and then cause it suddenly to collapse and disappear or to burn up.[2]

Defects and merits of Hippolytus' exposure of magic and of magic itself.

This exposition of the frauds of the magicians by Hippolytus is rather broken and incoherent, at least in the form in which his text has reached us.[3] Also we do not have much more faith in some of the methods by which he says the feats of magic are really done than he has in the ways by which the magicians claim to perform them. But while his notions of the chemical action of certain substances and of the occult virtue of others may be incorrect, the note-

[1] Since writing this sentence I have found an article by Diels on the discovery of alcohol in *Societas Regia Scientiarum, Abhandl. Philos.-Hist. Classe,* Berlin, 1913, in which he argues from this passage in Hippolytus that the discovery was made in the Alexandrian period and that it reached western Europe again only through the Arabs about the twelfth century, since alcohol is not mentioned in the older Schlettstadt version of the *Mappae clavicula.* If this be so, Adelard of Bath was perhaps the first to introduce it from the Arabs or the orient, although Diels does not say so.

[2] *Refutation of All Heresies,* IV, 29-41.

[3] In some places the text is illegible.

worthy point is that he endeavors to explain magic either as a deception or as employing natural substances and forces to simulate supernatural action, and that his exposure of magic devices leaves no place for the action of demons. Moreover, we see that magic fraud involves chemical experiment and considerable knowledge or error in the field of natural science. Under the guise or tyranny of magic experimental science is at work.

Hippolytus' sources.

The question then arises whether Hippolytus himself discovered these tricks of the magicians or whether he is simply copying his explanations of them from some previous work. An examination of the earlier chapters of his fourth book is sufficient to solve the question. His arguments against the practice of the Chaldean astrologers of predicting man's life from his horoscope at the time of his birth are drawn from the pages of the sceptical philosopher, Sextus Empiricus, whom he follows so closely that his editors are able to rectify his text by reference to the parallel passage in Sextus. We are therefore probably safe in assuming, especially in view of the resemblances to the *Alexander* of Lucian which have already been noted, that Hippolytus' attack on magic is also largely indebted to some classical work, possibly to that very treatise against magic by Celsus to which both Origen and Lucian refer, or perhaps to some account of apparatus with which to work marvels like Hero's *Pneumatics*.

Justin Martyr and others on the witch of Endor.

Turning back now to the subject of the witch of Endor, we find that some of the church fathers agree with Origen rather than Tertullian that the witch really invoked Samuel. Before Origen's time Justin Martyr in *The Dialogue with Trypho* [1] had mentioned as a proof of the immortality of the soul "the fact that the soul of Samuel was called up by the witch, as Saul demanded." Huet, who edited the writings of Origen, lists other Christian authors [2] who agreed

[1] Cap. 105.

[2] Leo Allatius "in syntagmate" *De engastrimytho*, cap. 7; Sulpicius Severus, *Historia sacra*, liber I; Anastasius Antiochenus, Ὁδηγός, quaest., 112; "et eorum quos laudat Bellarminus liber IV *de Christo*, cap. 11."

with Origen on this question, and further informs us that the ancient rabbis were wont to say that a soul invoked within a year after its death as Samuel's was, would be seen by the ventriloquist but not heard, and heard by the person consulting it but not seen, an observation which suggests that Saul was deceived by ventriloquism, while by others present the ghost would be neither seen nor heard.

Gregory of Nyssa and Eustathius concerning the ventriloquist.

Two ecclesiastics of the fourth century composed special treatises upon the ventriloquist or witch of Endor in which they took the opposite view from that of Origen. The briefer of these two treatises is by Gregory of Nyssa [1] who states, without mentioning Origen by name, that some previous writers have contended that Samuel was truly invoked by magic with divine permission in order that he might see his mistake in having called Saul the enemy of ventriloquists. But Gregory believes that Samuel was already in paradise and hence could not be invoked from the infernal regions; but that it was a demon from the infernal regions who predicted to Saul, "To-morrow you and Jonathan shall be with me." The longer treatise of Eustathius of Antioch is a direct answer to Origen's argument as its title, *Concerning the Ventriloquist against Origen,*[2] indicates. Eustathius holds that it was illegal to consult ventriloquists in view of Saul's own previous action against them and other prohibitions in Scripture, and that Origen's remarks are to be deplored as tending to encourage simple men to resort to arts of divination. Eustathius contends that the witch did not invoke Samuel but only made Saul think that she did, and that Saul himself did not see Samuel. Pharaoh's magicians similarly deceived the imagination with shadows and specters when they pretended to turn rods into snakes and water into blood. Eustathius does not agree with Origen that Samuel was in hell. He holds that the predictions made by the pseudo-Samuel were not impossible for a demon to make, and indeed were not strictly accurate,

[1] Περὶ τῆς ἐγγαστριμύθου, PG, XLV, 107-14.

[2] Migne, PG, XVIII, 613-74.

since Saul did not die the very next day but the day after it, and since not only Jonathan but his three sons were slain with him.[1] Furthermore, David was already so prominent in public affairs that a demon might easily guess that he would succeed Saul.

Gregory of Nyssa *Against Fate.*

Gregory of Nyssa also composed a treatise, entitled *Against Fate,*[2] in the form of a disputation between a pagan philosopher and himself at Constantinople in 382 A. D. His opponent holds that the life of man is determined by the constellations at his nativity, upon whose decree even conversion to Christianity would thus be made dependent. Gregory assumes the position of one hitherto ignorant of the principles of the art of astrology, of which the philosopher has to inform him, but on general grounds it seems very unlikely that he really was as ignorant as this of such a widespread superstition. Furthermore, he is sufficiently read in the subject to incorporate some of Bardesanes' arguments, of whose treatise both Gregory's title and dialogue form are reminiscent. Some of Gregory's reasoning, however, might well be that of a tyro and is scarcely worth elaborating here.

Astrology and the birth of Christ.

When the writer of the Gospel according to Matthew included the story of the wise men from the east who had seen the star, there can be little or no doubt that he inserted it and that it had been formulated in the first place, not merely in order to satisfy the ordinary, unlearned reader with portents connected with the birth of Jesus, but to secure the appearance of support for the kingship of Jesus from that art or science of astrology which so many persons then held in high esteem. To an age whose sublimest science was star-gazing it would seem fitting and almost inevitable that God should have announced the coming of the Prince of Peace in this manner, and the account in the Gospel of Matthew is in a sense an attempt to present the birth of Christ in a way to comply with the most searching tests of contem-

[1] The King James version, First Samuel, XXVIII, 19, reads, "and to morrow shalt thou and thy sons be with me," instead of "thou and Jonathan."

[2] Migne, PG, XII, 143-74.

porary science. But the early Christians were relatively rude and unlettered, and this effort to construct a royal horoscope for Jesus is a crude and faulty one from the astrological standpoint. For this, however, the author of the Gospel and not the art of astrology is obviously responsible. As a result, however, of the Gnostic reaction against astrological fatalism or of an orthodox Christian opposition to both Gnostics and astrologers, most of the early fathers of the church denied that this passage implied any recognition of the truth of astrology and attempted to explain away its obvious meaning. In doing this they often made the crude and imperfect astrology of the Gospel a criterion for criticizing the art of astrology itself.

Chrysostom on the star of the Magi.

Of patristic commentaries upon the passage in the Gospel of Matthew dealing with the Magi and the star of Bethlehem one of the fullest and most frequently cited by medieval writers is that attributed to Chrysostom. I say "attributed," because in addition to his genuine sixth homily upon Matthew [1] there was generally ascribed to Chrysostom in the middle ages another homily which is extant only in Latin [2] and has been thought to be the work of some Arian. The famous St. John Chrysostom was born at Antioch about 347 A. D. and there studied rhetoric under the noted sophist Libanius. From 398 to 404 he held the office of patriarch of Constantinople; then he was exiled to Cappadocia where he died in 407. One detail of his boyhood may be noted because of its connection with magic. When he was a lad, the tyrants in the city became suspicious of plots against them and sent soldiers to search for books of magic and sorcery. One of the men who was arrested and put to death had tried to rid himself of the damaging possession of a book of magic by throwing it into the river. Chrysostom and a playmate later unsuspectingly fished an object out of the water which turned out to be this very book, and

[1] Migne, PG, LVI, 61, *et seq.*

[2] Migne, PG, LVI, 637, *et seq.* *Homily* II, "Opus imperfectum in Matthaeum quod Chrysostomi nomine circumfertur." *Ibid.*, 602, *et seq.*, for opinions of various past writers as to its authenticity.

when a soldier happened to pass by just then, they were very frightened lest he should see what they had and they should be severely punished for it.[1]

Sixth homily on Matthew.

In his sixth homily upon Matthew Chrysostom recognizes the difficulties presented by the Scriptural account of the Magi and the star, and approaches the task of expounding it with prayers to God for aid. Some, he informs us, take the passage as an admission of the truth of astrology. It is this opinion which he is concerned to refute. He argues that it is not the function of astronomy to learn from the stars who are being born but merely to predict from the hour of birth what is going to happen, which seems a quite fallacious distinction upon his part. He also criticizes the Magi for calling Jesus the king of the Jews, when as Christ told Pilate His kingdom was not of this world. He further criticizes them for coming to Christ's birthplace when they might have known that it would cause difficulties with Herod, the existing king, and for coming, making trouble, and then immediately going back home again. But these shortcomings would seem to be those of the Scriptural narrative rather than of the art of astrology, although of course Chrysostom is trying to make the point that the Magi had not foreseen what would happen to themselves. He further argues that the star of Bethlehem was not like other stars nor even a star at all,[2] as was proved by its peculiar itinerary, its shining by day, its rare intelligence in hiding itself at the right time, and its miraculous ability in standing over the head of the child. Chrysostom therefore con-

[1] Migne, PG, LX, 274-5, in the 38th homily on the Book of Acts.

[2] On the other hand, D. Friedrich Münter, *Der Stern der Weisen: Untersuchungen über das Geburtsjahr Christi*, Kopenhagen, 1827, adopted the astrological theory that the star of Bethlehem was really a major conjunction of Saturn and Jupiter in Pisces, which Jewish tradition, too, seems to have regarded as the sign of the Messiah, and that therefore Jesus was born in 6 B. C. This view had already been advanced by Kepler, but recent writers seem to prefer a conjunction in Aries: see H. G. Voigt, *Die Geschichte Jesu und die Astrologie*, Leipzig, 1911; Kritzinger, *Der Stern der Weisen*, Gütersloh, 1911; von Oefele, *Die Angaben der Berliner Planetentafel P8279 verglichen mit der Geburtsgeschichte Christi im Berichte des Matthäus*, Berlin, 1903, in *Mitteil. d. Vorderasiatischen Gesellschaft.*

cludes that some invisible virtue put on the form of a star. He thinks that the star appeared to the Magi as a reflection upon the Jews, who had rejected prophet after prophet, whereas the apparition of a single star was sufficient to bring barbarian Magi to the feet of Christ. At the same time he believes that God especially favored the Magi in vouchsafing them a star, a sign to which they were accustomed, as the mode of announcement. Thus he comes dangerously near to admitting tacitly what he has just been denying, namely, that the stars are signs of the future and that there is something in the art of astrology. In short, the star appeared to the Magi because they as astrologers would comprehend its meaning. Chrysostom denies this openly and does his best to think up arguments against it, but he cannot rid his subconscious thought of the idea.

The spurious homily.

The other homily ascribed to Chrysostom repeats some of the points made in the genuine homily, but adds others. The preacher has read somewhere, perhaps in Origen where we have already met the suggestion, that the Magi had learned that the star would appear from the books of the diviner Balaam, "whose divination is also put into the Old Testament: 'A star shall arise from Jacob and a man shall come forth from Israel, and he shall rule all nations.'" But the preacher does not state why it is any better to have such a prediction made by a diviner than by an astrologer. The preacher has also heard some cite a writing, which is not surely authentic but yet is not destructive to the Faith and rather pleasing, to the effect that in the extreme east on the shores of the ocean live a people who possess a writing inscribed with the name of Seth and dealing with the appearance of this star and the gifts to be offered. This writing was handed down from father to son through successive generations, and twelve of the most studious men of their number were chosen to watch for the coming of the star, and whenever one died, another was chosen in his place. They were called Magi in their language because they glorified God silently. Every year after the threshing of the harvest they climbed a mountain to a cave with delightful springs shaded by carefully selected trees. There they washed themselves and for three days in silence prayed and praised God. Finally one year the star appeared in the form of a little child with the likeness of a cross above it; and it spoke with them and taught them and instructed them to set out for Judea.[1] When they had set out, it went before them for two years, during which time food and drink were never lacking in their wallets. On their return they worshiped and glorified God more sedulously than ever and preached to their people. Finally, after the resurrection, the apostle Thomas visited that region and they were baptized by him and were made his assistant preachers. This tale is indeed pleasing enough, and it saves the Magi from all imputation of magic arts and employment of demons and even denies that they were astrologers. But as a device to escape the natural inference from the Gospel story that the birth of Christ was announced by the stars and in a way which astronomers could comprehend it is certainly far-fetched, and shows how Christian theologians were put to it to find a way out of the difficulty. The homily goes on to advance some of the usual arguments against astrology, such as that the stars cannot cause evil, that the human will is free, and that a science of individual horoscopes cannot account for all men worshiping idols before Christ and abandoning idolatry and other ancient customs thereafter, or for the perishing in the deluge of all men except the family of Noah, or for national customs such as circumcision among the Jews and incest among the Persians. Here we again probably see the influence of Bardesanes.

We have already noted that Origen seems to have been the first of the fathers to state the number of the Magi as

[1] Mâle, *Religious Art in France*, 1913, p. 208, was not able to trace the legend that the star of the Magi appeared with the face of a child beyond *The Golden Legend* compiled by James of Voragine in the thirteenth century. We shall, however, find it mentioned in the twelfth century by Abelard, who derived it from this spurious homily of Chrysostom.

Number, names, and home of the Magi.

three, whereas the homily just considered implies that there were twelve of them. Their representation in art as three in number did not become general until the fourth century,[1] while the depiction of them as kings was also a gradual and, according to Kehrer, later growth.[2] Bouché-Leclercq, citing an earlier monograph,[3] states that the royalty of the Magi was invented towards the sixth century to show the fulfillment of Old Testament prophecies,[4] and that Bede is the first who knows their names. But Mâle says, "Their mysterious names are first found in a Greek chronicle of the beginning of the sixth century translated into Latin by a Merovingian monk," and are "Bithisarea, Melichior, Gathaspa." [5] The provenance of the Magi was variously stated by the Christian fathers: [6] Arabia according to Justin Martyr, Epiphanius, and Tertullian or Pseudo-Tertullian; Persia according to Clement of Alexandria, Basil, and Cyril; Persia or Chaldea according to Chrysostom and Diodorus of Tarsus; Chaldea according to Jerome and Augustine and the philosopher Chalcidius in his commentary upon Plato's *Timaeus*.[7] The homily which we were just considering gave the impression that they came from India.

In the middle ages the Magi appeared in liturgical drama as well as in art. An early instance is a tenth century lectionary from Compiègne, now preserved at Paris,[8] where

[1] They are twice so represented on the elaborately carved Christian sarcophagus in the museum at Syracuse, Sicily, where also the manger, ox, and ass are shown (compare note 4 below).

[2] Hugo Kehrer, *Die Heiligen drei Könige in Litteratur und Kunst*, Leipzig, 1908, 2 vols. An earlier work on the three Magi is Inchofer, *Tres Magi Evangelici*, Rome, 1639.

[3] J. C. Thilo, *Eusebii Alexandrini oratio* Περὶ ἀστρονόμων *(praemissa de magis et stella quaestione) e Cod. Reg. Par. primum edita*, Progr. Halae, 1834.

[4] A. Bouché-Leclercq, *L'Astrologie grecque*, 1899, p. 611, "La royauté des Mages fut inventée (vers le VIe siècle), comme la crèche (*sic!* see Luke, II, 12 and 16), le bœuf et l'âne pour montrer l'accomplissement des prophéties."

[5] *Religious Art in France*, 1913, p. 214 note, following, I presume, Kehrer's work, as he does on p. 213.

[6] For detailed references see Münter, *Der Stern der Weisen*, 1827, p. 15; and Bouché-Leclercq, 1899, p. 611, where they are stated somewhat differently.

[7] *Comm. in Platonis Timaeum*, II, vi, 125; quoted by Münter (1827), pp. 27-8.

[8] BN 16819, fol. 49r. Corpus Christi 134, early 12th century, fol. 1 v., has a brief "Magorum trium qui Domino Infanti aurum obtulere nomina et descriptio."

Liturgical drama of the Magi: *The Three Kings of Cologne.*

after homilies by various fathers there is added in a hand only slightly later the liturgical drama of the adoration of the Magi. In the later middle ages there came into existence the *History* or *Deeds of the Three Kings of Cologne,* as the Magi came to be called from the supposed translation of their relics to that city. Their bodies were said to have been brought by the empress Helena from India to Constantinople, whence they were transferred to Milan, and after its destruction by Barbarossa, to Cologne. This "fabulous narration," as it has well been entitled,[1] also has much to say of the miracles of the apostle Thomas in India and of Prester John, to whom we shall devote a later chapter. It asserts that the three kings reached Jerusalem on the thirteenth day after Christ's birth by a miraculously rapid transit by day and by night of themselves and their armies to the marvel of the inhabitants of the towns through which they passed, or rather, flew.[2] After they had returned home and had successively migrated to Christ above, another apparition of a star marked this fact.[3] The treatise exists in many manuscripts[4] and was printed more than once before 1500.

[1] Cotton Galba E, VIII, 15th century, fols. 3-28, Fabulosa narratio de tribus magis qui Christum adorarunt sive de tribus regibus Coloniensibus.

[2] Cap. 12 in the 1478 edition.

[3] *Ibid.*, cap. 34.

[4] At Munich all the following MSS are 15th century: CLM 18621, fol. 135, *Liber trium regum,* fol. 215, *Legenda trium regum excerpta ex praecedenti;* 19544, fols. 314-49, and 26688, fols. 157-92, *Laudes et gesta trium regum,* etc.; 21627, fols. 212-31, *Historia de tribus regibus;* 23839, fols. 112-37, and 24571, fols. 50-104, *Gesta trium regum;* 25073, fols. 260-83, *de nativitate domini et de tribus regibus.* At Berlin MSS 799 and 800, both of the 15th century, have the *Gesta trium regum* ascribed to John of Hildesheim. So Wolfenbüttel 3266, anno 1461. The printed edition of 1478 in 46 chapters and about 30 folios is also ascribed to John of Hildesheim. We read on the binding, "Ioannis Hildeshemensis Liber de trium regum translatione." The Incipit is: "Reverendissimo in Christo patri ac domino domino florencio de weuelkouen divina providencia monasteriensis ecclesie episcopo dignissimo." The colophon is: "Liber de gestis ac trina beatissimorum trium regum translacione . . . per me Johannem guldenschoff de moguncia." Some other MSS, also of the 15th century, are: Vatic. Palat. Lat. 859, de gestis et translationibus trium regum, and at Oxford, University College 33, Liber collectus de gestis et translationibus sanctorum trium regum de Colonia; Laud Misc., 658, The history of the three kings of Cologne, in forty-one chapters with a preface. It is thus seen that the number of

Another homily on the Magi.

Finally we may note the contents of the homily on the Magi which immediately precedes the liturgical drama concerning them in the above mentioned tenth century lectionary.[1] The Magi are said to have come on the thirteenth day of Christ's nativity. That they came from the Orient was fitting since they sought one of whom it had been written, *Ecce vir oriens.* It was also fitting that Christ's coming should be announced to shepherds of Israel by a rational angel, to Gentile Magi by an irrational star. This star appeared neither in the starry heaven nor on earth but in the air; it had not existed before and ceased to exist after it had fulfilled its function. Although he has just said that the star appeared in the air and not in the sky, the preacher now adds that when a new man was born in the world it was fitting that a new star should appear in the sky. He also, in pointing out how all the elements recognized that their Creator had come into the world, states that the sky sent a star, the sea allowed Him to walk upon it, the sun was darkened, stones were broken and the earth quaked when He died.

Priscillianists answered.

Since the heretics known as Priscillianists have adduced the star at Christ's birth to prove that every man is born under the fates of the stars, the preacher endeavors to answer them. He holds that since the star came to where Jesus lay He controlled it rather than vice versa. Then follow the usual arguments against genethlialogy that many men born under the sign Aquarius are not fishermen, that sons of serfs are born at the same time as princes, and the

chapters varies. Coxe's catalogue of the Laud MSS states that the Latin original was printed at Cologne in quarto in 1481, and that it is very different from the version printed by Wynkyn de Worde. "The Story of the Magi," in Bodleian (Bernard) 2325, covers only folio 68. At Amiens is a MS which the catalogue dates in the 14th century and ascribes to John of Hildesheim, and its Incipit is practically that of the printed edition: Amiens 481, fols. 1-58, "Reverendissimo in Christo Patri ac domino domino Florentino de Wovellonem (*sic*) divina providencia Monasteriensis ecclesie episcopo dignissimo. Cum venerandissimorum trium Magorum, ymo verius trium Regum." The work ends in the MS with the words, ". . . summi Regis servant legem incole Colonie. Amen. Explicit hystoria."

[1] BN 16819, 10th century, fols. 46r-49r.

case of Jacob and Esau. The star was merely a sign to the Magi and by its twinkling illuminated their minds to seek the new-born babe. It seems scarcely consistent that a star which the preacher has called irrational should illuminate minds.

Number and race of the Magi again.

The homily goes on to say that opinions differ as to who the Magi were and whence they came. Owing to the prophecy that the kings of Tarsus and the isles offer presents, the kings of the Arabs and Sheba bring gifts, some regard Tarsus, Arabia, and Sheba as the homes of the Magi. Others call them Persians or Chaldeans, since Chaldeans are skilled in astronomy. Others say that they were descendants of Balaam. At any rate they were the first Gentiles to seek Christ and they are well said to have been three, symbolizing faith in the Trinity, the three virtues, faith, hope and charity, the three safeguards against evil thoughts, words and works, and the three Gentile contributions to the Faith of physics, ethics, and logic, or natural, moral, and rational philosophy. The preacher then indulges in further allegorical interpretation anent Herod and what was typified by the gifts of the Magi.[1]

[1] Marco Polo (I, 13-14, ed. Yule and Cordier, 1903, vol. I, 78-81), who located the Magi in Saba, Persia, recounts further legends concerning them and their gifts. See also F. W. K. Müller, *Uigurica*, I, i, *Die Anbetung der Magier, ein Christliches Bruchstück,* Berlin, 1908.

CHAPTER XXI

CHRISTIANITY AND NATURAL SCIENCE: BASIL, EPIPHANIUS, AND THE PHYSIOLOGUS

Lactantius not a fair example—Commentaries on the Biblical account of creation—Date and delivery of Basil's *Hexaemeron*—The *Hexaemeron* of Ambrose—Basil's medieval influence—Science and religion—Scientific curiosity of Basil's audience—Allusions to amusements—Conflicts with Greek science—Agreement with Greek science—Qualification of the Scriptural account of creation—The four elements and four qualities—Enthusiasm for nature as God's work—Sin and nature—Habits of animals—Marvels of nature—Spontaneous generation—Lack of scientific scepticism—Sun worship and astrology—Permanence of species—Final impression from the *Hexaemeron*—*The Medicine Chest* of Epiphanius—Gems in the high priest's breastplate—Some other gems—The so-called *Physiologus;* problem of its origin—Does the title apply to any one particular treatise?—And to what sort of a treatise?—Medieval art shows almost no symbolic influence of the *Physiologus*—*Physiologus* was more natural scientist than allegorist.

Lactantius not a fair example.

THE opposition of early Christian thought to natural science has been rather unduly exaggerated. For instance, Lactantius, one of the least favorable to Greek philosophy and natural science of the fathers, should hardly be cited as typical of early Christian attitude in such matters. Nor does his opposition impress one as weighty.[1] He ridicules the theory of the Antipodes,[2] which he perhaps understands

[1] Beazley, *Dawn of Modern Geography*, I, 274, says, "Augustine and Chrysostom felt and spoke in the same way, though in more measured language, and nearly all early Christian writers who touched upon the matter did so to echo the voice of authorities so unquestioned." But I cannot agree with this statement. He goes on to imply that a majority of the fathers, like Cosmas Indicopleustes, attacked the belief in the sphericity of the earth; but here, too, I wonder if he is not following Letronne, *Des Opinions Cosmographiques des Pères,* without having examined the citations. Certainly no such attitude is found in Basil's *Hexaemeron*, Hom. 3 and 9 as the citation implies. I have not seen Marinelli, *La geographia e i Padri della Chiesa, estratto dal Bollettino della Società geografica italiana,* anno 1882, pp. 11-15.

[2] *Divin. Instit.,* III, 24.

none too well, asking if anyone can be so inept as to think that there are men whose feet are above their heads, although he knows very well that Greek science teaches that all weights fall towards the center of the earth, and that consequently if the feet are nearer the center of the earth that they must be below the head. He continues, however, to insist that the philosophers are either very stupid, or just joking, or arguing for the sake of arguing, and he declares that he could show by many arguments that the heaven cannot possibly be lower than the earth—which no one has asserted except himself—if it were not already time to close his third book and begin the fourth. Apparently Lactantius is the one who is arguing for the sake of arguing, or just joking, or else very stupid, and I fear it is the last. But other Christian fathers were less dense, and we already have heard the cultured pagan Plutarch scoff at the notion of a spherical earth and of antipodes. We may grant, however, that the ecclesiastical writers of the Roman Empire and early medieval period normally treat of spiritual rather than material themes and discuss them in a religious rather than a scientific manner.

Commentaries on the Biblical account of creation.

But in the commentaries upon the books of the Bible which the fathers multiplied so voluminously it was necessary for them, if they began their labors with *Genesis,* to deal at the very start in the first verses of the first book of the Bible with an explanation of nature which at several points was in disagreement with the accepted theories of Greek philosophy and ancient science. Such comment upon the opening verses of *Genesis* sometimes developed into a separate treatise called *Hexaemeron* from the works of the six days of creation which it discussed. Of the various treatises of this type the *Hexaemeron* of Basil [1] seems to have been both the best [2] and the most influential, and will be considered by us as an example of Christian attitude towards

[1] Migne, PG, vol. 29; PN, vol. 8.

[2] Duhem (1914) II, 394, however, prefers Gregory of Nyssa's work as "à la fois plus sobre, plus concis, et plus philosophique. . . ."

the natural science and, to some extent, the superstition of the ancient world.

Date and delivery of Basil's *Hexaemeron.*

Basil died on the first day of January, 379 A. D., and was born about 329. When or where the nine homilies which compose his *Hexaemeron* were preached is not known, but from an allusion to his bodily infirmity in the seventh homily and his forgetfulness the next day in Homily VIII we might infer that it was late in life. To all appearances these sermons were taken down and have reached us just as they were delivered to the people, to whose daily life Basil frequently adverts. The sermons were delivered early in the morning before the artisans in the audience went to their work and again at the close of the day and before the evening meal, since Basil sometimes speaks of the approach of darkness surprising him and of its consequently being time to stop.[1] One of the surest indications either that the sermons were delivered extemporaneously, or that Basil was repeating with variations to suit the occasion and present audience sermons which he had delivered so often as to have practically memorized, occurs in the eighth homily where he starts to discuss land animals, forgetting that the last day he did not get to birds, but is presently brought to a realization of his omission by the actions of his audience and, after a pause and an apology, makes a fresh start upon birds. The *Hexaemeron* was highly praised by Basil's contemporaries and was regarded as the best of his works by later Byzantine literary collectors and critics.

The *Hexaemeron* of Ambrose.

Basil's work, however, was not the first of its kind, as Hippolytus and Origen, at least, are known to have earlier composed similar treatises, and still earlier in the treatise

[1] Homily I was delivered in the morning, II in the evening; III was in the morning and speaks of a coming evening address. At the close of Homily VII Basil urges his hearers to talk over at their evening meal what they have heard this morning and this evening. If we regard Homily VI as the morning address referred to, we shall have Homily V left to cover an entire day. Homily VI, however, is the longest of the nine. In any case Homily VIII is clearly preached in the morning, and IX at evening.

of Theophilus *To Autolycus* we find a few chapters [1] devoted to the six days of creation. In one of his letters Jerome states that "Ambrose recently so compiled the *Hexaemeron* of Origen that he rather followed the views of Hippolytus and Basil." [2] This Latin work of Ambrose is extant and seems to me to follow Basil very closely. At times the order of presentation is slightly varied and the work of Ambrose is longer, but this is due to its more verbose rhetoric and greater indulgence in Biblical quotation, and not to the introduction of new ideas. The Benedictine editors of Ambrose admit that he has taken a great deal from Basil but deny that he has servilely imitated him.[3] But a striking instance of such servile imitation is seen in Ambrose's duplicating even Basil's mistake in omitting to discuss birds and then apologizing for it, reminding one of the Chinese workman who made all the new dinner plates with a crack and a toothpick stuck in it, like the old broken plate which he had been given as a model. It is true that Ambrose does not first discuss land animals for a page as Basil did, but makes his apology more immediately. The opening words of the eighth sermon in the twelfth chapter of his fifth book are, "And after he had remained silent for a moment, again resuming his discourse, he said . . ." Then comes his apology, expressed in different terms from Basil's and to the effect that in his previous discourse upon fishes he became so immersed in the depths of the sea as to forget all about birds. Thus the incident which in Basil had every appearance of a natural mistake, in Ambrose has all the earmarks of an affected imitation. It is barely possible, however, that Origen made the original mistake and that Basil and Ambrose have both imitated him in it. On the other hand, we are told that the *Hexaemerons* of Origen

[1] Bk. II, caps. 10-17.

[2] *Epistola 65, ad Pammachium.* Augustine's *De Genesi ad litteram*, which Cassiodorus (*Institutes*, I, 1) esteemed above the commentaries of Basil and Ambrose upon Genesis, is a somewhat similar work, but, after a briefer treatment of the work of creation, continues to comment on the text up to Adam's expulsion from Paradise.

[3] Migne, PL, 14, 131-2. The most recent edition of the *Hexaemeron* of Ambrose is by C. Schenkl. Vienna, 1896.

and Basil differed fundamentally in this respect, that Origen indulged to a great extent in allegorical interpretation of the Mosaic account of creation,[1] while Basil declares that he "takes all in the literal sense," is "not ashamed of the Gospel," and "admits the common sense of the Scriptures." [2]

Basil's medieval influence.

At any rate, Basil's *Hexaemeron* seems to have supplanted all such previous treatises in Greek, while its western influence is shown not only by Ambrose's imitation of it so soon after its production, but by Latin translations of it by Eustathius Afer in the fifth, and perhaps by Dionysius Exiguus in the sixth century. Medieval manuscripts of it are fairly numerous and sometimes of early date,[3] and include an Anglo-Saxon epitome ascribed to Aelfric in the Bodleian Library. Bartholomew of England [4] in the thirteenth century quotes "Rabanus who uses the words of Basil in the *Hexaemeron*" for a description of the empyrean heaven which I have been unable to find in the works of

[1] Fialon, *Étude sur St. Basile,* 1869, p. 296.

[2] Homily IX.

[3] For example, in the catalogue, published in 1744, of MSS in the then Royal Library at Paris there are listed five copies of Eustathius' Latin translation, dating from the ninth to the fourteenth century—2200, 4; 1701, 1; 1702, 1; 1787A, 2; 2633, 1; and fifteen copies of the *Hexaemeron* of Ambrose—1718; 1702, 2; 1719 to 1727 inclusive; 2387, 4; 2637 and 2638.

I have not noted what MSS of the *Hexaemerons* of Basil and Ambrose are found in the British Museum and Bodleian libraries. Some other medieval copies of Basil's in Latin translation are: BN 12134, 9th century Lombard hand; Vendôme 122, 11th century, fols. 1 v-60; Soissons 121, 12th century, fol. 97, Eustathius' prologue and a part of his translation; Grenoble 258, 12th century, fols. 1-45, "Eustathii translatio. . . ."

The *Hexaemeron* of Ambrose, since written originally in Latin, is naturally found oftener. The oldest MS is said to be CU Corpus Christi 193, large Lombard script of the 8th century which closely resembles BN 3836. Other MSS are: BN 11624, 11th century; BN 12135, 9th century; BN 12136, 12-13th century; BN 13336, 11th century; BN 14847, 12th century, fol. 163; BN nouv. acq. 490, 12th century; Vatican 269-273 inclusive, 10-15th centuries; Alençon 10, 12th century; Vendôme 129, 12th century, fols. 48-126; Semur, 10, 12th century; Chartres 63, 10-11th century, fols. 3-46; Orléans 35, 11th century; Orléans 192, 7th century, part of the first two books only; Amiens fonds Lescalopier 30, 12th century; le Mans 15, 11th century; Brussels 1782, 10th century; CLM 2549, 12th century; CLM 3728, 10th century; CLM 6258, 10th century; CLM 13079, 12th century; CLM 14399, 12th century; Novara 40, 12th century; and many other MSS of later date in these and other libraries.

[4] *De proprietatibus rerum,* VIII, 4.

either Rabanus or Basil. Bede, in a similar, though much abbreviated, work of his own, states that while many have said many things concerning the beginning of the *Book of Genesis,* the chief authorities, so far as he has been able to discover, are Basil of Caesarea, whom Eustathius translated from Greek into Latin, Ambrose of Milan, and Augustine, bishop of Hippo. These works, however, were so long and expensive that only the rich could afford to purchase them and so profound that only the learned could read and understand them. Bede had accordingly been requested to compose a brief rendition of them, which he does partly in his own words, partly in theirs.[1]

Science and religion.

The general tenor of Basil's treatise may be described as follows. He accepts the literal sense of the first chapter of *Genesis* as a correct account of the universe, and, when he finds Greek philosophy and science in disagreement with the Biblical narrative, inveighs against the futilities and follies and conflicting theories and excessive elaborations of the philosophers. On such occasions the simple statements of Scripture are sufficient for him. "Upon the essence of the heavens we are contented with what Isaiah says. . . . In the same way, as concerns the earth, let us resolve not to torment ourselves by trying to find out its essence. . . . At all events let us prefer the simplicity of faith to the demonstrations of reason." [2] These three quotations illustrate his attitude at such times. But at all other times he is apt to follow Greek science rather implicitly, accepting without question its hypothesis of four elements and four qualities, and taking all his details about birds, beasts, and fish from the same source.

Scientific curiosity of Basil's audience.

Moreover, while Basil may affirm that the edification of the church is his sole aim and interest, it is evident that his audience are possessed by a lively scientific curiosity,

[1] Bede, *Hexaemeron, sive libri quatuor in principium Genesis usque ad nativitatem Isaac et electionem Ismaelis,* in Migne, PL, 91, 9-100. Bede originally intended to carry his work only to the expulsion of Adam from Paradise, but subsequently added three more books.

[2] Homilies I, VIII, and X.

and that they wish to hear a great deal more about natural phenomena than Isaiah or any other Biblical author has to offer them. "What trouble you have given me in my previous discourses," exclaims Basil in his fourth homily, "by asking me why the earth was invisible, why all bodies are naturally endued with color, and why all color comes under the sense of sight? And perhaps my reason did not appear sufficient to you. . . . Perhaps you will ask me new questions." Basil gratifies this curiosity concerning the world of nature with many details not mentioned in the Bible but drawn from such works as Aristotle's *Meteorology* and *History of Animals*. This scientific curiosity displayed by Basil's hearers is the more interesting in that artisans who had to labor for their daily bread appear to have made up a large element in his audience.[1] It is perhaps on their account that Basil often speaks of God as the supreme artisan or artificer or artist,[2] or calls their attention to "the vast and varied workshop of divine creation," [3] and makes other flattering allusions to arts which support life or produce enduring work, and to waterways and sea trade.[4] He also seems to have a sincere appreciation of the arts and admiration of beauty, which he twice defines.[5]

Allusions to amusements.

At the risk of digression, it is perhaps worth noting further that Basil's hearers seem to have been very familiar with, not to say fond of, the amusements common in the cities of the Roman Empire. Twice he opens his sermons with allusions to the athletes of the circus and actors of the theater,[6] apparently as the surest way of quickly catching the attention of his audience, while on a third occasion, in concluding his morning address on what appears to have been a holiday, he remarks that if he had dismissed them earlier, some would have spent the rest of the day gambling with dice, and that "the longer I keep you, the longer you are out of the way of mischief." [7] He also alludes to the

[1] Homily III, 1 and 10.
[2] I, 7; III, 5 and 10.
[3] IV, 1.
[4] I, 7; III, 5; IV, 3, 4, and 7; VI, 9; VII, 6.
[5] II, 7; III, 10.
[6] IV, 1; VI, 1.
[7] VIII, 8.

spinning of tops and to what was apparently the game of push-ball.[1]

Conflicts with Greek science.

Taking up the contents of the *Hexaemeron* more in detail, we may first note those points upon which Basil supports the statements of the Bible against Greek science and philosophy. He of course insists that the universe was created by God and is not co-existent, much less identical, with Him.[2] He also denies that the form of the world alone is due to God and that matter is of separate origin.[3] Nor will he accept the arguments of the philosophers who "would rather lose their tongues" than admit that there is more than one heaven. Basil is ready to believe not merely in a second, but a third heaven, such as the apostle Paul speaks of being rapt to. He regards a plurality of heavens as no more difficult to credit than the seven concentric spheres of the planets, and as much more probable than the philosophic theory of the music of the spheres which he decries as "ingenious frivolity, the untruth of which is evident from the first word." [4] He also defends the statement of Scripture that there are waters above the firmament, not only against the doctrines of ancient astronomy,[5] but also against "certain writers in the church," among whom he probably has Origen in mind, who interpret the passage figuratively and assert that the waters stand for "spiritual and incorporeal powers," those above the firmament representing good angels and those below the firmament standing for evil demons. "Let us reject these theories as we would the interpretations of dreams and old-wives' tales." [6]

In connection with Basil's defense of the plurality of the heavens it may be noted that R. H. Charles presents evidence to show "that speculations or definitely formulated views on the plurality of the heavens were rife in the very cradle of Christendom and throughout its entire development," and that "the prevailing view was that of the seven-

[1] Homily V, 10; IX, 2
[2] I, 3.
[3] II, 1.
[4] III, 3.
[5] II, 4, *et seq.*
[6] III, 9.

fold division of the heavens."[1] He fails, however, to discriminate between the doctrine of Greek philosophy that the universe was one, although the circles of the planets are seven, and the plurality of the heavens, which Basil insists that the philosophers deny; and very probably the Jewish and early Christian notions of successive heavens full of angels and spirits developed from the spheres of the planets. Among the various early heresies described by the fathers are also found, of course, many allusions to these seven spheres or heavens. The disciples of Valentinus, for example, according to Irenaeus and Epiphanius, "affirm that these seven heavens are intelligent and speak of them as angels . . . and declare that Paradise, situated above the third heaven, is a powerful angel."[2]

Agreement with Greek science.

On the other hand, we may note some points where Basil is in accord with Greek science. He warns his hearers not to "be surprised that the world never falls; it occupies the center of the universe, its natural place."[3] He advances numerous proofs of the immense size of the sun and moon.[4] He accepts the hypothesis of four elements but abstains from passing judgment upon the question of a fifth element of which the heavens and celestial bodies may be composed.[5] He thinks that "it needs not the space of a moment for light to pass through" the ether.[6]

Qualification of the Scriptural account of creation.

Moreover, Basil finds it necessary to qualify some of the statements in the first chapter of *Genesis*. He interprets the command, "Let the waters under the heaven be gathered together unto one place," to apply only to the sea or ocean, which he contends is one body of water, and not to pools and lakes,[7] recognizing that otherwise "our explanation of the creation of the world may appear contrary to experience, because it is evident that all the waters did not flow together in one place." In this connection he

[1] Charles, *The Book of the Secrets of Enoch*, Introduction, pp. xxxi, xxxix.

[2] Irenaeus, I, 5; Epiphanius, ed. Petavius 186AB.

[3] Homily I, 10.

[4] VI, 9-11.

[5] I, 11.

[6] II, 7.

[7] IV, 2-4.

states that "although some authorities think that the Hyrcanian and Caspian Seas are enclosed in their own boundaries, if we are to believe the geographers, they communicate with each other and together discharge themselves into the Great Sea." He speaks of "the vast ocean, so dreaded by navigators, which surrounds the isle of Britain and western Spain." [1] Later he contends that "sea water is the source of all the moisture of the earth." [2] He has also to meet the following objection made to the eleventh and twelfth verses of the first chapter of *Genesis:* "How then, they say, can Scripture describe all the plants of the earth as seed-bearing, when the reed, couch-grass, mint, crocus, garlic, and the flowering rush and countless other species produce no seed? To this we reply that many vegetables have their seminal virtue in the lower part and in the roots." [3]

The four elements and four qualities.

Basil regards the words of *Genesis,* "God called the dry land earth," as a recognition of the fact that drought is the primal property of earth, as humidity is of air; cold, of water; and heat, of fire. He adds, however, that "our eyes and senses can find nothing which is completely singular, simple, and pure. Earth is at the same time dry and cold; water, cold and moist; air, moist and warm; fire, warm and dry." [4] Indeed, as he has already stated in the previous homily, the mixture of elements in actual objects is even more intricate than this last sentence might seem to indicate. Every element is in every other, and we not only do not perceive with our senses any pure elements but not even any compounds of two elements only.[5]

Enthusiasm for nature as God's work.

Basil is alive to the absorbing interest of the world of nature and to the marvelous intricacies of natural science. He tells his hearers that as "anyone not knowing a town is taken by the hand and led through it," so he will guide them "through the mysterious marvels of this great city of the universe." [6] As he had said in the preceding homily, "A

[1] Homily IV, 4.
[2] IV, 6.
[3] V, 2.
[4] IV, 5.
[5] III, 4.
[6] VI, 1.

single plant, a blade of grass is sufficient to occupy all your intelligence in the contemplation of the skill which produced it."[1] He sees "great wisdom in small things."[2] Thus by the argument from design he is apt to work back from nature to the Creator, so that his enthusiasm cannot be regarded as purely scientific. Going a step farther than Galen's argument from design, he contends that "not a single thing has been created without reason; not a single thing is useless."[3]

Sin and nature.

Basil also cherishes the notion, which we have already found both in pagan and Christian writers, that human sin leaves its stain or has its effect upon nature. The rose was without thorns before the fall of man, and their addition to its beauty serves to remind us that "sorrow is very near to pleasure."[4]

Habits of animals.

Basil discusses the habits of animals largely in order to draw moral lessons from them for human beings and he has several passages in the style supposed to be characteristic of the *Physiologus.* But he also refers in a number of places to the ability of animals to find remedies with which to cure themselves of ailments and injuries, or to their power of divining the future. The sea-urchin foretells storms; sheep and goats discern danger by instinct alone. The starling eats hemlock and digests it "before its chill can attack the vital parts"; and the quail is able to feed on hellebore. The wounded bear nurses himself, filling his wounds with mullein, an astringent plant; "the fox heals his wounds with droppings from the pine tree"; the tortoise counteracts the venom of the vipers it has eaten by means of the herb marjoram; and "the serpent heals sore eyes by eating fennel."[5]

Marvels of nature.

Indeed, far from being led by his acquaintance with Greek science into doubting the marvelous, Basil finds "in nature a thousand reasons for believing in the marvelous."[6] He is ready to ascribe astounding powers to animals, and

[1] Homily V, 3.
[2] V, 9.
[3] V, 4.
[4] V, 6.
[5] VII, 5; IX, 3.
[6] VIII, 6.

believes, like Pliny, that "the greatest vessels, sailing with full sails, are easily stopped by a tiny fish." [1] He tells us that nature endowed the lion with such loud and forceful vocal organs "that often much swifter animals are caught by his roaring alone." [2] He also repeats in charming style the familiar story of the halcyon days. The halcyon lays its eggs along the shore in mid-winter when violent winds dash the waves against the land. Yet winds are hushed and waves are calm during the seven days that the halcyon sits, and then, after its young are hatched and in need of food, "God in his munificence grants another seven days to this tiny animal. All sailors know this and call these days halcyon days." [3]

Spontaneous generation.

Like most ancient scientists, Basil believes that some animals are spontaneously generated. "Many birds have no need of union with males to conceive," a circumstance which should make it easy for us to believe in the Virgin birth of Christ.[4] Grasshoppers and other nameless insects and sometimes frogs and mice are "born from the earth itself," and "mud alone produces eels," [5] a theory not much more amazing than the assertion of modern biologists that eels spawn only in the Mediterranean Sea. Basil states that "in the environs of Thebes in Egypt after abundant rain in hot weather the country is covered with field mice," but without noting that abundant rain in upper Egypt in hot weather would itself be in the nature of a miracle.

Lack of scientific scepticism.

Basil is less sceptical than Apollonius of Tyana in regard to the birth of lions and of vipers, repeating unquestioningly the statement that the viper gnaws its way out of its mother's womb, and that the lioness bears only one whelp because it tears her with its claws.[6] Of purely scientific scepticism there is, indeed, little in the *Hexaemeron*. Basil does, however, question one of the powers ascribed to magicians, and this is his only mention of the magic

[1] Homily VII, 6.
[2] IX, 3.
[3] VIII, 5. See also Aristotle, *History of Animals*, V, 8.
[4] Homily VIII, 6.
[5] IX, 2.
IX, 5.

art. Discussing the immense size of the moon and its great influence upon terrestrial nature, he declares ridiculous the old-wives' tales which have been circulated everywhere that magic incantations "can remove the moon from its place and make it descend to the earth." [1]

Sun worship and astrology.

Sun worship still existed in Basil's time and he hails the fact that the sun was not created until the fourth day, after both light and vegetation were in existence, as a severe blow to those who reverence the sun as the source of life.[2] However, he does "not pretend to be able to separate light from the body of the sun." [3] Theophilus in his earlier discussion of creation had stated, perhaps copying Philo Judaeus, that the luminaries were not created until the fourth day, "because God, who possesses foreknowledge, knew the follies of the vain philosophers, that they were going to say, that the things which grow on earth are produced from the heavenly bodies"—which is, indeed, a fundamental hyopthesis of astrology—"so as to exclude God. In order, therefore, that the truth might be obvious, the plants and seeds were produced prior to the heavenly bodies, for what is posterior cannot produce that which is prior." [4] Basil does not make this point against the rule of inferior creation by the heavenly bodies, but in a succeeding homily he feels it necessary to devote several paragraphs [5] to refutation of the "vain science" of casting nativities, which some persons have justified by the words of God concerning sun, moon, and stars in the first chapter of *Genesis,* "And let them be for signs." Basil questions if it be possible to determine the exact instant of birth, declares that to attribute to the constellations and signs of the zodiac the characteristics of animals is to subject them to external influences, and defends human free will in much the usual fashion. He is ready, however, to grant that "the variations of the moon do not take place without exerting great influence upon the organization of animals and of all living

[1] Homily VI, 11.
[2] V, 1.
[3] VI, 3.
[4] *Ad Autolycum,* II, 15.
[5] Homily VI, 5-7.

things," and that the moon makes "all nature participate in her changes." [1]

Permanence of species.

Basil's utterances concerning the world of nature are not always consistent. In describing the creation of vegetation he asserts that species are unchanging, affirming that "all which sprang from the earth in the first bringing forth is kept the same to our time, thanks to the constant reproduction of kind." [2] Yet a few paragraphs later we find him saying, "It has been observed that pines, cut down or even submitted to the action of fire, are changed into a forest of oaks." [3] Nevertheless in the last homily he again asserts that "nature, once put in motion by divine command, . . . keeps up the succession of kinds through resemblance to the last. Nature always makes a horse succeed to a horse, a lion to a lion, an eagle to an eagle, and preserving each animal by these uninterrupted successions she transmits it to the end of all things. Animals do not see their peculiarities destroyed or effaced by any length of time; their nature, as though it had just been constituted, follows the course of ages forever young." [4]

Final impression from the *Hexaemeron.*

Concerning Basil in conclusion we may say that while he can scarcely be called much of a scientist, he is a pretty good scientist for a preacher. His knowledge of, and errors concerning, the world of nature will probably compare quite as well with the science of his day as those of most modern sermons will with the science of our days. His occasional flings at Greek philosophy are probably not to be taken too seriously. But what interests us rather more

[1] Homily VI, 10.

[2] V, 2.

[3] V, 7. But perhaps he simply means that oaks will grow where pines used to.

Tertullian, *De pallio,* cap. 2, dwelling on the law of change, speaks of the washing down of soil from mountains, the alluvial formation by rivers, and of sea-shells on mountain tops as a proof that the whole earth was once covered by water. He seems to have in mind a gradual process of geological evolution rather than Noah's flood, and Sir James Frazer states that Isidore of Seville is the first he knows of the many writers who have appealed "to fossil shells imbedded in remote mountains as witnesses to the truth of the Noachian tradition,"—*Origines,* XIII, 22, cited by J. G. Frazer, *Folk-Lore in the Old Testament* (1918), I, 159, who cites the passage in Tertullian at pp. 338-9.

[4] Homily IX, 2.

than Basil's attitude is that of his audience, curious concerning nature. Just as it is evident that many of them go to theaters and circuses, or play with dice, despite Basil's denunciation of the immoral songs of the stage and the evils of gambling; just so, we suspect, it was the attractive morsels of Greek astronomy, botany, and zoology which he offered them that induced them to come and listen further to his argument from design and his moral lessons based upon these natural phenomena. Nor were they likely to observe his censure of incantations and nativities more closely than his condemnation of theater and gaming. It would be rash to infer that they always practiced what he preached. By the same token, even if the church fathers had opposed scientific investigation—and it hardly appears that they did—they would probably have been no more successful in checking it than they were in checking the commerce of Constantinople, although "S. Ambrose regards the gains of merchants as for the most part fraudulent, and S. Chrysostom's language has been generally appealed to in a similar sense." [1]

The Medicine Chest of Epiphanius.

The same recognition of an interest in nature on the part of his audience and the same appeal to their scientific curiosity, which we have seen in Basil's sermons, is shown by Epiphanius of Cyprus (315-403) writing in 374-375 A. D.[2] He calls his work against heresies the *Panarion,* or "Medicine Chest," his idea being to provide antidotes and healing herbs in the form of salubrious doctrine against the venom of heretics whose enigmas he compares to the bites of serpents or wild beasts. This metaphor is more or less adhered to throughout the work, and particular heresies are compared to the asp, basilisk, dipsas,[3] buprestis,[4] lizard, dog-fish or shark, mole, centipede, scorpion, and various

[1] Cunningham, *Christian Opinion on Usury,* p. 9.

[2] Twice in the course of the *Panarion* (Dindorf, I, 280, and II, 428; Petavius, 2D and 404A) he gives the year of the reign of Valentinian and Valens, namely, the eleventh and the twelfth.

[3] Lucian's *De dipsadibus* will be recalled; see also Pliny, NH, XXIII, 80; Lucan, *Pharsalia,* IX, 719.

[4] Pliny, NH, XXIII, 18; XXX, 10.

vipers. We are further told of substances that drive away serpents, such as the herbs *dictamnon, abrotonum,* and *libanotis,* the gum *storax,*[1] and the stone *gagates.* As his authorities in such matters Epiphanius states that he uses Nicander for the natures of beasts and reptiles, and for roots and plants Dioscorides, Pamphilus, Mithridates the king, Callisthenes and Philo, Iolaos the Bithynian, Heracleides of Tarentum, and a number of other names.[2]

Gems in the high priest's breastplate.

If in his *Panarion* Epiphanius makes use of ancient botany, medicine, and zoology for purposes of comparison, in his treatise on the twelve gems in the breastplate of the Hebrew high priest[3] he perhaps gives an excuse and sets the fashion for the Christian medieval *Lapidaries.* This work was probably composed after the *Panarion,* and in the opinion of Fogginius even later than 392 A. D.[4] This treatise probably was better known in the middle ages than the *Panarion,* since the fullest version of it extant is the old Latin one, while the Greek text which has survived seems only a very brief epitome. The Greek version, however, embodies a good deal of what is said concerning the gems themselves and their virtues, but omits entirely the long effort to identify each of the twelve stones with one of the twelve tribes of Israel, which is left unfinished even in the Latin version. Epiphanius shows himself rather chary in regard to such virtues attributed to gems as to calm storms, make men pacific, and confer the power of divination. He does not go so far as to omit them entirely, but he usually qualifies them as the assertion of "those who construct fables" or "those who believe fables." It is without any such qualification, however, that he declares that the topaz,[5] when ground on a physician's grindstone, although red itself, emits a white milky fluid, and, moreover,

[1] Pliny, NH, XXV, 53; XXI, 92; XIX, 62; XII, 40 and 55.

[2] Dindorf, II, 450; Petavius, 422C.

[3] *Liber de XII gemmis rationalis summi sacerdotis Hebraeorum,* published in Dindorf's edition of the *Opera* of Epiphanius, vol. IV, pp. 141-248, with the preface and notes of Fogginius, and both the Latin and Greek versions.

[4] *Ibid.,* 160-62.

[5] P. 174.

that as many vessels as one wishes may be filled with this fluid without changing the appearance or shape or lessening the weight of the stone. Skilled physicians also attribute to this liquid a healing effect in eye troubles, in hydrophobia, and in the case of those who have gone mad from eating grape-fish.

Some other gems.

Epiphanius mentions a few other gems than those in the high priest's breastplate. Among these is the stone hyacinth [1] which, when placed upon live coals, extinguishes them without injury to itself and which is also beneficial to women in childbirth, and drives away phantasms. Certain varieties of it are found in the north among the barbarous Scythians. The gems lie at the bottom of a deep valley which is inaccessible to men because walled in completely by mountains, and moreover from the summits one cannot see into the valley because of a dark mist which covers it. How men ever became cognizant of the fact that there are gems there may well be wondered but is a point which Epiphanius does not take into consideration. He simply tells us that when men are sent to obtain some of these stones, they skin sheep and hurl the carcasses into the valley where some of the gems adhere to the flesh. The odor of the raw meat then attracts the eagles, whose keener sight is perhaps able to penetrate the mist, although Epiphanius does not say so, and they carry the carrion to their nests in the mountains. The men watch where the eagles have taken the meat and go there and find the gems which have been brought out with it. In the middle ages we find this same story in a slightly different form told of Alexander the Great on his expedition to India. Epiphanius has one thing to tell of India himself in connection with gems, which is that a temple of Father Liber (Bacchus) is located there which is said to have three hundred and sixty-five steps,—all of sapphire.[2]

[1] Pp. 190-91. [2] *Ibid.*, 184.

The so-called *Physiologus*: problem of its origin.

The problem of an early Christian work entitled *Physiologus* is no easy one, although much has been written concerning it [1] and more has been taken for granted. For instance, one often meets such wild and sweeping statement as that "the name Physiologus" was "given to a cyclopedia of what was known and imagined about earth, sea, sky, birds, beasts, and fishes, which for a thousand years was the authoritative source of information on these matters and was translated into every European tongue." [2] My later treatment of medieval science will make patent the inaccuracy of such a statement. But to return to the problem of the origin of Physiologus. The original Greek text,[3] which some would put back in the first half of the second century of our era, if it ever existed, is now lost, and its previous existence and character are inferred from numerous apparent citations of it, possible extracts from it, and what are taken to be imitations, abbreviations, amplifications, adaptations, and translations of it in other languages and of later date. Thus we have versions or fragments in Armenian,[4] Syriac,[5]

[1] Pitra, *Spicilegium Solesmense*, Paris, 1855, III, xlvii-lxxx. K. Ahrens, *Zur Geschichte des sogenannten Physiologus*, 1885. M. F. Mann, *Bestiaire Divin de Guillaume Le Clerc*. Heilbronn, 1888, pp. 16-33, "Entstehung des Physiologus und seine Entwicklung im Abendlande." F. Lauchert, *Geschichte des Physiologus*, Strassburg, 1889. E. Peters, *Der griechische Physiologus und seine orientalischen Uebersetzungen*, Berlin, 1898. M. Goldstaub, *Der Physiologus und seine Weiterbildung, besonders in der lateinischen und in der byzantinischen Litteratur*, in *Philologus*, Suppl. Bd. VIII (1898-1901), 337-404. Also in *Verhandl. d. 41 Versammlung deutscher Philologen u. Schulmänner in München*, Leipzig (1892), pp. 212-21. V. Schultze, *Der Physiologus in der kirchlichen Kunst des Mittelalters*, in *Christliches Kunstblatt*, XXXIX (1897), 49-55. J. Strzygowski, *Der Bilderkreis des griechischen Physiologus*, in *Byz. Zeitsch.* Ergänzungsheft, I (1899). E. P. Evans, *Animal Symbolism in Ecclesiastical Architecture*, 1896, is disappointing, being mainly compiled from secondary sources and having little to say on ecclesiastical architecture.

[2] EB, 11th ed., "Arthropoda."

[3] Lauchert (1889), pp. 229-79, attempts a critical edition of the Greek text.

[4] Pitra (1855), III, 374-90; French translation in Cahier, *Nouveaux mélanges* (1874), I, 117, *et seq.*

[5] O. G. Tychsen, *Physiologus Syrus*, 1795; from an incomplete Vatican MS. Land, *Otia Syriaca*, p. 31, *et seq.*, or in *Anecdota Syriaca*, IV, 115, *et seq.*, gives the complete text with a Latin translation.

Ethiopian,[1] and Arabic; [2] a Greek text from medieval manuscripts, mostly of late date; [3] various Latin versions in numerous manuscripts from the eighth century on; [4] in Old High German a prose translation of about 1000 A. D. and a poetical version later in the same language; [5] and Bestiaries such as those of Philip of Thaon [6] and William

[1] Hommel, *Die aethiopische Uebersetzung des Physiologus,* Leipzig, 1877. A bit of it was translated by Pitra (1855), III, 416-7.

[2] Land, *Otia Syriaca,* p. 137, *et seq.,* with Latin translation. A fragment in Pitra (1855), III, 535.

[3] Pitra (1855), III, 338-73, used MSS from the 13th to 15th century. The earliest known illuminated copies are of 1100 A. D. and later: see Dalton, *Byzantine Art and Archaeology,* Oxford, 1911, pp. 481-2.

[4] The oldest Latin MSS seem to be two of the 8th and 9th centuries at Berne. Edited by Mai, *Classici auctores,* Rome, 1835, VII, 585-96, and more completely by Pitra (1855), III, 418; also by G. Heider, in *Archiv f. Kunde österreich. Geschichtsquellen,* Vienna, 1850, II, 545; Cahier et Martin, *Mélanges d'archéologie,* Paris, II (1851), 85ff., III (1853), 203ff., IV (1856), 55ff. Cahier, *Nouveaux mélanges* (1874), p. 106ff.

Mann (1888), pp. 37-73, prints the Latin text which he regards as William le Clerc's source from Royal 2-C-XII, and gives a list of other MSS of Latin Bestiaries in English libraries.

Other medieval Latin Bestiaries have been printed in the works of Hildebert of Tours or Le Mans (Migne, PL, 171, 1217-24: really this poem concerning only twelve animals is by Theobald, who was perhaps abbot at Monte Cassino, 1022-1035, and it was printed under the name of Theobald before 1500,—see the volume numbered IA.12367 in the British Museum and entitled, *Phisiologus Theobaldi Episcopi de naturis duodecim animalium.* Indeed, it was printed at least nine times under his name,—see Hain, 15467-75): and in the works of Hugh of St. Victor (Migne, PL, 177, 9-164, *De bestiis et aliis rebus libri quatuor*). Both of these versions occur in numerous MSS, as does a third version which opens with citation of the remark of Jacob in blessing his sons, "Judah is a lion's whelp." The author then cites *Physiologus* as usual concerning the three natures of the lion. See Wolfenbüttel 4435, 11th century, fols. 159-68v, Liber bestiarum. "De leone rege bestiarum et animalium (est) etenim iacob benedicens iudam ait Catulus leonis iuda. De leone. Leo tres naturas habet." Laud. Misc. 247, 12th century, fol. 140-, . . . caps. 36, praevia tabula . . . Tit. "De tribus naturis leonis." Incip. "Bestiarium seu animalium regis; etenim Jacob benedicens filium suum Udam ait Catulus leonis Judas filius meus quis suscitabit eum; Fisiologus dicit, Tres res naturales habere leonem. . . ." Library of Dukes of Burgundy 10074, 10th century, "Etenim Jacob benedicens." CLM 19648, 15th century, fols. 180-95, "Igitur Jacob benedicens." CLM 23787, 15th century, fols. 12-20, "Igitur Jacob benedicens." CU Trinity 884, 13th century in a fine hand, with 107 English miniatures, fol. 89-, "Et enim iacob benedicens filium suum iudam ait catulus leonis est iudas filius meus"; this MS ends imperfectly.

[5] Printed by Lauchert (1889), pp. 280-99.

[6] Max F. Mann, *Der Physiologus des Philipp von Thaon und seine Quellen,* Halle, 1884, 53 pp.

the Clerk [1] in the Romance languages [2] and other vernaculars.[3] The *Physiologus* has been thought to have originated in Alexandria because of its use of the Egyptian names for the months and because Clement of Alexandria and Origen are supposed to have made use of it. But it is difficult to determine whether the church fathers drew passages concerning animals and nature from some such work or whether it was a collection of passages from their writings upon such themes. Ahrens, who thought he found the original form of the work in a Syriac *Book of the Things of Nature,*[4] regarded Origen as its author. In a medical manuscript at Vienna is a *Physiologus* in Greek ascribed to Epiphanius of Cyprus,[5] of whom we have just been treating, while we hear that Pope Gelasius at a synod of 496 condemned as apocryphal a *Physiologus* which was written by heretics and ascribed to Ambrose,[6] who so closely duplicated the *Hexaemeron* of Basil. A work on the natures of animals is also attributed to John Chrysostom.[7] I am not sure whether

[1] Mann, *Bestiaire Divin de Guillaume Le Clerc,* Heilbronn, 1888, in *Französische Studien,* VI, 2, pp. 201-306. Most recent edition by Robert, Leipzig, 1890.

[2] Besides the two foregoing see Goldstaub und Wendriner, *Ein tosco-venez. Bestiarius,* Halle, 1892. Magliabech. IV, 63, 13th century, mutilated, 53 fols., bestiario moralizato, in Italian prose. E. Monaci, *Rendiconti dell' Accad. dei Lincei, Classe di scienze morali, storiche e filol.,* vol. V, fasc. 10 and 12, has edited a Bestiario in 64 sonetti on as many animals from a private MS at "Gubbio nell' archivio degli avvocati Pietro e Oderisi Lucarelli," MS 25, fols. 112-27. See also M. Garver and K. McKenzie, *Il Bestiario Toscano secondo la lezione dei codice di Parigi e di Roma,* in *Studi romanzi,* Rome, 1912; McKenzie, *Unpublished Manuscripts of Italian Bestiaries,* in *Modern Language Publications,* XX (1905), 2; and Garver, "Some Supplementary Italian Bestiary Chapters," in *Romanic Review,* XI (1920), 308-27.

[3] For instance, A. S. Cook, *The Old English Elene, Phoenix, and Physiologus,* Yale University Press, 364 pp., 1919.

[4] K. Ahrens, *Das "Buch der Naturgegenstände,"* 1892.

[5] *Cod. Vind. Med.* 29, τοῦ ἁγίου Ἐπιφανίου ἐπισκόπου Κύπρου περὶ τῆς λέξεως πάντων τῶν ζώων φυσιολόγος. In the edition of Ponce de Leon, Rome, 1587, there are twenty animals described, and the symbolic interpretation is very short compared to later versions. Heider (1850), p. 543, regarded this as the oldest version and as extant in complete form.

[6] Mansi, *Concil.,* VIII, 151, "Liber Physiologus ab hereticis conscriptus et beati Ambrosii nomine presignatus apocryphus."

[7] Heider (1850), II, 541-82, "Physiologus nach einer Handschrift des XI Jahrhunderts": the text opens at p. 552, "Incipiunt Dicta Johannis Chrysostomi de naturis bestiarum." Lauchert used another MS, Vienna 303, 14th century, fol. 124v-, which was

a *Physiologus* ascribed to John the Scot in a tenth century Latin manuscript is the same work.[1]

Does the title apply to any one particular treatise?

The *Physiologus* is commonly described as a symbolic bestiary, in which the characteristics and properties of animals are accompanied by Christian allegories and instruction. Some have almost gone so far as to hold that any passages of this sort are evidence of an author's having employed the *Physiologus,* which some have held influenced the middle ages more than any other book except the Bible. But Pitra's point is well taken that the *Physiologus* is one thing and the allegorical interpretation thereof another. In the case of the discordant versions or fragments which he gathered and published from different manuscripts, centuries, and languages, he noted one common feature, that the allegorical interpretation was sharply separated from the extracts from *Physiologus* and sometimes omitted entirely. This is what one would naturally expect since a *physiologus* is a natural scientist on whose statements concerning this or that the allegorical interpretation is presumably based and added thereto. But this suggests another difficulty in identifying *Physiologus* as a single work. The abbreviations for the word in medieval manuscripts are very easily confused with those for philosophers or *phisici* (physical scientists), and just as medieval writers often cite what the philosophers say or the *phisici* say without having reference to any particular book, so may they not cite what *physiologi* or even *physiologus* says without having any particular writer in mind? In the *De bestiis* ascribed to

considerably different and was furthermore combined with the Physiologus of Theobald. An earlier MS than either of the foregoing is CLM 19417, 9th century, fols. 29-71, Liber Sancti Johannis episcopi regiae urbis Constantinopoli . . . Crisostomi quem de naturis animalium ordinavit. Another Vienna MS is 2511, 14th century, fols. 135-40, "Incipiunt dicta Johannis Chrysostomi de naturis animalium et primo de leone . . . / . . . Sic erit et scriba doctus in regno celorum qui profert de thesauro suo noua et uetera. Expliciunt dicta Johannis Crisostomi." A Paris MS of the same is BN 2780, 13th century, 14, Sancti Ioannis Chrysostomi liber qui physiologus appellatur.

[1] Additional 11,035, Johannis Scottigenae Phisiologiae liber. In the same MS are Macrobius' *Dream of Scipio* and the poems of Prudentius.

Hugh of St. Victor of the twelfth century *physici* are cited [1] as well as *Physiologus*. When Albertus Magnus states in the thirteenth century in his work on minerals that the *physiologi* have assigned very different causes for the marvelous occult virtue in stones, he evidently simply alludes to the opinions of scientists in general and has no such work or works as the so-called *Physiologus* in mind.[2] This is also clearly the case in a fragment from the introduction to a Latin translation from the Arabic of some treatise on the astrolabe, in which we find *phisiologi* cited as astronomical authorities.[3] Furthermore, even in works which deal with the natures of animals and which either have the word *Physiologus* in their titles or cite it now and then in the course of their texts, there exists such diversity that it becomes fairly evident not only that it is impossible to deduce from them the list of animals treated in the original *Physiologus* or the details which it gave concerning each, but also that it is highly probable that the title *Physiologus* has been applied to different treatises which did not necessarily have a common origin. Or at least the greatest liberties were taken with the original text and title,[4] so that the word *Physiologus* came to apply less to any particular book, author, or authority than to almost any treatment of animals in a certain style.

And to what sort of a treatise?

But of what style? It has too often been assumed that theology dominated all medieval thought and that natural science was employed only for purposes of religious symbolism. Of this general assumption the *Physiologus* has been seized upon as an apt illustration and it has been represented as a symbolic bestiary which influenced the middle ages more than any other book except the Bible [5] and whose allegories accounted for the animal sculpture of the Gothic

[1] *De bestiis et aliis rebus*, II, 1 (Migne, PL 177, 57). "Physici denique dicunt quinque naturales res sive naturas habere leonem. . . ."

[2] *Mineral.*, II, i, 1 (ed. Borgnet, V, 24).

[3] Bubnov (1899), p. 372.

[4] Thus even Lauchert (1899), p. 105, admits that Bartholomew of England, the thirteenth century Latin encyclopedist, cites *Physiologus* for much which does not come from *Physiologus*.

[5] Goldstaub (1899-1901), p. 341.

cathedrals and the strange or familiar beasts in the borders of the Bayeux Tapestry, the margins of illuminated manuscripts, and so on and so forth.

Medieval art shows almost no symbolic influence of the *Physiologus.*

The more recent scientific study of medieval art has largely dissipated this latter notion. It has become evident that in the main medieval men represented animals in art because they were fond of animals, not because they were fond of allegories. Their art was natural, not symbolic. They were, says Mâle, "craftsmen who delighted in nature for its own sake, sometimes lovingly copying the living forms, sometimes playing with them, combining and contorting them as they were led by their own caprice." St. Bernard, although "the prince of allegorists," saw no sense in the animal sculptures in Romanesque cloisters and inveighed against them. In short, with the exception of the symbols of the four evangelists, "there are few cases in which it is permissible to assign symbolic meaning to animal forms," and it is "evident that the fauna and flora of medieval art, natural or fantastic, have in most cases a value that is purely decorative." "To sum up," concludes Mâle, "we are of the opinion that the Bestiaries of which we hear so much from the archaeologists had no real influence on art until their substance passed into Honorius of Autun's book (*Speculum ecclesiae,* c. 1090-1120) and from that book into sermons. I have searched in vain (with but two exceptions) for representations of the hedgehog, beaver, tiger, and other animals which figure in the Bestiaries but which are not mentioned by Honorius." [1]

Physiologus was more natural scientist than allegorist.

These assertions concerning medieval art hold true also to a large extent of medieval literature and medieval science, although they were perhaps less natural and original than it and more dependent on past tradition and authority. But medieval men, as we shall see, studied nature from scientific curiosity and not in search for spiritual allegories, and even Goldstaub recognizes that by the thirteenth century the

[1] This and the preceding quotations in the paragraph are from Mâle (1913), pp. 48, 35, 49, 45.

scientific zoology of Aristotle submerged that of the *Physiologus* in writers like Thomas of Cantimpré and Albertus Magnus who, although they may still embody portions of the *Physiologus,* divest it of its characteristic religious elements.[1] But were its characteristic elements ever religious? Were they not always scientific or pseudo-scientific? Ahrens holds that the title was taken from Aristotle in the first place, and that Pliny was the chief source for the contents. The allegories do not appear in such early texts as the Syriac version or the fragments preserved in the Latin Glossary of Ansileubus. Not even the introductory scriptural texts appear in the Greek version ascribed to Epiphanius. Moreover, in the Bestiaries where the allegorical applications are included, it is for the natures of the animals, the supposedly scientific facts on which the symbolism is based, and for these alone that *Physiologus* is cited in the text. Thus the symbolism would appear to be somewhat adventitious, while the pseudo-science is constant. It is obvious that the allegorical applications cannot do without the supposed facts concerning animals; on the other hand, the supposedly scientific information can and does frequently dispense with the allegories. We do not know who was responsible for the allegorical interpretations in the first instance. Hommel would carry the origin of their symbolism back of the Christian era to the animal worship of Persia, India, and Egypt.[2] But we are assured over and over again that Natural Scientist or *Physiologus* vouches for the statements concerning the natures of animals. Thus the symbolic significance of the literature that has been grouped under the title *Physiologus* has been exaggerated, while the respect for and interest in natural science to which it testifies have too often been lost sight of.

[1] Goldstaub (1899-1901), pp. 350-1. The same statement could be made with equal truth of Vincent of Beauvais and Bartholomew of England.

[2] Hommel (1877), pp. xii, xv.

CHAPTER XXII

AUGUSTINE ON MAGIC AND ASTROLOGY

Date and influence of Augustine—Christianity and magic—Censure of magic and theurgy as well as *Goetia*—Magic due to demons—Marvels wrought by magic—Cannot be equalled by most Christians—Miracles of heretics—Theory of demons—Limitations to the power of magic—Its fantastic character—Samuel and the witch of Endor—Natural marvels—Relation between magic and science—Superstitions akin to magic—Survival of pagan superstition among the laity—Augustine's attack upon astrology—Fate and free will—Argument from twins—Defense of the astrologers—Elections—Are animals and plants under the stars?—Failure to disprove the control of nature by the stars—Natural divination and prophetic visions—The star at Christ's birth—Nature of the stars—Orosius on the Priscillianists and Origenists—Augustine's letter—Attitude toward astronomy—Perfect numbers.

Date and influence of Augustine.

THE utterances of Augustine concerning magic and astrology have been reserved for separate treatment in this chapter, partly because of his late date, 354 to 430 A. D., partly because of the voluminousness of his writings, but especially because of his approach to and influence upon the thought of the middle ages. It is, moreover, in his epoch-making book, *The City of God*, which better than any other single event marks, or at least sums up, the transition from classical to medieval civilization, from the life of the ancient city to that of the medieval church, that he descants with especial fulness upon magic, demons, and astrology, although he often also refers to these themes in his other treatises, which we shall cite as well. I separate the words, magic and astrology, here because Augustine, like most of the fathers, does so. Of Augustine's discussion of the Biblical account of creation in his *Confessions* and *De Genesi ad litteram* I shall not treat, having already presented Basil's *Hexaemeron* as an example of this type of work and of the Christian attitude toward natural science.[1] But later in treating of medieval writers on nature I may have occasion to point out certain passages in which they may have been influenced by Augustine.

Christianity and magic.

Even though writing in the fifth century Augustine still finds it necessary to defend Christ against those who imagine that He has converted peoples to Himself by means of the magic art.[2] And he tells us of books of magic which are ascribed to Christ Himself or to the apostles Peter and Paul.[3] In reply to such charges or assertions he insists that Christians have nothing to do with magic, and that their miracles "were wrought by simple confidence and devout faith, not by incantations and spells compounded by an art of depraved curiosity." [4] And he brings the counter-charge against Roman religion that King Numa, its founder, learned its secrets and sacred rites by means of hydromancy or necromancy.[5] He admits, however, that condemnation of magic and legislation against it had begun before Christianity.[6]

Magic and theurgy censured as well as *Goetia*.

Augustine uniformly speaks of magic with censure and several times adverts to "the crimes of magicians." [7] He speaks, however, of *goetia* or sorcery as "a more detestable name" than *magia* and of "theurgy" as "an honorable name." He also states that some persons draw a distinction between the *malefici* or sorcerers or practitioners of *goetia*, whom they call truly guilty of illicit arts and deserving of condemnation, and those who practice theurgy, whom they call praiseworthy. Porphyry, for instance, had

[1] Duhem, II (1914), 314, seems to me to have over-estimated the significance of *Confessions,* V, 5, and *De Genesi ad litteram,* I, 19, in saying, "L'assurance avec laquelle les Basile, les Grégoire de Nysse, les Ambroise, les Jean Chrysostome opposaient aux enseignements de la Physique profane les naïves assertions de leur science puérile contristait fort l'Évêque de Hippone." There is nothing, I think, to indicate that Augustine had these men or men of their stamp in mind, and I doubt if his scientific attainments were superior to Basil's.

[2] *De consensu Evangelistarum,* I, 11; in Migne, PL 34, 1049-50.

[3] *Ibid.,* I, 9-10.

[4] *De civitate Dei,* X, 9; PL vol. 41.

[5] *Ibid.,* VII, 34-35; and see Arnobius, *Against the Heathen,* V, 1, for Augustine's probable source.

[6] *De civ. Dei,* VIII, 19.

[7] *Ibid.,* VIII, 18, 19, 26; IX, 1.

stated that theurgy was useful to purge the soul and prepare it to receive spirits and to see God. Augustine, however, holds that in other passages Porphyry condemned theurgy, and in any case he himself refuses to sanction it.[1] He stoutly denies that "souls are purged and reconciled to God through sacrilegious likenesses and impious curiosity and magic consecrations." [2] Very possibly Augustine would have classed as improper theurgy some of the use of powerful names described by Origen.

Magic due to demons.

At any rate Augustine declares that theurgists and sorcerers alike "are entangled in the deceitful rites of demons who may masquerade under the names of angels." [3] For it is to demons that Augustine, like most of our Christian writers, attributes both the origin and the success of magic. The demons are enticed by men to work marvels, not by offerings of food, as if they were animals, but by symbols which conform to the individual taste of each as a spirit, namely, various stones, plants, trees, animals, incantations, and ceremonies,[4]—a good brief summary of the materials and methods of magic. Augustine believes that the spirits had first to instruct men what rites to perform and by what names to call them in order to summon them.

Marvels wrought by magic.

But when once the demons have revealed their secrets, henceforth the charms of the magic art have efficacy. Of the marvels worked by means of magic Augustine has little doubt; to deny them would indeed in his opinion be to deny the truth of the Scriptures, to whose accounts of Pharaoh's magicians,[5] the witch of Endor, and the Magi and the star, he adverts many times in his various works. If actors in the theater and performers in spectacles are able by art and exercise to display astounding alterations in the appearance of their earthly bodies, why may not the demons with

[1] *De civ. Dei,* X, 9-10.

[2] *De trinitate,* IV, 11; in Migne, PL 42, 897.

[3] *De civ. Dei,* X, 9.

[4] *De civ. Dei,* XXI, 6.

[5] In Grenoble 208, 12th century, containing works of Augustine, there is listed separately at fol. 54v, "De magis Pharaonis," to which the MSS catalogue adds, "et de CLIII piscibus." Probably it is an extract from one of Augustine's longer works as it covers only one leaf.

their aerial bodies produce marvelous changes in elementary substances or by occult influence construct phantom images to delude human senses?[1] Augustine even grants that the magicians are able to terrify the inferior spirits into obedience to their commands by adjuring them by the names of superior spirits, and thereby with divine permission "to exhibit to the eye of sense certain results which seem great and marvelous to men who through weakness of the flesh are incapable of beholding things eternal." He does not regard this as inconsistent with the assertion of Jesus that Satan cannot cast out Satan, since while it may be that thus demons are expelled from sick bodies, the evil one thereby only the more surely takes possession of the soul.[2]

Cannot be equalled by most Christians.

Augustine further grants that magicians, although stained with crime, can at present work miracles which most Christians and even most saints cannot perform. For this, however, he finds Scriptural precedent. Pharaoh's magicians performed feats which none of the Children of Israel could equal except Moses who excelled them by divine aid. Augustine, like earlier fathers, usually fails to mention Aaron in this connection.[3] This superiority of magicians to most Christians in working marvels Augustine believes is divinely ordained so that Christians may remain humble and practice works of justice rather than seek to perform miracles. Magicians seek their own glory; the saints strive only for the glory of God. And the more marvelous are the feats of magic, the more Christians should shun them; the greater the power of the demons, the closer Christians should cling to that Mediator who alone can raise men from the lowest depths.[4]

Miracles of heretics.

Like Origen, Augustine further distinguishes the miracles wrought by heretics both from magic and from the miracles of true Christians. He holds that every soul in

[1] *De trinitate,* IV, 11.

[2] *De diversis quaestionibus,* cap. 79; Migne, PL 40, 92-3.

[3] See also *De cataclysmo* (perhaps spurious), cap. 5, Migne, PL 40, 696; and *Sermo VIII,* PL 38, 74. *Sermo XC,* PL 38, 562, however, speaks of "Moyses et Aaron."

[4] *De civ. Dei,* XXI, 6; XVIII, 18.

part controls itself and exercises as it were a private jurisdiction, in part is subject to the laws of the universe just as any citizen is amenable to public jurisdiction. Therefore magicians perform their marvels by private contracts with demons; good Christians perform theirs by public justice; bad Christians perform theirs by the appearance or signs of public justice.[1] This view would seem to indicate that God, like the demons, regards the signs alone and not the character and purpose of the performer, so that Christian miracles, if they can be duplicated by heretics, would appear to be largely a matter of procedure and art, like magic.

Theory of demons.

For his theory of demons and their characteristics Augustine seems largely indebted to Apuleius, whom he cites in several chapters of the eighth and ninth books of *The City of God.* In his separate treatise, *The Divination of Demons,*[2] he explains their ability to predict the future and to perform marvels by the keenness of their sense, their rapidity of movement, their long experience of nature and life, and the subtlety of their aerial bodies. This last quality enables them to penetrate human bodies or affect the thoughts of men without men being aware of their presence. Augustine, however, of course does not believe that the world of nature is completely under the control of the demons. God alone created it and He still governs it, and the demons are able to do only as much as He permits.[1]

Limitations to the power of magic.

There were, for example, some things which Pharaoh's magicians could not do and in which Moses clearly excelled them. They were able to change their rods into snakes but his snake devoured theirs. How the magicians got their rods back, if at all, neither Augustine nor the Book of Exodus informs us. But whether with or without their magic wands, they were still able to duplicate one or two of the plagues sent upon Egypt. Augustine explains that neither they nor the demons who helped them really created snakes and frogs, but that there are certain seeds of life

[1] *De diversis quaestionibus,* cap. 79; *De doctrina Christiana,* II, 20, in Migne, PL 34, 50.

[2] Migne, PL 40, 581-92.

[1] *De trinitate,* III, 8; PL, 42 875.

hidden away in the elemental bodies of this world of which they made use. But their magic failed them when it came to the reproduction of minute insects.[1] Augustine furthermore has some hesitation about accepting the stories of magic transformations of men into animals, which he represents as current in his own day as well as in times past, so that certain female inn-keepers in Italy are said to transform travelers into beasts of burden by a magic potion administered in the cheese, just as Circe transformed the companions of Ulysses and as Apuleius says happened to himself in the book that he wrote under the title, *The Golden Ass.* These stories, in Augustine's opinion, "are either false or such uncommon occurrences that they are justly discredited." [2] He does not believe that demons can truly transform the human body into the limbs and lineaments of beasts, but the strange personal experiences of reliable persons have convinced him that men are deceived by dreams, hallucinations, and fantastic images.

Its fantastic character

Thus, as we have already seen over and over again, the fantastic and deceptive character of magic is dimly realized. Usually, however, when Augustine represents "the powers of the air" as deceiving men by magic, the deceit consists merely in the magicians' imagining that they are working the marvels which are really performed by demons, or in men being lured into subjection to Satan and to their ultimate and eternal damnation through the attractions of the magic art.[3]

Samuel and the witch of Endor.

Augustine twice responded to questions concerning the witch of Endor's apparent invocation of the spirit of Sam-

[1] *De trinitate,* III, 7-8. It seems strange to me that they should have failed on minute insects who in ancient and medieval science are often represented as produced by spontaneous generation. The Talmudists also, however, state that the Egyptians were unable to duplicate the plague of lice, as their art did not extend to things smaller than a barleycorn.

[2] *De civitate Dei,* XVIII, 22. In commenting on Genesis (PL 34, 445) he speaks even more harshly of "that absurd and harmful notion of the changing of souls and of men into beasts, or of beasts into men"; but perhaps he has reference to the doctrine of transmigration of souls rather than to magic transformations.

[3] *Confessions,* X, 42, in PL vol. 32.

uel, repeating in his *De octo Dulcitii quaestionibus* [1] what he had already said in *De diversis quaestionibus ad Simplicianum.*[2] In certain respects Augustine's treatment of the problem differs from those which we have previously examined. What, he asks, if the impure spirit which possessed the *pythonissa* was able to raise the very soul of Samuel from the dead? Is it not much more strange that Satan was allowed to converse personally with God concerning the tempting of Job, and to raise the very Christ aloft upon a pinnacle of the temple? Why then may not the soul of Samuel have appeared to Saul, not unwillingly and coerced by magic power but voluntarily under some hidden divine dispensation? Augustine, however, also thinks it possible that the soul of Samuel did not appear but was impersonated by some phantasm and imaginary illusion made by diabolical machinations. He can see no deceit in the Scripture's calling such a phantom Samuel, since we are accustomed to call paintings, statues, and images seen in dreams by the names of the actual persons whom they represent. Nor does it trouble him that the spirit of Samuel or pretended spirit predicted truly to Saul, for demons have a limited power of that sort. Thus they recognized Christ when the Jews knew Him not, and the damsel possessed of a spirit of divination in *The Acts* testified to Paul's divine mission. Augustine leaves, however, as beyond the limits of his time and strength the further problem whether the human soul after death can be so evoked by magic incantations that it is not only seen but recognized by the living. In his answer to Dulcitius he further calls attention to the passage in *Ecclesiasticus* (XLVI, 23) where Samuel is praised as prophesying from the dead. And if this passage be rejected because the book is not in the Hebrew canon, what shall we say of Moses who appeared to the living long after his death?

Natural marvels.

Augustine had some acquaintance with ancient natural science and in one passage rehearses a number of natural marvels which are found in the pages of Pliny and Solinus

[1] Quaest. VI; PL 40, 162-5. [2] II, 3; PL 40, 142-4.

in order to show pagans their inconsistency in accepting such wonders and yet remaining incredulous in regard to analogous phenomena mentioned in the Bible. So Augustine rehearses the strange properties of the magnet; asserts that adamant can be broken neither by steel nor fire but only by application of the blood of a goat; tells of Cappadocian mares who conceive from the wind; and hails the ability of the salamander to live in the midst of flames as a token that the bodies of sinners can subsist in hell fire. Augustine also admits "the virtue of stones and other objects and the craft of men who employ these in marvelous ways." [1] He denies, however, that the Marsi who charm snakes by their incantations are really understood by the serpents. There is some diabolical force behind their magic, as when Satan spoke to Eve through the serpent.[2]

Relation between magic and science.

Once at least, however, Augustine associates science and magic. In his *Confessions,* after speaking of sensual pleasure he also censures "the vain and curious desire of investigation" through the senses, which is "palliated under the name of knowledge and science." This is apt to lead one not only into scrutinizing secrets of nature which are beyond one and which it does one no good to know and which men want to know just for the sake of knowledge, but also "into searching through magic arts into the confines of perverse science." [3]

Superstitions akin to magic.

Of this dangerous borderland between magic and science Augustine has more to say in some chapters of his *Christian Doctrine.*[4] After mentioning as prime instances of human superstition idolatry, other false religions, and the magic arts, he next lists the books of soothsayers (*aruspices*) and augurs as of the same class, "though seemingly a more permissible vanity." In his *Confessions,*[5] however, he tells of a soothsayer who offered not only to consult the future for him, but to insure him success in a poetical contest in

[1] *De civitate Dei,* XXI, 4-6; PL 41, 712-6.

[2] *De Genesi ad litteram,* XI, 28-9; PL 34, 444-5.

[3] *Confessions,* X, 35; in PL vol. 32.

[4] II, 20 and 29.

[5] IV, 2-3.

which he was to engage in the theater. The incident is a good illustration of the fact that prediction of the future and attempting to influence events go naturally together, and that arts of divination cannot be separated either in theory or practice from magic arts. In the *Christian Doctrine* Augustine is inclined further to put in the same class all use of invocations, incantations, and characters, which he regards as signs implying pacts with evil spirits, and the use of which in working cures he asserts is condemned by the medical profession. He is also suspicious of ligatures and suspensions, and states that it is one thing to say, "If you drink the juice of this herb, your stomach will not ache," and is another thing to say, "If you suspend this herb from the neck, your stomach will not ache. For in one case a healing application is worthy of approval, in the other a superstitious signification is to be censured." Augustine recognizes, however, that such ligatures and suspensions are called "by the milder name of natural remedies (*physica*)"; and if they are applied without incantations or characters, possibly they may heal the body naturally by mere attachment, in which case it is lawful to employ them. But they may involve some signal to demons, in which case the more efficacious they are, the more a Christian should avoid them.

Survival of pagan superstition among the laity.

The same attitude toward superstitious medicine is shown in a sermon attributed to Augustine but probably spurious.[1] Here a tempter is represented as coming to the sick man and saying, "If you had only employed that enchanter, you would be well now; if you would attach these characters to your body, you could recover your health." Or another comes and says, "Send your girdle to that diviner; he will measure and scrutinize it and tell you what to do and whether you can recover. Or a third visitor may recommend someone who is skilled in fumigation. The preacher warns his hearers not to succumb to such advice or they will be sacrificing to the devil; whereas if they refuse

[1] PL 39, 2268-72.

such treatment and die, it will be a glorious martyr's death. The preacher, however, is not over-sanguine that his advice will be heeded, as he has often before admonished his hearers against pagan superstitions, and yet reports keep coming to him that some are continuing such practices. He therefore "warns them again and again" to forsake all diviners, *aruspices,* enchanters, phylacteries, augury, and observance of days, or they will lose all benefit of the sacrament of baptism and will be eternally damned unless they perform a vast amount of penance. The observance of days other than the Lord's Day is here condemned on the ground that God made the other six days without distinction. In another supposititious sermon [1] the practice of diligently observing on which day of the week to set out on a journey is censured as equivalent to worshiping the planets, or rather the pagan gods whose names they bear and who are said here to have originally been bad men and women who lived at the time that the Children of Israel were in Egypt. The preacher is even opposed to naming the days of the week after such persons or planets and exhorts his hearers to speak simply of the first day, second day, and so on.

Augustine's attack upon astrology.

Nor will Augustine, to return to his remarks in the *Christian Doctrine,*[2] exempt "from this genus of pernicious superstition those who are called *genethliaci* from their consideration of natal days and now are also popularly termed *mathematici.*" He holds that they enslave human free will by predicting a man's character and life from the stars, and that their art is a presumptuous and fallacious human invention, and that if their predictions come true, this is due either to chance or to demons who wish to confirm mankind in its error.[3] In his youth, when a follower of the Manichean sect, Augustine had been a believer in astrology and thereby "sacrificed himself to demons" at the same time that, owing to his Manichean scruples against animal sacrifice, he refused to employ a *haruspex.*[4] Perhaps on this account he

[1] *Sermo CXXX,* PL 39, 2004-5.
[2] II, 21-3; PL 34, 51-3.
[3] *De civitate Dei,* V, 7.
[4] *Confessions,* VII, 6.

felt the more bound to warn his readers against astrology in his old age. He often attacks the casters of horoscopes in his works and especially in the opening chapters of the fifth book of *The City of God,* on which we may center our attention as being a rather more elaborate discussion than the other passages and including almost all the arguments which he advances elsewhere. These arguments are not original with him, but his presentation of them was perhaps better known in the middle ages than any other.[1]

Fate and free will.

The objection to astrology as fatalistic does not come with the best grace from Augustine, the great advocate of divine prescience and of predestination, and in his discussion in *The City of God* he is forced to recognize this fact. He holds that the world is not governed by chance or by fate, a word which for most men means the force of the constellations, but by divine providence. He starts to accuse the astrologers of attributing to the spotless stars, or to the God whose orders the stars obediently execute, the causing of human sin and evil; but then recognizes that the astrologers will answer that the stars simply signify and in no way cause evil, just as God foresees but does not compel human sinfulness.

Argument from twins.

Thus thwarted in his attempt to show that the astrologers enslave the human will, although in other passages he still gives us to understand that they do,[2] Augustine adopts another line of argument, that from twins, an old favorite, which he twists first one way and then another, proposing to the astrologers a series of dilemmas as he finds them likely to escape from each preceding one. He seems to have been much impressed by the thought that at the same instant and hence with the same horoscope persons were born whose subsequent lives and characters were different. He brings forward Esau and Jacob as examples, and states that he himself has known of twins of dissimilar sex and

[1] Unless otherwise noted, the ensuing arguments are found in *The City of God*, V, 1-7.

[2] *De Genesi ad litteram*, II, 17; PL 34, 278. *De diversis quaestionibus*, cap. 45; PL 40, 28-9. *Epistola* 246; PL 33, 1061. *Sermo* 109; PL 38, 1027.

life. Moreover, he tells us in his *Confessions* that he was finally induced to abandon his study of the books of the astrologers, from which the arguments of "Vindicianus, a keen old man, and of Nebridius, a youth of remarkable intellect," had failed to win him, by hearing from another youth that his father, a man of wealth and rank, had been born at precisely the same moment as a certain wretched slave on the estate.[1]

Defense of the astrologers.

But the astrologers reply that even twins are not born at precisely the same instant and do not have the same horoscope, but are born under different constellations, so rapidly do the heavens revolve, as the astrologer Nigidius Figulus neatly illustrated by striking a rapidly revolving potter's wheel two successive blows as quickly as he could in what appeared to be the same spot. But when the wheel was stopped and examined, the two marks were found to be far apart. Augustine's counter argument is that if astrologers must take into account such small intervals of time, their observations and predictions can never attain sufficient accuracy to insure correct prediction; and that if so brief an instant of time is sufficient to alter the horoscope totally, then twins should not be as much alike as they are nor have as much in common as they do,—for instance, falling ill and recovering simultaneously. To this the astrologers are likely to respond that twins are alike because conceived at the same instant, but somewhat dissimilar in their life because of the difference in their times of birth. Augustine retorts that if two persons conceived simultaneously in the same womb may be born at different times and have different fates after birth, he sees no reason why persons who are born of different mothers at the same instant with the same horoscope may not die at different dates and lead different lives. But he does not recognize that very likely the astrologers would agree with him in this, since they often held that the influence of the stars was received variously by matter. He also asks why a certain sage is

[1] *Confessions*, IV. 2-3.

said to have selected a certain hour for intercourse with his wife in order to beget a marvelous son—possibly an inaccurate allusion to the story of Nectanebus [1]—unless the hour of conception controls the hour of birth, and consequently twins conceived together must have the same horoscope. He also objects that if twins fall sick at the same time because of their simultaneous conception, they should not be of opposite sex as sometimes happens.

Elections.

With this Augustine turns from the case of twins to urge the inconsistency of the astrological doctrine of elections, suggested by the story of the sage who chose the favorable moment for intercourse with his wife. He holds that this practice of choosing favorable times is inconsistent with the belief in nativities which are supposed to have determined and predicted the individual's fate already. He also inquires why men choose certain days for setting out trees and shrubs or breeding animals, if men alone are subject to the constellations.

Are animals and plants under the stars?

This last clause indicates how exclusively Augustine's attacks are directed against the prediction of man's life from the stars, and how little he has to say regarding the stars' control of the world of nature in general. He now goes on to consider this latter possibility, but interprets it too in the narrow sense of horoscope-casting, and as implying that every herb and beast must have its fate absolutely determined by the constellations at its moment of birth. This appears, however, to have been a widespread belief then, since he tells us that men are accustomed to test the skill of astrologers by submitting to them the horoscopes of dumb animals, and that the best astrologers are able not only to recognize that the reported constellations mark the birth of a beast rather than that of a human being, but also to state whether it was a horse, cow, dog, or sheep. Nevertheless, Augustine feels that he has reduced the art of casting horoscopes to an absurdity, as he feels sure that beasts and plants which are so numerous must frequently be born

[1] See below, chapter 24.

at precisely the same instant as human beings. Furthermore, it is plain that crops which are sown and ripen simultaneously meet with very diverse fates in the end. Augustine thinks that by this argument he will force the astrologers to say that men alone are subject to the stars, and then he will triumphantly ask how this can be, when God has endowed man alone of all creatures with free will. Having thus argued more or less in a circle, Augustine regains the point from which he had started, or rather, retreated.

Failure to disprove the control of nature by the stars.

Augustine cannot then be said to have advanced any telling arguments against some sort of control of inferior nature by the motions and influence of the heavenly bodies. He leaves the fundamental hypothesis of astrology unrebutted. His attention is concentrated upon genethlialogy, the superstition that the time and place of birth and nothing else determine with mathematical certainty and mechanical rigidity the entirety of one's life. This seems nevertheless to have been a superstition which was very much alive in his time, which he felt he must take pains repeatedly to refute, and to which he himself had once been in bondage. But he could not have studied the books of the astrologers very deeply, as he ascribes views to them which many of them did not hold. Also he seems never to have read the *Tetrabiblos* of Ptolemy. His attack upon and criticism of astrology was therefore narrow, partial, and inadequate, and did not prevent medieval men from devoting themselves to that subject, although they might cite his objections against ascribing to the constellations an influence subversive of human free will. But he cannot be said to have admitted the control of the stars over the world of nature. Apparently the most that he was willing to concede was that it was not absurd to say that the influence of the stars might produce changes in material things, as in the varying seasons of the year caused by the sun's course and the alternating augmentation and diminution of tides and shell-fish due, as he supposed, to the moon's phases. He concludes his discussion of the subject in *The City of God* by saying that, all things considered, if the astrologers make many marvelously true predictions, they do so by the aid and inspiration of the demons and not by the art of noting and inspecting horoscopes, which has no sound basis.

Natural divination and prophetic visions.

In another work Augustine tells of some young men who, while traveling, as a boyish prank pretended to be astrologers and either by mere chance or by natural and innate power of divination hit upon the truth in the predictions which they supposed that they were inventing. In the same context he proceeds to discuss in a credulous way the possibility of marvelous prophetic visions, concerning which he tells one or two other tall tales from his personal experience. He is, however, doubtful how far the human soul itself possesses the power of divination, which he is inclined to attribute rather to spirits, good or bad. But owing to Satan's ability in disguising himself as an angel of light it is often very difficult to tell to which sort of spirit to ascribe the vision in question.[1]

The star at Christ's birth.

In Augustine's time there were those who held that Christ Himself had been "born under the decree of the stars," because of the statement in the Gospel according to Matthew that the Magi had seen His star in the east. Of this matter Augustine treats in several of his works.[2] He denies that this would be true even if other men were subject to the fatal influence of the stars, which he denies as usual on the ground of free will. He contends that the star was not one of the planets or constellations but a special creation, since it did not keep to a regular course or orbit, but came to where the child lay. But how did the Magi know that it was the star of Christ when they saw it in the east, unless by astrology? Augustine can only suggest that this was revealed to them by spirits, whether good or bad he does not know.[3] Augustine further affirms that the star did not

[1] *De Genesi ad litteram,* XII, 22 and 17 and 12; PL 34, 472-3, 467-9, 464-5. See also the marvelous divinations of Albicerius recounted in *Contra Academicos,* I, 6; PL 32, 914-5.

[2] *Sermones* 199 and 374; PL 38, 1027-8, and 39, 1666. *Contra Faustum,* II, 15; PL 42, 212.

[3] In *Quaestiones ex Novo Testamento,* Quaest. 63, PL 35, 2258, which is probably a spurious

cause Christ to live a marvelous life, but Christ caused the star to make its marvelous appearance. "For, when born of a mother, He showed earth a new star in the sky, Who, when born of the Father, formed both heaven and earth." And, "when He is born, new light is revealed in a star; when He dies, old light is veiled in the sun." But these rhetorical flourishes and antitheses seem to attest rather than dispute the significance of celestial phenomena, so that Augustine cannot be said to have answered the astrological contention anent Christ's birth very satisfactorily.

Nature of the stars.

The problem of the nature of the stars is one which Augustine prefers to leave unsolved, although it comes up several times in his writings.[1] Whether they are simply lucid bodies without sense or intelligence, as some think; or have happy intellectual souls of their own, as Plato taught; whether they are to be classed with the Seats, Dominions, Principalities, and Powers of whom the apostle speaks; and whether they are ruled and animated by spirits: all these are questions which Augustine puts, but concerning whose answers he feels uncertain. His fullest discussion of the matter is in a letter against the Priscillianists to which we now come.

Orosius on the Priscillianists and Origenists.

An interchange of letters between Augustine and his Spanish disciple Orosius deals with the error of the Priscillianists and Origenists.[2] Nothing is said to convict them of magic, which was, however, the charge on which Pris-

work but was cited as Augustine's by Thomas Aquinas (*Summa*, III, 36, v), Balaam is said to have warned the Magi to watch for the star. It is also asserted, however, that "these Chaldean Magi watched the course of the stars, not from malevolence, but curiosity concerning nature" (*Hi Magi chaldaei non malevolentia astrorum cursum sed rerum curiositate speculabantur*).

[1] *Enchiridion, sive de fide, spe, et charitate,* I, 58; PL 40, 259-60. *De civitate Dei,* XIII, 16; PL 41, 388. *De Genesi ad litteram,* II, 18; PL 34, 279-80.

[2] *Orosii ad Augustinum Consultatio sive Commonitorium de errore Priscillianistarum et Origenistarum,* PL 31, 1211-22; also in G. Schepss (1889), in CSEL XVIII. *Augustini ad Orosium contra Priscillianistas et Origenistas,* PL 41, 669, *et seq.* Augustine also discusses the Priscillianists in *Epistle* 237, PL 33, 1034, *et seq.*, where he makes no charge either of magic or astrology against them.

cillian was put to death, but astrological tenets are ascribed to them. Orosius states that Priscillian taught that the soul was born of God and instructed by angels, but that it descended through certain circles of the heavens and was caught by evil principalities and thrust into different bodies; and that it remained subject to *Mathesis* or the laws of astrology until Christ set it free by His passion on the cross. Like the astrologers, continues Orosius, Priscillian associated the signs of the zodiac with the different members of the human body, Aries and the head, Taurus and the neck, and so on;[1] and he also taught that the names of the patriarchs of the twelve tribes were "members of the soul," Reuben in the head, Judah in the breast, Levi in the heart, and so on. Orosius adds that the Origenists regard the sun, moon, and stars not as elemental luminaries but as rational powers; and we have seen that Origen himself did so.

Augustine's letter.

Augustine in his reply states that we can see that the sun, moon, and stars are celestial bodies, but not that they are animated. He agrees firmly with Paul that there are Seats, Dominions, Principalities, and Powers in the heavens, "but I do not know what they are or what the difference is between them." On the whole, Augustine is inclined to regard this state of ignorance as a blissful one. He is somewhat troubled by the verses in the Book of Job, "How shall man be just in the sight of God, or how shall one born of woman purify himself? If He commands the moon and it does not shine, and if the stars are not pure before Him, how much more is man rottenness and the son of man a worm?" From this passage the Priscillianists infer that the stars have a rational spirit and are not free from sin, yet are placed in the heaven because their fault is less than that of sinful mankind. Origen too had argued, "If the stars are living and rational beings, there will undoubtedly appear among them both an advance and a falling back. For the language of Job, 'the stars are not clean in His

[1] This charge was later repeated by St. Leo, *Epistola XV;* see Withington, *History of Medicine,* 1894, p. 178; but the offense would seem a trivial one in any case.

sight,' seems to me to convey some such idea."[1] Augustine evades this difficulty by questioning whether this passage is to be received as of divine authority, since it is uttered by one of Job's comforters and not by Job himself, of whom alone it is said that he had not sinned with his lips against God.

Attitude towards astronomy.

So set is Augustine against astrology that he even holds that Christians may well leave the subject of astronomy alone, "because it is related to the most pernicious error of those who utter a fatuous fatalism," although he recognizes that there is nothing superstitious in predicting the future positions of the stars themselves from knowledge of their past movements. But except that to know the course of the moon is useful in determining the date of Easter, knowledge of the stars is of little or no help in interpreting the divine Scriptures.[2] In another passage Augustine is somewhat perturbed by the assertion of astronomers that there are many stars equal to or greater than the sun in size, but which seem smaller because they are farther off,—an assertion which seems to conflict with the statement of Genesis that in creating the sun and moon "God made two great lights." Augustine, however, does not stop to contest the point at length but leaves it with the excuse that Christians have many better and more serious matters to occupy their time than such subtle investigations concerning the relative magnitude of the stars and the intervals of space between them.[3]

Perfect numbers.

Augustine himself, however, was not above occupying his readers' time with discussion of the occult significance of numbers, towards belief in which he shows himself inclined. Six was a perfect number in his estimation, since God had created the world in six days, although He might have taken less or more time; and the Psalmist made no idle remark in saying that the Deity had ordered all things ac-

[1] *De principiis,* I, 7.

[2] *De doctrina Christiana,* II, 29, in Migne, 34, 57.

[3] *De Genesi ad litteram,* II, 16, in Migne, 34, 277.

cording to measure, number, and weight. Also six is the first number which can be obtained from adding together its factors: one, two, and three. Augustine was going on to say that seven was also a perfect number, when he checked himself lest he digress at too great length and seem "too eager to display his smattering of science." Hence he merely added that one indication of seven's perfection was its composition of the first complete odd number, three, and the first complete even number, four.[1] It is therefore not surprising to find ascribed to Augustine a sermon on the correspondence between the ten plagues of Egypt and the ten commandments which opens by remarking that it is not without cause that the number of precepts in God's law is the same as the number of plagues with which Egypt was afflicted.[2]

[1] *De civitate Dei*, XI, 30-31. He says about the same things concerning six and seven in *De Genesi ad litteram*, IV, 2.

[2] *Sermo supposititius* 21, in Migne, PL XXXIX, 1783, "De convenientia decem preceptorum et decem plagarum Egypti. Non est sine causa, fratres dilectissimi, quod preceptorum legis Dei numerus cum numero plagarum quibus Aegyptus percutitur exaequari videtur."

CHAPTER XXIII

THE FUSION OF PAGAN AND CHRISTIAN THOUGHT IN THE FOURTH AND FIFTH CENTURIES

Need of qualifying the patristic attitude—Plan of this chapter—Julius Firmicus Maternus—Date of the *Mathesis*—Are the attitudes in Firmicus' two works incompatible?—*De errore* is not unfavorable to astrology—Attitude of both works to the emperors—Religious attitude of the *Mathesis*—An astrologer's prayer—Christian objections to astrology met—Astrology proved experimentally—Information to be gained from the third and fourth books—Religion and magic; exorcists—Divination—Magic as a branch of learning—Interest in science—Diseases in antiquity—Place of Firmicus in the history of astrology—Libanius accused of magic—Declamation against a magician—Faith of Libanius in divination—Magic and astrology in Pseudo-Quintilian declamations—Fusion of Christianity and paganism in Synesius of Cyrene—His career—His interest in science—Belief in occult sympathy between natural objects—Synesius on divination and astrology—Synesius as an alchemist—Macrobius on number, dreams, and stars—Martianus Capella—Absence of astrology—Orders of spirits—*The Celestial Hierarchy* of Dionysius the Areopagite.

Need of qualifying the patristic attitude.

IN reading the writings of the Christian fathers one is impressed by the fact that their tone is almost invariably that of the preacher. In estimating therefore the practical effect of their utterances it is well to remember that these are counsels of perfection which were probably often not realized even by those who gave utterance to them. This is not to accuse the fathers of being pharisaical, but to suggest that as both clerics and apologists they were professionally bound to take up an irreproachable position morally and dogmatically. Basil has shown us that the audience who listened to his sermons were still under the spell of Roman amusements, dice, theater, and arena. And the average lay Christian mind was probably more easy-going in its attitude toward magic and superstition than Augustine. Not merely laymen, moreover, but Christian clergy and apologists of the declining Roman Empire might still hold to divination and astrology. It was a time, as has often been remarked, of religious syncretism, of fusion of pagan and Christian thought, when it is not always easy to tell whether the author of an extant writing is Christian or Neo-Platonist or both. Mr. Gwatkin states that "the surface thought" of Constantine's time, "Christian as well as heathen, tended to a vague monotheism which looked on Christ and the sun as almost equally good symbols of the Supreme." [1] Others believed that astrology was the truth back of all religions.[2]

Plan of this chapter.

In this chapter we shall therefore consider some writers of the fourth and fifth century who attest the existence of magic and astrology then, the influence of paganism on Christianity and of Christianity on paganism, and the fusion of Neo-Platonism, Christianity, and astrological theory. This, indeed, we have already done to some extent, as our previous chapters on Neo-Platonism and on the Christian fathers have carried us more or less into those centuries. But now as an offset to Augustine we take up other writers who have not yet been treated: Firmicus, the Latin Christian apologist and the astrologer of the mid-fourth century; Libanius, the Greek sophist of the same century; Macrobius and Synesius, Neo-Platonists writing respectively in Latin and Greek at the beginning of the fifth century, and of whom one was a Christian bishop; and probably in the same century the discussion of spirits by Martianus Capella in Latin and the Pseudo-Dionysius the Areopagite in Greek. Except for Libanius and Synesius, these authors were very influential in medieval Latin learning and might serve as well for an introduction to our following book on *The Early Middle Ages* as for a conclusion to this.

[1] *Cambridge Medieval History*, I, 9.

[2] The Greek work, *Hermippus or Concerning Astrology*, however, can no longer be regarded as an example of Christian belief in astrology at this period, since F. Boll, *Heidelberger Akad. Sitzb.*, 1912, No. 18, has shown it to be a fourteenth century work of John Katrarios, who makes use of a Greek translation of Albumasar.

Julius Firmicus Maternus

Julius Firmicus Maternus [1] flourished during the reigns of Constantine the Great and his sons. Sicily was his native land; he was of senatorial rank and very well educated for his time, showing interest in natural philosophy, literature, and rhetoric. Two works are extant under his name: one, *On the Error of Profane Religions,*[2] is addressed to Constantius and Constans, 340-350 A. D., and urges them to eradicate pagan cults. The other, *Mathesis,*[3] is a work of astrology written at the request of a similarly cultured friend, Lollianus or Mavortius, who is spoken of in the preface as *ordinario consuli designato,*[4] an office which we know that he held in 355 A. D. The writing of two such works by one man has long given critics pause, and is a splendid warning against taking anything for granted in our study of the past. Not long ago the general opinion was that there must have been two different authors by the name of Firmicus. This very unlikely theory has now been universally abandoned, as unmistakable similarities in style and wording have been noted in the two works. But it is still maintained that "there is no question but that he was a pagan when he wrote his astrological book." [5] This involves two considerations, whether the attitude expressed in

[1] For bibliography see F. Boll's "Firmicus" in PW. It does not include my article written subsequently on "A Roman Astrologer as a Historical Source: Julius Firmicus Maternus," in *Classical Philology,* VIII, No. 4, pp. 415-35, October, 1913. For bibliography see also Kroll et Skutsch, II, xxxiv.

[2] The edition of *De errore profanarum religionum* by K. Ziegler, Leipzig, 1907, is more critical than that in Migne, PL.

[3] *Iulii Firmici Materni Matheseos Libri VIII,* ed. W. Kroll et F. Skutsch, *Fasciculus prior libros IV priores et quinti prooemium continens,* Leipzig, 1897; *Fasciculus alter libros IV posteriores cum praefatione et indicibus continens,* 1913. My references will be by page and line to this text, unless otherwise noted. Earlier editions, which I used for the later books before 1913, are the *editio princeps, Julius Firmicus de nativitatibus, . . . Impressum Venetiis per Symonem papiensem dictum bivilaqua, 1497 die 13 Iunii,* cxv fols.; the Aldine edition of 1499 containing apparent interpolations, *Julii Firmici Astronomicorum libri octo integri et emendati ex Scythicis oris ad nos nuper allati . . .*"; and the Basel editions of 1533 and 1551 by M. Pruckner which reproduce the Aldine text. See Kroll et Skutsch, II, xxxiii, for another reproduction of the Aldine text, printed in 1503, and p. xxviii for a partial edition of books 3-5 of the *Mathesis* in 1488 and 1494 in *Opus Astrolabii plani . . . a Iohanne Angeli.*

[4] Kroll et Skutsch, I, 3, 27.

[5] Boll in PW, VI, 2365.

the two works is really incompatible and whether the *Mathesis* was written before or after the *De errore.*

Date of the *Mathesis.*

Mommsen contended that "it is beyond doubt" [1] that the *Mathesis* was written between 334 and 337 A. D., relying chiefly upon several apparent mentions of Constantine the Great as still living. The names, Constantine and Constantius are frequently confused in the sources, however,[2] and even while the words, *"Constantinum maximum principem et huius invictissimos liberos, domines et Caesares nostros,"* seem to refer unmistakably to Constantine, it must be remembered that they occur in a prayer to the planets and to the supreme God that Constantine and his children may "rule over our posterity and the posterity of our posterity through infinite succession of ages." As this is simply equivalent to expressing a hope that the dynasty may never become extinct, it is scarcely proof positive that Constantine the Great was still living when Firmicus published his book. On the other hand, to maintain the early date Mommsen was forced to treat the mention of Lollianus as *ordinario consuli designato* as mere prophetic flattery or as an appointment held up by Constantius for eighteen years. We know that Firmicus addressed the *De errore* to Constantius and Constans, probably between 345 and 350; we know that Lollianus was city prefect of Rome in 342, *consul ordinarius* in 355, and praetorian prefect in the following year; whereas we know nothing certainly of either of them before 337. Furthermore Firmicus explicitly states that the writing of the *Mathesis* has been long delayed,[3] and when the promise to compose it was first made, it is evident that neither he nor Lollianus was a young man. Lollianus was already *consularis* of Campania and according to inscriptions had

[1] *Hermes,* XXIX, 468-72. The treatise could not have been composed before 334 since Firmicus (I, 13, 23) refers to an eclipse in the consulship of Optatus and Paulinus which occurred in that year.

[2] For instance, at I, 37, 25, *"Constantinus scilicet maximus divi Constantini filius,"* might as well be rendered, "Constantius, son of Constantine," as "Constantine, son of Constantius."

[3] I, 1, 3, "Olim tibi hos libellos, Mavorti decus nostrum, me dicaturum esse promiseram verum diu me inconstantia verecundiae retardavit."

previously held a number of other offices; while still in this position Lollianus had frequently to spur his friend on to the task which Firmicus as frequently "gave up in despair." Then Lollianus became Count of all the Orient and continued his importunities. Finally, after Lollianus has become proconsul and ordinary consul elect, Firmicus completes the work and presents it to him. Meanwhile Firmicus himself—who had formerly "resisted with unbending confidence and firmness" factious and wicked and avaricious men, "who from fear of law-suits seemed terrible to the unfortunate"; and who "with liberal mind, despising forensic gains, to men in trouble . . . displayed a pure and faithful defense in the courts of law," by which upright conduct he incurred much enmity and danger; [1]—has retired from the sordid sphere of law courts and forum to spend his leisure with the divine men of old of Egypt and Babylon and to purify his spirit by contemplation of the everlasting stars and of the God who works through them. Yet we are asked to believe—if we accept a date before 337 for the *Mathesis*—not merely that he writes a vehement invective against "profane religions" a decade later, but also that twenty years after Lollianus is still a vigorous administrator.[2] It is possible, but seems unlikely.

Are the attitudes in Firmicus' two works incompatible?

Certainly the date of the *Mathesis* should be determined without any assumption as to what Firmicus' religion was when he wrote it. For, if we regard his attitudes in *Mathesis* and *De errore* as incompatible, it will be as difficult to explain how he could write the *De errore* after having composed the *Mathesis* as *vice versa*. After the steadfast affirmation of astrological principles in the *Mathesis* it is no easier to explain the fierce spirit of intolerance toward paganism in the *De errore* than it is after the mention of Christ in the *De errore* to explain the omission of that name in the *Mathesis*. But are the two works really incompatible? My answer is, No. The divergences are such as may be ex-

[1] I, 195-6.

[2] Ammianus Marcellinus, XVI, 8, 5, "iubetur Mavortius, tunc praefectus praetorio, vir sublimis constantiae, crimen acri inquisitione spectari."

plained by the different character of the two works and the different circumstances under which they were written. *De errore* is an impassioned polemic very possibly delivered as an oration before the emperors; *Mathesis* is a learned compilation on a pseudo-scientific subject composed at leisure for a friend with the help of previous treatises on the subject. Why should Firmicus mention Christ in the *Mathesis?* Does Boethius, after nearly two centuries more of Christian growth and although he wrote a work on the Trinity, mention Christ in *The Consolation of Philosophy?* Some apparent petty inconsistencies there may be between Firmicus' two works, but if we accept a host of contradictions in Constantine the Great, the first Christian emperor, why balk at some inconsistency in a writer who urges Constantine's children against profane cults? On the other hand, there are some striking correspondences between the *De errore* and *Mathesis.*

De errore is not unfavorable to astrology.

It is noteworthy in the first place that in the *De errore* Firmicus does not attack astrology. But if he had been converted to Christianity since writing the *Mathesis* and had abandoned the astrological doctrine there expounded, would he have failed to attack the error of that art like Augustine who testified that he had once believed in nativities? It is therefore obvious that Firmicus does not regard astrology as an error even at the time when he is penning the *De errore* as a Christian apologist. Moreover, his view of nature in the *De errore* is quite in accord with that of the astrologer, and he manifests the respect for natural science or *physica ratio* which one would expect from the author of the *Mathesis.* Thus we find him criticizing certain pagan cults as sharply for their incorrect physical notions as he does others for travestying Christian mysteries. In its opening chapters certain oriental religions are criticized for exalting each some one of the four elements above the others, and for neglecting that superior control of the world of terrestrial nature in which both Christian and astrologer confided. Another argument against pagan worships is that they include human and immoral elements which cannot be explained as based upon natural law [1] and the rule of that supreme God or "God the fabricator," "who composed all things by the orderly method of divine workmanship,"—phrases which, as Ziegler has shown,[2] occur both in the *De errore* and *Mathesis*. Furthermore, in the *De errore* Firmicus' allusions to the planets, which include a representation of the Sun making a reproachful address to certain pagans,[3] indicate that he regarded the stars as of immense importance in the administration of the universe.

Attitude of both works to the emperors.

It is also worth remarking that in both works Firmicus sets the emperors above the rest of mankind and closely associates them with the celestial bodies and "the supreme God." If in *Mathesis* he prays for the perpetuation of the line of Constantine and forbids astrologers to make predictions concerning the emperor on the ground that his fate is not subject to the stars but directly to the supreme God, "and inasmuch as the whole surface of the earth is subject to the emperor, he too is reckoned in the number of those gods whom the principal divinity has established to perform and preserve all things": [4]—if he says this in *Mathesis*, in *De errore* he repeatedly addresses the emperors as "most holy" [5] and in one passage says, "You now, O Constantius and Constans, most holy emperors, and the virtue of your venerated faith must be implored. It is erected above men and, separated from earthly frailty, joins in alliance with things celestial and in all its acts so far as it can follows the will of the supreme God. . . . Your felicity is joined with God's virtue, with Christ fighting at your side you have triumphed on behalf of human safety." [6]

Religious attitude of the *Mathesis*.

If the author of *De errore* is not unfavorable to astrology the author of the *Mathesis* is strongly inclined towards mon-

[1] Ziegler, p. 7, "Physica ratio quam dicis, alio genere celetur"; p. 9, "quod dicant physica ratione conpositum."

[2] Ziegler, p. 5.

[3] Ziegler, p. 23.

[4] Kroll et Skutsch, I, 86, 12-21.

[5] Ziegler, pp. 15, 38, 39, 64, 67, 81, 82, "sacratissimi imperatores"; pp. 31, 40, "sacrosancti principes"; p. 65, "sanctarum aurium vestrarum."

[6] Ziegler, pp. 53-4.

otheism and decidedly religious. He indignantly repels the accusation that astrology, which teaches that "all our acts are arranged by the divine courses of the stars," draws men away "from the cult of the gods and of religions." "We cause the gods to be feared and worshiped, we demonstrate their might and majesty." [1] The passage just quoted and some others are suggestive of polytheism, and Firmicus frequently speaks of the planets as "gods." Probably in this he is reproducing the phraseology and reflecting the attitude of the astrological works which he uses as his authorities and which belong to the period of the pagan past. His *apotelesmata,* too, or predictions of nativities for various horoscopes, give little or no indication of being especially adapted to a Christian society, although in some other respects they fit his own age.[2] But while the work contains a considerable residue of paganism, its prevailing conception of deity is one supreme God, the rector of the planets, "who composed all things by the arrangement of everlasting law," [3] and who made man the microcosm from the four elements.[4] He is prayed to thus:

An astrologer's prayer.

"But lest my words be bereft of divine aid and the envy of some hateful man impugn them by hostile attacks, whoever thou art, God, who continuest day after day the course of the heavens in rapid rotation, who perpetuatest the mobile agitation of ocean's tides, who strengthenest earth's solidity in the immovable strength of its foundations, who refreshest with night's sleep the toil of our earthly bodies, who when our strength is renewed returnest the grace of sweetest light, who stirrest all the substance of thy work by the salutary breath of the winds, who pourest forth the waves of streams and fountains in tireless force, who revolvest the varied seasons by sure periods of days: sole Governor and Prince of all, sole Emperor and Lord, whom all the celestial forces

[1] Kroll et Skutsch, I, 17-18.

[2] See my "A Roman Astrologer as a Historical Source," *Classical Philology,* VIII, 415-35, especially p. 421.

[3] I, 16, 20, "Summo illi ac rectori deo, qui omnia perpetua legis dispositione composuit. . . ."

[4] I, 16, 14; I, 57, 2; I, 90, 11, to 91, 10.

serve, whose will is the substance of perfect work, by whose faultless laws all nature is forever adorned and regulated; thou Father alike and Mother of every thing, thou bound to thyself, Father and Son, by one bond of relationship; to Thee we extend suppliant hands, Thee with trembling supplication we venerate; grant us grace to attempt the explanation of the courses of thy stars; thine is the power that somehow impels us to that interpretation. With a mind pure and separated from all earthly thoughts and purged from every stain of sin we have written these books for thy Romans." [1] Doubtless these words might have been written by a Neo-Platonist or a pagan, but it also seems likely that they were penned by a Christian astrologer.

Christian objections to astrology met.

Firmicus provides not only for divine government of the universe and creation of the world and man, but also for prayer to God and for human free will,[2] since by the divinity of the soul we are able to resist in some measure the decrees of the stars. He also holds that human laws and moral standards are not rendered of no avail by the force of the stars but are very useful to the soul in its struggle by the power of the divine mind against the vices of the body.[3] Indeed, not only is the astrologer himself urged at considerable length to lead a pure, upright, and unselfish life, but "to show the right way of living to sinful men, so that, reformed by your teaching, they may be freed from the errors of their past life." [4] The human soul is also immortal, a spark of that same divine mind which through the stars exerts its influence upon terrestrial bodies.[5] All this may be consistent or not both with itself and with the art of astrology, but it meets the chief objections that Christians might make and had made to the art.

Astrology proved experimentally.

These and other objections to the art of nativities are the theme to which the first of the eight books of the *Ma-*

[1] I, 280, 2-28.

[2] Besides the prayer just quoted, see I, 18, 10-13. See also the long prayer at the end of the first book to the planets and supreme God for the successful continuance of the dynasty of Constantine.

[3] I, 18, 25-9.

[4] I, 85-89 (Book II, chapter 30).

[5] I, 17, 2-23.

thesis is devoted. Firmicus points out that some of the other objections to astrology do not correctly state the doctrines of that art; others he admits are ingenious arguments which sound well on paper but he insists that if the opponents of astrology, instead of protesting that the influence of the stars at a given instant is incalculable, would put the matter to the test experimentally,[1] they would soon be convinced of the truth of astrologers' predictions, although he grants that unskilful astrologers sometimes give wrong responses. But he insists that persons who have not tested astrology experimentally are unfit to pass upon its merits.[2] He affirms that the human spirit which has discovered so many other sciences and to which so much of divinity and religion has been revealed is capable also of casting horoscopes, and that astrological prediction is a relatively easy task compared to the mapping out of the whole heavens and courses of the stars which the *mathematici* have already performed so successfully.[3] And he does not see why anyone persists in denying the power of fate in human affairs when all about him he can see the innocent suffering and the guilty escaping; the best men such as Socrates, Plato, and Pythagoras meeting an ill fate; and unprincipled persons like Alcibiades and Sulla prospering.[4]

Information to be gained from the third and fourth books.

The remaining seven books of the *Mathesis* are given over to the art of horoscope casting. The second book consists chiefly of preliminary directions, but the others state what men will be born under various constellations. Of these the last four books are extant only in manuscripts of the fifteenth and sixteenth centuries, while the first four are found in manuscripts going back to the eleventh century. Moreover, although books five to eight cover more pages than books three and four, they do not supply so many details or so satisfactory a picture of human society in their predictions. These divergences, which are mainly ones of omission, do not invalidate the results which we gain from

[1] I, 10, 3-.
[2] I, 11, 7-.
[3] Book I, Chapter 4 (I, 11-15).
[4] Book I, Chapter 7 (I, 19-30).

an analysis of the third and fourth books, but do raise the question whether the later books, especially the fifth and sixth, are genuine. In them the wording becomes vaguer, little knowledge is shown of conditions at the time that Firmicus wrote, the predictions are more sensational and rhetorical. Only the latter part of the eighth book carries the conviction of reality that books three and four do. These two books are both independent units and through their predictions of the future supply a general picture of human society, presumably that of Firmicus' own time or not long before. One naturally assumes that those matters to which Firmicus devotes most space and emphasis are the prominent features of his age. Let us see what his picture is of religion, divination, the occult science and magic, natural science and medicine.[1]

Religion and magic; exorcists.

To religion Firmicus gives less space than to politics. There are no clear references to Christianity, but there are few allusions to any particular cults. Firmicus, however, indicates the existence of many cults, speaking five times of the heads of religions, and characterizing men as "those who regard all religions and gods with a certain trepidation," "those devoted to certain religions," "those who cherish the greatest religions," and so on. Temples,[2] priests, and divination [3] are the three features of religion that he mentions most. Magic and religion are closely associated in his predictions, for instance, "temple priests ever famed in magic lore." Sacred or religious literatures and persons devoted to them are mentioned thrice, while in a fourth passage we

[1] For a fuller exposition of this quantitative method of source-analysis and the results obtained thereby see Thorndike (1913), pp. 415-35.

[2] Temple-robbers, 5; servile or ignoble employ in temples, 5; spending one's time in temples, 4; builders of temples, 3; beneficiaries of temples, 3; temple guards, 2; *neocori*, 3; and so on, making 35 references to temples in all. It is perhaps worth remarking that H. O. Taylor, *The Classical Heritage*, 1901, p. 80, notes that Synesius about 400 A.D. speaks of the Christian churches at Constantinople as "temples."

[3] Chief priests, 5; priests, 9; of provinces, 1; priestess, 1; priests of Cybele (*archigalli*), 3; *Asiarchae*, 1; priest of some great goddess, 1; illicit rites, 1. There are 27 passages concerning divination.

hear of men "investigating the secrets of all religions and of heaven itself." Other interesting descriptions [1] are of those who "stay in temples in an unkempt state and always walk abroad thus, and never cut their hair, and who would announce something to men as if said by the gods, such as are wont to be in temples, who are accustomed to predict the future"; and of "men terrible to the gods and who despise all kinds of perjuries. Moreover, they will be terrible to all demons, and at their approach the wicked spirits of demons flee; and they free men who are thus troubled, not by force of words but by their mere appearing; and however violent the demon may be who shakes the body and spirit of man, whether he be aerial or terrestrial or infernal, he flees at the bidding of this sort of man and fears his precepts with a certain veneration. These are they who are called exorcists by the people." Religious games and contests are mentioned four times: the carving, consecrating, adoring, and clothing of images of the gods, twice each; porters at religious ceremonies, thrice; hymn singers, twice; pipe-players once. Five passages represent persons professionally engaged in religion as growing rich thereby.

Divination.

We are told that men "predict the future either by the divinity of their own minds or by the admonition of the gods or from oracles or by the venerable discipline of some art." [2] Augurs, *aruspices,* interpreters of dreams, *mathematici* (astrologers), diviners, and prophets are mentioned. Once Firmicus alludes to false divination but he usually implies that it is a valid art.

Magic as a branch of learning.

From religion and divination we easily pass to the occult arts and sciences, and thence to learning and literature in general, from which occult learning is scarcely distinguished in the *Mathesis.* Magicians or magic arts are mentioned no less than seven times in varied relations with religion, philosophy, medicine, and astronomy or astrology, showing that magic was not invariably regarded as evil in that age, and

[1] Kroll et Skutsch, I, 148, 8 and 123, 4.

[2] Kroll et Skutsch, I, 201, 6.

that it was confused and intermingled with the arts and philosophy as well as with the religion of the times.[1] There are a number of other allusions to secret and illicit arts or writings; these, however, appear to be more unfavorably regarded and probably largely consist of witchcraft and poisoning.

Interest in science.

The evidence of the *Mathesis* suggests that the civilization of declining Rome was at least not conscious of the intellectual decadence and lack of scientific interest so generally imputed to it. We find three descriptions of intellectual pioneers who learn what no master has ever taught them, and one other instance of men who pretend to do so. We also hear of "those learning much and knowing all, also inventors," and of those "learning everything," and "desiring to learn the secrets of all arts." This curiosity, it is true, seems to be largely devoted to occult science, but it also seems plain that mathematics and medicine were important factors in fourth-century culture as well as the rhetorical studies whose rôle has perhaps been overestimated. Let us compare the statistics. Oratory is mentioned eighteen times, and it is to be noted that literary attainments and learning as well as mere eloquence are regarded as essential in an orator. Men of letters other than orators are found in six passages, and poets in only three. A passage reading "philologists or those skilled in laborious letters" suggests that four instances of the phrase *difficiles litterae* should perhaps be classed under linguistic rather than occult studies. There are four allusions to grammarians and two to masters of grammar, as against one description of "contentious, con-

[1] Cumont says (*Oriental Religions in Roman Paganism,* p. 188): "But the ancients expressly distinguished 'magic,' which was always under suspicion and disapproved of, from the legitimate and honorable art for which the name 'theurgy' was invented." This distinction was made by Porphyry and others, and is alluded to by Augustine in the *City of God,* but it is to be noted that Firmicus does not use the word "theurgy." Cumont also states (p. 179) that in the last period of paganism the name philosopher was finally applied to all adepts in occult science. But in Firmicus, while magic and philosophy are associated in two passages, there are five other allusions to magic and three separate mentions of philosophers.

tradictory dialecticians, professing that they know what no teaching has acquainted them with, mischievous fellows, but unable to do any effective thinking.'"[1] On the other hand, there are fourteen allusions to astronomy and astrology (not including the *mathematici* already listed under divination), three to geometry, and six to other varieties of mathematics.[2] Philosophers are mentioned five times; practitioners of medicine, eleven times;[3] surgeons, once; and botanists, twice. These professions seem to be well paid and are spoken of in complimentary terms.

Diseases in antiquity.

Death, injury, and disease loom up large in Firmicus' prospectus for the human race, making us realize the benefits of nineteenth-century medicine as well as of modern peace.[4] No less than 174 passages deal with disease and many of them list two or more ills. Mental disorders are mentioned in 37 places;[5] physical deformities in six. Other specific ailments mentioned are as follows: blindness and eye troubles, 10; deafness and ear troubles, 5; impediments of speech, 4; baldness, 1; foul odors, 1; dyspeptics, 4; other stomach complaints, 7; dysentery, 2; liver trouble, 1; jaundice, 1; dropsy, 5; spleen disorders, 1; gonorrhoea, 2; other diseases of the urinary bladder and private parts, 6; consumption and lung troubles, 6; hemorrhages, 6; apoplexy, 3; spasms, 5; ills attributed to bad or excessive humors, 12; leprosy and other skin diseases, 6; ague, 1; fever, 1; pains in various parts of the body, 6; internal pains and hidden diseases, 9; diseases of women, 5. There remain a large number of vague allusions to ill-health: 21 to debility, 12 to languor, 3 to invalids, and 49 other passages. Only eight passages allude to the cure of disease. Among the methods suggested are cauterizing, incantations, ordinary remedies,

[1] Kroll et Skutsch, I, 161, 26.

[2] *Computus*, 3; *calculus*, 2; and "those who excel at numbers," 1.

[3] Including two mentions of court physicians (*archiatri*). See *Codex Theod.*, Lib. XIII, Tit. 3, *passim*, for their position.

[4] I leave this sentence as I wrote it in 1913.

[5] *Aestus animi*, 5; insanity, 13; lunatics, 10; epileptics, 8; melancholia, 3; inflammation of the brain (*frenetici*), 4; delirium, dementia, demoniacs, alienation, and madness, one or two each; vague allusions to mental ills and injuries, 5.

and seeking divine aid, which last is mentioned most often. The eleven references to medical practitioners should, however, be recalled here. The predictions as to length of life are inadequate to the drawing of conclusions on that point.

Place of Firmicus in the history of astrology.

Firmicus regards his work as a new contribution so far as the Latin-speaking world is concerned.[1] Not that there had not been previous writing in Latin on the subject. Fronto "had written predictions very accurately," but "as if he were addressing persons already perfect and skilled in the art, and without first instructing in the elements and practice of the art." [2] Firmicus supplies this essential preliminary instruction, which hardly anyone of the Latins had given, and corrects Fronto's faulty presentation of *antiscia,* in which he followed Hipparchus, by the correcter method of Navigius (Nigidius?) and Ptolemy.[3] Firmicus gives no systematic account of his authorities [4] but occasionally cites them for some particular point and in general professes to follow not only the Greeks but the divine men of Egypt and Babylon, chief among whom seem to be Nechepso and Petosiris and the Hermetic works to or by Aesculapius and Hanubius. An Abram or Abraham is also cited several times. But Firmicus also gives the *Sphaera Barbarica,* "unknown to all the Romans and to many Greeks," and which escaped the notice even of Petosiris and Nechepso.[5] Firmicus himself is named by no ancient author [6] but was well known in the eleventh and twelfth centuries, as we shall see. In the *Mathesis* he cites two previous astrological treatises of his

[1] In his last chapter he says, "Take then, my dear Mavortius, what I promised you with extreme trepidation of spirit, these seven books composed conformably to the order and number of the seven planets. For the first book deals only with the defense of the art; but in the other books we have transmitted to the Romans the discipline of a new work," (II, 360, 10-15). And in the introduction to the fifth book he writes, "We have written these books for your Romans lest, when every other art and science had been translated, this task should seem to remain unattempted by Roman genius," (I, 280, 28-30).

[2] I, 41, 7 and 15; I, 40, 9-11.

[3] I, 41, 5 and 11; I, 40, 8.

[4] They are listed by Kroll et Skutsch, II, 362, *Index auctorum.*

[5] II, 294, 12-21.

[6] Kroll et Skutsch, II, p. iii.

own [1] and expresses his intention of composing another work in twelve books on the subject of *Myriogenesis*.[2] The astrologer Hephaestion of Thebes, who wrote later in the fourth century, seems also to have been a Christian, so that Firmicus was not a solitary case or an anomaly.[3]

Libanius accused of magic.

The writings of Libanius, 314-391 A. D., the sophist and rhetorician, throw some light on the relations between magic and learning in the fourth century, show that sorcery and divination were actually practiced, and largely duplicate impressions already received from Apuleius, Apollonius, and Galen, and a Christian like John Chrysostom as well as just now from Firmicus. Libanius tells us how Bemarchius, a rival of his at Athens, who would have poisoned him if he could, instead circulated reports that he (Bemarchius) was the victim of enchantments, and that Libanius had consulted against him an astrologer who was able to control the stars, so that he could confer benefits upon one man and work sorcery against another. This incidentally is another good illustration of how easily astrology passed from mere prediction of the future to operative magic, and of the essential unity of all magic arts. The mob was aroused against Libanius and a praetor who tried to protect him was ousted and another installed at daybreak who was ready to put Libanius to death. Torture was prepared and Libanius was advised to leave Athens, if he did not wish to die there, and took the advice and left.[4]

Declamation against a magician.

Among the declamations of Libanius is one against a magician,[5] supposed to have been delivered under the following circumstances. The city was afflicted with a pesti-

[1] I, 258, 10, "in singulari libro, quem de domino geniturae et chronocratore ad Murinum nostrum scripsimus"; II, 229, 23, "ex eo libro qui de fine vitae a nobis scriptus est."

[2] II, 18, 24; II, 283, 19.

[3] Engelbrecht, *Hephästion von Theben und sein astrologisches Compendium*, Vienna, 1887.

[4] *De vita sua*, in *Libanii sophistae praeludia oratoria LXXII declamationes XLV et dissertationes morales, Federicus Morellus regius interpres e MSS maxime reg. bibliothecae nunc primum edidit idemque Latine vertit . . . ad Henricum IV regem Christianissimum*, Paris, 1606, II, 15-18.

[5] *Magi accusatio, Ibid.*, I, 898-911.

lence and finally sent an embassy to the Delphic oracle to learn how to escape the scourge. Apollo replied that they must sacrifice the son of one of the inhabitants who should be determined by lot, and the lot fell to the son of a magician. The father then offered to stay the plague by means of his magic art, if they would agree to spare his son. Against this proposal Libanius argues, urging the people to carry out their original decision and not to anger the Delphic god by violating his oracle, whose reliability is attested by "long time and much experience and common testimony." He declares that magic is an evil art, and that magicians make no one happy but many wretched, ruining homes, bringing disaster to persons who have never harmed them, and disturbing even the spirits of the dead. He also censures the magician for not having offered to save the city from the plague before, and expresses some scepticism as to his magic power, asking why he did not prevent the fatal lot from falling to his son, or why he does not save him now by causing him to vanish from sight, or vouchsafe some other unmistakable sign of his magic power. It appears that the magician had asked a delay, saying that he must wait for the moon before he could operate against the plague. Libanius points out that meanwhile the citizens are perishing and that fulfillment of Apollo's oracle will bring instant relief. It would seem, however, that some of the citizens had more faith in the magician than in the god, which supports the oft-made general assertion that the magic arts waxed as pagan religion and its superstitious observances waned. Libanius concludes his oration or imaginary oration with the cutting and heartless witticism that the magician can lose his son more easily than can anyone else, since he will of course still be able to invoke his spirit from the dead.

Faith of Libanius in divination.

Libanius' own faith in divination is not only suggested by the attitude toward the Delphic oracle in the foregoing declamation but is attested by two passages in his autobiography. His great-great-grandfather had so excelled in *mantike* that he foresaw that his children would die by steel, although they would be handsome and great and good speakers. It also was rumored that a celebrated sophist had predicted many things concerning Libanius himself, which Libanius assures us had since come to pass.[1]

Magic and astrology in the pseudo-Quintilian declamations.

Of the same type as Libanius' declamation against the magician is the fourth pseudo-Quintilian declamation in Latin concerning an astrologer's prediction, which we shall later in the twelfth century find Bernard Silvester enlarging upon in his poem entitled *Mathematicus*. In another of the pseudo-Quintilian declamations the word *experimentum* is used of a magician's feat. "O harsh and cruel magician, O manufacturer of our tears, I would that you had not given so great an experiment! We are angry at you, yet we must cajole you. While you imprison the ghost, we know that you alone can evoke it." [2]

Fusion of Christianity and paganism in Synesius of Cyrene.

That more than fifty years after Firmicus adherence to Christianity might be combined with trust in divination of the future, occult science, and magical invocation of spirits, and with various other pagan and Neo-Platonic beliefs, is well illustrated by the case of Synesius of Cyrene,[3] a fellow-African and contemporary of Augustine. Synesius, however, traced his descent from the Heracleidae, wrote in Greek, and displayed a Hellenism unusual for his time,[4] and,

[1] *De vita sua, Opera,* II, 2-3.

[2] X, 196, 11, *De sepulcro incantato.*

[3] My citations of Synesius' works, unless otherwise noted, are from the edition: *Synesii Cyrenaei Quae Extant Opera Omnia,* ed. J. G. Krabinger, Landshut, 1850, vol. I, which has alone appeared. The older edition of Petavius with Latin translation is reprinted in Migne PG, vol. 66, 1021-1756. For a French translation, with several introductory essays, see H. Druon, *Œuvres de Synésius,* Paris, 1878. The *Letters* and *Hymns* have often been published separately. For this and other further bibliography see Christ, *Gesch. d. griech. Litt.,* 1913, II, ii, 1167-71, where, however, no note is taken of Berthelot's discussion of Synesius as a reputed author of alchemistic treatises.

Some works on Synesius are: H. Druon, *Études sur la vie et les œuvres de Synésius,* Paris, 1859; R. Volkmann, *Synesius von Cyrene,* Berlin, 1869; W. S. Crawford, *Synesius the Hellene,* London, 1901; G. Grützmacher, *Synesios von Kyrene,* Leipzig, 1913. In periodicals: F. X. Kraus in *Theol. Quartalschrift,* 1865 and 1866; O. Seeck, in *Philologus,* 1893.

[4] See Crawford, *op. cit.,* and monographs listed in Christ, *op. cit.,* p. 1168, notes 4 and 8.

while he did not find the Athens of his day entirely to his taste, continued the philosophical and rhetorical traditions of the sophists of the Roman Empire, like Libanius of whom we have just spoken. His extant letters show that Hypatia was numbered among his friends and had been his teacher at the Neo-Platonic and mathematical school of Alexandria. Hypatia was murdered by the fanatical Christian mob of that city in 415. But very different was the attitude of the people of Ptolemais to the like-minded Synesius. A few years before they had elected him bishop! [1] Moreover, he distinctly stipulated [2] that he should not renounce his wife and family nor his philosophical opinions, which seem to have involved a sceptical attitude towards miracles and the resurrection, and a belief in the eternity of the world and pre-existence of the soul rather than in creation,[3] in addition to the views which we are about to set forth. It has been observed also that his doctrine of the Trinity is more Neo-Platonic than Christian.[4]

Career of Synesius.

The dates of Synesius' birth and death are uncertain. He seems to have been born about 370. His last dateable letter appears to be written in 412, but some give the date of his death as late as 430. Others contend that he did not live to hear of Hypatia's murder. Before he was made bishop he had been to Constantinople on a mission to the emperor to secure alleviation of the oppressive taxation in Cyrene. He had lived in Athens and Alexandria as a student, and in Cyrene on his country estate. Here, if in his fondness for books and philosophy he constituted a survival of the past, in his fondness for the chase and dogs and horses and his repulsion of an invasion of Libyan marauders he was the forerunner of many a medieval feudal

[1] The date is variously stated as 411, 406, or 410.

[2] A. J. Kleffner, *Synesius von Cyrene . . . und sein angeblicher Vorbehalt bei seiner Wahl und Weihe zum Bischof von Ptolemais,* Paderborn, 1901. H. Koch, *Synesius von Cyrene bei seiner Wahl und Weihe zum Bischof,* in *Hist. Jahrb.*, XXIII (1902), pp. 751-74.

[3] Christ, *op. cit.*, p. 1168, note 1.

[4] *Ibid.*, p. 1170, citing K. Prächter, in *Genethliakon für C. Robert,* 1910, p. 244, *et seq.*

bishop. And after he became bishop, he launched an excommunication against the tyrannical prefect Andronicus.

His interest in science.

But our particular interest is less in his political and more purely literary activities than in his taste for mathematics and science. He knew some medicine and was well acquainted with geometry and astronomy. He believed himself to be the inventor of an astrolabe and of a hydroscope.

Belief in occult sympathies between natural objects.

With this interest in natural and mathematical science went an interest in occult science and divination. His belief that the universe was a unit and all its parts closely correlated not only led him to maintain, like Seneca, that whatever had a cause was a sign of some future event, or to hold with Plotinus that in any and every object the sage might discern the future of every other, and that the birds themselves, if endowed with sufficient intelligence, would be able to predict the future by observing the movements of human bipeds.[1] It led him also to the conclusion that the various parts of the universe were more than passive mirrors in which one might see the future of the other parts; that they further exerted, by virtue of the magic sympathy which united all parts of the universe, a potent active influence over other objects and occurrences. The wise man might not only predict the future; he might, to a great extent, control it. "For it must be, I think, that of this whole, so joined in sympathy and in agreement, the parts are closely connected as if members of a single body. And does not this explain the spells of the magi? For things, besides being signs of each other, have magic power over each other. The wise man, then, is he who knows the relationships of the parts of the universe. For he draws one object under his control by means of another object, holding what is at hand as a pledge for what is far away, and working through sounds and material substances and forms." [2] Synesius explained that plants

[1] Περὶ ἐνυπνίων (*On dreams*), ch. 2.

[2] Περὶ ἐνυπνίων (*On Dreams*), ch. 3. Ἔδει γάρ, οἶμαι, τοῦ παντὸς τούτου συμπαθοῦς τε ὄντος καὶ σύμπνου τὰ μέρη προσήκειν ἀλλήλοις, ἅτε ἑνὸς ὅλου τὰ μέλη τυγχάνοντα. Καὶ μή ποτε αἱ μάγων ἴυγγες αὗται; καὶ γὰρ θέλγεται παρ' ἀλλήλων, ὥσπερ σημαίνεται· καὶ σοφὸς ὁ εἰδὼς τὴν τῶν μερῶν τοῦ κόσμου συγγένειαν. Ἕλκει γὰρ ἄλλο δι' ἄλλον, ἔχων ἐνέχυρα παρόντα τῶν πλεῖστον ἀπόντων, καὶ φωνὰς, καὶ ὕλας

and stones are related by bonds of occult sympathy to the gods who are within the universe and who form a part of it, that plants and stones have magic power over these gods, and that one may by means of such material substances attract those deities.[1] He evidently believed that it was quite legitimate to control the processes of nature by invoking demons.

Synesius on divination and astrology.

The devotion of Synesius to divination has been already implied. He regarded it as among the noblest of human pursuits.[2] Dreams, on which he wrote a treatise, he viewed as significant and very useful events. They aided him, he wrote, in his every-day life, and had upon one occasion saved him from magic devices against his life.[3] Warned by a dream that he would have a son, he wrote a treatise for the child before it was born.[4] Of course, he had faith in astrology. The stars were well-nigh ever present in his thought. In his *Praise of Baldness* he characterized comets as fatal omens, as harbingers of the worst public disasters.[5] In *On Providence* he explained the supposed fact that history repeats itself by the periodical return to their former positions of the stars which govern our life.[6] In *On the Gift of an Astrolabe* he declared that "astronomy" besides being itself a noble science, prepared men for the diviner mysteries of theology.[7]

Synesius as an alchemist.

Finally, he held the view common among students of magic that knowledge should be esoteric; that its mysteries and marvels should be confined to the few fitted to receive them and that they should be expressed in language incomprehensible to the vulgar crowd.[8] It is perhaps on this

καὶ σχήματα . . . . . Evidently Synesius did not regard the magi as mere imposters.

[1] Περὶ ἐνυπνίων, ch. 3. Καὶ δὴ καὶ θεῷ τινὶ τῶν εἴσω τοῦ κόσμου λίθος ἐνθένδε καὶ βοτάνη προσήκει, οἷς ὁμοιοπαθῶν εἴκει τῇ φύσει καὶ γοητεύεται. In his *Praise of Baldness* (Φαλάκρας ἐγκώμιον), ch. 10, Synesius tells how the Egyptians attract demons by magic influences.

[2] Περὶ ἐνυπνίων, ch. 1. Αὗται μὲν ἀποδείξεις ἔστων τοῦ μαντείαν ἐν τοῖς ἀρίστοις εἶναι τῶν ἐπιτηδευομένων ἀνθρώποις.

[3] *Ibid.*, ch. 18.

[4] Δίων ἢ περὶ τῆς κατ' αὑτὸν διαγωγῆς.

[5] Φαλάκρας ἐγκώμιον, ch. 10.

[6] Αἰγύπτιοι ἢ περὶ προνοίας, bk. ii, ch. 7.

[7] Πρὸς Παιόνιον περὶ τοῦ δώρου, ch. 5.

[8] Δίων, ch. 7. Περὶ ἐνυπνίων, ch. 4. Ἐπιστολαί, 4, 49, and 142.

account that one of the oldest extant treatises of Greek alchemy is ascribed to him. Berthelot, however, accepted it as his, stating that "there is nothing surprising in Synesius' having really written on alchemy." [1]

Macrobius on number, dreams, and stars.

Synesius influenced the Byzantine period but probably not the western medieval world. But the Commentary of Macrobius on *The Dream of Scipio* by Cicero is one of the treatises most frequently encountered in early medieval Latin manuscripts. In the twelfth century Abelard made frequent reference to Macrobius and called him "no mean philosopher"; in the thirteenth Aquinas cited him as an authority for the doctrines of Neo-Platonism.[2] Macrobius himself affirmed that Vergil contained practically all necessary knowledge [3] and that Cicero's *Dream of Scipio* was a work second to none and contained the entire substance of philosophy.[4] Macrobius believed that numbers possess occult power. He dilated at considerable length upon every number from one to eight, emphasizing the perfection and far-reaching significance of each. He held the Pythagorean doctrine that the world-soul consists of number, that number rules the harmony of the celestial bodies, and that from the music of the spheres we derive the numerical values proper to musical consonance.[5] His opinion was that dreams and other striking occurrences will reveal an occult meaning to the careful investigator.[6] As for astrology, he regarded the stars as signs but not causes of future events, just as birds by their flight or song reveal matters of which they themselves are ignorant.[7] So the sun and other planets, though in a way divine, are but material bodies, and it is not from them but from the world-soul (pure mind), whence they too come, that the human spirit takes its origin.[8] In

[1] On Synesius as an alchemist see Berthelot (1885), pp. 65, 188-90; (1889), p. ix.

[2] T. R. Glover, *Life and Letters in the Fourth Century A. D.*, Cambridge, 1901, p. 187, note 1.

[3] *Saturnalia*, I, xvi, 12.

[4] *Commentary on the Dream of Scipio*, II, 17, "Universa philosophiae integritas"; ed. Nisard, Paris, 1883.

[5] *Ibid.*, I, 5-6; II, 1-2.

[6] *Ibid.*, I, 7.

[7] *Ibid.*, I, 19.

[8] *Ibid.*, I, 14.

his sole other extant work, the *Saturnalia,* Macrobius displays some belief in occult virtues in natural objects, as when Disaurius the physician answers such questions as why a copper knife stuck in game prevents decay.[1]

Martianus Capella.

The medieval vogue of the fifth century work of Martianus Capella, *The Nuptials of Philology and Mercury, and the Seven Liberal Arts,*[2] has been too frequently demonstrated to require further emphasis here, although it is still a puzzle just why a monastic Christian world should have selected for a text book in the liberal arts a work which contained so much pagan mythology, to say nothing of a marriage ceremony. Nor need we repeat its fulsome allegorical plot and meager learned content. Cassiodorus tells us that the author was a native of Madaura, the birth-place of Apuleius, in North Africa, and he appears to be a Neo-Platonist who has much to say of the sky, stars, and old pagan gods, often, however, by way of brief and vague poetical allusion.

Absence of astrology.

Of astrology there is very little trace in Capella's work. In a discussion of perfect numbers in the second book the number seven evokes allusion to the fatal courses of the stars and their influence upon the formation of the child in the womb; but the eighth book, which is devoted to the theme of astronomy as one of the liberal arts, is limited to a purely astronomical description of the heavens.

Orders of spirits.

The chief thing for us to note in the work is the account of the various orders of spiritual beings and their respective location in reference to the heavenly bodies.[3] Juno leads the virgin Philology to the aerial citadels and there instructs her in the multiplicity of diverse powers. From highest ether to the solar circle are beings of a fiery and flaming substance. These are the celestial gods who prepare the secrets of occult causes. They are pure and impassive and immortal and have little or no direct relation with mankind. Be-

[1] Glover (1901), p. 178.

[2] *De nuptiis philologiae et mercurii et de septem artibus liberalibus libri novem, Lugduni apud haeredes Simonis Vincentii,* 1539; ed. U. F. Kopp, Frankfurt, 1836; ed. F. Eyssenhardt, Leipzig, 1866.

[3] It occurs toward the close of the second book.

tween sun and moon come spirits who have especial charge of soothsaying, dreams, prodigies, omens, and divination from entrails and auguries. They often utter warning voices or admonish those who consult their oracles by the course of the stars or the hurling of thunderbolts. To this class belong the Genii associated with individual mortals and angels "who announce secret thoughts to the superior power." All these the Greeks call demons. Their splendor is less lucid than that of the celestials, but their bodies are not sufficiently corporeal to enable men to see them. Lares and purer human souls after death also come under this category. Between moon and earth the spirits subdivide into three classes. In the upper atmosphere are demi-gods. "These have celestial souls and holy minds and are begotten in human form to the profit of the whole world." Such were Hercules, Ammon, Dionysus, Osiris, Isis, Triptolemus, and Asclepius. Others of this class become sibyls and seers. From mid-air to the mountain-tops are found heroes and Manes. Finally the earth itself is inhabited by a long-lived race of dwellers in woods and groves, in fountains and lakes and streams, called Pans, Fauns, satyrs, Silvani, nymphs, and by other names. They finally die as men do, but possess great power of foresight and of inflicting injury.[1] It is evident that Capella's spiritual world is one well fitted for astrology, divination, and magic.

The Celestial Hierarchy of Dionysius the Areopagite.

Very different are the orders of spirits described in *The Celestial Hierarchy,* supposed to be the work of Dionysius the Areopagite, where are set forth nine orders of spirits in three groups of three each: Seraphim, Cherubim, and Thrones; Dominions, Virtues, and Powers; Princes, Archangels, and Angels. The threefold division reminds us of Capella, but there the resemblance ceases. The pseudo-Dionysius takes all his suggestions from the Old and New Testaments, rather than from classical mythology and such previous classifications of spirits as that of Apuleius. And

[1] In Kopp's edition pp. 202-23 are almost entirely taken up with notes setting forth other passages in the classics concerning such spirits.

while his starting from such verses of the Bible as "Every good gift and every perfect gift is from above, descending from the Father of lights," and "Jesus Christ the true light that lighteth every man that cometh into the world," and his using such phrases as "archifotic Father" and "thearchic ray," lead us to expect some Gnostic-like scheme of association of the spirits with the various heavens and celestial bodies, in fact he throughout speaks of the spirits solely as celestial and deiform and hypercosmic *minds,* and unspeakable and sacred enigmas of whose invisibility, transcendence, infinity, and incomprehensibility any description can be merely symbolic and figurative. Their functions seem to consist chiefly in contemplation of the deity or their superior orders and illumination of man and their inferior orders. They are not specifically associated by Dionysius with the celestial bodies, much less with any terrestrial objects, and so his account lays no foundation for magic and astrology, unless as its transcendent mysticism might pique some curious person to attempt some very immaterial variety of theurgy and sublimated theosophy. Although the Pseudo-Dionysius wrote in Greek,[1] his work was made available for the Latin middle ages by the translation of John the Scot in the ninth century.[2]

[1] Greek text in Migne, PG 3, 119-370.
[2] Migne, PL 122, 1037-70.

BOOK III. THE EARLY MIDDLE AGES

CHAPTER XXIV

THE STORY OF NECTANEBUS
OR
THE ALEXANDER LEGEND IN THE EARLY MIDDLE AGES [1]

The *Pseudo-Callisthenes*—Its unhistoric character—Julius Valerius —Oriental versions—Medieval epitomes of Julius Valerius—Letters of Alexander—Leo's *Historia de praeliis*—Medieval metamorphosis of ancient tradition—Survival of magical and scientific features—Who was Nectanebus?—A scientific key-note—Magic of Nectanebus—Nectanebus as an astrologer—A magic dream—Lucian on Olympias and the serpent —More dream-sending; magic transformation—An omen interpreted—The birth of Alexander—The death of Nectanebus—The Amazons and Gymnosophists—*The Letter to Aristotle.*

The Pseudo-Callisthenes.

THE oldest version of the legend or romance of Alexander is naturally believed to have been written in the Greek language but is thought to have been produced in Egypt at Alexandria. But the Greek manuscripts of the story are

[1] The following bibliography includes the editions of the texts concerned and the chief critical researches in the field. A. Ausfeld, *Zur Kritik des griechischen Alexanderromans; Untersuchungen über die unechten Teile der ältesten Ueberlieferung,* Karlsruhe, 1894. A. Ausfeld and W. Kroll, *Der griechische Alexanderroman,* Leipzig, 1907. H. Becker, *Die Brahmannen in der Alexandersage,* Königsberg, 1889, 34 pp. E. A. W. Budge, *History of Alexander the Great,* Cambridge University Press, 1889; the Syriac version of the *Pseudo-Callisthenes* edited from five MSS, with an English translation and notes. E. A. W. Budge, *The Life and Exploits of Alexander the Great,* Cambridge University Press, 1896; Ethiopic Histories of Alexander by the Pseudo-Callisthenes and other writers. D. Carrarioli, *La leggenda di Alessandro Magno,* 1892. G. G. Cillié, *De Iulii Valerii epitoma Oxoniensi,* Strasburg, 1905. G. Favre, *Recherches sur les histoires fabuleuses d'Alexandre le Grand,* in *Mélanges d'hist. litt.,* II (1856), 5-184. Ethé, *Alexanders Zug zur Lebensquelle im Lande der Finsterniss,* in *Atti dell' Accademia di Monaco,* 1871. B. Kübler, *Julius Valerius; Res gestae Alexandri Macedonis,* Leipzig, 1888 (see pp. xxv-xxvi for further bibliography). Levi, *La légende d'Alexandre dans le Talmud,* in *Revue des Études juives,* I (1880), 293-300. Meusel, *Pseudo-Callisthenes nach der Leidener Handschrift herausgegeben,* Leipzig, 1871. M. P. H. Meyer, *Alexandre le Grand dans la littérature française du moyen âge,* 2 vols., Paris,

all of the medieval or Renaissance period; indeed, none of them antedates the eleventh or twelfth century. Furthermore, they differ very considerably in content and arrangement, so that the problem of distinguishing or recovering the original text of the *Pseudo-Callisthenes,* as the work is commonly called, and of dating it, is one with which various scholars have grappled. It has been held that the original Greek text which lies back of the later versions was written not later than 200 A. D. But Basil, writing in Greek in the fourth century and well-versed in Greek culture, is apparently unfamiliar with the story of Nectanebus, since he says, "Without doubt there has never been a king who has taken measures to have his son born under the star of royalty." [1] Fortunately we are less interested in the original version than in the medieval development of the tradition. It should, however, perhaps be premised that certain features of the Alexander legend may be detected in embryo in Plutarch's *Life* of him.

1886. C. Müller, *Scriptores rerum Alexandri Magni,* Firmin-Didot, Paris, 1846 and 1877 (bound with Arrian, ed. Fr. Dübner); the first edition of the Greek text of the *Pseudo-Callisthenes* from three Paris MSS, also Julius Valerius, etc. Noeldeke, *Beiträge zur Geschichte des Alexanderromans, Denkschriften der Kaiserlichen Akademie der Wissenschaften in Wien, Philos. Hist. Classe,* vol. 38, Vienna, 1890; Budge says of this work, "Professor Noeldeke discusses in his characteristic masterly manner the Greek, Syriac, Hebrew, Persian, and Arabic versions, and ably shows how each is related to the other, and how certain variations in the narrative have arisen. No other writer before him was able to control, by knowledge at first hand, the statements of both the Aryan and Semitic versions; his work is therefore of unique value." *Padmuthiun Acheksandri Maketonazwui, I Wenedig i dparani serbuin Chazaru,* Hami, 1842; the Armenian version published by the Mechitarists, Venice, 1842. F. Pfister, *Kleine Texte zum Alexanderroman,* Heidelberg, 1910; *Sammlung vulgärlateinischer Texte herausg. v. W. Heraeus u. H. Morf, 4 Heft.* Spiegel, *Die Alexandersage bei den Orientalen,* Leipzig, 1851. Vogelstein, *Adnotationes quaedam ex litteris orientalibus petitae quae de Alexandro Magno circumferuntur,* Warsaw, 1865. A. Westermann, *De Callisthene Olynthio et Pseudo-Callisthene Commentatio,* 1838-1842. J. Zacher, *Pseudo-Callisthenes: Forschungen zur Kritik und Geschichte der ältesten Aufzeichnung der Alexandersage,* Halle, 1867 (see pp. 2-3 for further bibliography of works written before 1851). J. Zacher, *Julii Valerii Epitome, zum ersten mal herausgegeben,* Halle, 1867.

[1] *Hexaemeron,* VI, 7. On the other hand, Augustine, *De civitate dei,* V, 6-7, alludes to the sage who selected a certain hour for intercourse with his wife in order that he might beget a marvelous son.

The true Callisthenes was a historian who accompanied Alexander upon his Asiatic campaigns but then offended the conqueror by opposing his adoption of oriental dress, absolutism, and deification, and was therefore cast into prison on a charge of treason, and there died in 328 B. C. either from ill treatment or disease.[1] Since Callisthenes was also a relative and pupil of Aristotle, his name was an excellent one upon which to father the romance. However, the oldest Latin version of it professes to employ a Greek text by one Aesopus, possibly because Aesop's fables accompany the story of Alexander in some of the manuscripts. Yet other versions cite an Onesicritus,[2] and the *Pseudo-Callisthenes* has also been attributed to Antisthenes, Aristotle, and Arrian.

Its unhistoric character.

Perhaps no better single illustration of the totally unhistorical and romantic character of the *Pseudo-Callisthenes* can be given than the perversion of Alexander's line of march in most of the Greek and all of the Latin versions. He is represented as first proceeding to Italy and receiving royal honors at Rome; then he goes to Carthage and reaches the shrine of Ammon by traversing Libya; next he passes through Egypt into Syria and destroys Tyre, after which he crosses Arabia and has his first battle with Darius. Presently he is found back in Greece sacking Thebes and dealing with Corinth, Athens, and Sparta. Then his Asiatic conquests are resumed.

Julius Valerius.

The oldest Latin version of the Alexander romance is the *Res gestae Alexandri Macedonis* of Julius Valerius. Who he was and when he lived are matters still veiled in obscurity; but it is customary to place him in the early fourth century on the basis of Zacher's contention that the *Res gestae* is copied in certain portions of the *Itinerarium Alexandri,* which was written during the years 340-345 A. D. This

[1] Seneca in the *Natural Questions* (VI, 23) called the death of Callisthenes "the eternal crime" of Alexander which all his military victories and conquests could not outweigh,—a passage which did not keep Nero from forcing Seneca to commit suicide.

[2] Reitzenstein, *Poimandres,* Leipzig, 1904, pp. 308-309.

dating would also serve to explain why Basil, writing in Greek before 379, had never heard of a king who had taken steps to have his son born under the star of royalty, while Augustine, writing in Latin between 413 and 426, mentions the story of a sage who selected a certain hour for intercourse with his wife in order that he might beget a marvelous son. This would also suggest that the Latin version was older than the Greek, as in fact the extant manuscripts of it are. The oldest manuscript of Valerius, however, is a badly damaged palimpsest of the seventh century at Turin. Other manuscripts are one at Milan of the tenth century and another at Paris dating about 1200.[1] The text of Valerius differs considerably from the Greek *Pseudo-Callisthenes* and was to undergo further alteration in later medieval Latin versions.

Oriental versions.

Before speaking of these we may mention other oriental versions of the story. An Armenian text dates from the fifth century. A Syriac version, which dates from the seventh or eighth century and was "much read by the Nestorians," was itself derived from an earlier Persian rendering. It seems to make use of both the Greek Pseudo-Callisthenes and Julius Valerius since it includes incidents from either which are not found in the other. And it omits a considerable section of the Greek version besides adding episodes which are not found in it, although contained in Julius Valerius. We hear further of Arabic and Hebrew versions of the romance, while manuscripts of recent date supply an Ethiopic version of the *Pseudo-Callisthenes* of unknown authorship and date, together with other Ethiopic histories and romances of Alexander. These are based partly upon Arabic and Jewish works but take great liberties with their sources in making alterations to suit a Christian audience, omitting for example, as Budge points out, Alexander's vic-

[1] *Res gestae* of Alexander of Macedon, contained in three MSS of the Royal Library in the British Museum, dating according to the catalogue from the eleventh and twelfth centuries: Royal 13-A-I, Royal 12-C-IV, and Royal 15-C-VI, are not the full text of Julius Valerius, but the epitome of which I shall soon speak.

tory in the chariot race, and transforming Philip and Alexander into Christian martyrs, or the Greek gods into patriarchs and prophets like Enoch and Elijah. Even the Greek version did not remain unaltered in the Byzantine period when two recensions in prose and two more in verse are distinguished. Indeed, none of the Greek manuscripts of the work antedates the eleventh or twelfth century, they differ greatly, and some of them ascribe the romance to Alexander himself.

Medieval epitomes of Julius Valerius.

Such variations in the eastern versions of the story of Alexander illustrate how the middle ages made the classical heritage their own and prepare us for similar alterations in the Latin account current in western Europe. The work of Julius Valerius, though written in the rhetorical style characteristic of the declining Roman Empire and composed almost on the verge of the middle ages, was to undergo further alterations to adapt it more closely to medieval taste and use. By the ninth century, if not earlier, two epitomes of it had been made, and, beginning with that century, manuscripts of the shorter of these epitomes become far more numerous than those of the original Valerius.[1]

Letters of Alexander.

Two sections of the Alexander legend were omitted in the Epitome, not because medieval men had lost interest in them but because they had become so fond of them as to enlarge upon them and issue them as distinct works. They often, however, accompany the Epitome in the manuscripts. One of these was the Letter of Alexander to Aristotle on the Marvels of India.[2] It is longer than the corresponding

[1] The longer epitome is known from an Oxford MS, Corpus Christi MS 82, and was believed by Meyer to be intermediary between Valerius and the other briefer epitome. Cillié, however, tries to prove the shorter epitome to be the older.

[2] *Alexandri Magni Epistola ad Aristotelem de mirabilibus Indiae*, first printed with *Synesii Epistolae, graece; adcedunt aliorum Epistolae*, Venice, 1499; then Bologna, 1501; Basel, 1517; Paris, 1520, fols. 102v-14v, following the Pseudo-Aristotle, *Secret of Secrets;* etc. These early printed editions give the oldest Latin text, dating back as we have seen to at least 800.

Some MSS of the same version are:

BM Royal 13-A-I, fols. 51v-78r, a beautifully clear MS of the late 11th century with clubbed strokes. The Epistola is preceded by the

chapter of Valerius[1] where a letter of Alexander to Aristotle is quoted and also differs from any known Greek text. The fact that reference is made to it in the longer Epitome leads to the conclusion that the Letter is older. This would also seem to be the case with the other work, a short series of letters interchanged between Alexander and Dindimus, the king of the Brahmans, since the Epitome omits the two chapters of Valerius which tell of Alexander's interview with the Brahmans. It is believed that Alcuin, who died in 804, in one of his letters to Charlemagne speaks of sending these epistles exchanged between Alexander and Dindimus along with the equally apocryphal correspondence of the apostle Paul and the philosopher Seneca. No such letters are found in the *Pseudo-Callisthenes,* for the ten chapters on the Brahmans found in one Greek codex are interpolated from the treatise of Palladius, likewise in the form of a correspondence.[2] Julius Valerius does not even mention Dindimus, but a third epistolary discussion of the Brahmans exists in Latin, *De moribus Brachmannorum,* ascribed to St. Ambrose.[3]

Epitome of Valerius and followed by the correspondence with Dindimus.

Royal 12-C-IV, 12th century.

Royal 15-C-VI, 12th century.

Cotton Nero D VIII, fol. 169.

Sloane 1619, 13th century, fols. 12-17.

Arundel 242, 15th century, fols. 160-83.

BL Laud. Misc. 247, 12th century, fol. 186; preceded at fol. 171 by the "Ortus vita et obitus Alexandri Macedonis," and followed at fol. 196v by the letter to Dindimus.

BN MSS 2874, 4126, 4877, 4880, 5062, 6121, 6365, 6503, 6831, 7561, 8518, 8521A, *Epistola de itinere et situ Indiae;* 8607, *Epistolae eius nomine scriptae;* and 2695A, 6186, 6365, 6385, 6811, 6831, 8501A, for *Responsio ad Dindimum.*

CLM 11319, 13th century, fol. 88, *Alexandri epistola ad Aristotelem de rebus in India gestis,* preceded at fol. 72 by the *Epitome* and followed at fol. 97 by the *Dindimus.*

In the library of Eton College an imperfect copy of the *Epistola* follows *Orosius* in a MS of the early 13th century, 133, BL 4, 6, fols. 85r-87.

A somewhat different and later version of the *Letter to Aristotle* was published in 1910 at Heidelberg by Friedrich Pfister from a Bamberg MS of the 11th century, together with *Palladius* and the correspondence with Dindimus. Pfister believed all these to be translations from the Greek.

An Anglo-Saxon version of the *Letter to Aristotle* was edited by Cockayne in 1861 (see T. Wright, RS 34; xxvii).

[1] III, 17.

[2] First published by Joachim Camerarius about 1571.

[3] Published with *Palladius* by Sir Edward Bisse in 1665; MSS are numerous.

Leo, an archpriest of Naples, who went to Constantinople about 941-944 on an embassy for two dukes of Campania, John and Marinus, brought back with him a *History containing the conflicts and victories of Alexander the Great, King of Macedon*. Later Duke John, who was fond of science, had Leo translate this work from Greek into Latin, in which tongue it is entitled *Historia de praeliis*. We learn these facts from its prologue which is found only in the oldest extant manuscript, a Bamberg codex of the eleventh century,[1] and in a manuscript of the twelfth or thirteenth century at Munich. The location of these two manuscripts suggests that the work was early carried from Italy to Germany, lands then connected in the Holy Roman Empire. Of the *De praeliis* apart from the prologue there came to be many copies, but most of them date from the later middle ages, and the importance of the work as a source for the vernacular romances of Alexander has been somewhat overestimated, since Meyer has shown that no manuscript of it is found in France until the thirteenth century and since the manuscripts of the Epitome are far more numerous.[2]

Leo's *Historia de praeliis.*

In the foregoing observations we may seem to have digressed too far from our main theme of science and magic into the domain of literary history. But the development of the Alexander legend, which happens to have been traced more thoroughly than perhaps any other one thread in the medieval metamorphosis of ancient tradition, throws light at least by analogy upon many matters in which we are interested: the state of medieval manuscript material, the continuity and yet the alteration of ancient culture during the early middle ages, the process of translation from the Greek which went on even then, and the varying rapidity or slowness with which books circulated and ideas permeated.

Medieval metamorphosis of ancient tradition.

[1] From this same MS Pfister published the *Letter to Aristotle* and other treatises mentioned above.

[2] Its influence would therefore seem to have been upon the later prose romances and not upon French vernacular poetry. Known at first only in Italy and Germany, its popularity became general in western Europe toward the close of the middle ages.

Survival of magical and scientific features.

Moreover, the story of Alexander, especially as adapted by the middle ages, contained a large amount of magic and science, more especially the former. The Epitome might omit a great deal else, but it kept intact the opening portion of the *Pseudo-Callisthenes* and of Julius Valerius concerning the adventures of Nectanebus, the sage and magician from Egypt, the astrologer and the natural father of Alexander. Indeed, the titles in some manuscripts suggest that Nectanebus came to rival Alexander for medieval readers as the hero of the story. Thus we find a *History of Alexander, King of Macedon, and of Nectanebo, King of Egypt,*[1] or an account *Of the Life and Deeds of Neptanabus, astronomer of Egypt,*[2] or a Latin metrical version by "Uilikinus" or Aretinus Quilichinus of Spoleto in 1236 entitled, *The History of the Science of the Egyptians and of Neptanabus their king who afterwards was the true father of Alexander.*[3]

Who was Nectanebus?

Pliny in the *Natural History* describes the obelisk of Necthebis, king of Egypt, whom he places five centuries before Alexander the Great.[4] Plutarch, however, in his life of Agesilaus and Nepos in his life of Chabrias mention a Nectanebus II who struggled against Persia for the throne of Egypt about 361 B. C. and later was forced to flee to Ethiopia. In the Alexander romance, however, it is to Macedon that Nectanebus retreats. A Nectabis is listed as a magician along with Ostanes, Typhon, Dardanus, Damigeron, and Berenice, by Tertullian, writing about 200 A. D.[5] As a matter of fact, in the Thirtieth Dynasty were two kings named respectively Nektanebes or *Nekht-Har-ehbēt,* who ruled 378 to 361 B. C., and Nektanebos or *Nekhte-nebof,* who ruled 358 to 341 B. C. Both have left considerable

[1] Harleian 527, fols. 47-56.

[2] Amplon. Quarto 12, fols. 200-201; presumably it includes only those chapters concerned with Nectanebus.

[3] CUL 1429 (Gg. I, 34), 14th century, No. 5, 35 fols. Also in CU Trinity 1041, 14th century, fols. 200v-212v, "De Nectanabo mago quomodo magnum genuerit Alexandrum. Egipti sapientes. . . ."

[4] NH XXXVI, 14 and 19.

[5] *De anima,* cap. 57, in Migne, PL II, 792.

buildings.[1] It is the latter who was forced by the Persians to flee to Ethiopia nine years before Alexander conquered Egypt and who is the hero of our story. The stele of Metternich is covered with magical formulae ascribed to Nectanebo.[2]

A scientific key-note.

A note suggestive of both natural science and occult science is struck by the opening passage of the Latin epitomes and of the oldest Greek manuscript; the first page of Julius Valerius is missing and has to be supplied from the epitomes. The first words are "The Egyptian sages," and the first sentence describes their scientific ability in measuring the earth and in tracing the revolutions of the heavens and numbering the stars. "And of them all Nectanabus is recognized to have been the most prudent . . . for the elements of the universe obeyed him." In the opening sentences of the oldest Greek version and of the Ethiopic version even more emphasis is laid than in the Epitomes upon the learning of the Egyptians in general and of Nectanebus in particular, and of the close connection of that learning with astrology and magic.[3] We read, "Now there lived in the land of Egypt a king who was called Bektanis, and he was a famous magician and a sage, and he was deeply learned in the wisdom of the Egyptians. And he had more knowledge than all the wise men who knew what was in the depths of the Nile and in the abysses, and who were skilled in the knowledge of the stars and of their seasons and in the knowledge of the astrolabe and in the casting of nativities. . . . And by his learning and by his observations of the stars Nectanebus was able to predict what would befall anyone who was about to be born." [4] In one Latin manuscript of the fifteenth century the *History of Alexander the Great* begins with the

[1] The former built a Temple of Isis, now a heap of ruins, at Behbit el-Hagar and a colonnade to the Temple of Hibis in the oasis of Khirgeh; and his name appears upon a gate in the Temple of Mont at Karnak. Besides the Vestibule of Nektanebos at Philae there is a court of Nektanebos before the Temple of the Eighteenth Dynasty at Medinet Habu.

[2] Berthelot (1885), pp. 29-30.

[3] The Syriac version, on the contrary, emphasizes this point less.

[4] Budge's translation of the Ethiopic version.

sentence, "Books tell us how powerful the race of the Egyptians were in mathematics and the magic art." [1]

Magic of Nectanebus.

Next we are told, and the account is practically the same in all the versions of the story, how by means of his basin filled with water, his wax images of ships and men, his rod or wand of ebony, and the incantations with which he addressed the gods above and below, Nectanebus had been hitherto able to destroy all the armies and to sink all the fleets that had come against him. But when one day he found his magic unavailing to save him, he shaved his head and beard and fled to Macedon, where in linen garb he plied the trade of an astrologer.

Nectanebus as an astrologer.

In this he soon became so celebrated that the fame of his predictions reached the ears of the queen Olympias, who consulted him during an absence of Philip. When she asked Nectanebus by means of what art he divined the future so truthfully, he answered that there were many varieties of divination. Julius Valerius and the Latin epitomes mention specifically only interpreters of dreams and astrologers, but the Greek, Syriac, and Ethiopic versions give more elaborate lists of various kinds of diviners.[2] Nectanebus next produced an astrological tablet adorned with gold and ivory and with each planet and the horoscope represented by a different stone or metal. With the aid of this he read the

[1] CLM 215, fols. 176-94, "Egiptiorum gentem in mathematica magica quam in arte fuisse valentem littere tradunt."

[2] *Pseudo-Callisthenes*, I, 4, "casters of horoscopes, readers of signs, interpreters of dreams, ventriloquists, augurs, genethlialogists, the so-called magi to whom divination is an open book." Budge, Syriac version, p. 4, "The interpreters of dreams are of many kinds and the knowers of signs, those who understand divination, Chaldean augurs and casters of nativities; the Greeks call the signs of the zodiac 'sorcerers'; and others are counters of the stars. As for me, all of these are in my hands and I myself am an Egyptian prophet, a magus, and a counter of the stars." Budge, *Ethiopic Histories*, p. 11, "Then Nectanebus answered and said unto her, 'Yea. Those who have knowledge of the orbs of heaven are of many kinds. Some are interpreters of dreams, and some have knowledge of what shall happen in the future, and some understand omens, and some cast nativities, and there are besides all those who know magic and who are renowned because they are learned in their art, and some are skilled in the motion of the stars of heaven: but I have full knowledge of all these things.'"

queen's horoscope and told her that she would have a son by the God Ammon and would be forewarned soon to that effect in a dream. Olympias replied that if such a dream came to her, she would no longer employ Nectanebus as a *magus* but honor him as a god.

A magic dream.

Nectanebus thereupon sought for herbs useful to command dreams, plucked them, and pressed a syrup out of them. He placed a wax image of the queen inscribed with her name upon a little couch, lighted lamps, and poured his syrup over the wax figure, muttering a secret and efficacious incantation the while. By this means he brought it about that the queen would dream or think she dreamed whatever he said to the wax image of her. Later Nectanebus himself played the part of the god Ammon, announcing his coming beforehand to Olympias by making by his "science" a dragon which glided into her presence.

Lucian on Olympias and the serpent.

Lucian of Samosata in the second century tells us that it was a common story in his time that Olympias had lain with a serpent before giving birth to Alexander. He suggests as the explanation of how this tale originated the fact that at Pella in Macedonia there is a breed of large serpents, "so tame and gentle that women make pets of them, children take them to bed, they will let you tread on them, have no objection to being squeezed, and will draw milk from the breasts like infants. . . . It was doubtless one of these that was her bedfellow." [1] As is apt to be the case in ancient efforts to give a natural explanation of what purports to be miraculous or supernatural, Lucian's biology is only slightly less incredible than Nectanebus's magic transformations.

More dream-sending: magic transformation.

As the queen became pregnant, "Nectanebus consecrated a hawk and told it to go to Philip," who was still absent, "to stand by him through the night and to instruct him in a dream as it was ordered." [2] The vision in question was ex-

[1] From Fowler's translation of *Alexander: the False Prophet.* See also Plutarch's *Alexander.*

[2] The Syriac and Ethiopic versions are somewhat more detailed as to the magic by which Philip's dream was produced. Budge, Syriac version, p. 8, "Then Nectanebus . . . brought a hawk and muttered over it his charms and made it fly away with a small quantity of a drug, and that night

plained by an interpreter of dreams to Philip as signifying that his wife would have a son by the god Ammon. Nevertheless Philip was somewhat suspicious and hastened to bring his wars to a close and hurry home. Nectanebus, however, rendering himself invisible by means of the magic art, continued to deceive both king and queen. Once he terrified the court by appearing again in the form of a huge hissing serpent, but put his head in Olympias's lap and then kissed her. Thereupon he turned from a serpent into an eagle and flew away. Philip was then really convinced that his wife's lover was the god Ammon.

An omen interpreted.

Before the birth of Alexander the following omen befell Philip. As he sat absorbed in thought in a place where there were many birds flying about, one of them laid an egg in his lap. It rolled to the ground, the shell broke, and a snake issued forth. It circled about the egg-shell but when it tried to reenter the shell was prevented by death. When Antiphon, the interpreter of omens, was consulted concerning this portent, he said that it signified that a son should be born who would conquer the world but die before he could regain his native land.

The birth of Alexander.

The day of Olympias's delivery now approached and Nectanebus, in his office of astrologer, stood by her side to tell her when the favorable moment had arrived for the birth of her child. Once he urged her to wait, since a child born at that moment would be a slave and a captive. Again he bade her restrain herself, for at that moment an effeminate would be born. At last the favorable instant came for the birth of a world conqueror, and Alexander was born amid an earthquake, thunder, and lightning. In this case, therefore, the moment of birth is regarded as controlling the destiny. Many astrologers, however, considered the moment of conception as of greater importance; we have already

it shewed Philip a dream." Budge, *Ethiopic Histories*, p. 21, "Then Nectanebus took a swift bird and muttered over it certain charms and names, and . . . in one day and one night it traversed many lands and countries and seas, and it came to Philip by night and stopped. And it came to pass at that very hour . . . that Philip saw a marvelous dream."

heard Augustine tell of the sage who chose a certain hour for intercourse with his wife in order to beget a marvelous son; and in the thirteenth century Albertus Magnus, in his treatise on animals, informs us that "Nectanebus, the natural father of Alexander, in having intercourse with his mother Olympias, observed the time when the sun was entering Leo and Saturn was in Taurus, since he wished his son to receive the form and power of those planets." [1]

The death of Nectanebus.

The death of Nectanebus was as closely in accord with the stars as was the birth of Alexander. At the age of twelve Alexander found Nectanebus in consultation with Olympias and, attracted by his astrological tablet, made him promise to show him the stars at night. Then as Nectanebus walked along star-gazing, Alexander pushed him into a steep pit which they chanced to pass, and Nectanebus lay there with a broken neck. When he asked Alexander the reason for his act, the boy replied that it was in order to convince him of the futility of his art, since he gazed at the stars unmindful of what threatened him from the ground. But Nectanebus rebuts this revised version of the maid servant's taunt to Thales by telling Alexander that he had been forewarned by the stars that he should be killed by his own son, and by revealing to Alexander the secret of his birth.[2]

In concluding the story of Nectanebus it is perhaps worth while to emphasize the fact that the epitomes and Julius Valerius often use the word *magus* of Nectanebus as an astrologer and that in general magic, astrology, and divination are indissolubly connected.

[1] In another place, however, Albert calls Philip Alexander's father, *De causis et proprietatibus elementorum et planetarum,* II, ii, 1.

[2] The story is better told in the Syriac version (Budge, 14-17), where Alexander does not push Nectanebus into the pit until after he has asked the astrologer if he knows his own fate and has been told that Nectanebus is to be slain by his own son. Alexander then attempts to foil fate by pushing Nectanebus into the pit, but only fulfills it. In the Ethiopic version Nectanebus is represented as educating Alexander from his seventh year on in "philosophy and letters and the working of magic and the stars and their seasons." Aristotle becomes Alexander's tutor only after the death of Nectanebus. Aristotle, too, is represented as an adept in astrology, amulets, and the use of magic wax images. (Budge, *Ethiopic Histories,* pp. 31, xlv).

The Amazons and Gymnosophists.

Some account is given both in Julius Valerius and the longer epitome of Alexander's exchange of letters with the Amazons and of questions which he put to the Gymnosophists of India (i. e. the Brahmans) and their replies. Neither of these promising themes, however, results in the introduction of any magic or occult science. We also find in the *Stromata* of Clement of Alexandria [1] a list of ten questions which Alexander propounded to ten of the Gymnosophists of India and their ingenious answers given under pain of death if their responses proved unsatisfactory.

The letter to Aristotle.

Nor does Alexander's letter to Aristotle on the marvels of India reveal many specific instances of superstition that are at all interesting. For the most part it recounts his marches, the sufferings of his army from thirst, combats with wild beasts, serpents, and hippopotamuses, and the treasures which he captured. Alexander states that "in former letters I informed you about the eclipse of the sun and moon and the constancy of the stars and the signs of the air." [2] He tells now, however, of a place where there are two trees of the sun and moon, speaking Indian and Greek, one masculine and the other feminine, from which one may learn what the future has in store for good or evil. As to this Alexander was inclined to be incredulous, but the natives swore that it was true, and his companions urged him "not to be defrauded of the experience of so great a thing." Accordingly he made his way to the spot despite the innumerable beasts and snakes which beset his path. Chastity was essential in order to approach the trees, and he also had to lay aside his rings, royal robes, and shoes. The sun tree then told him at dawn that he would never see home or his mother and sisters again. At eventide the moon tree added that he would die at Babylon.[3] The third and final response,

[1] VI, 4.

[2] Royal 13-A-I, fol. 53v.

[3] In CU Trinity 1446 (1250 A. D.) *The Romance of Alexander* in French verse by Eustache (or Thomas) of Kent, among 152 pictures listed by James (III, 483-91) are two representing the hero's colloquy with the moon tree (fol. 31r). Marco Polo also tells of these marvelous trees. And see Roux de Rochelle, "Notice sur l'Arbre du Soleil, ou Arbre Sec, décrit dans la relation

vouchsafed by the sun tree, was that his death would be from poison, but the name of the poisoner the oracular tree refused to divulge lest Alexander try to kill him first and thus cheat the three Fates. Alexander has consequently had to content himself, as he informs Aristotle in the closing sentence of his letter, with building a monument to perpetuate his name among all mortals.[1]

Of other spurious treatises ascribed to Alexander in the middle ages, works of alchemy and works of astrology, we shall treat in a later chapter on the Pseudo-Aristotle.

des voyages de Marco Polo," in *Bulletin de la Société de géographie*, série 3, III (1845), 187-94.

[1] For the *Letter to Aristotle* I have employed the Paris, 1520 edition and Royal 13-A-I, which follow the early Latin version. As stated above, Pfister's edition (Heidelberg, 1910) gives a later version probably translated from the Greek.

CHAPTER XXV

POST-CLASSICAL MEDICINE

Three representatives of post-classical medicine—Bibliographical note—Medical compendiums: Oribasius and Paul of Aegina—Aëtius of Amida—How superstitious are Aëtius and Alexander of Tralles?—Compound medicines—Aëtius merely reproduces the superstition of Galen—Occult science mixed with some scepticism—Alexander of Tralles—Originality of his work—His medieval influence—His personal experience—Extent of his superstition—*Physica*—Occult virtue of substances applied externally—Other things used as ligatures and amulets—Astrology and sculpture of rings—Incantations—Conjuration of an herb—Medieval version seems less superstitious than the original text—Marcellus: date and identity—"Marcellus Empiricus"—Superstitious character of his medicine—Preparation of goat's blood—A rabbit's foot—Magic transfer of disease—Pliny and Marcellus compared on green lizards as eye-cures—More lizardry—Use of stones and an herb—Right and left: number—Incantations and characters—The art of medicine survives the barbarian invasions.

Three representatives of post-classical medicine.

IN this chapter as representatives of post-classical medicine and its influence upon medieval Latin medicine we shall consider three writers whose works date from the close of the fourth to the middle of the sixth century, Marcellus of Bordeaux or Marcellus Empiricus, Aëtius of Amida in Mesopotamia, and Alexander of Tralles in Asia Minor.[1] They have just been mentioned in their chronological order,

[1] There appears to have been no complete edition of Aëtius in Greek. The first eight of his sixteen books were printed at Venice in 1534, and the ninth at Leipzig in 1757, but for the entire sixteen books one must use the Latin translation of Cornarius, Basel, 1542, etc., which I have read in Stephanus, *Medicae artis principes,* 1567.

Recent editions of portions of Aëtius are: Αετιου λογος δωδεκατος πρωτον νυν εκδοθεις ὑπο Γεωργιου Α. Κωστομοιρου, pp. 112, 131, Paris, 1892.

Die Augenheilkunde des Aëtius aus Amida, Griechisch und deutsch herausg. von J. Hirschberg, pp. xi, 204, Leipzig, 1899.

Aetii sermo sextidecimus et ultimus (Αετιου περι των εν μητρᾳ παθων etc.). Erstens aus HSS veröffentl. mit Abbildungen, etc., v. S. Zervòs, pp. k', 172, Leipzig, 1901.

Αετιου Αμιδινου Λογος δεκατος πεμπτος, ed. S. Zerbos, 1909, in Επιστημονικη Εταιρεια, Αθηνα, vol. 21.

My references to Alexander of Tralles are both to the text of Stephanus (1567) and the more

but although Marcellus antedates the other two by a full century, we shall consider him last, since he wrote in Latin while they wrote in Greek, and since he includes Celtic words and probably Celtic folk-lore, and since he seems to have

recent edition by Theodor Puschmann, *Alexander von Tralles, Originaltext und Übersetzung nebst einer einleitenden Abhandlung*, Vienna, 1878-9, 2 vols. This gives a more critical text than any previous edition, but unfortunately Puschmann adopted still another arrangement into books than those of the MSS and previous editions, and also in my opinion did not make a sufficient study of the Latin MSS. His introduction contains information concerning Alexander's life and the MSS and previous editions of his works.

A valuable earlier study on Alexander was that of E. Milward, published in 1733 under the title, *A Letter to the Honourable Sir Hans Sloane Bart., etc.*, and in 1734 as *Trallianus Reviviscens*, 229 pp. Milward was preparing an edition of Alexander of Tralles, but it was never published. His estimate of Alexander's position in the history of medicine furnishes an incidental picture of interest of the state of medicine in his own time, the early eighteenth century.

The old Latin translation of Alexander of Tralles was the first to be printed at Lyons, 1504, *Alexandri yatros practica cum expositione glose interlinearis Jacobi de Partibus et (Simonis) Januensis in margine posite;* also Pavia, 1520 and Venice 1522. Next appeared a very free Latin translation by Torinus in 1533 and 1541, *Paraphrases in libros omnes Alexandri Tralliani.* The Greek text of Alexander was first printed by Stephanus (Robert Étienne) in 1548 (ed. J. Goupyl). The Latin translation by Guinther of Andernach, which is included in Stephanus (1567), first appeared in 1549, Strasburg, and was reprinted a number of times.

Another work by Puschmann may also be noted: *Nachträge zu Alexander Trallianus. Fragmente aus Philumenus und Philagrius nebst einer bisher noch ungedruckten Abhandlung über Augenkrankheiten*, Berlin, 1886, in *Berliner Studien f. class. Philol. und Archaeol.*, V, 2; 188 pp., in which he segregates as fragments of Philumenus and Philagrius portions of the text of Alexander as found in the Latin MSS.

My references for the *De medicamentis* of Marcellus apply to Helmreich's edition of 1889 in the Teubner series. This edition is based on a single MS of the ninth century at Laon which Helmreich followed Valentin Rose in regarding as the sole extant codex of the work. As a result Rose indulged in ingenious theories to explain how the *editio princeps* by Ianus Cornarius, Basel, 1536, included the prefatory letter and other preliminary material not found in the Laon MS, whose first leaves and some others are missing.

But as a matter of fact BN 6880, a clear and beautifully written MS of the ninth century, contains the *De medicamentis* entire with all the preliminary letters. Moreover, it is evident that the *editio princeps* was printed directly from this MS, which contains not only notes by Cornarius but the marks of the compositors.

The text of the edition of 1536 was reproduced in the medical collections of Aldus, *Medici antiqui*, Venice, 1547, and Stephanus, *Medicae artis principes*, 1567.

Jacob Grimm, *Über Marcellus Burdigalensis*, in *Abhandl. d. kgl. Akad. d. Wiss. z. Berlin* (1847), pp. 429-60, discusses the evidence for placing Marcellus under the older Theodosius, lists the Celtic

been a native of Gaul, if not of Bordeaux,[1] and thus is geographically closer to the scene of medieval Latin learning. Aëtius and Alexander have the closer connection not only with the eastern and Greek world but also with the past classical medicine of Galen and so will provide a better point of departure. Presumably from the places and periods in which they lived, all three of our authors were Christians, but it must be said that the chief evidence of Christianity in their works is the use of Christian or Hebrew proper names in incantations, and there are some analogous relics of pagan superstition.

Medical compendiums: Oribasius and Paul of Aegina.

As Tribonian and Justinian boiled down the voluminous legal literature of Rome into one *Digest,* so there was a similar tendency to reduce the past medical writings of the Greeks into one compendious work. Paul of Aegina, writing in the seventh century, observes in his preface[2] that it is not right, when lawyers who usually have plenty of time to reflect over their cases have handy summaries of their subject to which they can refer, that physicians whose cases often require immediate action should not also have some

words and expressions found in the *De medicamentis,* and also one hundred specimens of its folk-lore and magic. This article was reprinted in *Kleinere Schriften,* II (1865), 114-51, where it is followed at pp. 152-72 by a supplementary paper, *Über die Marcellischen Formeln,* likewise reprinted from the Academy Proceedings for 1855, pp. 51-68.

The magic of Marcellus was further treated of by R. Heim, *De rebus magicis Marcelli medici,* in *Schedae philol. Hermanno Usener oblatae* (1891), pp. 119-37, where he adds *nova magica ex Marcelli libris collata* which Grimm had omitted.

[1] Marcellus is often called of Bordeaux, notably in Grimm's article, *Über Marcellus Burdigalensis,* 1847; also by C. W. King, *The Gnostics and their Remains,* 1887, p. 219; and by J. G. Frazer, *The Golden Bough,* I, 23; but there seems to be no definite proof that he was from that city.

Jules Combarieu, *La musique et la magie,* 1909, p. 87, says in reference to the following incantation recommended by Marcellus, *tetunc resonco bregan gresso,* "Je remarque en passant qu'il faut frotter l'œil en disant ce *carmen,* et que dans le patois du Midi, *brégua* ou *brége,* signifie frotter. Marcellus, si je ne me trompe, était de Bordeaux."

Grimm, however (1847), p. 455, interpreted *bregan* as "lies"—"breigan gen. pl. von breag lüge," and the whole line as in modern Irish *teith uainn cre soin go breigan greasa* ("fleuch von uns staub hinnen zu der lügen genossen!").

[2] Stephanus (1567), I, 347, *et seq.* For an English translation of the text see F. Adams, *The Seven Books of Paulus Aegineta,* London, 1844-1847.

convenient handbook, and the more so since many of them are called upon to exercise their profession not in large cities with easy access to libraries, but in the country, in desert places, or on shipboard. Oribasius, friend and physician of the emperor Julian, 361-363 A. D., had made such a compendium by that emperor's order. In this he embodied so much of Galen's teachings that he became known as "the ape of Galen," [1] although he also used more recent writers. But Paul of Aegina regarded this work of Oribasius as too bulky, since it originally comprised seventy-two books although only twenty-five are now extant, and so essayed a briefer compilation of his own. Two centuries ago, however, Friend and Milward protested against regarding Paul, Aëtius, and Alexander as mere compilers and maintained that they "were really men of great learning and experience" [2] who "have described distempers which were omitted before; taught a new method of treating old ones; given an account of new medicines, both simple and compound; and made large additions to the practice of surgery." [3] Puschmann more recently states that Paul's compendium was "composed with great originality and independence" and is of great value "particularly in its surgical sections." [4] After Paul, however, the Byzantine medical writers, such as Palladius, Theophilus, Stephen of Alexandria, Nonus, and Psellus, were of an inferior caliber.[5] With Paul's work, however, we are not now further concerned, nor with that of Oribasius, but with the somewhat similar compendiums of Aëtius and Alexander which lie chronologically between these other two. It is Aëtius and Alexander whom Payne accuses of "introducing into classical medicine the magical elements derived from the East" [6] and whom we

[1] *Simia Galieni,* according to Guinther in his translation of Alexander of Tralles, Stephanus (1567), I, 131.

[2] Milward (1733), 9-11.

[3] John Friend (or Freind), *History of Physick* (1725), I, 297.

[4] Puschmann, *History of Medical Education,* 1891, p. 153.

[5] Milward (1733), p. 11.

[6] J. F. Payne, *English Medicine in Anglo-Saxon Times,* 1904, pp. 102-8.

might therefore expect to possess an especial interest for our investigation.

Aëtius of Amida.

Of the life and personality of Aëtius we know very little, but inasmuch as he mentions St. Cyril, archbishop of Alexandria, and Peter the Archiater, a physician of Theodoric, while he himself is cited by Alexander of Tralles, he seems to have lived at the end of the fifth and beginning of the sixth century.[1] And since Alexander cites him only in his book on fevers which seems to have been composed after the rest of his work, it seems probable that Aëtius was almost contemporary with him and wrote in the sixth rather than the fifth century. His *Tetrabiblos*—each of the four books subdivides into four sections and often these are spoken of as sixteen books—occupies a middle position not only in time but in length between the works of Oribasius and Paul, and resembles the latter in making a great deal of use of the former. Aëtius' extracts from the older writers are shorter than those of Oribasius, however, and he also differs from him in combining several authorities in a single chapter, the method usually adopted by the medieval Latin encyclopedists. It has been noted that the wording of the original authorities was often preserved in the oldest medieval manuscripts of Aëtius, until the copyists of the time of the Italian Renaissance began to touch up the style in accordance with their erroneous notions of what constituted classical Greek.[2] It may also be said that these systematically arranged handbooks of Oribasius, Aëtius, and the rest, where one could find what one was looking after, were far superior in systematic and orderly presentation to the discursive works of Galen which, like many other classical writings, often seem rambling and without any particular plan.[3] This more logical, if somewhat cut-and-

[1] Milward (1733), p. 19; Puschmann (1878), I, 104.

[2] Ch. Daremberg, *Histoire des Sciences Médicales*, Paris, 1870, I, 242.

[3] This general impression received from reading many classical and medieval works I was glad to find confirmed by Milward (1733), p. 29, in the particular case of Alexander of Tralles, of whom he writes: "As our author's stile is excellent, so likewise is his method, and there is no respect in which he is more distinguished from the other

dried method, was also to be a virtue of medieval Latin learning. Whether Aëtius directly influenced the Latin middle ages is doubtful, since no early Latin translation of him seems to be known.[1] The work of Oribasius, however, exists in Latin translation in manuscripts of the seventh century as well as in others of the ninth and twelfth.[2]

How superstitious are Aëtius and Alexander?

The works of Aëtius and Alexander of Tralles do not impress me as containing an unusually large amount of superstitious medicine. Much less am I inclined to agree with Payne that they are responsible for the introduction into classical medicine of magical elements derived from the east. These elements, whether derived from the orient any more than any other feature of classical civilization or not, at any rate had been a prominent feature of classical medicine long before the days of Aëtius and Alexander, as Pliny's review of medicine before his time abundantly proved and as is also shown by the extraordinary virtues which Pliny himself, his contemporary Dioscorides, and even the great Galen attributed to medicinal simples.

Compound medicines.

It is true that Aëtius and Alexander abound in recipes for elaborate medical compounds composed of numerous ingredients. Of such concoctions one example must suffice, a plaster which Aëtius recommends for tumors, hard lumps, and gout. "Of the terebinth-tree, of the stone of Asia, of bitumen three hundred and sixty drams each; of washing-soda (*spumae nitri*), calf-fat, wax, laurel berries, ammonia, and thyme three hundred and forty drams each; of the stone pyrites and quick-lime one hundred and twenty drams each; of the ashes of asps which have been burned alive one

Greek writers in physick than in this. The works of Hippocrates, Galen, and indeed of all of them except it be Aretaeus are not only very voluminous but put together with little or no order, as is evident enough to all such as have been conversant with them."

[1] Daremberg (1870), I, 258-9, said that a mass of MSS in a score of European libraries contained as yet unidentified Latin translations of Greek medical writers.

[2] BN 10233, 7th century uncial; BN nouv. acq. 1619, 7-8th century, demi-uncial; BN 9332, 9th century, fol. 1-, Oribasii synopsis medica; CLM 23535, 12th century, fols. 72 and 112. V. Rose, *Soranus*, 1882, pp. iv-v, speaks of a sixth century Latin version of *Oribasius*.

hundred and forty drams; of old oil two pounds. First liquefy the oil and wax, then the bitumen, which should have first been pulverized. Add to these the fat, and presently the ammonia and terebinth; and when these are taken off the fire mix in the lime and stone of Asia, then the laurel berries and washing-soda, and finally after the medicament has cooled sprinkle the ashes of asps upon it." [1] Such concoctions are to a large extent borrowed by Aëtius, Alexander, and Marcellus from earlier writers. Moreover, while Pliny had excluded such compounds from the pages of his *Natural History,* he had also made it abundantly evident that they were already in general use by his time, and they are to be found in great numbers in the works of Galen who cites many from preceding writers.

Aëtius merely reproduces the superstition of Galen.

Indeed, it was from Galen himself and not from the east that Aëtius at least derived his most strikingly superstitious passages. This was accidentally and convincingly proven by my own experience. It so happened that I wrote an account of the passages in the *Tetrabiblos* of Aëtius before I had read extensively in Galen's works. When I came to do so, I found that almost every passage that I had selected to illustrate the superstitious side of Aëtius was contained in Galen: for example, the use as an amulet of a green jasper suspended from the neck by a thread so as to touch the abdomen; [2] the story of the reapers who found the dead viper in their wine and cured instead of killing the sufferer from elephantiasis to whom they gave the wine to drink; [3] the tale of his preceptor who roasted river crabs to an ash in a red copper dish in August during dog-days on the eighteenth day of the moon, and administered the powder daily for forty days to persons bitten by mad dogs.[4] Such

[1] *Tetrabiblos,* IV, iii, 15.

[2] *Ibid.,* I, iv, 9, where Galen is not cited, and III, i, 9, where Galen is cited. In Galen, *De simplicibus,* IX, ii, 19 (Kühn, XII, 207).

[3] *Ibid.,* I, ii, 170, where Galen is not cited; *De simplicibus,* XI, i, 1 (Kühn, XII, 311-4).

[4] *Tetrabiblos* I, ii, 175; Kühn XII, 356-9. Galen is not cited in this, nor in any of the following passages from the *Tetrabiblos* listed in the notes, unless this is expressly stated.

passages are usually repeated by Aëtius in such a way as to lead the reader to think them his own experiences, a fact which warns us not to accept the assertions of ancient and medieval authors that they have experienced this or that at their face value, and which makes us wonder if Friend and Milward were not too generous in regarding Aëtius at least as more than a compiler. He also repeats some of Galen's general observations anent experience as that the virtues of simples are best discovered thus, and that he will not discuss all plants but only those "of which we have information by experience."[1] He further reproduces Galen's attitude of mingled credulity and scepticism concerning the basilisk, combining the two passages into one;[2] also Galen's questioning the efficacy of incantations and telling of having seen a scorpion killed by the mere spittle of a fasting man without any incantation.[3] Like Galen again, he omits all injurious medicaments and expresses the opinion that men who spread the knowledge of such drugs do more harm than actual poisoners who perhaps cause but a single death.[4] Like Galen he announces his intention to omit all "abominable and detestable recipes and those which are prohibited by law," mentioning as instances the eating of human flesh and drinking urine or *menses muliebres*.[5] But also like Galen, he devotes several chapters to the virtues of human and animal excrement, especially recommending that of dogs after they have been fed on bones for two days.[6] Somewhat similar to Galen's recommendation to fill cavities in the teeth with roasted earthworms is the recipe of Aëtius for painless extraction of teeth "without iron." The tooth must first be thoroughly scraped or the gum cut loose about it, and then sprinkled with the ashes of earthworms. "Therefore use this remedy with confidence, for it has already often

[1] *Tetrabiblos* at the beginning, pp. 6-7 in Stephanus (1567).

[2] *Tetrabiblos* IV, i, 33; Kühn XIV, 233, and XII, 250-1.

[3] *Tetrabiblos* I, ii, 109; Kühn XII, 288.

[4] *Tetrabiblos* I, ii, 84; Kühn XII, 253.

[5] *Tetrabiblos* I, ii, 84; Kühn XII, 248, 284-5.

[6] *Tetrabiblos* I, ii, 111; Kühn XII, 291-3.

been celebrated as a mystery."[1] Such use of earthworms continued a feature of medieval dentistry.

Occult science mixed with some scepticism.

Of my original selections from Aëtius very few are now left, and it is not unlikely that they too might be found somewhere in Galen's works if one looked long enough. Aëtius asserts that drinking bitumen or asphalt in water will prevent hydrophobia from developing,[2] and recommends for wounds inflicted by sea serpents an application of lead with a slice of the serpent itself.[3] He takes the following prescription from Oribasius. To cure impotency anoint the big toe of the right foot with oil in which the pulverized ashes of a lizard have been mixed. To check the operation of this powerful stimulant one has merely to wash off the ointment from the toe.[4] On the other hand, an instance of a sceptical tendency is the citation of the view of Posidonius that the so-called *incubus* is not a demon but a disease akin to epilepsy and insanity and marked by suffocation, loss of voice, heaviness, and immobility.[5] It may also be noted that in discussing the medicinal virtues of the beaver's testicles Aëtius does not include the story of its biting them off in order to escape its hunters.[6] He does, however, cite several authorities, Piso, Menelbus, Simonides, Aristodemus, and Pherecydes for instances of the remarkable powers of certain animals in discovering the presence of poisons and preserving themselves and their owners from this danger: a partridge who made a great noise and fuss whenever any medicament or poison was being prepared in the house; a pet eagle who would attack anyone in the house who even plotted such a thing; a peacock who would go to the place where the dose had been prepared and raise

[1] *Tetrabiblos* II, iv, 34; Kühn XII, 860. Perhaps a closer correspondence than this could be found. In his preceding 33rd chapter, headed *Curatio erosorum dentium ex Galeno,* Aëtius includes use of the tooth of a dead dog pulverized in vinegar, which is to be held in the mouth, or filling the ear next the tooth with "fumigated earthworms" or with oil in which earthworms have been cooked.

[2] *Tetrabiblos* I, ii, 49.

[3] *Tetrabiblos* IV, i, 39.

[4] *Tetrabiblos* III, iii, 35.

[5] *Tetrabiblos* II, ii, 12. Marcellus, cap. 20 (p. 188) also speaks of "those who often think that they are made sport of by an incubus."

[6] *Tetrabiblos,* I, ii, 177.

a clamor, or upset the receptacle containing the potion, or dig up a charm, if it had been buried underground; and a pet ichneumon and parrot who were endowed with very similar gifts.[1] Aëtius shows a slight tendency in the direction of astrological medicine, giving a list of "times ordained by God" for the risings and settings of various stars, since these affect the air and winds, and since "the bodies of persons in good health, and much more so those of the sick, are altered according to the state of the air." [2] But on the whole, of our three authors, Aëtius seems to contain the smallest proportional amount of superstitious medicine and occult science.

Alexander of Tralles.

Alexander of Tralles was the son of a physician and, according to the Byzantine historian, Agathias,[3] the youngest of a group of five distinguished brothers, including Anthemius of Tralles, architect of St. Sophia at Constantinople, and Metrodorus the grammarian, whom Justinian summoned also to his court. Alexander had visited Italy, Gaul, and Spain as well as all parts of Greece [4] before settling down in old age, when he could no longer engage in active medical practice,[5] to the composition of his *magnum opus* in twelve books beginning with the head, eyes, and ears, and ending with gout and fever. Aside from his citation of Aëtius in the book on fevers, the latest writer named by Alexander is Jacobus Psychrestus, physician to Leo the Great about 474.[6] It seems rather strange that Alexander says nothing of the pestilence of 542.[7]

Originality of his work.

Alexander embodied the results of his own practice to a much greater extent than Oribasius and Aëtius. His book is more a record of his own medical observations and experiences than a compilation from past writings, a fact recog-

[1] *Tetrabiblos,* IV, i, 86.

[2] *Tetrabiblos* I, iii, 164. This passage was printed separately in the *Uranologion* of D. Petavius, Paris, 1630 and 1703.

[3] Agathias, *De imperio et rebus gestis Justiniani,* Paris, 1860, p. 149.

[4] Milward (1733), p. 17, "he travel'd through Greece, Gaul, Spain, and several other places whose mention we find up and down in his works."

[5] Puschmann (1878), I, 288, *διὸ καὶ γέρων λοιπὸν πειθαρχῶ καὶ κάμνειν οὐκέτι δυνάμενος . . .*

[6] Milward (1733), p. 25.

[7] Puschmann (1878), I, 83.

nized in the first edition which entitled it *Practica,* and "though he pays a due deference to the ancients, yet he is so far from putting an implicit faith in what they have advanced that he very often dissents from their doctrines." [1] Puschmann regarded him as the first doctor for a long time who had done any original thinking,[2] and esteemed his pathology as highly as his therapeutics had been esteemed by his sixteenth century translator, Guinther of Andernach.[3] Friend wrote of him in the early eighteenth century, "His method is extremely rational and just and after all our discoveries and improvements in physick scarce anything can be added to it." [4] Alexander seems to have been a practitioner of much resource and ingenuity, stopping hemorrhage of the nose by blowing down or fuzz up the nostrils through a hollow reed, and directing patients, a thousand years before the discovery of the Eustachian tube, to sneeze with mouth and nose stopped up in order to dislodge a foreign object from the ear.[5] According to Milward, Alexander was the first Greek medical writer to mention rhubarb and tape-worms, and the first practitioner to open the jugular veins.[6] Indeed, Alexander advises blood-letting a great deal, but Milward, whose age still approved of that practice, notes that he was "no ways addicted to those superstitious rules of opening this or that vein in particular cases which several of the ancients and some even among the moderns have been so very fond of." [7] Finally, Alexander's concise and orderly method of presentation compares favorably with that of the classical medical writers.

His medieval influence.

Alexander's book traveled west, as its author had done, and was current in a free and abbreviated Latin translation from an early date.[8] In fact, it was from the Latin version

[1] Milward (1733), p. 27.
[2] Puschmann (1891), 152-3.
[3] Stephanus (1567), I, 131.
[4] Friend (1725), I, 106.
[5] Milward (1733), pp. 65-6, 57 *et seq.*
[6] *Ibid.,* pp. 104, 92-3, 71.
[7] *Ibid.,* pp. 48-9.
[8] See V. Rose, *Hermes,* VIII, 39; *Anecdota,* II, 108. I presume that BN 9332, 9th century, fol. 139, "Alexandri hiatrosofiste therapeut(i)con" (libri tres) is the free Latin translation in a

that the work was translated into Hebrew and Syriac.[1] Not only are Latin manuscripts of Alexander's work as a whole or of extracts from it [2] found from the ninth century on, while printed editions in Latin were numerous through the sixteenth century, but it was much used and cited by medieval writers such as Constantinus Africanus, Gariopontus,[3] and Gilbert of England.[4] It is not, however, always safe to assume that citations of *Alexander*

Paris MS of the ninth century alluded to by Daremberg (1870), I, 258-9. Puschmann (1878) I, 91-2, in a blind and inadequate account of the Latin MSS, does not mention it, but lists a Monte Cassino codex (97) of the 9-10th century and an Angers MS of the 10-11th century. He also alludes to a MS at Chartres without giving any number or date for it, but probably has reference to Chartres 342, 12th century, fols. 1-139, "Libri tres Alexandri Yatros." He alludes to BN 6881 and 6882, both 13th century, libri tres de morbis et de morborum curatione; but not to CLM 344, 12-13th century, fols. 1-60, libri III de medicina,—integra versio Latina Lugduni a. 1504 edita. Other MSS are: Gonville and Caius 400, early 13th century, fols. 4v-83v, "Inc. Alexander yatros sophista"; Royal 12-B-XVI, late 13th century, fol. 113, Practica Alexandri.

It will be noted that the text in all these Latin MSS is in only three books, but it follows the same order as the twelve books. It is also, at least in the edition of 1504, not as abbreviated as one might infer from Rose. Rather the later editors, Albanus Torinus and Guinther of Andernach, seem to have taken greater liberties with, and made unwarranted additions to Alexander's text. At the same time the early Latin text treats of some topics such as toothache which are not included in Puschmann's Greek text, and also includes (II, 79-103, and 104-50) treatments of diseases of the abdomen and spleen for which there seems to be no genuine Greek text and which Puschmann, *Nachträge*, 1886, has published separately as fragments of Philumenus and Philagrius, medical writers of the first and fourth centuries. His chief reason seems to be that cap. 79 is entitled, *De reumate ventris filominis*, and cap. 104, *Ad splenem philogrius*, while cap. 151 is headed, *Causa que est ydropicie alexandri*. These passages are, however, found in the Latin MSS of Alexander's work from the first, and the use of Romance words by the unknown Latin translator indicates that the translation was made in the early medieval period, — Puschmann (1886), p. 12.

[1] Puschmann (1878), I, 91.

[2] As in Vendôme 109, 11th century, fol. 1, Mulsa Alexandri (Tralliani), fol. 68v, "De reuma ventris, de libro Alexandri" (not here ascribed, it will be noted, to Philumenus), fol. 71, "De secundo libro Alexandri de cura nefreticorum." The *Mulsa Alexandri* is found also in two other 11th century MSS of the same library: Vendôme 172, fol. 1, and 175, fol. 2.

In Royal 12-E-XX, 12th century, fols. 146v-151v, "Incipit liber dietarum diversarum medicorum, hoc est Alexandri et aliorum." This extract, made up of a number of Alexander's chapters on the diet suitable in different ailments, is often found in the MSS, as here, with the Pseudo-Pliny and was printed as its fifth book in 1509 and 1516.

[3] Puschmann (1878), I, 97.

[4] Milward (1773), p. 179.

medicus, encountered in thirteenth century writers on the nature of things like Thomas of Cantimpré and Bartholomew of England, have reference to Alexander of Tralles, since a treatise on fevers is also ascribed to Alexander of Aphrodisias,[1] while a work on the pulse and urine in fevers is thought to be by some medieval Alexander.[2] And medical treatises are sometimes ascribed even to Alexander the Great of Macedon in the medieval manuscripts.[3]

His personal experience.

We have already said that Alexander is no mere compiler but embodies the results of his own observation and experience during a long period of travel and medical practice. He frequently asserts that he has tested this or that for himself, or that the prescription in question has been "approved by long use and experience," [4] so that it is not surprising that we find the name Alexander still associated with medical "experiments" in manuscripts dating from the twelfth to fifteenth centuries.[5] One of his cures for epilepsy he learned "from a rustic in Tuscany" (*Thuscia?*) but afterwards often employed with success himself.[6] "It is a marvelous and exceptional medicine which you will communicate to no one," concludes Alexan-

[1] Thus in Vendôme 109 (see note 2, p. 577) besides the extracts from Alexander of Tralles we find at fol. 58, "Alexander (Aphrodisiensis) amicus veritatis in tertio libro suo ubi de febribus commemorat." The Arabs seem to have confused these two Alexanders: see Steinschneider (1862), p. 61; Puschmann (1878), I, 94-5.

[2] See the discussion by Choulant in *Janus* (1845), p. 52, and Henschel in De Renzi (1852-9) II, 11, of a 12th century MS at Breslau, "Liber Alexandri de agnoscendis febribus et pulsibus et urinis"; also Puschmann (1878) I, 105-6, concerning BN Greek MS 2316, which seems to be a late Greek translation of it,—another instance that a Greek text is not necessarily the original.

[3] Corpus Christi 189, 11-12th century, fols. 1-5, "Antidotum pigra magni Alexandri Macedonii quod facit stomaticis epilenticis." Steinschneider, cited by Puschmann (1878) I, 106, has also noted the attribution in Hebrew MSS to Alexander the Great of a work on fever, urine, and pulse, presumably identical with that mentioned in the foregoing note.

[4] Stephanus (1567) I, 176, 204, 216, 225; and Puschmann, II, 575, are a few specimens.

[5] Amplon. Quarto 204, 12-13th century, fols. 90-5, Experimentorum Alexandri medici collectio succincta. Digby 79, 13th century, fols. 180-92v, "Alexandrina experimenta de libro percompendiose extractata meliora ut nobis visum est ad singulas egritudines." Additional 34111, 15th century, fol. 77, "Experimenta Alexandri," in English.

[6] Stephanus I, 156; Puschmann II, 563.

der, a rather surprising prohibition in view of the fact that it was a popular remedy to begin with. Folk-lore, however, is often supposed to be kept secret. Another general rule which holds true in Alexander's case is that these empirical remedies are apt to be the most superstitious, and conversely that marvels are apt to be supported by solemn assurance of their experimental testing.

Extent of his superstition.

Two centuries ago Milward wrote of Alexander of Tralles, "But there is another objection to our author's character which I cannot pretend to say much in defence of, and that is, his being addicted to charms and amulets. It is very surprising that one who discovers so much judgment in other matters should show so much weakness in this." [1] Alexander certainly devotes more space to superstition relatively to the length of his book than Aëtius does and also is hospitable to a wider range of more or less magical notions and practices. One notices, however, in his book that the treatment of certain diseases, such as epilepsy, colic, gout, and quartan fever, is more likely to involve magical and astrological procedure than that of other ailments such as earache and disorder of the spleen. This is also apt to be the case with other ancient and medieval medical works. But it is doubtful if the distinction can be sharply drawn that magic was resorted to more in those diseases which seemed most mysterious and incurable.

Physica.

The chief circumstance which renders some parts of Alexander's work more superstitious than others is that he sometimes, after concluding the usual medical description of the disease and prescriptions for it, adds a list of what he calls physical or natural medicines (φυσικά), which are for the most part ligatures and suspensions but involve also the employment of incantations and engraved images or characters. Apparently he calls these remedies *physica,* because they supposedly act by some peculiar property or occult virtue of the substance which is bound on

[1] Milward (1733), p. 168.

or suspended and constitute a sort of natural magic. Alexander explains that "since some cannot observe a diet nor endure medicine, they compel us in the case of gout to employ physical remedies and ligatures; and in order that the well-trained physician may be instructed in every side of his art and able to help all sick persons in every way, I come to this subject." [1] This rather apologetic tone and the fact that he separates the *physica* from his other remedies show that he regards them as not quite on the same level with normal medical procedure. He goes on to say, however, that although there are many of these "physical" remedies which are efficacious, he will write down only those proved true by long use. In discussing fevers he again justifies the inclusion of *physica* in much the same way and says that those now mentioned were learned by him during a long-extended practice and experience.[2] It is to be noted that some of these chapters on physical ligatures do not appear in the Latin version in three books, at least as it was printed in 1504.

Occult virtue of substances applied externally.

One ligature which is "quite celebrated and approved by many" and which instantly lessens the pain of ulcers in the feet, makes use of muscles from a wild ass, a wild boar, and a stork, binding the right muscles about the patient's right foot and the left muscles about the left foot. Some persons, however, do not intertwine the muscles of the stork with the others but put them separately into the skin of a sea-calf. Also they take care to bind the other muscles about the patient's feet when the moon is in the west or in a sterile sign and approaching Saturn. Others bind on the tendons and claws of a vulture, or the feet of a hare who should remain alive.[3] Alexander seems to regard the carcass of the ass as especially remedial in the case of epilepsy. In Spain he learned to use the skull of an ass reduced to ashes and he recommends employing the forehead and brain of an

[1] Stephanus I, 312; Puschmann II, 579.

[2] Stephanus I, 345, see also 296 and 339; Puschmann I, 407, 437.

[3] Stephanus I, 312; Puschmann II, 579.

ass as amulets.[1] A suspension for quartan fever consists of a live beetle firmly fastened on the outside of a red linen cloth and hung about the neck. "This is true and often tested by experience," Alexander assures us. Also excellent for this purpose are hairs from a goat's cheek or a green lizard combined with clippings of the patient's finger nails and toe nails. It is confirmed by the testimony of all "natural" physicians that the blood *qui primus a virgine fuerit excretus* is naturally hostile to quartan fever. Even if the girl is not chaste, the blood will be efficacious, if applied to the patient's right hand or arm.[2] Alexander knew a man who treated quartan fever by giving an undergarment of the patient to a woman in childbirth to wear, after which the patient wore it again and was cured "miraculously by some antipathy and occult influence." [3]

Other things used as ligatures and in amulets.

The materials employed in Alexander's therapeutics are sometimes those which we associate especially with magic arts, such as the hair and nail-parings already mentioned. Against epilepsy he employs nails from a cross or wrecked ship, or the blood-stained shirt of a gladiator or criminal who has been slain. The nails are bound to the patient's arm; the shirt is burned and the patient given the ashes in wine seven times. The use of a nail from a cross is a method ascribed to Asclepiades. Other materials recommended by Alexander against gout and epilepsy include the herb night-shade, the stones magnet and aetites, blood of a swallow and urine of a boy, chameleons in varied forms, and the stones found in dissected swallows of which we have heard before and shall hear yet again. For Alexander these stones are black and white, but he states that they are not found in all young swallows but are said to appear only in the first-born, so that one often has to dissect a great many birds before one finds any. In these passages on *Physica* Alexander cites such authors of magical reputation

[1] Stephanus I, 156; Puschmann I, 565.

[2] Stephanus I, 345; Puschmann I, 437.

[3] Καὶ θαυμαστῶς ὅπως ἀντιπαθείᾳ τινὶ καὶ λόγῳ ἀρρήτῳ.

as Ostanes and Democritus, and tells how the latter suffered in youth from epilepsy until an oracle from Delphi instructed him to make use of the worms in goats' brains. When a goat sneezes violently, some of these worms are expelled into his nostrils, whence they should be carefully extracted in a cloth without allowing them to touch the ground. Either one or three of them should then be worn about the epileptic's neck wrapped in the thin skin of a black sheep.[1]

Astrology and sculpture of rings.

One passage has already been cited where astrological conditions were observed. Alexander sometimes prescribes the day of the month upon which things shall be done; an oil, for instance, is to be prepared on the fifth of March.[2] In one place Alexander advises engraving upon a copper die a lion, a half-moon, a star, and the name of the beast. This is to be worn enclosed in a gold ring upon the fourth finger.[3] That the lion may not stand for a sign of the zodiac is suggested by another instruction concerning an engraved stone to be set in a gold ring, and which is to be carved with a figure of Hercules suffocating a lion.[4] For gout, however, one writes a verse of Homer on a copper plate when the moon is in Libra or Leo.[5] For colic one inscribes upon an iron ring with an octangular circumference a charm beginning, "Flee, flee, colic." [6]

Incantations.

The employment of such incantations is expressly justified by Alexander, who maintains that even "the most divine" Galen, who once thought that incantations were of no avail, came after a long time and much experience to be convinced that they were of great efficacy. Alexander then quotes from a treatise which is not extant but which he asserts is a work by Galen entitled, *On medical treatment in Homer*.[7] "So some think that incantations are like old-wives' tales and so I thought for a long while, but in process

[1] For the passages in this paragraph see Stephanus I, 156-7, 313; Puschmann I, 561, 567-73.

[2] Stephanus I, 312.

[3] Stephanus I, 281; Puschmann II, 475.

[4] Stephanus I, 296; Puschmann II, 377.

[5] Stephanus I, 313.

[6] Stephanus I, 296; Puschmann II, 377.

[7] Stephanus I, 281; Puschmann II, 475.

of time from perfectly plain instances I have become persuaded that there is force in them, for I have experienced their aid in the case of persons stung by scorpions. And no less in the case of bones stuck in the throat, which were straightway expelled by an incantation." Alexander himself thereupon continues, "If such is the testimony of divinest Galen and many other ancients, what prevents us too from communicating to you those which we have learned from experience and which we have received from trustworthy friends?"

Conjuration of an herb.

Both incantations and observance of astrological conditions play an important part in the instructions given by Alexander for digging and plucking with imprecations an herb to be used in the treatment of fluxions of hands or feet. "When the moon is in Aquarius under Pisces, dig before sunset, not touching the root. After digging with two fingers of the left hand, namely, the thumb and middle finger, say, 'I address you, I address you, sacred herb. I summon you to-morrow to the house of Philia to stay the fluxion of feet and hands of this man or this woman. But I adjure you by the great name, Iaoth, Sabaoth, God who established the earth and fixed the sea abounding in fluid floods, who desiccated Lot's wife and made her a statue of salt, receive the spirit of thy mother earth and its powers, and dry up this fluxion of feet or of hands of this man or woman.' On the morrow ere sunrise, taking the bone of some dead animal, dig up the root, and holding it say, 'I adjure you by the sacred names, Iaoth, Sabaoth, Adonai, Eloi,' and sprinkle a pinch of salt on that root, saying, 'As this salt is not increased, so be not the ailment of this man or of this woman.' Then bind one end of the root to the patient, taking care that it is not moist, and suspend the rest of it over the fire for 360 days." [1] The mention of mother earth in this charm perhaps indicates an ultimate pagan origin, but the allusions to one God, and to incidents in the Old Testament, and the use of names of spirits show Jewish

[1] Stephanus I, 314; Puschmann II, 585.

or Christian influence, while the number 360 perhaps points to the Gnostics.

Medieval version seems less superstitious than the original text.

While in conformity with the character of our investigation we have emphasized those passages in Alexander which are suggestive of magic and its methods, it should be said that many of the passages which we have cited are apparently [1] not found in the medieval Latin versions which seem to omit many, although not all, of the chapters devoted to physical ligatures. Here then apparently is a case where the early medieval translator and adapter, instead of retaining and emphasizing the superstition of the past, has largely purged his text of it. But we have next to consider a Latin work, written apparently about the year 400 A. D. and known to us through two manuscripts of the ninth century, in which magic is far more rampant than in any version of Alexander of Tralles. Judging, however, from the small number of extant manuscripts, it was less influential through the medieval period than was Alexander's book.

Marcellus: date and identity.

The *De medicamentis* opens in one of the two extant manuscripts with a dedicatory letter from "Marcellus, an illustrious man of the main office of Theodosius the Elder (?)" to his sons.[2] This ascription is generally accepted as genuine, and Grimm believed this to be the same Marcellus as the physician who is gratefully mentioned, together with his sons, then mere infants, in the letters of Libanius, whose severe headaches Marcellus had alleviated, and as the *Marcellus magister officiorum* who is mentioned twice in the Theodosian Code under the year 395. The date of the *De medicamentis* may be further fixed from its including "a singular remedy for spleen which the patriarch Gamaliel recently revealed from proved experiments." This

[1] If the MSS, which I have not examined, agree with the 1504 edition.

[2] Both in BN 6880 and the edition of Basel, 1536, "Marcellus vir inluster ex magno officio Theodosii Sen. filiis suis salutem d(icit)." In the MS, however, a later hand has written above the now faded line an incorrect copy in which "Theodosii Sen." is replaced by "theodosiensi." Helmreich (1889), on the other hand, has replaced "ex magno officio" by "ex magistro officio." It is perhaps open to doubt whether the "Sen." goes with "Theodosii" or "Marcellus."

Gamaliel was Jewish patriarch at Constantinople from some time before 395 on to 415 or later. The question, however, of Marcellus' authorship is complicated by the fact that he is twice cited in the work itself. One of these passages concerns an "oxyporium which Nero used for the digestion, which Marcellus the eminent physician revealed, which we too have tested in practice." [1] This sounds as if some later person had had a hand in the work as it has reached us, since Marcellus himself would scarcely have cited another person of the same name without some distinguishing epithet. Furthermore Aëtius cites a Marcellus for a passage which does not appear in the *De medicamentis* concerning wolfish or canine insanity, in which men imagine themselves to be wolves or dogs and act like them during the night in the month of February. But the *De medicamentis* as a whole is of the character promised by Marcellus in the introductory letter to his sons and so may be taken as his work.

"Marcellus Empiricus."

The empiricism which we have already noted in Alexander of Tralles becomes most pronounced and most extreme in Marcellus, who indeed is often called Marcellus Empiricus on this account, and many of whose chapter and other headings [2] terminate with these words descriptive of their contents, "various rational and natural remedies learned by experience" (*remedia rationabilia et physica diversa de experimentis*). In his preface, too, he speaks of his book not as *De medicamentis* but as *De empiricis.* He has, it is true, utilized "the old authorities of the medical art set down in the Latin language," and likewise more recent writers and "the works of studious men" who were not especially trained in medicine; but he also includes what he has learned from hearsay or from personal experience, and "even remedies chanced upon by rustics and the populace and simples which they have tested by experience." One prescription, which he characterizes as efficacious beyond human hope and incapable of being satisfactorily

[1] Cap. 20 (1889), p. 204.

[2] In BN 6880 there are other headings written in capitals than those which mark the openings of the 36 chapters.

lauded, he purchased from an old-wife of Africa who cured many at Rome by it, while the author himself has employed it in the cure of "several persons neither of humble rank nor unknown, whose names it is superfluous to mention." This remedy is a concoction of such things as ashes of deer-horn, nine grains of white pepper, a little myrrh, and an African snail pounded shell and all while still alive in a mortar and then mixed with Falernian wine. Very detailed and explicit directions are given as to its preparation and administration, including an instruction to drink the dose facing towards the east.[1] In another passage Marcellus says of certain compounds, "If there is any faith, both I myself have always found them by experience to be useful remedies and I can state that others are of the same mind; and I will add this, that other medicines can not compare to this liniment, which in similar cases several of my friends, whom I trust as I do myself, have affirmed on oath they have found by experience a remarkable cure."[2] Of an eye-remedy he remarks, "And that we may believe the author of this remedy from experience, he states that after he had been blind for twelve years it restored his sight within twenty days."[3] Marcellus also frequently couples marvelousness with experimentation, saying, "You will experience a wonderful remedy." In one passage he uses the word "experiment as a verb rather than as a noun, coining a new expression, *experimentatum remedium,*[4] but his commonest expressions are *de experimento* or *de experimentis, expertum,* and *experieris* or *experietur.*[5] Some of his "experi-

[1] Cap. 29 (1889), pp. 304-6.

[2] Cap. 35 (1889), p. 361.

[3] Cap. 8 (1889), p. 80.

[4] Cap. 5 (1889), p. 49.

[5] For such mentions of experience and experiment see the following passages in the 1889 edition, numbers referring to page and line: 31, 7; 34, 3; 35, 14; 44, 2; 53, 1; 58, 21; 64, 34; 65, 30; 66, 26; 72, 22; 73, 7; 74, 2; 77, 9; 80, 28; 81, 29; 89, 3 and 29; 96, 14 and 31; 102, 27; 120, 32; 123, 15; 129, 21; 133, 10; 145, 33; 148, 25; 149, 26; 160, 18; 176, 5; 178, 25; 186, 15; 190, 20; 192, 31; 211, 1; 222, 18; 224, 31; 230, 3; 235, 15; 236, 14; 239, 8 and 26; 242, 8 and 23; 248, 20; 256, 9; 258, 5; 264, 21; 276, 35; 281, 19 and 27; 282, 15; 308, 21; 312, 6 and 19 and 22; 314, 25; 326, 28; 327, 13; 334, 29; 343, 23; 351, 23 and 25; 353, 4; 354, 19; 356, 6; 362, 32; 370, 22 and 37.

ences" really are purposive experiments, as where one discovers whether a tumor is scrofulous by applying an earthworm to it. Then put the worm on a leaf and if the tumor was scrofulous, the worm will turn into earth.[1] The following experiment indicates that sufferers from spleen should drink in vinegar the root or dried leaves of the tamarisk. Give tamarisk to a pig to eat for nine days, then kill the animal and you will find it without a spleen.[2]

Superstitious character of his medicine.

As Marcellus appeals the most to experience, so he is by far the most given to superstition and folk-lore of our three authors. Practically his entire work is of the character of the passages devoted to *Physica* by Alexander of Tralles. He indulges in no medical theory, he does not diagnose diseases, nor prescribe a regimen of health in the form of bathing, diet, and exercise. His work is wholly composed of medicaments and for the most part empirical ones. Besides the elaborate compounds which were so frequent in Aëtius and Alexander, he is extremely addicted to absurd rigmarole and all sorts of superstitious practices in the application or administration of medicinal simples. His pharmacy includes not only herbs and gems, to which he attributes occult virtue and which he sometimes directs to have engraven with characters and figures, such as SSS or a dragon surrounded with seven rays[3]—the emblem of the Agathodaemon, but also all kinds of animals, reptiles, and parts of the same, after the fashion of Pliny's medicine. He is constantly calling into requisition such things as the ashes of a mole, the blood of a bat, the brains of a mouse, the gall of a hyena, the hoofs of a live ass, the liver of a wolf, woman's milk, sea-hares, a white spider with very long legs, and centipedes or multipedes, especially the variety that rolls up into a ball when touched. But it is scarcely feasible to separate Marcellus' materials from his procedure, so we will begin to consider them together in some prescriptions where animals play the leading part.

[1] Cap. 15 (1889), p. 146.

[2] Cap. 23 (1889), p. 239.

[3] Caps. 20 and 24 (1889), pp. 208 and 244.

Preparation of goat's blood.

For those suffering from stone is recommended a remedy prepared in the following fashion. In August shut up in a dry place for three days a goat, preferably a wild one who is one year old, and feed him on nothing but laurel and give him no water to drink; finally on the third day, which should fall on a Thursday or Sunday, kill him. Both the person who kills the goat and the patient should be chaste and pure. Cut the goat's throat and collect his blood—it is best if the blood is collected by naked boys—and burn it to an ash in an earthen pot. After combining it with various herbs and drugs, there are further directions to follow as to how it may best be administered to the patient. Marcellus, by the way, affirms that adamant can be broken only by goat's blood.[1]

A rabbit's foot.

The following prescription involves the familiar superstition that a rabbit's foot is lucky: "Cut off the foot of a live rabbit and take hairs from under its belly and let it go. Of those hairs or wool make a strong thread and with it bind the rabbit's foot to the body of the patient and you will find a marvelous remedy. But the remedy will be even more efficacious, so that it is hardly credible, if by chance you find that bone, namely, the rabbit's ankle-bone, in the dung of a wolf, which you should guard so that it neither touches the earth nor is touched by woman. Nor should any woman touch that thread made of the rabbit's wool." Marcellus further recommends that in releasing the rabbit after taking its wool you should say, "Flee, flee, little rabbit, and take the pain away with you." [2]

Magic transfer of disease.

Of such magical transfer of disease to other animals or objects there are a number of examples. Toothache may be stopped by standing on the ground under the open sky and spitting in a frog's mouth and asking it to take the toothache away with it and then releasing it.[3] Even consumptives who seem certain to die and who labor continually with an unbearable cough, may be cured by giving them

[1] Cap. 26 (1889), pp. 264-6.

[2] Cap. 29 (1889), p. 311; and see cap. 28, p. 298.

[3] Cap. 12, p. 123.

to drink for three days the saliva or foam of a horse. "You will indeed cure the patient without delay, but the horse will die suddenly." [1] Splenetic persons are benefited by imposing anyone of three kinds of fish upon the spleen and then replacing the fish alive in the sea.[2] Warts may be got rid of by rubbing them with something the moment you see a star falling in the sky; but if you rub them with your bare hand, you will simply transfer them to it.[3] Another superstition connected with falling stars which Marcellus records is that one will be free from sore eyes for as many years as he can count numbers while a star is falling.[4] The first time you hear or see a swallow, hasten silently to a spring or well and anoint your eyes with the water and pray God that you may not have sore eyes that year, and the swallows will bear away all pain from your eyes.[5] With slight variations the same procedure may be employed to prevent toothache. In this case you fill your mouth with water, rub your teeth with the middle fingers of both hands, and say, "Swallow, I say to you, as this will not again be in my beak, so may my teeth not ache all year long." [6] Marcellus advises anyone whose nose is stuffed up to blow it on a piece of parchment, and, folding this up like a letter, cast it into the public way,[7]—which would very likely spread the germs, if not take away the cold.

Pliny and Marcellus compared on green lizards as eye cures.

In his preface Marcellus refers to Pliny as one of his authorities and many of his quaint animal remedies will be found substantially duplicated in the *Natural History*. Both, for example, state that one can stop one's nose from running by kissing a mule.[8] Marcellus, however, adds much from other sources or of his own. This may be illustrated by comparing their accounts of the use of lizards to cure eye diseases.[9] Marcellus omits the following portion of Pliny's account: "Some shut up a green lizard in a new earthen pot,

[1] Cap. 16, p. 166.
[2] Cap. 23, p. 238.
[3] Cap. 34, p. 357.
[4] Cap. 8, p. 69.
[5] Cap. 8, p. 66.
[6] Cap. 12, p. 125.
[7] Cap. 10, p. 113.
[8] Cap. 10, p. 112; NH 30, 11.
[9] Cap. 8, p. 68; NH 29, 38.

and they mark the little stones called *cinaedia,* which are bound on for tumors of the groin, with nine signs and take out one daily. On the ninth day they let the lizard go, and keep the pebbles for pains of the eyes." Pliny next proceeds: "Others put earth under a green lizard that has been blinded and shut it up in a glass vase with rings of solid iron or gold. When through the glass the lizard is seen to have recovered its sight, it is released and the rings are used for sore eyes." This recipe is in Marcellus who, however, words it differently and adds that the lizard must be blinded with a copper needle, that the rings may be of silver, electrum, or copper, that the vase must be carefully sealed and opened on the fifth or seventh day following, and that one should not only wear the rings afterwards on one's fingers but also frequently apply them to one's eyes and strengthen the sight by looking through them. He further cautions to leave the vase in a clean grassy spot, to collect the rings only after the lizard has departed, to catch the lizard in the first place on a Thursday in September between the nineteenth and twenty-fifth day of the moon, and to have the operation performed by a very pure and chaste man. Marcellus also states that an amulet made either of the eyes of the said lizard enclosed in a lead bull or gold coin, or of its blood caught on clean wool and wrapped in purple cloth will effectually prevent eye diseases. Meanwhile Pliny for his part has gone on to tell how efficacious the ashes of green lizards are.

More lizardry.

Marcellus employs green lizards in other connections which are not paralleled in Pliny. To stay colic one binds about the patient three times with an incantation a string with which a copper needle has been threaded and drawn through a lizard's eyes, after which the reptile is released at the same point where it was captured.[1] In another passage Marcellus recommends the drawing by a silver needle of threads of nine different colors other than black or white through the eyes of a new-born puppy before they open and

[1] Cap. 29, p. 313.

ita ut per anum eius exeant, after which the puppy is to be thrown into the river.[1] But to return to our lizards. For those suffering from liver complaint the liver of a lizard is to be extracted with the point of a reed and bound in purple or black cloth to the patient's right side or suspended from his arm, while the lizard is to be dismissed alive with these words, "Lo, I send you away alive; see to it that no one whom I touch henceforth has liver complaint." [2] To insure a wife's fidelity one touches her with the tip of a lizard's tail which has been cut off by the left hand.[3] Here again the lizard is released but apparently is not expected to survive for long, since one is instructed to "hold the tail shut in the palm of the same hand until it dies." In a fourth example the lizard is neither mutilated nor released but hung in the doorway of a splenetic's bedroom where it will touch his head and left hand as he comes and goes.[4]

Use of stones and an herb.

One or two other prescriptions may be added where the procedure is connected with herbs or stones rather than with animals. On entering a city one is advised to pick up some of the pebbles lying in the road before the city gate, stating that they are being collected for headache. Then bind one of them on the head and throw the others behind your back without looking around.[5] A certain herb must be gathered on Thursday in a waning moon. When it is administered in drink, the recipient must take it standing and facing the east. He receives the cup from the right hand and then, in order not to look back, returns it to the left to him who gave it. Only these two persons should touch the drink.[6]

Right and left: number.

Right and left, as just illustrated, are much observed in Marcellus' medicine. When a tooth aches on the left side of the mouth, a hot cooked dried bean is applied to the right elbow for three days, a process which is reversed

[1] Cap. 29, p. 314. Pliny has a similar procedure with a frog and a reed.

[2] Cap. 22, p. 230.

[3] Cap. 33, p. 347, "mulierem verendaque eius dum cum ea cois tange."

[4] Cap. 23, p. 239.

[5] Cap. 1, p. 34.

[6] Cap. 25, p. 247.

if the tooth is on the right side.[1] The following exercise recommended for a stiff neck would seem to stand more chance of success than most of Marcellus' prescriptions. While fasting the patient should spit on his right hand and rub his right thigh, and then do the same with his left hand and thigh. Thrice repeated this is warranted to work an immediate cure.[2] A ring worn on the middle finger of the left hand is said to stop hiccough.[3] The power of the planets or of mere number is indicated in the advice, given several times, to make seven knots in a string.[4] Once instructions are given to make as many knots as there are letters in the patient's name.[5]

Incantations and characters.

Incantations and characters, as has already been incidentally illustrated, abound in Marcellus' pages. Some are in Greek, some in Latin, some perhaps in Celtic; many, as we have seen, are coherent statements, commands, or requests; many others are to all appearance a jargon of meaningless words, like the jingle, *Argidam, margidam, sturgidam,*[6] which is to be repeated seven times on Tuesday and Thursday in a waning moon to cure toothache. Marcellus well calls one of these *carmen idioticum.*[7] For stomach and intestinal troubles he recommends pressing the abdomen with the left thumb and saying, "Adam, bedam, alam, betur, alem, botum." This is to be repeated nine times, then one touches the earth with the same thumb and spits, then says the charm nine more times, and again for a third series of nine, touching the ground and spitting nine times also. *Alabanda, alabandi, alambo* is another incantation, variously repeated thrice with hands clasped above and below the abdomen. Yet another consists in rubbing the abdomen with the left thumb and two little fingers and saying, "A tree stood in the middle of the sea and there hung an urn full of human intestines; three virgins went

[1] Cap. 12, p. 126.
[2] Cap. 18, p. 178.
[3] Cap. 17, p. 176.
[4] Cap. 32, pp. 337, 338, 340.
[5] Cap. 8, p. 70.
[6] Cap. 12, p. 123.
[7] Cap. 36, p. 379. Marcellus employs the phrase, of course, to indicate a private or personal incantation, and as a matter of fact it is somewhat less absurd than a number of others.

around it, two make it fast, one revolves it." As you repeat this thrice, you touch the ground thrice and spit, but if the charm is for veterinary purposes, for the words "human intestines" should be substituted "the intestines of mules" or horses or asses as the case may be.[1] The following is a specimen of the characters prescribed by Marcellus: [2]

Λ Ψ Μ Θ Κ Ι Α
Λ Ψ Μ Θ Κ Ι Α
Λ Ψ Μ Θ Κ Ι Α

The art of medicine survives the barbarian invasions.

It is perhaps worth while to point out in concluding this chapter that apparently at no time during the period of barbarian invasions and early medieval centuries did medical practice or literature cease entirely in the west. We have seen that there is reason to suspect that portions of the work ascribed to Marcellus may be contributions of the centuries following him, and that there were early medieval Latin translations of the works of Oribasius and Alexander of Tralles. Furthermore, the laws of the German kingdoms, the allusions of contemporary chroniclers and men of letters, the advice of Gregory the Great to a sick archbishop to seek medical assistance, and many other bits of evidence [3] show that physicians were fairly numerous and in good repute, and that medieval Christians at no time depended entirely upon the healing virtues of relics of the saints or other miraculous powers credited to the church or divine answer to prayer.

[1] Cap. 28, p. 301.

[2] Cap. 29, p. 310. For further instances of incantations and characters in the *De medicamentis* see page 110, lines 18-27; 111, 26-33; 112, 29 - 113, 2; 116, 8-11; 133, 18-22, 26-31; 139, 17-26; 142, 19-26; 149, 4-11; 151, 18-33; 152, 9-14, 19-24; 180, 1-3; 220, 11-20; 221, 2-6; 223, 15-18; 241, 1-6, 14-22; 244, 26-28; 248, 16-19; 260, 22-24; 295, 18-22; 333, 9-15; 382, 16-18.

[3] Daremberg (1870) I, 257-8.

CHAPTER XXVI

PSEUDO-LITERATURE IN NATURAL SCIENCE OF THE EARLY MIDDLE AGES

General character—*Medicine of Pliny*—*Herbarium of Apuleius*—Specimens of its occult science—A "Precantation of all herbs"—Other treatises accompanying the *Herbarium*—*Cosmography of Aethicus*—Its medieval influence—Character of the work—Its attitude to marvels—The *Geoponica*—Magic and astrology therein—Dioscorides—Textual history of the *De materia medica*—Alterations made in the Greek text—Dioscorides little known to Latins before the middle ages—Partial versions in Latin—*De herbis femininis*—The fuller Latin versions—Peter of Abano's account of the medieval versions—Pseudo-Dioscorides on stones—Conclusions from the textual history of Dioscorides—Macer on herbs; its great currency—Problem of date and author—Virtues ascribed to herbs—*Experiments of Macer.*

General character.

A CLASS of writings which seems to have been very characteristic of the waning culture of the declining Roman Empire and the scanty erudition of the early medieval period were the brief epitomes of, or disorderly collections of fragments from, the writers of the classical period. Such works often passed under the name of some famous author of the previous period and sometimes are more or less based upon his writings. Most of the works in the field of natural science are of such derivative or pseudo-authorship: the *Medicine* of the Pseudo-Pliny, the *Herbarium* of the Pseudo-Apuleius, the geographical work ascribed to Aethicus, the *Geoponica,* the treatises on herbs attributed to Macer and Dioscorides. Indeed, the whole textual history of the latter's *De materia medica* is so full of vicissitudes and uncertainties that I have postponed its treatment until this chapter. The names of the actual compilers or abbreviators of these works are usually unknown and it is also usually impossible to date them with any approach to accuracy.

Roughly speaking of them as a whole, they may be said to have gradually taken on their present form at almost any time between the third and tenth centuries. In the case of these works of natural science at least, it is not quite fair to class them all as brief epitomes or disorderly collections. In some we see an obvious attempt to rearrange the old materials in a form more convenient for present use. In others to the stage of abbreviation from ancient authors has succeeded another stage of later additions from other sources.

Medicine of Pliny.

The *Medicina*, or *Art of Medicine,* of the Pseudo-Pliny [1] consists of three books in which medical passages, drawn from Pliny's *Natural History,* are rearranged according to diseases instead of, as in the genuine Pliny, by simples. The first two books deal with diseases of the human body in descending order from top to toe and from headache to gout, a favorite arrangement throughout the course of medieval medicine. The last book then considers afflictions which are not necessarily connected with any particular part of the body, such as wounds and fevers. Thus this compilation attests Pliny's medieval influence and the practical use made of his work, while it of course continues much of his medical magic and superstition. The compiler's rearrangement is an essential one, if the medical recommendations of the *Natural History* were to be made available for ready reference. In this case, therefore, the epitomizer has rather improved upon than disordered the arrangement of the original. This compilation is believed to have been used by Marcellus Empiricus, and a *Letter of Plinius Secundus to his friends about medicine,* which Marcellus gives along with other medical epistles, is thought to be the preface of the abbreviator, who in that case depicts himself as composing his volume so that his friends and himself when traveling may avoid the payment of exorbitant fees asked by strange physicians. If we can regard everything in the

[1] *Plinii Secundi Iunioris de medicina libri tres,* ed. V. Rose, Lipsiae, 1875. V. Rose, "Ueber die Medicina Plinii," in *Hermes,* VIII (1874) 19-66.

work of Marcellus as we have it as having been written by 400, the *Medicine of Pliny* must have been written during the declining Roman Empire. The manuscripts used by Rose in his edition were of the tenth and twelfth centuries. There is also a later version of the *Medicine of Pliny* in five books,[1] of which the two last are entirely new additions, the fifth being an extract from the old Latin translation of Alexander of Tralles. And in the first three books the earlier Pseudo-Pliny has been worked over with additions. The Pseudo-Pliny is also embodied with alterations and accompanied by some prayers and incantations in a tenth century manuscript at St. Gall.[2]

The Herbarium of Apuleius.

Several works besides the six commonly regarded as genuine [3] were attributed to Apuleius in the middle ages, grammatical [4] and rhetorical [5] treatises, the Hermetic *Asclepius,*[6] a treatise on physiognomy,[7] and the very widespread *Sphere of Life and Death,* of which we shall treat in another chapter.[8] We shall now consider the *Herbarium of Apuleius,*[9] the one of his spurious works, which has most to do with the world of nature, and, with the exception of the brief *Sphere,* the one which occurs most often in the manuscripts. The *Herbarium* was first printed about 1480 by the physician of Pope Sixtus IV

[1] *C. Plinii Secundi Medicina,* ed. Thomas Pighinuccius, Rome, 1509.

[2] Codex St. Gall 751; described by V. Rose, *Hermes,* VIII, 48-55; *Anecdota* II, 106.

[3] For the list of his six genuine works see above p. 222.

[4] *De nota aspirationis* and *De diphthongis,* ed. Osann, Darmstadt, 1826, with *De orthographia,* a forgery by a sixteenth century humanist.

[5] Περὶ ἑρμηνείας, sometimes printed as the third book of the *De dogmate Platonis.* Some scholars, however, regard it as genuine, and there are a number of MSS of it from the 9th, 10th, and 11th centuries. See Schanz (1905), 127-8.

[6] See above p. 290.

[7] See Schanz (1905), 139-40.

[8] See below p. 683. Schanz fails to mention it among the apocryphal works of Apuleius.

[9] H. Köbert, *De Pseudo-Apulei herbarum medicaminibus,* Bayreuth, 1888. Schanz (1905) 138, mentions only continental MSS, although there are numerous MSS of it in the British Museum and Bodleian libraries, some of which have been used and others described by O. Cockayne in his edition of the *Herbarium* and the other treatises accompanying it in his *Leechdoms, Wortcunning, and Starcraft of Early England,* Vol. I (1864) in RS XXXV. Nor does Schanz note Cockayne's book.

from a manuscript at Monte Cassino, and then, after various other editions, was included in 1547 in the collection of ancient Latin medical writers issued by the Aldine Press. We are told, however, that with the close of the fifteenth century the Apuleius began to be superseded by German herbals. The medieval manuscripts of the *Herbarium* are often noteworthy for their illuminations of the herbs in vivid colors. Those of the mandragora root are especially interesting, showing it as a man standing on the back of a dog or a human form with leaves growing on the head and led by a dog chained to his waist.[1] The oldest manuscripts are of the sixth century, and there are some in Anglo-Saxon, but as one would expect, the work underwent many additions and alterations, and different manuscripts of it vary considerably. The author is usually spoken of as Apuleius the Platonist and is sometimes said to have received his work from the centaur Chiron, the master of Achilles, and from Esculapius.[2]

Specimens of its occult science.

In the *Herbarium* the plants are listed and described and their virtues, especially medicinal, stated. Usually the names for each herb in several languages or regions are given—Latin, Greek, Punic, Biblical (by the Prophets),

[1] See Sloane 1975, a vellum MS of the 12th or early 13th century written in fine large letters and beautifully illuminated; Ashmole 1431, end of 11th century, and 1462, 13th century, fol. 45r. Harleian 4986, Apuleii Platonici de medicamentis cum figuris pictis, is another early illuminated English MS. Cockayne I, lxxxii, does not date it, but the MSS catalogue lists it as tenth century. In CU Trinity 1152, 14th century, James (III, 162-3) estimates the number of colored drawings as between 800 and 1000; he describes only a few. Singer (1921) reproduces a number of such illuminations from MSS of the *Herbarium* and of Dioscorides.

[2] Lucca 236, 9-10th century, "Herbarium Apuleii Platonici quem accepit a Chironi magistro Achillis et ab Escolapio explicit feliciter." In Cotton Vitellius C-III, early 11th century, in Anglo-Saxon, although the title reads, "The Herbarium of Apuleius the Platonist which he received from Esculapius and Chiron the centaur, the master of Achilles," a full page painting shows Plato and Chiron receiving the volume from Aesculapius (Cockayne, I, lxxxviii). And Sloane 1975 and Harleian 1585 speak of the *Herbarium* as "Liber Platonis Apoliensis." In a 15th century MS (Rawlinson C-328, fol. 113v-, Incipit de herbis Galieni Apolei et Ciceronis) Galen and Cicero, who perhaps replace Chiron and Aesculapius, are associated with Apuleius as authors.

Egyptian, Syrian, Gallic, Dacian, Spanish, Phrygian, Tuscan. By no means all of these are listed in every case, however. The virtues of the herbs often operate in an occult manner, or procedure suggestive of magic is involved in collecting or applying them. Often diseases are cured merely by holding an herb in the hand, wearing it with a string about the neck, or placing it behind one ear, or wearing it in a ring. Lunatics, for example, are treated by binding an herb about the neck with red cloth when the moon is waxing in the sign of the bull or the first part of the scorpion. Not only does observance of astrology assist the medicinal application of herbs; plants are in turn of assistance in the pursuit of astrology. To learn under the rule of what star you are, be in a state of purity, pluck the herb Montaster, keep it in a bit of clean linen until you find a whole grain of wheat in a loaf of bread, then place this with the herb under your pillow and pray to the seven planets to reveal your guardian star to you in your sleep. Indeed prayers and incantations are frequently employed and in one case must be repeated nine times. Sometimes the herb itself is addressed, as in the conjuration, "Herb Erystion, I implore you to aid me and cheerfully afford me all your virtues and cure and make whole all those ills which Aesculapius and Chiron the centaur, masters of medicine, healed by means of you." Sometimes the earth is conjured as in the prayer beginning, "Holy goddess Earth." Such prayers are scarcely consonant with Christianity and in some manuscripts have been omitted and replaced by the Lord's Prayer or other Christian forms, or left in with their wording slightly altered to avoid paganism.[1] Personal purity and clean clothing are often en-

[1] Daremberg (1853), 11-12, said that the pagan incantations were preserved intact in a number of MSS at Oxford and Cambridge. Conjurations of herbs are not limited to the Pseudo-Apuleius in medieval MSS but sometimes occur singly as in Perugia 736, 13th century, where at fol. 267 a 14th century hand has added a passage in Latin which may be translated: "In the name of Christ, Amen. I conjure you, herb, that I may conquer by lord Peter etc. by moon and stars etc. and may you conquer all my enemies, pontiff and priests and all laymen and all women and all lawyers

joined upon those gathering the herbs and such instructions are added as to mark the circle about the plant with gold, silver, ivory, the tooth of a wild boar, and the horn of a bull, or to fill the hole with honeyed fruits. Some herbs protect their bearers from all serpents or even from all evils. Others, like asparagus if you use a dry root of it to sprinkle the patient with spring water, break the spell of witchcraft. Asparagus is also beneficial for toothache and wonderfully relieves a tumor or bladder trouble, if it is boiled in water and drunk by the patient fasting for seven days and also used in bathing for a number of days. But one must be careful not to go out in the cold during this time nor to take cold drinks.[1]

A "Precantation of All Herbs."

In some manuscripts a "Precantation of all herbs" is placed at the beginning of the treatise.[2] It prescribes such procedure as holding a mirror over the herb before plucking it before sunrise under a waning moon. The person plucking the herb and uttering the incantation must be barefoot, ungirded, chaste, and wear no ring. The plant is adjured not only "by the living God" and "the holy name of God, Sabaoth," but also by Seia, the Roman goddess of sowing, and by "GS," which presumably stands for *Gaia Seia,* an expression which is once written out in full. Some meaningless words are also repeated.

Other treatises accompanying the *Herbarium.*

The *Herbarium* is often accompanied in the manuscripts by other treatises on herbs ascribed to Dioscorides and Macer, of which we shall speak presently; by a work on the medicinal properties of animals, or more particularly of quadrupeds, by Sextus Papirius Placidus [3] Actor [4]—an

who are against me etc." In Sloane 1571, 15th century, fols. 1-6, at the close of fragments of a Latin-English dictionary of herbs is a Latin prayer entitled, *Benedictio omnium herbarum.*

[1] The above passages are from Sloane 1975 and the edition of 1547.

[2] Ashmole 1431, 11th century, fol. 3r, "In nomine domini incipit herboralium apuleii platonis quod accepit ascolapio et chirone centauro magistro. Lege feliciter. Precantatio omnium herbarum ad singulas curas." CU Trinity 1152, 14th century, fol. 1. Gonville and Caius 345, 14th century, fol. 89v.

[3] Or Papyriensis Placitus.

[4] Perhaps merely for "auctor." ed. Fabricius, Bibl. Graec. XIII, 395-423, *Sexti Placiti liber de medicina ex animalibus.*

otherwise quite unknown personage;[1] by a "letter concerning a little beast" from the king of Egypt or Aesculapius to the emperor Octavian Augustus;[2] and by introductory letters, such as we find prefaced to the *De medicamentis* of Marcellus Empiricus, of "Hippocrates to his Moecenas"[3] and "Antonius Musus to Moecenas Agrippa." The epistle of the Egyptian king or Aesculapius to Augustus, however, really forms the introduction or opening chapter to the treatise of Sextus Papirius Placidus on the medicinal properties of animals, and after the little beast or quadruped called *mela* or *taxo*[4] follow fast the stag, serpent, fox, hare, scorpion, and so forth. As for the *taxo,* Augustus is told that by means of it he can protect himself from sorcerers, avoid defections in his army, and preserve his troops from the pestilence which the barbarians bring; and the city of Rome from both pestilences and fires. To this end a lustration should be performed with its flesh, and it should then be buried at the city gates. One way to appropriate its virtue is to extract its large teeth, repeating a jargon of strange words the while.

Cosmography of Aethicus.

Another characteristic product of declining antique learning and of early medieval effort is found in the field of geography in the *Cosmography* of Aethicus Istricus, translated into Latin by the priest Jerome (*Hieronymus Presbyter*). The oldest manuscript is one of the eighth

[1] In Montpellier 277, 15th century, "Liber Sesti platonis de animalibus," perhaps because the Apuleius of the *Herbarium* is called a Platonist. In Digby 43, late 14th century, fol. 15, "Liber Septiplanti Papiensis de bestiis et avibus medicinalis." In Rawlinson C-328, 15th century, fol. 128, "Incipit liber Papiriensis ex animalibus ex avibus." The work is sometimes found in juxtaposition with a somewhat similar "Liber medicinalis de secretis Galieni," concerning which see below, chapter 64, II, 761.

[2] V. Rose (1875) 337-8 suggests that this is a fragment from a fuller work of Aesculapius to Augustus cited by Thomas of Cantimpré, Albertus Magnus, and Vincent of Beauvais. See also Peter of Abano, *De venenis,* cap. 5, "in epistola Esculapii philosophi ad Octavianum." But perhaps these writers refer to the entire work of Sextus Papirius.

[3] Ed. Ruellius, with Scribonius Largus, Paris, 1529.

[4] In a later medieval vocabulary *taxus* is given as a synonym for the animal called *camaleon: Alphita,* ed. Daremberg from BN 6954 and 6957 in De Renzi, *Collectio Salernitana,* III, 272-322.

century in the British Museum,[1] where it is also found in several other fairly early manuscripts [2] in the respectable company of Vitruvius, Vegetius, Sallust, and Suetonius,[3] as well as with the more congenial work of Solinus. This *Cosmographia* was not printed until 1852, when it was edited at Paris by M. d'Avezac and again in 1854 at Leipzig by M. H. Wuttke. It is an entirely different work from what had hitherto been repeatedly printed as the *Cosmography* of Aethicus but is really to be identified with fragments of Julian Honorius and Orosius. The Latin translator of our treatise had been identified in the middle ages with St. Jerome, the church father, and Wuttke still ascribed it to him, but Bunbury protested against this,[4] and Mommsen placed our treatise not earlier than the seventh century.[5]

Its medieval influence

Bunbury added, however, that the *Cosmography* "appears to have been much read in the middle ages, and is therefore not without literary interest." The apparent greatness of the names on the title page seems to have given the middle ages an exaggerated notion of the work's importance. Aethicus himself is spoken of as from Istria and according to the *Explicit* of at least one manuscript [6] was a Scythian, but this does not mean that his attitude towards learning was that of a Hun, for the same *Explicit* goes on to inform us that he was of noble lineage and, if I correctly

[1] Cotton Vespasian B, X, *6.

[2] Harleian 3859, called tenth century in the Harleian catalogue which is often incorrect in its dating, but 11th or 12th century by d'Avezac, Mommsen in his edition of Solinus, and Beazley, *Dawn of Geography,* I, 523. Royal 15-B-II and 15-C-IV, both of the 12th century. For other MSS at Paris, Leyden, and Rome see Beazley, *op. cit.*

[3] But after all is Suetonius any more respectable a historian than Aethicus and Solinus are geographers?

[4] Bunbury, *History of Ancient Geography,* II, Appendix: "How M. Wuttke can attach any value to such a production is to me quite incomprehensible; still more that he should ascribe the translation to the great ecclesiastical writer," Jerome. Bunbury believed that the work was not earlier than the seventh century. Beazley, *Dawn of Geography,* I, 355-63, is of the same opinion.

[5] In his edition of Solinus, p. xxvii, he contends that certain passages which Wuttke pointed out as common to Aethicus and Solinus are borrowed by Aethicus from Isidore who died in 636.

[6] Harleian 3859.

interpret the faulty syntax of its Latin, that from him the ethical philosophy of other sages drew its origins. Somewhat later Roger Bacon said in discussing faults in the study of theology in his day, "From the authorities of the philosophers whom the saints cite I shall abstain, except that I will strengthen the utterances of Ethicus the astronomer and Alchimus the philosopher by the authority of the blessed Jerome, since no one could credit that they had said so many marvelous things about Christ and the angels and demons and men who are to be glorified or damned unless Jerome or some other saint proved that they had said so." [1]

Character of the work.

As Bacon's words indicate, Christian influence is manifest in the *Cosmography,* although, as they also indicate, the original Aethicus is not supposed to have been a Christian, but, as one manuscript informs us, an Academic philosopher.[2] Oriental influence, too, is perhaps shown in flights of poetical language and unrestrained imagination, in a number of allusions to Alexander the Great, and in an extraordinary ignorance of early Roman history which leads the author to tell how Romulus invaded Pannonia and fought against the Lacedaemonians. "How great carnage," he exclaims, "in Lacedaemonia, Noricum and Pannonia, Istria and Albania, northern regions near my home, first at the hands of the Romans and the tyrant Numitor, then under the brothers Romulus and Remus, and later under the first Tarquin, the Proud." The author eulogizes Athens as well as Alexander, and mentions a people called *Turchi,* but whether or not he has Turks in mind would be hard to say.

Its attitude to marvels.

As we have it, the *Cosmography* cites both the Ethicus and the Alchimus to whom Roger Bacon referred. Indeed, our treatise does not pretend to be the original work of Ae-

[1] Steele, *Opera hactenus inedita,* 1905, Fasc. I, pp. 1-2.

[2] CUL 213, 14th century, fols. 103v-14, "Qui hunc librum legit intelligat Ethicum philosophum non omnia dixisse que hic scripta sunt, set Solinus (so James, but *Jeronimus* in d'Avezac, p. 237) qui eum transtulit sententias veritati consonas ex libro eiusdem excerpsit et easdem testimonias scripture nostre confirmavit. Non enim erat iste philosophus Christianus sed Ethnicus et professione Achademicus."

thicus, which it repeatedly cites, but is apparently the work of some epitomizer or abbreviator who intersperses remarks and comments of his own, and, according to one manuscript, makes the statements of Aethicus conform to Christian Scripture. From the volumes of the original work he makes only a few excerpts, professing to omit what is unheard of or unknown or seems too formidable, and including only with hesitancy a few bits concerning unknown races on the testimony of hearsay. The enigmas of Aethicus and other philosophers often give our abbreviator pause, and he regards as incredible the story of Aethicus that the Amazons nurse young minotaurs and centaurs who fight for them in return. Aethicus also tells of the wonderful armor of the Amazons which they treat with bitumen and the blood of their own offspring. In Crete Aethicus found herbs unknown in other lands which ward off famine. Very beautiful gems are mentioned, including those extracted from the brains of immense dragons and basilisks, but little is said of their virtues, occult or otherwise. Indeed, the amount either of specific information or specific misinformation in the book is very scanty. It deals largely in uncouth rhetoric, glittering generalities, and obscure allusion anent the wanderings of Aethicus over the face of the earth and the strange marvels which he encountered in distant lands. He is described as well versed in astrology and as reproving the astrologers of Scythia(?) and Mantua(?), and one passage vaguely speaks of the stars as signs of the present and future; but otherwise the abbreviator gives little evidence of knowledge of the subject, although Roger Bacon[1] cited *Ethicus Astronomicus in Cosmographia* as one of his authorities when discussing the question of Jesus Christ's nativity and its relation to the stars, and although Pico della Mirandola ranked the *Cosmography* as one of the most absurd of astrological works.[2] As for magic, in one passage *malefici* and *magi* are censured along with idolaters, and the

[1] Bridges I, 267-8.

[2] Cited by d'Avezac, pp. 257 and 267.

author presently speaks of vain characters and superstitious doctrines. But elsewhere a magician (*Pirronius magus*) is named as the inventor of ships and discoverer of purple. On the whole, in its loose and hazy way the *Cosmography* not only is romantic and religious enough to appeal to medieval readers, it also is of a character to offer encouragement, if not data, to a later and more detailed interest in alchemy, occult virtues, astrology, and magic.

The *Geoponica.*

Upon the subject of agriculture in the early middle ages we have the collection known as the *Geoponica*. It properly belongs to Byzantine literature and perhaps had little direct influence upon western Europe. Nevertheless at least a portion of it upon vineyards was translated into Latin by Burgundio of Pisa in the twelfth century.[1] In any case as the "only formal treatise on Greek agriculture" extant it is a rather important historical source; it also is a good specimen of early medieval compilations from classical works; and in its inclusion of superstitious and magical details it is probably roughly representative of the period, whether in east or west. In the form which we now possess it was published about 950 A. D. and dedicated to the Byzantine emperor, Constantine VII or Porphyrygennetos. But this issue was perhaps little more than an abbreviated revision of the work of Cassianus Bassus of the sixth century, whose introductory words to his son are still given at the beginning of the seventh book. Cassianus is believed in his turn to have been especially indebted to two fourth century writers, Vindanius Anatolius of Beirut, whose agricultural teaching was of a sober and rational sort, and Didymus of Alexandria, who was more given to superstition and magic.[2]

Magic and astrology therein.

Nevertheless, magic and astrology find no place in the index to the most recent edition of the work.[3] A survey, however, of the text itself reveals some indications of the

[1] Vienna 2272, 14th century, fol. 92, De vindemiis a Burgundione translatus: Pars Geoponicorum.

[2] Such is the view set forth in PW *Geoponica*.

[3] H. Beckh, *Geoponica sive Cassiani Bassi scholastici de re rustica eclogae*, Lipsiae, Teubner, 1895. PW criticizes this edition as "*leider völlig verfehlten.*" Its preface lists the earlier editions.

presence of both. The very first of its twenty books deals with astrological prediction of the weather and cites some spurious work or works by Zoroaster a great deal. In later books, too, Zoroaster is sometimes cited for semi-astrological advice, such as guarding wine jars against sun or moonbeams when opening them, or testing seed by exposing it to the rays of the dog-star.[1] Zoroaster is also used as an authority on the sympathy and antipathy existing between natural objects.[2] Damigeron and Democritus are other names cited which are suggestive of the occult and magical.[3] There are not, however, many cases of extreme superstition in the *Geoponica*. Something is said of the marvelous properties of gems, of the effect of a hyena's shadow falling upon a dog by moonlight, and how dogs will not attack a person who holds a hyena's tongue in his hand.[4] Incantations of a sort are occasionally recommended.[5] To keep wine from turning sour one is directed to write the divine words, "Taste and see that the Lord is good" upon the wine-jar.[6] Another passage advises a person who finds himself in a place full of fleas to cry, "Ouch! Ouch!" and then they will not bite him.[7]

Dioscorides.

Perhaps the chief ancient work on pharmacology was the *De materia medica* or Περὶ ὕλης ἰατρικῆς of Pedanius Dioscorides of Anazarba. Galen, as we have seen, found things to criticize in it but nevertheless made great use of it in his own work on medicinal simples. Dioscorides of course had his previous sources but seems to have surpassed them in fulness and orderliness of arrangement. Of the man himself his preface tells us all that we know, and his dedication shows that he probably wrote during the reign of Nero. He was born in Cilicia near Tarsus, he had traveled in many lands as a soldier, and his work was based

[1] *Geoponica,* VII, 5; II, 15.
[2] VII, 11; XV, 1.
[3] I, 12; VII, 13; etc.
[4] XV, 1.
[5] R. Heim, *Incantamenta magica graeca latina,* in *Jahrb. f. class. Philologie,* Suppl. Bd. 19, Leipzig, 1893, pp. 463-576, drew from the *Geoponica* 13 out of his total of 245 instances of incantations from Greek and Latin literature.
[6] VII, 14.
[7] XIII, 15.

partly upon personal observation and experience as well as previous books.

Textual history of the *De materia medica.*

Dioscorides' influence continued and even increased as time went on; but if future centuries were deeply influenced by his book, it was also seriously affected by them, for it seems to have been subjected to a long series of repeated abbreviations and omissions, additions and interpolations, changes in form and in order. Thus all sorts of versions of what was called Dioscorides came into being, but which in some cases can hardly be regarded as more than compilations from all the favorite pharmacies of the time, in which the genuine Dioscorides constituted but a remnant or a core. Thus most early printed editions of what purports to be the *De materia medica* must be handled with great caution, and it may perhaps be doubted if even the latest effort of Max Wellmann to recover the original Greek text has been entirely successful.[1] Of the five books regarded as genuine and original the first dealt with spices, salves, and oils; the second, with parts of animals and animal products like milk and honey, with grains, vegetables, and pot-herbs. Other plants and roots were considered in the third and fourth books, while the last dealt with wines and minerals.[2]

Alterations made in the Greek text.

Whether we now possess Dioscorides' original text or not, at any rate the oldest Greek manuscripts do not contain it, but only that portion dealing with herbs. Moreover, this has been rearranged in alphabetical order and has been adapted to fit a set of pictures of plants which were perhaps taken over from the work of Crateuas, one of Dioscorides' chief sources. Such is the famous early sixth century illuminated manuscript made for Juliana Anicia, daughter of the emperor Olybrius (472 A. D.) and wife of the consul

[1] The first two volumes, published at Berlin in 1907, 1906, covered the first four of the five genuine books. A previous attempt was K. Sprengel's edition in vols. 25-26 of C. J. Kühn's *Medici Graeci*, Leipzig, 1829. On the textual history and problems see further Wellman's articles: "Dioskurides" in Pauly-Wissowa, and in *Hermes*, XXXIII, (1898) 360ff.

[2] *Περὶ βοτανῶν, περὶ ζῴων παντοίων, περὶ παντοίων ἐλαίων, περὶ ὕλης δένδρων, περὶ οἴνων καὶ λίθων*, is another order suggested.

Areobindus (about 512 A. D.).[1] The alphabetical rearrangement of the Greek text of Dioscorides was made at some time between Galen and Oribasius, who cites from it in the fourth century. Not only were the five books of the genuine *De materia medica* interpolated, but additional spurious books were added "On Harmful Drugs" and "On Poisons." [2] The work on medicinal simples attributed to Dioscorides is extant in no manuscript earlier than the fourteenth century and some versions of it are much more interpolated than others. As Galen does not cite it while Oribasius and Aëtius do use it, it is assumed that it was com-

[1] The MS is said by Singer (1921) 60, to have now been removed from Vienna to St. Mark's Library at Venice; it was procured from Constantinople in 1555 for the future Emperor Maximilian II (1564-1576). A photographic copy was published in 1906 in the Leiden Collection, *Codices Graeci et Latini*, by A. W. Sijthoff, with an introduction by A. von Premerstein, C. Wessely, and J. Mantuani (C. Wessely, *Codex Anciae Iulianae*, etc., 1906). See also A. v. Premerstein in the Austrian *Jahrbuch* (1903) XXIV, 105ff.

I have examined the fac-simile of this MS and found the large but faded and partially obliterated illuminations which precede the text rather disappointing after having read the description of them in Dalton's *Byzantine* Art, (1911) 460-61, which, however, I presume is accurate and so reproduce here. These large illuminations include a portrait of Juliana Anicia, an ornamental peacock with tail spread, groups of doctors engaged in medical discussions, and Dioscorides himself seated writing, and again seated on a folding stool receiving the herb mandragora (which, of course, was a medieval favorite) from a female figure personifying Discovery (Εὕρησις), "while in the foreground a dog dies in agony," presumably from the fatal effects of the herb. There are rough reproductions of this last picture in Woltmann and Woermann, *History of Painting*, I, 192-3, and Singer (1921) 62. When the text proper begins the illuminations are confined to medicinal plants.

Other early Greek manuscripts are the *Codex Neapolitanus*, formerly at Vienna, now at St. Mark's, Venice, an eighth century palimpsest from Bobbio, and a Paris codex, (BN Greek 2179) of the ninth century. An Arabic translation from the Greek seems to have been made about 850; a century later the Byzantine emperor sent a Greek manuscript of Dioscorides to the caliph in Spain.

For the full text of the *De materia medica* we are dependent on MSS of the 11th, 12th, 13th and later centuries.

[2] *Περὶ δηλητηρίων φαρμάκων* and *περὶ ἰοβόλων*, edited by Sprengel in Kühn (1830), XXVI, as was the *Περὶ εὐποριστων ἁπλῶν τε καὶ συνθέτων φαρμάκων*. The *Περὶ φαρμάκων ἐμπειρίας*. ("Experimental Pharmacy"), of which a Latin version, *Alphabetum empiricum, sive Dioscoridis et Stephani Atheniensis . . . de remediis expertis,* was edited by C. Wolf, Zürich, 1581, is an alphabetical arrangement by diseases ascribed to Dioscorides and Stephen of Athens (and other writers).

posed in the third or early fourth century with a forged dedication to a contemporary of Dioscorides, but that it made considerable use of the genuine Dioscorides, to which it bore much the same relation as the *Medicina Plinii* did to the *Historia Naturalis*. Later, however, some Byzantine compiler of the eleventh, twelfth, or thirteenth century introduced a great deal of new material from Galen's genuine and spurious works in that field and from John of Damascus.[1]

Dioscorides little known to Latins before the middle ages.

What more especially concern us are the medieval Latin versions of Dioscorides. As a matter of fact, although the *De materia medica* was from the start highly regarded and widely used by Greek physicians, it seems to have been little known to Latin writers until the verge of the medieval period. Gargilius Martialis, a Roman writer on agriculture in the third century of our era, was the only old Latin author to cite Dioscorides, which he did, however, no less than eighteen times in his *Medicinae ex oleribus et pomis*. This has led to the suggestion that he was perhaps responsible for the first Latin translation or version of Dioscorides; but it seems unlikely that the work had been put into Latin as early as his time, since it is not cited again by a Latin writer until the sixth century and is not used by such medical authors as Serenus Sammonicus, Cassius Felix, Theodorus Priscianus, and Marcellus Empiricus.

Partial versions in Latin.

But at least a portion of Dioscorides seems to have been translated into Latin by the time of Cassiodorus, who, writing in the first half of the sixth century, states that those who cannot read Greek may consult the *Herbarium Dioscoridis*.[2] This naturally suggests a version limited to medicinal plants like the early Greek text in the manuscript of Juliana Anicia. This impression is confirmed by the preface to some early Latin version of Dioscorides, which Rose discovered in one of the manuscripts of the *Herbarium of*

[1] Max Wellmann, *Die Schrift des Dioskurides* Περὶ ἁπλῶν φαρμάκων, 1914, and col. 1140 of his article "Dioskurides" in Pauly-Wissowa.

[2] *De inst. div. lit.* cap. 31.

Apuleius in the British Museum.[1] This preface implies that the translation which it introduced was limited to the botanical books of Dioscorides and states that it was accompanied by illustrations of herbs.

De herbis femininis.

Based upon this partial translation rather than identical with it is believed to have been the *De herbis femininis*,[2] which was ascribed to Dioscorides in the middle ages and which often accompanies the *Herbarium* of the Pseudo-Apuleius in the manuscripts. In this case the herbs of the Pseudo-Apuleius are sometimes called masculine, but as a matter of fact only a minority of those in the Pseudo-Dioscorides seem to be distinctly feminine. Of seventy-one plants Kaestner classed fifteen or sixteen as feminine, while in only thirty cases are they prescribed for female complaints. Rose dated this work before Isidore of Seville by whom he believed it was used.[3] It seems to combine a free Latin translation of excerpts from the genuine Dioscorides with numerous additions from other sources.

The fuller Latin versions.

Besides such abbreviated and interpolated Latin versions or perversions of Dioscorides, there was also in existence in the early middle ages a literal translation of all five books

[1] V. Rose in Hermes VIII, 38A. Harleian 4986, fol. 44v, ". . . marcelline libellum botanicon ex dioscoridis libris in latinum sermonem conversum in quo depicte sunt herbarum figure ad te misi . . ."

[2] Heinrich Kaetsner, *Kritisches und Exegetisches zu Pseudo-Dioskorides de herbis femininis*, Regensburg, 1896; text in *Hermes* XXXI (1896) 578-636. Singer (1921) 68, gives as the earliest MS, Rome Barberini IX, 29, of 9th century. Some other MSS are: BN 12995, 9th century; Additional 8928, 11th century, fol. 62v-; Ashmole 1431, end of 11th century, fols. 31v-43, "Incipit liber Dioscoridis ex herbis feminis"; Sloane 1975, 12th or early 13th century, fols. 49v-73; Harleian 1585, 12th century, fol. 79-; Harleian 5294, 12th century; Turin K-IV-3, 12th century, #5, "Incipit liber dioscoridis medicine ex herbis femininis numero LXXI . . / . . Liber medicine dioscoridis de herbis femininis et masculinis explicit feliciter."

In Vienna 5371, 15th century, fols. 121v-124v, is a briefer Latin treatise ascribed to Dioscordes, which begins with the herb *aristologia* and mentions silk (*sericum*) at its close. I have not seen the MS but from the title, *Quid pro quo*, and the fact that the writer dedicates it to his uncle, one might fancy that it was a work written by Adelard of Bath's nephew in return for the *Natural Questions* of his uncle. (See below, chapter 36).

[3] *Hermes* VIII, 38, comparing *Etymologies* XVII, 93, with cap. 30 of the *De herbis femininis*.

of the *De materia medica.* It is full of Latinisms and barbarisms but otherwise reproduces the complete and genuine Dioscorides, or is supposed to do so. Rose and Wellmann [1] say that it was current from the sixth century on, and the few extant manuscripts of it date from the early medieval period.[2] One reason for this seems to be that this literal translation was replaced by another Latin version which in a Bamberg manuscript [3] is ascribed to Constantinus Africanus, the medical translator and writer of the eleventh century. In this version the items are arranged alphabetically, and additions are embodied from other sources. This version apparently became much better known than the earlier literal translation and has been called "the most widely disseminated handbook of pharmacy of the whole later middle ages." [4] It is stated by Rose to be identical with the "Dyascorides," upon which Peter of Abano lectured and commented about 1300 and which was printed at Colle in 1478 and again at Lyons in 1512.[5]

Peter of Abano's account of the medieval versions.

Peter of Abano tells us in his preface [6] that in his time there were current two different versions, although both had the same preface. One of these was in five books with a great many short chapters, so short in fact that often the treatment of a single thing was scattered over several chapters. This version was rare in Latin. The other version contained fewer but longer chapters with material added from Galen, Pliny, and other writers. This version was

[1] *Anecdota graeca et graecolatina,* Berlin, 1864, II, 115 and 119; Hermes VIII, 38; Wellmann (1906), p. xxi.

[2] BN 9332, 8th century; CLM 337, 9-10th century from Monte Cassino; ed. T. M. Auracher et H. Stadler, in *Rom. Forsch.* I, 49-105; X, 181-247 and 368-446; XI, 1-121; XII, 161-243.

[3] Cod. Bam. L-III-9.

[4] PW "Dioskurides." A fairly early MS is CU Jesus 44, 12-13th century, fols. 17-145r, "diascorides per modum alphabeti de virtutibus herbarum et compositione olerum." I have not seen it but, if correctly dated, it and Bologna University Library 378, 12th century, which is said to differ from the printed editions, are too early to be Peter of Abano's version.

[5] *Explicit dyascorides quem petrus paduanensis legendo corexit et exponendo quae utiliora sunt in lucem deduxit,* Colle, 1478. *Dioscorides digestus alphabetico ordine additis annotatiunculis brevibus et tractatu aquarum,* Lugduni, 1512. And see Chap. 70, Appendix II.

[6] I have read it in BN 6820, fol. 1r, as well as in the 1478 edition.

arranged alphabetically. It was this version which *Aggregator*[1] had followed and imitated, but sometimes there were chapters in either "Dyascorides" which were missing in *Aggregator*. Peter had also seen an alphabetical version of Dioscorides in Greek.

Pseudo-Dioscorides on stones.

There seems also to have been current, at least in the later middle ages, a Pseudo-Dioscorides on stones, drawn in part, like the *Feminine Herbs*, from the genuine *De materia medica*, whose discussion of the virtues of stones is incredible enough.[2] This *Dioscorides on Stones* is cited by Arnold of Saxony and Bartholomew of England in the thirteenth century, and portions at least of the work are extant in manuscripts at Erfurt and Montpellier.[3] A work on physical ligatures is ascribed to Dioscorides in a late manuscript,[4] but is really a collection of items from various authors since Dioscorides on the marvelous virtues of animals, herbs, and stones, especially when bound on the body, held in the hand, or worn around the neck.

Conclusions from the textual history of Dioscorides.

The history of the medieval versions of Dioscorides, even in the brief and incomplete outline given here, is instructive, showing us in general the vicissitudes to which the transmission of the text of any ancient author may have been subjected, but more especially proving that the middle ages, whether Latin or Byzantine, were ready to take great liberties with ancient authorities and to adapt them to their own taste and requirements. And indeed, why should they not rearrange and make additions to their

[1] A work by Serapion which Simon Cordo of Genoa translated from Arabic into Latin with the help of Abraham, a Jew of Tortosa. Serapion states at the beginning that his work is a combination of Dioscorides and of the work of Galen on medicinal simples. *Aggregator* was printed in 1479, *Liber Serapionis aggregatus in medicinis simplicibus. Translatio Symonis Ianuensis interprete Abraam iudeo tortuosiensi de arabico in latinum.*

[2] Ruska (1912), p. 5, says that Dioscorides, V, 84-133, among other things describes "eine ganze Reihe von höchst zweifelhaften Steinen mit unglaublichen Wirkungen die in den Arabischen Arzneimittelverzeichnissen und Steinbüchern niederkehren."

[3] Amplon. Folio 41, fols. 36-7; Montpellier 277, caps. 46-67 of the treatise entitled, *Liber aristotelis de lapidibus preciosis secundum verba sapientium antiquorum.*

[4] Sloane 3848, 17th century, fols. 36-40.

Dioscorides? After all it was a compilation to begin with. But the case of Dioscorides has also taught us that we do not have to wait until the medieval period for the appearance of new versions of an ancient author.

Macer on herbs; its great currency.

With the possible exception of the *Herbarium* of the Pseudo-Apuleius, probably the best known single and distinct treatment of the virtues of herbs produced during the middle ages was the poem *De viribus herbarum* which circulated under the name of Macer Floridus.[1] It was often cited by the medieval encyclopedists and other writers on nature and medicine in the twelfth and thirteenth centuries.[2] It is found in an Anglo-Saxon version [3] and was even translated into Danish in the early thirteenth century.[4] Manuscripts of it are very numerous [5] and there are many early printed editions.[6] Even as recently as the first half of the nineteenth century a historian of medicine and natural science, in the preface of his edition of Macer, stated as one argument for the modern study of medieval medicine that much might be learned from writings of that period concerning the virtues of herbs.[7]

Problem of date and author

The poem was certainly not written by the classical poet, Aemilius Macer, who was a friend of Vergil and Ovid, and whose descriptions of plants, birds, and reptiles are cited by Pliny in his *Natural History* and also preserved in some extracts by the grammarians. Proof of this is that our

[1] *Macer Floridus de viribus herbarum una cum Walafridi Strabonis, Othonis Cremonensis et Ioannis Folcz carminibus similis argumenti,* ed. Ludovicus Choulant, 1832.

[2] V. Rose himself corrected (*Hermes,* VIII, 330-1) the strange statement which he had made (*Hermes,* VIII, 63) that the name "Macer" is not found in connection with this work until MSS of the 14th and 15th centuries. Both the treatise and the name are frequent in the earlier MSS.

[3] Cotton, Vitellius C, III.

[4] The Dane, Harpestreng, who died in 1244, translated and commented upon the poem; published by Christian Molbech, Copenhagen, 1826.

[5] There are a large number in the MSS collections of the British Museum alone. Some said to be of the 12th century are Harleian 4346, and at Erfurt Amplon. Octavo 62a and 62b.

[6] See the British Museum catalogue of printed books. I have used besides Choulant's text of 1832 an illustrated octavo edition probably of 1489. The poem also appears in medical collections such as *Medici antiqui omnes,* Aldus, Venice, 1547, fols. 223-46.

[7] Choulant (1832) Preface.

poem cites Pliny; in fact, it cites him more frequently than any other author. It also cites Galen six times, Dioscorides four, and as late an author as Oribasius twice.[1] But Oribasius is not the latest author cited since Walafrid Strabo is also used.[2] Strabo was born about 806, became abbot of Reichenau in 842, and died in 849. In his *Hortulus,* a poem dedicated to Grimoald, the abbot of St. Gall, he described twenty-three herbs in 444 hexameters.[3] Indeed Stadler holds that the Pseudo-Macer uses the *De gradibus* of Constantinus Africanus who did not die until 1087.[4] The true author of our poem ascribed to Macer is said on the authority of certain manuscripts to have been an Odo of Meung on the Loire, apparently the same town as the birthplace of Jean Clopinel or de Meun, the learned author of the latter portion of *The Romance of the Rose.* Choulant, however, did not regard this as sufficiently proved, and Stadler has recently noted that some manuscripts ascribe the poem to a physician, Odo of Verona; and others to the Cistercian, Odo of Morimont, who died in 1161.[5] In any case, unless the mentions of Strabo are later interpolations, the author must be regarded as post-Carolingian, while he cannot be later than the eleventh century in view of a remark of Sigebertus Gemblacensis in 1112,[6] the Anglo-Saxon version, the many twelfth century manuscripts, and the frequent use of his poem in the *Regimen Salernitanum.*[7]. Although Macer seems a pseudonym to begin with, the original poem, consisting of 2269 lines in which 77 herbs are discussed, is sometimes accompanied by additional lines regarded as spurious.[8]

[1] Choulant (1832) *Prolegomena ad Macrum,* p. 14.

[2] See the description of *Ligusticum,* lines 900-6.

[3] Often printed: ed. F. A. Reuss, Würzburg, 1834; in Migne PL 114, 1119-30.

[4] H. Stadler, *Die Quellen des Macer Floridus,* in Sudhoff (1909).

[5] Stadler, *op. cit.;* Choulant (1832), p. 4.

[6] "Macer scripsit metrico stilo librum. de viribus herbarum,"—Stadler (1909), 65.

[7] It was, however, a good deal subject to later interpolation.

[8] Choulant (1832) adds as *Macri spuria* 487 lines concerning twenty herbs.

In Vienna 3207, 15th century, fols. 1-50, Macer Floridus, De viribus herbarum; fols. 50-52, Pseudo-Macer, De animalibus et lignis.

Virtues ascribed to herbs.

Our poet does not appear to have much of his own to offer on the subject of the virtues of herbs. When he does not cite his authority by name, he usually qualifies the statement made by a vaguer "they say" or "it is said." He does not connect certain herbs with certain stars or otherwise introduce anything that can be called astrological. He repeats Pliny's statement of the powers ascribed to vervain by the *magi,* such as to gain one's desires, win the friendship of the powerful, and dispel disease and fever. Pliny had spoken of the *magi* as "raving about this herb"; our poet says:

"Although potent Nature can grant such virtues,
Yet they really seem to us idle old-wives' tales."[1]

Nevertheless he himself about fifteen lines before had said of the vervain:

"If, holding this herb in the hand, you ask the patient,
'Say, brother, how are you?' and the patient answers, 'Well,'
He will live; but if he says 'Ill,' there is no hope of safety."[2]

Our poet not only thus associates with herbs the virtue of divination, but is guilty of sympathetic magic when he believes that the ancients learned by experience that *Dragontea* or snake-weed dispels poisons, wards off snakes, and is good for snake-bite from observing the similarity between the spotted rind of the herb and the skin of a snake.[3] Odo or Macer repeats Galen's story of curing an epileptic boy by suspending a root of peony about his neck,[4] and later asserts the same virtue for the herb *pyrethrum.*[5] Even more magical is the ceremony for curing toothache which he takes from Pliny and which consists in digging up the herb *Senecion* without use of iron, touching the aching tooth with it three times, and then replacing the plant in the place where it came from so that it will grow again.[6] Pliny is also cited

[1] Lines 1901-2, *Quae, quamvis natura potens concedere posset Vana tamen nobis et anilia iure videntur.*

[2] Lines 1881-3, *Hanc herbam gestando manu si queris ab egro Dic frater quid agis? bene si responderit eger, Vivet, si vero male, spes est nulla salutis.*

[3] Herb 54, lines 1728-.

[4] Herb 49, lines 1617-27.

[5] Herb 67, lines 2095-.

[6] Herb 51, lines 1685-9.

concerning the swallow's restoring the sight of its young by swallow-wort.[1] Our poet also repeats such beliefs as that the herb *Buglossa* preserves the memory,[2] or that the smoke of *Aristochia* dispels demons and exhilarates infants.[3] If the hives are anointed with the juice of the herb *Barrocus,* the bees will not desert them; while carrying that plant with one is a protection against the stings of bees, wasps, and spiders.[4] Among the virtues most frequently attributed to herbs are expelling or killing worms, curing pestiferous bites or poisons, and provoking urine or vomiting. On the whole, "Macer" contains only a moderate amount of superstition, although rather more proportionally than Walafrid Strabo.

Experiments of Macer.

Although Odo or Macer seems to make no original contribution to botany, cites authorities frequently, and speaks often of the ancients or men of old, he also at least once cites "experts" [5] and we have also seen his belief that the ancients had tested the virtues of plants by experience. This rather slight experimental character of the work is further emphasized in some manuscripts of it, where the title is "Experiments of Macer" and the matter seems to have been re-arranged under diseases instead of by herbs.[6]

[1] Herb 52.

[2] Herb 34, lines 1135-8.

[3] Herb 41, lines 1421-2.

[4] Herb 50, lines 1641-63.

[5] Herb 69, *Cyminum,* lines 2118-9, "Hoc orthopnoicis miram praestare medelam Experti dicunt cum pusce saepius haustum."

[6] Vienna 2532, 12th century, fols. 106-17, "Experimenta Macri. Ad dolorem capitis. Accipe balsamum et instilla . . . / . . . adde sucum celidonie et superpone vulneribus."

Arundel 295, 14th century, fols. 222-33, "Experimenta Macri collecta sub certis capitulis a Gotefrido."

CHAPTER XXVII

OTHER EARLY MEDIEVAL LEARNING: BOETHIUS, ISIDORE, BEDE, GREGORY THE GREAT

Aridity of early medieval learning—Historic importance of *The Consolation of Philosophy*—Medieval reading—Influence of the works of Boethius—His relation to antiquity and middle ages—Attitude to the stars—Fate and free will—Music of the stars and universe—Isidore of Seville—Method of the *Etymologies*—Its sources—Natural marvels—Isidore is rather less hospitable to superstition than Pliny—Portents—Words and numbers—History of magic—Definition of magic—Future influence of Isidore's account of magic—Attitude to astrology—In the *De natura rerum*—Bede's scanty science—Bede's *De natura rerum*—Divination by thunder—Riddles of Aldhelm—Gregory's *Dialogues*—Signs and wonders wrought by saints—More monkish miracles—A monastic snake-charmer—Basilius the magician—A demon salad—Incantations in Old Irish—The *Fili*.

Aridity of early medieval learning.

THE erudite fortitude of students of the Merovingian period commands our admiration, but sometimes inclines us to wonder whether anyone without a somewhat dry-as-dust constitution could penetrate far or tarry long in the desert of early medieval Latin learning without perishing of intellectual thirst. As a rule the writings of the time show no originality whatever, and least of all any scientific investigation; they are of value merely as an indication of what past books men still read and what parts of past science they still possessed some interest in. Under the same category of condemnation may be placed most of the Carolingian period so far as our investigation is concerned. We shall therefore traverse rapidly this period of sparse scientific productivity and shall be doing it ample justice, if from its meager list of writers we select for consideration Boethius of Italy at the opening of the sixth century and Gregory the Great at its close, Isidore of Spain at the opening of the seventh century, and Bede in England at the beginning of the eighth century, with some brief allusion to the riddles of Aldhelm, bishop of Sherborne, and to Old Irish literature. We should gain little or nothing by adding to the list Alcuin at the close of the eighth century and Rabanus Maurus in the ninth century, although it may be noted now that later medieval writers cite Rabanus for statements which I have failed to find in his printed works. In general it may be said that the writers whom we shall consider are those during the period who are most cited by the later medieval authors.

Historic importance of *The Consolation of Philosophy.*

Of the distinguished family and political career of Boethius who lived from about 480 to 524 A. D., and his final exile, imprisonment, and execution by Theodoric the East Goth, we need scarcely speak here. Our concern is with his little book, *The Consolation of Philosophy,* one of those memorable writings which, like *The City of God* of Augustine, stand out as historical landmarks and seem to have been written on the right subject by the right man at the most dramatic moment. The timely appearance of such works, produced in both these cases not under the stimulus of triumphant victory but the sting of bitter defeat, is nevertheless perhaps less surprising than is their subsequent preservation and enormous influence. We often are alternately amused and amazed by the mistakes concerning historical and chronological detail found in medieval writers. Yet medieval readers showed considerable appreciation of the course of history, of its fundamental tendencies, and of its crucial moments by the works which they included in their meager libraries.

Medieval reading.

But were medieval libraries as meager as we are wont to assume? Bede and Alcuin both tell of the existence of sizeable libraries in England,[1] and Cassiodorus urged those monks whose duty it was to tend the sick to read a number of standard medical works.[2] I sometimes wonder if too

[1] R. L. Poole, *Medieval Thought,* 1884, pp. 19, 21.
[2] Migne, PL 70, 1146.

much attention has not been given to medieval writing and too little to medieval reading, of which so much medieval writing, in Latin at least, is little more than a reflection. We get their image, faint perhaps and partial; but they had the real object. It has been assumed by some modern scholars that medieval writers had usually not read the works, especially of classical antiquity, which they profess to cite and quote, but relied largely upon anthologies and *florilegia*. In the case of various later medieval authors we shall have occasion to discuss this question further. For the present I may say that in going through the catalogues of collections of medieval manuscripts I have noticed few *florilegia* or anthologies from the classics in medieval Latin manuscripts,—perhaps Byzantine ones from Greek literature are more common—and few indeed compared to the number of manuscripts of the old Latin writers themselves. We owe the very preservation of the Latin classics to medieval scribes who copied them in the ninth and tenth centuries; why deny that they read them? Latin *florilegia* of any sort do not exist in impressive numbers, but other kinds are as often met with as are those from classic poets or prose writers, for instance, selections from the church fathers themselves. On the whole, the impression I have received is that those authors included in *florilegia,* commonplace books, and other manuscripts made up of miscellaneous extracts, were likewise the authors most read *in toto.* I am therefore inclined to regard the *florilegia* as a proof that the authors included were read rather than that they were not. But from extant Latin manuscripts one gets the impression that the whole matter of *florilegia* is of very slight importance, and that the theory hitherto based upon them is a survival of the prejudice of the classical renaissance against "the dark ages."

Influence of the works of Boethius.

At any rate, however scanty medieval libraries may have been, they were apt to include a copy of *The Consolation of Philosophy,* and however little read some of their volumes may have been, its pages were certainly well thumbed. Lists of its commentators, translators, and imitators, and other indications of its vast medieval influence may be found in Peiper's edition.[1] Other writings of Boethius were also well known in the middle ages and increased his reputation then. His translations and commentaries upon the Aristotelian logical treatises [2] are of course of great importance in the history of medieval scholasticism. His translations and adaptations of Greek treatises in arithmetic, geometry, and music occupy a similar place in the history of medieval mathematical studies.[3] Indeed, his treatise on music is said to have "continued to be the staple requisite for the musical degree at Oxford until far into the eighteenth century." [4] The work on the Trinity and some other theological tracts, attributed to Boethius by Cassiodorus and through the middle ages, are now again accepted as genuine by modern scholars and place Boethius' Christianity beyond question.[5]

His relation to antiquity and middle ages.

Boethius has often been regarded as a last representative of Roman statesmanship and of classical civilization. His defense of Roman provincials against the greed of the Goths, his stand even unto death against Theodoric on behalf of the rights of the Roman senate and people, his preservation through translation of the learned treatises of expiring an-

[1] *Anicii Manlii Severini Boetii Philosophiae Consolationis Libri quinque,* ed. R. Peiper, Lipsiae, 1871, pp. xxxix-xlvi, li-lxvii. See also Manitius (1911), pp. 33-5.

It was by seeking comfort in *The Consolation of Philosophy* after the death of Beatrice that Dante was led into a new world of literature, science, and philosophy, as he tells us in his *Convivio;* cited by Orr (1913), p. 1.

[2] Manitius (1911), pp. 29-32.

[3] *Ibid.,* 26-8. At the time I went through the various catalogues of MSS in the British Museum item by item it was not my intention to include Boethius in this investigation, and I am therefore unable to say whether the Museum has MSS which may throw further light upon the problems connected with the mathematical treatises ascribed to Boethius. Manitius mentions no English MSS in this connection, but there are likely to be some at London, Oxford, or Cambridge.

[4] *Boethius' Consolation of Philosophy,* translated from the Latin by George Colville, 1556; ed. with Introduction by E. B. Box, London, 1897, p. xviii.

[5] Manitius (1911) pp. 35-6; Usener, *Anecdota Holderi,* Bonn, 1877, pp. 48-59; E. K. Rand, *Der dem Boethius zugeschriebene Traktat De fide catholica,* 1901. The *De fide catholica,* however, is not mentioned by Cassiodorus and is regarded as spurious.

tiquity, and the almost classical Latin style and numerous allusions to pagan mythology of *The Consolation of Philosophy*:—all these combine to support this view. But the middle ages also made Boethius their own, and several points may be noted in which *The Consolation of Philosophy* in particular foreshadowed their attitude and profoundly influenced them. Both a Christian and a classicist, both a theologian and a philosopher, Boethius set a standard which subsequent thought was to follow for a long time. The very form of his work, a dialogue part in prose and part in verse, remained a medieval favorite. And the fact that this sixth century author of a work on the Trinity consoled his last hours with a work in which Christ and the Trinity are not mentioned, but where Phoebus is often named and where Philosophy is the author's sole interlocutor:—this fact, combined with Boethius' great medieval popularity, gave perpetual license to those medieval writers who chose to discuss philosophy and theology as separate subjects and from distinct points of view. The great medieval influence of Aristotle and Plato, and in particular of the latter's *Timaeus*, also is already manifest in *The Consolation of Philosophy*. Aristotle, it is true, appears to be incorrectly credited by Boethius with the assertion that the eye of the lynx can see through solid objects,[1] but this ascription of spurious statements to the Stagirite also corresponds to the attribution of entire spurious treatises to him later in the middle ages.

Attitude to the stars.

Of the ways in which *The Consolation of Philosophy* influenced medieval thought that which is most germane to our investigation is its attitude toward the stars and the problem of fate and free will. The heavenly bodies are apparently ever present in Boethius' thought in this work, and especially in the poetical interludes he keeps mentioning Phoebus, the moon, the universe, the sky, and the starry constellations. *Per ardua ad astra* was a true saying for those last days in which he solaced his disgrace and pain with philosophy. It is by contemplation of the heavens

[1] *De consol. philos.*, III, 8, 21.

that he raises his thought to lofty philosophic reflection; his mind may don swift wings and fly far above earthly things

> "Until it reaches starry mansions
> And joins paths with Phoebus." [1]

He loves to think of God as ruling the universe by perpetual reason and certain order, as sowing stars in the sky, as binding the elements by number, as Himself immovable, yet revolving the spheres and decreeing natural events in a fixed series.[2] The attitude is like that of the *Timaeus* and Aristotle's *Metaphysics,* closely associating astronomy and theology, favorable to belief in astrology, in support of which later scholastic writers cite Boethius.

Fate and free will.

We may further note the main points in Boethius' argument concerning fate and free will, providence and predestination,[3] which was often cited by later writers. He declares that all generation and change and movement proceed from the divine mind or Providence,[4] while fate is the regular arrangement inherent in movable objects by which divine providence is realized.[5] Fate may be exercised through spirits, angelic or daemonic, through the soul or through the aid of all nature or "by the celestial motion of the stars." [6] It is with the last that Boethius seems most inclined to identify *fati series mobilis.* "That series moves sky and stars, harmonizes the elements one with another, and transforms them from one to another." More than that, "It constrains human fortunes in an indissoluble chain of causes, which, since it starts from the decree of immovable Providence, must needs itself also be immutable." [7] Boethius, however, does not believe in a complete fatalism, astrological or otherwise. He holds that nothing escapes

[1] *De consol. philos.,* IV, 1.
[2] *Ibid.,* III, 9, 1; III, 12, 14; III, 9, 10; III, 12, 99; II, 8, 13.
[3] *Ibid.,* IV, 6, 10, "In hac enim de providentiae simplicitate, de fati serie, de repentinis casibus, de cognitione ac praedestinatione divina, de arbitrii libertate quaeri solet." To the ensuing argument are devoted the sixth and seventh chapters of Book IV and all of Book V.
[4] *Ibid.,* IV, 6, 21.
[5] *Ibid.,* IV, 6, 30.
[6] *Ibid.,* IV, 6, 48.
[7] *Ibid.,* IV, 6, 77.

divine providence, to which there is no distinction between past, present, and future.[1] As the human reason can conceive universals, although sense and imagination are able to deal only with particulars, so the divine mind can foresee the future as well as the present. But there are some things which are under divine providence but which are not subject to fate.[2] Divine providence imposes no fatal necessity upon the human will, which is free to choose its course.[3] The world of nature, however, existing without will or reason of its own, conforms absolutely to the fatal series provided for it. As for chance, Boethius agrees with Aristotle's *Physics* that there is really no such thing, but that what is commonly ascribed to chance really results from an unexpected coincidence of causes, as when a man plowing a field finds a treasure which another has buried there.[4] Thus Boethius maintains the co-existence of the fatal series expressed in the stars, divine providence, and human free will, a thesis likely to reassure Christians inclined to astrology who had been somewhat disturbed by the fulminations of the fathers against the *genethliaci,* just as his constant rhapsodizing over the stars and heavens would lead them to regard the science of the stars as second only to divine worship. Indeed, his position was the usual one in the subsequent middle ages.

Music of the stars and universe.

The stars also come into Boethius' treatise on music, where one of the three varieties of music is described as mundane, where the music of the spheres is declared to exist although inaudible to us, and where each planet is connected with a musical chord. Plato is quoted as having said, not in vain, that the world soul is compounded of musical harmony, and it is affirmed that the four different and contrary elements could never be united in one system unless some harmony joined them.[5]

[1] *De consol philos.,* V, 4-6.

[2] *Ibid.,* IV, 6, 58.

[3] *Ibid.,* V, 2-3 and 6, 110, "tametsi nullam naturae habeat necessitatem atqui deus ea futura quae ex arbitrii libertate proveniunt praesentia contuetur."

[4] *Ibid.,* V, 1.

[5] *De musica libri quinque,* I, 1-2 and 27; in Migne, PL 63, 1167-1300.

Isidore of Seville.

Isidore was born about 560 or 570, became bishop of Seville in 599 or 600, and died in the year 636. Although mention should perhaps be made of his briefer *De natura rerum,*[1] a treatise dedicated to King Sisebut who reigned from 612 to 620, Isidore's chief work from our standpoint is the *Etymologiae.*[2] His friend, bishop Braulio, writing after Isidore's death, says that he had left unfinished the copy of this work which he made at his request, but this was apparently a second edition, since in a letter written to Isidore probably in 630, Braulio speaks of copies as already in circulation, although he describes their text as corrupt and abbreviated. But apparently the work had been composed seven years before this.[3] The *Etymologies* was undoubtedly a work of great importance and influence in the middle ages, but one should not be led, as some writers have been, into exaggerated praise of Isidore's erudition on this account.[4] For the work's importance consists chiefly in showing how scanty was the knowledge of the early middle ages. Its influence also would seem not to have been entirely beneficial, since writers continued to cite it as an authority as late as the thirteenth century, when it might have been expected to have outlived its usefulness. We suspect that it proved too handy and convenient and tended to encourage intellectual laziness and stagnation more than any anthology of literary quotations did. Arevalus listed ten

[1] Migne, PL 83, 963-1018. In Harleian 3099, 1134 A. D., the *Etymologies* at fols. 1-154, are followed by the *De natura rerum,* the last chapter of which (fol. 164v) is numbered 42 instead of 48 as in Migne. But up to chapter 27, *Utrum sidera animam habeant,* the division into chapters seems the same as in the printed text.

[2] Migne, PL 82, 73-728, a reprint of the edition of Arevalus, Rome, 1796. Large portions of the *Etymologies* have been translated into English with an introduction of some seventy pages by E. Brehaut, *An Encyclopedist of the Dark Ages; Isidore of Seville,* 1912, in *Columbia University Studies in History,* etc., vol. 48, pp. 1-274. For Isidorean bibliography see pp. 17, 22-3, 46-7 of Brehaut's introduction.

[3] Manitius (1911), pp. 60-61; Brehaut (1912), p. 34.

[4] To say, for example, that "so hospitable an attitude toward profane learning as Isidore displayed . . . was never surpassed throughout the middle ages" (Brehaut, p. 31), is unfair to many later writers, as our discussion of the natural science of the twelfth and thirteenth centuries will show.

printed editions of it before 1527, showing that it was as popular in the time of the Renaissance as in the middle ages.

Method of the *Etymologies.*

The *Etymologies* is little more than a dictionary, in which words are not listed alphabetically but under subjects with an average of from one to a half dozen lines of derivation and definition for each term. The method is, as Brehaut well says, "to treat each subject by . . . defining the terms belonging to it." [1] Pursuing this method, Isidore treats of various arts and sciences, human interests and natural phenomena: the seven liberal arts, medicine, and law; chronology and bibliography; the church, religion, and theology; the state and family, physiology, zoology, botany, mineralogy, geography, and astronomy; architecture and agriculture; war and sport; arms and armor; ships and costume and various utensils of domestic life. Such is the classification which later medieval writers were to adopt or adapt rather than the arrangement followed in Pliny's *Natural History.* Isidore's association of words and definitions under topics makes an approach, at least, to the articles of encyclopedias: sometimes there is a brief discussion of the general topic before the particular terms and names are considered; sometimes there are chronological tables, family trees, or lists of signs and abbreviations. In short, Isidore forms a connecting link between Pliny and the encyclopedists of the thirteenth century.

Its sources.

In a prefatory word to Braulio Isidore describes the *Etymologies* as a collection made from his recollection and notes of old authors,[2] of whom he cites a large number in the course of the work. It has been suspected that some of these writers were known to Isidore only at second or third hand; at any rate he has not made a very discriminating selection from their works and he has been accused more than once of not clearly understanding what he tried to abridge. On the other hand, Isidore seems to me to display a notable

[1] Brehaut (1912), p. 34.

[2] Migne, PL 82, 73, "Opus de origine quarumdam rerum, ex veteris lectionis recordatione collectum, atque ita in quibusdam locis adnotatum, sicut exstat conscriptum stylo maiorum."

power of brief generalization, of terse expression and telling use of words. We should not have to go back to the middle ages for textbook writers who have written more and said less. This power of condensed expression probably accounts for Isidore's being so much cited. Many of the derivations proposed for words are so patently absurd that we would fain ascribe them to Isidore's own perverse ingenuity, but it is doubtful if he possessed even that much originality, and they are probably all taken from classical grammarians such as Varro.[1] Isidore, however, still displays a considerable knowledge of the Greek language. And again it may be said in excuse of Isidore and his sources that the absurd etymologies are usually proposed in the case of words whose derivation is still problematic.

In the passages dealing with natural phenomena and science Isidore borrows chiefly from Pliny and Solinus, sometimes from Dioscorides, giving us a faint adumbration of their much fuller confusion of science and superstition. Occasionally bits of information or misinformation are borrowed through the medium of the church fathers. A work of Galen, for instance, is cited[2] through the letter of Jerome to Furia against widows remarrying. Galen, indeed, is seldom mentioned by Isidore who draws his unusually brief fourth book on medicine chiefly from Caelius Aurelianus.[3]

Natural marvels.

In his treatment of things in nature Isidore seldom gives their medicinal properties as Pliny does, and this reduces correspondingly the amount of space devoted to marvelous virtues. Indeed, of the twenty books of the *Etymologies* but one is devoted to animals other than man, one to vegetation which is combined in the same book with agriculture, and one to metals and minerals. The book on animals is the longest and is subdivided under the topics of domestic

[1] See, for example, *Etymol.*, VIII, 7, 3, "Vates a vi mentis appellatos, Varro auctor est."

[2] *Etymol.*, XX, 2, 37.

[3] Cassiodorus, however, urged the monks of the sixth century who cared for the sick to read Hippocrates and Galen as well as Dioscorides and Caelius Aurelianus; Brehaut (1912), p. 87, note, citing PL 70, 1146, in the *De instit. divin. litterarum.*

animals, wild beasts, minute animals, serpents, worms, fish, birds, and minute flying creatures. Isidore also tends to ascribe more marvelous virtues to animals than to plants or stones. From Pliny and Solinus are repeated the tales of the basilisk, echeneis, and the like,[1] while Augustine's *Commentary on the Psalms* is cited for the story of the asp resisting the incantations of its charmers by laying one ear to the ground and stopping up the other ear with the end of its tail.[2] On the other hand, Isidore omits Pliny's superstitious assertions concerning the river tortoise and gives only his criticism that the statement that ships move more slowly if they have the foot of a tortoise aboard is incredible.[3] Even in the books on minerals and vegetation we still hear of animal marvels: [4] how the coloring matter, cinnabar, is composed of the blood shed by the dragon in its death struggle with the elephant, how the fiercest bulls grow tame under the Egyptian fig-tree, how swallows restore the sight of their young with the swallow-wort, or of the use of fennel and rue by the snake and weasel respectively, the former tasting fennel to enable him to shed his old skin, and the latter eating rue to make him immune from venom in fighting the snake. All these items, too, are from Pliny.

Isidore is rather less hospitable to superstition than Pliny.

But on the whole I should estimate that Isidore contains less superstitious matter even proportionally to his meager content than Pliny does in connection with the virtues of animals, plants, and stones. In discussing plants he says nothing of ceremonial plucking of them and he contains practically no traces of agricultural magic. He describes as a superstition of the Gentiles the notion that the herb *scylla*, suspended whole at the threshold, drives away all evils.[5] He mentions the use of mandragora as an anaesthetic in surgical operations, and remarks that its root is of human form, but says nothing of its applications in magic.[6] In his discussion of stones he repeats after Pliny and So-

[1] *Etymol.*, XII, 4, 6 and 6, 34.

[2] *Ibid.*, XII, 4, 12.

[3] *Ibid.*, XII, 6, 56.

[4] *Ibid.*, XVII, 7, 17 and 9, 36; XIX, 17, 8.

[5] *Ibid.*, XVII, 9, 85.

[6] *Ibid.*, XVII, 9, 30.

linus the marvelous virtues ascribed to a number of them, but follows Pliny's method of making the magicians responsible for these assertions or of inserting a word of caution such as "if this is to be believed" with each statement. Finally he introduces together a number of cases of marvelous powers ascribed to stones with the introduction, "There are certain gems employed by the Gentiles in their superstitions." [1]

Portents.

Isidore lists a number of mythical monsters as well as cases of portentous births in the third chapter, *De portentis,* of his eleventh book. He there affirms that God sometimes wishes to signify future events by means of monstrous births as well as by dreams and oracles, and declares that this "has been proved by numerous experiences." [2]

Words and numbers.

Brehaut is impressed by Isidore's "confidence in words," which he thinks "really amounted to a belief, strong though perhaps somewhat inarticulate, that words were transcendental entities." [3] Isidore's faith in the power of words does not seem, however, to have led him to recommend the use of any incantations; he was content with etymologies and allegorical interpretation. He was also a great believer in the mystic significance of numbers and wrote a separate treatise upon those numbers which occur in the sacred Scriptures. In the *Etymologies,* too, he more than once dwells upon the perfection of certain numbers. We have already heard how perfect most of the numbers up to twelve are, but this is our first opportunity to hear the Pythagorean method applied to the number twenty-two. However, Isidore is not the first to do this; he is, indeed, simply quoting one of the fathers, Epiphanius.[4] "The *modius* is so-called because it is of perfect mode. For this measure contains forty-four pounds, that is, twenty-two *sextarii.* And the reason for this number is that in the beginning God performed twenty-two works. For on the first day He made

[1] *Etymol.,* XVI, 15, 21-26.

[2] *Ibid.,* XI, 3, 4, "quod plurimis etiam experimentis probatum est."

[3] Brehaut (1912), p. 3.

[4] *Etymol.,* XVI, 26, 10, from Epiphanius, *Liber de ponderibus et mensuris.*

seven works, namely, unformed matter, angels, light, the upper heavens, earth, water, and air. On the second day only one work, the firmament. On the third day four things: the seas, seeds, grass, and trees. On the fourth day three things: sun and moon and stars. On the fifth day three: fish and aquatic reptiles and flying creatures. On the sixth day four: beasts, domestic animals, land reptiles, and man. And all twenty-two kinds were made in six days.[1] And there are twenty-two generations from Adam to Jacob. . . . And twenty-two books of the Old Testament. . . . And there are twenty-two letters from which the doctrine of the divine law is composed. Therefore in accordance with these examples the *modius* of twenty-two *sextarii* was established by Moses following the measure of sacred law. And although various peoples have added something to or ignorantly subtracted something from its weight, it is divinely preserved among the Hebrews for such reasons." With such mental magic and pious "arithmetic," as Isidore's friend Braulio called it, might the Christian attempt to sate the inherited thirst within him for the operative magic and pagan divination in which his conscience and church no longer allowed him to indulge.

History of magic.

Isidore's chapter on the *Magi* or magicians, which occurs in his eighth book on the church and divers sects, is a notable one, of whose great future influence we shall presently speak. His own borrowing here is only in small part from Pliny's famous passage on the same theme. On such a subject Isidore naturally has recourse mainly to Christian writers: Augustine, Jerome, Lactantius, Tertullian. From the occasional similarity of his wording to these authors it seems fairly certain that his account is a patchwork from their works, and the context is too Christian to have been drawn *in toto* from some Roman encyclopedist now lost to us. Perhaps the most noteworthy point about Isidore's chapter is that he has made magic and magicians the general and

[1] Hence, presumably, the *sextarii*, from *sex*.

inclusive head under which he presently lists various other minor occult arts and their practitioners for separate definition. But first we have a longer discussion, though long only by comparison, of magic in general. Its history is sketched; Zoroaster and Democritus, as in Pliny, are mentioned as its founders, but it is not forgotten that the bad angels were really responsible for its dissemination. From the first Isidore identifies magic and divination; after stating that the magic arts abounded among the Assyrians, he quotes a passage from Lucan which speaks of the prevalence of liver divination, augury, divination from thunder, and astrology in Assyria. Also the magic arts are said to have prevailed over the whole world for many centuries through their prediction of the future and invocation of the dead. Brief allusion is further made to Moses and Pharaoh's magicians, to the invocation of Samuel by the witch of Endor, to Circe and the comrades of Ulysses, and to several other passages in classical literature anent magic.

Next comes a formal definition of the *Magi*. They are "those who are popularly called *malefici* or sorcerers on account of the magnitude (a characteristic bit of derivation) of their crimes. They agitate the elements, disturb men's minds, and slay merely by force of incantation without any poisoned draught. Hence Lucan writes, 'The mind, though polluted by no venom of poisoned draught, perishes by enchantment.' [1] For, summoning demons, they dare to work their magic so that anyone may kill his enemies by evil arts. They also use blood and victims and sometimes corpses." After this very unfavorable, although sufficiently credulous, definition of magic, which is represented as seeking the worst ends by the worst means, Isidore goes on to list and briefly define a number of subordinate or kindred occult arts. First come necromancers; then hydromancy, geomancy, aeromancy, and pyromancy; next diviners, those employing incantations, *arioli, aruspices,* augurs, *auspices, pythones,* astrologers and their cognates, the *genethliaci* and *mathe-*

Definition of magic.

[1] "*Mens hausti nulla sanie polluta veneni*
Incantata perit . . ."

matici, who as Isidore notes are spoken of in the Gospel as *Magi,* and *horoscopi.* "*Sortilegi* are those who profess the science of divination under the pretended guise of religion through certain devices called *sortes sanctorum* and predict by inspection of certain scriptures." *Salisatores* are those who predict from the jerks of their limbs. To this list of magic arts Isidore adds in the words of Augustine all ligatures and suspensions, incantations and characters, which the art of medicine condemns and which are simply the work of the devil. With mention of the origin of augury among the Phrygians, the discovery of *praestigium* which deceives the eye by Mercury, and the revelation of *aruspicina* by Tagus to the Etruscans, Isidore closes the chapter. Some of its items will be found again in his *De differentiis verborum,*[1] listed under the appropriate letters of the alphabet. It may also be noted that he briefly treats of transformations worked by magic in the fourth chapter of the eleventh book of the *Etymologies.*

Future influence of Isidore's account of magic.

We turn to the future influence of this account of magic which seems to have been first patched together by Isidore. Juiceless as it is, it seems to have become a sort of stock or stereotyped treatment of the subject with succeeding Christian writers down into the twelfth century. Somewhat altered by omission of poetical quotations or the insertion of transitional sentences, it was otherwise copied almost word for word by Rabanus Maurus (about 784 to 856), in his *De consanguineorum nuptiis et de magorum praestigiis falsisque divinationibus tractatus,* and by Burchard of Worms and Ivo of Chartres (died 1115) in their respective collections of *Decreta,* while Hincmar of Rheims in his *De divortio Lotharii et Tetbergae* copied it with more omissions.[2] It was also in substance retained in the *Decretum* of

[1] Migne, PL 83, 9.

[2] For Rabanus' account see Migne, PL 110, 1097-1110; Burchard, PL 140, 839 *et seq.;* Ivo, PL 161, 760 *et seq.;* Hincmar, PL 125, 716-29. Moreover, Burchard continues to follow Rabanus word for word for some ten columns after the conclusion of their mutual excerpt from Isidore, while Ivo is identical with Burchard for fifteen more columns. In "Some Medieval Conceptions of Magic," *The Monist*

Gratian in the twelfth century, when, too, Hugh of St. Victor probably made use of it and John of Salisbury made it the basis of his fuller discussion of the subject. Isidore's account of magic, like his discussion of many other topics, sounds as if he had ceased thinking on the subject, and it must have meant still less to those who copied it. John of Salisbury is the first of them to put any life into the subject and give us any assurance that such arts were still practiced in his day. We have, however, other evidence that magic continued to be practiced in the interval. And such practices as the *sortes sanctorum,* though included in Isi-

January, 1915, XXV, 107-39, I stated (p. 109, note 2) that I thought that I was the first to point out the identity of these four accounts with Isidore's.

Since then, however, I have noticed that Manitius (1911), p. 299, notes the identity of Rabanus with Isidore, "Dass Hraban sich auch sonst ganz an Isidor anlehnt, beweist er in der Schrift *De consanguineorum nuptiis* im Abschnitt *de magicis artibus* (Migne, 109, 1097ff.) der aus *Etym.* 8, 9 stammt." Also Mr. C. C. I. Webb, in his 1909 edition of the *Polycraticus* notes John of Salisbury's borrowings from Isidore and Ivo of Chartres. Finally, J. Hansen, *Zauberwahn, Inquisition, und Hexenprozess im Mittelalter,* 1900, at p. 49 notes that Isidore's sketch of the history of magic keeps recurring in medieval writings, at p. 71 the dependence of Rabanus and Hincmar upon Isidore, and perhaps he somewhere notes the identity with the foregoing of the accounts of magic in Burchard and the other decretalists, but in the absence of an index to his volume I do not find such a passage. At p. 128, however, he notes that John of Salisbury's description of magic is in part taken word for word from Isidore and Rabanus.

Professor Hamilton, in one of his papers on *Storm-Making Springs,* which appeared at about the same time as my article (*Romanic Review,* V, 3, 1914; but, owing probably to war conditions, this issue did not actually appear until after the number of *The Monist* containing my article), came near noting the same thing when he spoke (p. 225) of Isidore's chapter as "quoted at length" by Gratian—who seems to me, however, to give the substance of Isidore's chapter rather than his exact wording—and further noted that four lines of Latin which he quoted were found alike in Rabanus, Hincmar, Ivo, and the *Polycraticus* of John of Salisbury.

In my article I also stated: "Professor Burr, in a note to his paper on 'The Literature of Witchcraft' (*American Historical Association Papers,* IV (1890), p. 241) has described the accounts of Rabanus and Hincmar but without explicitly noting their close resemblance, although he characterizes Rabanus' article as 'mainly compiled.'" Professor Burr subsequently wrote to me, "That I did not mention the relation in my old paper on "The Literature of Witchcraft" was partly because they borrowed from other sources as well and partly because Isidore is himself a compiler. I hoped to come back to the matter in a more careful study of the whole genesis of these stock passages."

dore's stock definition of magic, were probably not generally regarded as reprehensible.[1]

Attitude to astrology.

Isidore's repetition of the views of the fathers concerning demons is so brief and trite [2] that we need not further notice it, but turn to his attitude toward astrology. We have just heard him associate astrologers with practitioners of the magic arts, but in his third book in discussing the *quadrivium* he states that astrology is only partly superstitious and partly a natural science. The superstitious variety is that pursued by the *mathematici* who augur the future from the stars, assign the parts of the soul and body to the signs of the zodiac, and try to predict the nativities and characters of men from the course of the stars. Such superstitions "are without doubt contrary to our faith; Christians should so ignore them that they shall not even appear to have been written." *Mathesis,* or the attempt to predict future events from the stars, is denounced, according to Isidore, "not only by doctors of the Christian religion but also of the Gentiles,—Plato, Aristotle, and others." Isidore also states that there is a distinction between astronomy and astrology, but what it is, especially between astronomy and natural astrology, he fails to elucidate.[3]

In the *De natura rerum.*

In the preface to his *De natura rerum,* which deals chiefly with astronomical and meteorological phenomena, Isidore asserts that "it is not superstitious science to know the nature of these things, if only they are considered from the standpoint of sane and sober doctrine." He also states that his treatise is a brief sketch of what has been written by the men of old and especially in the works of Catholics. In it some of the stock questions which gave difficulty to Christian scientists are briefly discussed, for instance, "Concerning the waters which are above the heavens," and "Whether the stars have souls?" [4] Isidore rejects as "ab-

[1] See below, chapter 60 on Aquinas.

[2] *Etymol.,* VIII, 11, 15-17; *Differentiarum,* II, 14.

[3] Indeed, *Differentiarum,* II, 39, he defines astrology as he had astronomy in *Etymol.,* III, 27. In *Etymol.,* III, 25, he ascribes the invention of astronomy to the Egyptians and that of astrology to the Chaldeans.

[4] Caps. 14 and 27.

surd fictions" imagined by the stupidity of the Gentiles their naming the days of the week from the planets, "because by the same they thought that some effect was produced in themselves, saying that from the sun they received the spirit, from the moon the body, from Mercury speech and wisdom, from Venus pleasure, from Mars ardor, from Jupiter temperance, from Saturn slowness." [1] Yet later in the same treatise we find him saying that everything in nature grows and increases according to the waxing and waning of the moon.[2] Moreover, he calls Saturn a cold star and explains that the planets are called *errantia,* not because they wander themselves but because they cause men to err.[3] He also describes man as a microcosm.[4] Like most ecclesiastical writers, no matter how hostile they may be to astrologers, he is ready to assert that comets signify political revolutions, wars, and pestilences.[5] In the *Etymologies* he not only attributes racial and temperamental differences among the peoples of different regions to "force of the star" [6] and "diversity of the sky," [7] phrases which seem to imply astrological influence rather than the mere influence of climate in our sense. He also encourages astrological medicine when he says that the doctor should know astronomy, since human bodies change with the qualities of the stars and the change of times.[8] Isidore might as well have taken the planets as signs in the astrological sense as have ascribed to them the absurd allegorical significance in passages of Scripture that he did. He states that the moon is sometimes to be taken as a symbol of this world, sometimes as the church, which is illuminated by Christ as the moon receives its light from the sun, and which has seven meritorious graces corresponding to the seven forms of the moon.[9]

[1] *De nat. rer.,* III, 4; PL 83, 968.

[2] *Ibid.,* XIX, 2.

[3] *Ibid.,* XXII, 2-3.

[4] *Ibid.,* IX, 1-2.

[5] *Ibid.,* XXVI, 15; *Etymol.,* III, 71, 16.

[6] *Etymol.,* XIV, 5, "vim sideris."

[7] *Ibid.,* IX, 2, "secundum diversitatem enim coeli."

[8] *Ibid.,* IV, 13, 4.

[9] *De nat. rerum,* XVIII, 5-7.

Bede's scanty science.

The scientific acquisitions of Bede have too often been referred to in exaggerated terms. Sharon Turner said of him, "He collected and taught more natural truths with fewer errors than any Roman book on the same subjects had accomplished. Thus his work displays an advance, not a retrogradation of human knowledge; and from its judicious selection and concentration of the best natural philosophy of the Roman Empire it does high credit to the Anglo-Saxon good sense."[1] Dr. R. L. Poole more moderately says of Bede, "He shows an extent of knowledge in classical literature and natural science entirely unrivalled in his own day and probably not surpassed for many generations to come."[2] Bede perhaps knew more natural science than anyone else of his time, but if so, the others must have known practically nothing; his knowledge can in no sense be called extensive. As a matter of fact, we have evidence that his extremely brief and elementary treatises in this field were not full enough to satisfy even his contemporaries. In the preface to his *De temporum ratione*[3] he says that previously he had composed two treatises, *De natura rerum* and *De ratione temporum,* in brief style as he thought fitting for pupils, but that when he began to teach them to some of the brethren, they objected that they were reduced to a much briefer form than they wished, especially the *De temporibus,* which Bede now proceeds to revise and amplify. It is noteworthy that in order to fulfill the monks' desire for a fuller treatment of the subject he found it necessary to do some further reading in the fathers. In addition to Bede's own statement of his aim, the frequency with which we find manuscripts of early date[4] of the *De natura rerum* and

[1] *History of the Anglo-Saxons,* III, 403.

[2] *Illustrations of the History of Medieval Thought,* 1884, p. 20; p. 18 in 1920 edition.

[3] Migne, PL 90, 293-4.

[4] A few MSS, chiefly from France, earlier than the 12th century, are: BN 5543, 9th century; BN 15685, 9th century; BN nouv. acq. 1612, 1615, and 1632, all 9th or 10th century; Amiens 222, 9th century; Cambrai 925, 9th century; Ivrea 3, 9th century; Ivrea 6, 10th century; Berlin 128, 8-9th century; Berlin 130, 9-10th century; CLM 18158, 11th century; CLM 21557, 11th century.

I have not noted the MSS of

De temporibus suggests that they were employed as textbooks in the monastic schools of the early middle ages. As the Carolingian poet expressed it,

> *Beda dei famulus nostri didasculus evi*
> *Falce pia sophie veterum sata lata peragrans.*

Bede's *De natura rerum.*

Of Bede's *Hexaemeron* we spoke in an earlier chapter. His chief extant genuine scientific treatise is the aforesaid *De natura rerum,*[1] a very curtailed discussion of astronomy and meteorology. It is very similar to Isidore's treatise of the same title, but is even briefer, omitting for the most part the mention of authorities and the Biblical quotations and allegorical applications which make up a considerable portion of Isidore's brief work. One of the few authorities whom Bede does cite is Pliny in a discussion of the circles of the planets.[2] Like Isidore he accepts comets as signs of war and political change, of tempests and pestilence.[3] He also states that the air is inhabited by evil spirits who there await the worse torments of the day of judgment.[4] In his Biblical commentaries Bede briefly echoes some of the views of the fathers concerning magic and demons, for instance, in his treatment of the witch of Endor.[5]

Divination by thunder.

Bede also translated into Latin a treatise on divination from thunder, perhaps from the works of the sixth century Greek writer, John Lydus. In the preface to Herefridus, at whose request he had undertaken the translation, he speaks of it as a laborious and dangerous task, sure to expose him to the attacks of the invidious and detractors who will perhaps insinuate that he is possessed of an evil spirit or is a practitioner of magic. The three chapters of the treatise give the significance of thunder for the four points of the compass, the twelve months of the year, and the seven days of the week. For instance, if thunder arises in the east,

Bede in the British Museum and Bodleian collections.

[1] PL 90, 187-278; the text occupies but a small portion of these columns.

[2] *Ibid.*, Cap. 14.

[3] *Ibid.*, Cap. 24.

[4] *Ibid.*, Cap. 25.

[5] *In Samuelem prophetam allegorica expositio,* IV, 7; PL 91, 701.

according to the traditions of subtle philosophers there will be in the course of that year copious effusion of human blood. Each signification is introduced with some bombastic phraseology concerning the agile genius or sagacious investigation of the philosophers who discovered it.[1] Other tracts on divination which were attributed to Bede are probably spurious and will for the most part be considered later in connection with other treatises of the same sort.[2]

Riddles of Aldhelm.

Some interest in and knowledge of natural science is displayed in the metrical riddles[3] of St. Aldhelm, abbot of Malmesbury and bishop of Sherborne, who died in 709, "the first Englishman who cultivated classical learning with any success and the first of whom any literary remains are preserved." Most of them are concerned with animals, such as silkworms, peacock, salamander, bee, swan, lion, ostrich, dove, fish, basilisk, camel, eagle, taxo, beaver, weasel, swallow, cat, crow, unicorn, minotaur, Scylla, and elephant; or with herbs and trees, such as heliotrope, pepper, nettles, hellebore, and palm; or with minerals, such as salt, adamant, and magnet; or with terrestrial and celestial phenomena, such as earth, wind, cloud, rainbow, moon, Pleiades, Arcturus, Lucifer, and night. There is a close resemblance between some of these riddles and a score of citations from an Adhelmus made in the thirteenth century by Thomas of Cantimpré in his *De natura rerum*.[4] Pitra,[5] however, suggested

[1] *De tonitruis libellus ad Herefridum*, PL 90, 609-14.

[2] See below, chapter 29.

[3] The *Aenigmatum Liber* forms a part of the *Liber de septenario et de metris* in Aldhelm's works as edited by Giles, Oxford, 1844, and reprinted in Migne, PL 89, 183-99.

[4] Cantimpré's citations of Adhelmus seem almost certainly drawn from the *Aenigmata* in the cases of *Leo, ciconia, hirundinus, nycticorax, salamander, luligo* (or, *loligo*), *perna, dranguntia lapis* (*natrix*), *myrmicoleon, colossus,* and *molossus*. On the other hand, the citations concerning *onocentaur* do not correspond to the riddle *De monocero sive unicorni;* the two accounts of Scylla are different; and I do not find *cacus* or *onager* or harpy or siren or locust or the Indian ants larger than foxes in the *Riddles* as edited by Giles.

The passages in which Thomas of Cantimpré cites Adhelmus are printed together by Pitra (1855) III, 425-7.

[5] Pitra (1855) III, xxvi. Only in the case of the salamander does Pitra say, "Thomas huc adduxit Adhelmi Shirbrunensis aenigma de Salamandra vatemque a philosopho clare distinxit."

that the Adhelmus cited by Thomas of Cantimpré was a brother of John the Scot of the ninth century.

Gregory's *Dialogues*.

The total lack of originality and the extremely abbreviated character of the infrequent scientific writing in the west is not, however, a fair example of the total thought and writing of early medieval Latin Christendom. When we turn to the lives of the saints, to the miracles recorded of contemporary monks and missionaries, we find that in the field of its own supreme interests the pious imagination of the time could display considerable inventiveness and was by no means satisfied with brief compendiums from the Bible and earlier Fathers. Here too the superstition and credulity, which had been held back by fear of paganism in the case of natural and occult science, ran luxuriant riot. Such literature lies rather outside the strict field of this investigation, but it is so characteristic of the Christian thought of the period that we may consider one prominent specimen, the *Dialogues* of Gregory the Great,[1] pope from 590 to 604. We shall sufficiently illustrate the nature of this farrago of pious folk-lore by a résumé of the contents of the opening pages of the first of its four books. We need not dwell upon the importance of Gregory in the history of the papacy, of monasticism, and of patristic literature, further than to emphasize the point that so distinguished, influential, and for his times great, a man should have been capable of writing such a book. Similar citations which might be multiplied from other authors of the period could not add much force to this one impressive instance of the naïve pious credulity and superstition of the best Christian minds of that age. Not only were the *Dialogues* well known throughout the medieval period in the Latin reading world, but they were translated into Greek at an early date and in 779 from that language into Arabic, while King Alfred made an Anglo-Saxon translation of the Latin in the closing ninth century.

[1] I have used the text in Migne, PL vol. 77.

Signs and wonders wrought by saints.

In the *Dialogues* Gregory narrates to Peter the Deacon some of the virtues, signs, and marvelous works of saintly men in Italy which he has learned either by personal experience or indirectly from the statements of good and trustworthy witnesses. The first story is of Honoratus, the son of a *colonus* on a villa in Samnium. When the lad evinced his piety by abstaining from meat at a banquet given by his parents, they ridiculed him, declaring that he would find no fish to eat in those mountains. But when the servant presently went out to draw some water, he poured a fish out of the pitcher upon his return which provided the boy with enough food for the entire day. Subsequently the lad was given his freedom and founded a monastery on the spot. Still later he saved this monastery from an impending avalanche by frequent calling upon the name of Christ and use of the sign of the cross. By these means he stopped the landslide in mid-course and the rocks may still be seen looking as if they were sure to fall.

More monkish miracles.

A tale follows of Goths who stole a monk's horse, but found themselves unable to force their own horses to cross the next river to which they came until they had restored his horse to the monk. In another case where Franks came to plunder this same monk, he remained invisible to them. This same monk was a disciple of the afore-mentioned Honoratus and once raised a woman's child from the dead by placing upon its breast an old shoe of his master which he cherished as a souvenir. Thus he contrived to satisfy the mother's pleading and at the same time preserve his own modesty and humility. Gregory does not doubt that the woman's faith also contributed to the miracle. Gregory adds, however, that he thinks the virtue of patience greater than signs and miracles and tells another story of the same monk to illustrate that virtue.

A monastic snake-charmer.

We may pass on, however, to the third chapter which contains a story of the gardener of a monastery who set a snake to catch a thief who had made depredations upon the garden, adjuring the snake as follows: "In the name of

Jesus I command you to guard this approach and not permit the thief to enter here." The serpent obediently stretched its length across the path, and when the gardener returned later, he found the thief hanging head first from the hedge, in which his foot had caught as he was climbing over it and had been surprised by the sight of the serpent. The monk of course then freely gave the thief what he had come to steal, but also of course gave him a brief moral lecture which was perhaps less welcome.

Basilius the magician.

After a brief account of a miraculous release from sexual passion Gregory comes to a tale of Basilius the magician. This is the same man concerning whose arrest and trial on the charge of practicing magic and sinister arts we find directions given in two of the letters of Cassiodorus.[1] According to Gregory he took refuge with the aid of a bishop in a monastery, although the abbot saw something diabolical about him from the very start. Soon a virgin who was under the charge of the monastery became so infatuated with Basilius as to call publicly for him, declaring that she should die unless he came to her aid. The abbot then expelled him from the monastery, on which occasion Basilius confessed that he had often by his magic arts suspended the monastery in mid-air but that he had never been able to injure anyone who was in it. This is more detailed information concerning the nature of Basilius' magic than Cassiodorus gives us. Gregory further adds that not long after Basilius was burned to death at Rome by the zeal of the Christian people.

A demon salad.

A female servant of this same monastery once ate a lettuce in the garden without making the sign of the cross first, and became possessed of a demon straightway. When the abbot was summoned, the demon attempted to excuse himself, exclaiming, "What have I done? what have I done? I was just sitting on a lettuce when she came along and ate me." The abbot nevertheless indignantly proceeded to drive the evil spirit out of his serf.

[1] *Variarum* IV, *Epist.* 22-23, Migne, PL 69, 624-25.

Such are a few specimens of the monkish magic that was considered perfectly legitimate and rapturously admired at the same time that men like Basilius were burned at the stake on charges of magic by the zealous Christian populace.

Incantations in Old Irish.

We may add a word at this point concerning Old Irish literature [1] which, as it has reached us, is almost entirely religious in character,[2] produced and preserved by the Christian clergy. Yet we find a number of traces of magic in these remains of Celtic learning and literature during the dark ages. Indeed, the sole document in the Irish language which is ascribed to St. Patrick is a *Hymn* or incantation in which he invokes the Trinity and the powers of nature to aid him against the enchantments of women, smiths, and wizards. By repeating this rhythmical formula Patrick and his companions are said to have become invisible to King Loigaire and his Druids. The spell is perhaps as old as Patrick's time. Three other incantations for urinary disease, sore eyes, and to extract a thorn are contained in the Stowe Missal. An Irish manuscript of the eighth or ninth century in the monastery of St. Gall has four spells for similar purposes and another is found in a ninth century codex preserved in Carinthia.

The *Fili*.

The Irish had their *Fili* corresponding somewhat to the Druids of Gaul or Britain. They were perhaps less closely connected with heathen rites, since the church seems to have been less opposed to them than to the Druids. They were poets and learned men, and a large part of their learning, at least originally, seems to have consisted of magic and divination. There are many instances in Irish literature of their disfiguring the faces of their enemies by raising blotches upon them by the power of words which they uttered. St. Patrick forbade two of their three methods of divination.

[1] I derive the following facts from E. C. Quiggin, "Irish Literature," in EB V, 622 *et seq.*, where further bibliography is given.

[2] "The Gaelic medical MSS, whether preserved in Ireland, Scotland, or elsewhere, . . . are all, or nearly all, of foreign origin":—Mackinnon, in the *International Congress of Medicine*, London, 1913, p. 413.

CHAPTER XXVIII

ARABIC OCCULT SCIENCE OF THE NINTH CENTURY

Plan of the chapter—Works of Alkindi—*On Stellar Rays,* or *The Theory of the Magic Art*—Radiation of occult force from the stars—Magic power of words—Problem of prayer—Figures, characters, and sacrifice—Experiment and magic—Alkindi's medieval influence—Divination by visions and dreams—Weather prediction—Alkindi as an astrologer—Alkindi on conjunctions—Alkindi and alchemy—Astrological works of Albumasar—The *Experiments* of Albumasar—*Albumasar in Sadan*—*Book of Rains*—Costa ben Luca's translation of Hero's *Mechanica*—Latin versions of his *Epistle concerning Incantation*—Form of the epistle—Incantations directly affect the mind alone—Men imagine themselves bewitched—How are amulets effective?—Citations from the lapidary of the Pseudo-Aristotle—From Galen and Dioscorides—Occult virtue—*On the Difference between Soul and Spirit*—The nature of *spiritus*—Thought explained physiologically—Views of other medieval writers—Thebit ben Corat—The Sabians—Thebit's Relations to Sabianism—Thebit as encyclopedist, philosopher, astronomer—His occult science—Astrological and magic images—Life of Rasis—His 232 works—Charlatans discussed—His interest in natural science—Rasis and alchemy—Titles suggestive of astrology and magic—Conclusion.

Plan of the chapter.

IN this chapter we shall consider a number of learned men who wrote in Arabic or other oriental languages in the ninth and early tenth century: Alkindi, Albumasar, Costa ben Luca, Thebit ben Corat, and Rasis—to mention for the present only the brief and convenient form of their names by which they were commonly designated in medieval Latin learning. Not all of these men were Mohammedans; not one was an Arab, strictly speaking; but they lived under Mohammedan rule and wrote in Arabic. We shall note especially those of their works which deal with occult science and which were plainly influential upon the later medieval Latin learning. Indeed, most of the works of which we shall treat seem to be extant only in Latin translation. This chapter aims at no exhaustive treatment of Arabic science and magic in the ninth century, but merely, by presenting a few prominent examples, to give some idea of it and of its influence upon the middle ages. In subsequent chapters we shall have occasion to mention many other such medieval translations from Arabic and other oriental languages.

Works of Alkindi.

One of the great names in the history of Arabic learning is that of Alkindi (Ya'kûb ibn Ishâk ibn Sabbâh al-Kindî), who died about 850 or 873 A. D.[1] Comparatively few of his writings have come to us, however, although some two hundred titles prove that he covered the whole field of knowledge in his own day. He translated the works of Aristotle and other Greeks into Arabic, and wrote upon philosophy, politics, mathematics, medicine, music, astronomy, and astrology, discriminating little between science and superstition in his enthusiasm for extensive knowledge. The first treatise of his to appear in print was an astrological one on weather prediction in Latin translation.[2] In 1875 Loth printed an Arabic text of his treatise on the theory of conjunctions. More recently Nagy has edited Latin versions of some of his philosophical opuscula, and Björnbo has published an optical treatise by him entitled *De spectaculis*.

On Stellar Rays, or *The Theory of the Magic Art.*

In a manuscript of the closing fourteenth century are contained several sets of errors of Aristotle and various Arabs, also others condemned at Paris in 1348 and 1363, at Oxford in 1376, and so on. Among these are listed the

[1] G. Flügel, *Alkindi, genannt der Philosoph der Araber, ein Vorbild seiner Zeit*, Leipzig, 1857.

F. Dieterici, *Die Naturanschauung und Naturphilosophie der Araber im zehnten Jahrhundert*, Berlin, 1861.

O. Loth, *Al-Kindi als Astrolog.* in *Morgenländische Forschungen. Festschrift für Fleischer*, Leipzig, 1875, pp. 263-309.

A. Nagy, *Die philosophischen Abhandlungen des Al-Kindis*, 1897 in *Beiträge z. Gesch. d. Philos. d. Mittelalt.*, II, 5.

A. A. Björnbo and S. Vogl, *Alkindi, Tideus, und Pseudo-Euclid, Drei Optische Werke*, Leipzig, 1911, in *Abhandl. z. Gesch. d. Math. Wiss.*, XXVI, 3.

For further bibliography see the last-named work and Steinschneider (1905) 23-4, 47, (1906) 31-33.

The Apology of Al Kindy (Sir Wm. Muir, London, 1882) is a defense of Christianity by another writer of about the same time.

[2] *Astrorum iudicis Alkindi, Gaphar de pluviis imbribus et ventis ac aeris mutatione, ex officina Petri Liechtenstein: Venetiis, 1507.*

Errors of Alkindi in the Magic Art.[1] The allusion is to a treatise by Alkindi, variously styled *The Theory of the Magic Art* or *On Stellar Rays,* which is found in Latin version in a number of medieval manuscripts,[2] but which has never been published or described at all fully.

Radiation of occult force from the stars.

Alkindi begins the treatise by asserting the astrological doctrine of radiation of occult influence from the stars. The diversity of objects in nature depends upon two things, the diversity of matter and the varying influence exerted by the rays from the stars. Each star has its own peculiar force and certain objects are especially under its influence, while the movement of the stars to new positions and "the collision of their rays" produce such an infinite variety of combinations that no two things in this world are ever found alike in all respects. The stars, however, are not

[1] Amplon. Quarto 151, fols. 17-19.

[2] In the 1412 catalogue of Amplonius, Math. 48 was "Theorica Alkindi de radiis stellicis seu arcium magicarum vel de phisicis ligaturis"; and at present Amplon. Quarto 349, 14th century, fols. 47v, 65v, 66r-v, 16r-v, 29r, contains "Liber Alkindi de radiis Omnes homines qui sensibilia / Explicit theorica artis magis (*sic*). Explicit Alkindi de radiis stellicis."

Harleian 13, 13th century, given by John of London to St. Augustine's Abbey, Canterbury (#1166, James, 330-1), fols. 166-74, "de radiis stellicis Omnes homines qui sensibilia / explicit Theoria Artis Magice Alkindi."

Digby 91, 16th century, fols. 66-80, Alkindus de radiis stellarum, "Omnes homines qui sensibilia sensu percipiunt. . . ."

Digby 183, end 14th century, fols. 38-45.

Selden supra 76 (Bernard 3464), fols. 47r-60v, "Incipit theoreita artium magicarum. Capitulum de origine scientie. Omnes homines qui sensibilia sensu percipiunt. . . ."; Selden 3467, #4.

Canon. Misc. 370, fols. 240-59, "Explicit theoria magice artis sive libellus Alkindi de radiis stellatis anno per me Theod. scriptus Domini 1484. . . ."

Rawlinson C-117, 15th century (according to Macray, but since the MS once belonged to John of London it is more likely to be 13th century), fols. 157-69, "Incipit theorica Alkindi et est de causis reddendis circa operationes karacterum et conjurationes et suffumigationes et ceteris huiusmodi quae pertinent ad artem magicam. 'Omnes homines qui sensibilia.' . . ."

BN nouv. acq. 616, 1442 A.D., Liber Jacobi Alchindi de radiis.

CU Trinity 936 (R. 15, 17) 17th century, Alkyndus de Radiis.

Ste. Geneviève 2240, 17th century, fol. 32 (?)—since the treatise is listed between two others which begin at fols. 68 and 112, respectively — "Alkyndus de radiis; de virtute verborum."

Steinschneider (1906), 32, has already listed four of these MSS, but was mistaken in thinking Cotton Appendix VI, fols. 63v-70r, "Explicit Iacob alkindi de theorica planetarum," the same treatise as *The Theory of the Magic Art.*

the only objects which emit rays; everything in the world of the elements radiates force, too. Fire, color, and sound are examples of this. The science of physics considers the action of objects upon one another by contact, but the sages know of a more occult interaction of remote objects suggested by the power of the magnet and the reflection of an image in a mirror. All such emanations, however, are in the last analysis caused by the celestial harmony, which governs by necessity all the changes in this world. Thus the men of old, by experiments and by close scrutiny of the secrets of both superior and inferior nature and of the disposition of the sky, came to comprehend many hidden things in the world of nature and were able to discover the names of those who had committed theft and adultery.

The border-line between science and magic.

Alkindi has thus prepared the reader's mind for the consideration of phenomena beyond the realm of ordinary physical action. At the same time he has approached the occult by arguing on the analogy of natural phenomena and he has laid down as a fundamental scientific premise what we now regard as a superstition of astrologers. In other words, he is not unaware of a difference in method and character between physics and astrology, between science and superstition, yet he tries to formulate a scientific basis for what is really a belief in magic.

Magic power of words.

Although Alkindi does not, as I recall, use the word magic, he next argues in favor of what is commonly called the magic power of words. He affirms that the human imagination can form concepts and then emit rays which will affect exterior objects just as would the thing itself whose image the mind has conceived. Muscular movement and speech are the two channels by which the mind's conceptions can be transformed into action. Frequent experiments have proven clearly the potency of words when uttered in exact accordance with imagination and intention, and when accompanied by due solemnity, firm faith, and strong desire. The effect produced by words and voices is heightened if they are uttered under favorable astrological conditions.

Some go best with Saturn, others wtih the planet Jupiter, some with one sign of the zodiac and others with another. The four elements are variously affected by different voices; some voices, for instance, affect fire most powerfully. Some especially stir trees or some one kind of tree. Thus by words motion is started, accelerated, or impeded; animal life is generated or destroyed; images are made to appear in mirrors; flames and lightnings are produced; and other feats and illusions are performed which seem marvelous to the mob.

Problem of prayer.

Alkindi even ventures to touch upon the subject of prayer. He states that the rays emitted by the human mind and voice become the more efficacious in moving matter, if the speaker has fixed his mind upon and names God or some powerful angel. Human ignorance of the harmony of nature also often necessitates appeal to a higher power in order to attain good and to avoid evil. Faith, and observance of the proper time and place and attendant circumstances have their bearing, however, upon the success or failure of prayer as well as of other utterances. And there are some authorities who would exclude spiritual influence entirely in such matters and who believe that words and images and prayers as well as herbs and gems are completely under the universal control exercised by the stars.

Figures, characters, and sacrifice.

The treatise concludes by discussing the virtues of figures, characters, images, and sacrifices in much the same way as it has treated of the power of words. We are assured that "The sages have proved by frequent experiments that figures and characters inscribed by the hand of man on various materials with intention and due solemnity of place and time and other circumstances have the effect of motion upon external objects." Every such figure emits rays having the peculiar virtue which has been impressed upon it by the stars and signs. There are characters which can be employed to cure disease or to induce it in men or animals. Images constructed in conformity with the constellations emit rays having something of the virtue of the celestial harmony. Alkindi also defends the practice of animal sacrifice. Whether God or spirits are placated thereby or not, none the less the sacrifice is efficacious, if made with human intent and due solemnity and in accordance with the celestial harmony. The star and sign which are dominant when any voluntary act of this sort is begun, rule that work to its finish. The material and forms employed should be appropriate to the constellation, or the effect produced will be discordant and perverted.

Experiment and magic.

It will have been noted that Alkindi more than once asserts that his conclusions have been demonstrated experimentally. Thus we have one more example of the connection, supposititious or real, between magic and experimental method.

Alkindi's medieval influence.

The doctrine here set forth by Alkindi of the radiation of force and his explanation of magic by astrology were both to be very influential conceptions in Latin medieval learning. We shall find Roger Bacon, for example, repeating the same views in almost the same language concerning stellar rays and the power of words, and it is appropriate that in two manuscripts his utterances are placed together with those of Alkindi.[1]

Divination by visions and dreams.

Alkindi's treatise *De somno et visione,* as we have it in the Latin translation by Gerard of Cremona,[2] accepts clairvoyance and divination by dreams as true and asks why we see some things before they happen, why we see other things which require interpretation before they reveal the future, and why at other times we foresee the contrary of what is to be.[3] His answer is that the mind or soul has innate

[1] In Digby 91 Roger Bacon on Perspective is followed by Alkindi on the rays of the stars, while in Digby 183 a marginal note to Alkindi's treatise reads "Nota hoc quod est extractum de libro Rogeri Bakun de celo et mundo, capitulo de numero celorum," and following the work of Alkindi we have Bacon on the retardation of old age and perhaps also *de radiis solaribus.*

[2] Edited by Nagy (1897). A MS of the late 12th or early 13th century which Nagy fails to note is Digby 40, fols. 15v-25, de somno et visionibus.

[3] Nagy, p. 18, "Quare autem videamus quasdam res antequam sint? et quare videamus res cum

natural knowledge of these things, and that "it is itself the seat of all species sensible and rational." Vision is when the soul dismisses the senses and employs thought, and the formative or imaginative virtue of the mind is more active in sleep, the sensitive faculties when one is awake.

Weather prediction.

While by some persons, at least, opinions of Alkindi in his *Theory of the Magic Art* were regarded as erroneous, Albertus Magnus in his *Speculum astronomiae* listed among works on judicial astrology with which he thought that the church could find no fault "a book of Alchindi" which opened with the words *Rogatus fui.*[1] This is a work on weather prediction which still exists in a number of manuscripts [2] and was printed in 1507 at Venice, and in 1540 at Paris, together with a treatise on the same theme by Albumasar, of whom we shall say more presently.[3]

interpretatione significantes res antequam sint? et quare videamus res facientes nos videre contrarium earum?"

[1] Spec. astron. cap. 7. More fully the Incipit is, "Rogatus fui quod manifestem consilia philosophorum. . . ."

[2] Digby 68, 14th century, fols. 124-35, Liber Alkindii de impressionibus terre et aeris accidentibus. CU Clare College 15 (Kk. 4, 2), c. 1280, fols. 8-13, "In nomine dei et eius laude Epistola Alkindi de rebus aeribus et pluviis cum sermone aggregato et utili de arabico in latinum translata."

Steinschneider (1906) 32 gives the title as *De impressionibus aeris,* and suggests that it is the same as a *De pluviis* or *De nubibus,* which seems to be the case, as they have the same Incipit—Steinschneider (1905) 13—as does a *De imbribus* in Digby 176, 14th century, fols. 61-63. Steinschneider also suggested that BN 7332, *De impressionibus planetarum* was probably the same treatise; and this is shown to be true by the Explicit of Alkindi's treatise in another MS, Cotton Appendix VI, fol. 63v, "Explicit liber de impressionibus planetarum secundum iacobum alkindi." See also BN 7316, 7328, 7440, 7482.

The opening words of an anonymous *Tractatus de meteorologia* in Vienna 2385, 13th century, fols. 46-49, show that it is the Alkindi. A very similar treatise on weather prediction, *De subradiis planetarum* or *De pluviis,* is ascribed to Haly and exists in three Digby MSS (67, fol. 12v; 93, fol. 183v; 147, fol. 117v) and in some other MSS noted by Steinschneider. It belongs, I suspect, together with a brief *Haly de dispositione aeris* (Digby 92, fol. 5) which Steinschneider listed separately.

[3] Some notion of the number of these astrological treatises on the weather may be had from the following group of them in a single MS.

Vienna 2436, 14th century,
fols. 134-6, "Finitur Hermanni liber de ymbribus et pluviis"
136-8, Iohannes Hispalensis, Tractatus de mutatione aeris
139, Haomar de pluviis
139-40, Idem de qualitate aeris et temporum
140, de pluvia, fulgure, tonitruis et vento

Alkindi as an astrologer.

A majority, indeed, of the works by Alkindi extant in Latin translation are astrological.[1] Several were translated by Gerard of Cremona, and one or two by John of Spain and Robert of Chester.[2] Geomancies are attributed to Alkindi in manuscripts at Munich.[3] Loth notes concerning Alkindi's astrology what we have already found to be the case in his theories of radiation and magic art and of divination by dreams; namely, that while he believes in astrology unconditionally, he tries to pursue it as a science in a scientific way, observing mathematical method and physical laws—as they seemed to him—while he attacked the vulgar superstitions which were popularly regarded as astrology.

Alkindi on conjunctions.

The astrological treatise by Alkindi, of which Loth edited the Arabic text, is a letter on the duration of the empire of the Arabs. This bit of political prediction was, as far as Loth knew, the first instance of the theory of conjunctions in Arabian astrology. The theory was that lesser conjunctions of the planets, which occur every twenty years, middling conjunctions which come every two hundred and forty years, and great conjunctions which occur only every nine hundred and sixty years, exert a great influence not only upon the world of nature but upon political and religious events, and, especially the great conjunctions, open new periods in history. Thus, as Loth says, the conjunction is for the macrocosmos what the horoscope is for man the microcosmos; the one forecasts the fate of

140-1, Dorochius, De hora pluvie et ventorum caloris et frigoris
141, Idem, De hora pluvie
141-2, Alkindus, alias Dorochius, De aeris qualitatibus
142, Idem, De imbribus
143, Jergis, De pluviis
198, 206, Iacobus Alkindus, Liber de significationibus planetarum et eorum naturis, alias de pluviis."

[1] Their titles are listed by Steinschneider (1906) 99; 31-3. We may note BN 6978, 14th century, Incipit epistola Alkindi Achalis de Baldac philosophi de futurorum scientia; Corpus Christi 254, fol. 191, "de aspectibus"—a fragment from a 14th century MSS.

[2] MSS of Robert's translation of Alkindi's *Judgments* are numerous in the Bodleian library: Digby 91, fol. 80-; Ashmole 179; 209; 369; 434; and extracts from it in other MSS. It opens, "Quamquam post Euclidem."

[3] CLM 392, 15th century, fol. 80-; 489, 16th century, fols, 207-21.

the individual; the other, that of society. Loth knew of no Latin translation of Alkindi's letter, and medieval writers in Latin cite Albumasar usually as their authority on the subject of conjunctions. But Loth held that Albumasar, who was a pupil of Alkindi, merely developed and popularized the astrological theories of his master, and Loth showed that Albumasar embodied our letter on the duration of the Arabian empire in large part in his work *On Great Conjunctions* without mentioning Alkindi as his authority.

Alkindi and alchemy.

Although a believer in astrology to the point of magic, and not unacquainted with metals as his work *On the Properties of Swords* shows, Alkindi regarded the art of alchemy as a deception and the pretended transmutation of other metals into gold as false.[1] He affirmed this especially in his treatise entitled, *The Deceits of the Alchemists,* but also in his other writings.[2]

Astrological works of Albumasar.

Something further should be said concerning the astrological treatises of Albumasar (Abu Ma'shar Ja'far ben Muhammad al-Balkhî) whence also his briefer appellations, Japhar and Dja'far. He died in 886 and has been called the most celebrated of all the ninth century Bagdad astrologers, although he has also been accused of plagiarism, as we have seen. In 1489 at Augsburg Erhard Ratdolt published three of his works, the *Greater Introduction to Astronomy* in eight books, the *Flowers*—which Roger Bacon cites as severely condemning physicians who do not study astrology[3]—and the eight books concerning great conjunctions and revolutions of the years. Of these the *Introduction* was translated both by John of Spain and Hermann of Dalmatia, but the former translation, although found in many manuscripts, remains unprinted. The *Flores* is found in numerous manuscripts and was reprinted in 1495. The

[1] O. Loth (1875), pp. 271-2; at 280-2 he gives the Latin of the passage in question from Albumasar, following the Arabic of Alkindi at 273-9.

[2] E. Wiedemann in *Journal f. praktische Chemie,* 1907, p. 73. *et seq.;* cited by Lippmann (1919) p. 399.

[3] Bridges, *Opus Maius,* I, 262, note.

work on conjunctions and revolutions was printed again in 1515 and also exists in many manuscripts.[1] A French translation which Hagins the Jew, working for Henri Bate of Malines, made in 1273 of "Le livre des revolutions de siècle," of whose six chapters he translated only four,[2] probably applied to a part of this work.

The *Experiments* of Albumasar.

Albertus Magnus in the *Speculum astronomiae,* in listing irreproachable works of astronomy and astrology, mentions a "Book of Experiments" by Albumasar instead of the Conjunctions and Revolutions along with his *Flowers* and *Introduction.*[3] This book of experiments by Albumasar is often met with in the manuscripts. It is a different and shorter work than that in eight parts on Conjunctions, but itself

[1] Steinschneider (1905), p. 47.

[2] HL 21, 499-503.

[3] *Spec. astron.* cap. 6. He gives the Incipit of the *Experiments* of Albumasar as "Scito horam introitus" which serves to identify it with the following:

Amplon. Quarto 365, 12th century, fols. 1-18, liber experimentorum.

Ashmole 369-V, 13th century, fols. 103-23v, ". . . . incipit liber in revolutione annorum mundi. Perfectus est liber experimentorum . . ."

Ashmole 393, 15th century, fol. 95v, "Item Albumasar de revolutionibus annorum mundi sive de experimentis . . ."

BN 16204, 13th century, pp. 302-333, "Revolutio annorum mundi . . . Perfectus est liber experimentorum Albumasar . . ."

Arsenal 880, 15th century, fol. 1-.

Arsenal 1036, 14th century, fol. 104v.

Dijon 1045, 15th century, fol. 81-.

Other MSS containing *Experiments* of Albumasar but where I am not sure of the wording of the Incipit are:

Laud. Misc. 594, 14-15th century, fol. 123-, Liber experimentorum.

Harleian 1, fols. 31-41, de experimentis in revolutione annorum mundi.

CLM 51, 1487, and 1503.

Vienna 2436, 14th century, following John of Spain's translation of the *Introductorium magnum* at fols. 1-85 and a *Liber magnarum coniunctionum* at fols. 144-98, comes at fol. 242, "Liber experimentorum seu Capitula stellarum oblata regi magno Sarracenorum ab Albumasore." The Incipit here is "Dispositio est ut dicam ab ariete sic initium" but the treatise is incomplete.

In some MS at Oxford which I cannot now identify the *Flores* of Albumasar close with the statement that the book of Experiments will follow. A different hand then adds "The following work is Albumazar on the revolutions of years," while a third hand adds the explanation, "And according to some authorities it and the book of experiments are one," which is the case.

In some MSS, however, another treatise on revolutions accompanies the *Experiments.* In Amplon. Quarto 365 it is followed at fols. 18-27 by *Sentencie de revolucione annorum,* while in Laud. Misc. 594 it is preceded at fol. 106 by *Liber Albumasar de revolutionibus annorum collectus a floribus antiquorum philosophorum,* which is the same as the *Flores.*

deals with the subject of revolutions. It is not, however, to be confused with still another work by Albumasar on revolutions as connected with nativities.[1]

Albumasar in Sadan.

Another work on astrology with which the name of Albumasar is connected is cited by medieval writers, notably Peter of Abano,[2] as *Albumasar in Sadan* (or Sadam), and is also found in Latin manuscripts where it is also called "Excerpts from the Secrets of Albumasar." [3] Steinschneider regarded the Latin translation as a shortened or incomplete version of an Arabic original entitled *al-Mudsakaret,* or *Memorabilia* by Abu Sa'id Schâdsân, who wrote down the answers of his teacher to his questions.[4] There is also a Greek text, entitled *Mysteries,* which differs considerably from the Latin and of which Sadan perhaps made use.[5] The Latin version might be described as a miscellaneous collection of astrological teachings, anecdotes, and actual cases of Albumasar gathered up by his disciples and somewhat resembling Luther's *Table-Talk* in form.

Book of rains.

We have already alluded to the treatise on weather prediction by Albumasar which was printed with a similar

[1] The distinction between these various works is made quite clear in BN 16204, 13th century, where at pp. 1-183 is John of Spain's translation of the *Liber introductorius maior* in eight parts; at 183-302 the *Conjunctions,* also in eight parts; at 302-333 the *Revolutio annorum mundi* or *Liber experimentorum;* at 333-353 the *Flores,* and at 353-369 the *De revolutione annorum in revolutione nativitatum,* which opens *"Omne tempus breve est operandi . . ."* At the same time the Explicit of this treatise bears witness to the ease with which these works of Albumasar are confused, for it was at first written, *"Explicit liber albumasar de revolutione annorum mundi,"* and some other hand has crossed out this last word and substituted *"nativitatis."*

[2] *Conciliator,* Diff. 156.

[3] Laud. Misc. 594, 14-15th century, fols. 137-41, Liber Sadan, sive Albumasar in Sadan. "Dixit Sadan, Audivi Albumayar dicentem quod omnis vita viventium post Deum est sol et luna / Expliciunt excerpta de secretis Albumasar."

Cat. cod. astrol. Graec. V, i, 142, quotes from a 15th century MS, "Expliciunt excerpta de secretis Albumasaris per Sadan discipulum cuius (eius?) et vocatur liber Albumasaris in Sadan."

The treatise, according to Steinschneider (1906), 36-8, is also found in Amplon. Quarto 352.

CLM 826, 14th century, written and illuminated in Bohemia, fols. 27-33, Tractatus de nativitatibus, "Dixit Zadan: audivi Albumazar dicentem . . ."

[4] Steinschneider (1906), 36-38.

[5] *Cat. cod. astrol. Graec.* V, i, 142. In Vienna MS 10583, 15th century, 99 fols., we find a "de revolutionibus nativitatum" by Albumasar "greco in latinum."

work by Alkindi in 1507 and 1540, and also often accompanies it in the manuscripts. In this "book of rains according to the Indians"[1] Albumasar is variously disguised under the names of Gaphar, Jafar, and Iafar and is called an Indian, Egyptian, or Babylonian.[2] In his Latin translation of it Hugo Sanctellensis tells his patron, the "antistes Michael" that the treatise was written by Gaphar, an ancient astrologer of India, and has since been abbreviated by a Tillemus or Cilenius or Cylenius Mercurius.[3] To Japhar is also attributed a *Minor Isagoga* to astronomy in seven lectures or *sermones,* which Adelard of Bath is said to have translated from the Arabic.[4]

Costa ben Luca's translation of Hero's *Mechanica.*

We turn next to Costa ben Luca, or Qustá ibn Lūqá, of Baalbek, and especially to his treatise *On Physical Ligatures,* or more fully, *The Epistle concerning Incantations, Adjurations, and Suspensions from the Neck.* The scientific importance of Costa ben Luca may be seen from the circumstance that the *Mechanica* of Hero of Alexandria, of which the Greek text is for the most part lost, has been preserved in the Arabic translation which Costa prepared in 862-866

[1] BN 7316, 15th century, #13, liber imbrium secundos Indos . . . authore Jafar; so too BN 7329, 15th century, #6; BN 7316 #16, de mutatione temporum secundum Indos, seems, however, to be another anonymous treatise on the same subject. Perhaps the following, although not so listed in the catalogue, is by Albumasar.

Digby 194, fol. 147v- "Sapientes Indi de pluviis indicant secundum lunam, considerantes ipsius mansiones / quum dominus aspectus aspicit dominum vel est ei conjunctus."

[2] Corpus Christi 233, 13-15th century, fol. 122- "Japhar philosophi et astrologi Aegyptii. Cum multa et varia de nubium congregatione precepta Indorum traxit auctoritas . . ."

Cod. Cantab. Ii-I-13, "Incipit liber Gaphar de temporis mutatione qui dicitur Geazar Babiloniensis. Universa astronomiae iudicia prout Indorum . . ."

[3] The text printed in 1507 and 1540 is Hugo's translation. So is Bodleian 463 (Bernard 2456) 14th century, fols. 20r-24r, "Incipit liber imbrium editum a Iafar astrologo et a lenio et mercurio (Cilenio Mercurio) correcto." See also Savile 15 (Bernard 6561), Liber imbrium ab antiquo Indorum astrologo nomine Jafar editus, deinde a Cylenio Mercurio abbreviatus.

[4] Digby 68, 14th century, fol. 116- "Ysagoga minor Japharis mathematici in astronomiam per Adhelardum Bathoniencem ex Arabico sumpta. Quicunque philosophie scienciam altiorem studio constanti inquireris . . ."

Sloane 2030, fols. 83-86v, according to Haskins in EHR (1913), but my notes, which it is now too late to verify, suggest that it is a fragment occupying less than a page at fol. 87.

for the caliph al-Musta. Several manuscripts of this Arabic text are still extant at Cairo, Constantinople, Leyden, and London, and it has been twice printed.[1]

Latin versions of his *Epistle concerning Incantation, etc.*

The work in which we are more especially interested has also been printed in editions of the works of Galen, of Constantinus Africanus, of Arnald of Villanova, and of Henry Cornelius Agrippa.[2] The treatise is also attributed to Rasis in the library at Montpellier.[3] Its inclusion among Galen's works is a manifest error; in the edition of Agrippa it is appended as *The Letter of an Unknown Author (Epistola incerti authoris)*; while Arnald is represented as translating the work from Greek—a language of which he was ignorant—into Latin. He could read Arabic, however, and perhaps rendered the treatise from that language.[4] But it had certainly been translated before his time, the end of the thirteenth century, and presumably by Constantinus Africanus, c1015-1087, since it not merely appears in his printed works but is found together with an imperfect copy of his *Pantegni* in a manuscript of the twelfth century.[5] In a fifteenth century manuscript Unayn or Honein ben Ishak is named as the author of our treatise, but this seems to be a mistake.[6] Albertus Magnus in the middle of the thirteenth century cites our treatise both in his *Vegetables and Plants*,[7] where he alludes to "the books of incantations of Hermes the philosopher and of Costa ben Luca the philosopher, and the books of physical ligatures," and in his *Minerals*,[8] where

[1] By Carra de Vaux in *Journal asiatique, 9e série*, I, 386, II, 152, 420, with a French translation; and by Nix, Leipzig, 1900, with a German translation, also printed separately in 1894.

[2] Galen, ed. Chart. X, 571; Constantinus Africanus, ed. Basel, 1536, pp. 317-21; Arnald of Villanova, *Opera*, Lyons, 1532, fol. 295, and also in other editions of his works; H. C. Agrippa, *Occult Philosophy*, Lyons, 1600, pp. 637-40.

[3] HL XXVIII, 78-9.

[4] *Idem.*

[5] Additional 22719, 12th century, fol. 200v, "Quesivisti fili karissime de incantatione adjuratione colli suspensione . . ." In view of this and the citations of the work by Albertus Magnus who wrote before Arnald of Villanova, I cannot agree with Steinschneider (1905), pp. 6 and 12, in denying that Constantinus translated the work and in ascribing the translation exclusively to Arnald.

[6] Florence II, III, 214, 15th century, fols. 72-4, "Liber Unayn de incantatione. Quesisti fili karissime . . ."

[7] *De vegetabilibus*, V, ii, 6.

[8] *Mineral.* II, ii, 7, and II, iii, 6.

the *Liber de ligaturis physicis,* as he calls it, is the source whence he has borrowed statements concerning gems ascribed to Aristotle and Dioscorides.

Form of the epistle.

Our treatise is in the form of a reply by Costa ben Luca to someone whom he addresses as "dearest son" and who has asked him what validity there is in incantations, adjurations, and suspensions from one's neck, and what the books of the Greeks and Indians have to say upon these matters. The wording of Costa's epistle varies considerably in the printed editions owing probably to careless interpretation of the manuscripts or careless copying by the earlier scribes, but its general tenor is the same.

Incantations directly affect the mind alone.

Costa first affirms that all the ancients have agreed that the virtue of the mind affects the state of the body. Galen in particular is cited as to the effect of passions upon health and the advisability of the physician's cheering the minds of gloomy patients even by resort to deception to a limited extent, if it seems necessary. A perfect mind generally goes with a perfect body and an imperfect mind with an imperfect body, as is seen in the case of children, old men, and women, or in the inhabitants of the intemperate zones, either torrid Ethiopia or the frozen north. Here one text specifies Scotland (*Scotie*); another, *Schytie,* which is perhaps intended for Scythia. Costa therefore argues that if anyone believes that an incantation will help him, he will at least be benefited by his own confidence. And if a person is constantly afraid that incantations may be directed against him, he may easily fret himself into a fever. This, Costa thinks, was what Socrates had in mind when he described incantations as "words deceiving rational souls by their interpretation or by the fear they produce or by despair." According to Albertus Magnus, who embodies a good deal of Costa's *Epistle* in his *Minerals,* Socrates said more fully that incantations, or perhaps better, enchantments, were made in four ways, namely, by suspending or binding on objects, by imprecations or adjurations, by characters, and by images; and that they dement rational souls so that they fall into fear and despair or rise to joy and confidence; and that through these accidents of the mind bodies are altered either in the direction of health or of chronic infirmity.[1] Costa states that the medical men of India believe that incantations and adjurations are beneficial. But he says nothing to indicate that they, much less the Greeks or himself, have faith in the efficacy of incantations or words to work changes in matter *per se* or directly, nor does he say anything to indicate that demons may be summoned and given orders by this method. Perhaps his discussion of incantations is a trifle constrained and not sufficiently outspoken, but it is moderate and scientific and shows a fair degree of scepticism for that period, especially when we compare it with Alkindi's attitude towards the power of words.

Men imagine themselves bewitched.

Costa ben Luca's attitude towards sorcery seems the same as towards incantations. He concludes his discussion of this point by a story of "a certain great noble of our country" who had convinced himself that he had been bewitched and consequently became impotent. After vainly endeavoring to convince him that this was simply due to his imagination, Costa decided that there was nothing to do but humor him in his delusion. He therefore showed him a passage in *The Book of Cleopatra* which prescribed as an aphrodisiac the anointing of the entire body with the gall of a crow mixed with sesame.[2] The noble followed the prescription and had so much faith in it that his imaginary complaint disappeared.

How are amulets effective?

Finally Costa considers the question of the validity of amulets, or ligatures and suspensions, which we have heard Socrates class with incantations, adjurations, characters, and images. Costa says that he has read in many works by the ancients that objects suspended from the neck are potent not through their natural, but their occult properties. He will not deny that this may be so, but is inclined as before

[1] *Mineral.* II, iii, 6 (ed. Borgnet, V, 55-6).

[2] I am not certain as to this word: it is *sizamelon* in one text, *sesameleon* in another.

to attribute the result rather to the comforting effect which such things have upon one's mind. He proceeds, however, to list a number of suspensions recommended by ancient writers.

Citations from the lapidary of the Pseudo-Aristotle.

First he cites from "Aristotle in the Book of Stones," a spurious treatise of which we shall have more to say in the chapter on Aristotle in the middle ages, a number of examples of the marvelous powers of gems worn suspended from the neck or set in a ring upon the finger. One augments the flow of saliva, another checks the flow of blood. The stone hyacinth enables its bearer to pass safely through a pestilent region, and makes him honored in men's thoughts and procures the granting of his petitions by rulers. The emerald wards off epilepsy, "wherefore we often prescribe to nobles that their children should wear this stone hung about the neck lest they incur this infirmity."

From Galen and Dioscorides.

Costa also cites some recommendations of ligatures and suspensions from Galen, such as curing stomach-ache by suspending coral about the neck or abdomen, or the dung of wolves who have eaten bones, which should preferably be bound on with a thread made from the wool of a sheep eaten by that wolf. To Dioscorides are attributed such amulets as the teeth of a mad dog who has bit a man, which will safeguard their wearer from ever being so bitten—and it would be somewhat of a coincidence, if he were—and the seed of wild saffron which, held in the hand or worn about the neck, is good for the stings of scorpions. The Indians are cited for what is a recipe rather than an amulet: *stercum elephantinum cum melle mixtum et in vulva mulieris positum numquam permittit concipere.* And some say that a woman who spits thrice in a frog's mouth will not conceive for a year. A number of other examples are given without mention of any particular authority. Some of them, indeed, are very familiar and could be found in many authors, and we shall meet them in other contexts.

Occult virtue.

Costa concludes by saying that he himself has not tested these statements extracted from the works of the ancients, but that neither will he deny them, since there exist in nature many strange phenomena and inexplicable forces. We would not believe that the magnet attracts iron, if we had not seen it. Similarly lead breaks adamant which iron cannot break. There is a stone which no furnace can consume and a fish which paralyzes the hand of the person catching it. These strange properties act in some subtle and mighty fashion which is not perceptible to our senses and which we cannot account for by reasoning.[1] But it is noteworthy that as in discussing incantations Costa said nothing of demons, so he fails to ascribe occult virtue to the influence of the stars.

On the Difference between Soul and Spirit.

Another treatise by Costa ben Luca, *On the Difference between Soul and Spirit*,[2] has little to do with occult science, but gives too good a glimpse of medieval notions in the field of physiological psychology to pass it by. It was translated into Latin by John of Spain for Archbishop Raymond of Toledo in the twelfth century,[3] and is found in many manuscripts, often together with the works of Aristotle.[4] Probably by a confusion of the names Costa ben Luca and Constantinus [5] it was printed among the latter's works,[6]

[1] "Quorum enim actio ex proprietate est non rationibus, unde sic comprehendi non potest. Rationibus enim tantum comprehenduntur que sensibus subministrantur. Aliquando ergo quedam substantie habent proprietatem ratione incomprehensibilem propter sui subtilitatem et sensibus non subministratum propter altitudinem sui magnam." I doubt if these last three words refer to the influence of the stars.

[2] *Liber de differentia spiritus et animae*, or *De differentia inter animam et spiritum*. The prologue opens: "Interrogasti me—honoret te Deus!—de differentia . . ."

[3] Steinschneider (1866), p. 404; (1905), p. 43, "wovon ich das Original in Gotha 1158 erkannte."

[4] So in Corpus Christi 114, late 13th century, fol. 229, and at Paris in the following MSS of the 13th or 14th century mostly: BN 6319, #11; 6322, #11; 6323, #6; 6323A; 6325, #17; 6567A; 6569; 8247; 16082; 16083; 16088; 16142; 16490.

[5] Specific illustrations of such confusions between the two names in the MSS are: BN 6296, 14th century, #15, ". . . authore filio Lucae Medici Constabolo"; Brussels, Library of Dukes of Burgundy 2784, 12th century, "Constaben"; Sloane 2454, late 13th century, "Liber differentiae inter animam et spiritum quem Constantinus Luce amico suo scriptori Regis edidit."

[6] Constantinus Africanus, *Opera*, Basel, 1536, pp. 307-17, "Qui voluerit scire differentiam, que est inter duas res . . . / . . . Hec igitur de differentiis spiritus et anime tibi dicta sufficiant, valeto." Edited more recently by S Barach, Innsbruck, 1878, pp. 120-39.

and indeed we find very similar views in his *Pantegni* [1] and in his treatise *On Melancholy*. The work has also been ascribed to Augustine,[2] Isaac,[3] Avicenna,[4] Alexander Neckam, Thomas of Cantimpré, and Albertus Magnus.[5] A different work with a similar title and somewhat similar contents is the *De spiritu et anima,* which is printed with the works of Augustine [6] but which cites such later authors as Boethius, Isidore, Bede, Alcuin, St. Bernard, and Hugh of St. Victor, to whom also it has been attributed.[7] Thomas Aquinas called it the work of an anonymous Cistercian.[8] But to return to our treatise.

The nature of *spiritus.*

Costa ben Luca has, as we have hinted, some diverting passages in the fields of physiological psychology. He believes in the existence of *spiritus,* which is not spirit in one of our senses of that word, but "a subtle body," unlike the soul which is incorporeal. This subtle *spiritus* perishes when separated from the body and it operates most of the vital processes of the body such as breathing and the pulse, sensation and movement. The two former processes are operated by *spiritus* "arising from the heart and borne in the pulsating veins to vivify the body." The two latter processes are caused by *spiritus* which arises from the brain and operates through the nerves. Thus *spiritus* is the cause of life in the body and it leaves this mortal frame with our dying gasp. The clearer and more subtle this *spiritus* is, the more readily it lends itself to mental processes, while the more perfect the human body, the more perfect the *spiritus* and the human mind. Hence the intellectual powers of children and women are inferior, and the same is true of races subjected to excessive heat or cold like the Ethiopians or Slavs.

[1] *Theorica,* III, 12.

[2] Corpus Christi 154, late 13th century, pp. 356-74, ascribed to Augustine in both Titulus and Explicit.

[3] S. Marco 179, 14th century, fols. 57-9, 83, Liber Ysaac de differentia spiritus et animae.

[4] CU Gonville and Caius 109, 13th century, fols. 1-6v, "Avicenna de differencia spiritus et anime."

[5] So says Coxe, anent Corpus Christi 114, and Steinschneider (1905), p. 43.

[6] Migne, PL 40, 779-832.

[7] By Trithemius; but earlier so cited by Vincent of Beauvais (PL 40, 779-80). See also Exon. 23, 13th century, fol. 196v.

[8] Migne, PL 40, 779-80.

Here we have the same views repeated as in the *Epistle concerning Incantation.* Some physicians and philosophers think that there are two vessels in the heart and that there is more *spiritus* than blood in the left hand vessel and more blood than *spiritus* in the right hand vessel. The *spiritus* in the brain becomes more subtle and apt to receive the virtues of the soul by its passage from one cavity of the brain to another. The less subtle *spiritus* the brain uses for the five senses; Costa speaks of "hollow nerves" from the brain to the eye through which the *spiritus* passes for the purpose of vision. The most subtle *spiritus* is employed in the higher mental processes such as imagination, memory, and reason.

Thought explained physiologically.

Costa ben Luca gives an amusing explanation of how these processes take place in the brain. The opening between the anterior and posterior ventricles of the brain is closed by a sort of valve which he describes as "a particle of the body of the brain similar to a worm." When a man is in the act of recalling something to memory, this valve opens and the *spiritus* passes from the anterior to the posterior cavity. Moreover, the speed with which this valve works or responds differs in different brains, and this fact explains why some men are of slow memory and why others answer a question so much sooner. The habit of inclining the head when deep in cogitation is also to be explained as tending to open this valve. However, the relative subtlety of the *spiritus* is another important factor in intellectual ability.

Views of other medieval writers.

Other medieval writers differed somewhat from these views of Costa ben Luca as to the nature of *spiritus* and the cavities of the brain. For instance, Constantinus Africanus in his treatise *On Melancholy* states that the *spiritus* of the brain is called the rational soul, which is inconsistent with the distinction drawn between soul and spirit in the other treatise. In the eleventh century both Constantinus in his *Pantegni* and *Anatomy* or *De humana natura*,[1] and Petrocellus the Salernitan in his *Practica;*[2] in the twelfth century both Hildegard of Bingen[3] and the Pseudo-Augustinian *Liber de spiritu et anima;*[4] in the thirteenth century both Bartholomew of England, who seems to cite Johannitius (Hunain ibn Ishak) on this point,[5] and Vincent of Beauvais agree that the brain has three main cavities. The first is phantastic, from which the senses are controlled, where the sensations are registered, and where the process of imagination goes on. The middle cell is logical or rational, and there the forms received from the senses and imagination are examined and judged. The third cell retains such forms as pass this examination and so is the seat of memory.[6] The Pseudo-Augustine, however, represents it further as the source of motor activity. Constantinus and Vincent of Beauvais, who quotes him in the thirteenth century, further distinguish the phantastic cavity as hot and dry, the logical cell as cold and moist, and the seat of memory as cold and dry. Moreover, the phantastic cell which multiplies forms contains a great deal of *spiritus* and very little medulla, while the cell of memory which retains the smaller number of forms selected by reason contains much medulla and little *spiritus*. Thus the general point of view of these other authors resembles that of Costa ben Luca despite the divergence from him in details. They perhaps also owe something to Augustine, who in his genuine works speaks of the three cells of the brain but makes the

[1] Both passages were excerpted by Vincent of Beauvais, *Speculum naturale*, XXIX, 41.

[2] De Renzi (1852-9) IV, 189; Petrocellus is very brief on the cells of the brain.

[3] Singer (1917), pp. 45 and 51, has noted that Hildegard's description of the brain as divided into three chambers is anteceded by the *Liber de humana natura* of Constantinus, and contained "in the writings of St. Augustine."

[4] PL 40, 795, cap. 22.

[5] *De proprietatibus rerum*, III, 10 and 16; V, 3.

[6] Similarly E. G. Browne (1921), p. 123, writing of Arabian medicine and Avicenna, says, "Corresponding with the five external senses, taste, touch, hearing, smelling, and seeing, are the five internal senses, of which the first and second, the compound sense (or 'sensus communis') and the imagination, are located in the anterior ventricle of the brain; the third and fourth, the co-ordinating and emotional faculties, in the mid-brain; and the fifth, the memory, in the hind-brain." Galen had somewhat similar ideas.

hind-brain the center of motor activity, and the mid-brain the seat of memory.[1]

Thebit ben Corat.

Thabit ibn Kurrah ibn Marwan ibn Karaya ibn Ibrahim ibn Marinos ibn Salamanos (Abu Al Hasan) Al Harrani or Thabit ben Corrah ben Zahrun el Harrani, or Tabit ibn Qorra ibn Merwan, Abu'l-Hasan, el-Harrani, or Thabit ben Qorrah or Thabit ibn Qurra, or Tabit ibn Korrah, or Thabit ben Korra, as he is variously designated by modern scholars;[2] or Thebit ben Corat, or Thebith ben Corath, or Thebit filius Core, or Thebites filius Chori, also Tabith, Tebith, Thabit, Thebeth, Thebyth, and Benchorac, ben corach, etc., as we find it in the medieval Latin versions—Thebit ben Corat seems the prevalent medieval spelling and so will be adopted here—was born at Harran in Mesopotamia about 836, spent much of his life at Bagdad, and lived until about 901.[3] He wrote in Arabic as well as Syriac, but was not a Mohammedan, and Roger Bacon alludes to him as "the supreme philosopher among all Christians, who has added in many respects, speculative as well as practical, to the work of Ptolemy."[4] As a matter of fact, he was a heathen or pagan, a member of the sect of Sabians, whose chief seat was at his birth-place, Harran.

The Sabians.

The Sabians appear to have continued the paganism and astrology of Babylonia, but also to have accepted the Agathodaemon and Hermes of Egypt,[5] and to have had relations with Gnosticism and Neo-Platonism. They seem to have laid especial stress upon the spirits of the planets,[6] to whom they made prayers, sacrifices, and suffumigations,[7] while days on which the planets reached their culminating-

[1] *De Genesi ad litteram,* VII, 18 (PL 34, 364).

[2] The fullest treatment of him will be found in D. A. Chwolson, *Die Ssabier und der Ssabismus,* Petrograd, 1856, 2 vols., *passim.* For a list of his works see Steinschneider. *Zeitschrift f. Math.,* XVIII, 331-38.

[3] There is some difficulty with these dates or their Arabic equivalents, because we are not certain whether the length of his life is given in lunar or solar years: see Chwolson, I, 532-3, 547-8.

[4] Bridges, I, 394.

[5] Carra de Vaux, *Avicenne,* Paris, 1900, p. 68.

[6] Chwolson, II, 406, 422, 431, 440, 453, 610, 703.

[7] *Ibid.,* I, 741; II, 7, 258, 386, 677, etc.

points were celebrated as festivals.[1] They observed the houses and stations of the planets, their risings and settings, conjunctions and oppositions, and rule over certain hours of the day and night.[2] Some planets were masculine, others feminine; some lucky, others unlucky; [3] they were related to different metals; [4] the different members of the human body were placed under different signs of the zodiac; [5] and in general each planet had its own appropriate figures and forms, and ruled over certain climates, regions, and things [6] in nature. Most of this, however, is astrological commonplace whether of pagans, Mohammedans, or Christians. Nor were the Sabians peculiar in associating intellectual substances or spirits with the planets.[7] It was only in worshiping these and denying the existence of one God and in their practice of sacrificial divination that they could be distinguished as heathen or pagan. However, they seem to have devoted a rather unusual amount of attention to astrology and other forms of magic such as oracular heads,[8] magic knots and figures,[9] and seal-rings carved with peculiar animal figures. These last they often buried with the dead for a time in order to increase their virtue.[10]

Thebit's relations to Sabianism.

Thebit, at any rate, seems to have prided himself upon being a descendant of pagan antiquity. In a passage praising his native town he said, "We are the heirs and posterity of heathenism," [11] and he described with veneration a ruined Greek temple at Antioch.[12] He had, however, some religious disagreement with the Sabians of Harran and was finally forced to leave.[13] He met a philosopher who took him to Bagdad where he became one of the Caliph's astronomers [14] and founded there a Sabian community to his own taste.

[1] Chwolson, II, 386-97, 500, 525, 530, 676.

[2] *Ibid.*, I, 737.

[3] *Ibid.*, II, 30, 373.

[4] *Ibid.*, II, 411, 658, 839.

[5] *Ibid.*, II, 253.

[6] *Ibid.*, I, 738.

[7] *Ibid.*, I, 733-4.

[8] *Ibid.*, II, 19, 148, 150.

[9] *Ibid.*, II, 21, 138-9.

[10] *Ibid.*, I, 526; II, 141.

[11] Quoted by Bishop Gregory Bar-hebraeus in his *Syrian Chronicle:* Chwolson, I, 177-80.

[12] Chwolson, I, 195; II, 623.

[13] *Ibid.*, I, 482-3.

[14] Again there seems to be uncertainty as to dates, since the Arabic sources name a caliph who was not contemporary with the philosopher in question: Chwolson, I, 548-9.

His numerous religious writings show the value which he attached to various Sabian usages and rites: ceremonials at burials, hours of prayer, rules of purity and impurity and concerning the animals to be sacrificed, readings in honor of the different planets.[1]

Thebit as encyclopedist, philosopher, astronomer.

Thebit was a writer of encyclopedic range and translated from the Greek [2] into Arabic or Syriac such authors as Apollonius, Archimedes, Aristotle, Euclid, Hippocrates, and Galen. He "was famed above all as a philosopher," [3] but most of his philosophical works are lost, but some geometrical treatises by him are extant, and a work on weights appears in Latin translation.[4] A group of four astronomical treatises by him also occurs with fair frequency in medieval manuscripts.[5] On the basis of these specimens of his astronomy Delambre was not moved to assign him any great place in the history of the science; [6] Chwolson objects that they are too brief to do him justice,[7] but they are probably the cream of his own contributions to the subject or the middle ages would not have translated and preserved them so sedulously.

His occult science.

Whatever Thebit's contributions to positive knowledge may or may not have been, there is no dispute as to the fact that he was given to occult science and even superstition. His attitude towards alchemy, indeed, is doubtful, as a work of alchemy is ascribed to him in one manuscript of

[1] Chwolson, I, 485. Chwolson perhaps lays himself open a little to the charge of arguing in a circle, since Thebit's writings are his main source concerning Sabianism.

[2] *Ibid.*, I, 553-64, for a list of his translations of, extracts from, and commentaries upon Greek works.

[3] *Ibid.*, I, 484.

[4] BN 10260, 16th century, "Incipit liber Karastoni de ponderibus . . . / . . . editus a Thebit filio Core." Also in BN 7377B, 14-15th century, #3; 7424, 14th century, #6; Vienna 5203, 15th century, fols. 172-80. For other MSS see Björnbo (1911) 140.

[5] Harleian 13, fol. 118- Thebit de motu octave spere; fol. 120v- Liber Thebith ben Corath de his qui indigent expositione antequam legitur Almagestum; 123- Liber Thebit de ymaginatione spere et circulorum eius diversorum; 124v- Liber Thebith de quantitatibus stellarum et planetarum.

Also in Harl. 3647, #11-14; Tanner 192, 14th century, fol. 103-; BN 7195, 14th century, #12-15; Magliabech. XI-117, 14th century; CUL 1767 (Ii. III, 3) 1276 A. D., fols. 86-96; and many other MSS.

[6] Delambre (1819) 73.

[7] Chwolson, I, 551.

the fourteenth century and some notes against the art in another.[1] But of his adhesion to astrology there is no doubt,[2] and Chwolson notes his interest in the mystic power of letters and magic combinations of them.[3] But the one outstanding example of his occult science is his treatise on images, which seems to have been a favorite with the Latin middle ages, since it appears to have been translated into Latin twice, by Adelard of Bath [4] and by John of Seville,[5]

[1] BN 6514, #10, *Thebit de alchymia;* Amplon. Quarto 312, written before 1323 A.D., fol. 29, *Notule Thebith contra alchimiam.*

[2] A work on judgments is ascribed to him in a Munich MS, CLM 588, 14th century, fol. 189- *Thebites de iudiciis;* followed by, 220- *Liber iudicialis Ptolomei,* 233- *Libellus de iudiciis,* and 238- *Modus iudicandi.* The treatise on fifteen stars, fifteen herbs, and fifteen stones, which as we have seen is usually ascribed to Hermes or Enoch, is attributed to Thebit in at least one MS, BN 7337, page 129-.

[3] I, 551.

[4] Lyons 328, fols. 70-74, Liber prestigiorum Thebidis (Elbidis) secundum Ptolemeum et Hermetem per Adhelardum bathoniensem translatus, opening, "Quicunque geometria atque philosopia peritus astronomiae expers fuerit ociosus est." In this MS the treatise closes with the words, "ut prestigiorum artifex facultate non decidat." This seems to be the only MS known where the translation is ascribed to Adelard of Bath. It seems to have once been part of Avranches 235, 12th century, where the same title is listed in the table of contents. Haskins, in EHR (1911) 495, fails to identify the work, calling it "a treatise on horoscopes." It is to be noted, however, that Albertus Magnus in listing bad necromantic books on images in the *Speculum astronomiae* (cap. xi, Borgnet, X, 641) gives the same Incipit for a *liber praestigiorum* by Hermes, "Qui geometriae aut philosophiae peritus, expers astronomiae fuerit . . ." Undoubtedly the two were the same.

[5] Of John of Seville's translation the MSS are more numerous. The following will serve as a representative. Royal 12-C-XVIII, 14th century, fols. 10v-12r, "Dixit thebyth bencorat et dixit aristoteles qui philosophiam et geometriam exercet et omnem scientiam legit et ab astronomia vacuus fuerit erit occupatus et vacuus quod dignior geometria et altior philosophia est ymaginum scientia. / Explicit tractatus de imaginibus Thebith Bencorath translatus a Iohanne Hyspalensi atque Limiensi in Limia ex Arabico in Latinum. Sit laus deo maximo."

This is the version cited by Michael Scot in his *Liber Introductorius* (Bodleian 266, fol. 200) where he gives the Incipit, "Dixerunt enim thebith benchorath et aristoteles quod si quis philosophiam . . . ," etc., substantially as above.

But now comes a good joke on Albertus, who has listed among good astronomical books of images (*Speculum astronomiae,* cap. xi, Borgnet, p. 642) the work of "Thebith eben chorath" opening "Dixit A. qui philosophiam . . ." which of course is that just mentioned. Thus he condemns one translation of the same book and approves the other; is he perhaps having some fun at the expense of the opponents of both astrology and necromancy?

It will be noted that it is Aristotle, rather than Hermes or Ptolemy, who is cited at the start in John of Seville's translation. I

since the manuscripts of it are numerous,[1] and it also was printed,[2] and since Thebit is cited as an authority on the subject of images by such medieval writers as Roger Bacon, Albertus Magnus,[3] the author of *Picatrix*,[4] Peter of Abano,[5] and Cecco d'Ascoli.[6]

Astrological and magic images.

The work begins by emphasizing the need of a knowledge of astronomy in order to perform feats of magic (*praestigia*). The images described are astronomical or astrological and must be constructed under prescribed constellations in order to fulfill the end sought. Often, however, they are human forms rather than astronomical figures. It is not necessary to engrave them upon gems; Thebit expressly states that the material of which they are made or

therefore am uncertain whether Chwolson has our treatise in mind, when he speaks of Thebit's commenting upon "eine pseudohermetische Schrift über Talismane u.s.w." In the printed text of 1559 Aristotle and Ptolemy are cited in the first paragraph, but in the MSS Aristotle is cited twice.

[1] Some other MSS differ slightly from the foregoing in their opening words, but perhaps not enough to suggest a third translation:

Ashmole 346, 16th century, fols. 113-15v, "Incipit liber de ymaginibus secundum Thebit. In nomine pii et misericordis Dei. Dixit Thebit qui geometrie aut philosophie expers fuerit."

Bodleian 463 (Bernard 2456), written in Spain, 14th century, fols. 75r-75v, "Dixit thebit bencorat Ar. qui legit phylosophiam et geumetriam et omnem scientiam et alienus fuerit ab astronomia erit impeditus vel occupatus."

The following MSS ascribe the translation to John of Spain and have the usual opening words, "Dixit Thebit ben Corat, Dixit Aristoteles, qui philosophiam, etc."

Digby 194, 15th century, fol. 145v-.

S. Marco XI-102, 14th century, fols. 150-53.

Berlin 963, 15th century, fol. 140- "Dixit thebit ben corach Cum volueris operari de ymaginibus," but then at fol. 199, with the usual Incipit.

Harleian 80 has the first part missing but ends, fol. 76r, like John's translation.

Still other MSS are:

Harleian 3647, 13th century.

Sloane 3846, fols. 86v-93; 3847; and 3883, fols. 87-93: all three 17th century.

Amplon. Quarto 174, 14th century, fols. 120-1.

BN 7282, 15th century, #4, interprete Joanne Hispalensi.

Berlin 964, 15th century, fols. 213-5.

Vienna 2378, 14th century, fols. 41-63.

CLM 27, 14-15th century, fols. 71-77; 59, 15th century, fols. 239-43.

Florence II-iii-214, 15th century, fols. 1-4, "Incipit liber Thebit Benchorac de scientia omigarum et imaginum. (D) ixit Aristottiles qui."

[2] *De tribus imaginibus magicis*, Frankfurt, 1559.

[3] *Mineral.* II, iii, 3.

[4] Magliabech. XX-20, fol. 12r; Sloane 1305, fol. 19r.

[5] *Conciliator*, Diff. X., fol. 16GH, in ed. Venice, 1526.

[6] *Commentary on the Sphere*, cap. 3.

upon which they are engraved is unimportant, and that lead or tin or bronze or gold or silver or wax or mud or anything you please will do. The essential thing and "the perfection of mastery" is careful conformity to astrological conditions. This science of images is indeed, as Aristotle and Ptolemy have testified, the acme of astrology. Nevertheless, after the image has been properly constructed, there is usually some non-astrological ceremony to be executed in connection with it which savors of magic. Often the image is to be buried, not however in a grave as in the case of the ancient curses upon lead tablets, but in the house of someone concerned. Once two images are to be placed facing each other and wrapped in a clean cloth before burying them. Instructions are also given as to the direction in which the person burying the image should face. Also forms of words are prescribed which are to be repeated as the image is buried. Once the name of the person whom it is desired to injure is to be written with "names of hate on the back of the image." Among the objects supposed to be achieved by such images are driving off scorpions, destroying a given region, causing misfortunes to happen to others, recovery of stolen objects, success in business or politics, protection from possible injury at the hands of the king, or the causing of an enemy's death by bringing him into disfavor with the monarch. The treatise closes, at least in the printed text, with an admission of its essentially magic character by saying, "And this is what God the highest wished to reveal to his servants concerning magic, that His name may be honored and praised and ever exalted through the ages." But no mention is made of demons, unless an instruction to name one image "by a famous name" alludes to some spirit.

We shall now conclude the present survey with some account of Rasis and his writings, with the exception of a number of books of experiments ascribed to him, but which it is impossible to separate from those ascribed to Galen and other authors, and of which we shall treat later under the head of such experimental literature.

Life of Rasis.

The full name of Rasis or Rhazes was Abu Bakr Muhammad ibn Zakariya ar-Razi,[1] the last word indicating his birthplace in Persia. The date of his birth is uncertain, perhaps about 850. He died in 923 or 924.[2] For the facts of his life we are dependent upon two Arabic writers of the thirteenth century [3] who do little except tell one "good" story after another about him, or quote his famous sayings, most of which sound as if culled from the works of Galen. When about thirty years of age Rasis came to Bagdad and is said to have been attracted to the study of medicine by hearing how an inflamed and swollen forearm which gave great pain was marvelously cured by the application of an herb, which came to be called "the vivifier of the world." In the early years of the tenth century Rasis served as physician in the hospital at Bagdad. According to Withington he has been called "the first and most original of the great Moslem physicians." He also was interested in philosophy and alchemy, as his writings will show.

His 232 works.

There has come down to us a list of some 232 works ascribed to Rasis.[4] Some of them are probably merely different wordings of the same title, others are very likely chapters repeated from his longer works, but at any rate they serve to give us some idea of his interests and the

[1] Also given as Muhammad ibn Zakariya (Abu Bakr) ar-Razi and Abu Bekr Mohammed ben Zachariah.

[2] Withington in his *Medical History,* 1894, gives the date as 932, perhaps by a misprint.

[3] Ibn Abi Usaibi'a (1203-1269, himself a physician and son of an oculist) "Sources of Information concerning Classes of Physicians," compiled at Damascus, 1245-1246, ed. by Müller, Cairo, 1882; and Ibn Khallikan (1211-1282), "Obituaries of Men of Note," written between 1256 and 1274.

For these titles and most of the general account of the life and works of Rasis which follows I am indebted to G. S. A. Ranking's "The Life and Works of Rhazes," pp. 237-68, in *Transactions of the Seventeenth International Congress of Medicine, Section XXIII,* London, 1913.

[4] The list is reproduced by Ranking (1913) in Arabic and Latin, largely on the basis of a MS at the University of Glasgow, which contains a Latin translation by a Greek priest, who died in 1729, of the Arabic work of Usaibi'a, or part of it, mentioned in the previous note: Hunterian Library, MS 44, fols. 1-19v.

ground he covered, although of course some may be incorrectly attributed to him. Editions of the Latin translations of some of his chief medical works were printed before the end of the fifteenth century at Milan in 1481 and Bergamo in 1497.[1] These contain the famous *Liber Almansoris* or *Liber El-Mansuri dictus* with its ten subordinate treatises: (1) introduction to medicine and discussion of human anatomy, (2) the doctrine of temperaments and humors and a discussion of the art of physiognomy,[2] with a chapter on how to select slaves, (3) diet and drugs, (4) hygiene, (5) cosmetics, (6) rules of health and medicines for travelers, (7) surgery or "the art of binding up broken bones and concerning wounds and ulcers," (8) poisons, (9) treatment of diseases from head to foot, (10) fevers. Following this in both editions come his works on Divisions, on diseases of the joints, on the diseases of children, and his Aphorisms or six books of medicinal secrets. Other writings by Rasis found in one or both of the printed editions are a brief treatise on Surgery, Cautery, and Leeches,[3] the book of Synonyms, the table of antidotes, and some others which we shall have occasion to mention later. His treatise on the pestilence or on smallpox and measles was printed many times from the fifteenth to sixteenth century.

Charlatans discussed.

In the list of 232 titles are three works which all seem to bear on the same point and are perhaps different descriptions of one treatise, or else show that this was a favorite theme with Rasis. The idea in all three seems to be that no physician is perfect or can cure all diseases of all patients,

[1] I have examined both these editions at the British Museum; Withington does not mention them in his *History of Medicine*, but cites editions of the *Continens,* Venice, 1542, and *Opera Parva,* 1510, and a modern edition (1858) by the Sydenham Society of *On the Small Pox and Measles.* The pages are not numbered in the edition of 1481, so that I shall not be able to give exact references to them.

[2] This was sometimes reproduced separately: see Wolfenbüttel 2885, 15th century, fol. 1, Phisonomia Rasis, fol. 2, Phisonomia Aristetelis, Rasis et Philomenis, summorum magistrorum in philosophia.

[3] It occupies but a little over three pages in the 1481 edition. Since in the middle of the treatise we read "Magister rasis fecit cauterizari quidem artheticum . . . ," etc., it is perhaps by a disciple rather than Rasis himself.

that this is why many persons go to charlatans, and why sometimes quacks, old-wives, and popular practice succeed in certain cases where the most learned doctors have failed.[1]

His interest in natural science.

Other titles show that Rasis was interested in natural science and not merely in the practice of medicine. Besides what would appear to have been a general treatise entitled, *Opinions concerning Natural Things,* he wrote on optics, holding that vision was not by rays sent forth from the eye, and discussing some of the figures in the work on optics ascribed to Euclid. In a letter he inquired into the reason for the creation of wild beasts and venomous reptiles; and in a third treatise wrote of the magnet's attraction for iron and of vacuums.[2] His interest in natural philosophy of a rather theoretical sort is indicated by an *Explanation of the book of Plutarch or commentary on the book of Timaeus.*[3] Other titles attest his experimental tendency.[4]

Rasis and alchemy.

Eight titles deal with alchemy [5] and show that Rasis regarded transmutation as possible. One is a reply to Alkindi who held the opposite opinion.[6] None of these writings seem to be extant in Arabic, however, and the Latin works of alchemy ascribed to Rasis are generally regarded as spurious. The thirteenth century encyclopedist, Vincent

[1] 79, *Dissertatio de causis quae plerorumque hominum animos a praestantissimis ad viliores quosque medicos solent deflectere.*

124, *Liber, Quod medicus acutus non sit ille qui possit omnes curare morbos quoniam hoc non est in hominum potestate . . . ,*

125, *Epistola, Quod artifex omnibus numeris absolutus in quacumque arte non existat nedum in medicina speciatim: et de causa cur imperiti medici, vulgus, et etiam mulieres in civitatibus, foeliciores sint in sanandis quibusdam morbis quam viri doctissimi et de excusatione medici hoc propter.*

There appears to be a German translation by Steinschneider of this work by Rasis on the success of quacks and charlatans in *Virchow's Archiv f. Pathologische Anatomie,* XXXVI, 570-86.

[2] Ranking (1913), #180, 15, 138, 163.

[3] *Ibid.,* #137; also 145, *Supplementum libris Plutarchi.*

[4] *Ibid.* #126, *Liber, De probatis et experientia compertis in arte medica; per modum syntagmatis est digestus.* #205, *Liber, Quod in morbis qui determinari atque explicari non possunt oporteat ut medicus sit assiduus apud aegrotantem et debeat uti experimentis ad illos cognoscendos. Et de medici fluctatione.*

[5] *Ibid.* #25, 26, 32-35, 38, 40. I should guess that 201, *Arcanum arcanorum de sapientia,* was the same as 35, *Arcanum arcanorum.*

[6] *Ibid.* #40, *Responsio ad philosophum el-Kendi eo quod artem al-Chymi in impossibili posuerit.*

of Beauvais, made a number of citations from the treatise *De salibus et aluminibus* attributed to Rasis, but Berthelot [1] regarded this work as later than Rasis and it is not found among our eight titles. The *Lumen luminis,* which is ascribed to Rasis [2] and seems to have been translated by Michael Scot [3] in the early thirteenth century, is also mainly devoted to these two substances, salts and alums. A *Book of Seventy* is ascribed to Rasis as well as to Geber. Berthelot was inclined to think that a *Book of Secrets* perhaps went back to Rasis. At least some good stories are told by Arabic chroniclers of Rasis' connection with alchemy. One is to the effect that he abandoned the art as a result of a sound beating to which the caliph subjected him when he failed to transmute metals at order. Another states that in preparing the elixir he injured his eyes with its vapors and was cured by a physician who charged him a fee of five hundred *dinars.* Rasis paid the doctor's bill, but, remarking that at last he had discovered the true alchemy and the best art of making gold, devoted the remainder of his life to the study and practice of medicine.[4]

Titles suggestive of astrology and magic.

Rasis also wrote treatises on mathematics and the stars but it is not always easy to infer their contents from the titles which have alone reached us or to tell when *mathematica* means astrology. In one work he seems to have shown the excellence and utility of *mathematica,* but to have confuted those who extolled it beyond measure.[5] In a letter he denied that the rising and setting of the sun and other planets was because of the earth's motion and held that it was due to the movement of the celestial orb.[6] In another letter he discussed the opinion of natural philosophers concerning the sciences of the stars and whether or

[1] Berthelot (1893), I, 68 and 286-7. On the alchemy of Rasis see further in this same volume the chapter, *L'Alchimie de Rasis et du Pseudo-Aristote.*

[2] BN 6514 and 7156.

[3] Riccardian 119, fol. 35v, "Incipit liber luminis luminum translatus a magistro michahele scotto philosopho." Printed by J. Wood Brown (1897), p. 240 *et seq.*

[4] Lippmann (1919), p. 400, citing the *Biographies* of Albaihaqi (1105-1169).

[5] Ranking, #8.

[6] *Ibid.* #107.

not the stars were living beings.[1] Rasis also discussed the difference between dreams from which the future can be forecast and other dreams.[2] The title, *Of exorcisms, fascinations, and incantations,* under which, according to Negri's Latin translation Rasis discussed the causes and cures of diseases by these methods and magic arts, should, in Ranking's opinion, be more accurately translated as *The Book of Divisions and Branches.*[3] A work *On the Necessity of Prayer* is also included in the list of 232 works ascribed to Rasis,[4] while a Lapidary produced for Wenzel II of Bohemia (1278-1305) cites Rasis *On the virtues of words and characters.*[5]

Conclusion.

Herewith we conclude our present survey of Arabian occult science especially in the ninth century, although in the following chapters we shall frequently encounter its influence. We have found the occult science closely associated with natural science and difficult to sever from it. In the authors and works reviewed we have found both scepticism and superstition, both rationalism and empiricism. But perhaps the most impressive point is that even superstition pretends to be or attempts to be scientific.

[1] Ranking, #134. Other titles in mathematics and astronomy are: 73, *Liber de sphaeris et mensuris compendiosis;* 128, *De septem planetis et de sapientia;* 155, *De quadrato in mathesi epistola;* also 109 and 110.

[2] *Ibid.* #13.

[3] *Ibid.* #51.

[4] *Ibid.* #158, *De necessitate precationis.*

[5] Printed as the Lapidary of Aristotle, Merseburg, 1473, p. 2.

CHAPTER XXIX

LATIN ASTROLOGY AND DIVINATION: ESPECIALLY IN THE NINTH, TENTH, AND ELEVENTH CENTURIES

Astrology in Gaul before the twelfth century—Figures of astrological medicine—The divine quaternities of Raoul Glaber—Celestial portents and other marvels—An eleventh century calendar—Astrology and divination in ecclesiastical *compoti*—Notker on the mystic date of Easter—Prediction from the Kalends of January—Other divination by the day of the week—Divination by the day of the moon—Authorship of moon-books—Spheres of life and death: in Greek—Medieval Latin versions—Survival of such methods in medical practice of about 1400—Egyptian days—Their history—Medieval attempts to explain them—Other perilous days—Firmicus read by an archbishop of York—Relation of Latin astrology to Arabic—Appendix I. Some manuscripts of the Sphere of Pythagoras or Apuleius—Appendix II. Egyptian days in early medieval manuscripts.

Astrology in Gaul before the twelfth century.

ASTROLOGY had continued to flourish in Gaul in the last declining days of the Roman Empire, despite the strictures of Christian writers and clergy,[1] and it was one of the first subjects to revive after the darkness of the Merovingian period. Two centuries ago Goujet in a treatise on the state of the sciences in France from the death of Charlemagne to that of King Robert noted that from the reign of Charlemagne astronomy continued to be increasingly studied. "The councils in their decrees, the bishops in their statutes, the kings in their capitularies, expressly recommended the study of it to the clergy." [2] With the study of astronomy naturally developed a belief in astrology. According to the *Histoire Littéraire de la France* it became quite the fashion during the reign of Louis the

[1] See De la Ville de Mirmont, *L'Astrologie chez les Gallo-Romains,* Bordeaux, 1904; also published in *Revue des Études anciennes,* 1902, p. 115-; 1903, p. 255-; 1906, p. 128-.

[2] Goujet (1737), p. 50; cited by C. Jourdain (1838), pp. 28-9.

Pious, Charlemagne's successor, when we are told that there was no great lord but had his own astrologer. Adalmus, before he became abbot of Castres, wasted much time upon this pseudo-science, and Rabanus Maurus showed tendencies in that direction. In the tenth century such celestial phenomena as comets and eclipses were feared as sinister portents, and men resorted to enchantments, auguries, and other forms of divination.[1] A brief treatise in a manuscript of the ninth century in the Vatican library also develops the thesis that comets signify disasters.[2] In the eleventh century Engelbert, a monk of Liège, and Odo, teacher at Tournai, were devoted to the study of the stars; and Gilbert Maminot, bishop of Lisieux, and for a time chaplain and physician to William the Conqueror, would rather spend his nights in star-gazing than in sleep. "But what was the outcome of all this toil and study?" inquires the *Histoire Littéraire* and replies to its own question, "The making of some wretched astrologers and not a single true astronomer!" [3]

Figures of astrological medicine.

These words were written nearly two hundred years ago, but such a recent investigation of manuscripts in French libraries as that of Wickersheimer on figures illustrative of astrological medicine from the ninth, tenth, and eleventh centuries has on the whole confirmed the importance of astrology in the meager learning of that time.[4] The manuscripts in English libraries, I have found, tell a similar story. Of the human figures marked with the twelve signs of the zodiac, which become so common in the manuscripts by the fourteenth century, and in which the head rests upon the

[1] HL IV, 274-5; V, 182-3; VI, 9-10.

[2] Palat. Lat. 487, fol. 40, opening, "Nouo et insolito siderum ortu infausta quaedam uel tristitia potius quam laeta uel prospera miseris uentura significari mortalibus pene omnia ueterum aestimauit auctoritas."

[3] HL VII, 137.

[4] Ernest Wickersheimer, *Figures médico-astrologiques des neuvième, dixième et onzième siècles*, in *Transactions of the Seventeenth International Congress of Medicine, Section XXIII, History of Medicine*, London, 1913, p. 313 *et seq.* I have not seen A. Fischer *Aberglaube unter den Angelsachsen*, Meiningen, 1891, or M. Förster, *Die Kleinlitteratur des Aberglaubens im Altenglischen*, in *Archiv. f. d. Studium d. Neuer. Sprachen*, vol. 110, pp. 346-58.

Ram, the feet on Pisces, while the intervening members of the body are marked by their respective signs,—of these Wickersheimer found none before the twelfth century. But in a medical manuscript of the eleventh century the twelve signs with their names and the names of the parts of the human body to which they apply are grouped about a half figure of Christ, who has His right hand raised to bless, while about His head is a halo or sun-disk with twelve rays.[1] Less favorable to astrology is the accompanying legend, "According to the ravings of the philosophers the twelve signs are thus denoted." On the page following the text describes the twelve signs "according to the Gentiles." Schemes in which the world, the year, and man were associated, and where are shown the four elements, four seasons, four humors, four temperaments, four ages, four cardinal points, and four winds, are frequently found in extant manuscripts of the ninth, tenth, and eleventh centuries.[2]

The divine quaternities of Raoul Glaber.

Such association reminds one of the opening of the chronicle of Raoul Glaber, written in the eleventh century, "Since we are to treat of events in the four quarters of the earth, it will be well to touch first upon the power of divine and abstract quaternity." There are four elements, he gives us to understand, four virtues and four senses. There are four Gospels and they have their relation to the four elements. Matthew, dealing with Christ's incarnation, corresponds to earth; Mark to water, since it emphasizes baptism; Luke to air, because it is the longest Gospel;

[1] Charles Singer, *Studies in the History and Method of Science,* Oxford, 1917, Plate XV, opposite p. 40, reproduces this illumination. The MS, BN 7028, seems to have once belonged to the abbey of St. Hilary at Poitiers.

[2] Besides those in France mentioned by Wickersheimer may be noted two of the tenth century at Munich: CLM 18629, fol. 105, "Tabula cosmica cum nominibus ventorum, germanicorum quoque"; CLM 18764, fols. 79-80, "Schema de genitura mundi." Also Vatic. Lat. 645, 9th century, fol. 66, Ventorum imagines et in circulo Adam in medio ferarum; fol. 66v, Planetarum figura. This same MS contains a conjuration written in a later hand of the eleventh or twelfth century: fol. 4v, "In nomine patris. . . . Tres angeli ambulaverunt in monte. . . ."

For such an astrological diagram in an Arabic work of the tenth century see E. G. Browne (1921), 117-8.

and John to fire or ether as the most spiritual. In like manner can be associated with the four cardinal virtues those four famous rivers which had their sources in Paradise: Phison and prudence, Geon and temperance, the Tigris and fortitude, the Euphrates and justice. Finally the ages of the world are found to be four by Raoul, instead of the six eras corresponding to the days of creation which we find in Isidore, Bede, and other medieval historians; and these four ages also relate to the four virtues. The days of Abel, Enoch, and Noah were days of prudence; but on leaving Noah we have temperance marking the age of Abraham and the patriarchs; fortitude is the feature of the time of Moses and the prophets; while justice characterizes the period since the incarnation of the Word.

Celestial portents and other marvels.

The faith of Raoul and his contemporaries in the mystic significance of numbers, if not also in astrology, and the fact that they were constantly on the lookout for portents and prodigies, are further attested by the stress laid in his chronicle upon the thousandth anniversaries of Christ's birth and of His passion. Says Raoul, "After the multiplicity of prodigies which, although some came a little before and some a trifle afterwards, happened in the world around the thousandth year of Christ the Lord, there were many industrious men of sagacious mind who prophesied that there would be others not inferior to these in the thousandth year of our Lord's passion." That they were not mistaken in this premonition he shows later by several chapters, including an account of the eclipse of the sun in that year. Like many another medieval historian, Raoul is careful to note the appearance of comets—in the Bayeux tapestry of the same century one marks the death of Edward the Confessor; Raoul also believes that if a living person is visited by spirits, either good or evil, it is a sign of his approaching death; he holds the usual view that demons may sometimes work marvels by divine permission, and tells of a magician-impostor whom he saw work miracles upon pseudo-relics.

But from the superstition of medieval chroniclers we must turn back to astrological manuscripts proper.

An eleventh century calendar.

An eleventh century calendar at Amiens [1] reveals both a simple form of astrological medicine and a belief in some peculiar significance of the number seven, whether as a sacred or an astrological number. At the head of each month are brief instructions as to what herbs to use during that month, as to bleeding and bathing, and what disease may most easily be cured then.[2] In the same manuscript one miniature shows someone striking seven bells with a hammer, perhaps as notes in a scale, and another miniature represents a seven-branched candlestick, of which the branches are respectively labeled, "Spirit of piety, Spirit of fortitude, Spirit of intellect, Spirit of wisdom, Spirit of prudence, Spirit of science, Spirit of the fear of God." [3]

Astrology and divination in ecclesiastical *Compoti.*

Indeed works of astrology and divination are especially likely to be found in the same manuscripts with ecclesiastical calendars and *computi*. *Computus* or *compotus,* as one manuscript states, was "the science considering times." [4] For example, in a brief *compotus* of the ninth century [5] a divining sphere of Pythagoras occurs twice, and we have also a moon book, an account of the Egyptian days, and a method of divination from winds. In a twelfth century manuscript,[6] sandwiched in between calendars and reckonings of Easter and eclipses and Bede's work *On the Natures of Things,* are a sphere of divination, an account of Egyptian days, a method of divination from thunder, and a portion of a work on judicial astrology beginning with the eleventh chapter which tells how to determine whether anyone will be poor or rich by inspection of the planet in his nativity.[7]

[1] Amiens, fonds Lescalopier, 2, 11th century, fols. 1-12.

[2] For instance, for February, "Bibe agrimoniam et apii semen; oculos turbulentos sanare debes": for March, "Merum dulce primum bibe, assum balneum usita, sanguinem non minuas, ruta et levestico utere."

[3] *Ibid.*, fols. 11 and 19.

[4] Pembroke 278, early 14th century, fol. 25, "Compotus est sciencia considerans tempora."

[5] BN nouv. acq. 1616, 14 leaves.

[6] BN 7299A.

[7] BN 7299A, fols. 35v, 37v, 56r.

Notker on the mystic date of Easter.

The very dating of Easter itself might be the occasion for indulging in mystic speculation of a semi-astrological nature. Thus Notker Labeo, c 950-1022, the well-known monk of St. Gall,[1] in a treatise to his disciple Erkenhard on four questions of *compotus*,[2] states that the principal problem, with which all others are connected, is that of the date of Easter. He gives the time as in the first full moon after the vernal equinox, but adds that this is because of a certain mystery. For if there were no mystery connected with the date of Easter, and it merely celebrated like other festivals the memory of an event which once happened, there is no doubt but that it would occur every year without variation upon the twenty-seventh of March, which was the day of the Lord's resurrection. But as after the vernal equinox the days grow longer than the nights, and as at the full of the moon its splendor is revolved on high, so we should overcome the darkness of sin by the light of piety and faith and turn our minds from earthly to celestial things, if we wish to celebrate Easter worthily.

Prediction from the Kalends of January.

But let us consider in more detail the methods of divination found in such manuscripts. Simplest of all perhaps are predictions as to the character of the ensuing year according to the day of the week upon which the first of January falls. For example, "If the kalends of January shall be on the Lord's day, the winter will be good and mild and warm, the spring windy, and the summer dry. Good vintage, increasing flocks; honey will be abundant; the old men will die; and peace will be made." [3] In some

[1] Notker is especially famed for his translations with learned commentaries from Latin into German, of which five are extant, namely: *The Consolation of Philosophy* of Boethius, *The Marriage of Mercury and Philology* of Martianus Capella, the *Psalter*, and Aristotle, *De categoriis* and *De interpretatione:* see Piper, *Die Schriften Notkers,* Freiburg, 1882-1883, vols. I-III.

[2] BN nouv. acq. 229, fols. 10v-14v. *Notker erkenhardo discipulo de IIII questionibus compoti.* It seems not to have been printed.

[3] Cotton Tiberius A, III, a MS written in various hands before the Norman conquest, partly in Latin and partly in Anglo-Saxon, and containing among other things the Colloquy of Aelfric Our item occurs at fol. 34r in

manuscripts these predictions concerning the weather, crops, wars, and king for the ensuing year are called *Supputatio Esdrae* or signs which God revealed to the prophet Esdras.[1] In another manuscript [2] the weather for winter and summer is predicted according to the day of the week upon which Christmas falls and Lent begins. Christmas of course was sometimes regarded as the first day of the new year and in any case it falls on the same day of the week as the following first of January. In a ninth century manuscript [3] predictions for the ensuing year are made according as there is wind in the night on Christmas eve and the eleven nights following. For instance, "If there is wind in the night on the night of the natal day of our Lord Jesus Christ, in

Latin with an Anglo-Saxon interlinear version, and at fol. 39v in Anglo-Saxon only.

Cotton Titus D, XXVI, 10th century, fols. 10v-11v, gives a slightly different version for some days of the week.

[1] Harleian 3017, 10th century, fols. 63r-64v, CLM 6382, 11th century, fol. 42, Supputatio Esdrae; Incipit, "Kal. Jan. si fuerint dominico die hiems bona erit."

Vatican, Palat. Lat. 235, 10-11th century, fol. 39, "Subputatio quam subputavit Esdras in templo Hierusalem," opening, "Si in prima feria fuerint kl. Ianuarii hiemps bona erit."

Also found in Egerton 821, fol. 1r, which is of the twelfth century and adds a more elaborate method of divination according to what planet rules the first hour of the first night of January and which of its 28 mansions the moon is in.

CLM 9921, 12th century, fol. 1, is a calendar with verses beginning, "Jani prima dies et septima fine timetur."

[2] Sloane 475, this portion perhaps 11th century, fol. 217r. Other MSS of later date than the period we are now considering are: Harleian 2258, fol. 191, "prognostica a die nativitatis Domini a luna et somniis petita," predictions from Christmas, the moon, and dreams. CUL 1338, 15th century, fol. 65v, Prognostications derived from the day on which Christmas falls (in Latin); fol. 74v, Prognostications drawn from the day of the week on which the year commences. CU Trinity 1109, 14th century, fol. 148, "Prognostica anni sequentis ex die natalium Domini."

[3] BN nouv. acq. 1616, 9th century, fol. 12v. Similar later MSS are:

Digby 86, 13th century, fols. 32-4, Prognosticatio ex vento in nocte Natalis Domini, and fols. 40v-41r, "Les singnes del jour de Nouel," predictions in French according to the day of the week on which Christmas falls.

Digby 88, 15th century, fol. 77, "Howe all ye yere ys rewlyde by the day that Christemas day fallythe on," and fol. 40r, "Prognostication from the sight of the sun on Christmas and the ten days following" (Prognosticatio ex visione solis in die Natalis Domini et in decem diebus subsequentibus), and fol. 75, a poem of prognostications for Christmas day. This same MS contains a large number of other brief anonymous treatises in the fields of astrology and divination.

that year kings and pontiffs will perish," and "If on twelfth night there shall be wind, kings will perish in war."

Other divination by the day of the week

Divination from thunder is another form of judicial astrology, if it may so be called, found in these early manuscripts. Perhaps the simplest variety of it is according to the day of the week on which thunder is heard.[1] Predictions were also made according to the month in which thunder was heard,[2] or the direction from which it was heard.[3] It may be recalled that the three chapters of Bede's translation of some work on divination from thunder had been respectively devoted to these three methods by the direction from which the thunder is heard, the month, and the day of the week. Nativities of infants are also given according to the day of the week on which they are born, and further taking into account whether the hour of birth is diurnal or nocturnal.[4] It is also regarded as important to note upon which day of the week the new moon occurs,[5] and we are further informed of the various hours of the days of the week when it is advisable to perform bloodletting.[6] In a method of divination according to the day of the week and the letters in the boy's or girl's name the Lord's day is assigned the number thirteen, the day "of the moon" eighteen, and that "of Mars" fifteen.[7] Since

[1] Titus D, XXVI, fol. 9v. Tiberius A, III, fols. 38r and 35r. Cockayne, *Leechdoms* etc., III, 150-295, in RS vol. 35, published this and a number of other extracts from Tiberius A, III, and other early English MSS.

Vienna 2245, 12th century, fols. 59r-69v are devoted to various prognostications, beginning with, "Three days are to be observed above all others," and ending with, "Thunder at dawn signifies the birth of a king." A dream book by Daniel follows at fols. 69v-75r.

[2] Vatican Palat. Lat. 235, fol. 40, "In mense Ianuario si tonitru fuerit." In Egerton 821, 12th century, the significance of thunder is given according to the twelve signs of the zodiac, and we are told of what the Egyptians write, and of famine in Babylon. In CUL 1687, 13-14th century, fols. 68v-69r, Latin verses containing prognostications concerning thunder are followed by "a list of the number of quarters of flour, beer, etc., used in the year *at the monastery*" and by "a note on the symbolism of the pastoral staff."

[3] Combined with the method by the day of the week in BN 7299A, 12th century, fol. 37v.

[4] Tiberius A, III, fol. 63r; Vatican Palat. Lat. 235, fol. 40.

[5] Tiberius A, III, fol. 38v.

[6] Sloane 475, fol. 135v.

[7] Sloane 475, fol. 133r. The method is almost identical with that of the spheres of life and death, of which we shall speak

the days of the week bore the names of the planets, it was not strange that they should have been credited with something of the virtues of the stars.

Divination by the day of the moon.

A commoner method of divination and one more nearly approaching approved astrological doctrine was that by the day of the month or moon. Briefest of such moon-books is that which merely designates each of the thirty days as favorable or unfavorable.[1] We also find a *Lunarium* for the sick, stating the patient's prospects from the day of the moon on which he contracted his illness; [2] a work ascribed to "Saint Daniel" on nativities by the day of the moon; [3] and an equally brief interpretation of dreams upon the same basis.[4] Or all these matters may be considered in the same treatise and each of them somewhat more fully, and we may be told whether the day is a good one on which to buy and sell, to board a ship, to enter a city, to operate upon a patient, to send children off to school, to breed animals, to build an aqueduct or mill, or whether it is best to

presently. In CU Trinity 987, *The Canterbury Psalter,* about 1150 A. D., the value assigned *Dies Solis* is 24.

[1] Vatic. Palat. Lat. 235, fol. 40, "De lunae observatione: Luna I omnibus rebus agendis utilis."

Tiberius A, III, fol. 63r, where, however, such parts of the day as morning and evening are further distinguished.

Vatic. Palat. Lat. 485, 9th century, fol. 15v, "Ad sanguinem minuendum," merely states which days of the moon are favorable or unfavorable for blood-letting.

St. John's 17, 1110 A. D., fol. 4, Luna quibus diebus bona est et quibus non; fol. 154v, a table of lucky and unlucky numbers.

[2] Harleian 3017, fol. 58v; the Incipit states that it is by the same author as the preceding Sphere of Pythagoras and Apuleius.

Titus D, XXVI, fol. 8.

Cotton Caligula A, XV, 10th century, fol. 121v, Latin and Anglo-Saxon.

Egerton 821, fol. 32r, is a twelfth century instance.

The method seems combined or confused with the Egyptian days in Vatic. Palat. Lat. 485, 9th century, fol. 13v, "Dies aegyptiaci. Signa in quibus aegrotus an periclitare aut evadere non potest," but opening, "Luna I. qui ceciderit in infirmitatem difficile euadit."

[3] Harleian 3017, fol. 58v, "Incipit lunarium sancti danihel de nativitate infantium. Luna I qui fuerit natus vitalis erit; Luna II, mediocris erit . . . Luna IIII, tractator regum erit . . . Luna XII, religiosus erit . . . Luna XXX, negotias multas tractabit."

Tiberius A, III, fols. 63r and 34v.

Titus D, XXVI, fols. 7v and 6v.

[4] Tiberius A, III, fol. 33v. Titus D, XXVI, fol. 9r. CLM 6382, 11th century, fol. 42, De somni ueris uel mendosis quidam incipiunt in aetatibus lunae exploratis.

abstain on it from most business. Also such predictions as that the boy born on that day will be illustrious, astute, wise, and lettered; that he will encounter danger on the water, but will live to old age if he escapes; while the girl born on the same day will be "chaste, benign, good-looking, and pleasing to men." That anyone who takes to his bed on that day will suffer a long sickness, but that it is a favorable day for blood-letting, and that one should not worry about dreams he has then, since they possess no significance either for good or evil. Also what chance there is of recovering articles stolen on that day.[1] In later manuscripts at least it is further stated that certain Biblical characters were born on this day or that day of the moon: Adam on the first, Eve on the second, Cain on the third, Abel on the fourth, and so on.[2]

[1] Tiberius A, III, fols. 30v-33v, "Finiunt somnia danielis prophete."

Sloane 475, fols. 211-6, is almost identical, but I believe does not mention Daniel as its author.

Vatic. Palat. Lat. 235, fol. 39v.

BN nouv. acq. 1616, 9th century, is roughly similar but names no author and does not distinguish the fates of boys and girls. It usually states whether slaves who run away and thieves who steal on the day in question will be caught or escape. It opens and closes thus: "Luna prima qui incenditur in ipsa sanabitur et bona et in omnibus dare et accipere et nubere et navigare in mare et vendere et emere et omnis quicumque fugerit in ipsa aut servus aut liber non poterit sed capitur aut qui incendit incendio sanabitur (presumably an allusion to the medical practice of cauterization) et qui natus fuerit vitalis erit . . . / . . . Luna XXX bona est ambulare in piscatione et qui fugit post multos annos revertitur in loco suo et qui natus fuerit dives erit et honoratissimus erit et qui incadit aut manducet aut non vivet periculo mortis habebit."

Titus D, XXVII, fols. 22-25r, "judicia de diebus quibusdam cuiusque mensis"; fols. 27-9, "argumentum lunare, quando et qualiter observentur tempora ad res agendas."

Of the twelfth century, Vienna 2532, fols. 55-9, "Luna I. Hec dies omnibus egrotantibus utilis est . . . / . . . Puer natus negotia multa sectabit."

[2] Sloane 2461, end of 13th century, fols. 62-4. No Biblical character is mentioned for the fifth and sixth days, but we are told that on the seventh day of the moon Abel was slain by Cain.

BN 3660A, 16th century, fols. 53r-57r, ascribes the birth of Nebuchadnezzar to the fifth day, leaves the sixth blank, has Abel slain on the seventh, Methusaleh born on the eighth, Lamech on the ninth, and so on.

Egerton 821, 12th century, fol. 12r, "Natus est Samuel propheta. . . ."

Digby 88, 15th century, fol. 62r, has English verses beginning:

"God made Adam the fyrst day
of the moone,
And the second day Eve good
dedis to doone."

A similar poem occurs at fol. 64 of the same MS and in Ashmole 189, fol. 213v.

Authorship of moon-books.

In the early manuscripts moon-books are anonymous or ascribed to Daniel, but in later medieval manuscripts other authors are named. The name of Adam is coupled with that of Daniel in both of two rather elaborate moon-books in a fourteenth century manuscript,[1] where Adam is said to have worked out these " lunations" "by true experience." A fifteenth century one is attributed to a philosopher, astrologer, and physician named Edris,[2] perhaps the Esdras of the method of divination by the kalends of January rather than the Arab Edrisi. It briefly predicts from the relation of the moon to the twelve signs whether patients will recover and captives escape. In a sixteenth century manuscript at Paris are "Significations of the days of the moon which the most excellent astronomer Bezogar revealed to his disciples and transmitted to them as a very great secret and most precious gift." [3] But such an ascription is rather obviously a late fiction.

Spheres of life and death: in Greek.

Determining the fate of the patient from the day of the moon upon which his illness was incurred enters also into certain spheres of life and death which were much employed in the early middle ages. But in these the number of the day of the moon is combined with a second number obtained by a numerical evaluation of the letters forming the patient's name. This method came down from the ancient Greek-speaking world, as in a "Sphere of Democritus, prognostic of life and death" found in a Leyden papyrus,[4] while the very similar *Sphere of Petosiris*, the

[1] Ashmole 361, mid 14th century, fols. 156v-158v, "Iste sunt lunaciones quas Adam primus homo disposuit secundum veram experientiam quam etiam suis filiis tradidit et quam maxime Abel et ceteris de posteritate ad quos etiam concordavit Daniel propheta . . ."; fol. 159, "Modo agitur de numero lune ad videndum que sit bona vel que mala et usum istarum lunacionum invenerunt Adam et Daniel propheta."

[2] Canon. Misc. 517, fol. 35r, "Incipit scientia edita ab edri philosopho astrologo et medico."

[3] BN 3660A, fols. 53r-57r. In the catalogue of Ashburnham MSS at Florence the name of Giovannino di Graziano is connected with a moon-book in Ashburnham 130, 13-15th century, fols. 25-6, "Luna prima Adam natus fuit. . . ." But perhaps this name should go only with some prognostications, exorcisms, and recipes which occur at the close of the predictions for the thirty days of the moon.

[4] Ed. Leemans, 1833-1885.

mythical Egyptian astrologer, is variously dated by W. Kroll from the second century before Christ, by E. Riess from the first century before Christ, and by F. Boll in the first century of our era.[1] The so-called "Sphere" is really only a wheel of fortune, circle, or other plane figure divided into compartments where different numbers are grouped under such headings as "Life" and "Death." Having calculated the value of a person's name by adding together the Greek numerals represented by its component letters, and having further added in the day of the moon, one divides the sum by some given divisor and looks for the quotient in the compartments. This method of divination was also employed in regard to fugitive slaves and the outcome of gladiatorial combats.[2]

Medieval Latin versions.

In the medieval Latin versions of these Spheres of life and death the numerical value of the Greek letters was naturally usually lost and arbitrary numerical equivalents were assigned to the Roman letters or some other method of calculation was substituted. The *Sphere of Petosiris* was perpetuated in the form of a letter by him to Nechepso, king of Egypt.[3] But more common than this in manuscripts of the ninth, tenth, and eleventh centuries was the Sphere of life and death of Apuleius or Pythagoras or both[4] which replaced that of Democritus. Like it, it consisted of the numbers from one to thirty arranged in six compartments, three above a line each containing six numbers, and three below the line having four each. John of Salisbury, in the twelfth century, presumably refers to

[1] Bouché-Leclercq (1899), 537-42; (1879-1882), I, 258-65. Berthelot, *Alchimistes grecs* (1888), I, 86-90. K. Sudhoff (1902), pp. 4-6.

[2] Arundel 319, 13th century, fol. 2r, Versus de faustis vel infaustis nominibus pugnantium, is a medieval Latin example.

[3] Printed among treatises of dubious or spurious authorship with Bede's works, Migne, PL 90, 963-6; and more recently in Riess' edition of the fragments of Nechepso and Petosiris (*Philologus,* Suppl. VI, 1891-1893, pp. 382-3) from Cod. Laur. XXXVIII, 24, 9-10th century, fol. 174v. Wickersheimer (1913), pp. 315-7, notes BN 17868, 10th century, fol. 13. For other MSS see Appendix I to this chapter.

[4] Printed by Paul Lehmann, *Apuleiusfragmente, Hermes* XLIX (1914), 612-20. For a list of some MSS of it see Appendix I at the close of this chapter.

it when he speaks of divination or lot-casting "by inspection of the so-called Pythagorean table";[1] and it continues to be found with great frequency in the manuscripts of subsequent centuries.[2] It is not to be confused, however, with the *Prenostica Pitagorice,* a more elaborate, although somewhat similar, method of divination by means of geomantic tables, of which we shall treat later in the chapter on Bernard Silvester. A Sphere ascribed to St. Donatus in a twelfth century manuscript includes instructions how to determine the sign of the zodiac under which a person was born by computing the difference between his name and his mother's name. If this amounts to four letters, he was born under the fourth sign, and so on.[3]

Survival of such methods in medical practice of about 1400.

The survival of such superstitious methods of divination into the later middle ages is attested not only by the frequent recurrence of the *Sphere of Apuleius* and the divinations from the kalends of January in manuscripts of the later centuries, but by the medical notebook, written in middle English, of John Crophill, who practiced medicine in Suffolk under Henry IV.[4] Besides a record of his patients and the sums of money due from them, rules of dieting and blood-letting for the twelve months of the year, and his "more regular and masterly observations upon Urin," his notes include a treatise on astrological medicine which, in the sarcastic language of the old catalogue of the Harleian Manuscripts, concludes "with a masterpiece of art, namely, a tretys or chapter of 'Calculation to know

[1] *Polycraticus* I, 13, ed. Webb, I, 54. Mr. Webb in a note refers to an article in a German periodical (K. Gillert, *Neues Archiv d. Gesellschaft f. ältere deutsche Geschichtskunde,* V, 254) concerning a MS of the *Sphere of Pythagoras* preserved at Petrograd, but says nothing of the MSS in the British Museum listed in Appendix I to this chapter,—a good illustration of the unnecessary obsequiousness of English towards German scholarship which has frequently prevailed in the past.

[2] A few of them will be found listed in Appendix I to this chapter.

[3] Egerton 821, 12th century, fol. 15r, "Hec est spera quod fecit sanctus Donatus. Quicumque egrotare incipit. . . ." It is followed on the next page by the usual figure for the *Sphere of Apuleius.*

[4] Harleian 1735; the passages referred to in the following account occur at fols. 36v, 41, 43, 29, 44v, 40, and 39v respectively.

what thou wilt,' and this by observation of persons' names." The notebook also contains "Oracular Answers prepared beforehand by this great Doctor for those of both Sexes who shall come to consult him in the momentous affair of Matrimony; according to the several Months of the year wherein they should apply themselves." Further contents are an incantation in Latin for women in child-birth, and "The names of the 12 signs with such marks as shew that this John Crophill was a dabbler in Geomancy."

Egyptian days.

Brief lists of "Egyptian Days" are of rather common occurrence in both Latin and Anglo-Saxon manuscripts of the ninth, tenth, and succeeding centuries.[1] Often it is merely stated what days of the year they are; sometimes it is simply added that the doctor should not bleed the patient upon them. As early as a ninth century manuscript,[2] however, we are further warned not to take a walk or plant or carry on a lawsuit or do any work upon these days. And under no circumstances, no matter what the seeming necessity, is it permitted to bleed man or beast on these days. Two Egyptian days are then listed for each month, one reckoned as so many days from the beginning and the other as so many days before the close of the month. Eleven days is the farthest removed that any Egyptian day is from the first of the month and twelve the most from the close, so that they never fall in the middle of a month nor on the very first or last day. Our ninth century manuscript then mentions three of these days in April, August, and December as especially dangerous. Whoever falls ill or receives a potion on them is sure to die soon. Whoever, male or female, is born on one of them will die an evil and painful death. "And if one drinks water on those three days, he will die within forty days." The account then closes with the statement that on the Egyptian days the people of Egypt were cursed with Pharaoh. In another ninth century manu-

[1] See Appendix II to this chapter for a list of MSS other than those mentioned in the following notes.

[2] BN nouv. acq. 1616, 9th century, fol. 12r.

script a bare list of the Egyptian days is followed by a somewhat similar account of the three which must be observed with especial care.[1] In a calendar of saints' days in this same manuscript only the third of March and the third of July are marked *dies egiptiagus.*[2] Egyptian days are also marked in the calendar of Marianus Scotus, the well-known chronicler and chronologist.[3] A somewhat different account in a twelfth century manuscript states that "these are the days which God sent without mercy." It also, however, lists two of them for each month and distinguishes the three in April, August, and December as especially dangerous.[4]

Their history.

There seems to be no doubt that these Egyptian days were a relic of the unlucky days in the ancient Egyptian calendar,[5] of which we learn from several papyri, although of course the ancient Egyptians were also accustomed to distinguish further the three divisions of each day as lucky or unlucky. The Egyptian days are noted in official calendars of the Roman Empire about 354 A. D., and in the *Fasti Philocaliani* there are twenty-five in all, of which three fall in January. In the middle ages, as has already been illustrated, there were usually but twenty-four, two to each month.[6] They were mentioned in the *Life of Proclus* by Marinus, and both Ambrose and Augustine testified that many Christians still had faith in them.[7] Indeed, they passed into the ecclesiastical calendar, as the Franciscan, Bartholomew of England, states in the thirteenth century.[8]

[1] Digby 63, end of 9th century, fol. 36.

[2] *Ibid.*, fols. 40-5.

[3] CU Trinity 1369, 11th century, fol. 1v.

[4] BN 7299A, 12th century, fol. 37v.

[5] For further information on this point see Budge, *Egyptian Magic*, 1899, pp. 225-8; Webster, *Rest Days*, 1916, pp. 295-7.

[6] Webster (1916), pp. 300-301, however, speaks of 30 in a 14th century MS, 32 in an English MS of Henry VI's reign, and 31 in another 15th century MS.

[7] Cited by Bouché-Leclercq, *L'Astrologie grecque*, 1899, pp. 485-6, 623.

[8] *De proprietatibus rerum*, 1488, Lindelbach, Heidelberg, IX, 20. This is not to say, however, that they always appear in medieval calendars; I did not find them in any of the 14th and 15th century calendars from Apulia and Iapygia published by G. M. Giovene, *Kalendaria vetera*, Naples, 1828. His calendars consist of little save saints' days, although

Medieval attempts to explain them.

By that time the notion had become prevalent that they were anniversaries of the days upon which God afflicted Egypt with plagues, as our citations from the manuscripts have shown. Bartholomew, indeed, is at pains to explain that the days are placed in the church calendar, "not because one should omit anything upon them more than upon other days, but in order that God's miracles may be recalled to memory." The circumstance that there are twenty-four days does not embarrass him; he simply explains that this proves that God sent more plagues upon Egypt than the ten which are especially famed. Our citations from earlier manuscripts have shown that most people would not agree with Bartholomew that nothing should be omitted on these days. Moreover, other explanations of their origin had been already given in the middle ages than that from the plagues of Egypt. Honorius of Autun stated in the twelfth century that they were called Egyptian days because they had been discovered by the Egyptians, and since Egypt means dark,[1] they are called *tenebrosi,* because they are declared to bring the incautious to the shadows of death.[2] The Dominican, Vincent of Beauvais,[3] who probably wrote his encyclopedia soon after that of Bartholomew, did not find the discrepancy between ten plagues and twenty-four days so easy to explain away. He states that of the two Egyptian days in each month one comes near the beginning and the other near the close, as we have already learned. He adds that some call them lucky days, while others say that the astrologers of Egypt discovered that they were unlucky. Yet another explanation of their origin is that on these days the Egyptians were accustomed to sacrifice to demons with their own blood, a circumstance which would not seem to recommend them for inclusion in the ecclesiastical calendar. Bernard Gordon, a medical writer at the

in some of them the beginning of dog-days is marked and when the sun enters each sign of the zodiac.

[1] "Black earth" was the name given by the Egyptians to their country.

[2] *Imago mundi,* II, 109.

[3] *Speculum naturale,* XVI, 83, printed by Anth. Koburger, Nürnberg, 1485.

end of the thirteenth century, reverts to the position that the Egyptian days were in memory of the plagues in Egypt. He declares that there is no sense in the prohibition of blood-letting upon these days, since they have no astrological significance, but are the anniversaries of miracles worked by special providence.[1] Gilbert of England, earlier in the thirteenth century, had advised against bleeding on Egyptian days, if the moon was then influenced by any evil planet.[2]

Other perilous days.

On the other hand, not only did the twenty-four Egyptian days and the three in April, August, and December which were considered especially dangerous, continue to be listed in the fourteenth and fifteenth century manuscripts, but imitations of them appeared. Thus in a fourteenth century manuscript we read of forty perilous days which should be observed with the utmost care and which Greek masters have tested by experience;[3] while in a second manuscript of the closing medieval period appear fifty-eight dangerous days "according to the Arabs."[4] Of the Greek days only twenty-nine are actually listed, seven in January, three in February, and so on, omitting the months of July and August entirely, which perhaps should contain the missing eleven days.[5] The Arabic days vary in number per month from seven in March, which is the first month listed, to three in February. "And there are four other days and nights according to Bede on which no one is ever born or con-

[1] HL 25, 329. My impression is that some medieval astronomers also denied to these Egyptian days any astrological importance, since they always came upon the same days of the months without reference to the phases of the moon or courses of the other planets: but I cannot put my hand on such passages.

[2] And is approvingly cited to that effect by Arnald of Villanova, *Regulae generales curationis morborum. Doctrina IV*.

[3] Ashmole 361, mid 14th century, fols. 158v-159.

[4] BN 7337, 14-15th century, p. 75. Ad-Damîrî states in his zoological lexicon, (ed. A. S. G. Jayaker, 1906, I, 134) that Mohammed is reported to have said, "Be cautious of twelve days in the year, because they are such as cause the loss of property and bring on disgrace or dishonor."

[5] M. Hamilton, *Greek Saints and Their Festivals*, 1910, p. 187, states that "in all parts of (modern) Greece on certain days of August and March it is considered necessary to abstain from particular kinds of work in order to avoid disaster."

ceived, and if by chance a male is conceived or born, its body will never be freed from putridity." [1]

Firmicus read by an archbishop of York.

That astrological knowledge in England, at least soon after the Norman conquest, was not limited to such meager and simple treatises as the moon-books described above from Anglo-Saxon manuscripts, is seen from the closing incident in the career of Gerard, a learned and eloquent man, bishop of Hereford under William Rufus and archbishop of York under Henry I, whom he supported in the investiture struggle with Anselm and the pope. The story goes that Gerard, who had been feeling slightly indisposed, lay down to rest and enjoy the fresh air and fragrance of the flowers in a garden near his palace, asking his chaplains to leave him for a while. On their return after dinner they found him dead, and beneath the cushion upon which his head rested was a copy of the astrological work of Julius Firmicus Maternus. Gerard had not been popular with the inhabitants of York, and when his corpse was brought back to town, boys stoned the bier and the canons refused it burial within the cathedral, which, however, his successor granted. "His enemies," we are told, "interpreted his death, without the rites of the church, as a divine judgment for his addiction to magical and forbidden arts." At any rate the story shows that the work of Firmicus was well known by this time; it is from the eleventh century that the oldest manuscripts of it date; and we suspect that some of his enemies were rather hypocritical in the horror which they expressed at a bishop's reading such a book. "Too independent a thinker for his contemporaries," writes Miss Bateson, "his opponents held up their hands in horror that an astrological

[1] Mention may perhaps be made in this connection of the "Tobias nights," three nights of abstinence which newly wedded couples were sometimes accustomed to observe in the middle ages in order to defeat the demons. The practice is mentioned in the Vulgate, but not in most ancient versions of the *Book of Tobit*. In 1409 the citizens of Abbeville won a lawsuit with the bishop of Amiens who claimed the right to grant dispensations from the observance of the Tobias nights and required that fees be paid him for that purpose. See J. G. Frazer (1918), I, 498-520, where analogous practices of primitive tribes are listed.

work by Julius Firmicus Maternus should be found under his pillow when he died."[1] The style of Firmicus is much imitated by the anonymous author of *The Laws of Henry I* and another legal work entitled *Quadripartitus* written in 1114. F. Liebermann states that the author was in the service of archbishop Gerard aforesaid.[2]

Relation of Latin astrology to Arabic.

Charles Jourdain once made the generalization that before the translation of the *Quadripartite* of Ptolemy and the works of the Arabian astrologers into Latin in the twelfth century, astrology had little hold among men of learning in western Europe.[3] An even more erroneous assertion was that in Burckhardt's *Die Kultur der Renaissance in Italien* that "at the beginning of the thirteenth century" the superstition of astrology "suddenly appeared in the foreground of Italian life."[4] Even Jourdain's assertion the entire present chapter tends to disprove, but since it has been quoted with approval by a subsequent writer on the thirteenth century,[5] we may deal with it a little farther. The reason which Jourdain added in support of his generalization was that before the translations from the Arabic "those who cultivated astrology had no other guides than Censorinus, Manilius, and Julius Firmicus, who might indeed seduce a few

[1] Bateson, *Medieval England,* 1904, p. 72; I have in the main followed the fuller account in DNB "Gerard," from which the previous quotation is taken. William of Malmesbury, *Gesta Pontificum Anglorum,* III, 118 (ed. N. E. S. A. Hamilton, RS, vol. 52, 1870) does not say definitely that the book found under Gerard's pillow was Firmicus. Also he says nothing of boys stoning the bier or of Gerard's enemies interpreting his death as a divine judgment, and in his autograph copy of the *Gesta Pontificum* he afterwards erased the statements that rumor accused Gerard of many crimes and lusts, and that he was said to practice sorcery because he read Julius Firmicus on the sly before the mid-day hours, and that people say that a book of curious arts was found beneath his pillow when he died. This, the late medieval chroniclers say, was Firmicus: see Ranulf Higden, ed. Lumby, VII, 420, and Knyghton, ed. Twysden, X, SS., 2375.

[2] *Firmicus Maternus,* ed. Kroll et Skutsch, II (1913), p. iv; and F. Liebermann, ed. *Quadripartitus,* Halle, 1892, p. 36, and *Die Gesetze der Angelsachsen,* Halle, 1903-1906, I, 548.

[3] C. Jourdain, *Nicolas Oresme et les astrologues à la cour de Charles V,* in *Revue des Questions Historiques,* 1875, p. 136.

[4] English translation, ed. of 1898, p. 508.

[5] N. Valois (1880), p. 305.

isolated dreamers but did not have enough weight to convince philosophers. Ptolemy and the Arabs, on the contrary, appeared as masters of a regular science having its own principles and method." This sounds as if Jourdain had not read Firmicus who gives a more elaborate presentation of the art of astrology than the elementary *Quadripartite* of Ptolemy. It is true that Ptolemy had a great scientific reputation from his other writings, but Manilius is a poet of no small merit, and there would be no reason why an age which accepted Ovid and Vergil as authorities concerning nature and regarded such works as *De vetula* and the *Secret of Secrets* as genuine works of Ovid and Aristotle, should draw delicate distinctions between Firmicus and Albumasar or Manilius and Alkindi. It was because reading Firmicus and even practicing the cruder modes of divination which we have described had already aroused an interest in astrology that other works in the field were sought out and translated. Moreover, there is an even more cogent objection to Jourdain's generalization which will be developed in the following chapter, and it is that the taking over of Arabic astrology had already begun long before the twelfth century. We have, indeed, in the present chapter told only half the story of astrology in the tenth and eleventh centuries, and must now turn back to Gerbert and the introduction of Arabic astrology.

APPENDIX I

SOME MANUSCRIPTS OF THE SPHERE OF PYTHAGORAS OR APULEIUS

Besides the copies noted by Wickersheimer (1913) in French manuscripts from the ninth to the eleventh centuries, such as Laon 407, Orléans 276, and BN nouv. acq. 1616, where in fact it occurs twice: at fol. 7v, "Ratio spere phytagor philosophi quem epulegus descripsit," and at fol. 14r, "Ratio pitagere de infirmis,"—the following may be listed.

BN 5239, 10th century, ♯ 12.

Harleian 3017, 10th century, fol. 58r, "Ratio spherae Pythagorae philosophi quam Apuleius descripsit."

Cotton Tiberius C, VI, 11th century, fol. 6v, Imagines vitae et mortis quarum utraque rotulum tenet longum literis et numeris quae ad sphaeram Apuleii ad latera adscriptis, cum versibus pagina circumscriptis. The figures are of *Vita* with halo, robes, and angelic face, and of *Mors,* who wears only a pair of drawers, whose ribs show through his flesh, and who has wings like a demon. One has to turn the page upside down in order to read some of it.

CU Trinity 1369, 11th century, fol. 1r, just before the Calendar of Marianus Scotus, "Racio spere pytagorice quam apuleius descripsit."

Chartres 113, 9th century, fol. 99, following works by Alcuin, "Spera Apuleii Platonis."

Ivrea 19, 10th century, ♯ 5, De spera Putagorae.

CLM 22307, 10-11th century, fol. 194, Ratio sphaerae Phitagoreae philosophi quam Apulegius descripsit, "Petosiris philosophus Micipso regi salutem . . .", where it would seem to be confused with the letter of Petosiris to Nechepso.

Vatican Palat. Lat. 176, 10th century, fol. 162v, "Eulogii ratio sperae Pitagorae philosophi," in a MS containing works of Jerome, Augustine, and Ambrose.

Vatican Urb. Lat. 290, 11-13th century, fol. 2v, Ratio spere Pitagoras quam Apuleius descripsit; fol. 3, Petosiris Micipso regi salutem.

I suspect that the following would also prove upon examination to be one of these Spheres of life and death.

CLM 18629, 10th century, fol. 95, Characteres literarum secretarum, item incantationes. Alphabetum Graecorum et numeri per tabulam dispositi; fol. 106, Tractatus de literis alphabeti (mysticus).

Vatican Palat. Lat. 485, 9th century, fol. 14, Litterae graecae cum interpretatione alphabetica et numerica.

Vatican 644, 10-11th century, fol. 16v.

Of the numerous occurrences of the *Sphere of Pythagoras* or of Apuleius in MSS later than the eleventh century I have noted only a few examples.

Vienna 2532, 12th century, fols. 1-2, Tractatus astrologicus de divinando exitu morborum e positionibus lune et de sphere Pythagore.

Vatican 642, 12th century, fol. 82, a somewhat different mode of divination, by which one tells what another is thinking or is holding in his hand, is attributed to Bede.

Madrid 10016, early 13th century, fol. 3, "spera de morte vel vita"; fol. 85v, the letter of Petosiris to Nechepso. It is interesting to note that this MS originally belonged to an English Cluniac monastery: Haskins, EHR (1915), p. 65.

BN 7486, 14th century, fol. 66v, "Canon supra rotam Pictagore," opens, "Pictagoras is said to have written thus to Nasurius, king of the Chaldees;" then at fol. 67r comes "The Sphere of Pictagoras the philosopher which Epuleus Platonicus briefly described;" which is followed at fol. 68r by a long treatise ascribed to Ptolemy, *Exortatio ad artem prescientie ptholomei regis egypti,* in which various questions are answered by numerical and alphabetical calculations and one is also by the same method referred to nativities arranged under the 28 mansions of the moon.

CU Trinity 1109, 14th century, fol. 15, Spera apulei et platonici; fol. 20, "Ratio spere pictagis philosophe quod apollonius scripsit;" fol. 392, S(p)era Fortune.

Digby, 58, 14th century, fol. 1v, "Spera philosophorum."

Bodleian 26 (Bernard 1871), 13-14th century, fols. 207 and 216v.

Bodleian 177 (Bernard 2072), late 14th century, ♯ 1, Pythagorae sphaera quam Apuleius exaravit ut scias an aeger convalescat; ♯ 14, fol. 22r, Apuleii Platonici Sphaera de vita et morte et de omnibus negotiis quae inquirere volueris.

Amplon. Quarto 380, 14th century, at the close of a Geomancy by Abdallah, "Spera Apuley de vita et morte vel de omnibus negociis de quibus scire volueris; sic facias. . . ."

Additional 15236, 13-14th century, fol. 108, "Spera (Pictagore) de vita et morte sive de re alia quacunque secundum Apuleium."

Harleian 5311, 15th century, folder i, "Spera Apullei."

S. Marco XI, 111, 16th century, ascribes a wheel of life and death to "Bede the presbyter," and another to Apollonius and Pythagoras.

APPENDIX II

EGYPTIAN DAYS IN EARLY MEDIEVAL MANUSCRIPTS

The following citations could probably be greatly multiplied.

BN nouv. acq. 1616, 9th century, fol. 12r.

Digby 63, end of 9th century, Anglo-Saxon minuscule, fol. 36, "Dies Egipciachi."

Berlin 131 (Phillips 1869, Trier), 9th century, fol. 12r.

Lucca 236, about 900 A. D., on its last 3 leaves are Egyptian days and a dream-book; described by Giacosa (1901), p. 349.

Harleian 3017, 10th century, fol. 59r, De diebus Egiptiacis qui mali sunt in anno circulo. The catalogue dates this MS as 920 A. D. but at fol. 66r the date is given as DCCClxii or DCCCClxii (962 A. D.)—a letter seems to have been erased which probably was the fourth C.

Harleian 3271, 10th century (?), fol. 121, Versus ad dies Egyptiacas inveniendas. See also Baehrens, *Poet. lat. min.* V, 354-6; Mommsen CIL I, 411.

Sloane 475, this portion of the MS 10-11th century, fol. 216v, Versus de significatione dierum mensis, opening, "Tenebrae Aegyptus Grecos sermone vocantur. . . ."

Additional 22398, 10th century, fol. 104.

Cotton Caligula A, XV, written mostly in Gaul before 1000 A. D., fol. 126, a list of lucky and unlucky days for medical purposes, in Anglo-Saxon.

Cotton Titus D, XXVI, 10th century, fol. 3v.

Cotton Vitellius A, XII, fol. 39v.

Cotton Vitellius C, VIII, in Anglo-Saxon, fol. 23, de tribus anni diebus Aegyptiacis.

CU Trinity 945, early 11th century, fol. 37.

CU Trinity 1369, 11th century (perhaps 1086 A. D.) fol. 1v.

Vatican 644, 10-11th century, fol. 77r, versus duodecim de diebus aegyptiis, and a fragment "de tribus diebus aegyptiis."

Dijon 448, 10-12th century, fol. 88, Calendrier, avec jours égyptiaques ajoutés; fol. 191, "De Egyptiacis diebus." Bede's *De temporibus* and *De natura rerum* occur twice in this MS and at fol. 181 is an incantation for use in fevers.

Harleian 1585 and Sloane 1975, where the Egyptian days are found with the *Herbarium* of Apuleius, are both 12th century but probably copied from earlier MSS.

So in Chalons-sur-Marne 7, 13th century, fol. 41, verses on the Egyptian days occur with the *Ars calculatoria* of Helpericus of Auxerre who wrote in the ninth century.

I have usually not noted the occurrence of the Egyptian days in later manuscripts. A few exceptions are:

BN 7299A, 12th century, fol. 37r.

CLM 23390, 12-13th century, the last item is, "Verses concerning the twelve signs and the Egyptian days." The previous contents were mainly religious.

Cambrai 195, fol. 208; 229, fol. 56; 829, fol. 54; all three MSS of the 12th century.

Cambrai 861, early 13th century, fol. 56.

Sloane 2461, end of 13th century, fols. 62r-64v.

The verses concerning the ten plagues of Egypt contained in CLM 18629, 10th century, fol. 93, and ascribed by the catalogue to Eugenius Toletanus have, I presume, no connection with the Egyptian days. Such proved to be the case with BN 16216, 13th century, fol. 251v, de decem plagis Egyptiorum et de vii diebus, although from the fact that it follows "Precepta Pithagore" I suspected before examining it that it might have something to do with divination. But not even the Pythagorean precepts have in this case.

CHAPTER XXX

GERBERT AND THE INTRODUCTION OF ARABIC ASTROLOGY

Arabic influence in early manuscripts—A preface and twenty-one chapters on the astrolabe—Are they parts of one work?—Their relation to Gerbert and the Arabic—Hermann's *De mensura astrolabii*—Attitude towards astrology in the preface—Question of Gerbert's attitude towards astrology—His posthumous reputation as a magician—An anonymous astronomical treatise; its possible relation to Gerbert—Contents of its first two books—Attitude towards astrology—The fourth book—Citations: Arabic names—*Mathematica* of Alchandrus or Alhandreus—An account of its contents—Astrological doctrine—Nativities and name-calculations—Interrogations and more name-calculations—Alchandrus or Alhandreus not the same as Alexander—Alkandrinus or Alchandrinus on nativities according to the mansions of the moon—Albandinus—Geomancy of Alkardianus or Alchandianus—An anonymous treatise or fragment of the tenth century.

Arabic influence in early manuscripts.

THE usual view has been that western Latin learning was not affected by Arabic science until the twelfth or even the thirteenth century. We shall see in other chapters that the translations of the Aristotelian books of natural philosophy were current rather earlier than has been recognized, that in medicine a period of Neo-Latin Salernitan tradition can scarcely be distinguished from one of Arabic influence, and that in chemistry owing to the misinterpretation of the date of Robert of Chester's translation of the book of Morienus Romanus—in which Robert says that the Latin world does not yet know what alchemy is—Berthelot in his history of medieval alchemy placed the introduction of Arabic influence half a century too late. In the present chapter we shall see that the voluminous work of translation of Arabic astrologers which went on in the twelfth century—and to which another chapter will later be devoted—was preceded in the eleventh and even tenth centuries by numerous signs of Arabic influence in works of astronomy and astrology and also by translations of Arabic authors. I was somewhat startled when I first found works by Arabic authors and use of astronomical terminology drawn from the Arabic in a manuscript of the eleventh century in the British Museum [1] and Wickersheimer was similarly surprised at the traces of Arabic influence in a similar but still earlier manuscript of the tenth century at Paris.[2] Bubnov, however, had already noted this Paris manuscript as a proof that Arabic books were being translated into Latin in Gerbert's time,[3] and one of Gerbert's letters, written in 984 to a Lupitus of Barcelona (*Lupito Barchinonensi*), asking him to send Gerbert a book on "astrology" which he had translated, points in the same direction. In the present chapter we shall discuss the contents of the early manuscripts just mentioned and of some others which seem to have some connection either with Gerbert or the introduction of Arabic astrology into Latin learning.

A preface and twenty-one chapters on the astrolabe.

In an eleventh century manuscript at Munich [4] the astrological work of Firmicus is preceded by writings in a different hand upon the astrolabe. One of these, in its present state an anonymous fragment, is a stilted and florid introduction to a translation from the Arabic of a work on the astrolabe.[5] Another is a treatise on the astrolabe in twenty-one chapters and containing many Arabic names.[6]

[1] Additional 17,808, a narrow folio in vellum with all the treatises written in the same large, plain hand with few abbreviations. A considerable part of the MS is occupied by the work on music of Guido of Arezzo (c. 995-1050). This MS is not noted by Wickersheimer or by Bubnov, although it includes treatises on the abacus and the astrolabe which are perhaps by Gerbert.

[2] BN 17,868, from the chapter of Notre Dame of Paris, 21 leaves. Wickersheimer (1913), 321-3, states that it has all the marks of the writing of the tenth century: Delisle so dated it. Bubnov (1899), LXVII, regards fols. 14r *et seq.* as by a slightly older hand than the first portion.

[3] Bubnov (1899), 124-6, note.

[4] CLM 560, described in Bubnov, *Gerberti opera mathematica*, 1899, p. xli.

[5] *Ibid.*, fols. 16r-19, Fragmentum libelli de astrolabio a quodam ex Arabico versi. Incipit, "Ad intimas summe phylosophie disciplinas et sublimia ipsius perfectionis archisteria." Printed by Bubnov (1899), pp. 370-75.

[6] Incipit "Quicumque astronomiam peritiam disciplinae"; the printed editions insert a *discere* after *astronomiam,* but it has not been there in the MSS which I have seen and is not needed. Printed by Pez, *Thesaurus Anecdotorum Noviss.* III, ii, 109-30,

Bubnov lists three other copies of the introductory fragment, and they are all in manuscripts where the second treatise is also included;[1] it, however, is often found in other manuscripts where the anonymous fragment does not appear, and it must be admitted that its omission is no great loss.

Are they parts of one work?

Although the fragment precedes the other treatise in only one manuscript mentioned by Bubnov, there is reason to think that they belong together, since both are concerned with the *Wazzalcora* or planisphere or *astrolapsus* of Ptolemy, and since the plan outlined by the writer of the introduction is followed in the treatise of twenty-one chapters except that it ends incompletely. Bubnov recognized this, yet did not unite them as a single work.[2] In 984 Gerbert wrote to a *Lupito Barchinonensi* asking Lupitus to send him a work on "astrology" which Lupitus had translated.[3] If Lupitus was of Barcelona, his translation was probably from the Arabic, and as such translations were presumably not common in the tenth century, it is natural to wonder if he may not be the above-mentioned anonymous translator. This Bubnov suggested in the case of the introductory fragment,[4] but the treatise in twenty-one chapters he placed among the doubtful works of Gerbert,[5] because a monastic catalogue composed before 1084 speaks of a work of Gerbert on the astrolabe, while six manuscripts of the

(1721) and incorrectly ascribed by him to Hermannus Contractus, because it often occurs in the MSS together with another treatise on the astrolabe by a "Herimannus Christi pauperum peripsima et philosophiae tyronum asello imo limace tardior assecla." Of this last we shall have more to say presently. The edition of Pez reappears in Migne, PL vol. 143. Bubnov (1899), 114-47, gives a new edition, and at pp. 109-13 a list of the MSS of the work, in which, however, he fails to note the following: and they are also absent from his general index of 153 codices at pp. xvii-xc. BM Additional MS 17808, 11th century, fols. 73v-79r, under the title as in other MSS of "Regulae ex libris Ptolomei regis de compositione astrolapsus." Yet Bubnov says, p. cxvi, "Catalogues of Additional MSS (omnia volumina inspexi, quae ante a. 1895 edita sunt)." BM Egerton 823, 12th century, fol. 4r. BN 7412, 12th and 13th centuries, fols. 1-9, "Waztalkora sive tract. de utilitatibus astrolabii." Professor D. B. Macdonald suggests that *Waztalkora* is for *rasmu-l-kura*, "the describing of the sphere in lines."

[1] (1899), p. 370.

[2] (1899), p. 374.

[3] Ep. 24.

[4] (1899), p. 370.

[5] P. 109.

twelfth and thirteenth centuries, although none earlier to his knowledge, ascribe this very treatise of twenty-one chapters to Gerbert. Bubnov believed that whoever the author of the treatise in twenty-one chapters was, he had utilized the full work of the anonymous translator. But this seems a rather unnecessary refinement. For what has become of that translation? Why is only its wordy and rhetorical preface extant? If the writer of the twenty-one chapters destroyed its text after plagiarizing it, why did he not also make away with the preface? It seems more plausible that the twenty-one chapters are the original translation from the Arabic, and that many makers of manuscripts have copied it alone and omitted the wordy and rather worthless preface of the translator. If, as Bubnov suggested, the treatise in twenty-one chapters is Gerbert's revision and polishing up of Lupitus' translation,[1] why did he not prefix a new introduction of his own? And why should anyone try to polish up the style of so rhetorical a writer as he who penned the extant anonymous introduction?

Their relation to Gerbert and the Arabic.

If we accept this anonymous introduction as the preface to the twenty-one chapters, Gerbert would be the most likely person to ascribe both to, unless we argue that he could not make a translation from the Arabic and that his letter asking to see a translation from the Arabic by Lupitus is a proof of this. If Gerbert is not the author, Lupitus would perhaps be the next most likely person, but the hint contained in Gerbert's letter is all that points to Lupitus, and indeed the only mention that we have of him. If the translator is some third unknown person, at least he is not later than the eleventh century. If, on the other hand, we regard the introduction of the translator and the twenty-one chapters as by different persons, who perhaps had no connection with each other, and Gerbert's letter of 984 as having nothing to do with either, we have the more evidence of an early and widespread interest in astronomy and

[1] Bubnov (1899), 370 . . . "Hoc opusculum ex Arabico versum ad manum habuit, retractavit dicendique genere expolivit."

knowledge of Arabic in the western Latin learned world.

Hermann's *De mensura astrolabii.*

One reason why the treatise on the astrolabe in twenty-one chapters is so seldom found in the manuscripts preceded by the introduction of the translator may be that it is more often found with and preceded by another treatise on the astrolabe, sometimes entitled *De mensura astrolabii,* and attributed to a Hermann who modestly calls himself "the offscouring of Christ's poor and the butt of mere tyros in philosophy." [1] This treatise tells how to construct an astrolabe, thus filling in the deficiency left by the incomplete ending of the treatise in twenty-one chapters, which fails to carry out fully this last item in the plan of the introductory fragment. A note in one manuscript, reproduced in part by Macray in his catalogue of the Digby Manuscripts in the Bodleian Library, states that the treatise in twenty-one chapters is by Gerbert and that when a certain Berengarius read it, he found it told how to exercise the art but not to make the instrument and asked Hermann to tell him how to make one. Hermann therefore composed the work in question, dedicated it to Berengarius, and prefixed it to Gerbert's treatise.[2] Of late there has been a tendency to identify this Hermann with Hermann of Dalmatia, the twelfth century translator from the Arabic,[3] rather than with Hermann the Lame, the chronicler, who died in 1054, but if Bubnov is correct in dating two manuscripts [4] containing

[1] Printed by Pez. *Thesaur. Anecdot. Noviss.* III, ii, 95-106. "Herimannus Christi pauperum peripsima et philosophiae tyronum asello imo limace tardior assecla." The MSS are numerous.

[2] Digby 174, fol. 210v; also noted by Bubnov (1899), p. 113. Hermann's dedicatory prologue, however, does not give his friend's name in full, but reads in this MS, "B. amico suo."

[3] See Clerval, *Hermann le Dalmate,* Paris, 1891, in *Compte rendu du Congrès scientifique international des catholiques, Sciences Historiques,* 163-9. Also, I believe, published separately as *Hermann le Dalmate et les premières traductions latines des traités arabes d'astronomie au moyen âge,* Paris, Picard, 1891, 11 pp. Clerval adduced only one MS in support of his contention and took up the untenable position that Arabic astronomy was unknown in Latin until the twelfth century. He also did not distinguish between the different works on the astrolabe.

[4] Munich CLM 14836, fols. 16v-24r. BM Royal 15-B-IX, fol. 51r-: in both cases followed by the treatise of twenty-one chapters.

Hermann's treatise on the astrolabe in the eleventh century, they could not be the work of Hermann the translator of the next century.[1] Moreover, in the thirteenth century the treatise seems to have been regarded as the work of Hermann the Lame.[2] The author's self-depreciatory description of himself is also a mark of Hermann the Lame, who in another treatise addressed to his friend Herrandus and discussing the length of a moon calls himself "of Christ's poor a vile abortion." [3]

Attitude towards astrology in the preface.

In the treatise of twenty-one chapters, which simply tells how to use the astrolabe, there is naturally no reference to judicial astrology. But in the introduction of the anonymous writer to his translation from the Arabic of a work on the astrolabe there is mention of the influence of the stars. Their "concord with all mundane creatures in all things" is regarded as established by "secret institution

[1] Professor Haskins has announced as in preparation an article on Hermann the translator which will perhaps solve the difficulties.

[2] In a Berlin manuscript of the twelfth century (Berlin 956, fol. 11) there is added a note in a thirteenth century hand recounting the legend that this Hermann was the son of a king and queen and that, his mother having been asked before his birth whether she would prefer a handsome and foolish son or a learned and shamefully ugly one and she having chosen the latter alternative, he was born hunchbacked and lame. It was from this MS of the treatise on the astrolabe that Pertz edited the legend in the *Monumenta Germaniae* (*Scriptores,* V, 267). Rose (1905), p. 1179, calls the writer of this note Berengar, too, asking anent the opening words of the note, "De isto hermanno legitur in historia," "Aus welcher *historia* hat der Schreiber (Berengarius) seine Fabeln?" The note at the close of the treatise in Digby 174, fol. 210v, gives a different version of the legend, stating that Hermann was a good man and dear to God and that one day an angel offered him his choice between bodily health without great wisdom and the greatest science with corporal infirmity. Hermann chose the latter and afterwards became a paralytic and gouty.

[3] This treatise, in which Hermann expresses amazement that Bede has so underestimated the duration of the moon, immediately precedes the one on the astrolabe in BN nouv. acq. 229, a German MS of the twelfth century, fols. 17r-19r (formerly pp. 265-269). After the treatise on the astrolabe follows a third work by Hermann, "de quodam horologio," fols. 25v-28r. Then follows the treatise in twenty-one chapters on the astrolabe.

These citations alone are sufficient to demonstrate the error of Clerval's assertion: (1891), 165. "On ne peut invoquer aucune preuve sérieuse en faveur d'Hermann Contract. Jacques de Bergame et Trithème . . . sont les premiers qui aient attribué au moine de Constance les traités en question."

of divinity and by natural law" and testified to by scientists.[1] Not only is the effect of the moon on tides adduced as usual as an example, but God is believed to have set the seal of His approval upon "this discipline," when He made miraculous use of the stars and heavens to mark the birth and passion of His Son. The writer, however, stigmatizes as a "frivolous superstition" the doctrine of the Chaldean *genethlialogi,* "who account for the entire life of man by astrological reasons" and "try to explain conceptions and nativities, character, prosperity and adversity from the courses of the stars." Something nevertheless is to be conceded to them, provided all things are recognized as under divine disposition. But their doctrine is an egg which is not to be sucked unless rid of the bad odors of error.[2] The translator urges the importance of a knowledge of astronomy in determining the date of church festivals and canonical hours. He cites Josephus concerning Abraham's instruction of the Egyptians in arithmetic and astronomy, but regards Ptolemy as the most illustrious of all astronomers and the astrolabe as the invention of his "divine mind." The translator wishes his readers to understand that he is offering them nothing new but only reviving the discoveries of the past, and that he is simply presenting what he finds in the Arabic.

[1] Bubnov (1899) 372. "Habet etiam ex divinitatis archana institutione et physica lata ratione cum omnibus mundanis creaturis concordiam in rebus omnibus, secundum phisiologos non parvam congruentiam. . . ." Bubnov unfortunately used only one of his four MSS in printing this text, and there often seems to be something wrong with it or with his punctuation. This criticism applies more especially to the passage quoted in the following footnote.

[2] *Ibid.,* "Et ut Chaldaicas reticeam gentilogias (*sic*) qui omnem humanam vitam astrologicis attribuunt rationationibus et quosdam constellationum effectus per xii signa disponunt, quique etiam conceptiones et nativitates, hominumque mores, prospera seu adversa ex cursu siderum explicare conantur. Quod illorum tamen frivolae superstitiositati concedendum est, dum omnia divinae dispositioni commendanda sint. Illud est ovum a nullo forbillandum (Bubnov suggests the reading *furcillandum* in parentheses, but *sorbillandum* seems to me the obvious reading), nisi prius foetidos inscitiae exhalaverit ructus et feces mundialium evomerit studiorum." The passage is rather incoherent as it stands, but I hope that I have correctly interpreted its meaning.

Question of Gerbert's attitude toward astrology.

If Gerbert could be shown to be the translator who wrote this introduction, it would be a more valuable bit of evidence as to his attitude toward astrology than anything that we have at present. His surely genuine mathematical works, as edited by Bubnov, consist solely of a short geometry and a few of his letters in which mathematical topics, mainly the abacus, are touched upon. His contemporary and disciple, the historian Richer, tells in the well-known passage [1] how Borellus, "the duke of Hither Spain," took Gerbert as a youth from the monastery at Aurillac in Auvergne back with him across the Pyrenees and entrusted his education to Hatto, bishop of Vich, in the north-eastern part of the peninsula. Whether Gerbert studied Arabic or not Richer does not state. Since he is still described as *adolescens* when the duke and bishop take him with them to Italy and leave him there with the pope, one would infer that he probably had not engaged in the work of translation from the Arabic. Another almost contemporary writer, alluding very briefly to Gerbert, makes him visit Cordova, but is perhaps mistaken.[2] Richer does, however, state that Berbert especially studied *mathesis,* a word which, as various medieval writers inform us, may mean either mathematics or divination. Apparently Richer uses it in the former sense, for later he mentions only Gerbert's achievements in arithmetic, geometry, music, and astronomy.[3] But Robert, king of France, 987-1031, whose teacher Gerbert had been, seems to refer to him as "that master Neptanebus" in some verses,[4] a name which certainly suggests an astrologer, as well as an instructor of royalty, if not also a magician.

His posthumous reputation as a magician.

But Gerbert's reputation for magic seems to start with William of Malmesbury in the first half of the twelfth century, who makes him flee by night from his monastery to Spain to study "astrology" and other arts with the Saracens,

[1] III, 43-45.

[2] Ademarus Cabannensis, who died about 1035 (Bubnov, 1899, 382-3). For Gerbert's sources in Barcelona see J. M. Burnam, "A Group of Spanish Manuscripts," in *Bulletin Hispanique, Annales de la Faculté des Lettres de Bordeaux,* XXII, 4, p. 329.

[3] III, 48-53.

[4] "Plurima me docuit Neptanebus ille magister" (Bubnov, 381).

until he came to surpass Julius Firmicus in his knowledge of fate. There too, according to William of Malmesbury, "he learned what the song and flight of birds portend, to summon ghostly figures from the lower world, and whatever human curiosity has encompassed whether harmful or salutary." William then adds some more sober facts concerning Gerbert's mathematical achievements and associates.[1] Michael Scot in his *Introduction to Astrology* in the early thirteenth century speaks of a master *Gilbertus* who was the best nigromancer in France and whom the demons obeyed in all that he required of them day and night because of the great sacrifices which he offered and his prayers and fastings and magic books and great diversity of rings and candles. Having succeeded in borrowing an astrolabe for a short time he made the demons explain its purpose, how to operate it, and how to make another one. Later he reformed and became bishop of Ravenna and pope.[2] In a manuscript early in the thirteenth century is a statement that Gerbert became archbishop and pope by demon aid and had a spirit enclosed in a golden head whom he consulted as to knotty problems in composing his commentary on arithmetic. When the demon expounded a certain very difficult place badly, Gerbert skipped it, and hence that unexplained passage is called the *Saltus Gilberti.*[3]

An anonymous astronomical treatise; its possible relation to Gerbert.

In a manuscript in the Bodleian library which seems to have been written early in the twelfth century[4] is an astronomical treatise in four books which Macray suggested might be the *Liber de planetis et mundi climatibus* which Ethelwold, bishop of Winchester from 963 to 984, is said to have composed.[5] The present treatise indeed embodies

[1] *De rebus gestis regum Anglorum,* II, 167-8.

[2] Bodleian 266, fol. 25r.

[3] Bubnov (1899), 391. On Gerbert as a magician see further J. J. I. Döllinger, *Die Papst-Fabeln des Mittelalters,* Munich, 1863, pp. 155-59.

[4] Digby 83, quarto in skin, well written in large letters with few abbreviations and illustrated with many figures in red, 76 leaves. For the *Incipits* of the four books and their prologues see Macray's Catalogue of the Digby MSS.

[5] Another indication of mathematical activity in tenth century England is provided by some old verses in English in Royal 17-A-I, fols. 2v-3, which state that Euclid's geometry was introduced into England "Yn tyme of good

a *Letter of Ethelwold to Pope Gerbert* on squaring the circle.[1] It seems, however, that this letter on squaring the circle was really written by Adelbold, bishop of Utrecht from 1010 to 1027.[2] Adelbold speaks of himself in the letter as a young man [3] and of course wrote it before Gerbert's death in 1003, and very probably before Gerbert became Pope Silvester II in 999. But he could scarcely have written the letter early enough to have it included in a work written by Ethelwold who died in 984. Our astronomical treatise in four books is therefore not by Ethel-

kyng Adelstones day." Usually the first Latin translation of Euclid is supposed to have been that by Adelard of Bath in the early twelfth century. Halliwell (1839), 56.

[1] Digby 83, fol. 24, "Epistola Ethelwodi ad Girbertum papam. Domino summo pontifici et philosopho Girberto pape athelwoldus vite felicitatem. . . ." Gerbert of course did not become pope until long after Ethelwold's death, but this Titulus and Incipit are open to suspicion anyway, since if Gerbert had become pope he should have been addressed as Pope Silvester. The article on Ethelwold (DNB) states that "a treatise on the circle, said to have been written by him and addressed to Gerbert, afterwards Pope Silvester II, is in the Bodleian Library (1684, Bodl. MS. Digby 83, f. 24)." William of Malmesbury mentioned "Adelboldum episcopum, ut dicunt, Winterbrugensem" as the author of the letter to Gerbert, quoted by Bubnov (1899), 388.

[2] It has always been so printed: by Pez, Olleris, Curtze, and Bubnov, and seems to be ascribed to him in most MSS, for which and other evidence pointing to the bishop of Utrecht as author see Bubnov (1899), 300-309, 41-45, 384, etc. Bubnov, however, failed to note Digby 83 either in connection with this letter or at all in his long list of mathematical MSS (XVII-CXIX). It may therefore be well to note that the letter as given in Digby 83 differs considerably from the version printed by Bubnov. It in general omits epistolary amenities which do not bear directly on the mathematical question in hand, notably the entire first paragraph of Bubnov's text and the close of the second and third paragraphs. It also abbreviates portions of the fifth paragraph and the last sentence of the eighth and last paragraph. On the other hand after the first sentence of the fifth paragraph of Bubnov's text it inserts the following passage which seems to be missing in Bubnov's text of the letter: "Si quis ergo vult invenire quadraturam circuli dividat lineam in VII partes spatiumque unius septime partis semotim ponat. Deinde lineam in VII divisam in duo distribuat et spatium alterius duorum separatim ponat. Post hoc lineam in VII partitam triplicet cui triplicate spatium unius septime quod semoverat adiciat. Ipsa denique totam in IIII partiatur quarum quarta angulis directis per lineam quadrangulam metiatur. Ad ultimum sumpto spatio alterius duorum quod prius reposuerat deposito puncto in medio quadranguli eodem spatio circumducat circinum (circulum) et sic inveniet circuli quadraturam."

[3] Bubnov (1899), 41-42, "quod tantum virum quasi conscolasticum iuvenis convenio."

wold, unless the letter be a later interpolation, but it is possibly by Adelbold or by Gerbert.[1] Its opening words, "Quicumque mundane spere rationem et astrorum legem . . . ," are similar to those of the treatise on the uses of the astrolabe which has often been ascribed to Gerbert, "Quicumque astronomice peritiam discipline . . ." [2]

Contents of its first two books.

Our treatise then may be by Gerbert or it may be a specimen of the astronomy of the eleventh or early twelfth century. As it appears to be little known and never to have been published, it may be well to give a brief summary of its contents. An introductory paragraph outlines some of the chief points with which the treatise will be concerned, such as the twelve signs of the zodiac, their positions, "most varied qualities," the reasons for their names, and the diverse opinions of gentile philosophers and Catholics as to their significations; the four elements; and the seven planets. In the text which follows, these topics are considered in rather the reverse order to that in which they were named in the preface. After some discussion of "the founders of astronomy and the doctors of astrology," the first book is occupied with a description of the sphere or heavens. The second book is largely geographical, beginning with the question of the size of the earth, the zones, the ocean, and how to draw a T map. This geographical digression the author justifies in the prologue to his third book by the statement that often the position of the stars can be determined from the location of countries, and that

[1] Bubnov does not include it in his edition of the mathematical works of Gerbert, but as we have seen he was unaware of the existence of this MS, i.e., Digby 83.

[2] And also to the *Incipit* of a treatise in a tenth century MS at Paris, BN 17,868, fol. 14r, "Quicumque nosse desiderat legem astrorum. . . ." The treatise or fragment in this Paris MS seems to end at fol. 17r, or at least at fol. 17v, after which most of the few remaining leaves of the MS, which has only 21 leaves in all, are blank. There is some similarity of contents, but the Paris MS is more astrological. Possibly, however, it is a different part of, or rather extracts from the same work, since we shall see reasons for thinking that the text in Digby 83 is incomplete.

if the habitat of peoples is known one can more easily arrive at the effect of the stars.[1]

Attitude towards astrology.

This suggests that the author believes in astrological influence, and in the two following books he states a number of astrological doctrines, not, however, as his own convictions but as the opinions of the *genethliaci* or astrologers, or "those who will have it that prosperity and adversity in human life are due to these stars." [2] On the other hand, he seldom subjects the astrologers to any adverse criticism. Indeed, early in the third book, he states that the belief of the *genethliaci* that human wealth and honors, poverty and obscurity, depend upon the stars, pertains to another subject than that which he is at present discussing; namely, prognostication, concerning which he will treat fully in later chapters. But I cannot see that he fulfills this promise in the present manuscript, which seems to end rather abruptly,[3] so that possibly there is something missing. In the previous passage, however, he immediately proceeded to admit that the sun and moon greatly affect our life and to tell further how it is connected with the other five planets. In the star of Saturn the soul is said to busy itself especially with reasoning and intelligence, logic and theory. Jupiter is practical and represents the power of action. Mars signifies animosity; Venus, desire; Mercury, interpretation. Men have proved the moon's moist influence by sleeping out-of-doors and finding that more humor collected in their heads when they slept in the moon-light than when they did not.[4] After mentioning the twelve signs, "through which the aforesaid planets revolving exert varied influences, and even, according to the *genethliaci,* make a good man in some nativities and a bad man in others," [5] the author goes on to tell which

[1] At least such seems to me to be the meaning of the passage, fol. 21r, "Quippe cum aliquando per situm gentium ipsarum positionem stellarum demonstrati simus precognita populorum habitatione rei effectus ad faciliorem curret eventus."

[2] Fol. 22r.

[3] Fol. 76r, the closing words are, "Quod autem de elementis diximus idem de temporibus deque humoribus intellige sicut hec figura evidentissime designat." But the figure is not given.

[4] Fol. 27v.

[5] Fol. 31v, "per que predicti planete revoluti diversa in diver-

signs are masculine and which are feminine, to relate them to the four cardinal points and to the four elements, to define the twenty-eight mansions and their distribution among the twelve signs and seven planets,[1] and to tell how the planets differ in quality.[2] All this is providing at least the basis for astrological prediction.

The fourth book.

The fourth book of the treatise is mainly taken up with descriptions and figures of the constellations, concerning which the author often repeats the fables of antiquity. After discussing the six ages of the world, the author intended to insert a figure on what is the next to last page of the present text to show "the harmony of the elements, climates of the sky, times of the year, and humors of the human body," for, as he goes on to say, man is called a microcosm by the philosophers. This missing figure or figures would have been analogous to those which Wickersheimer investigated in the early medieval manuscripts in the libraries of France.

Citations: Arabic names.

Our author does not make many citations, but among them are Eratosthenes,[3] Aratus, Ptolemy, Macrobius, and Martianus Capella. Some of these authors are perhaps known to him only indirectly, and he seems to make use of Isidore and Pliny without mentioning them. He shows, however, an acquaintance with foreign languages, listing the seven heavens as "oleth, lothen, ethat, edim, eliyd, hachim, atarpha," and giving Greek, Hebrew, and "Saracen" names for the seven planets, as well as a "Similitudo," or corresponding metal, and "Interpretatio," or quality such as "Obscurus, Clarus, Igneus." [4] He also gives the Arabic names for the twenty-eight mansions into which the circle of the zodiac subdivides.[5] We now turn to another treatise, found in tenth and eleventh century manuscripts, in which Arabian influence is apparent.

sis possunt et etiam secundum genethliacos bonum quidam in quibusdam malum vero in quibusdam quidam nativitatibus hominem astruunt."

[1] Fol. 32r.
[2] Fol. 36r.
[3] Fol. 59r, "Herastotenes."
[4] Fol. 21r-v.
[5] Fol. 32r.

The *Mathematica* of Alchandrus or Alhandreus.

William of Malmesbury, writing in the first half of the twelfth century concerning Gerbert's studies in Spain, says, probably with a great deal of exaggeration, that Gerbert surpassed Ptolemy in his knowledge of the astrolabe, Alandraeus in his knowledge of the distances between the stars, and Julius Firmicus in his knowledge of fate.[1] It is rather remarkable that a work ascribed to Alhandreus or Alcandrus, "supreme astrologer," should be found in two manuscripts of the eleventh century [2] in both of which occurs also the work on the astrolabe which is perhaps by Gerbert, while in one is found also the *Mathesis* of Julius Firmicus Maternus. Alchadrinus or Archandrinus is cited in Michael Scot's long *Introduction to Astrology* as the author of a "book of fortune making mention of the three *facies* of the signs and the planets ruling in them," and Michael adds that a similar method of divination is employed in general among the Arabs and Indians as can be seen in the streets and alleys of Messina where "learned women" answer the questions of merchants.[3] Peter of Abano in his *Lucidator astronomiae*,[4] written in 1310, mentions Alchandrus as a successor of Hermes Trismegistus in the science of astronomy but as flourishing before the time of Nebuchadnezzar. Alchandrus was probably scarcely as ancient as that, but the treatise ascribed to him also exists in Latin in a manuscript of the tenth century,[5] and seems to be a translation from

[1] *De rebus gestis regum Anglorum*, II, 167.

[2] Addit. 17808, fols. 85v-99v, "Mathematica Alhandrei summi astrologi. Luna est frigide nature et argentei coloris / oculis descriptio talis subiciatur": and CLM 560, fols. 61-87, which I have not seen but which from the description in the catalogue is evidently the same treatise and has the same *Incipit*, although no author or title seems to be given.

[3] Bodleian 266, fol. 179v, "libellum fortune faciens mentionem de tribus faciebus signorum et planetis regnantibus in eisdem . . . mulieres docte."

[4] BN 2598, 15th century, fol. 108r.

[5] BN 17868, fols. 2r-12v. "Incipit liber Alchandrei" (Wickersheimer) or Alchandri (Bubnov) "philosophi. Luna est frigide nature et argentei coloris." In a passage of Addit. 17808, fol. 86v, where the years from the beginning of the world are being reckoned, the year of writing is apparently given as 1040 A. D., but the existence of the treatise in BN 17868 shows that it was written before 1000. Also there is something wrong with the passage mentioned in Addit. 17808—as is very apt to be the case with such figures in medieval MSS—for the number of years from the

the Arabic. In any case it is full of Arabic and Hebrew words, and professes to cite the opinions of Egyptians, Ishmaelites, and Chaldeans in general as well as those of Ascalu the Ishmaelite and Arfarfan or Argafalan or Argafalaus [1] the Chaldean in particular. Since the name Alchandrus or Alhandreus is found so far as I know in no historian or bibliographer of Arabian literature or learning,[2] we shall treat somewhat fully of the work and its author here.

An account of its contents.

The "Mathematic of Alhandreus, supreme astrologer," as it is entitled in one manuscript, opens somewhat abruptly with a terse statement of the qualities of the planets. Two estimates of the number of years between creation and the birth of Christ are then given, one "according to the Hebrews," the other "according to others." [3] There follow letters of the Greek alphabet with Roman numerals expressing their respective numerical values, perhaps for future reference in connection with some sphere of life or death. Next is considered the division of the zodiac into twelve signs for which Hebrew as well as Latin names are given. The movements of the planets through the signs are then discussed, and it is explained in the usual astrological style that Leo is the house of the sun, Cancer of the moon, while two signs are assigned to each of the other five planets. Every planet is erect in some one sign and falls in its opposite, and any planet is friendly to another in whose house it is erect and hostile to another in whose house it declines. Presently the author treats of "the order of the planets according to nature and their names according to the Hebrews," [4] and then of their sex and courses, which last leads to considerable

beginning of the world to the birth of Christ is given as 4970 and then the sum of the two as 6018 instead of 6010 years, while at fol. 85v other estimates are given of the number of years between the Creation and the Incarnation.

[1] The spellings of such proper names vary in the different MSS or even in the same one.

[2] Steinschneider (1905) 30, briefly notes "Alcandrinus," however. See below, p. 715 of the present chapter.

[3] Addit. 17808, fol. 85v; BN 17868, fol. 2r.

[4] Addit. 17808, fols. 86r-87r; BN 17868, fol. 3v.

digressions anent the solar and lunar calendars.[1] Then the twelve signs are related to the four "climates" and elements.

Astrological doctrine.

All this implies a favorable attitude to astrology, and the author has already expressed his conviction more than once that human affairs are disposed by the seven planets according to the will of God.[2] Since man like the world is composed of the four elements it is no false opinion which persuades us that under God's government human affairs are principally regulated by the celestial bodies.[3] To make this plainer the author proposes to insert an astrological figure "which Alexander of Macedon composed most diligently," and which presumably would have been of the microcosmus or Melothesia type, but the space for it remains blank in the manuscript. Next comes a paragraph on the sex of the signs and their rising and setting, and then lists of the hours of the day and night governed by the signs and by each planet for all the days of the week.[4]

Nativities and name-calculations.

Then we read, "These are the twenty-eight principal parts or stars (i.e. constellations) through which the fates of all are disposed and pronounced indubitably, future as well as present. Anyone may with diligence forecast goings and returnings, origins and endings, by the most agreeable aid of these horoscopes." [5] These twenty-eight parts are

[1] Addit. 17808, fols. 87v-88r.

[2] BN 17868, fol. 2r; Addit. 17808, fol. 85v; "Iuxta que quia omnia humana secundum nutum dei disponuntur per septem planetas que subter (subtus) feruntur eorum nobis potestas innuitur": BN 17868, fol. 3r; Addit. 17808, fol. 86v, "Per has autem vii planetas quia ut diximus et adhuc probabimus humana fata disponuntur regulam certam demus qua in quo signo queque sit pronoscatur." Only in a third passage does he attribute such views to the mathematici; Addit. 17808, fol. 88v, "Cum sint signa xii in zodiaco cumque iuxta mathematicos et secundum horum diversissimos potestates fata omnium ita volente sapientissimo domino disponantur. . . ."

[3] Addit. 17808, fol. 89r, "Que quum ita discernuntur non falsa opinio persuasit istis humana principaliter gubernante domino moderari cum itaque ut mundus homo unusquisque ex his iiii compaginetur elementis."

[4] Addit. 17808, fol. 89v. But the lists are left incomplete and a blank leaf, which is also left unnumbered, follows in the MS.

[5] BN 17868, fol. 5r: Addit. 17808, fol. 90r, "Hec sunt xxviii principales partes vel astra per que omnium fata disponuntur et indubitanter tam futura quam presentia prenuntiantur a quocumque itus reditus ortus occasus horum horoscoporum iocundissimo auxilio diligenter providentur."

of course the sub-divisions of the zodiac into mansions of the sun or moon which we have already encountered, and Arabic names are given for them beginning with *Alnait,* the first part of the sign Aries. First, however, we are instructed how to determine under which one of them anyone was born by a numerical calculation of the value of his name and that of his natural mother similar to that of the spheres of life and death except that it is based upon Hebrew instead of Greek letters.[1] Then follow statements of the sort of men who are born under each of the twenty-eight mansions, their physical, mental, and moral characteristics, and any especial marks upon the body,—either birth-marks or inflicted subsequently by such means as hot irons and dog-bite,—their health or sickness, term of life, and manner of death,—which in the case of Alnait, the first mansion, will be "by the machinations or imaginations of the magic arts." [2] Also the number of their children is roughly predicted.

Interrogations and more name-calculations.

Next is discussed the course of the planets through the signs, the houses of the planets, and their positions in the signs at creation.[3] The author then turns to the influence of the planets upon men and gives another method of numerical calculation of a man's name in order to determine which planet he is under.[4] Under the heading "Excerpts from the books of Alexander, the astrologer king," [5] directions are given for the recovery of lost or stolen articles and descriptions of the thief are provided for the hour of each planet. The letter of Argafalaus to Alexander instructs how to read men's secret thoughts as Plato the Philosopher used to do, and how to tell what is hidden in a person's hand by means of the hours of the planets.[6] After some fur-

[1] BN 17868, fol. 5v.

[2] BN 17868, fol. 6r.

[3] BN 17868, fol. 9r-; Addit. 17808, fols. 94v-95v.

[4] BN 17868, fol. 10r; Addit. 17808, fol. 96r.

[5] Addit. 17808, fol. 97r.

[6] Addit. 17808, fol. 97v. In BN 17868, fol. 11r, we read, "Explicit liber primus. Incipit liber secundus." And then begins the letter of Argafalaus with the words, "Regi macedonum Alexandro astrologo et universa philosophia perfectissimo Argafalaus servuus suus condicione et nacione ingenuus caldeus, professione vero secundus ab illo astrologus."

ther discussion of astrological interrogations the manuscript at the British Museum closes with the Breviary of Alhandreus, supreme astrologer,[1] for learning anything unknown by a method of computation from Hebrew and Arabic letters.

Alchandrus or Alhandreus not the same as Alexander.

Someone may wonder if the names Alhandreus and Alchandrus may not be mere corruptions of Alexander who is cited and quoted even more than has yet been indicated,[2] and if some careless head-line writer has not inserted the name *Alchandri* or *Alhandrei* instead of *Alexandri* in the *Titulus*. But this would leave the statements of William of Malmesbury and of Peter of Abano to be explained away. Or, if it is argued that the name of Alhandreus should be attached only to the Breviary, it must be remembered that in the earliest manuscript, which does not contain the Breviary, the treatise is none the less called the Book of Alchandreus. As a matter of fact there is found also in the manuscripts a "Mathematica Alexandri summi astrologi," [3] but while the title is the same, the contents are different from the "Mathematica Alhandrei summi astrologi."

However, the treatise itself is found together with the

[1] Addit. 17808, fol. 99r-v. This does not appear in BN 17868 which goes on to discuss various astrological influences of the 12 hours of the day and of the night. After this there is a space left blank in the middle of fol. 12v: then more is said concerning hours of the planets and interrogations until at the bottom of fol. 13r comes the letter of Phethosiris to Nechepso. But no definite ending is indicated either of the letter of Argafalaus or the Liber Secundus of Alchandrus.

In a MS now missing but listed in the late 15th century catalogue of the MSS in the library of St. Augustine's Abbey, Canterbury (No. 1172, James 332) was a "Breviarium alhandredi su'm astrologi et peritissimi de soia (scienda?) qualibet ignota nullo decrete." This was one of the MSS donated to the monastery by John of London.

BN 4161, 16th century, #5, Breviarium Alhandriae, summi Astrologi, de scientia qualiter ignota nullo indicante investigari possit.

[2] Addit. 17808, fol. 89r, "figuram quam super hac re Alexander Macedo composuit diligentissime posterius describemus"; fol. 95r, "Hinc Alexander macedo dicit eclipsin solis et lune certissima ratione colligi"; fol. 96r. "Aut iuxta alexandrum macedonem draco quasi octava planeta."

[3] Ashmole 369, late 13th century, fols. 77-84v. "Mathematica Alexandri summi astrologi. In exordio omnis creature herus huranicus inter cuncta sidera XII maluit signa fore . . . / . . . nam quod lineam designat eandem stellam occupat. Explicit." A further discussion of the contents of this work will be found below in Chapter 48, vol. II, p. 259.

Mathematica Alhandrei in a tenth century manuscript.[1] But no author is mentioned, and instead of *Mathematica* the title reads "Incipiunt proportiones cppfcfntfs knkstrprx indxstrkb," which may be deciphered as "Incipiunt proportiones competentes in astrorum industria." [2] Possibly therefore this treatise is a part of the work of Alchander, and the title *Mathematica Alexandri* is an error for *Mathematica Alhandrei.*

Alkandrinus or Alchandrinus on nativities according to the mansions of the moon.

Moreover, in later manuscripts we encounter authors with names very similar to Alchandrus and works by them of the same sort as that we have just considered. In a fifteenth century manuscript at Oxford we find ascribed to Alkandrinus an account of the types of men born in each of the twenty-eight mansions of the moon [3] such as we have seen formed a part of the *Mathematica Alhandrei.* And in a fifteenth century manuscript at Paris occurs under the name of Alchandrinus what seems to be a Christian revision of that same part of the *Mathematica Alhandrei.*[4] What appears to be another revision and working over of this same discussion of nativities according to the twenty-eight mansions of the moon [5] appeared in print a number of times in the sixteenth and seventeenth centuries and in

[1] BN 17868, fol. 17r. The Incipit is the same as in Ashmole 369. The work here seems to be incomplete, since after fol. 17v most of the remaining leaves of the MS (which has 21 fols. in all) are blank.

[2] The vowels being represented by the consonants following, a common medieval cipher.

[3] All Souls 81, 15th century, fols. 145v-164r. "Cum sint 28 mansiones lune. . . ." Coxe was mistaken in thinking that the work of Alkandrinus continued to fol. 188 and was in two parts, for at fol. 163r we read, "Expliciunt iudicia libri Alkandrini que sunt in divisione triplici 12 signorum que sunt apparencie per certa tempora super terram." Moreover, the seven chapters on the planets which follow end at fol. 183v ". . . finem fecimus. Completa fuit hec compilatio in conversione sancti pauli apostoli anno domini 1350 (1305?) vacante sede per mortem Benedicti undecimi cuius anima requiescat in pace. Amen." It would therefore seem that some compiler has made an extract from Alchandrus on the twenty-eight mansions.

[4] BN 10271, fols. 9r-52v, "Incipit liber alchandrini philosophi de nativitatibus hominum secundum compositionem duodecim signorum celi, quem reformavit quidem philosophus cristianus prout patet, quia in quibusdam differt iste liber ab antiquo primordiali. Primo facies arietis in homine sive in masculo. Alnaliet est prima facies arietis. . . ."

[5] Steinschneider (1905), 30.

French and English translations as well as Latin. The author's name in these printed editions is usually given as Arcandam, but the English edition of 1626 adds "or Alchandrin." [1]

Albandinus.

Two other manuscripts at Paris [2] contain under the name of Albandinus a "book of similitudes of the sons of Adam, fortunate and unfortunate, of life or death, according to nations, that is, their nativities according to the twelve signs." The treatise opens with a method of calculating a person's nativity from the letters in his own and his mother's name similar to that which occurs in the course of the *Mathematica Alhandrei*, but then applies it directly to the twelve signs rather than to the twenty-eight mansions of the moon. It also does not bother with the Hebrew alphabet but gives numerical equivalents directly for the Latin letters. Some treatise by Albandinus on sickness and the signs in a manuscript at Munich [3] is perhaps identical with the foregoing.

Geomancy of Alkardianus or Alchandiandus.

To an Alkardianus or Alchandiandus is ascribed a geomancy,[4] and since it also is arranged according to the twenty-eight divisions of the zodiac with 28 judges and 28 chapters each consisting of 28 lines in answer to as many questions, it would seem almost certain that it is by the same author who treated of the influences of the 28 houses or

[1] The *editio princeps* seems to be "Arcandam doctor peritissimus ac non vulgaris astrologus, de veritatibus et praedictionibus astrologiae et praecipuc nativitatum seu fatalis dispositionis vel diei cuiuscunque nati, nuper per Magistrum Richardum Roussat, canonicum Lingoniensem, artium et medicinae professorem, de confuso ac indistincto stilo non minus quam e tenebris in lucem aeditus, re cognitus, ac innumeris (ut pote passim) erratis expurgatus, ita ut per multa maxime necessaria et utilissima adiecerit atque adnotaverit modo eiusdem dexteritate praelo primo donatus." Paris, 1542.

The British Museum also contains another Latin edition of Paris, 1553; French editions of Rouen, 1584 and 1587, Lyons 1625; and English versions printed at London, 1626 (translated from the French), 1630, 1637, and 1670.

[2] BN 7349, 15th century, fol. 56r, seems only a fragment of the work; BN 7351, 14th century, takes up the various signs.

[3] CLM 527, 13-14th century, fols. 36-42, de physica signorum et supernascentium et aegrotantium.

[4] Addit. 15236, English hand of 13-14th century, fols. 130-52r, "libellus Alchandiandi." BN 7486, 14th century, "Incipit liber alkardiani phylosophi. Cum omne quod experitur sit experiendum propter se vel propter aliud. . . ."

facies of the twelve signs upon those born under them. Moreover, this Alkardianus or Alchandiandus states in his preface that he has composed certain books on the dispositions of the signs and the courses of the planets and on prediction of the future from them. "But since moderns always rejoice in brevity," he has added this handy and rapid geomantic means of answering questions and ascertaining the decrees of the stars. The 28 tables of 28 lines each of this Alkardianus or Alchandiandus are identical with one of the two such sets [1] commonly included in the *Experimentarius* [2] of Bernard Silvester, a work of geomancy which he is said to have translated from the Arabic.[3] He lived in the twelfth century and will be the subject of one of our later chapters.

An anonymous treatise or fragment of the tenth century.

It still remains to speak of a portion of our tenth century manuscript at Paris which begins, after the book of Alchandrus seems to have concluded, with the words, "Quicunque nosse desiderat legem astrorum . . ." [4] This *Incipit* is so similar to that of the twenty-one chapters on the astrolabe, "Quicumque astronomiam peritiam disciplinae . . ." and to that of the four books of astronomy, "Quicumque mundane spere rationem et astrorum," that one is tempted to imply some relation between them, and, in view of the tenth century date of the one at present in question, to connect it like the others with the name of Gerbert. Our present treatise or fragment of a treatise is largely astrological in character, "following for the present the wisdom of the *mathematici* who think that mundane affairs are carried on under the rule of the constellations." This refusal to accept personal responsibility for astrological doctrine is similar to the attitude of the author of the four books of

[1] The set in which the first line reads. "Tuum indumentum durabit tempore longo."

[2] Very probably this title was derived from the *Incipit* just given in note 4, p. 716.

[3] See Sloane 2472, 3554, 3857.

[4] BN 17868, fol. 14r-16v. The letter of Petosiris on the sphere of life and death at fol. 13r-v "Incipit epistola Phetosiri de sphaera" separates this treatise or fragment from the preceding *liber Alchandri philosophi*. Also this treatise is in a different and slightly older hand than fols. 2-13 are, or at least such was Bubnov's opinion (1899), 125, note.

astronomy, so that perhaps the present text is the missing fragment required to fulfil his promise to treat of the subject of prognostication in later chapters. If so it indulges in some repetition, as it goes into the relations existing between signs, planets, and elements, and gives the "Saracen" names [1] for the twenty-eight mansions of the moon. It includes a way to detect theft for each planet and a method of determining if a patient will recover by computation of the numerical value of the letters in his name. These features are suggestive of the *Mathematica* of Alchandrus.

[1] BN 17686, fol. 14v, "que sarraceni nuncupant ita."

CHAPTER XXXI

ANGLO-SAXON, SALERNITAN, AND OTHER LATIN MEDICINE IN MANUSCRIPTS FROM THE NINTH TO THE TWELFTH CENTURY

Plan of this chapter—Instances of early medieval additions to ancient medicine—*Leech-Book of Bald and Cild*—Magical procedure and incantations—A superstitious compound—Summary—Cauterization—Treatment of demoniacs—Incantations and characters—In a twelfth century manuscript—Magic with a split hazel rod—More incantations and the virtues of a vulture—*Lots of the saints*—Superstitious veterinary and medical practice—Two Paris manuscripts—Blood-letting—Resemblances to Egerton 821—Virtues of blood—Pious incantations and magical procedure—More superstitious veterinary practice—The School of Salerno—Was Salernitan medicine free from superstition?—The *Practica* of Petrocellus—Its sources—Fourfold origin of medicine—Therapeutics of Petrocellus—The *Regimen Salernitanum*—Its superstition—The *Practica* of Archimatthaeus—A Salernitan treatise of about 1200—The wives of Salerno.

Plan of this chapter.

In this chapter our purpose is to treat of early medieval medicine as distinct on the one hand from post-classical medicine, to which we have already devoted a chapter, and on the other hand from later medieval medicine as affected by translations from the Arabic and other oriental influence. Perhaps one of the outcomes of our discussion will be to suggest that any such distinctions cannot be at all sharply or chronologically drawn. However, the writings which we shall discuss now are contained mainly in manuscripts dating from the ninth to the twelfth century, although some of them may have been first composed at an earlier date than that of the manuscript in which they chance to be preserved. Some are in Anglo-Saxon; more, in Latin. Some it has been customary to classify under the caption of Salernitan. We shall postpone until the next chapter our consideration of Constantinus Africanus, although the dates of his life fall within the eleventh century, because he already at that early date represents the introduction of Arabic medicine to the western world.

Instances of early medieval additions to ancient medicine.

A good instance of the working over by men of the early medieval period of the medical writings of the late Roman period is provided by a manuscript of the ninth or tenth century at Berlin.[1] It now consists of a number of fragments whose original order can no longer be determined. These are made up of extracts from different sources or from other collections, but the collection also bears the mark of its last compiler who has introduced new remedies of his own and words derived from the vernacular of his day. Even extracts on fevers taken from the old Latin adaptation of Galen [2] are added to by some Christian physician, who introduces among other things some incantations, such as, "I adjure you, spots, that you go away and recede from and be destroyed from the eye of the servant of God." [3] The manuscript also comprises more than one tract on how dreams or the fate of the patient or child born can be foretold from the day of the moon.[4] Another tract [5] tells how God made the first man out of eight parts, of which the first was the mud of the earth and the last the light of the world. This would seem to be rather a novel departure from the usual four element theory but perhaps involves ancient Gnostic error. The author further argues that individual divergences of character depend upon the preponderance of one or another of the eight constituents of the body.

Leech-Book of Bald and Cild.

The Anglo-Saxon *Leech-Book of Bald and Cild* [6] has been called "the first medical treatise written in western Eu-

[1] Berlin 165 (Phillips 1790), 9-10th century. I have not seen the MS, but follow Rose's full description of it in his *Verzeichnis der lateinischen Handschriften,* I, 362-9.

[2] Cod. Casin. 97 Gal. I, 24-51.

[3] Berlin 165, fol. 88.

[4] *Ibid.,* fols. 40-2.

[5] *Ibid.,* fol. 39v.

[6] Edited with an English translation, which I employ in my quotations, by Rev. Oswald Cockayne in vol. II of his *Leechdoms, Wortcunning, and Starcraft of Early England,* in RS vol. 35, in 3 vols., London, 1864-1866. The relation of Bald and Cild to the work is indicated by the colophon at the close of the second book: "Bald habet hunc librum, Cild quem conscribere iussit,"—"Bald owns this book; Cild is the one he told to write (or copy?) it." The following third book is therefore presumably of other authorship.

rope which can be said to belong to modern history."[1] It was produced in the tenth century. However, it extracts a good deal from late Greek medical writers, such as Paul of Aegina and Alexander of Tralles, and cites Pliny, "the mickle leech," for the cure of baldness by application of dead bees burnt to ashes,[2] a remedy also found in the *Euporista* ascribed to Galen. On the whole, however, it uses parts of animals somewhat less than Pliny, although sometimes a powdered earthworm is recommended, or a man stung by an adder is to drink holy water in which a black snail has been washed, or the bite of a viper is to be smeared with earwax while thrice repeating "the prayer of Saint John."[3] And a man about to engage in combat is advised to eat swallow nestlings boiled in wine.[4] Herbs are as useful against a woman's tongue as birds against a foeman's steel, for we are told: "Against a woman's chatter; taste at night fasting a root of radish; that day the chatter cannot harm thee."[5] There are directions for plucking herbs similar to those in Pliny,[6] and the significance which he ascribed to cart ruts is paralleled by the injunction, after one has treated a venomous bite by striking five scarifications, one on the bite and four around it, to "throw the blood with a spoon silently over a wagon way."[7] Eight virtues of the stone agate are enumerated.[8]

Magical procedure and incantations.

Not only such occult virtues of animals, vegetables, and minerals, but also magical procedure and incantations abound in the work. In a prescription "for flying venom and every venomous swelling" butter is to be churned on a Friday from the milk of a "neat or hind all of one color," and a litany, paternoster, and incantation of strange words are to be repeated nine times each.[9] A great deal of superstitious use is made of such Christian symbols, names, and forms of prayer as the sign of the cross, the names of the four evan-

[1] J. F. Payne, *English Medicine in Anglo-Saxon Times*, 1904, p. 155.
[2] Book I, cap. 87.
[3] I, 45.
[4] I, 85.
[5] III, 47.
[6] I, 86.
[7] I, 68.
[8] II, 66.
[9] I, 45.

gelists, and masses, psalms, and exorcisms. Fear of witchcraft and enchantment is manifested, and the ills both of man and beast are frequently attributed to evil spirits. "A drink for a fiend-sick man" is on one occasion "to be drunk out of a church bell," with the accompaniment of much additional ecclesiastical hocus-pocus.[1] "If a horse is elf-shot, then take the knife of which the haft is horn of a fallow ox, and on which are three brass nails. Then write upon the horse's forehead Christ's mark, and on each of the limbs which thou may feel at. Then take the left ear; prick a hole in it in silence. This thou shalt do; then take a yerd, strike the horse on the back, then it will be whole. And write upon the horn of the knife these words, *Benedicite omnia opera domini dominum*. Be the elf what it may, this is mighty for him to amends." [2]

A superstitious compound.

Neither Bald and Cild nor their continuator shared Pliny's prejudice against compound medicines. In the third book by the continuator is described "a salve against the elfin race and nocturnal visitors, and for women with whom the devil hath carnal commerce." One takes the ewe hop plant, wormwood, bishopwort, lupin, ashthroat, henbane, harewort, viper's bugloss, heatherberry plants, cropleek, garlic, grains of hedgerife, githrife, and fennel. These herbs are put in a vessel and placed beneath the altar where nine masses are sung over them. They are then boiled in butter and mutton fat; much holy salt is added; the salve is strained through a cloth; and what remains of the worts is thrown into running water. The patient's forehead and eyes are to be smeared with this ointment and he is further to be censed with incense and signed often with the sign of the cross.[3]

Summary.

The "modern" character of Bald's and Cild's book cannot be said to have produced any diminution of superstition as against the writings of antiquity. But we do find native herbs introduced, also popular medicine, and probably a considerable amount of Teutonic and perhaps also Celtic folk-

[1] I, 63.
[2] II, 65
[3] III, 61.

lore, which, however, has been more or less Christianized. Indeed the connection between medicine and religion is remarkably close.

Cauterization.

The medicine of this period may be further illustrated by two Latin manuscripts of the eleventh century in the Sloane collection of the British Museum.[1] One contains a brief treatise which illustrates the common tendency at that time to employ cauterization not only for surgical purposes in connection with wounds, but as a medical means of giving relief to internal diseases and trivial complaints with which cauterization could have no connection. That the practice was very largely a superstition is further evident from the fact that one part of the body often was cauterized for a complaint in another or opposite portion or member. In the present example, under the alluring names of Apollonius and Galen as professed authors,[2] are presented a series of human figures showing where the cautery should be applied. These pictures of naked patients marked all over their anatomy with spots where the red-hot iron should be applied, or submitting with smiling or wry faces to its actual administration in the most tender places, are both amusing and, when we reflect that this useless pain was actually repeatedly inflicted through long centuries, pathetic.[3]

Treatment of demoniacs.

In a general and much longer work on diseases and their remedies which follows in the same manuscript and which is professedly compiled from Hippocrates, Galen, and Apollonius, the treatment prescribed for demoniacs,[4] who, it states, are in Greek called *epilemptici* (epileptics), includes among

[1] Sloane 475 (olim Fr. Bernard 116), 231 leaves, including two codices, one of the 12th century, which is also medical but with which we shall not deal at present, and the other of the 10th or 11th century and written in different hands. The MS is mutilated both at the beginning and the close.

Sloane 2839, 11th century, 112 leaves.

[2] Sloane 2839, fols. 1v-3, "Liber Cirrurgium Cauterium Apollonii et Galieni." James, *Western MSS in Trinity College*, Cambridge, III, 26-8, describes fifty drawings, chiefly of surgical operations, in MS 1044, early 13th century. By that date cauterization seems to have become less common.

[3] Professor T. W. Todd thinks that I am too severe upon the practice of cauterization, and that it may sometimes have served as a counter-irritant like mustard plasters and the blister.

[4] Sloane, 2839, fols. 79v-80v.

other things vaporization between the shoulder blades with various mixtures, scarification and bleeding, application of leeches to the "stomach where you ought not to operate with iron," [1] shaving and "imbrocating" [2] the scalp, and anointing the hands and feet with oil. Both our manuscripts contain recipes for expelling or routing demons.[3] For this purpose such substances are employed as the stone *gagates* and holy water, and elsewhere the usual confidence is reposed in the virtues of herbs and such parts of animals as the liver of a vulture.

Incantations and characters.

In one of the manuscripts is a treatise in which much use is made of incantations and characters. There are prayers to "Lord Jesus and Holy Mary" to heal the sick, while characters, sometimes engraved upon lead plates, are employed not only for medical purposes, but to prevent women from conceiving, to make fruit trees bear well, and against enemies.[4] Later on in the manuscript instructions for plucking a medicinal herb include facing east and reciting a paternoster.[5]

In a twelfth century manuscript.

The twelfth century portion of this same manuscript consists mainly of a long medical medley with no definitely marked beginning or ending but apparently originally in five books.[6] Towards its close occur a number of incantations and characters quite in the style of Marcellus Empiricus.[7] Indeed, "a marvelous charm" for toothache is an exact copy of his instructions to repeat seven times in a waning moon

[1] "Ad stomachum ubi ferro operare non oportes sansugias apponas."

[2] *Imbrocare.* I have not discovered exactly what it means.

[3] Sloane 475, fol. 224r; Sloane 2839, fol. 97r.

[4] Sloane 475, fol. 133, *et seq.*

[5] Sloane 475, fol. 224v.

[6] Sloane 475, fols. 1-124. At fol. 36r occurs the familiar pseudo-letter of Hippocrates to Antigonus; at fols. 8v-10r is a passage almost identical with that at the close of the *De medicamentis* of Marcellus, 1889, p. 382; an incantation from Marcellus is repeated at fol. 117v. At fol. 37r we read "Explicit Liber II. Incipit Liber Tertius ad ventris rigiditatem"; at fol. 60r, "Explicit liber tertius. Incipit Liber IIII"; at fol. 85r, "Incipit Liber V."

[7] See fol. 110r, "Cros, oros, comigeos, delig(c)ros, falicros, spolicros, splena mihi"; and fol. 114r, "Opas, nolipas, opium, nolimpium." Those who delight in ciphers will perhaps detect in the latter incantation a hidden allusion to opiates.

on Tuesday or Thursday an incantation beginning, "Aridam, margidam, sturgidam." [1] To make all his enemies fear him a man should gather the herb verbena on a Thursday, repeating seven times a formula in which the plant is personally addressed and the desire expressed to triumph over all foes as the verbena conquers winds and rains, hail and storms.[2] If here the influence of pagan religion is still present, many of the incantations are in Christian form and expressed in the name of God or the Father. To find a thief characters are employed together with the incantation, "Abraham bound, Isaac held, Jacob brought back to the house." [3] A charm against fever opens, "Christ was born and suffered; Christ Jesus rose from the dead and ascended unto heaven; Christ will come at the day of judgment. Christ says, According to your faith it shall be done." Then the sign of the cross is employed and "sacred words," which seem, however, to include not only Matthew, Mark, Luke, and John, but Maximianus, Dionysius, John, Serapion, and Constantinus. As we have to do with a twelfth century manuscript the last two names might be presumed to have reference to the medical writers of the eleventh century, but another manuscript which contains a similar incantation states that they are the names of the seven sleepers.[4] Our charm then continues "In the name of Christ" and with a prayer to God to free from sickness anyone who "bears this writing in Thy name." [5]

Magic with a split hazel rod.

In the same work occurs the earliest instance of which I am aware of the magical "experiment" with a split rod and an incantation, to which we shall hear William of Auvergne, Albertus Magnus, John of St. Amand, and Roger Bacon refer in the thirteenth century. A rod of four cubits length is to be cut with repetition of the Lord's Prayer. It is to be split, and the two halves are to be held apart at the

[1] Fol. 117v; see Marcellus (1889), p. 123, cap. 12.

[2] Fol. 111r.

[3] Fol. 111v.

[4] BN nouv. acq. 229, fol. 7v (once p. 246), "nomina septem sanctorum germanorum dormientium que sunt hec, Maximianus, Malchus, Martinianus, Constantinus, Dionisius, Iohannes, Serapion."

[5] Sloane 475, fol. 122v.

ends by two men. Then, making the sign of the cross, one should repeat the following incantation, "Ellum sat upon ella and held a green rod in his hand and said, Rod of green reunite again," [1] together with the Lord's Prayer until the two split halves bend together in the middle. One then seizes them in one's fist at the junction point, cuts off the rest of the rods, and makes magic use of the section remaining in one's grasp. [2]

More incantations and the virtues of a vulture.

Another manuscript of the twelfth century [3] contains many similar charms, incantations, prayers, and characters for healing purposes. One formula employed is, "Christ conquers, Christ reigns, Christ commands." In cases of miscarriage a drink of verbena is recommended and repetition of the following incantation with three Paternosters, "Saisa, laisa, relaisa, because so Saint Mary did when she bore the Son of God." Presently a paragraph opens with the assertion that the human race does not know how great virtue the vulture [4] possesses and how much it improves health. But certain ceremonial directions must be observed in making use of it. The bird should be killed in the very hour in which it is caught and with a sharp reed rather than a sword. Before beheading it, one should utter an incantation containing such names as Adonai and Abraam. Various healing virtues appertain to the different parts of its carcass, although here again there are instructions to be observed. The bones of its head should be bound in hyena skin; its eyes should be suspended from the neck in wolf's skin. Binding its wings on the left foot of a woman struggling in child-birth produces a quick delivery. One who wears its tongue will receive the adoration of all his enemies; if one has its heart bound in the skin of a lion or wolf, all demons will avoid one and robbers will only worship one. Its gall taken in quite a mixture cures epileptics

[1] "Ellum super ellam sedebat et virgam viridem in manu tenebat et dicebat, Virgam viridis reunitere in simul."

[2] Sloane 475, fol. 112v. Unintelligible letters follow.

[3] Egerton 821, 12th century, fols. 52v-60v.

[4] *Ibid.*, fol. 53v, *vultilis*, which I assume should be *vulturis* rather than *vituli*, or bull-calf.

and lunatics; its lung in another compound cures fevers; and so on.

Lots of the saints.

There follow *Sortes sanctorum,* introduced by a page and a half of prayers of this tenor, "In the name of our Lord Jesus Christ, we ask Father and Son and Holy Ghost, Three and One; we ask Saint Mary, the mother of our Lord Jesus Christ; we ask the nine orders of angels; we ask the whole chorus of patriarchs; we ask the whole chorus of apostles, martyrs, confessors, and virgins, and the whole chorus of God's faithful that they deign to reveal to us these lots which we seek, and that no seduction of the devil may deceive us." The treatise closes, "These are the lots of the saints which never fail; so ask God and obtain what you desire."

Superstitious veterinary and medical practice.

The next items in the manuscript are some cases of superstitious veterinary practice, with such pious incantations as "May God who saved the thief on the cross save this beast!" [1] and with instructions concerning the religious invocations and written characters to be employed in blessing the food and salt to be given to domestic animals in order to keep them in good health. Characters are also mentioned which will prevent the blood of a pig from flowing when it is slaughtered, provided they are bound upon the breast or are written on the knife with which the pig is to be stuck.[2] Holy water and bread that has been blessed are used for medical purposes and instructions are given on what days medicinal herbs should be gathered. The prayers employed are usually put in Christian form, but one for the cure of toothache has slipped by at least partially uncensored. It opens with the words "O lady Moon, free me. . . ." [3]

Two Paris manuscripts.

If we turn from medical manuscripts of the eleventh and twelfth centuries in the British Museum to those of the Bibliothèque Nationale, we find the same occurrence of superstitious passages. In an eleventh century codex which contains parts of the medical work of Celsus and the *De dina-*

[1] Egerton 821, fol. 57.
[2] *Ibid.,* fol. 58v.
[3] *Ibid.,* fol. 60r.

midis of Galen are also found prayers to God for the medicinal aid of the angel Raphael against the treacherous attacks of the demons, a work on the virtues of stones which has much to say of their marvelous properties, and figures and text concerning the twelve signs of the zodiac and twelve winds.[1] Much more superstitious, however, is an anonymous treatise occupying the first ten leaves of a twelfth century manuscript [2] which is apparently of German origin from the number of German words and phrases introduced near its close. This treatise is followed in the manuscript by the works of Notker, Hermann the Lame, and others on *computus* and the astrolabe.

Blood-letting.

After discussing the effect of food upon health, listing potions of herbs to be drunk in each month of the year,[3] treating of the veins and of the four winds, four seasons, and four humors, and the relations existing between the two last-named, the author enumerates the many advantages of blood-letting in a long passage which is worth quoting in part. "It contains the beginning of health, it makes the mind sincere, it aids the memory, it purges the brain, it reforms the bladder, it warms the marrow, it opens the hearing, it checks tears, it removes nausea, it benefits the stomach, it invites digestion, it evokes the voice, it builds up the sense, it moves the bowels, it enriches sleep, it removes anxiety, it nourishes good health . . ." : and so on. The operation of bleeding should not be performed on the tenth, fifteenth, twenty-fifth, or thirtieth day of the moon, nor should a potion be taken then. The Egyptian days and dog-days are to be similarly observed. The hours of the day when each humor predominates are then given.

[1] BN 7028, 11th century, fols. 136v, 140-3, 154r, and 156r.

[2] BN nouv. acq. 229, 12th century, fols. 1r-10r (once pp. 233-51), opening, "Rationem observationis vestre pietati secundum precepta doctorum medicinalium ut potui. . . ."

[3] BN nouv. acq. 229, fol. 2r. March is treated first and February last, while a similar discussion later in the same work (fols. 8r-9r, Quid unoquoque mense utendum quidve vitandum sit) begins with January.

Resemblances to Egerton 821.

There then is introduced rather abruptly an account of the medicinal virtues of the vulture almost identical with that in the British Museum manuscript. Once again, too, herbs are to be plucked with repetition of the Lord's Prayer.[1] The use of characters to prevent a slaughtered pig from bleeding is introduced somewhat otherwise than in the other manuscript. Having first recommended as a cure for human sufferers from flux of blood the binding about the abdomen of a parchment inscribed with the characters in question, the author adds, "And if you don't believe it, write them on a knife and kill a pig with it, and you will see no blood flow from the wound." [2]

Virtues of blood.

Considerable medicinal use is made of blood in this treatise. For cataract is recommended instilling in the eye the blood which flows from a certain worm (*oudehsam?*) when "you cut it in two near the tail." [3] To break the stone one employs goat's blood caught in a glass vessel in a waning moon and dried eight days in the sun together with the pulverized skin of a rabbit caught in a waning moon and roasted over marble. These are to be mixed in wine and given in the name of the Lord to the patient to drink while he is in the bath.[4] Another remedy consists of three drops of the milk of a woman nursing a male child given in a raw egg to the patient without his knowledge.[5]

Pious incantations and magical procedure.

The work abounds in characters and in incantations which consist either of seemingly meaningless words or of Biblical phrases and allusions. These are very much like those in the manuscripts already considered and are often accompanied by elaborate procedure. For example, the prayer, "O Lord, spare your servant N., so that chastised with deserved stripes he may rest in your mercy," is to be written on five holy wafers which are then to be placed on the five wounds of a figure of Christ on a crucifix. The patient is to approach barefoot, eat the wafers, and say:

[1] BN nouv. acq. 229, fol. 7

[2] Fol. 6r.

[3] Fol. 4v.

[4] Fols. 4v-5r.

[5] Fol. 7r.

"Almighty God, who saved all the human race, save me and free me from these fevers and from all my languors. By God Christ was announced, and Christ was born, and Christ was wrapped in swaddling clothes, and Christ was placed in a manger, and Christ was circumcised, and Christ was adored by the Magi, and Christ was baptized, and Christ was tempted, and Christ was betrayed, and Christ was flogged, and Christ was spat upon, and Christ was given gall and vinegar to drink, and Christ was pierced with a lance, and Christ was crucified, and Christ died, and Christ was buried, and Christ rose again, and Christ ascended unto heaven. In the name of the Father and of the Son and of the Holy Ghost, Jesus, rising from the synagogue, entered the house of Simon. Moreover, Simon's daughter was sick with a high fever. And they entreated Him on her behalf. And standing over her He commanded the fever and it departed." [1] To cure epilepsy an interesting combination of scriptural incantation and rather unusual magic procedure is recommended. Before the attack comes on, the words of the Gospel of Matthew, "Jesus was led by the spirit into the desert; and angels came and ministered unto Him," are to be written on a wooden tablet with some black substance which will wash off readily. Then, when the fit comes on, this writing is to be washed off into a vessel with still water and given to the patient to drink in the name of Father, Son, and Holy Ghost. "If you do this three times, God helping the patient will be cured." [2]

More superstitious veterinary practice.

Our manuscript further resembles Egerton 821 of the British Museum in containing remedies for beast as well as man. If a horse suffers from over-eating, one should learn his name and procure some hazel rods. Then one is to whisper in his right ear an incantation consisting of outlandish words accompanied by the Lord's Prayer, and is to bind his thighs and feet with the rods. This ceremony, too, is to be repeated thrice.[3]

[1] Fol. 7r-v.
[2] Fol. 7v.
[3] Fol. 9v.

The School of Salerno.

We now come to the consideration of treatises supposed to have been produced by the school of medicine at Salerno. But not only are the origins of the so-called School of Salerno "veiled in impenetrable obscurity," [1] much of its later history is scarcely less uncertain, and it is no easy matter to say what men and what writings may be properly called Salernitan, or when they lived or were composed. The manuscripts of Salernitan writings seem to have been found more frequently north of the Alps than in Italian libraries. It would perhaps be carrying scepticism too far to doubt if medicine developed much earlier or more rapidly at Salerno than elsewhere, since it seems certain that the town was famous for its physicians at an early date, and that we have medical writings of Salernitans produced in the early eleventh century. But one is inclined to view with some scepticism the assumption of historians of medicine[2] that the word Salernitan represents a separate body of doctrine, or of method in practice, which may be sharply distinguished from Arabic medicine or from later medieval medicine as affected by Arabic influence. Rather the medical literature and practice of Salerno is an integral and

[1] What is known of the School of Salerno has already been briefly indicated in English by H. Rashdall, *Universities of Europe in the Middle Ages,* 1895, I, 75-86, and T. Puschmann, *History of Medical Education,* English translation, London, 1891, pp. 197-211. The standard work on the subject is Salvatore De Renzi, *Collectio Salernitana,* in Italian with Latin texts, published at Naples in five volumes from 1852 to 1859. It contains a history of the School of Salerno by Renzi and various texts brought to light and dissertations discussing them by Renzi, Daremberg, Henschel, and others.

Unfortunately this publication proceeded by the unsystematic piecemeal and hand-to-mouth method, and new texts and discoveries were brought to the editor's attention during the process, so that the history of the school and the texts in the earlier volumes have to be supplemented and corrected by the fuller versions and dissertations in the later volumes. It is too bad that all the materials could not have been collected and more systematically arranged and collated before publication. Also some of the texts printed have but the remotest connection with Salerno, while others have nothing to do with medicine.

To this collection of materials some further additions have been made by P. Giacosa, *Magistri Salernitani nondum editi,* Turin, 1901.

For further bibliography see in the recent reprint of Harrington's English translation, *The School of Salerno* (1920), pp. 50-52.

[2] Notably Daremberg.

scarcely distinguishable part of medieval medicine as a whole. Many Salernitan treatises themselves belong to the later medieval period, and very few of them can be shown to antedate Constantinus Africanus, whose translations seem to mark the beginning of Arabic influence. And on the other hand there are equally early medieval medical treatises, such as those we have hitherto been considering, which are not Salernitan and yet show no sign of Arabic influence. Thus the word Salernitan cannot accurately be identified with a first period of medieval Latin medicine based upon early or Neo-Latin translations of Greek medical authors and upon independent medical practice. Such activity was not confined to Salerno. But if we so employ the word Salernitan for a moment, there seems no reason for thinking that such a development would be very different from the Arabic and Byzantine continuations of Greek medicine. A place so open to Saracen and Byzantine influence as the coast of southern Italy is hardly the spot where we should look for a totally distinct medical development, and the influence of Celtic and Teutonic folklore upon medical practice would presumably be more felt north of the Alps. And it is to Salerno that Constantinus Africanus, the earliest known importer of Arabic medicine, comes.

Was Salernitan medicine free from superstition?

The notion, too, that the Salernitan or early medieval Latin medical practice was sound and straightforward and sensible and free from the superstition with which the holders of this opinion represent Arabic and later medieval medicine as overburdened, is also probably illusory. We have already seen evidence of rather extreme superstition in early medieval Latin medicine which shows no trace of Arabic influence, and the medical practitioners of Salerno are sometimes represented in the sources as empiricists or old-wives. The place was peculiarly noted for its female practitioners, of whom more anon; and one of the earliest mentions of a physician of Salerno is the account in Richer's chronicle[1] of the mutual poisoning of two rival physicians in 946 A. D. Here the Salernitan is described as lacking in Latin book-knowledge and skilful from natural talent and much experience. He was the queen's favorite physician, but was worsted by another royal physician, Bishop Deroldus, in a debate which the king, Louis IV, instituted in order to find out "which of them knew more of the natures of things." The defeated Salernitan then "prepared sorcery" and tried to poison the bishop, who cured himself with theriac and secretly poisoned his rival in turn. The Salernitan was then reduced to the humiliating position of being forced to beseech the prelate to cure him, but in his case the theriac only drove the poison into his foot, which had to be amputated by a surgeon. This tale, be it true or not, suggests that there were good Latin physicians and surgeons outside of Salerno at an early date as well as that Salernitan medicine was far from being free from magic and empiricism.

The *Practica* of Petrocellus.

It is fairer, however, to judge Salerno by its own best written productions rather than by the stories of perhaps jealous northerners, and we may note Payne's comparison of the *Practica* of Petrocellus,[2] written probably in the early eleventh century, with the earlier *Leech-Book of Bald and Cild.* Selected recipes, it may first be said, were translated from the *Practica* into Anglo-Saxon.[3] Dr. Payne was impressed by "the complete freedom of the former from the magic and superstition which tainted the Anglo-Saxon and

[1] II, 59 (MG. SS. III, 600).

[2] S. de Renzi, *Collectio Salernitana,* IV, 185, *Practica Petroncelli,* perhaps from an imperfect copy; IV, 315, Sulle opere che vanno sotto il nome di Petroncello. Heeg, *Pseudodemocrit. Studien,* in *Abhandl. d. Berl. Akad.* (1913), p. 42, shows that what Renzi printed tentatively as the table of contents and an extract from the third book of the *Practica,* is not by Petrocellus but by the Pseudo-Democritus, and that one MS of it dates from the ninth or tenth century.

[3] Petrocellus, Περὶ διδάξεων, Eine Sammlung von Rezepten in englischer Sprache aus dem 11-12 Jahrhundert. Nach einer Handschrift des Britischen Museums herausg. v. M. Löweneck (in Anglo-Saxon and Latin), 1896, pp. viii, 57, Heft 12 in *Erlanger Beiträge z. englischen Philologie.* The treatise perhaps also contains selections from the *Passionarius* of Gariopontus. It had been published before in Cockayne, *Anglo-Saxon Leechdoms,* 1864-1866, III, 82-143.

all other European medicine of the time." Payne noted that the compounds of Petrocellus contained fewer ingredients, and regarded the Salernitan selection of drugs as "more intelligent." The Salernitan formulae are "clear, simple, and written on a uniform system which implies traditional skill and culture." [1] "The pharmacy is generally very simple; and, as might be expected, there is an entire absence of charms and superstitious rites." [2] Such simplicity, however, is at best a negative sort of virtue; and we wonder if this early specimen of the School of Salerno is free from elaborate superstition for the very reason that the work is simple and elementary. The less medicine, the less superstition perhaps. Moreover, superstition is not quite absent, since Payne himself quotes the following recipe: "For those who cannot see from sunrise to sunset. . . . This is the leechcraft which thereto belongeth. Take a kneecap of a buck [3] and roast it, and, when the roast sweats, then take the sweat and therewith smear the eyes, and after that let him eat the same roast; and then take fresh asses' dung and squeeze it, and smear the eyes therewith, and it will soon be better with them." [4]

Its sources.

Petrocellus is thought to have used Greek writings directly without the intermediary of Arabic versions.[5] He says in the introductory letter which opens the *Practica* that he reduces to brief form in the Latin language those "authors who have culled the dogmas of all cases from Greek places." [6] But these words might be taken to indicate that he has used Greek sources only indirectly, while the fact that the person to whom the work is addressed is called "dearest son" and "sweetest son" is rather in the style of Arabian and Hebrew medieval writers. He goes on to

[1] Payne (1904), pp. 155-6.

[2] *Ibid.*, p. 148.

[3] The Latin text reads, "liver of a hedgehog," and doubtless either would be equally efficacious.

[4] Quoted by Payne (1904), p. 152, from Cockayne's translation.

[5] Renzi (1852-9), IV, 185.

[6] Renzi, IV, 190, "Propterea fili karissime cum diuturno tempore de medicina tractassemus omnipotentis Dei nutu admonitus placuit ut ex grecis locis sectantes auctores omnium causarum dogmata in breviloquium latino sermone conscriberemus."

assure this person that everything in the work has been tested by experience and that nothing should be added to or subtracted from it.

Fourfold origin of medicine.

This introductory epistle also embodies an account of the origin of medicine which, while not exactly superstitious, is quite in the usual naïve and uncritical style so often employed by both ancient and medieval writers in treating of a distant past. Apollo and his son Esculapius, Asclepius and "Ypocras" are named as the four founders of the medical art. Apollo discovered *methoyca,* which presumably means methodism, but which Petrocellus proceeds to identify with surgery. Esculapius invented *empirica,* which is described as pharmacy rather than empiricism, although perhaps the distinction *is* slight. Asclepius founded *loyca,* which is probably meant for the dogmatic school. Hippocrates' contribution was *theoperica,* which may mean therapeutics but is further described as the prognostication or "prevision of diseases." It is in this same introductory epistle that Petrocellus makes the division of the brain into three cells of which we spoke in the chapter on Arabic occult science. Besides distinguishing the three cells as phantastic, logical, and mnemonic, he adds that good and evil are distinguished in the middle cell and that the soul is in the posterior one.

Therapeutics of Petrocellus.

In the *Practica* proper the method of Petrocellus is to take up one disease at a time, tell what the Greeks call it, and briefly describe it, sometimes listing its symptoms or causes, but devoting most of his space to such methods of curing it as diet and bleeding, simples and compounds. I saw no instance of astrological medicine nor of resort to amulets and incantations in the version published by Renzi from a twelfth century manuscript at Paris. But in a fragment of the work from a Milan manuscript where twenty-six lines are devoted to the treatment of epilepsy instead of but seven as in the other text,[1] one is advised to use antimony in the holy water "which the Greeks bless on

[1] For the two passages on epilepsy see Renzi, IV, pp. 235 and 293.

Epiphany" and to chant the Lord's Prayer three times. If this passage be a later addition, it shows that Petrocellus was less inclined to superstitious methods than others and that his injunction that nothing should be subtracted from or added to his work was not well observed. But in any case it illustrates my previous point that the more medicine, the more superstition. In twenty-six lines on epilepsy one is much more likely to find something superstitious than in seven. Indeed, the treatment of epilepsy was so generally superstitious that my recollection is that any account of it of any considerable length which I have seen in medieval writings contained some superstition. In fact, even if Petrocellus wrote the longer passage, he could be praised for having resorted to charms and formulae only in the case of that mysterious disease.

The *Regimen Salernitanum.*

The work most generally known as a characteristic product of the School of Salerno is the Latin poem [1] which opens with the line, "To the King of the English writes the whole School of Salerno." [2] This poem has been variously entitled *Schola Salernitana, Regimen Salernitanum,* and *Flos medicinae.* How much more influential and widespread it was than the *Practica* of Petrocellus may be seen from the fact that manuscripts of the text of the latter are rare, though the introductory letter is more common, and that it was first published by Renzi in the nineteenth century, whereas about one hundred manuscripts and two hundred and fifty printed editions of the poem have been found. It was known chiefly through the brief version of 362 verses, upon which Arnald of Villanova commented at the close of the thirteenth century, until as a result of the researches of Baudry de Balzac, Renzi, and Daremberg the number of lines was increased to 3526. This patchwork from many manuscripts can scarcely

[1] Renzi, I, 417-516, *Flos medicinae,* a text of 2130 lines; V, 1-104, the fuller text of 3526 lines; 113-72, Notice bibliographique; 385-406, Notes choisies de M. Baudry de Balzac au *Flos Sanitatis.*

[2] "Anglorum Regi scribit Schola tota Salerni." Some MSS have Francorum or Roberto instead of Anglorum.

be regarded as the work of any one author, time, or even school, and it may be seriously questioned how many of the verses really emanated from Salerno. Certainly it is not free from Arabic influence, since it cites Alfraganus as well as Ptolemy.[1] Pliny is used a great deal for the virtues of herbs. Much of it sounds like a late versification of commonplaces for mnemonic purposes. Sudhoff has recently pointed out that it was not generally known until the middle of the thirteenth century, before which time Frederick II, the cultured monarch, and Giles de Corbeil, the medical poet, appear unaware of its existence.[2]

Its superstition.

The brief version of the poem commented upon by Arnald of Villanova naturally contains only one-tenth of the superstition found in the fuller text which is ten times longer. In some respects this brief version might pass as a restrained, though quaint, early set of directions how to preserve health, to which later writers have added superstitious recipes. But as a matter of fact it is too superstitious for even one as hospitable to theories of occult influence as Arnald, who rejects as false and worthless [3] its assertion that the months of April, May, and September are lunar and that in them consequently fall the days upon which bleeding is prohibited. In the lines upon which Arnald comments marvelous properties are mentioned in the case of the plant rue, but the fuller text has many mentions of the occult virtues of herbs, stones, and animals. Almost at a glance we read that the urine of a dog or the blood of a mouse cures warts; that juice of betony should be gathered on the eve of St. John the Baptist, that rubbing the soles of the feet cures a stiff neck, and that pearls or the stone found in a crab's head are of equal virtue for heart trouble.[4] And not far away is a passage [5] on the virtue of the *Agnus Dei*,

[1] Lines 2692-3.

[2] K. Sudhoff, *Zum Regimen Sanitatis Salernitanum*, in *Archiv f. Gesch. d. Medizin*, VII (1914), 360, and IX (1915-1916), 1-9.

[3] Arnald de Villanova, *Opera*, Lyons, 1532, fol. 147v.

[4] Lines 1918-9, 1932-3, 1973-4, 1985, in Renzi's first text of 2130 lines; in the fuller version they are somewhat more widely separated: lines 3053, 3130, 3227, 3267.

[5] Lines 1845-55 or 2873-83.

made of balsam, pure wax, and the Chrism. It protects against lightning and the waves of the sea, aids women in child-birth, saves from sudden death, and in short from "every kind of evil." Astrology is by no means omitted from the *Regimen Salernitanum;* in fact Balzac seems to have taken the fact that verses were astrological in character as a sign that they belonged in the Salernitan collection.

The *Practica* of Archimatthaeus.

A third work which may be considered as an example of the medicine of Salerno is the *Practica* of Archimatthaeus which Renzi placed in the twelfth century and conjectured to be the work of Matthaeus Platearius the Elder.[1] One or two expressions, however, might be taken as indications that the writer is neither of early date nor himself a Salernitan. He speaks of curing pleurisy in a different way from the treatment recommended in the *Practica's* and tells how the Salernitans try to prevent their hair from falling out by reason of their pores opening too wide when they frequent the bath.[2] Renzi hailed this treatise with delight as "a true medical clinic," [3] since the author describes some twenty-two specific cases. He states at the beginning that he does not propose to write a systematic treatise or to deal with every variety of disease, but only with those in which he has learned new and better methods by experience, "and in which God has put the desired effect in my hand." [4] Through the work we encounter such phrases as *expertum est, aliud probatissimum,* "I tell you what I have proved," "We have tested this by experience and rejoiced at the result." These utterances seem really to refer to the writer's own experience and not to be copied from previous authors. The following is an example of his cases. "A certain lady incurred paralysis of the face during sleep after the bath," which he attributes to dissolution of humors which affected the muscles. First he bled the cephalic vein, hoping thereby to draw off somewhat the humors from the afflicted place.

[1] Renzi, V, 377-8.
[2] *Ibid.*, 372-3.
[3] *Ibid.*, 379-81.
[4] *Ibid.*, 350.

Then for three successive days he gave her "the potion of St. Paul with wine of a decoction of salvia and castoria which in part prevent dissolution, in part consume it." He also had her hold that wine in her mouth for a long time before swallowing it. At length he gave her a purgative with pills of yerapiga (*sacrum amarum*), mixed with golden pills. "Afterwards we injected pills of diacastoria into her nostrils and placed her near the fire. Finally we gave *opopira* (bread free from furfure) with the aforesaid wine, and so she was cured, only a certain tumor remained in her face and made her eye water. We anointed her face with golden unguent and the potion of St. Paul mixed together and the tumor disappeared; for the tears we gave golden Alexandrina and they were checked; and thus it was that this year in your presence we cured a certain paralytic." [1] Like Galen's accounts of his actual cases this makes us realize that all the gruesome mixtures of which we read in the books were actually forced upon patients, often several of them upon one poor sick person, and that medical practice was rather worse than medical theory. An interesting observation concerning the lot of the lower classes is let fall by our author when, in discussing involuntary emission of urine, he states that serfs and handmaids are especially subject to this ailment, since they go about ill-clad and with bare feet and become thoroughly chilled.[2]

A Salernitan treatise of about 1200.

Giacosa classed one of the treatises which he published as Salernitan because it was written in a Lombard or Monte Cassino hand of about 1200.[3] He described its contents as purely therapeutical and regarded its author as showing "a certain repugnance" to the popular remedies and superstitions recommended by other contemporary treatises. For

[1] Professor T. Wingate Todd comments upon this passage: "Of course this is *post hoc propter hoc*, but it is the typical history of a case of Bell's palsy occurring after a 'chill.'"

[2] Renzi, V, 371, "Involuntariam urine emissionem quidam patiebantur et adhuc multi patiuntur et maxime servi et ancille qui male induti et discalciati incedunt, unde frigiditate incensa vesica fit quasi paralitica cum urinam nequeat continere."

[3] Giacosa (1901), pp. 71-166.

this conclusion the chief evidence seems to be a passage where the author, after listing such means to prevent a woman from conceiving as binding her head with a red ribbon or holding the stone found in the head of an ass, says that he thinks that such remedies "operate more by faith than reason." [1] But he makes much use of parts of animals and of suffumigations, advising for example on the same page that after conception there should be fumigation with a root of mandragora or peony or the excrement of an ass mixed with flour, an operation which he characterizes as *expertissimum*. And on the preceding page, as Giacosa has noted, he recommends a procedure which is even more improbable than it is immoral, whereby patients who show themselves ungrateful to the physician after they have been cured may be made to suffer again.[2]

The wives of Salerno.

We promised to say something of the female practitioners of Salerno. Trotula is no longer believed to be a woman and we have to judge the women of Salerno mainly by what others say of them. In a commentary of a Master Bernard of Provence, who I suspect may be Bernard Gordon, the medical writer at Montpellier of the closing thirteenth century, are a number of practices attributed to the women of Salerno which Renzi has already brought together.[3] In these cases the practices are chiefly those employed by the women themselves in child-birth. We may note three from the list that savor strongly of magic. "The women of Salerno cook doves with the acorns which the doves eat; then they remove the acorns from the gizzard and eat them, whence the retentive virtue is much comforted." "When the women of Salerno fear abortion, they carry with them the pregnant stone," which our author explains is not the magnet. The other recipe had perhaps better remain untranslated: *Stercus asini comedunt mulieres Salernitanae in crispellis et dant viris suis ut melius retineant sperma et sic concipiant.* As we shall see in our chapter on

[1] Giacosa (1901), p. 146.
[2] *Ibid.*, p. 145.
[3] Renzi, V, 331-2.

Arnald of Villanova, another medical writer of the late thirteenth and early fourteenth century, he condemned the use of incantations in cases of child-birth by old-wives of Salerno but approved of a very similar procedure by which a priest had cured him of warts, and also mentioned favorably the cures wrought by female practitioners at Rome and Montpellier.

CHAPTER XXXII

CONSTANTINUS AFRICANUS: C. 1015-1087.

Reputation and influence—His studies in the Orient—His later life in Italy—His works were mainly translations—*Pantegni*—*Viaticum*—Other translations—*The book of degrees*—*On melancholy*—*On disorders of the stomach*—Medical works ascribed to Alfanus—Constantinus and experiment—"Experiments" involving incantations—Superstition comparatively rare in Constantinus—And of Greek rather than Arabic origin—Some signs of astrology and alchemy—Constantinus and the School of Salerno—*Liber aureus* and John Afflacius—Afflacius more superstitious than his master.

Reputation and influence.

CONSTANTINUS AFRICANUS will be here considered at perhaps greater length than his connection with the history either of magic or experimental science requires, but which his general importance in the history of medicine and the lack of any good treatment of him in English may justify.[1]

[1] Many of the works listed by Peter the Deacon and some others which he does not name have been printed under Constantinus' name, either in the edition of the works of Isaac issued at Lyons in 1515, or in the partial edition of the works of Constantinus printed at Basel in 1536 and 1539, or in an edition of Albucasis published at Basel in 1541.

An early MS containing several of Constantinus' works is Gonville and Caius 411, 12-13th century, fol. 1-, Viaticum, 69- de melancholia, 77v- de stomacho, 98v- de oblivione, 100r- de coitu, (no author is named for 109v- liber elefantie, 113- de modo medendi), 121- liber febrium, (169- de inamidarium Galieni).

The chief secondary investigations concerning Constantinus Africanus are:

Daremberg, *Notices et Extraits des Manuscrits Médicaux,* 1853, pp. 63-100, "Recherches sur un ouvrage qui a pour titre Zad el-Monçafir en arabe, Ephrodes en grec, Viatique en latin, et qui est attribué dans les textes arabes et grecs à Abou Djafar, et dans le texte latin à Constantin."

Puccinotti, *Storia della Medicina,* II, i, pp. 292-350, 1855, devoted several chapters to Constantinus and tried to defend him from the charge of plagiarism and to maintain that the *Viaticum* and some other works were original.

Steinschneider, *Constantinus Africanus und seine arabischen Quellen,* in Virchow's *Archiv für Pathologische Anatomie,* etc., Berlin, 1866, vol. 37, pp. 351-410. This should be supplemented by pp. 9-12 of his *Die europäischen Übersetzungen aus dem Arabischen* (1905).

Our discussion of him as an importer of Arabic medicine will also serve to support our attitude towards the School of Salerno. Daremberg wrote in 1853, "We owe a great debt of gratitude to Constantinus because he thus opened for Latin lands the treasures of the east and consequently those of Greece. He has received and he deserves from every point of view the title of restorer of medical literature in the west." [1] Daremberg proceeded to propose that a statue of Constantinus be erected in the center of the Gulf of Salerno or on the summit of Monte Cassino. Yet in 1870 he made the surprising assertion that "the voice of Constantinus towards the close of the eleventh century is an isolated voice and almost without an echo." [2] But as a matter of fact Constantinus was a much cited authority during the twelfth and thirteenth centuries in the works both of medicine and of natural science produced in Latin in western Europe, and his translations were cited under his own name rather than those of their original authors.[3]

His studies in the Orient.

A brief sketch of Constantinus' career and a list of his works[4] is twice supplied us by Peter the Deacon, who wrote in the next century,[5] and who treats of Constantinus both in the chronicle of Monte Cassino, which he continued to the year 1138,[6] and in his work on the illustrious men of Monte Cassino.[7] Peter tells that Constantinus was born

[1] *Notices et Extraits des Manuscrits Médicaux* (1853), p. 86.

[2] *Histoire des Sciences Médicales* (1870), I, 261.

[3] Indeed Daremberg said in 1853 (p. 85, note) "dans le moyen âge beaucoup d'auteurs citent volontiers Constantine comme une autorité."

[4] Perhaps through the fault of the printer the list of the writings of Constantinus given by Peter the Deacon is defective as reproduced in tabular form by Steinschneider (1866), pp. 353-4. Steinschneider also incorrectly speaks of Leo of Ostia as well as Peter the Deacon as a source for Constantinus (p. 352, "Die Schriften Constantins sind bekanntlich von seinen alten Biographen, Petrus Diaconus und Leo Ostiensis verzeichnet worden"), since Leo's portion of the *Chronicle* ends before Constantinus is mentioned.

[5] Peter was born about 1107 and was placed in the monastery of Monte Cassino by his parents in 1115. He became librarian. *Monumenta Germaniae, Scriptores,* VII, 562 and 565.

[6] *Chronica Mon. Casinensis,* Lib. III, auctore Petro, MG. SS. VII, 728-9; Muratori, *Scriptores,* IV, 455-6 (lib. III, cap. 35).

[7] *Petri Diaconi De viribus illustribus Casinensibus,* cap. 23, in Fabricius, *Bibl. Graec.,* XIII, 123.

at Carthage, by which he probably means Tunis, since Carthage was no longer in existence, but went to Babylon, by which Cairo is presumably designated, since Babylon had ages before been reduced to a dust heap,[1] to improve his education. His birth must have been in about 1015. There he is said to have studied grammar, dialectic, geometry, arithmetic, "mathematics," astronomy, and physics or medicine (*physica*). To this curriculum in the *Chronicle* Peter adds in the *Lives of Illustrious Men* the subjects of music and necromancy. When so little was said of spirits in the occult science of the Arabic authors of the ninth century whom we considered in an earlier chapter, it is rather a surprise to hear that Constantinus studied necromancy, but that subject is listed along with mathematical and natural sciences by Al-Farabi in his *De ortu scientiarum*,[2] and we shall find this classification reproduced by two western Christian scholars of the twelfth century.[3] The *mathematica* and astronomy which Constantinus studied very likely also included considerable astrology and divination. At any rate we are told that he not only pursued his studies among "the Chaldeans, Arabs, Persians, and Saracens," and was fully imbued with "all the arts of the Egyptians," but even, like Apollonius of Tyana, visited India and Ethiopia in his quest for learning. It was only after a lapse of thirty-nine or forty years that he returned to North Africa. Most modern secondary accounts here state that Constantinus was soon forced to flee from North Africa because of the jealousy of other physicians who accused him of magic,[4] or from fear that his fellow citizens would kill him as a wizard.

[1] Yet modern compilers and writers of encyclopedia articles invariably repeat "Carthage" and "Babylon."

[2] BN 14700, fol. 171v, cited by Baur (1903), who also notes parallel passages in Al-Gazel, *Phil. tr.* I, 1; and Avicenna, *De divis. philos.*, fol. 141.

[3] Gundissalinus and Daniel Morley. Al-Farabi's list of eight mathematical sciences, including "the science of spirits," was also reproduced by Vincent of Beauvais in the thirteenth century, *Speculum doctrinale*, XVI.

[4] Possibly there is some confusion with Galen's similar experience with the physicians of Rome, which Constantinus may have reproduced in some one of his translations of Galen in such a way as to lead the reader to consider it his own experience.

In view of his study of necromancy, this may well have been the case. Peter the Deacon, however, simply states that when the Africans saw him so fully instructed in the studies of all nations, they plotted to kill him,[1] and gives no further indication of their motives.

Constantinus secretly boarded ship and made his escape to Salerno, where he lived for some time in poverty, until a brother of the caliph (*regis Babiloniorum*) who chanced to come there recognized him, after which he was held in great honor by Duke Robert Guiscard. The secondary accounts say that he became Robert's confidential secretary and that he had previously occupied a similar position under the Byzantine emperor, Constantine Monomachos,[2] but of these matters again Peter the Deacon is silent. When Constantinus left the Norman court, it was to become a monk at Monte Cassino, where he remained until his death in 1087.

His later life in Italy.

[1] The words are the same both in the *Chronicle* and *Illustrious Men:* "quem cum vidissent Afri ita ad plenum omnibus (omnium?) gentium eruditum, cogitaverunt occidere eum."

[2] Pagel (1902), p. 644, "Vorher soll er kurze Zeit noch in Reggio, einer kleinen Stadt in der Nähe von Byzanz, als Protosekretär des Kaisers Constantinos Monomachos sich aufgehalten und das Reisehandbuch des Abu Dschafer übersetzt haben." But Pagel gives no source for this statement.

Apparently the notion is due to the fact that a Greek treatise entitled *Ephodia,* of which there are numerous MSS and which seems to be a translation of the same Arabic work as that upon which Constantinus based his *Viaticum,* speaks of a Constantine as its author who was protosecretary and lived at Reggio or Rhegium.

Daremberg (1853), p. 77, held that a Vatican MS of the *Ephodia* was of the tenth century and therefore this Greek translation could not be the work of Constantinus Africanus in the next century, but Steinschneider (1866), p. 392, only says, "Die griechische Uebersetzung des Viaticum soll bis in die Zeit Constantins hinaufreichen."

Another MS, Escorial &-II-9, 16th century, fol. 1-, contains a "Commeatus Peregrinantium" whose author is called "Ebrubat Zafar filio Elbazar," which perhaps designates Abu Jafar Ahmed Ibn-al-Jezzar, whom Daremberg and Steinschneider call the author of the Arabic original of the *Viaticum.* The work is said to have been translated into Greek "a Constantino Primo a secretis Regis," which suggests that Constantinus was perhaps first of the royal secretaries rather than of Reggio either in Norman Italy or near Byzantium. The translation from Greek into Latin is ascribed to Antonius Eparchus. The opening sentences of each book of this Latin version from the Greek by Eparchus differ in wording but agree in substance with those of the *Viaticum* of Constantinus Africanus, if we omit some transitional sentences in the latter.

In a work addressed to the archbishop of Salerno he speaks of himself as *Constantinus Africanus Cassinensis*[1] and Albertus Magnus cites him as *Constantinus Cassianensis.*[2] What purports to be a picture of Constantinus is preserved in a manuscript of the fifteenth century at Oxford.[3]

His works were mainly translations.

Peter the Deacon states both in the *Chronicle* and in the *Illustrious Men* that while at the monastery of Monte Cassino Constantinus Africanus "translated a great number of books from the languages of various peoples." Peter then lists the chief of these. It is interesting to note, in view of the fact that Constantinus in prefaces and introductions appears to claim some of the works as his own, and that he was accused of fraud and plagiarism by medieval writers who followed him as well as by modern investigators, that Peter the Deacon speaks of *all* his writings as translations from other languages. Peter does not, however, give us much information as to who the Greek or Arabic authorities were whom Constantine translated. It may be added that if Constantinus claimed for himself the credit for Latin versions which were essentially translations, he was merely continuing a practice of which Arabic authors themselves had been repeatedly guilty. Indeed, we are told that they sometimes even destroyed earlier works which they had copied in order to receive sole credit for ideas which were not their own.[4]

Pantegni.

The longest of Constantinus' translations and the one most often cited in the middle ages was the *Pantechni* or *Pantegni,* comprising ten books of theory and ten of prac-

[1] *Opera* (1536), p. 215.

[2] *De animalibus,* XXII, i, 1.

[3] Rawlinson C, 328, fol. 3. It is accompanied by the legend, "This is Constantinus, monk of Monte Cassino, who is as it were the fount of that science of long standing from the judgment of urines, and it has exhibited a true cure in all the diseases in this book and in many other books. To whom come women with urine that he may tell them what is the cause of the disease." The illumination shows Constantinus seated, holding a book on his knees with his left hand, while he raises his right hand and forefinger in didactic style. He wears the tonsure, has a beard but no mustache, and seems to be approached by one woman and two men carrying two jars of urine.

[4] See Margoliouth, *Avicenna,* 1913, p. 49.

tice as printed in 1515 with the works of Isaac,[1] although Peter the Deacon speaks of Constantinus' dividing the *Pantegni* into twelve books and then of a *Practica* which also consisted of twelve books. What is the ninth book of the *Practica* in this printed version is listed as a separate book on surgery by Peter in his *Illustrious Men,* although omitted from his list in the *Chronicle,* and was so printed in the 1536 edition of the works of Constantinus.[2] And the *Antidotarium* which Peter lists as a separate title is probably simply the tenth book of the *Practica* as printed with the works of Isaac.[3] The *Pantegni,* however, is not a translation of any work by Isaac, but an adaptation of the *Khitaab el Maleki,* or Royal Art of Medicine, of Ali Ibn Abbas. The preface of Constantinus [4] says nothing of Ali but tells the abbot Desiderius that, failing to find in the many works of the Latins or even in "our own writers, ancient and modern," such as Hippocrates, Galen, Oribasius, Paulus, and Alexander, exactly the sort of treatise desired, he has composed "this little work of our own" (*hoc nostrum opusculum*). But Stephen of Pisa, who also translated Ali into Latin in 1127,[5] accused Constantinus of having suppressed both the author's name and title of the book and of having made many omissions and changes of order both in preface and text but without really adding any new contributions of his own.[6] Stephen further justified his own translation by asserting that not only had the first part of *The Royal Art of Medicine* of Ali Ibn Abbas been "corrupted by the shrewd fraud of its translator," but also that the last and greater portion was missing in the version by

[1] Only the ten books of theory are printed in the 1539 edition of Constantinus.

[2] *Chirurgia,* at pp. 324-41.

[3] *Opera omnia ysaac* (1515), fol. 126v, "Liber decimus practice qui antidotarium dicitur in duas divisus partes."

Isaac Israeli is the subject of the first chapter in Husik (1916), who calls him (p. 2) "the first Jew, so far as we know, to devote himself to philosophical and scientific discussions."

[4] Daremberg (1853), pp. 82-5, gives the prefaces of Ali and Constantinus in parallel columns.

[5] Printed in 1492 with the works of Ali ben Abbas; Stephen's translation was made at Antioch in Syria.

[6] Steinschneider (1866), p. 359.

Constantinus.[1] Also Ferrarius said in his gloss to the *Universal Diets* of Isaac that Constantinus had completed the translation of only three books of the *Practica,* losing the rest in a shipwreck.[2] A third medieval writer, Giraldus Bituricensis, adds [3] that Constantinus substituted in its place the *Liber simplicis medicinae* and *Liber graduum,* and that it was Stephen of Pisa who translated the remainder of the work of Ali ben Abbas which is called the *Practica Pantegni et Stephanonis.* Stephen's translation is indeed different from the ten books of the *Practica* printed with the works of Isaac. From these facts and from an examination of the manuscripts of the *Practica* Rose concluded [4] that Constantinus wrote only its first two books [5] and the first part of the ninth, which is roughly the same as the *Surgery* published separately among Constantinus' works. The rest of this ninth book was translated into Latin at the time of the expedition to besiege Majorca, that is, in 1114-1115, by a John [6] who had recently been converted to Christianity [7] and whom Rose was inclined to identify with John Afflacius, "a disciple of Constantinus," of whom we shall have more

[1] "Ultimam et maiorem deesse sensi partem, alteram vero interpretis callida depravatam fraude."

[2] Amplon. Octavo 62.

[3] In his gloss to the *Viaticum* of Constantinus.

[4] *Berlin HSS Verzeichnis* (1905), pp. 1059-65, to whom I owe the preceding references to Ferrarius and Giraldus.

[5] Rose cites Bamberg L-iii-9. The two following MSS are perhaps also worth noting: The *Pantegni* as contained in CU Trinity 906, 12th century, finely written, fols. 1-141v, comprises only ten books. The first opens, "Cum totius generalitas tres principales partes habeat"; the tenth ends, "Unde acutum oportet habere sensum ad intelligendum. Explicit."

St. John's 85, close of 13th century, "Constantini africani Pantegnus in duas partes divisus quarum prima dicitur Theorica continens decem libros secunda dicitur Practica 33 capita continens," as a table of contents written in on the fly-leaf states. The ten books of theory end at fol. 100r, "Explicit prima pars pantegni scilicet de theorica. Incipit secunda pars scilicet practica et est primus liber de regimento sanitatis." This single book in 33 chapters on the preservation of health ends at fol. 116v, and at fol. 117r begins the *Liber divisionum* of Rasis.

[6] In Berlin 898, a 12th century MS of Stephen's translation of Ali's *Practica,* this ninth section by Constantinus and John is for some reason substituted for the corresponding book of Stephen.

[7] He calls himself, "iohannes quidam agarenus (Saracenus?) quondam, qui noviter ad fidem christiane religionis venerat cum rustico pisano belle filius ac professione medicus."

to say presently. Rose further held that this John completed the *Practica* [1] commonly ascribed to Constantinus with the exception of its tenth book which, as we have suggested, seems originally to have been a distinct *Antidotarium*. Different from the *Pantegni* is the *Compendium megategni Galeni* by Constantinus published with the works of Isaac, and the *Librum Tegni, Megategni, Microtegni* listed by Peter the Deacon.

Viaticum.

Perhaps the next best known and the most frequently printed [2] of Constantinus' translations or adaptations from the Arabic is his *Viaticum* which, as Peter the Deacon states, is divided into seven books. In the preface Constantinus states that the *Pantegni* was for more advanced students, this is a brief manual for others. He also adds that he appends his own name to it because there are persons who profit by the labors of others and, "when the work of someone else has come into their hands, furtively and like thieves inscribe their own names." Daremberg designated Abu Jafar Ahmed Ibn-al-Jezzar as author of the Arabic original of the *Viaticum*. Moses Ibn Tibbon, who made a Hebrew translation in 1259, criticized the Latin version of Constantinus as often abbreviated, obscure, and seriously altered in arrangement.[3] Constantinus seems to be alluded to in the *Ephodia* or Greek version of the same work.[4]

[1] The main objection to this theory is that Stephen of Pisa, translating in 1127, speaks as if the latter portion of Ali's work was still untranslated. Rose therefore holds that John had not yet published his translation, although we have seen that he completed the surgical section by 1115.

[2] In *Opera omnia ysaac*, Lyons, 1515, II, fols. 144-72, "Viaticum ysaac quod constantinus sibi attribuit"; in the Basel, 1536, edition of the works of Constantinus, pp. 1-167, under the title, "De morborum cognitione et curatione lib. vii"; in the Venice, 1505, edition of Gerardus de Solo (Bituricensis), "Commentum eiusdem super viatico cum textu"; and in the Lyons, 1511, edition of Rhazes, *Opera parva Albubetri*.

A fairly early but imperfect MS is CU Trinity 1064, 12-13th century.

Laud. Misc. 567, late 12th century, fol. 2, recognizes in its Titulus that the *Viaticum* is a translation, "Incipit Viaticum a Constantino in Latinam linguam translatam."

[3] Steinschneider (1866), 368-9.

[4] See above, page 745, note 2.

Other translations.

If neither the original of the *Pantegni* nor of the *Viaticum* is to be assigned to Isaac, Constantinus nevertheless did translate some of his works, namely, those on diets, urines, and fevers.[1] Moreover, Constantinus himself admits that these Latin works are translations, stating in the preface to the treatise on urines that, finding no satisfactory treatment of the subject in Latin, he turned to the Arabic language and translated the work which Isaac had compiled from the ancients. Constantinus also states that he translated the treatise on fevers from the Arabic. We have already seen that the alphabetical Latin version of Dioscorides which had most currency in the middle ages is ascribed in at least one manuscript to Constantinus. He also translated some treatises ascribed to Hippocrates and Galen, such as Galen's commentary on the *Aphorisms* and *Prognostics* of Hippocrates [2] and the *Tegni* of Galen. Constantinus has also been credited with translating works of Galen on the eyes, on diseases of women, and on human nature, but these are not genuine works of Galen.

The book of degrees.

In his list of the works which Constantinus translated from various languages.[3] Peter the Deacon includes *The book of degrees,* but it has not yet been discovered from what earlier author, if any, it is copied or adapted. The work is a development of Galen's doctrine that various

[1] In the 1515 edition of Isaac's works, I, 11-, 156-, and 203-. Peter the Deacon presumably refers to these three works in speaking of "Dietam ciborum. Librum febrium quem de Arabica lingua transtulit. Librum de urinis." Whether the two initial treatises in the 1515 edition of Isaac, dealing with definitions and the elements, were translated by Constantinus or by Gerard of Cremona is doubtful.

[2] See CLM 187, fol. 8; 168, fol. 23; 161, fol. 41; 270, fol. 10; 13034, fol. 49, for 13-14th century copies of Galen's commentary upon the *Aphorisms* of Hippocrates with a preface by Constantinus. University College Oxford 89, early 14th century, fol. 90, Incipiunt amphorismi Ypocratis cum commento domini Constantini Affricani montis Cassienensis monachi; fol. 155, Eiusdem Prognostica cum Galeni commento, eodem interprete; fols. 203-61, Eiusdem liber de regimine acutorum cum eiusdem commento eodem interprete.

[3] *De viris illustribus,* cap. 23, ". . . transtulit de diversis gentium linguis libros quamplurimos in quibus praecipue . . .": *Chronica,* Lib. III, ". . . transtulit de diversorum gentium linguis libros quamplurimos in quibus sunt hi praecipue. . . ."

medicinal simples are hot or cold, dry or moist, in varying degrees. Constantinus presupposes four gradations of this sort. Thus a food or medicine is hot in the first degree if its heating power is below that of the normal human body; if it is of the same temperature as the body, it ranks as of the second degree; if its heat is somewhat greater than that of the body, it is of the third degree; if its heat is extreme and unbearable, it is of the fourth degree. The rose is cold in the first degree, is dry towards the end of the second degree, while the violet is cold towards the end of the first degree and moist in the beginning of the second degree. Thus Constantinus distinguishes not only four degrees but a beginning, middle and end of each degree, and Peter the Deacon once gives the title of the work as *The book of twelve degrees.*[1] This interesting though crude beginning in the direction of scientific thermometry and hydrometry unfortunately rested upon incorrect assumptions as to the nature and causation of heat and moisture, and so was perhaps destined to do more harm than good.

On melancholy.

A glossary of herbs and species and a work on the pulse, which Peter the Deacon includes in both his lists of Constantinus' works or translations, do not seem to have been printed or identified as Constantinus'. On the other hand, the printed edition of the works of Constantinus includes treatises on melancholy and on the stomach [2] which are not mentioned in Peter's list. In a preface to the *De melancholia* which is not included in the printed edition [3] Constantinus Africanus speaks of himself as a monk of Monte Cassino and states that, while he has often touched on the disease of melancholy in the many medical books which he has added to the Latin language, he has decided also to write a separate brochure on the subject because it is an important malady and because it is especially prevalent "in these regions." "Therefore I have collected this booklet from

[1] "Librum duodecim graduum" in *De viris illus.*: in the *Chronicle*, "Liber graduum."

[2] Edition of Basel, 1536, at pp. 280-98 and 215-74 respectively.

[3] It is found in Laud. Misc. 567, late 12th century, fol. 51v.

many volumes of our adepts in this art." Whether the word "our" here refers to Greek or Arabic writers would be hard to say. Constantinus states that melancholy is a disease to which those are especially liable who are always intent on study and books of philosophy, "because of their scientific investigations and tiring their memories and grieving over the failure of their minds." This ailment also afflicts "those who lose their beloved possessions, such as their children and dearest friends or some precious thing which cannot be restored, as when scholars suddenly lose their books." Constantinus also describes the melancholy of "many religious persons who live lives to be revered, but fall into this disease from their fear of God and contemplation of the last judgment and desire of seeing the *summum bonum.* Such persons think of nothing and seek for nothing save to love and fear God alone, and they incur this complaint and become drunk as it were with their excessive anxiety and vanity." [1] Such passages would seem to describe Constantinus' own associates and environment, but they may possibly be a mere translation of some work of an earlier Christian Arab, such as Honein ben Ishak who translated or pretended to translate a number of works of Greek medicine into Arabic. In a later chapter [2] we shall find that Honein perhaps had something to do with another work called *The Secrets of Galen,* in which remedies for religious ascetics who have ruined their health by their austerities form a rather prominent feature.

On disorders of the stomach.

That the treatise on disorders of the stomach is Constantinus' own work is indicated by its preface, which is addressed to Alfanus, archbishop of Salerno from 1058 to 1087 and earlier a monk of Monte Cassino. Alfanus had himself translated Nemesius Περὶ φύσεως ἀνθρώπου [3] and was the center of a group of learned writers: the dialectician, Alberic the Deacon, the historian, Amatus of Salerno, and

[1] Edition of 1536, pp. 283-4.

[2] See below, Chapter 64.

[3] *Zeitsch. f. klass. Philol.* (1896), pp. 1098ff.

the mathematician and astronomer, Pandulf of Capua.[1] Constantinus states that he writes this treatise for Alfanus as a compensation for his recent failure to relieve a stomach-ache with which that prelate was afflicted. Such instances of self-confessed failure, be it noted in passing, are rare indeed in ancient and medieval medicine, and for this reason we are the more inclined to deal charitably with the charges of literary plagiarism which have been preferred against Constantinus. He goes on to say that he has sought with great care but in vain among ancient writings for any treatise devoted exclusively to the stomach, and has only succeeded in finding here and there scattered discussions which he now presumably combines in the present special treatise.

Medical works ascribed to Alfanus.

This archbishop Alfanus appears to have written on medicine himself, since *A treatise of Alfanus of Salerno concerning certain medical questions* was listed among the books at Christchurch, Canterbury about 1300.[2] Also a collection of recipes entitled, *Experiments of an archbishop of Salerno,* in a manuscript of the early twelfth century are very likely by him.[3] They follow a treatise on melancholy which does not, however, appear to be that of Constantinus Africanus.[4]

Constantinus and experiment.

Peter the Deacon's bibliography of the works of Constantinus includes a *De experimentis* which, if extant, has not been identified as Constantinus'. In such works of his as are available, however, we find a number of mentions of experience and its value. It is of course to be remembered that such expressions as "we state what we have tested and what our authorities have used," [5] and "we have had personal experience of the confection which we now mention," [6] may refer to the experience of the past authors

[1] J. A. Endres, *Petrus Damiani und die weltliche Wissenschaft,* 1910, p. 35, in *Beiträge,* VIII, 3.

[2] James (1903), p. 59, "Tractatus Alfani Salernitanus de quibusdam questionibus medicinalibus."

[3] CU Trinity 1365, early 12th century, fols. 155-162v, *Experimenta archiep. Salernitani.*

[4] Judging from its opening and closing words as given by James.

[5] *De coitu,* edition of 1536, p. 306.

[6] *Viaticum,* VI, 19.

whose works Constantinus is using or translating rather than to his own. In the *Pantegni* [1] "ancient medical writers" are divided into *experientes* and *rationabiles,* and we are told that the empirics declare that compound medicines can be discovered only in dreams and by chance, while the rationalists hold that these can be deduced from a knowledge of the virtues and qualities and accidents of bodies and diseases. This much is of course simply Galen over again. Constantinus occasionally gives medical "experiments," as in the case of "proved experiments to eject reptiles from the body," [2] or the placing of a live chicken on the place bitten by a mad dog. The chicken will then die while the man will be cured "beyond a doubt." [3] Such medical "experiments" by Constantinus were often cited by subsequent medieval writers.

"Experiments" involving incantations.

Incantations are involved in some of these "experiments." One approved experiment, we are told, consists in whispering in the ear of the patient the words, *Recede demon quia dee fanolcri precipiunt.* The effect of this procedure is that when the epileptic rises, after remaining like one dead for an hour, he will answer any question that may be put to him. Another experiment to cure epilepsy is frequently cited by subsequent medieval medical writers from Constantinus, and, while it may not have originated with him, is apparently of Christian rather than Greek or Mohammedan origin. If the epileptic has parents living, they are to take him to church on the day of the four seasons and have him hear mass on the sixth day and also on Saturday. When he comes again on Sunday the priest is to write down the passage in the Gospel where it says, "This kind is not cast out save by fasting and prayer." Presumably the epileptic is to wear this writing, in which case a sure cure is promised, "be he epileptic or lunatic or demoniac." But it is added that the charm will not work in the case of persons born of incestuous marriages.[4]

[1] *Practica,* X, 1; in Isaac, *Opera,* 1515, II, fol. 126.

[2] *Ibid.,* VII, 31; fol. 111r.

[3] *Ibid.,* IV, 37; fol. 96r.

[4] *Ibid.,* V, 17; fol. 99r.

Superstition comparatively rare in Constantinus.

But as a rule incantations and superstitious ceremony are comparatively rare in the works of Constantinus, which contain little to justify the charge of magic said to have been made against him in Africa or the charge of superstition made against the Arabic medicine which his writings so largely reflect. Also these superstitious passages seem limited to the treatment of certain ailments of a mysterious character like epilepsy and insanity, which, Constantinus says, the populace call *divinatio* and account for by possession by demons.[1] It is against epilepsy and phantasy that it is recommended to give a child to swallow before it has been weaned the brains of a goat drawn through a golden ring. And it is for epilepsy that we find such suspensions as hairs from an entirely white dog or the small red stones in swallows' gizzards, from which they must have been removed at midday. When Constantinus is treating of eye and ear troubles, or even of paralysis of the tongue and toothache, use of amulets is infrequent and there is only an occasional suggestion of marvelous virtue. Gout is treated with unguents and recipes but without the superstitious ligatures often found in medieval works of medicine. [2] Parts of animals are employed a good deal: thus if you anoint the entire body with lion fat, you will have no fear of serpents, and binding on the head the fresh lung of an ox is good for frenzy.[3] But Constantinus more often explains the action of things in nature from their four qualities of hot, cold, moist, and dry, than he does by assuming the existence of occult virtues.

And of Greek rather than Arabic origin.

It is also to be noted that those passages where Constantinus' medicine borders most closely upon magic are apt to be borrowed from, or at least credited to, Galen and Dioscorides. Neither Constantinus nor his Arabic authorities introduced most of these superstitious elements into medicine. In his work on degrees Constantinus repeats

[1] *De melancholia* (1536), p. 290.

[2] *Practica,* VIII, 40; ed. of 1515, fol. 118v.

[3] *Practica,* IV, 39, and V, 7; ed. of 1515, fols. 96r and 98r.

Galen's story of the boy who fell into an epileptic fit whenever the suspended peony was removed from his neck.[1] In the *Viaticum*[2] he ascribes the suspension of a white dog's hairs and the use of various other parts of animals for epileptics to Dioscorides, but they do not seem to be found in that author's extant works. Water in which blacksmiths have quenched their irons is another remedy prescribed for various disorders upon the authority of Dioscorides and Galen.[3] Theriac and *terra sigillata* are of course not forgotten. That there is a magnetic mountain on the shore of the Indian Ocean which draws all the iron nails out of passing ships, and that the magnet extracts arrows from wounds is stated on the authority of the *Lapidary of Aristotle,* a spurious work. Constantinus adds that Rufus says that the magnet comforts those afflicted with melancholy and removes their fears and suspicions.[4] However, it is without citation of other authors that Constantinus states that the plant *agnus castus* will mortify lust if it is merely suspended over the sleeper.[5]

Some signs of astrology and alchemy.

There is not a great deal of astrological medicine in the works of Constantinus Africanus. There are some allusions to the moon and dog-days,[6] Galen being twice cited to the effect that epilepsy in a waxing moon is a very moist disease, while in a waning moon it is very cold. In a chapter of the *Pantegni*[7] the relation of critical days to the course of the moon and also to the nature of number is discussed. In another passage of the same work[8] we read that if other remedies fail in the case of a patient who cannot hold his water while in bed, he should eat the bladder of a river fish for eight days while the moon is waxing and waning

[1] Ed. of 1536, p. 358; also in the *Viaticum,* I, 22; p. 20.

[2] *Viaticum,* I, 22; p. 21.

[3] *Viaticum,* VII, 13: *De gradibus* (1536), p. 377.

[4] According to Steinschneider (1866), p. 402, it is only from the citations of Constantinus that we know of a work by Rufus on melancholy. See especially *De melancholia* (1536), p. 285, "Invenimus Rufum clarissimum medicum de melancholia fecisse librum. . . ."

[5] *De gradibus* (1536), p. 378.

[6] Edition of 1536, pp. 20, 290, 356.

[7] *Theorica,* X, 9; ed. of 1515, fol. 54.

[8] *Practica,* VII, 59 (1515), fol. 114v.

and he will be freed from the complaint. But Hippocrates testifies that in old men the ailment is incurable. But the principal astrological passage that I have found in the works of Constantinus is that in *De humana natura*[1] where he traces the formation of the child in the womb and the influence of the planets upon the successive months of the process, and explains why children born in the seventh or ninth month live while those born in the eighth month die. This passage was cited by Vincent of Beauvais in his *Speculum naturale*.[2] Belief in alchemy is suggested when Constantinus repeats the assertion of some book on stones that lead would be silver except for its smell, its softness, and its inability to endure fire.[3]

Constantinus and the School of Salerno.

The relation of Constantinus Africanus to the School of Salerno has been the subject of much dispute and of divergent views. Some have held that Salerno's medical importance practically began with him; others have tried to maintain for Salernitan medicine a Neo-Latin character quite distinct from Constantinus' introduction of Arabic influence. From the fact that Constantinus passed from Salerno to Monte Cassino, where most, if not all, of his writing seems to have been done, it has been assumed that there was an intimate connection between the monks and the rise of a medical school at Salerno. On the other hand, Renzi and Rashdall have ridiculed the notion, declaring the distance and difficulty of communication between the two places to be an insurmountable difficulty. It must be remembered, however, that Constantinus himself both attended the archbishop of Salerno in a case of stomach trouble and sent a treatise on the subject to him afterwards. A strong personal influence by him upon the practice and still more upon the literature of Salernitan medicine is therefore not precluded, though his stay at Salerno may have been brief and his literary labor performed entirely

[1] Ed. of 1541, pp. 319-21.
[2] *Spec. nat.*, XVI, 49.
[3] *De gradibus* (1536), p. 360, "de quo Arabũ (Aristotle?) in libro de lapidibus intitulato."

at the monastery. In any case a Master John Afflacius, who is associated with other Salernitan writers in a compilation from their works, was a disciple of Constantinus and, as we are about to see, perhaps the author of some of the treatises which have been published under Constantinus' name. It certainly would seem that Constantinus and his disciple have as good a right to be called Salernitan as most of the authors included in Renzi's collection.

Liber aureus and John Afflacius.

In a medical manuscript which Henschel discovered at Breslau in 1837 [1] and which he regarded as a composition of the School of Salerno and dated in the twelfth century, he found in the case of two works compiled from various authors [2] that the passages ascribed to a Master John Afflacius, who was described as "a disciple of Constantinus," [3] were identical with passages in the *Liber aureus* or *De remediorum et aegritudinum cognitione* published as a work of Constantinus in the Basel edition of 1536. He also identified a *Liber urinarum* attributed to the same John Afflacius, disciple of Constantinus, in the Breslau manuscript with the *De urinis* which follows the *Liber aureus* in the printed edition of Constantinus' works. Thus either the pupil appropriated or completed and published the work of his master, or Constantinus had the same good fortune in having his own name attached to the compositions of his pupil [4] as in the case of the writings of his Arabic predecessors.

It may be further noted that the disciple seems to have been more superstitious than the master, for in one of the passages ascribed to Afflacius in the aforesaid compilation,

[1] *Manoscritto Salernitano dilucidato dal Prof. Henschel,* in Renzi (1853), II, 1-80, especially pp. 16, 41, 59.

[2] *De aegritudinum curatione tractatus,* Renzi, II, 81-386; *De febribus tractatus,* II, 737-68.

[3] The preface to Constantinus' translation of Isaac on fevers is addressed to his "dearest son, John": see Brussels, Library of Dukes of Burgundy 15489, 14th century, "Quoniam te karissime fili Iohanne"; Cambrai 914, 13-14th century; Cambrai 907, 14th century, fol. 1, Prefatio Constantini ad Johannem discipulum.

[4] However, in an Oxford MS the *Liber aureus* itself is ascribed to "John, son of Constantinus": Bodleian 2060, #1, Joannis filii Constantini de re medica liber aureus.

Afflacius more superstitious than his master.

after the correspondence with the *Liber aureus* has ceased, the text goes on to prescribe the suspension of goat's horn over one's head as a soporific and gives the following "prognostic of life or death." Smear the forehead of the patient from ear to ear with *musam eneam*. "If he sleeps, he will live; but if not, he will die; and this has been tested in acute fevers." Another method is to try if the patient's urine will mix with the milk of a woman who is suckling a male child. If it will, he will live. Another procedure to induce sleep is then given, which consists in reading the first verse of the Gospel of John nine times over the patient's head, placing beneath his head a missal or psalter and the names of the seven sleepers written on a scroll. This is not the first instance of such Christian magic that we have encountered in connection with the School of Salerno and we begin to suspect that it was rather characteristic. At any rate it was not uncommon in medieval medicine in general and was almost certainly introduced before Innocent III who in 1215 forbade ordeals and who frowned on other superstitious practices. Probably such Christian magic dates from a period before Arabic influence began to be felt. Thus again we have reason to doubt whether early medieval medicine or Salernitan medicine was less superstitious than Arabic medicine or than medieval medicine after the introduction of Arabic medicine. At least Constantinus Africanus who represents the introduction of translations from the Arabic is comparatively free from superstition.

CHAPTER XXXIII

TREATISES ON THE ARTS BEFORE THE INTRODUCTION OF ARABIC ALCHEMY

Latin treatises on the arts and colors—Progress of the arts even during the early middle ages—Scantiness of the sources—Character of Arabic alchemy—Different character of our Latin treatises—*Compositiones ad tingenda—Mappe Clavicula*—Some of its recipes—Question of symbolic nomenclature—Magical procedure with goats: in *Mappe Clavicula*—Similar passages in Heraclius—And Theophilus—A magic figure—Use of an incantation in tenth century alchemy—Experimental character of the work of Theophilus—How to make Spanish gold—The question of symbolic terminology again—Alchemy in the eleventh century—St. Dunstan and alchemy and magic—Introduction of Arabic alchemy in the twelfth century.

". . . campum latissimum diversarum artium perscrutari. . . ."

—Theophilus, Schedula, I, Praefatio.

Latin treatises on the arts and colors.

WE come to the consideration of several treatises dealing with colors and the arts and dating from about the eighth to the twelfth centuries and probably in part of earlier origin. These are the *Compositiones ad tingenda* in a manuscript of the eighth or ninth century, the *Mappe clavicula* found in part in a tenth century manuscript and more fully in one of the twelfth century, the poem of Heraclius on *The colors and arts of the Romans,* and the remarkable treatise of Theophilus *On diverse arts* in three books. [1] The

[1] Interest in such works was aroused by the almost simultaneous publication of R. Hendrie's English translation of Theophilus, London, 1847; the publication of the *Mappe clavicula* in a "Letter from Sir Thomas Phillipps to Albert Way" in *Archaeologia,* XXXII, 183-244, London, 1847; and the inclusion of Heraclius, *De coloribus et de artibus Romanorum,* in Mrs. Merrifield's *Ancient Practice of Painting,* London, 1849. Hendrie printed the Latin text of Theophilus with his translation. A. Ilg published a revised Latin text with a German translation in 1874, with a fuller account of the MSS.

oldest known manuscripts of Theophilus are of the twelfth century and he has been dated at the beginning of that century or end of the eleventh, and Heraclius, from whom he takes a number of his chapters, still earlier. But it scarcely seems that some of Theophilus' descriptions of ecclesiastical art would have been written before the twelfth century. Mrs. Merrifield regarded only the first two metrical books of *The colors and arts of the Romans* as the work of Heraclius, and the third book in prose as a later addition of the twelfth or thirteenth century and probably written by a Frenchman, whereas she believed that Heraclius wrote in southern Italy under Byzantine influence.[1] His poem sounds to me like an attempt to imitate Lucretius, while one also is inclined to associate it with the perhaps nearly contemporary poems in which the so-called Macer and Marbod recounted in verse form some of the properties of herbs and stones which they had learned from ancient writers.

Progress of the arts even during the early middle ages.

Berthelot regarded these treatises on the arts as proof that the knowledge of industrial and alchemical processes continued unbroken even in western Europe from Egypt to the middle ages, although he held that the theories of transmutation and the like reached the west only in the twelfth century through the Arabs.[2] Moreover, there is progress in the technical processes just as there was progress in Romanesque and Gothic art. New items and recipes appear in the lists. Even in the declining Roman Empire and earliest middle age we have evidence of new discoveries. The artificial fabrication of cinnabar becomes known at some time after Dioscorides and Pliny and before the eighth century.[3] The hydrostatic balance is described not only in the *Mappe clavicula* but in the *Carmen de ponderibus* of Priscian or of Q. Remnius Fannius Palaemo of the fourth

[1] Merrifield (1849), I, 166-74.

[2] Berthelot (1893), I, 29. He dated, however, Robert of Chester's translation of Morienus thirty-eight years too late in that century, mistaking the Spanish for the Christian era.

[3] *Ibid.*, p. 18.

or fifth century A. D.[1] Heraclius speaks more than once in his poem with admiration of the works of art of the Roman "kings" and people, and asks, "Who now is capable of investigating these arts, is able to reveal to us what those potent artificers of immense intellect discovered for themselves?" [2] However, his aim is to resurrect these arts; he assures the reader that he writes nothing which he has not first proved himself; [3] and he tells in particular how he discovered by close scrutiny of a piece of Roman glass that there was gold-leaf placed between two layers of glass, a work which he successfully imitated. [4] On the other hand, lead glazing, according to Alexandre Brongniart, director of the Sèvres manufactory, is not found in European pottery before the twelfth century, when it was applied in Pesaro about 1100 and is found on pottery in a tomb at Jumièges of about 1120.[5]

Scantiness of the sources.

During the early medieval centuries the Byzantine Empire, Syria and Egypt after they were conquered by the Arabs, the busy streets of Bagdad and Cordova, and Persia undoubtedly produced a far more flourishing activity in the fine arts and the industrial arts than was the case in backward western Christian Europe. Yet the surviving evidence for such activity is disappointing, and seems limited to some notices and allusions in Arabian and Jewish travelers and historians, and to the dust-heaps of ruined cities like Fostat, Rai, and Rakka. As the finest early specimens of Byzantine mosaics are preserved in Italy at Ravenna, so our Latin treatises concerning the arts are perhaps the best extant for the early medieval period up to the twelfth century.

[1] Berthelot (1893), I, 169.

[2] Merrifield (1849), I, 183. See also pp. 189-91.

[3] *Ibid.*, p. 183, "Nil tibi scribo equidem quod non prius ipse probassem."

[4] *Ibid.*, p. 187.

[5] *Traité des Arts Céramiques*, p. 304, cited by Merrifield, I, 177. This is not, however, to be regarded as the invention of lead glazing, since, as William Burton writes ("Ceramics" in EB, p. 706), "lead glazes were extensively used in Egypt and the nearer East in Ptolemaic times." He adds, "And it is significant that, though the Romans made singularly little use of glazes of any kind, the pottery that succeeded theirs, either in western Europe or in the Byzantine Empire, was generally covered with glazes rich in lead."

Character of Arabic alchemy.

A number of treatises on alchemy in Arabic have reached us but they, like the Byzantine, chiefly continue the fantastic mysticism and obscurity, the astrology and magic, of the ancient Greek alchemists. Thus in the *Book of Crates* we have a virgin priestess of the temple of Serapis at Alexandria, and the snake Ouroburos, also a vision of the seven heavens of the planets. The *Book of Alhabib* invokes Hermes Trismegistus and says that the sages have not revealed the secret of transmutation for fear of the anger of the demons. The *Book of Ostanes,* in which Andalusia is mentioned, has eighty-four different names for the philosopher's stone, and a fantastic dream concerning seven doors and three inscriptions in Egyptian, concerning the Persian Magi, and a citation from an Indian sage concerning the healing virtues of the urine of a white elephant. The *Book of Like Weights* of Geber states that the sage can discern the mixture of the four elements in animals, plants, and stones by astrology and many other signs involving varied superstition. His *Book of Sympathy* again emphasizes the seven planets as the key to alchemy and has much about the spirit in matter. His *Book on Quicksilver,* although it promises clarity, is the most mystic and incomprehensible of all. In it we read of raising the dead and of use of such liquids as "a divine water" and the milk of an uncorrupted virgin.[1]

Different character of our Latin treatises.

Our Latin treatises are as free from mysticism and obscurity, from dreams and visions, as they are from theoretical discussion. They are collections of recipes and directions which are supposed at least to be practical and which are written in a simple and straightforward style. They are not, however, taken together, by any means entirely free from astrological directions or belief in occult virtue or yet other superstition, and they include recipes for making

[1] For these works see Berthelot (1893), III, or Lippmann (1919), who follows him. I have not had access to E. Wiedemann, *Zur Chemie bei den Arabern,* in *Sitzungsberichte der physikalisch-medizinischen Societät in Erlangen,* XLIII (1911); and his *Die Alchemie bei den Arabern,* in *Journal für praktische Chemie,* LXXVI (1907), 85-87, 105-23.

gold. Of this there is least in the first treatise we have to consider.

Compositiones ad tingenda.

The *Compositiones ad tingenda,*[1] a treatise or collection of notes and recipes preserved in a manuscript dating from the time of Charlemagne, throws some light on the technical processes preserved in the Latin west in the early middle ages and on the amount of knowledge of natural phenomena preserved in connection with the arts,—applied science in other words. It tells how to color glass and make mosaics, and describes a glass furnace; how to dye skins and make parchment; how to make gold-leaf, gold-thread, silver-leaf and tin-leaf; how to give copper the color of gold; it gives various directions and preparations for painting and gilding; and a description of various minerals and herbs employed in the above processes. Much is repeated that is found already in Pliny and Dioscorides, or in Aristotle and the Greek alchemists. But several things are mentioned, at least so far as we know, for the first time, although Berthelot believed that the compiler of the *Compositiones ad tingenda* had copied them from earlier works, very probably Byzantine or late Roman, and not invented them himself. We find here the first mention of vitriol and of "bronze,"—a word apparently derived from Brundisium. *Amor aquae* is used for the first time for the scum formed on waters containing iron salts and other metals, and we also meet the first instance of the preparation of cinnabar by means of sulphur and mercury. The work contains very little superstition with the exception of one passage which Berthelot has already noted.[2] Once a stone is spoken of as having solar virtue; lead is distinguished as masculine and feminine; the gall of a tortoise is used in a composition for writing golden letters, and pig's blood

[1] The full title is "Compositiones ad tingenda musiva, pelles et alia, ad deaurandum ferrum, ad mineralia, ad chrysographiam, ad glutina quaedam conficienda, aliaque artium documenta." The MS, Bibliotheca capituli canonicorum Lucensium, Arm. I, Cod. L, was printed in Muratori, *Antiquitates Italicae,* II (1739), 364-87. It is described by Berthelot (1893), I, 7-22, whose comparison of it with previous treatises I follow.

[2] Berthelot (1888), I, 12, note.

is employed in another connection. But these are trifling signs of occult science.

More alchemistic in character is the *Mappe Clavicula,*[1] which, in its fuller twelfth century form, embodies the *Compositiones ad tingenda* in a different order,[2] and adds about twice as many more recipes for making gold, making colors, writing with gold, glues and various other matters, including building directions. Berthelot regarded two items instructing how to make images of the gods as signs of an ancient pagan origin for the work.[3] One of these items occurs in the twelfth century text, the other in the tenth century table of contents. On the other hand Berthelot believed that the twelfth century version contained the oldest directions for the distillation of alcohol.[4] The *Mappe Clavicula* adds a good deal that is of a superstitious character to the *Compositiones ad tingenda* which it includes, and at the same time lays considerable stress upon experimental method.

Mappe Clavicula.

It opens with a recipe "for making the best gold," the first of a long series. One of the ingredients in this case is "a bit of moon-earth, which the Greeks call *Affroselinum.*" The third recipe advises one to experiment at first with only a little of the compound in question, until one learns the process more thoroughly.[5] The ingredients for gold-making in the sixth recipe include the gall of a goat and of a bull, and saffron from Lycia or Arabia, which is to be pounded in a Theban mortar in the sun in dog-days. At the close of the fourteenth recipe, into which the gall of a bull again enters we have one of the injunctions to secrecy so dear to the alchemist: "Hide the sacred secret which should be transmitted to no one, nor give to

Some of its recipes.

[1] Text and some discussion thereof in *Archaeologia*, XXXII (1847), 183-244. Analyzed by Berthelot (1893), I, 23-65. On the Schlestadt MS of the 10th century, see Giry in *Bibliothèque de l'École des Hautes Études,* XXXV (1878), 209-27.

[2] See recipes 105-93.

[3] Berthelot (1893), I, 57.

[4] *Ibid.*, 61. Others, however, would trace the discovery of alcohol back to Hippolytus. See above, p. 468.

[5] "Accipies ad experimentum donec primitus discas non multum cum semel facias."

anyone the prophetic."[1] It is also implied that alchemy is a religious or divine art in the twentieth recipe where it is said that operators should concede all things to divine works. But such mystic allusions are infrequent as well as brief. In the same twentieth item gold is supposed to be made from a mixture of iron rust, magnet, foreign alum, myrrh, gold, and wine. It is also stated that those who will not credit the great utility that there is in humors are those who do not make demonstration for themselves, another instance of the experimental character of the work. The forty-first recipe states that gold may be dissolved in order to write with it by dipping it in the blood of an Indian dragon, placing it in a glass vessel, and surrounding it with coals. In the sixty-ninth item the blood of a dragon or of a cock is mixed with urine and the stone *celidonius*. The gall of a bull and the blood of a pig are used again in recipes sixty-eight and one hundred and twenty-eight.

Question of symbolic nomenclature.

It has sometimes been contended, chiefly by persons who did not realize how universal was the ascription of great virtue to the parts of animals in ancient and medieval science and their use as remedies in the medicine of the same periods, that they are not to be taken literally in alchemical recipes but are to be understood symbolically and are cryptic designations for common mineral substances. Thus Berthelot cites a passage from the Latin *De anima*, ascribed to Avicenna, which says, "I am going to tell you a secret: the eye of a man or bull or cow or deer signifies mercury," and so on.[2] But despite what Berthelot goes on to say about the "old prophetic nomenclature" of the Egyptians, I am inclined to think that such symbolism is mainly a refinement of later alchemists, and that originally most such expressions were intended literally. Certainly it would be impossible to explain all the medicinal use of parts of animals in Pliny's *Natural History* as either symbolic or derived from the Egyptian priests. Like the suggestion that Roger Bacon

[1] "Absconde sanctum et nulli tradendum secretum neque alicui dederis propheta."

[2] Berthelot (1893), I, 303-4.

wrote in cipher, the symbolic nomenclature theory is based on the assumption that the men of old concealed great secrets under an appearance of error. And where such cryptograms and symbols were employed, it was almost invariably done, we may be sure, with the object of impressing the reader with an exaggerated notion of the importance of what was written rather than because the writer really had any great discovery that he wished to conceal. That symbolic language was employed by alchemists, especially in the latest middle age and early modern centuries, is not to be questioned. The use of the names of the planets for the corresponding metals is a familiar example. But most such symbolic nomenclature is equally obvious, while there is no reason for not taking the use of parts of animals literally. Indeed, in many passages it must be so taken, as in a later item of the *Mappe Clavicula* [1] which has no concern with alchemy and where in order to poison an arrow for use in battle, we are instructed to dip it in the sweat from the right side of a horse between the hip-bones. The following experiments with goats also illustrate the great value set upon animal fluids and substances.

Magical procedure with goats in the *Mappe Clavicula.*

We are reminded of the directions given by Marcellus Empiricus for the preparation of goat's blood by a recipe for making figures of crystal which occurs near the close of the *Mappe Clavicula.*[2] A he-goat which has never indulged in sexual intercourse is to be shut up in a cask for three days until he has completely digested everything that he had in his belly. He is then to be fed on ivy for four days, at the end of which time he is to be slain and his blood mixed with his urine which is now collected from the cask. By soaking the crystal overnight in this mixture it can be moulded or carved at will. This experiment is immediately preceded by a somewhat similar procedure for cutting glass with steel.[3] The glass is to be softened and the steel is to be tempered by placing them either in the milk of a Saracen

[1] Item 265.

[2] Item 290.

[3] Item 289.

she-goat, who has been fed upon ivy and milked by scratching her udders with nettles, or in the lotion of a small girl of ruddy complexion, which must be taken before sunrise.

Similar passages in Heraclius.

Very similar passages are found in the works of Heraclius and Theophilus, the former of whom gives the following directions for glass engraving: "Oh! all you artists who wish to engrave glass correctly, now I will show you just as I myself have proven. I sought the fat worms which the plow turns up from the earth, and the useful art in such matters bade me at the same time seek vinegar and the hot blood of a huge he-goat, which I had taken pains to tie up under cover and to feed on strong ivy for a while. Next I mixed the worms and vinegar with the warm blood and anointed all the bright shining phial. This done, I tried to engrave the glass with the hard stone called pyrites."[1] In another passage Heraclius recommends the use of the urine and blood of a goat in engraving gems,[2] and he also states that the blood of a goat makes crystal easier to carve.[3]

And Theophilus.

Theophilus states that poets and artificers have greatly cherished the ivy, "because they recognized the occult powers which it contains within itself."[4] He also affirms that the blood of a goat makes crystal easier to carve, but he recommends the blood of a living goat two or three years old and repeated insertion of the crystal in an incision between the animal's breast and abdomen.[5] He also recommends a somewhat similar procedure to that of the *Mappe Clavicula* with a goat and a cask.[6] In this case the goat should be three years old, and after being bound for three

[1] *De coloribus et artibus Romanorum,* I, iv. I have somewhat altered Mrs. Merrifield's translation (I, 186).

[2] *Ibid.,* I, xi; Mrs. Merrifield (1849), I, 189-91.

[3] *Ibid.,* I, xii:
"Sed vim cristalli cruor antea temperet hirci
Sanguis enim facilem ferro facit hic adamantem."
Mrs. Merrifield (I, 194) has incorrectly rendered this passage, "But let the blood of a goat first temper it, for this blood makes the iron so hard that even adamant is soft compared to it." What Heraclius says is,
"But first let the blood of a he-goat temper the force of the crystal,
For this blood makes adamant soft to the iron."

[4] *Schedula diversarum artium,* III, 98.

[5] *Ibid.,* III, 94.

[6] *Ibid.,* III, 21.

days without food should be fed for two days on nothing but fern. The following night he should be shut up in a cask with holes in the bottom through which his urine can be collected in another vessel for two or three nights, when the goat may be released and the urine employed to temper iron tools. Or the urine of a small red-headed boy may be employed, as it is better for tempering than plain water. Indeed, both Theophilus and Heraclius make much use of parts of animals in the arts: various animals' teeth to shine and polish things with, horse dung mixed with clay, skins and bladders, saliva and ear-wax to polish niello, and so forth.

A magic figure.

Returning to the *Mappe Clavicula* we note the employment of a magic figure called *arragab,* which Berthelot thinks is a small lead image.[1] By means of it the flow of a spring may be stopped; a cup may be made either to retain or to empty its contents; if the cows drink first from the trough, there will be enough water for both the cows and the horses, but if the horses drink first, there will not be enough for either. The same figure enables one to fill a pitcher from a cask without diminishing the amount of liquid in the cask, or to construct a lamp which will produce phantoms. It also makes soldiers leave their camp without their spears and yet return with them. After this flight into the realm of magic we come back to a more plausibly physical basis for marvels in a description of four revolving hoops or circles within which a vessel may be revolved in any direction without spilling its contents.[2]

Use of an incantation in tenth century alchemy.

The passages which we have just noted in the *Mappe Clavicula* cannot be surely traced back earlier than the twelfth century version of it and do not appear in the table of contents which is preserved in the tenth century Schlestadt manuscript and which covers only a portion of the chapters of the twelfth century manuscript, but also some

[1] Berthelot (1893), I, 63. His French translation omits some of the Latin text as published in *Archaeologia,* cap. 288.

[2] "Cardan's concentric circles," according to Berthelot (1893), I, 64.

other chapters which are not extant. But that magic was not entirely absent from the earlier version to which this table of contents seems to apply is evidenced by the fact that one of the chapter headings dealing with the fabrication of gold mentions a prayer or incantation to be recited during the process.[1]

Experimental character of the work of Theophilus.

The great importance of the work of Theophilus in the history of art is too generally recognized to need elaboration here. Our purpose is rather to point out that in it information of great value is found side by side with a considerable amount of misguided natural theory and magical ceremony. The stress laid by Theophilus upon personal observation, experience, and experimental method should not, however, pass unnoticed. He has scrutinized the works of art in the church of St. Sophia one by one "with diligent experience," has tested everything by eye and hand, has as a "curious explorer" made all sorts of experiments, and appears to represent transparent stained glass as his own discovery or idea.[2] Nor is he the only experimenter; he also speaks of "modern workmen" who deceive many incautious persons by their imitation of the appearance of most precious Arabian gold which "is frequently found employed in the most ancient vases.[3]

How to make Spanish gold.

Theophilus, however, believes that other metals can really be transmuted into gold, and we may repeat his amusing account of how Spanish gold "is made from red copper

[1] Berthelot (1893), I, 55.

[2] II, prologus (closing passage). "Huius ergo imitator desiderans fore, apprehendi atrium agiaę Sophiae conspicorque cellulam diversorum colorum omnimodo varietate refertam et monstrantem singulorum utilitatem ac naturam. Quo mox inobservato pede ingressus, replevi armariolum cordis mei sufficienter ex omnibus, quae diligenti experientia sigillatim perscrutatus, cuncta visu manibusque probata satis lucide tuo studio commendavi absque invidia. Verum quoniam huiusmodi picturae usus perspicax non valet esse, quasi curiosus explorator omnibus modis elaboravi cognoscere, quo artis ingenio et colorum varietas opus decoraret, et lucem diei solisque radios non repelleret. Huic exercitio dans operam vitri naturam comprehendo, eiusque solius usu et varietate id effici posse considero, quod artificium, sicut visum et auditum didici, studio tuo indagare curavi." Ilg's Latin text (1874).

[3] III, 47.

and powdered basilisk and human blood and vinegar." "For the Gentiles, whose skill in this art is well known, create basilisks in this wise. They have an underground chamber completely walled in on all sides with stone, and with two windows so small as scarcely to admit any light. In this they put two cocks of twelve or fifteen years and give them plenty of food. These, when they have grown fat, from the heat of their fat have commerce together and lay eggs. As soon as the eggs are laid the cocks are ejected and toads are put in to sit on the eggs and are fed upon bread. When the eggs are hatched chicks come forth who look like young roosters, but after seven days they grow serpents' tails and would straightway burrow into the ground, were the chamber not paved with stone. Guarding against this, their masters have round brazen vessels of great amplitude, perforated on all sides, with narrow mouths, in which they put the chicks and close the mouths with copper covers and bury them underground, and the chicks are nourished for six months by the subtle earth which enters through the perforations. After this they uncover them and apply a strong fire until the beasts within are totally consumed. When this is over and it has cooled off, they remove and carefully pulverize them, adding a third part of the blood of a ruddy man, which blood is dried and powdered. Having compounded these two they temper them with strong vinegar in a clean vessel; then they take very thin plates of the purest red copper and spread this mixture over them on both sides and place them in the fire. And when they grow white hot, they take them out and quench and wash them in the same mixture, and this process they repeat until the mixture has eaten through the copper, and so obtain the weight and color of gold. This gold is suited for all operations." [1]

Mr. Hendrie held that Theophilus was here describing in symbolic language a process "for procuring pure gold by the means of the mineral acids;" and that "the toads of

[1] I have followed Ilg's rather than Hendrie's text; III, 48.

The question of symbolic terminology again.

Theophilus which hatch the eggs are probably fragments of the mineral salt, nitrate of potash; . . . the blood of a red man . . . probably a nitrate of ammonia; fine earth, a muriate of soda (common salt); the cocks, the sulphates of copper and iron; the eggs, gold ore; the hatched chickens, which require a stone pavement, sulphuric acid produced by burning these in a stone vessel, collecting the fumes. . . . The elements of nitro-muriatic acid are all here, the solvent for gold." [1] Mr. Hendrie leaves, however, a number of details unexplained and he admits that "Unfortunately each chemist appears to have varied the symbols in use." Certainly one would have to vary them in almost every case to make any sense out of such procedures as this of Theophilus. On the other hand, there is nothing very surprising in his procedure taken literally to one who is acquainted with the beliefs of ancient and medieval science and magic. And certainly Shakespeare's line concerning the precious jewel in the toad's head, which Hendrie quotes in this connection, is much more likely to be meant literally than to be the symbolic "jargon of the alchemist." Later we shall hear again from Alexander Neckam, in a passage which has no connection with alchemy, of the basilisk hatched by a toad from an egg laid by a cock, and we shall hear from Albertus Magnus of an experiment in which a toad's eye was proved superior in virtue to an emerald.

Alchemy in the eleventh century.

The treatises which we have been considering appear, at least for the most part, to antedate the Latin translations of works of alchemy from the Arabic, although it is possible that, just as the first translations of mathematical and astronomical works from the Arabic go back to the tenth century at least, so the reception of Arabic alchemy may have begun in a small way before the twelfth century. At any rate we find that in the eleventh century not only were Michael Psellus and other Byzantine scholars spreading the doctrines of alchemy, [2] but a scholium to Adam of Bremen

[1] Hendrie (1847), pp. 432-3.
[2] Ernst von Meyer, *History of Chemistry,* 1906.

records the presence at the court of Bishop Adalbert of Bremen of an alchemist in the person of a baptized Jew.[1]

St. Dunstan and alchemy and magic.

To St. Dunstan, the famous abbot of Glastonbury, archbishop of Canterbury, and statesman of the tenth century (924 or 925 to 988), is attributed a treatise on the philosopher's stone contained in a Corpus Christi manuscript of the fifteenth century at Oxford and printed at Cassel in 1649. No genuine works by him seem to be extant, however, but it is interesting to note that along with his reputation for learning and mechanical skill went the association of his name with magic. In his studious youth he was accused of magic, driven from court, and thrown into a muddy pond. His contemporary biographer also narrates how the devil appeared to him in various animal and other terrifying forms. His favorite studies were mathematics and music, and he was said to own a magic harp which played while hanging by itself on the wall.[2]

Introduction of Arabic alchemy in the twelfth century.

Berthelot has associated the introduction of Arabic alchemy into Christian western Europe with the Latin translation by Robert of Chester of *The Book of Morienus*, but incorrectly dated it in 1182 A. D.,[3] whereas the mention of that date in the manuscripts has reference to the Spanish era and denotes the year 1144 A. D.[4] The main reason for regarding Robert's translation as one of the earliest is that he remarks in his preface, "What alchemy is and what is its composition, your Latin world does not yet know truly." Of the work translated by Robert we shall treat more fully in a later chapter on *Hermetic Books in the Middle Ages*. Here we may further note the existence of a work

[1] Migne, PL 146, 583-4. Some accused the bishop of resort to magic arts: *Ibid.*, 606.

[2] W. Stubbs, in RS LXIII, p. cix. C. L. Barnes, *Science in Early England*, in Smithsonian Report for 1895, p. 732. Of the alchemy ascribed to Dunstan, Elias Ashmole remarked in his *Theatrum Chemicum Britannicum*, 1652, "He who shall have the happiness to meet with St. Dunstan's work *De occulta philosophia* . . . may therein read such stories as will make him amazed to think what stupendous and immense things are to be performed by virtue of the Philosopher's Mercury, of which a taste only and no more."

[3] Berthelot (1893), I, 234.

[4] Karpinski (1915), pp. 26-30; Haskins, EHR, XXX (1915), 62-5.

of alchemy in another twelfth century manuscript.[1] It is a brief work in four chapters and its superstitious character may be inferred from its opening instruction to "take four hundred hen's eggs laid in the month of March," and its citation of Artesius concerning divination by the reflection or refraction of the sun's rays or moon-beams in liquids or a mirror. Since the treatise bears the title *Alchamia,* it is probably safe to assume that it represents Arabic influence.

[1] Berlin 956, 12th century, "Hic incipit alchamia. Accipe CCCC ova gauline que generata sunt et facta in mense martii . . . / . . . ut recentiora sint semper et calidiora. Explicit alchamia." The titles of the last three chapters are, "de iiii ollis, de cognitione, de observatione stestarum." I have not seen the MS but follow Rose's description in the Berlin MSS catalogue.

CHAPTER XXXIV

MARBOD, BISHOP OF RENNES, 1035-1123

Career of Marbod—Relation of his *Liber lapidum* to the prose *Evax*—Problem of Marbod's sources—Influence of the *Liber lapidum* —Occult virtue of gems—*Liber lapidum* meant seriously—*De fato et genesi.*

"Nec dubium cuiquam debet falsumque videri
Quin sua sit gemmis divinitus insita virtus;
Ingens est herbis virtus data, maxima gemmis."
—*Marbod, Liber lapidum.*

Career of Marbod.

Of medieval Latin Lapidaries the earliest and what also seems to have been the classic on the subject of the marvelous properties of stones is the *Liber lapidum seu de gemmis* by Marbod, bishop of Rennes,[1] who lived from 1035 to 1123 and so had very likely completed this work before the close of the eleventh century. Indeed one manuscript of it seems to date from that century [2] and there are numerous twelfth century manuscripts. These early manuscripts bear his name and the style is the same as in his other writings. Born in the county of Anjou, Marbod attended the church

[1] I have used the edition of Marbod's poems in Migne, PL vol. 171, which also contains a life of Marbod. Two secondary accounts of Marbod are C. Ferry, *De Marbodi Rhedonensis Episcopi vita et carminibus*, Nemansi, 1877; L. V. E. Ernault, *Marbode, Évêque de Rennes, Sa vie et ses Œuvres*, in *Bull. et Mém. de la Société Archéologique du dept. d'Ille-et-Vilaine*, XX, 1-260, Rennes, 1889. See also V. Rose, *Aristoteles De Lapidibus und Arnoldus Saxo*, in *Zeitsch. f. deutsches Alterthum*, XVIII (1875), p. 321, *et seq.*; L. Pannier, *Les lapidaires français du moyen âge*, Paris, 1882. C. W. King, *The Natural History, Ancient and Modern, of Precious Stones and Gems*, London, 1865.

[2] CLM 23479, 11th century, fols. 4-10, Carmina de lapidibus eadem quae Marbodo tribuuntur sed alio ordine. Of CUL 768, 15th century, fols. 67-80, "Marbodi liber lapidum," the Catalogue says, "This Latin poem has been often printed but it does not appear that the editors have collated this MS. The order of the sections is different from all those of which Beckmann speaks in his edition (Göttingen, 1799), answering, however, most nearly to his own."

school there, became the schoolmaster himself from 1067 to 1081, during which time he probably composed the *Liber lapidum*, then served as archdeacon under three successive bishops, and finally himself became a bishop in 1096. He attended church councils in 1103 and 1104 and died in September, 1123, in an Angevin monastery, whose monks issued a eulogistic encyclical letter on that occasion, while two archdeacons celebrated his integrity, learning, and eloquence in admiring verse. Marbod's own productions are also in poetical form. It is interesting to note that despite his early date he was eulogized not as a lone man of letters in an uncultured age but as "the king of orators, although at that time all Gaul resounded with varied studies."

Relation of the *Liber lapidum* to the prose *Evax*.

The *Liber lapidum* is a Latin poem of 734 hexameters describing sixty stones. In the opening lines Marbod writes:

"Evax, king of the Arabs, is said to have written to Nero,
Who after Augustus ruled next in the city.[1]
How many the species of stones, what names, and what colors,
From what regions they came, and how great the power of each one."

Making use of this worthy book, Marbod has decided to compose a briefer account for himself and a few friends only, believing that he who popularizes mysteries lessens their majesty. As a result of this opening line and the fact that in some manuscripts Marbod's own name is not given, his poem is sometimes listed in the catalogues as the work of Evax.[2] There is also, however, extant a work in Latin

[1] The full name of Tiberius was, of course, Tiberius Claudius Nero Caesar.

[2] Library of Dukes of Burgundy 8890, 12th century, Evacis regis. BN 2621, 12th and 15th centuries, #6, Poemation de gemmis cuius author dicitur Evax, Rex Arabiae.

Montpellier 277, Liber lapidum preciosorum Evax rex Arabum.

Riccard. 1228, 12th century, fols. 41-54; Incipit prologus Evacis regis Arabie ad Neronem Tyberium de lapidibus. Incipit lapidarius Evacis habens nomina gemmarum lx.

BL Hatton 76 contains two letters of Evax, king of the Arabs, to Tiberius Caesar, on the virtues of stones, according to Cockayne (1864), I, xc and lxxxiv.

prose which opens, "Evax, king of Arabia, to the emperor Tiberius greeting." [1] But as this prose work is not much longer than Marbod's poem, and seems to be known only from a single manuscript of the fourteenth century, it is doubtful if it is the work which he professed to abbreviate. This prose work is also ascribed to Amigeron or Damigeron,[2] to whom we have already seen that the author of *Lithica* was supposed to be indebted and whose name was regarded as that of a famous magician. After alluding to the magnificent gifts which the emperor had sent to Evax by the centurion Lucinius Fronto and offering this book in return, the author of the prose version lists seven stones appropriate, not, strangely enough, to the seven planets, but to seven of the signs of the zodiac.[3] Fifty chapters are then devoted to as many stones, beginning with *Aetites,* which is twenty-fifth in Marbod's list, and ending with *Sardo,* while *Sardius* comes tenth in Marbod's poem. Marbod's own order, however, sometimes varies in the manuscripts.[4]

Problem of Marbod's sources.

King, and Rose after him, asserted [5] that despite Marbod's professed abridgement of a work which Evax was supposed to have presented to Tiberius, he drew largely from Isidore of Seville's *Etymologies.* Rose thought that some of the descriptions of stones were from Solinus, the rest from Isidore, but that the account of their virtues was from Evax. King also noted occasional extracts from the Orphic work, *Lithica,* which is not surprising in view of the fact that both Evax and the *Lithica* seem based on Damigeron. This question of sources and ultimate origins is, however, as usual of relatively little moment to our investigation. My own impression would be that in antiquity and the middle age

[1] Printed by J. B. Pitra, III (1855), 324-35.

[2] BN 7418, 14th century, fol. 116-, (D)amigeronis peritissimi de lapidibus. Since this is the sole MS known of the prose version (Rose, 1875, p. 326) and is of the 14th century, whereas we have numerous early MSS of Marbod's poem, it would seem that this may be derived from Marbod rather than even from the earlier and fuller work which he is supposed to have used.

[3] Namely, Leo, Cancer, Aries, Sagittarius, Taurus, Virgo, and Capricorn.

[4] See page 775, note 2.

[5] King (1865), p. 7; Rose (1875), p. 335.

there exists a sort of common fund of information and stock of beliefs concerning gems which naturally is drawn upon and appears in every individual treatise upon them. But the number of gems discussed and the order in which they are considered or classified varies with each new author, and there is apt to be a similar variation in the number of statements made concerning any particular stone and the way in which these are arranged. In fine, all ancient and medieval accounts of the natures and virtues of stones bear a general resemblance to one another which is more impressive than is the similarity between any two given accounts, and testify to a consensus of opinion and to a common learned tradition concerning gems which is more significant than the possible borrowings of individual authors from one another.

Influence of the *Liber lapidum.*

However, there seems to be little doubt that the poem of Marbod is itself an outstanding work among medieval accounts of precious stones, first because of the early date of its authorship, and second because of its late persistence and popularity, which is indicated by the fourteen editions that appeared after the invention of printing.[1] Its convenient form perhaps accounts to a considerable extent for its popularity. At any rate the manuscripts of it are numerous, and it was much used by subsequent writers of the twelfth and thirteenth centuries, although citations of *Lapidarius* cannot always be assumed to refer to Marbod. But at least the notions concerning gems which we find in his poem are a fair sample of what we should find in any Latin treatment of the same subject for several centuries to come. It is found also in a medieval French version.

Occult virtue of gems.

It does not make much difference where we begin or what stones we select from Marbod's list as examples, since the same sort of marvelous powers are ascribed to all of them. In his prologue Marbod describes the occult virtues of gems as those "whose hidden cause gives manifest effects." No one should doubt them or think them false,

[1] Ferry (1887), p. 69.

"since the virtue in gems is divinely implanted. Enormous virtue is given to herbs, but the greatest to gems."

Adamant, hard as it is, cracks when heated with goat's blood. It counteracts the action of the magnet. It is used in the magic arts and makes its bearer indomitable. It drives off nocturnal specters and idle dreams. It routs black venom, heals quarrels and contentions, cures the insane, and repels fierce foes.

Allectory, found inside cocks, slakes thirst. Milo overcame other athletes, and kings have won battles by its aid. It restores promptly those who have been banished, enables orators to speak with a flow of language, makes one welcome on every occasion, and endears a wife to her husband. It is advised to carry it concealed in the mouth.

The sapphire nourishes the body and preserves the limbs whole. Its bearer, who should be most chaste, cannot be harmed by fraud or envy and is unmoved by any terror. It leads those in bonds from prison. It placates God and makes Him favorable to prayers. It is good for peace-making and reconciliation. It is preferred to other gems in hydromancy, since prophetic responses can be obtained by it. As for medicinal qualities, it cools internal heat, checks perspiration, powdered and applied with milk it heals ulcers, cleanses the eyes, stops headache, and cures diseases of the tongue.

Gagates, worn as an amulet, benefits dropsy; diluted with water, it prevents loose teeth from falling out; fumigation with it is good for epileptics and it is thought to be hostile to demons; it remedies indigestion and constipation and overcomes magical illusions (*praestigia*) and evil incantations. Also

Per suffumigium mulieri menstrua reddit

. . . .

Et solet, ut perhibent, deprehendere virginitatem.
Praegnans potest aquam triduo qua mersus habetur
Quo vexabatur partum cito libera fundit.

Gagates burns when washed with water; is extinguished by anointing it with olive oil

The magnet is especially used in the illusions of magic. The great Deendor is said to have first used it, realizing that there was no more potent force in magic, and after him the famous witch Circe employed it. Among the Medes experience revealed still further virtues of the stone. It is used to test a wife's chastity while she is sleeping; if she is unchaste, she will fall out of bed when the gem is applied to her head. A burglar can commit theft unmolested by sprinkling it over hot coals and so driving away all the occupants of the house.

In the case of *Chelonitis* Marbod's account is very similar to that in Pliny's *Natural History,*[1] citing the Magi for the power of divination it bestows when carried under the tongue at certain times of the moon, according to whose phases its power varies. Of the gems hitherto described only in the case of adamant and gagates was there any resemblance between Marbod and Pliny and there only partial.

Pliny also briefly states that the stone *diadochos* resembles beryl, but does not have Marbod's statements that it is employed in water divination to show varied images of demons, "nor is there other stone stronger to evoke shades." But if by chance it comes in contact with a corpse, it loses its wonted force, since the stone is sacred and abhors dead bodies.[2]

Liber lapidum was meant to be taken seriously.

The vast powers, not only medicinal and physical, but of divination and magic, over the mind and affections, miraculous and supernatural, even over God, as in the statement that the sapphire can be employed to secure a more favorable answer to prayer, which Marbod assigns to gems without a sign of scruple or scepticism or disapproval on his part, have so shocked some moderns that suggestions have been made, in order to explain away the acceptance of talismanic powers of gems to such a degree by a Christian clergyman who became a bishop, that Marbod must have com-

[1] NH XXXVI, 56. Pliny, however, makes these statements about chelonia and not chelonitis which follows it.

[2] The stones which I have taken as examples are numbers 1, 3, 5, 18, 19, 39, and 57 respectively.

posed his poem when quite young and lived to repent it, or that he regarded it merely as a poetical flight and exercise, not as an exposition of scientific fact. But wherefore then was it not only widely read in the literary twelfth century but also widely cited as an authority in the scientific and equally Christian thirteenth century? No; everyone else took it precisely as Marbod meant it, as a serious statement of the marvelous powers which had been divinely implanted in gems. And why should not God be more easily reached through the instrumentality of gems, since He had endowed them with their marvelous virtues? Marbod affirms his own faith in the great virtues of gems not only at the beginning but the close of his poem, stating that while some have doubted the marvelous properties attributed to them, this has been due to the fact that so many imitation gems are made of glass, which deceive the unwary but of course lack the occult virtues of the genuine stones. If the stones are genuine and duly consecrated, the marvelous effects will without a doubt follow.

De fato et genesi.

Marbod's belief in the almost boundless talismanic virtues of gems is thrown into the higher relief by the fact that in another of his poems he makes an attack upon genethlialogy or the prediction of the entire life of the individual from the constellations at his birth. In *De fato et genesi* he writes against "the common notion" *(opinio vulgi)* that all things are ruled by fate, that the hour of nativity controls man's entire life, and the contention of the *mathematici* that the seven planets control not only the external forces with which man comes in contact but also human character. He objects to such a doctrine as that, when Venus and Mars appear in certain relations to the sun, the babe born under that constellation will be destined to commit incest and adultery in later life. He objects that such beliefs destroy all the foundations of morality, law, and future reward or punishment; contends that there are certain races which never commit adultery or crime, yet have the same seven planets; and argues that since Jews are all circumcised on the eighth day, they should all have the same horoscope. These are familiar contentions, at least as old as Bardesanes. Marbod declares further that the astrological writer, Firmicus, employs "infirm arguments," and that his own horoscope, taken according to Firmicus' methods and interpreted likewise, turned out to be false, "as I proved when once I dabbled in that art." This is interesting as showing that Gerard of York[1] was not the only bishop of the eleventh century who was acquainted with the work of Julius Firmicus Maternus, and that even opponents of astrology are apt to have once been dabblers in it. Marbod concludes his poem with this neat turn:

> "I thought I ought to write these lines briefly against genethlialogy.
> Nevertheless, that I may not seem to repel fate and horoscope utterly,
> I assert that my fate is the Word of the supreme Father,
> By Whom should all things be ruled and all men confess;
> And I say that the computation of my constellation is innate in me
> And the liberty by which I can tend whither I will.
> Therefore, if my will shall be in conjunction with reason
> In the sign of the Balances with Christ regarding me,
> All things will turn out prosperously for me here and everywhere :—
> This is the favorable horoscope of all Christ's followers."

[1] See above, chapter 29, page 689.

GENERAL INDEX

Names of men of learning will be found for the most part in the bibliographical index.

BIBLIOGRAPHICAL INDEX

Names of authors, editors, translators, publishers, etc., in Roman type. Titles and periodicals in italics. Leading passages in italics. Bibliographical abbreviations, such as EB, HL, PG, PL, are as a rule not indexed. In the abbreviated titles such opening words as *De* and *Liber* are omitted to facilitate aphabetical arrangement. In proper names *De* and *Von* are usually designated by *d.* and *v.*, and are treated as initials.

INDEX OF MANUSCRIPTS